Human Histology

Human Histology

A Text and Atlas for Physicians and Scientists

Bertalan Dudás M.D., Ph.D. Habil.

ELSEVIER

ACADEMIC PRESS
An imprint of Elsevier

To my sons,

Demian,
Dominik
and Nimrod.

Academic Press is an imprint of Elsevier
125 London Wall, London EC2Y 5AS, United Kingdom
525 B Street, Suite 1650, San Diego, CA 92101, United States
50 Hampshire Street, 5th Floor, Cambridge, MA 02139, United States
The Boulevard, Langford Lane, Kidlington, Oxford OX5 1GB, United Kingdom

ISBN: 978-0-323-91891-6

For information on all Academic Press publications
visit our website at https://www.elsevier.com/books-and-journals

Publisher: Stacy Masucci
Acquisition Editor: Patricia M. Osborn
Editorial Project Manager: Sam Young
Production Project Manager: Omer Mukthar
Cover Designer: Matthew Limbert

Typeset by STRAIVE, India

Contents

Preface vi

Abbreviations vii

Chapter 1: Basic Tissues 1

Chapter 2: Lymphatic System 35

Chapter 3: Cardiovascular System 54

Chapter 4: Integumentary System 66

Chapter 5: Digestive System I: Upper Alimentary Tract 83

Chapter 6: Digestive System II: Lower Alimentary Tract 118

Chapter 7: Digestive System III: Liver and Pancreas 175

Chapter 8: Respiratory System 190

Chapter 9: Urinary System 212

Chapter 10: Male Reproductive System 228

Chapter 11: Female Reproductive System 265

Chapter 12: Endocrine System 304

Chapter 13: Nervous System 332

Chapter 14: Sensory Organs 349

Acknowledgement 372

Index 373

Preface

Teaching histology in the medical school is a challenging task due to the shortness of time allocated to the classic morphological subjects. As our medical knowledge advances, the focus often shifts to the clinical field, with less and less time being available for the basic and system histology courses. Nevertheless, morphology represents a crucial foundation that is essential for a practicing physician. One way to embark upon this problem is to eliminate redundancies between the histology, physiology and biochemistry curricula; however, in the absence of concise, easy to read volumes, educators are often forced to provide their own notes. Many excellent histology books are guilty of these issues, since they contain extended physiological and biochemical details. While this information is certainly helpful for understanding the link between morphology and function, students simply cannot find the study time required for this approach, rendering these volumes extremely hard to use. At the same time, a practicing physician or a scientist may also be required to swiftly refresh histological knowledge learnt during their medical education without going through the evident - and therefore unnecessary - details.

An additional problem with many of the existing, otherwise excellent volumes is that they use animal data to illustrate microanatomy. While this practice may be inevitable in certain cases due to the significant *post mortem* time of the available human tissues, and therefore the difficulty in achieving proper fixation, human material should certainly be preferred in medical education.

The present volume is intended to address these issues. Compact but comprehensive histological descriptions of organs and tissues include only the essential functional information, and the text is coupled with full page micrographs with insets containing magnified details. Almost exclusively, micrographs of human tissues have been used in the book. Multiple samples of the same organs obtained from different patients, as well as from fetal and newborn subjects, have also been included to illustrate morphological variations. The vast majority of the sections are stained with haematoxylin & eosin (H&E) that is considered to be a standard staining procedure in histology; however, some additional specific staining methods have also been used to reveal delicate morphological details. The intention is for the material to include all of the necessary minutiae discoverable by light microscopy without elaborating on the ultrastructural details, and provide a simple, yet thorough description of the microanatomy of the tissues and organs. I hope that these aspirations will meet the requirements of medical students, physicians and researchers alike, and the book will prove to be a useful addition to the education of medical histology.

Bertalan Dudas MD

Bertalan Dudas M.D., Ph.D. Habil.
Erie, 2022 April

Abbreviations:

ac, anterior chamber of the eye
ad, adipocytes
afn, anterior funiculus
ahp, adenohypophysis
al, alveolus
ald, alveolar duct
als, alveolar sack
alv, alveolar bone
amn, amnion
ao, arteriole
apl, autonomic plexus
arm, arytenoid muscle
ary, arytenoid cartilage
as, apocrine sweat gland
asm, anal submucosal muscle
atz, anal transitional zone
bc, Bowman's capsule
bm, basilar membrane
bm, bone marrow
bmm, Bowman's membrane
böt, cuboidal cells of Böttcher
bpl, basal plate, decidua basalis
br, Brunner's gland
bt, bone trabecule
bv, blood vessel
cam, central adrenomedullary vein
cav, cavern
cc, cervical canal, uterus
ccl, lamina choriocapillaris
ccv, corpus cavernosum
cd, cochlear duct
cf, collagen fiber
cg, cardiac gland
chr, choroid of the eye
cib, ciliary body
cip, ciliary process
cl, crypts of Lieberkühn
cla, cuboidal cells of Claudius
clx, minor calyx
cm, cementum
cn, cochlear nerve
cor, cornea
cp, capsule
cpl, chorionic plate
cr, cremasteric muscle
cri, cricoid cartilage
crz, colorectal zone
ct, connective tissue
cuz, cutaneous zone
cv, central vein
cvg, cervical gland, uterus
cvl, chorionic villi
cx, cortex
cy, crypt
cys, cystic cavity
csk, circular skeletal muscle
csp, corpus spongiosum
dar, deferential artery
dct, distal convoluted tubule
ded, ductus epididymidis
dfs, deep (Buck's) fascia of penis
dh, dorsal horn
dip, dilator pupillae muscle
dm, dermis
dmm, Descemet's membrane
dn, dentin
dp, dermal papilla
dr, dura mater
drrm, dorsal ramus, spinal nerve
drrt, dorsal root
ds, dermal sheath
dt, duct
ed, epidermis
edc, efferent ductules, epididymis
eel, external elastic lamina
eem, extraembryonic mesoderm
egl, epiglottis
ej, ejaculatory duct
el, endosteal lamellae
elm, external limiting membrane (retina)
emt, endometrium
epi, epithelium
epn, epineurium
ers, external root sheath
es, early spermatid
esf, external spermatic fascia
esp, external anal sphincter
evl, epiglottic vallecula
exc, excretory duct
fc, follicular cells
fg, fundic glands
film, inner limiting membrane (retina)
flp, filiform papilla
fmb, fimbria
fnp, fungiform papilla
fpo, fibrous periosteum
gc, granulosa cells
gcl, glans clitoridis
gep, germinal epithelium
gfl, Graafian follicle
ggc, ganglion cell layer (retina)
ggl, ganglion cells
gl, glomerulus
glc, granulosa lutein cells
gp, gastric pit
grn, granular layer
gv, gingiva
hc, Haversian canal
hf, hair follicle
hl, haversian lamellae
hp, papilla of the hair follicle
hr, hair
hy, hyalin cartilage
ian, interarytenoid notch
ib, intralobular duct
ibr, inner border cells
ic, intercalated duct
icp, renal capsule, inner layer
id, interlobular duct
iel, internal elastic lamina
ihr, inner hair cells
il, islets of Langerhans
iln, inner longitudinal layer
ils, interlobar sulcus
im, inner muscularis externa
inl, inner nuclear layer (retina)
iph, inner phalangeal cells
ipl, inner pillar cells
ipl, inner plexiform layer (retina)
irs, internal root sheath
isf, internal spermatic fascia
isl, interstitial lamellae
isp, internal anal sphincter
it, inner spiral tunnel
lca, lat. cricoarytenoid muscle
lf, lymphatic follicle
lfd, lactiferous duct
lfn, lateral funiculus
lh, loop of Henle
lmn, labium minus
lnd, lymph node
lp, lamina propria
ls, late spermatid
lsk, longitudinal skeletal muscle
lut, corpus luteum
lv, lymph vessel
ly, lymphatic aggregation
ma, muscular artery
mc, medullary cords
mcg, prostatic mucosal glands
mcr, middle circular layer
md, medulla
me, tunica muscularis externa
mec, caecal muscularis externa

mei, ileal muscularis externa
mf, muscle fiber
mg, esophageal mucosal gland
mm, tunica muscularis mucosae
mmt, myometrium
mng, prostatic main glands
mol, molecular layer
mr, medullary rays of Ferrein
ms, medullary sinus
msp, mesosalpinx
mu, mucous acinus
mv, muscular venule
mx, matrix
nab, Nabothian follicle
nfl, layer of nerve fibers (retina)
nhp, neurohypophysis
nv, nerve
obr, outer border cells of Hansen
oc, oocyte
ocl, osteoclast
ocn, ossification center
ocp, renal capsule, outer layer
ocr, outer circular layer
od, odontoblast cells
ohr, outer hair cells
oln, outermost longitudinal layer
om, outer muscularis externa
onl, outer nuclear layer (retina)
ooc, orbicularis oculi muscle
oph, outer phalangeal cells
opl, outer pillar cells
opl, outer plexiform layer (retina)
opo, osteogenic periosteum
opt, optic nerve
os, osteon
osl, osseous spiral lamina
ot, outer spiral tunnel
oxc, oxyphil cells, parathyroid
pa, Pacinian corpuscle
pca, post. cricoarytenoid muscle
pch, perichondrium
pct, proximal convoluted tubule
pcu, pars cutanea
pcx, paracortex
pfn, posterior funiculus
pg, pyloric gland
pia, pia mater
pl, periosteal lamellae
pm, perimysium
pmt, perimetrium
pmu, pars mucosa
po, periodontal ligament
poc, posterior chamber of the eye
pp, pampiniform plexus
prc, principal cells, parathyroid
prn, perineurium
prp, prepuce (foreskin)
ps, primary spermatocyte
pt, portal vein
ptm, peritubular myoid cells
pu, dental pulp
pv, postcapillary venule
py, pyriform recess
ra, Rokitansky-Aschoff sinuses
rac, layer of rods & cones (retina)
rcl, renal column
rpl, renal pelvis
rpp, renal papilla
sas, subarachnoidal space
sb, sebaceous gland
sba, stratum basale
sc, subcutis
scl, sclera
sco, stratum corneum
se, tunica serosa
ser, Sertoli cell
sft, seminiferous tubules
sg, spermatogonia
sgr, stratum granulosum
sk, skeletal muscle
sl, salivary gland
slu, stratum lucidum
sm, tunica submucosa
smc, caecal submucosa
smg, prostatic submucosal glands
smi, ileal submucosa
smm, smooth muscle
sn, sinusoid
spf, spaces of Fontana
spg, spiral ganglion
spl, spiral ligament
spp, sphincter pupillae muscle
sr, serous acinus
ss, subcapsular sinus
ssp, stratum spinosum
st, striated duct
stn, superior thyroid notch
str, stroma
sty, scala tympani
sve, scala vestibuli
sw, sweat gland
ta, tunica adventitia
tal, tunica albuginea
tb, terminal bronchiolus
tc, tectorial membrane
teg, thyroepiglottic muscle
teg, thyroepiglottic muscle
tgl, thyroid gland
tha, thyroarytenoid muscle
the, theca externa
thf, thyroid follicles
thi, theca interna
thy, thyroid cartilage
ti, tunica intima
tm, tunica media
tpl, tarsal plate
tr, trabecule
tra, testicular artery
trc, tracheal cartilage
ts, trabecular sinus
uar, umbilical artery
ur, ureter
ut, prostatic utricle
uv, musculus uvulae
uvn, umbilical vein
vcf, vocal fold
vcl, vocal ligament
vcm, vocalis muscle
vd, vas deferens
ve, von Ebner's gland
vh, ventral horn
vl, villi
vm, vestibular membrane
vn, vein
vpr, vocal process
vrm, verumontanum
vs, venous sinus
vsf, vestibular fold
vtrm, ventral ramus, spinal nerve
vtrt, ventral root
vz, vermillion zone
whj, Wharton's jelly
wm, white matter
zc, zone of reserve cartilage
zfc, zona fasciculata
zgl, zona glomerulosa
zh, zone of hypertrophy
zo, zone of ossification
zp, zone of proliferation
zrt, zona reticularis

Chapter 1

Basic Tissues

Epithelium

Functionally, epithelium can be subdivided into two main types: covering (surface) epithelium and glandular epithelium. Surface epithelium covers mucous and serous membranes of tubular (hollow) organs as well as the outer surface of the body as epidermis. Structurally, surface epithelium can be composed of a single layer of epithelial cells (simple epithelium) or multiple layers of cells (stratified epithelium). Epithelial cells forming the simple epithelium can be squamous, with flattened nuclei, cuboidal with round-shaped centrally located nuclei and columnar with cylindrical nuclei. These cells form the top layer of stratified epithelia as well; consequently, they define the stratified squamous, stratified cuboidal and stratified columnar epithelia, while the underlying layers are usually composed of cuboidal (polygonal) cells. Stratified squamous epithelia can be keratinized, when the surface cells lose their nuclei and they become filled with keratin; in some cases, the keratinized layer (stratum corneum) retains some of the nuclei resulting in parakeratinized epithelium. The flattened superficial cells of the stratified squamous nonkeratinized epithelium do not show signs of keratinization. A special type of the stratified epithelia is the transitional epithelium or urothelium, due to its location in the urinary system; here, a single layer of large umbrella cells covers the surface. The appearance of the umbrella cells depends on the filling of the urinary tract; when the mucous membrane is stretched, these cells appear to be more flattened while the relaxed mucous membrane tends to form folds with barrel-shaped umbrella cells. Umbrella cells have round-shaped or slightly oval nuclei, cells with two nuclei are common. A specialized form of simple epithelia is the pseudostratified epithelium that is composed of columnar cells that all rest on the basement membrane; however, in contrast of the simple columnar epithelium, their nuclei are not aligned at the same level. The luminal surface of the columnar cells in simple, stratified, and pseudostratified epithelia is often covered by microvilli or cilia.

Glandular epithelia can be functionally categorized as endocrine glands and exocrine glands; the secretion is the former drains into the blood stream while the secretory product of the latter drains on the surface of the mucous membrane. Multicellular exocrine glands can be classified as tubular or alveolar (acinar) glands according to the shape of the base of the gland; they can also be simple (unbranched) or compound glands with branched ducts. The most common form of secretion of exocrine glands is the merocrine (eccrine) secretion when the cell does not lose substantial amounts of cytoplasm during the secretion; vesicles in the cytoplasm fuse with the luminal surface of the cells and empty their content there (exocytosis). Merocrine secretion is characteristic to the salivary glands, where the basal sac of the gland (acinus) forms serous (dark cells with central rounded nuclei) or mucous acini (light cells with more flattened, peripheral nuclei). Occasionally, a serous cap can be observed on the surface of the mucous acinus that represents mixed, seromucous acini with both serous and mucous cells. These serous demilunes are composed of serous cells that are squeezed from between the mucous ones that swell during the histological procedure. Goblet cells are mucous, typically unicellular glands that appear in columnar epithelia. The distended apical portion of these cells contains mucus and it is lined with short microvilli while the thinner basal part contains the nucleus. In contrast to the merocrine secretion, in apocrine secretion the entire apex of the columnar cell, filled with the secretory product, is discharged, and therefore, the shape of the secretory cell varies between columnar and cuboidal. This type of secretion is characteristic for the lactating mammary gland; apocrine sweat glands exhibit the characteristic features of apocrine secretion, but consecutive studies revealed that they secrete with merocrine mechanism. Finally, during holocrine secretion the entire secretory cell dies forming the secretory product itself; this type of secretion can be observed in sebaceous glands.

http://doi.org/10.1016/B978-0-323-91891-6.50001-5,

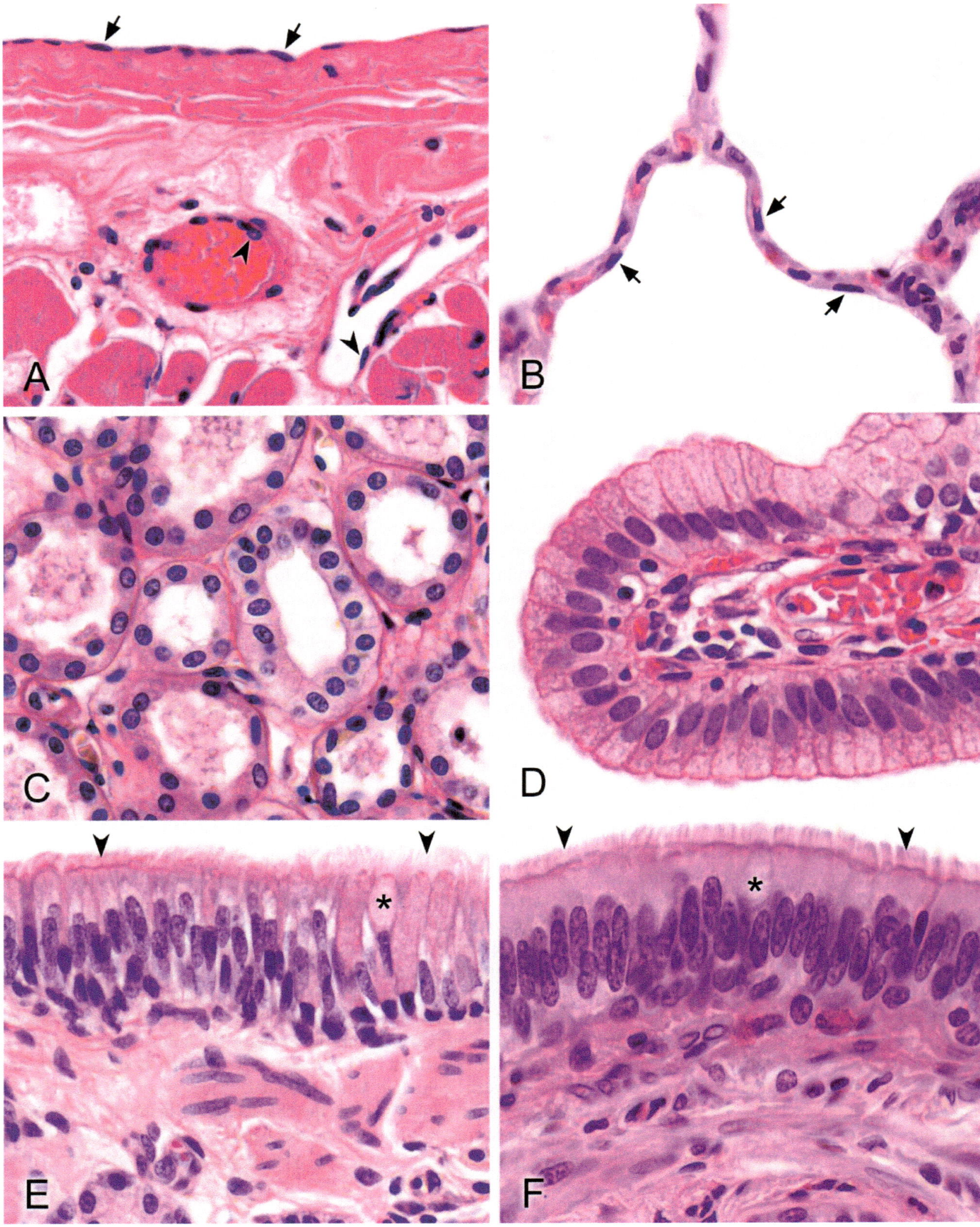

Figure 1. Simple epithelia. (A) Simple squamous epithelia: mesothelium (arrows) and endothelium (arrowheads). (B) Simple squamous epithelium (arrows) lining the lung alveoli. (C) Simple cuboidal epithelium (kidney). (D) Simple columnar epithelium (gallbladder). (E, F) pseudostratified columnar epithelium of the respiratory system. Columnar cells are covered with kinocilia (arrowheads). Mucous secreting goblet cells are dispersed between the columnar cells (asterisks).

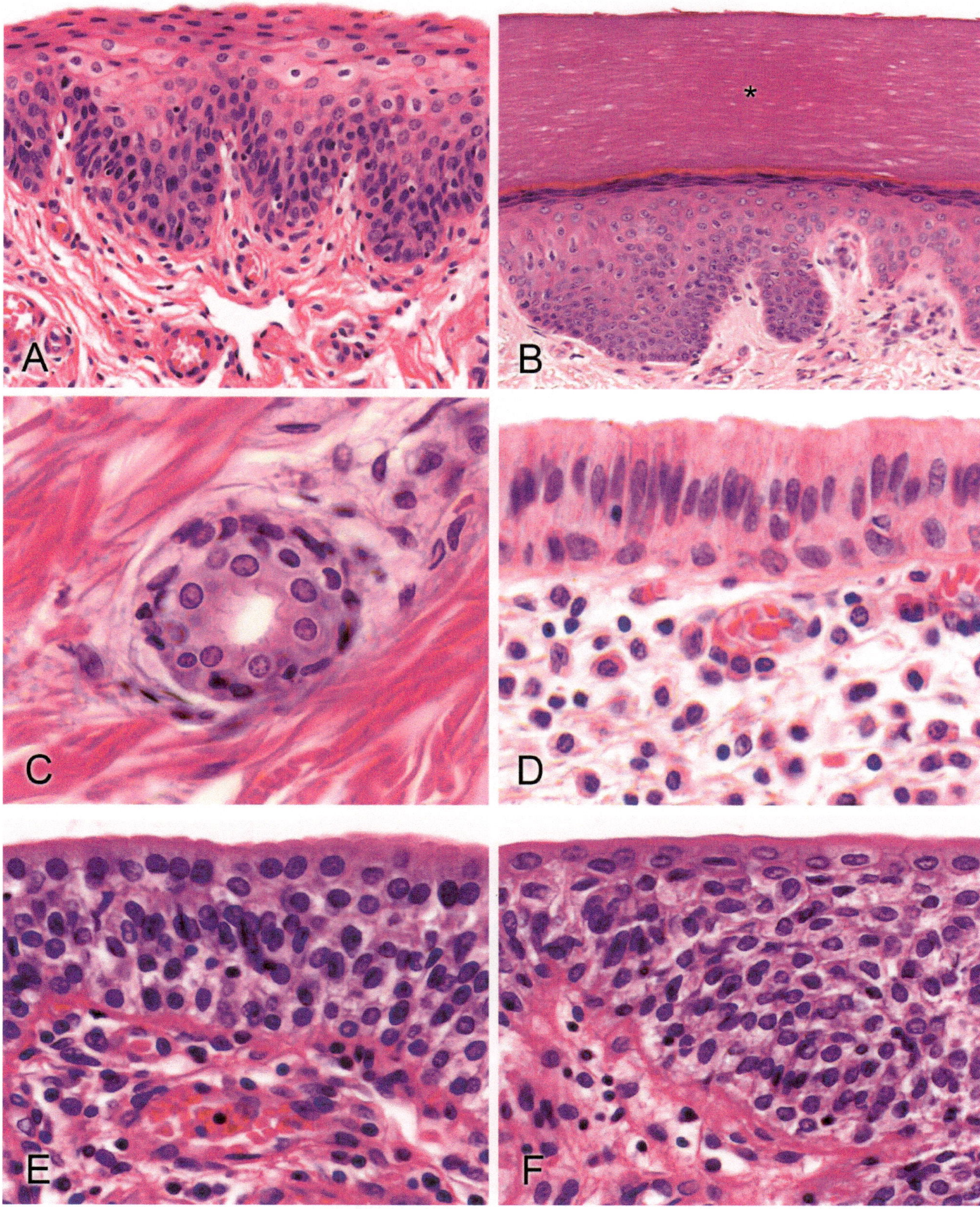

Figure 2. Stratified epithelia. (A) Nonkeratinized stratified squamous epithelium of the esophagus. (B) Keratinized stratified squamous epithelium (skin). Asterisk denotes the stratum corneum. (C) Stratified cuboidal epithelium (sweat gland duct). (D) Stratified columnar epithelium (excretory duct, salivary gland). (E, F) Transitional epithelium (urothelium) in relaxed (E) and stretched stages (F). Luminal surface is covered by a single layer of large umbrella cells.

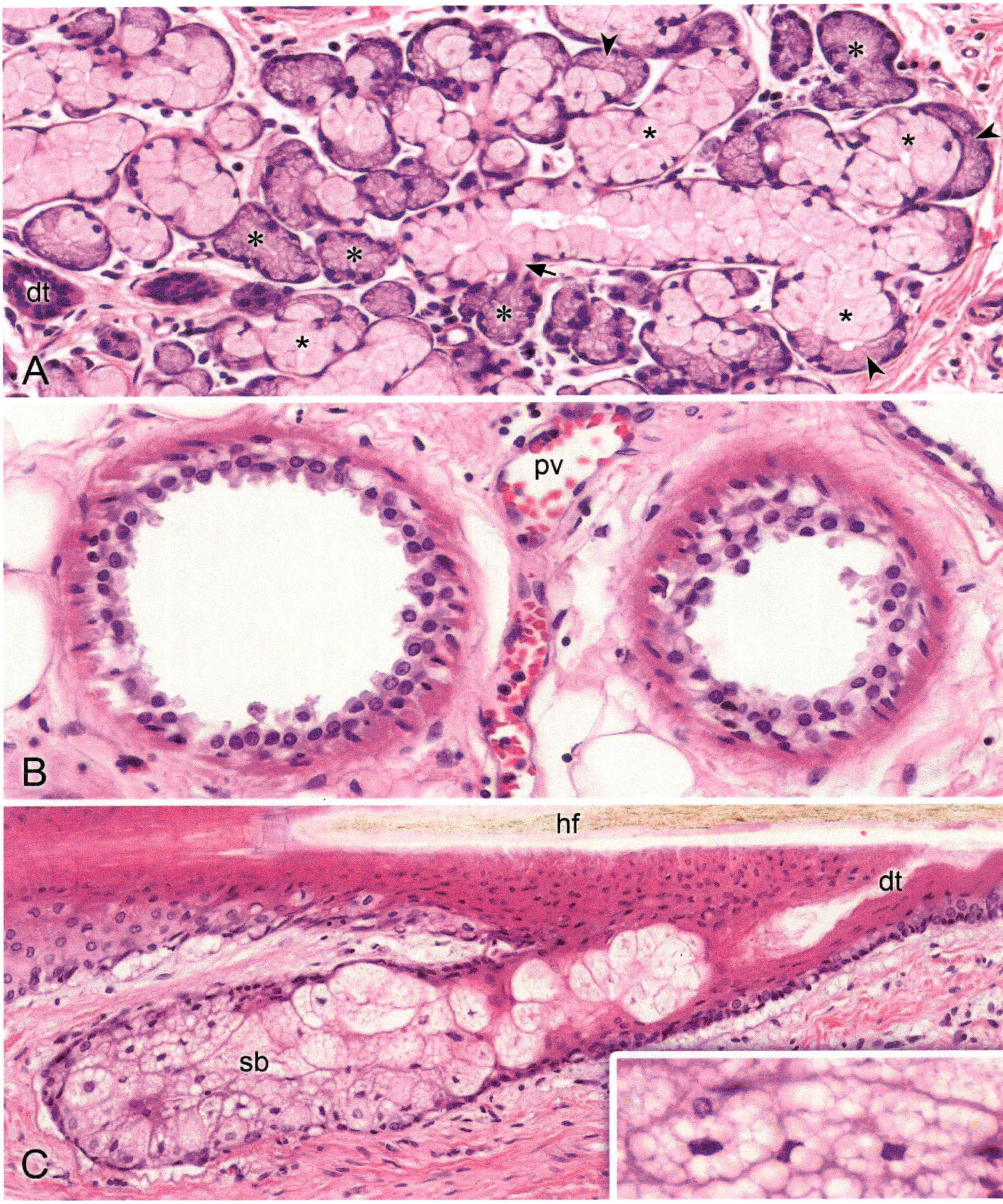

Figure 3. Glandular epithelia. (A) Salivary glands are merocrine glands formed by serous (six-pointed asterisks) and mucous acini (five-pointed asterisks) that drain to the surface through ducts (dt). Arrowheads denote serous demilunes that are serous caps on mucous acini and represent seromucous acini. Arrow points to an intercalated duct. (B) Although apocrine sweat glands have merocrine secretion, they exhibit typical apocrine morphology, when the apical parts of the secretory cells are sloughing off during the secretion. Vessels, including postcapillary venules (pv) can be observed in the surrounding connective tissue. (C) Holocrine sebaceous glands (sb) open into hair follicles (hf) via ducts (dt). Cells have shrunken nuclei and accumulate lipid droplets (inset); they eventually die forming the sebum.

Muscle tissue

Muscle tissue is capable of contraction. Skeletal muscle is composed of multinucleated muscle fibers that can be as long as the muscle itself and contain flattened nuclei under the sarcolemma (plasma membrane). The muscle fiber contains contractile myofibrils that exhibit dark A (anisotropic) and light I (isotropic) bands forming a cross striation along the muscle fiber. A thin connective tissue envelope called endomysium surrounds individual muscle fibers, while a thicker perimysium envelopes fiber bundles and the entire muscle is ensheathed by dense connective tissue epimysium. Muscle fibers in the skeletal muscle are innervated by nerve fibers that form motor end plates or neuromuscular junctions on the sarcolemma that operate with cholinergic neurotransmission.

Cardiac muscle forms the myocardium of the heart, and it is composed of branching cardiac muscle cells that form a complex network. The boundaries between the cells appear as intercalated discs. Cardiac muscle cells exhibit delicate cross striation and contain an axially located round-shaped or slightly oval nucleus; occasionally cells may contain two nuclei. Cardiac muscle is very vascular, with a rich capillary network between the cells.

Smooth muscle is composed of spindle-shaped cells that are densely packed against each other. Nuclei are spindle-shaped and pointy and they are located at the axis of the cells; consequently, the transverse section of the cells contains round-shaped cross sections of the nuclei. Smooth muscle is located mainly in the wall of the tubular organs in the gastrointestinal tract and in the urinary and respiratory systems as well as in the wall of vessels.

Myoepithelial cells are branching contractile elements that typically surround the base of the exocrine glands above the basement membrane of the secretory cells and facilitate the release of the secretory product. They contain elongated nuclei as well as cytokeratin filaments.

Connective tissue

Connective tissues have primarily structural (connecting) and space-filling roles. Loose connective tissue can be found in multiple organs as part of the mucous and serous membranes (lamina propria mucosae or serosae). It is composed of a loose mixture of cells and fibers embedded into a ground substance that contains a large amount of glycosaminoglycans. Transient cell types, including granulocytes, monocytes and lymphocytes migrate into the tissue from blood, while the permanent cell types are the primary structural and functional components of the loose connective tissue. These cells include fibroblasts and their less active form, the fibrocytes, that are spindle-shaped cells with multiple processes and oval (disc)-shaped nuclei as well as adipose cells (adipocytes) filled with a single droplet of lipid (white adipose tissue) or multiple lipid droplets (brown adipose tissue), mastocytes (mast cells, originating from German Mastzelle) with large basophilic granules and rounded nucleus that is not obscured by the granules, and pluripotential mesenchymal cells. Fibroblasts are responsible for the formation of collagen and elastic fibers of the connective tissues. These fibers stain red or pink with eosin; elastic fibers can also be stained specifically with orcein (brown) and resorcin (bluish gray).

A special type of loose connective tissue is the adipose tissue that is almost exclusively formed by white or brown adipocytes; the latter type is characteristic to newborns and responsible for rapid energy release.

Dense connective tissue is composed of closely packed type I collagen fiber bundles with tightly squeezed fibroblasts with rod-shaped nuclei between them. These fibers can run parallel to each other, forming dense regular connective tissue that can be observed intendons, ligaments, and capsules of organs. In the submucosa of the tubular organs and the dermis layer of the skin, type I collagen fibers are running in multiple directions, forming dense irregular connective tissue. A special type of connective tissue contains excess amount of elastic fibers that can be observed in the ligamentum flavum and the tunica media of the large elastic arteries (elastic connective tissue).

Mucous connective tissue is located in the umbilical cord (Wharton's jelly) and in the embryo (mesenchyme). Wharton's jelly is composed of loosely arranged network of fibroblasts and occasional mesenchymal cells embedded into a large amount of viscous, jelly like ground substance, while mesenchyme is mostly formed by star-shaped, pluripotential mesenchymal cells that can also be observed postnatally, primarily around small vessels.

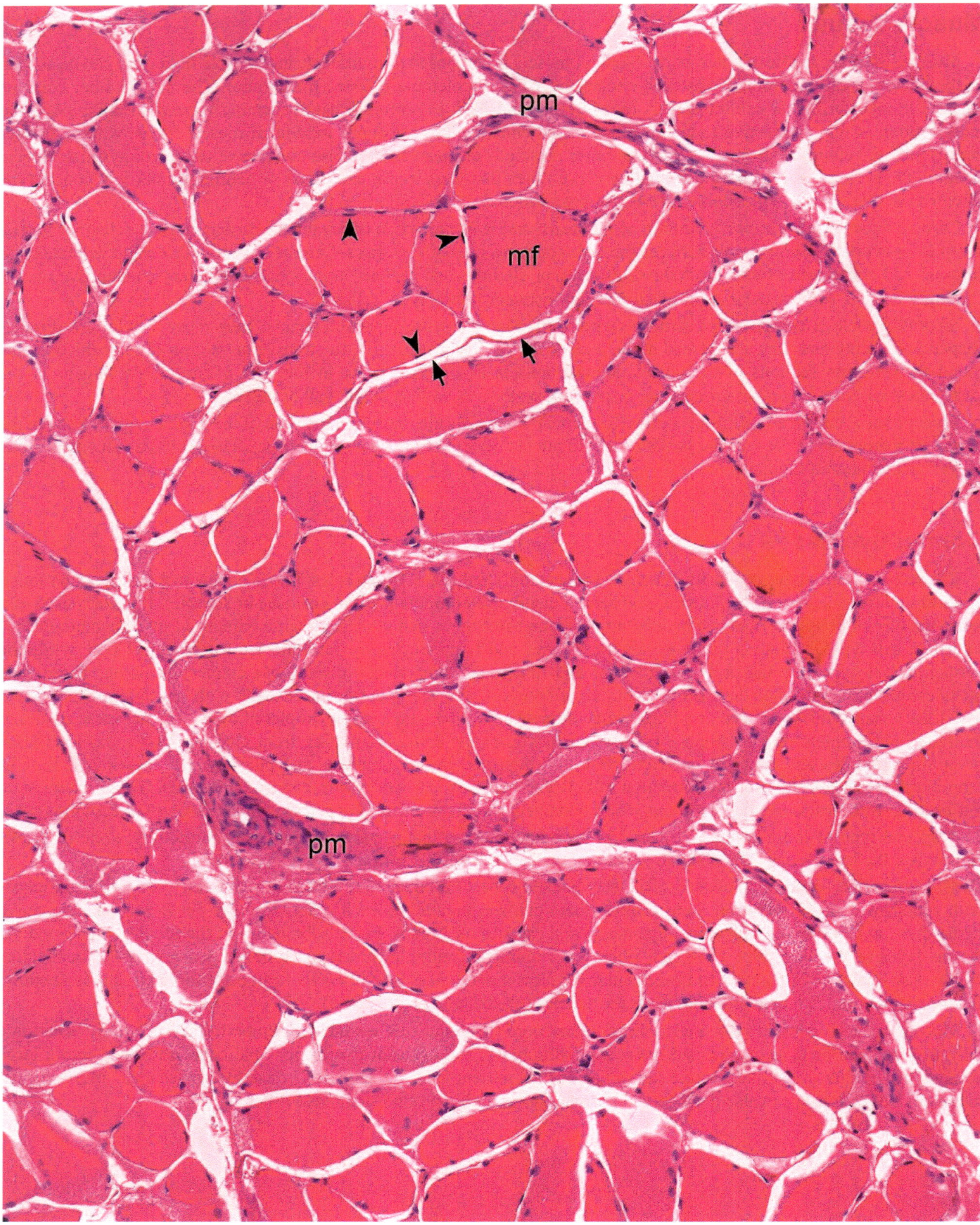

Figure 4. Skeletal muscle, cross section. Skeletal muscle is composed of multinucleated muscle fibers (mf) with peripheral flattened nuclei (arrowheads). Bundles of muscle fibers are surrounded by a connective tissue perimysium (pm), while individual fibers are ensheathed by endomysium (arrows).

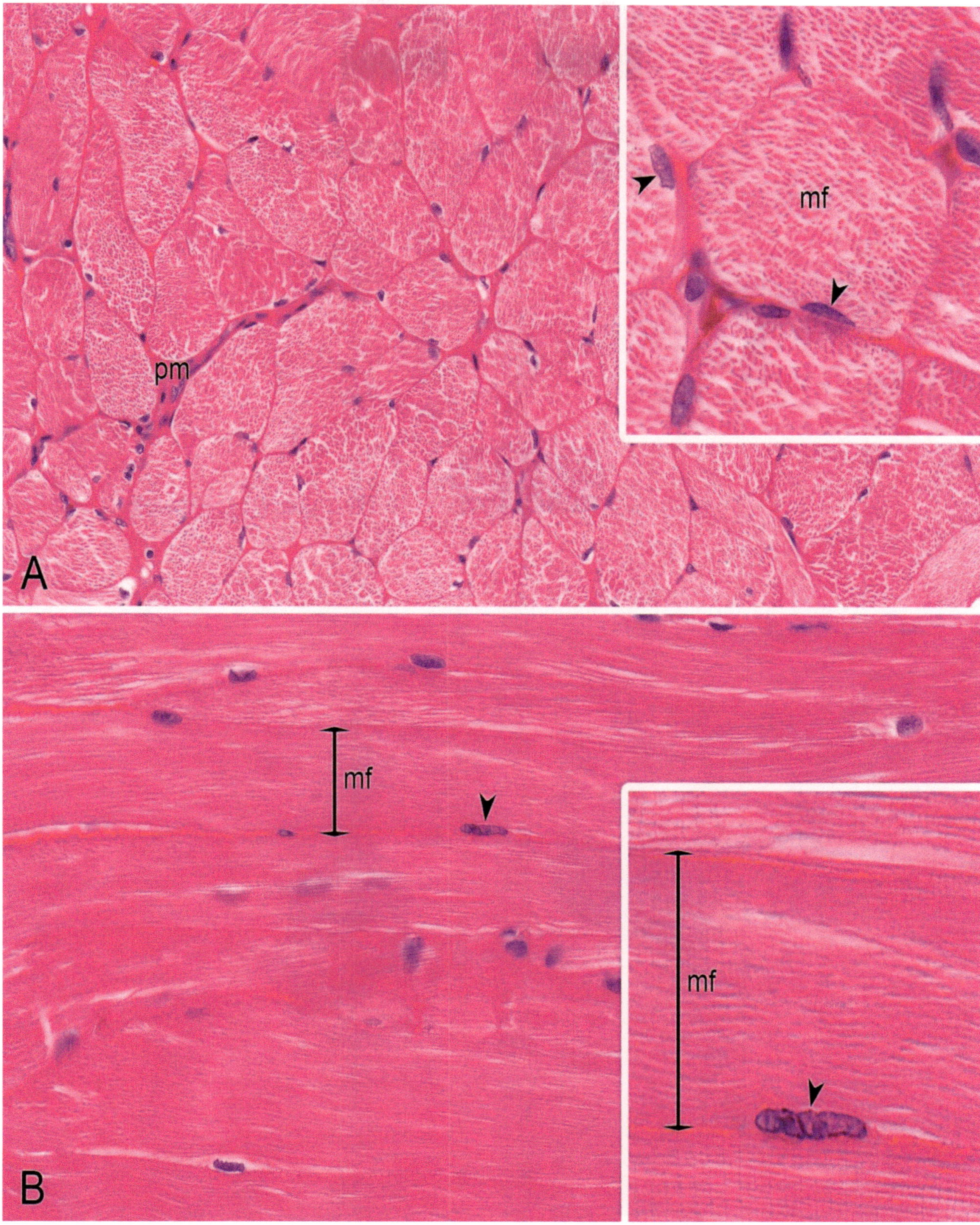

Figure 5. Skeletal muscle. (A) Cross section of the muscle fibers (mf) reveals the muscle fibrils inside the sarcoplasm (upper inset). Perimysium (pm) surrounds bundles of muscle fibers. (B) Longitudinal section of the muscle fibers (mf) reveals the cross striation (lower inset) that is made of periodic alternation of dark (A) and light (I) bands. Multinucleated muscle fibers contain peripheral, flattened nuclei (arrowheads).

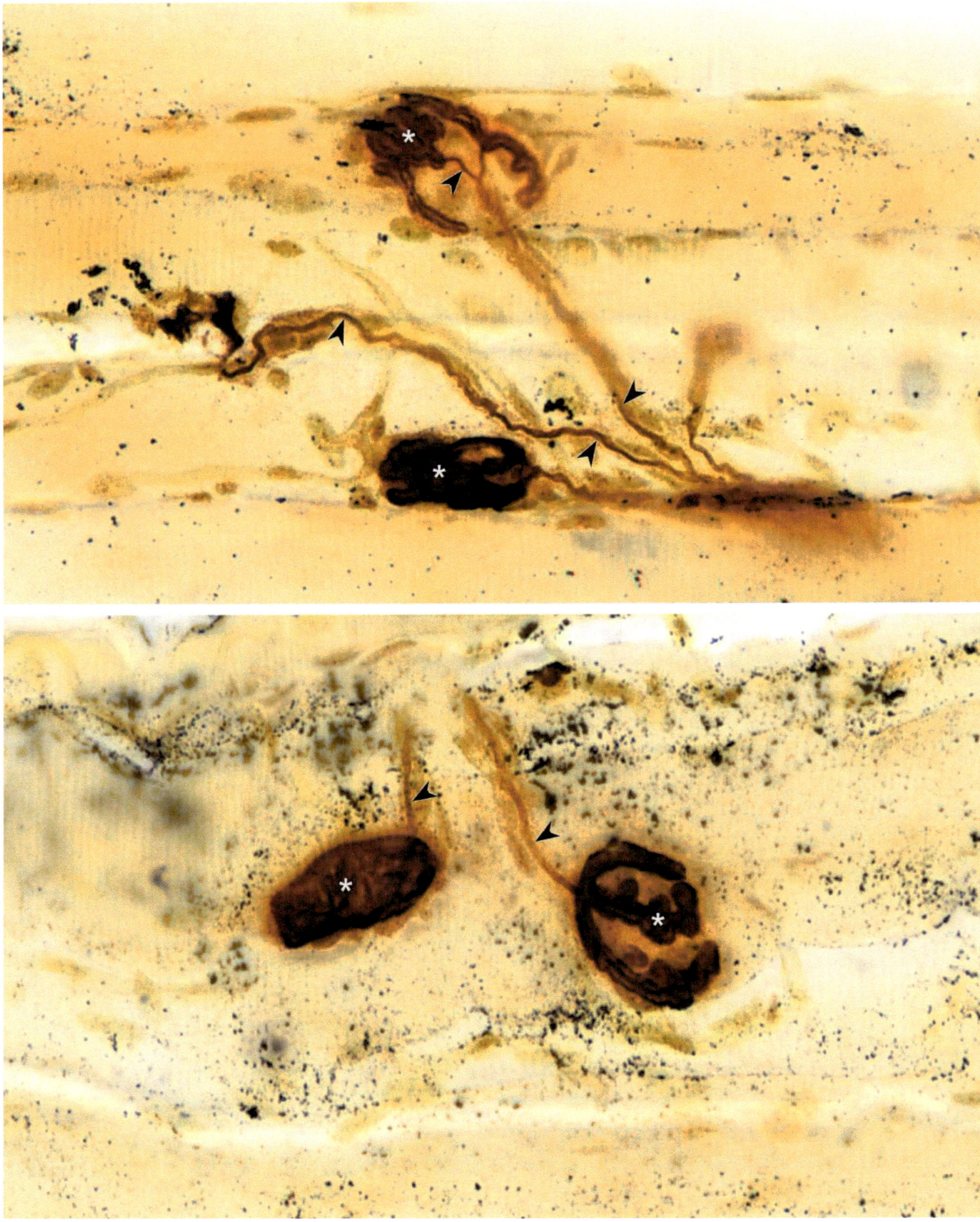

Figure 6. Neuromuscular junctions (motor end plates) of the skeletal muscle. Nerve fibers (arrowheads) terminate in motor end plates (asterisks) while making contact with the muscle fiber. Neuromuscular junctions are synapses working with cholinergic neurotransmission.

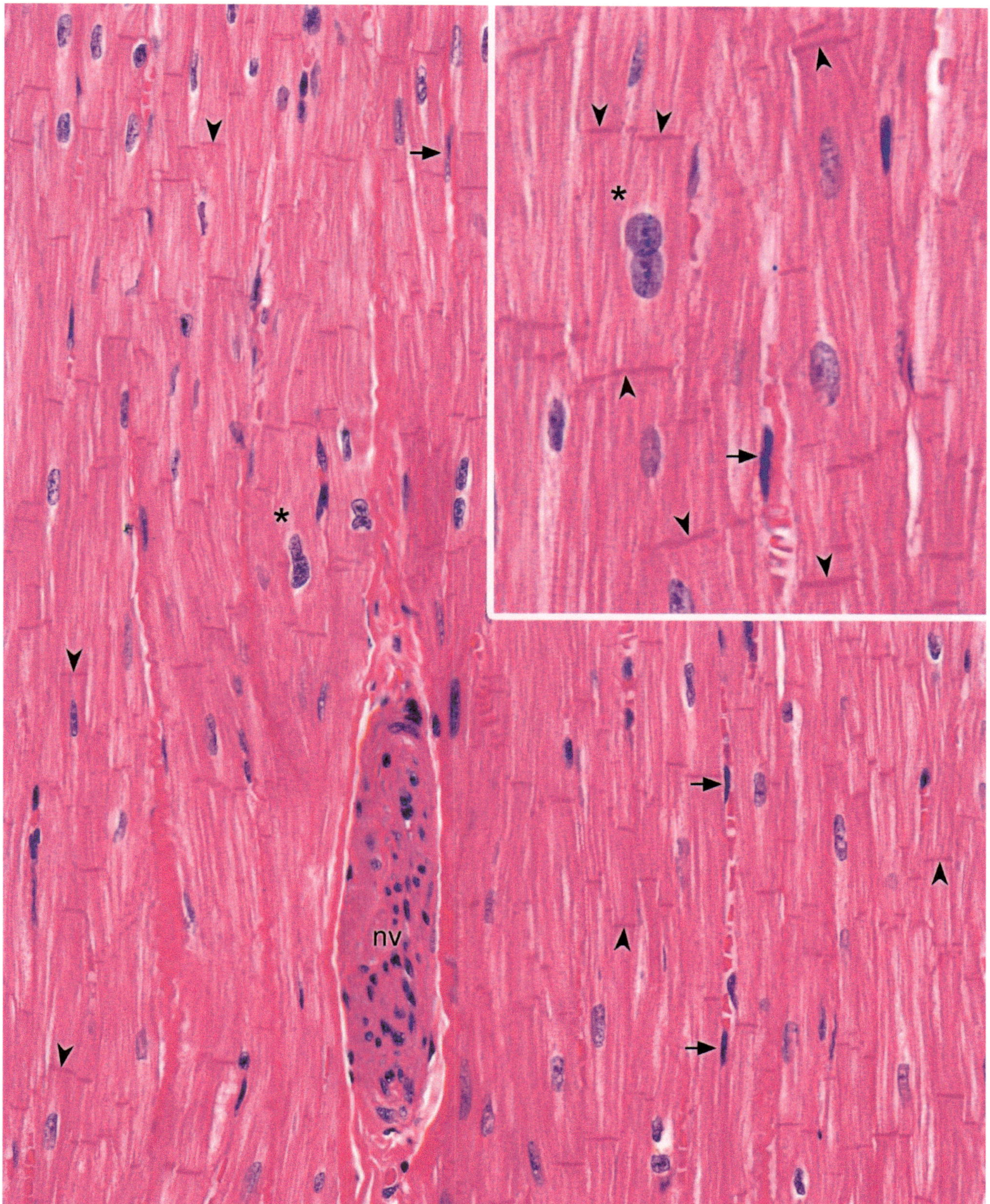

Figure 7. Cardiac muscle, longitudinal section. Branched cardiac muscle cells are connected by intercalated discs (arrowheads). Central nuclei are spherical or oval-shaped and many cells are binucleated (asterisks). Cells exhibit a delicate cross striation (inset). Cardiac muscle is vascular tissue with rich capillary network lined with endothelial cells (arrows). Occasional nerves (nv) can be observed between the cardiac muscle cells.

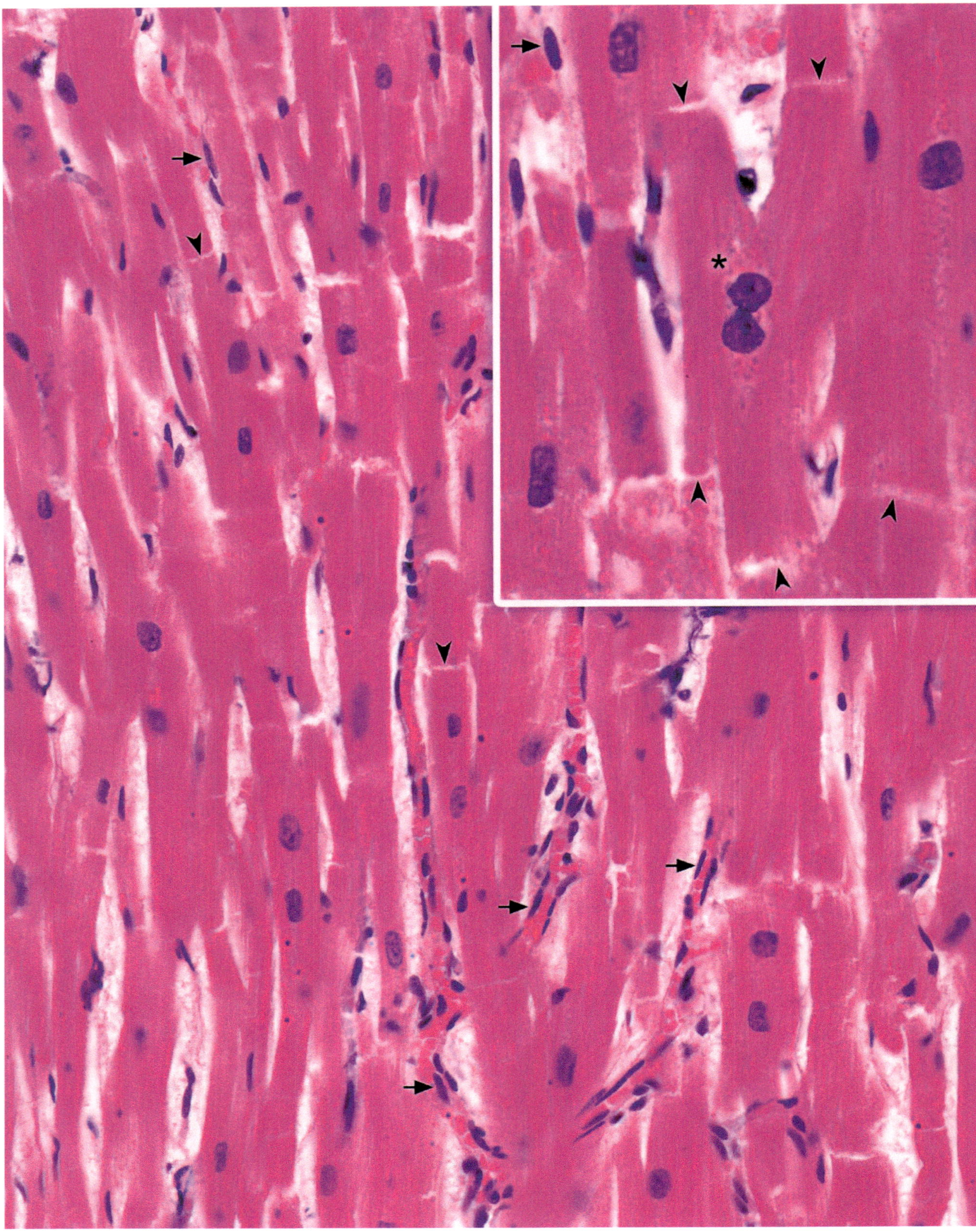

Figure 8. Cardiac muscle, longitudinal section. Cardiac muscle cells may be artificially separated along the intercalated discs (arrowheads) during the sectioning of the tissue. Cells are branched (inset), central nuclei are spherical or oval-shaped and many cells are binucleated (asterisk). Cells exhibit a delicate cross striation. Cardiac muscle contains numerous capillaries lined with endothelial cells (arrows).

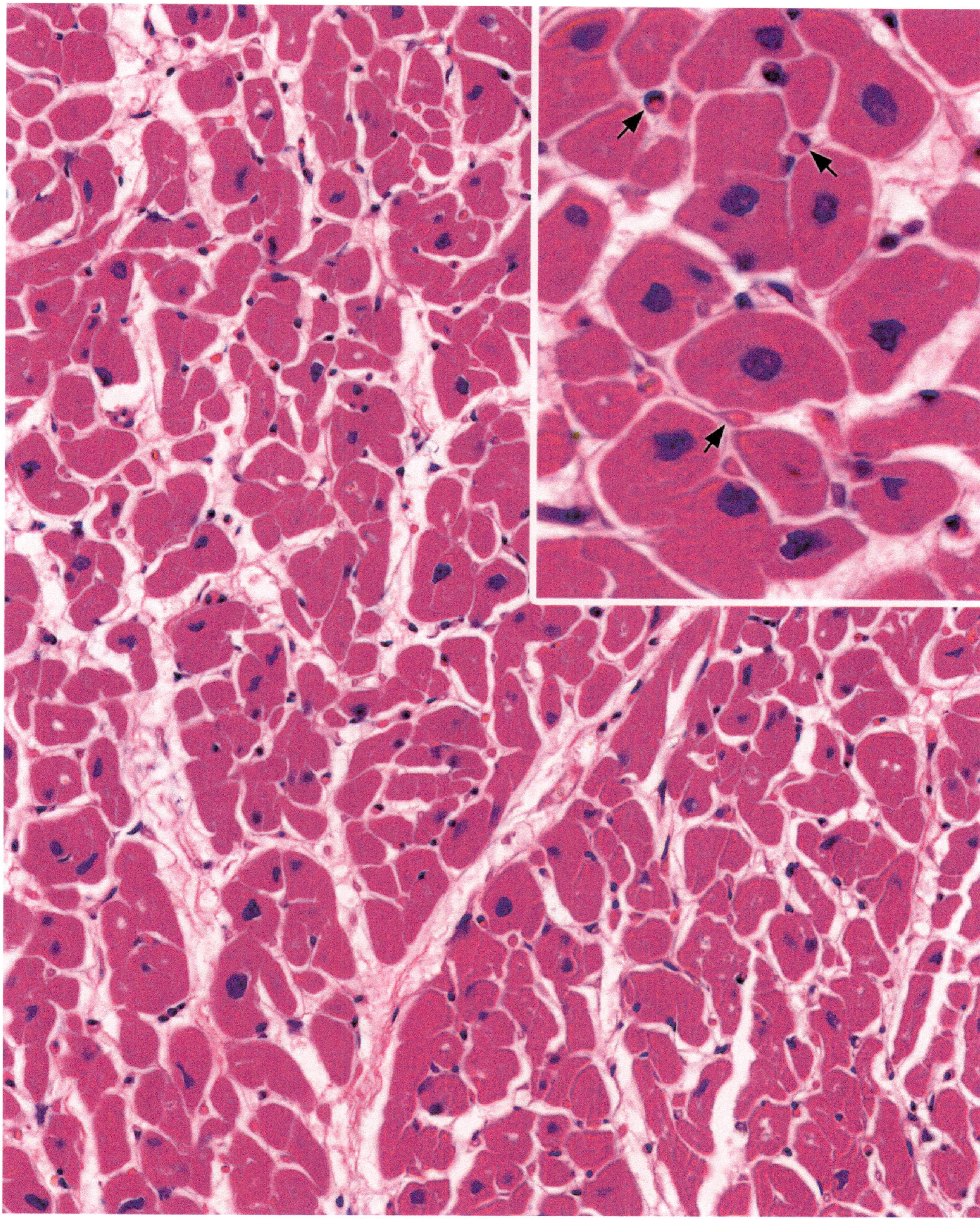

Figure 9. Cardiac muscle, cross section. In cross section, cardiac muscle cells exhibit centrally located spherical nuclei. Cells are ensheathed by delicate connective tissue envelope with rich capillary network (arrows).

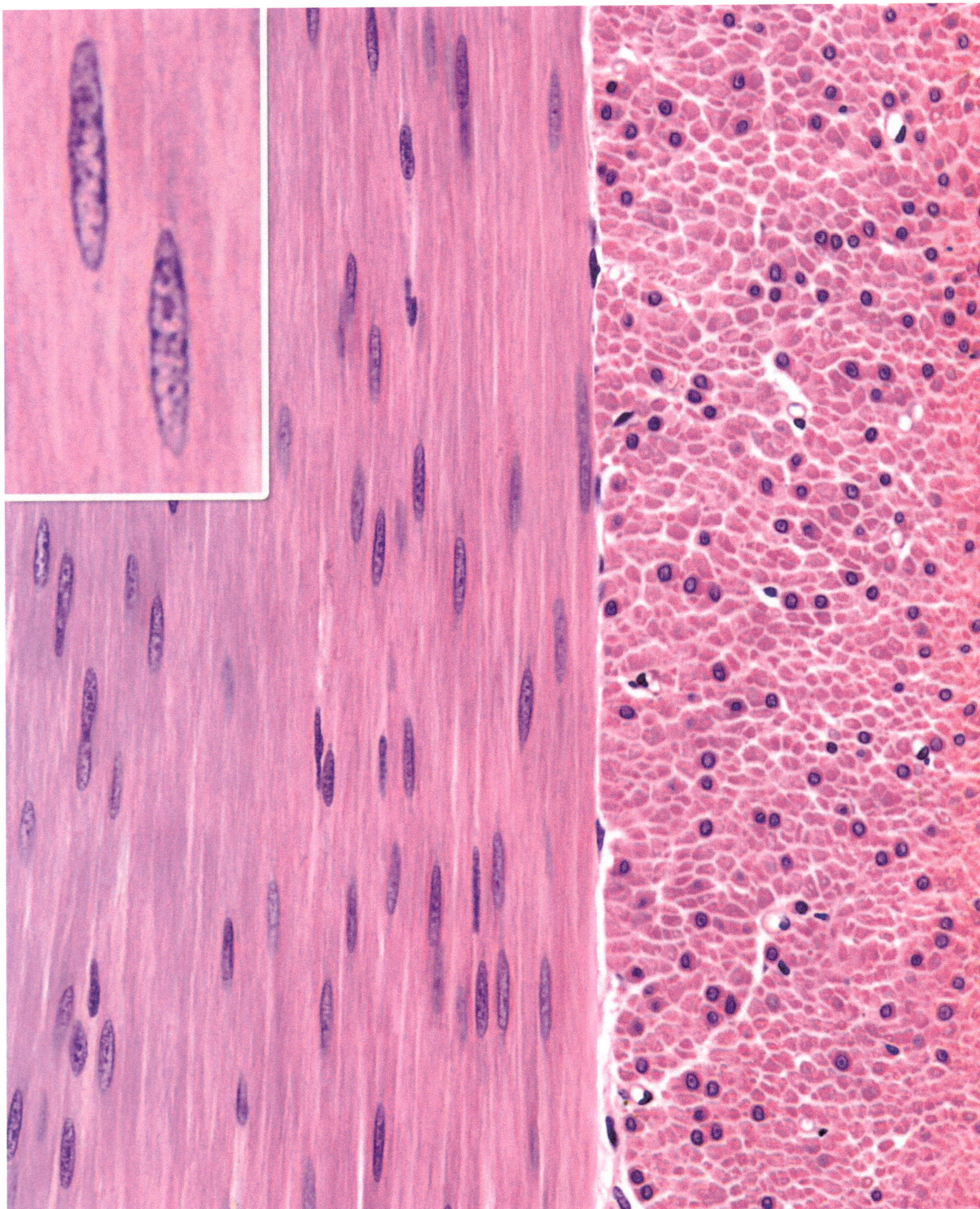

Figure 10. Smooth muscle, intestinal wall. Longitudinal section of the smooth muscle cells (left side) reveals centrally located, spindle-shaped nuclei with pointy endings. The same spindle-shaped muscle cells look entirely different in cross section on the right; elongated nuclei that are in the plane of the section appear to be spherical.

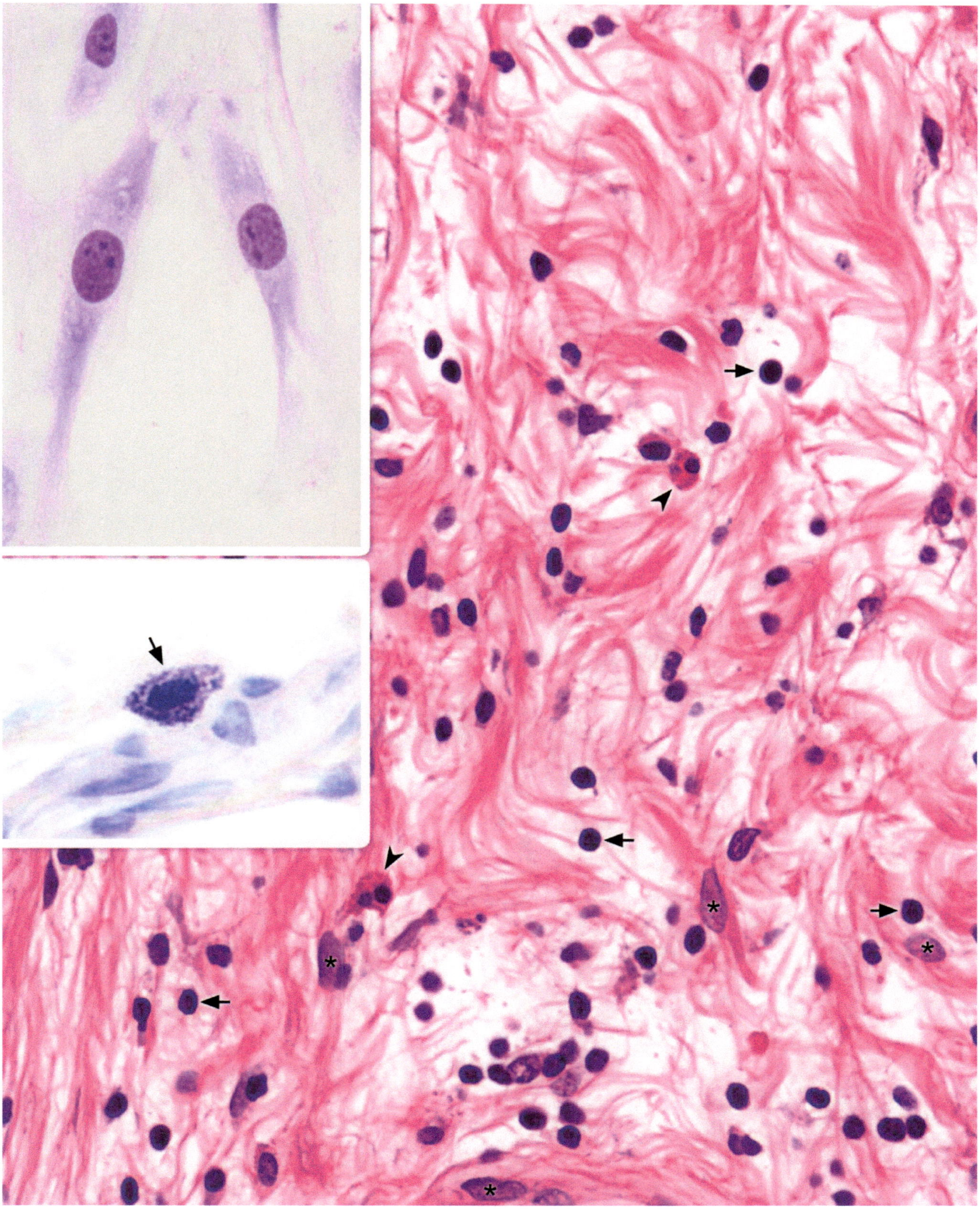

Figure 11. Loose connective tissue. Loose connective tissue is composed of a loose network of cells and collagen and elastic fibers. The dominant cell type is the fibroblast (upper inset) that have multiple processes and an elongated, oval-shaped nucleus when it is observed in loose connective tissue (asterisks). Other cell types include lymphocytes (arrows), eosinophils (arrowheads), mastocytes (lower inset, arrow) and various resident and transient cells.

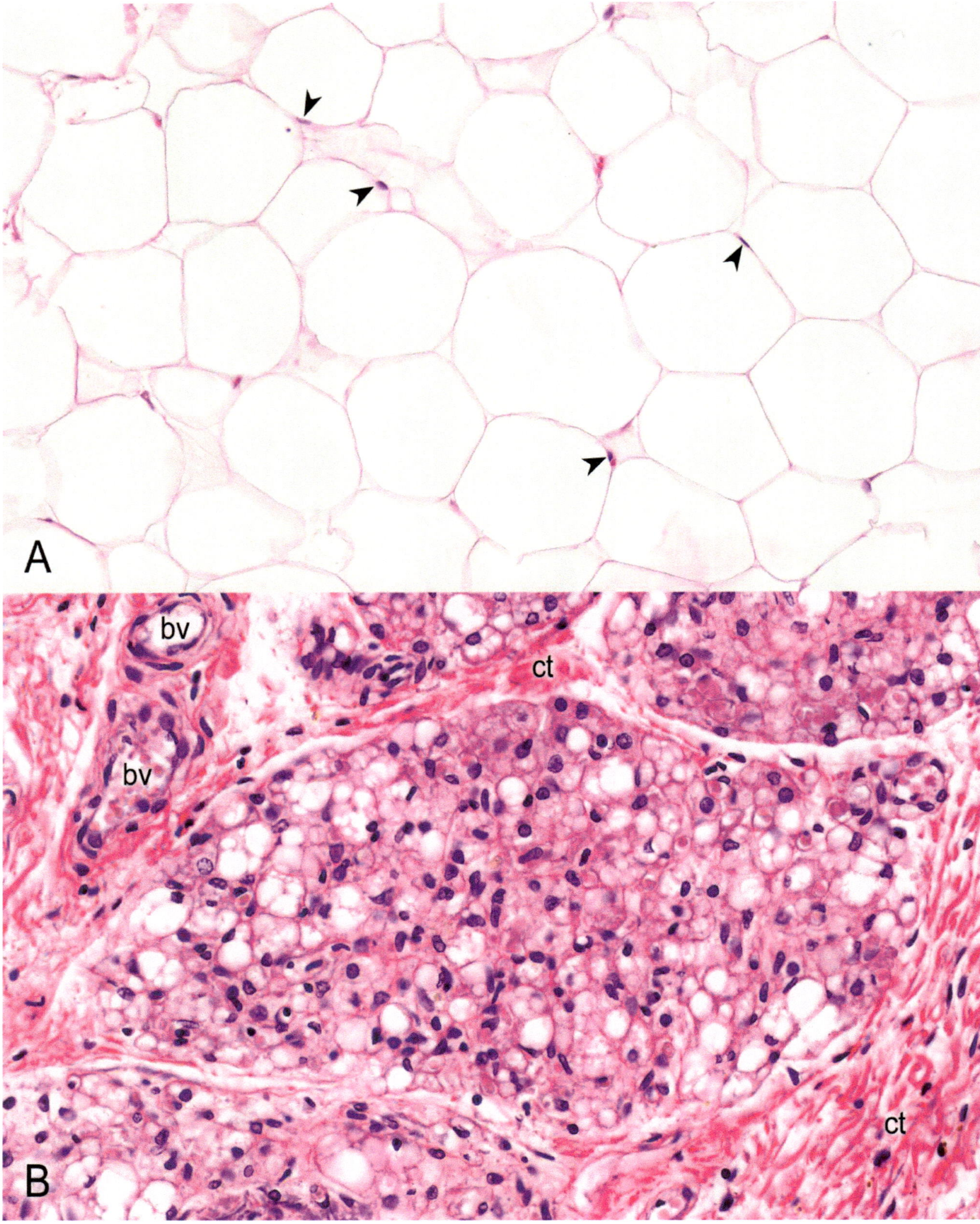

Figure 12. Adipose tissue. (A) White adipose tissue is formed by large adipocytes that contain a single lipid droplet that fills the entire cytoplasm. The cells appear to be empty because the lipid is dissolved during the common histological procedure. Nuclei are flattened and eccentric. (B) Brown adipose tissue is commonly located in the newborn, and it is very vascular; brown adipocytes contain multiple small lipid droplets and they are surrounded by connective tissue (ct) and blood vessels (bv).

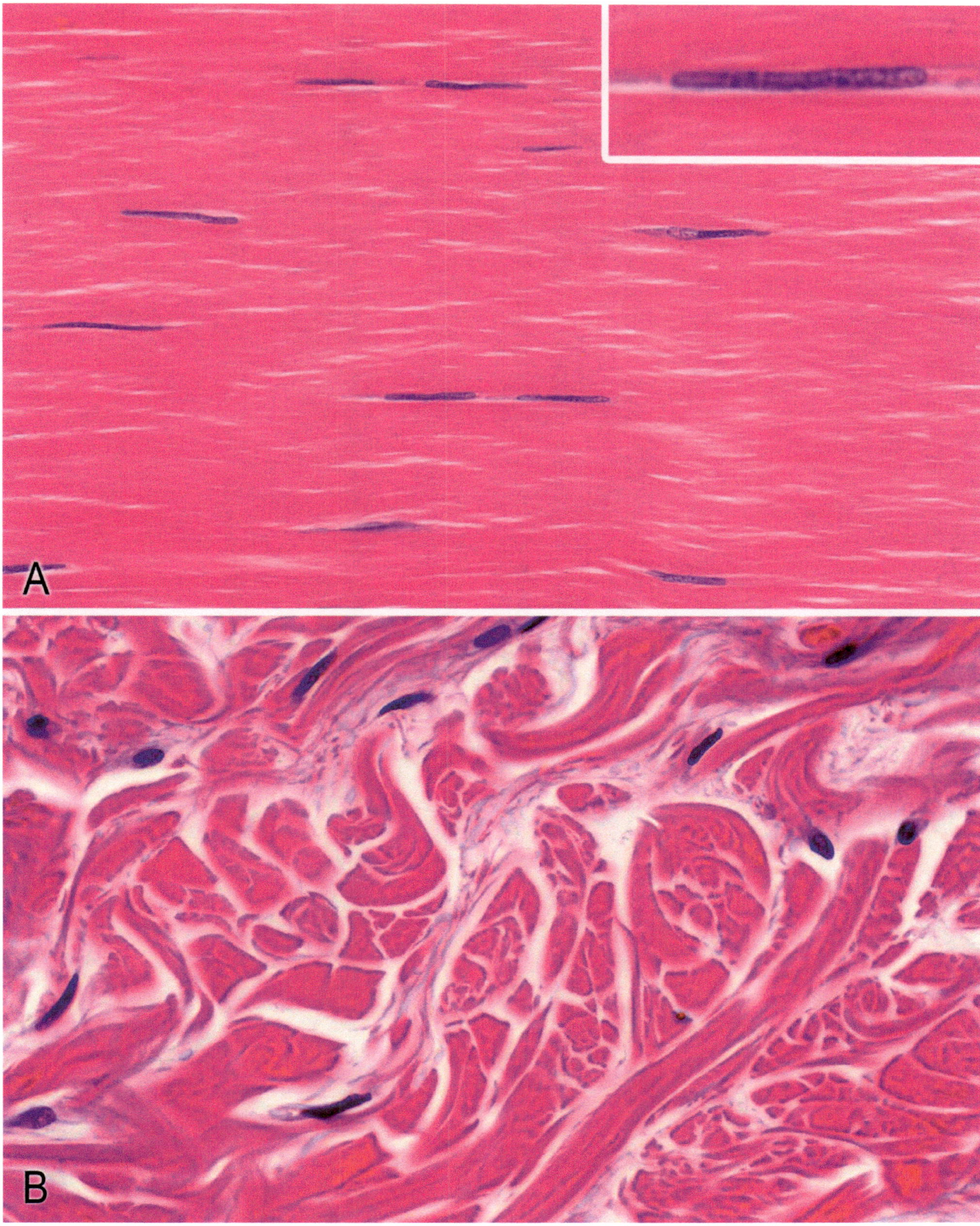

Figure 13. Dense connective tissue. (A) Dense regular connective tissue is composed of parallelly arranged, thick collagen type I fiber bundles and fibroblasts between with rod-shaped nuclei (inset). Dense regular connective forms tendons and ligaments. (B) Dense irregular connective tissue has the same constituents, but the collagen fibers are randomly oriented. Dense irregular connective tissue can be found in the tunica submucosa of the alimentary tract.

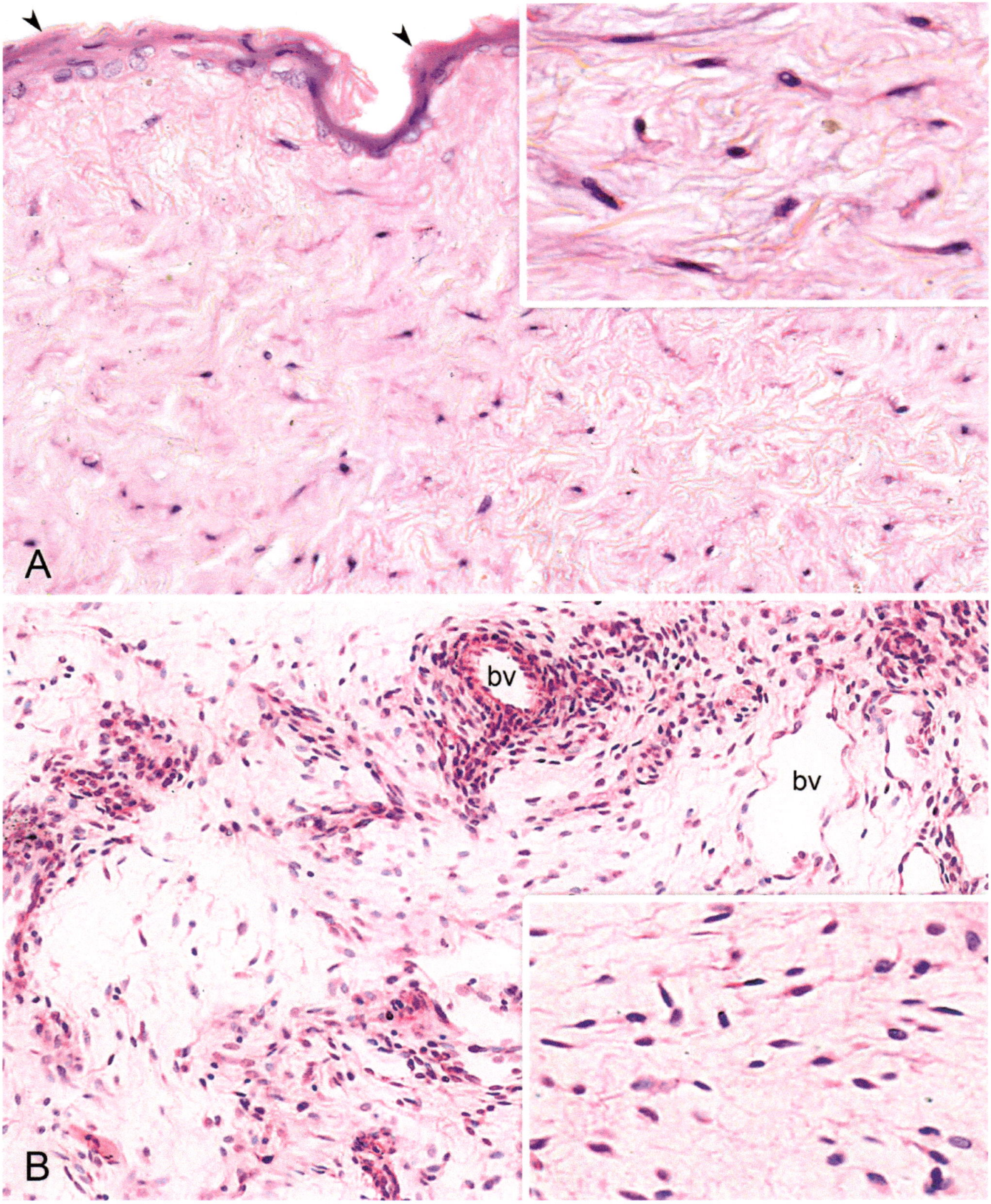

Figure 14. Embryonic connective tissue. (A) A relatively mature type of embryonic connective tissue is the mucous connective tissue or Wharton's jelly located in the umbilical cord that is covered with the amnion (arrowheads). Wharton's jelly is mostly composed of fibroblasts that resemble mesenchymal cells (upper inset) embedded into a gelatinous matrix with few fibers. (B) Mesenchyme can be found in the embryo often surrounding blood vessels (bv). It is less developed tissue than the Wharton's jelly and composed of mostly mesenchymal cells that possess long and thin processes (lower inset). Jelly-like ground substance with few thin fibers forms the extracellular matrix.

Cartilage

Cartilage is composed of chondrocytes, the primary cell types of cartilage; these are large, rounded cells, with a relatively small, round-shaped nucleus and pale cytoplasm. Chondrocytes often form clusters (isogenous groups) in mature cartilage that are surrounded by the ground matrix. According to the matrix, three types of cartilage can be distinguished. In hyaline cartilage the matrix is rigid and basophilic, containing type II collagen fibers that are completely obscured by the ground substance. Chondrocytes are sitting in small clusters in the cavities of the matrix (lacunae) and they are surrounded by a darker zone of matrix rich in glycosaminoglycans, the territorial matrix. The rest of the matrix between the isogenous groups is lighter and called interterritorial matrix. Hyalin cartilage can be observed in joint surfaces, in costal cartilage as well as in the larynx and trachea, and it also serves as a template for the developing bone. Isogenous groups of chondrocytes are less defined in elastic cartilage that contains delicate elastic fiber bundles observable with both eosin and orcein/resorcin staining and can be found in the epiglottis and the external ear (auricle). Fibrocartilage is the primary component of the intervertebral discs and the transitional zone between the tendon and the bone. Fibrocartilage contains of rows of chondrocytes between thick bundles of type I collagen stained with eosin; in this sense it resembles dense connective tissue, but instead if fibroblasts, it contains chondrocytes. Elastic cartilage and hyalin cartilage that does not cover joint surfaces are surrounded by perichondrium that has an outer dense connective tissue layer and an inner chondrogenic layer composed of mesenchymal cells. Since cartilage in human normally does not contain vessels, nutrition is provided by diffusion from vessels located in the perichondrium or at the periphery of the fibrocartilage.

Bone

Bone is a supportive tissue where the extracellular matrix is rigid, formed by hydroxyapatite crystals. The outer surface of the bone is covered by periosteum, and – similarly to the perichondrium – it is composed of an outer dense connective tissue layer and an inner osteogenic layer that contains osteoprogenitor cells that are responsible for the appositional growth of the bone. The inner surface of the bone facing the marrow cavity is also rich in osteoprogenitor cells that also line Haversian canals. The bone matrix is organized in lamellae that enclose spider-shaped osteocytes in the lacunae. Bone is continuously remodeled and therefore these lamellar systems are continuously changing shape and size. The primary unit of this rather dynamic system is the osteon (Haversian system) that is composed of concentrically arranged Haversian lamellae surrounding a single Haversian canal that carries vessels and nerves. The space between the cylindrical osteons is filled with the fragments of older Haversian systems that have been partially destroyed during the bone development; these are called interstitial lamellae. Finally, at the outer and inner surface of the bone shaft, under the periosteum and endosteum, several layers of outer and inner circumferential lamellae can be distinguished. Apart from the osteoprogenitor cells, and the spider-shaped osteocytes, there are two additional types of bone cells that can be observed primarily in the developing bone or during bone remodeling; multinucleated osteoclasts sitting in resorption bays are responsible for the bone resorption, and cuboidal osteoblasts forming dark basophilic lining on the developing bone trabeculae. The dark staining of these cells is due to the large amount of rough endoplasmic reticulum (rER) in their cytoplasm; indeed, these cells form the organic components of the bone matrix that can be observed under the basal surface of the cells as a lighter, more eosinophilic zone (osteoid) that later calcifies. During the bone development, osteoblasts are regularly enclosed into the developing lacunae, and they turn into osteocytes. This process can be observed during the bone formation that has two major types, the intramembranous and the endochondral ossification. Intramembranous ossification is characteristic for the fetal development of the neurocranium, and the process starts with a connective tissue plate, where mesenchymal cells turn into osteoprogenitor cells that in turn form osteoblasts responsible for the bone formation. Endochondral ossification can be observed in the epiphyseal plate during the longitudinal growth of the bone; here, hyalin cartilage (zone of reserved cartilage) is gradually replaced by bone at the ossification centers, exhibiting a characteristic layered structure. Chondrocytes first proliferate, forming tightly packed cell columns (zone of proliferation),and then they undergo hypertrophy exhibiting pycnotic nuclei (zone of hypertrophy), and eventually die and replaced by bone (zone of ossification) laid down by the osteoblasts on the interstitium among the dying chondrocytes.

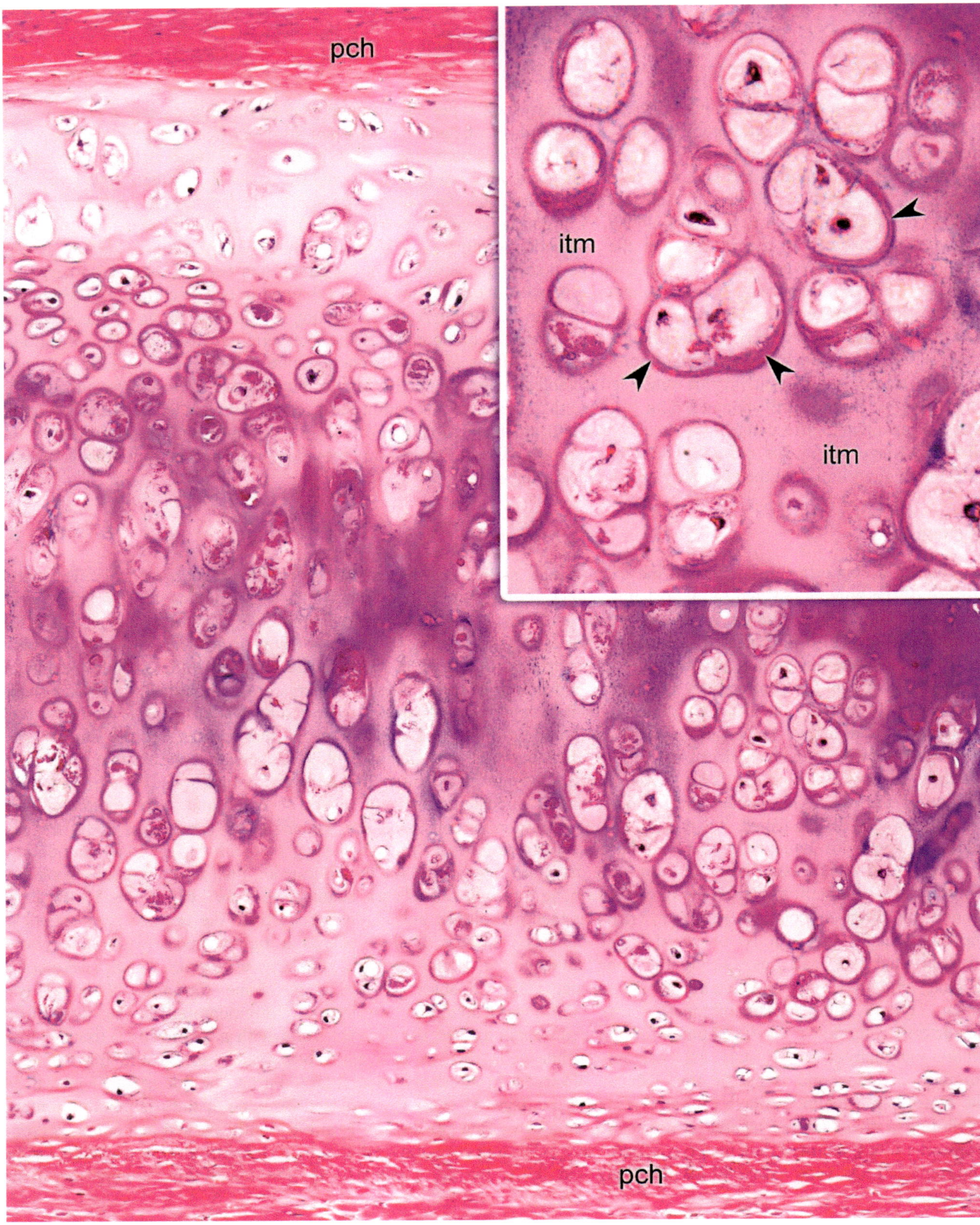

Figure 15. Hyalin cartilage. Mature hyalin cartilage is avascular tissue with rigid amorphous intercellular substance and a connective tissue coverage, the perichondrium (pch). Chondrocytes are large cells with pale cytoplasm and small spherical nucleus; the cells are arranged in isogenous groups, and they are located in the cavities of the matrix (lacunae) that is surrounded by a basophilic capsule, the territorial matrix (arrowheads). The lighter interterritorial matrix (itm) lies between the cell groups.

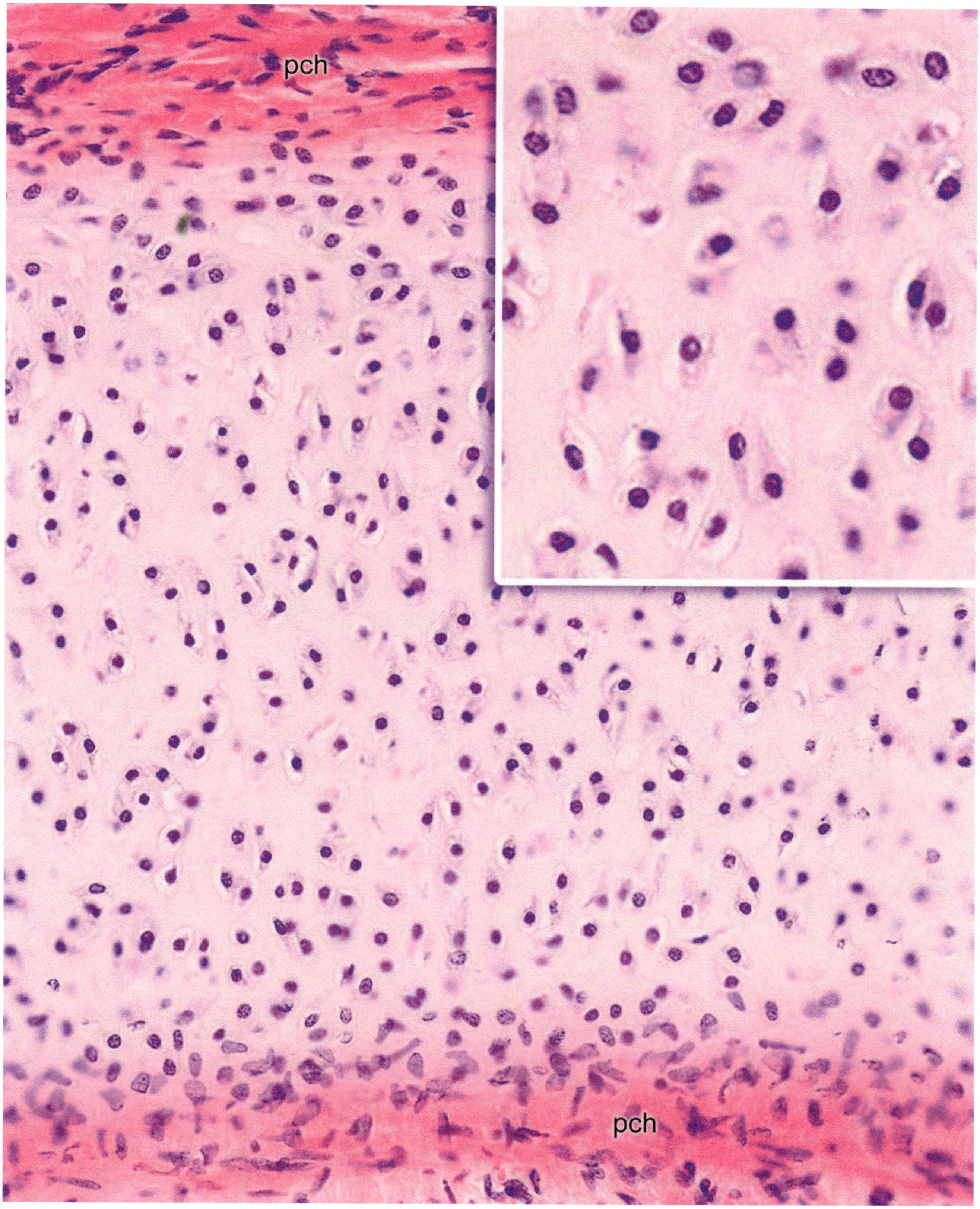

Figure 16. Hyalin cartilage. In immature hyalin cartilage the chondrocytes are smaller and the isogenous groups contain less cells. The amorphous intercellular matrix is rigid, and it is covered by the perichondrium (pch). Territorial and interterritorial matrix is less defined when compared to the mature hyalin cartilage.

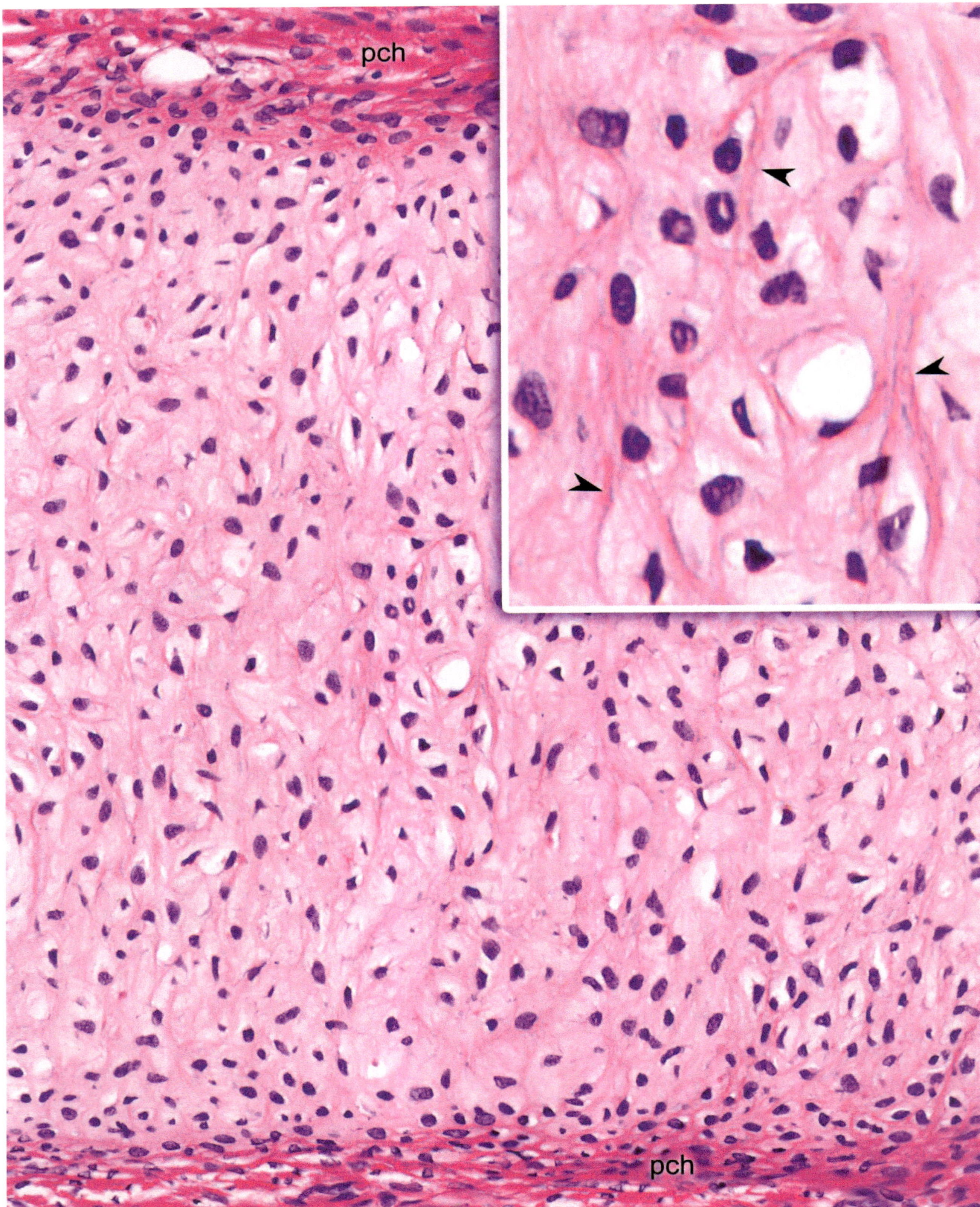

Figure 17. Elastic cartilage, H&E staining. The intercellular matrix of the elastic cartilage is flexible and contains thin, eosinophilic elastic fibers (arrowheads). The isogenous groups of the cells are poorly defined, distinct territorial and interterritorial matrix is not observable. Elastic cartilage is avascular, blood supply is provided by a connective tissue perichondrium (pch).

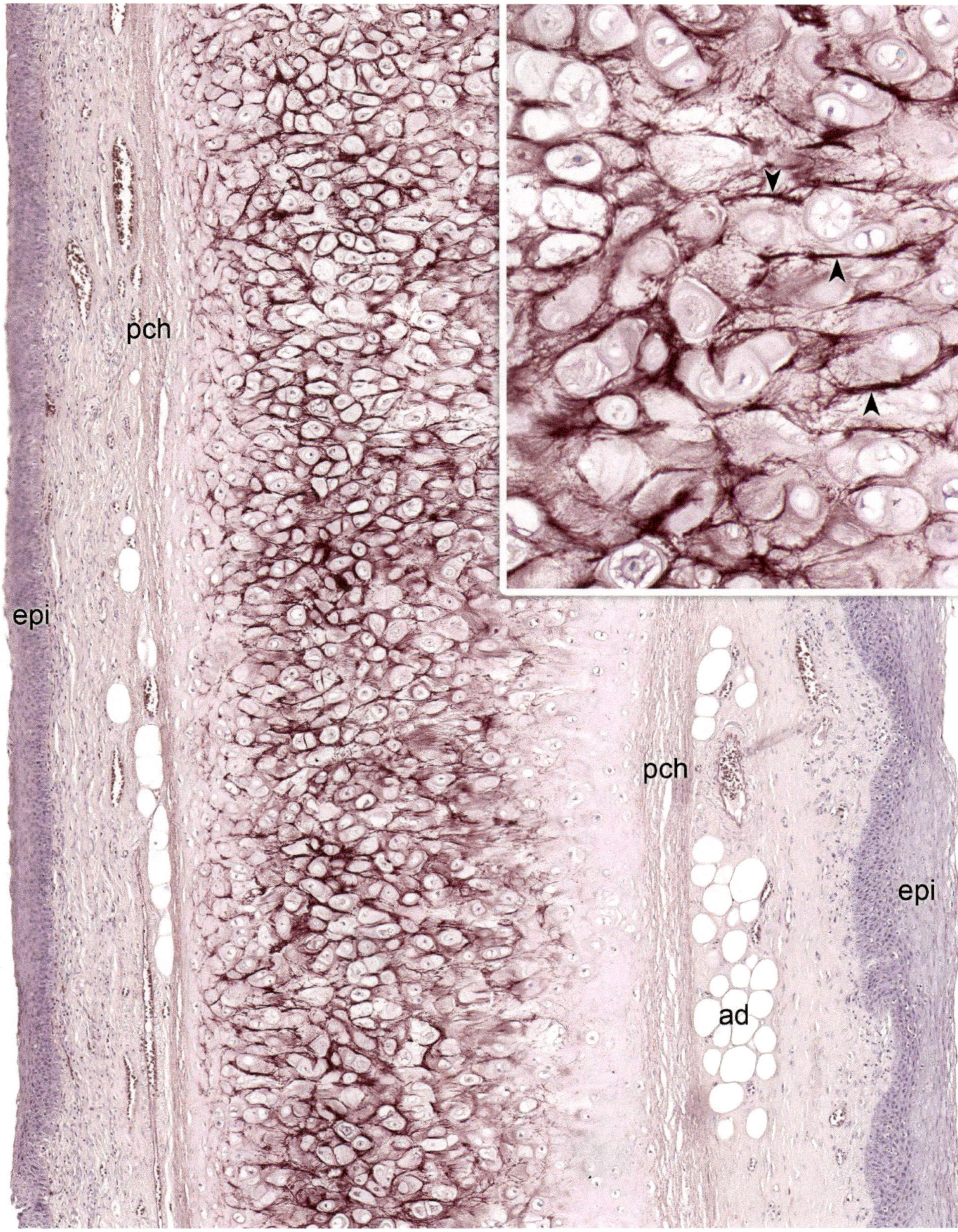

Figure 18. Elastic cartilage, orcein staining. Epiglottis contains an elastic cartilage core covered by mucous membrane that is composed of the epithelium (epi) and the underlying loose connective tissue with adipocytes (ad). Orcein stains the elastic fibers of the cartilage matrix brown (arrowheads). Elastic cartilage is covered by perichondrium (pch).

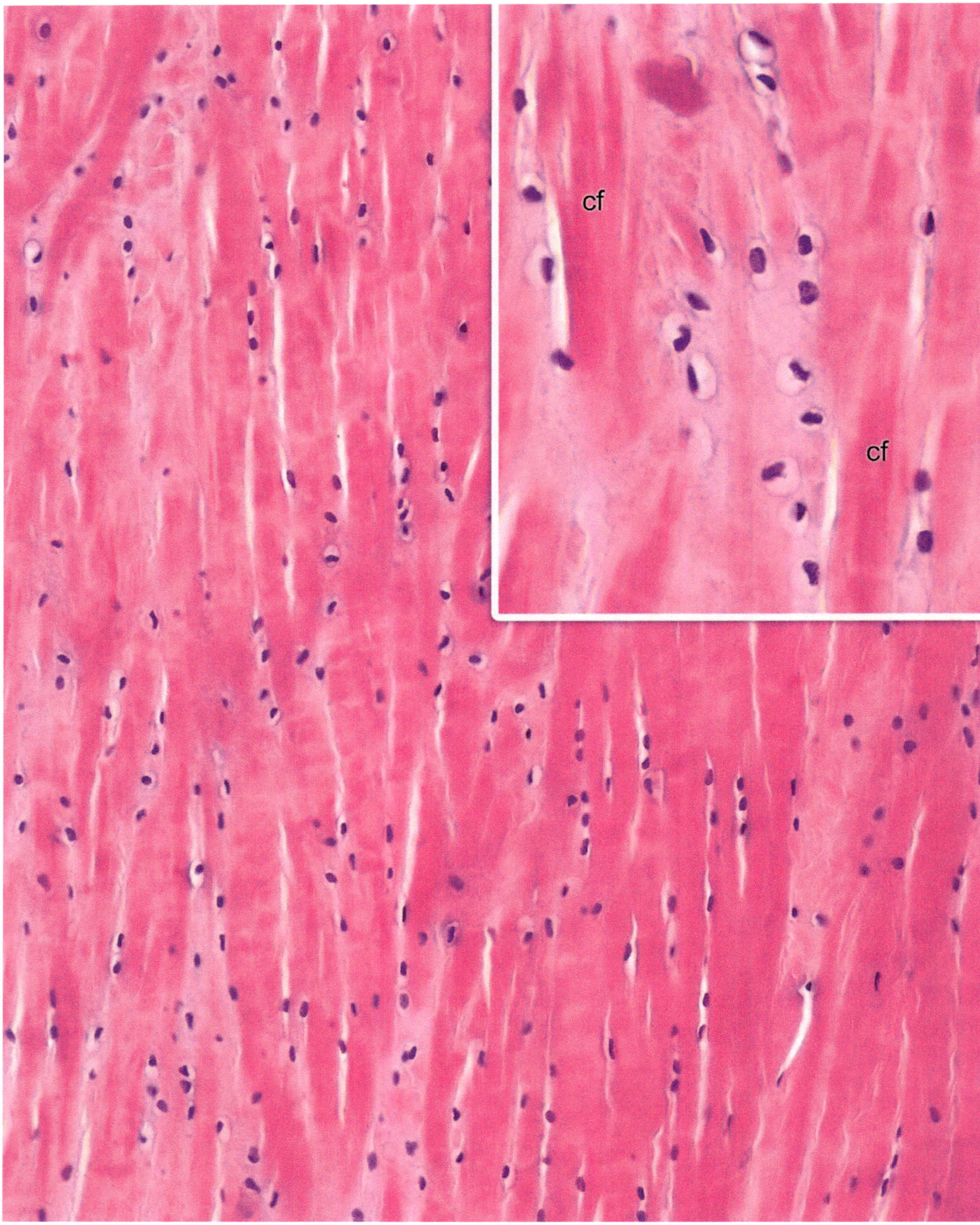

Figure 19. Fibrocartilage. The extracellular matrix of the fibrocartilage contains thick, eosinophilic, collagen type I bundles that are typically parallelly arranged. Chondrocytes are small and they are arranged in rows between the fibers. Fibrocartilage resembles dense regular connective tissue, but instead of the fibroblast, fibrocartilage contains chondrocytes. Fibrocartilage lacks perichondrium and it is located at the transition between tendon and bone; it also forms the annulus fibrosus of the intervertebral discs.

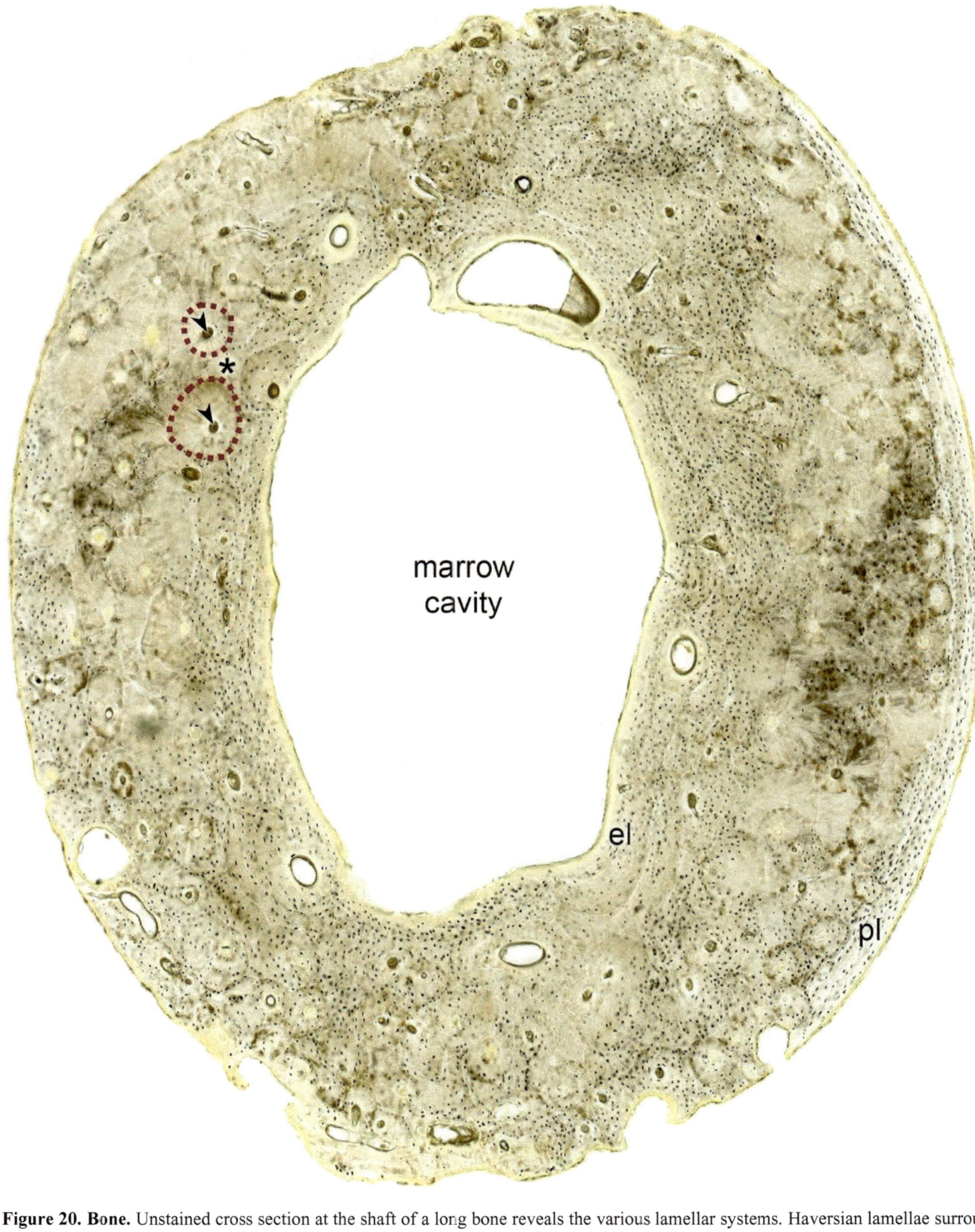

Figure 20. Bone. Unstained cross section at the shaft of a long bone reveals the various lamellar systems. Haversian lamellae surround the Haversian canals (arrowheads) forming Haversian systems (osteons) denoted by the dotted lines. Spaces between the osteons are filled by the interstitial lamellae (asterisk) that are fragments of older Haversian systems partially destroyed during the remodeling of the bone. Endosteal lamellae (el) line the marrow cavity while periosteal lamellae (pl) cover the external surface of the bone under the periosteum.

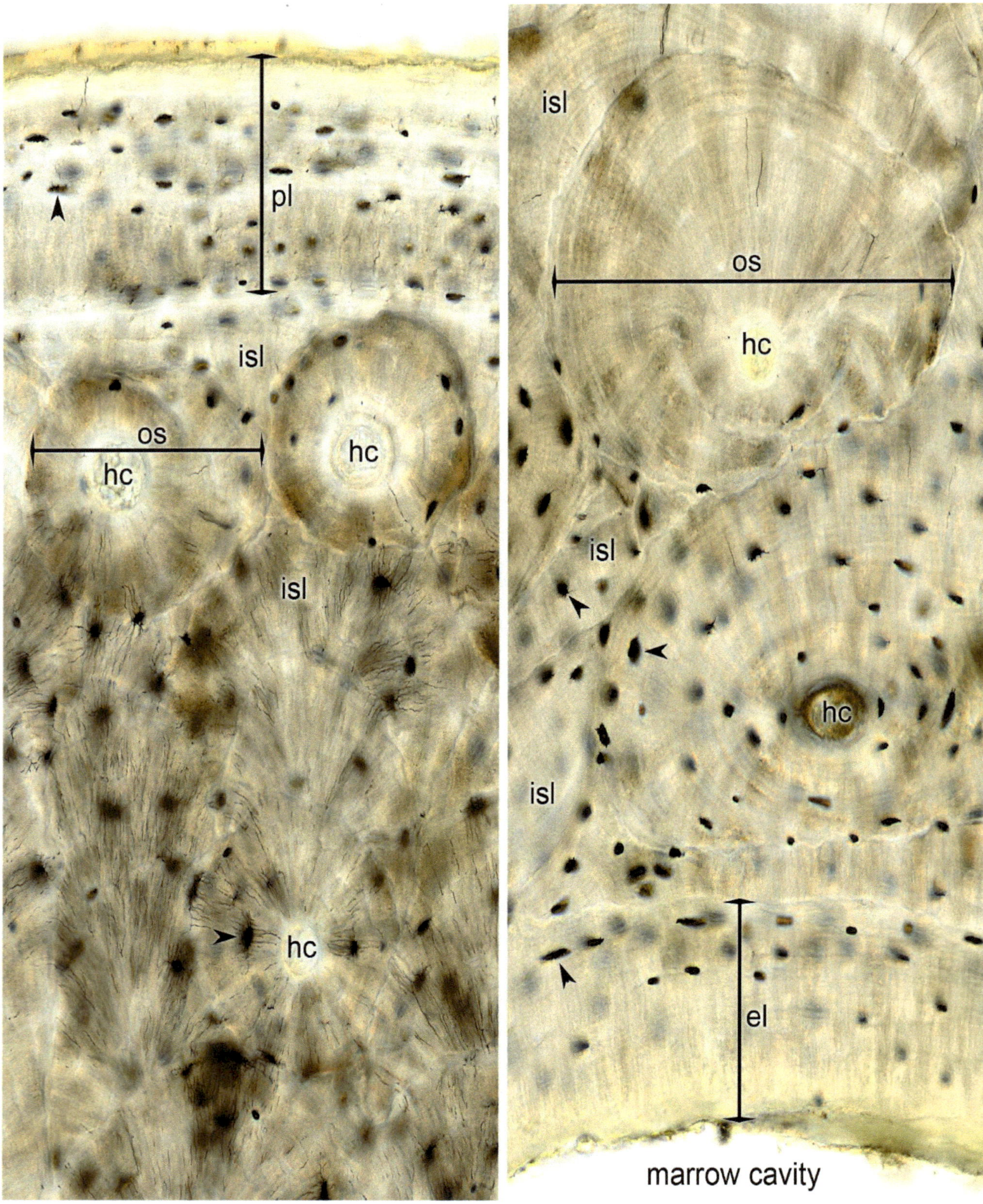

Figure 21. Bone. Unstained cross section at the outer (left) and inner segment (right) of the diaphysis of a long bone. Haversian lamellae surround the Haversian canals (hc) forming osteons or Haversian systems (os). Spaces between the osteons are filled by the interstitial lamellae (isl) that are fragments of older Haversian systems partially destroyed during the remodeling of the bone. Endosteal lamellae (el) line the marrow cavity while periosteal lamellae (pl) cover the external surface of the bone under the periosteum. Osteocytes (arrowheads) with multiple processes are located in the lacunae between the lamellae.

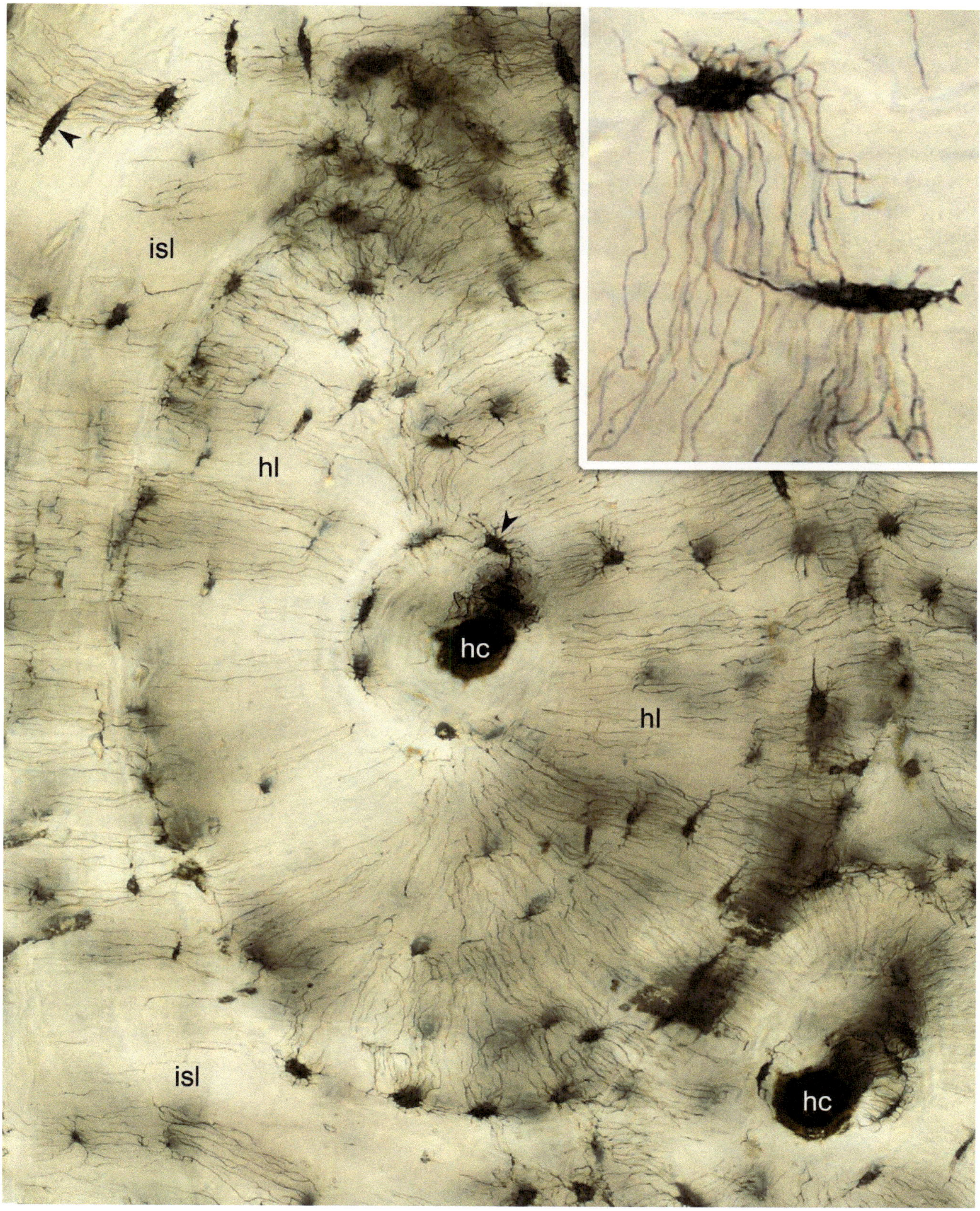

Figure 22. Bone. Unstained cross section of an osteon or Haversian system. Haversian lamellae (hl) surround the Haversian canals (hc) forming osteons or Haversian systems. Spaces between the osteons are filled by the interstitial lamellae (isl) that are fragments of older Haversian systems partially destroyed during the remodeling of the bone. Osteocytes (inset) are insectoid cells enclosed into the lacunae located between the lamellae. Osteocytic processes lie in the canaliculi piercing the lamellae.

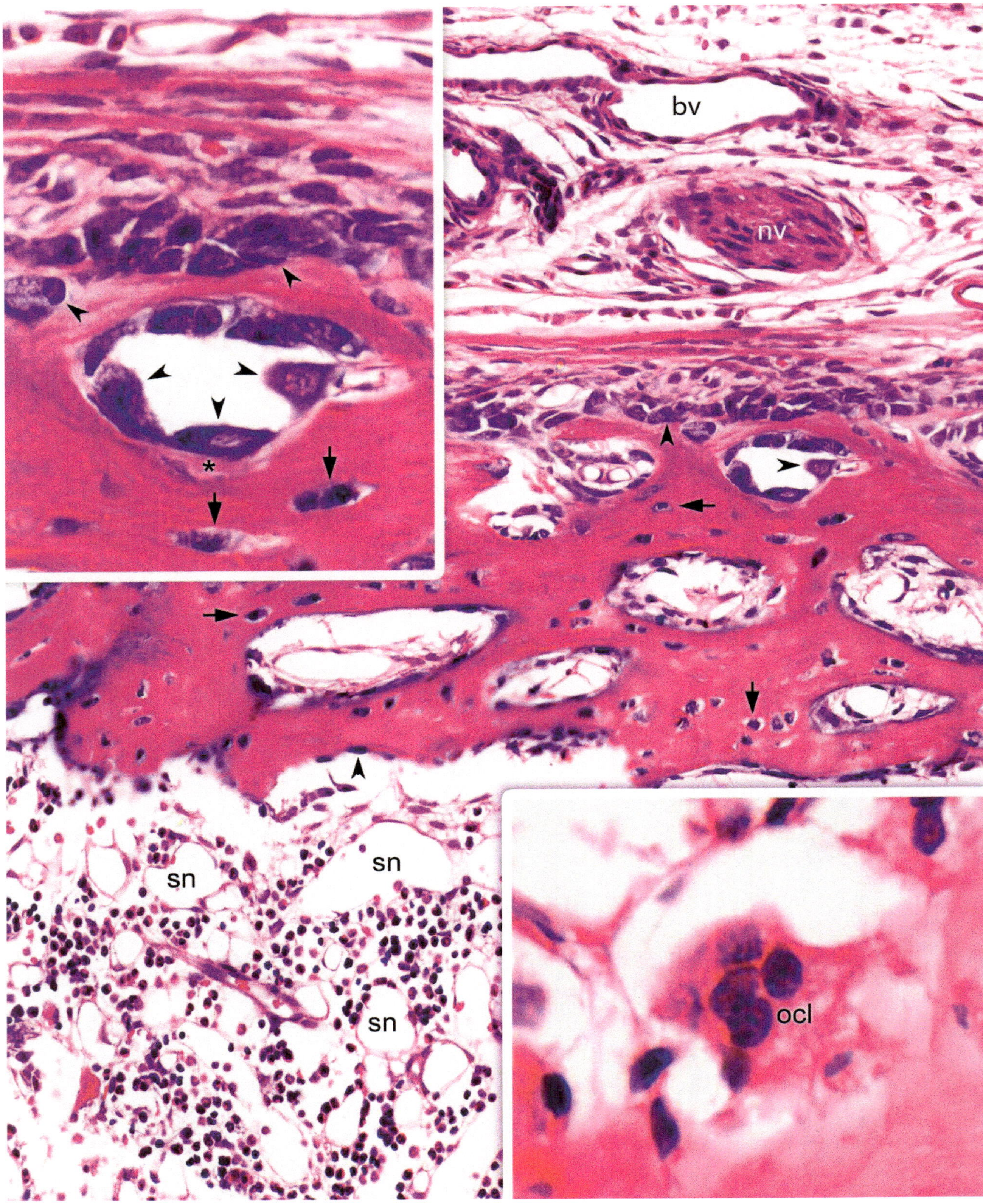

Figure 23. Bone and bone marrow. Bone is formed by osteoblasts (arrowheads, upper inset) that are typically cuboidal cells with spherical or oval nucleus and basophilic cytoplasm. Osteoblasts sit on a layer of osteoid (asterisk) that is not entirely mineralized matrix, and they are eventually enclosed into the developing bone matrix and become osteocytes (arrows). Osteoclasts (ocl) are large, multinucleated cells that dissolve the bone matrix (lower inset). Bone is surrounded by connective tissue with blood vessels (bv) and nerves (nv). Marrow cavity contains the bone marrow interlaced with sinusoids (sn).

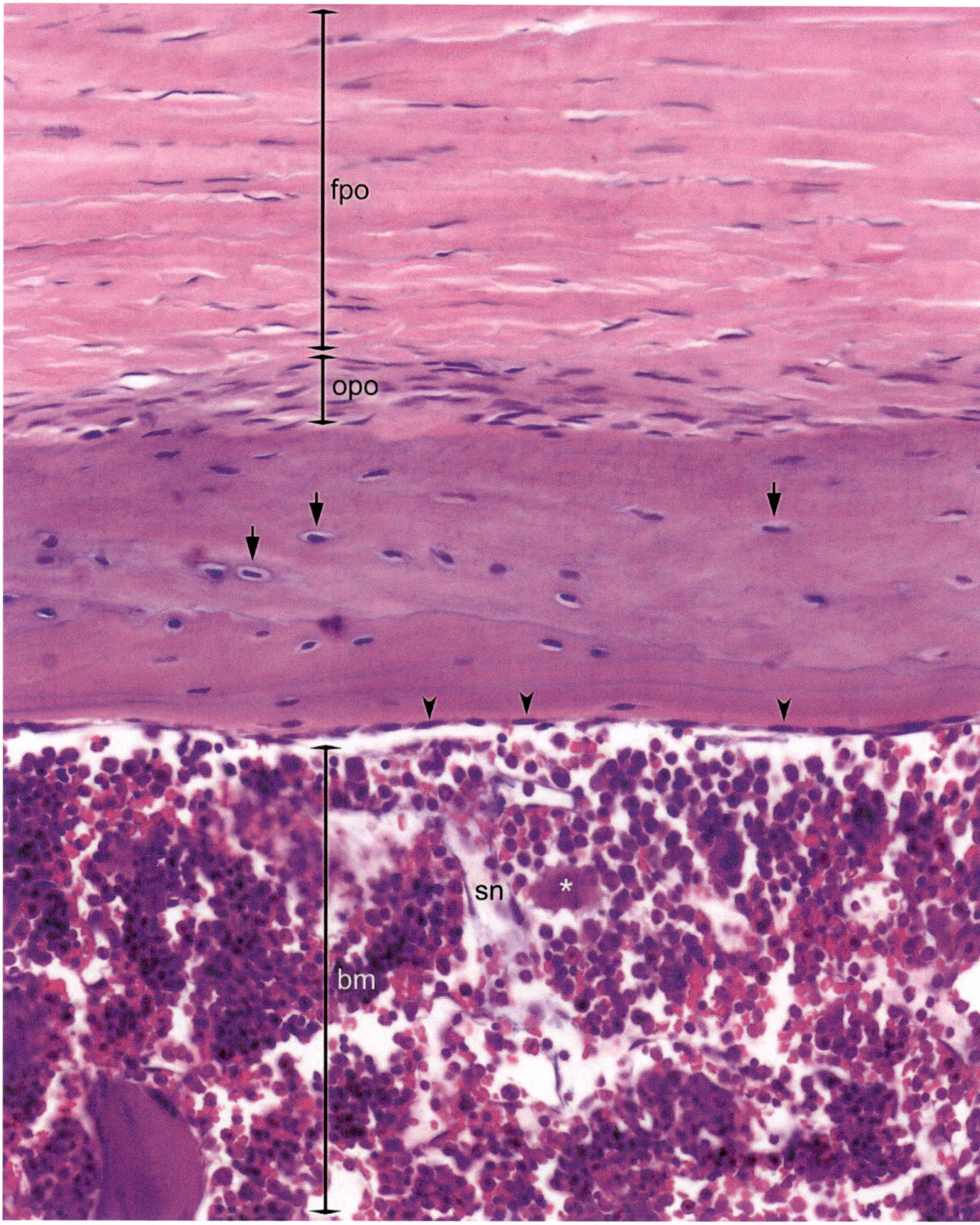

Figure 24. Bone and bone marrow. Osteocytes are enclosed into the bone matrix (arrows). Endosteum lines the marrow cavity and it is mostly composed of osteoprogenitor cells (arrowheads). Bone marrow (bm) contains sinusoids (sn) that are lined with discontinuous endothelium and large megakaryocytes (asterisk) that form the platelets (thrombocytes). Periosteum is composed of an outer fibrous layer (fpo) formed by dense connective tissue and an inner cellular osteogenic layer (opo) that is rich in osteoprogenitor cells..

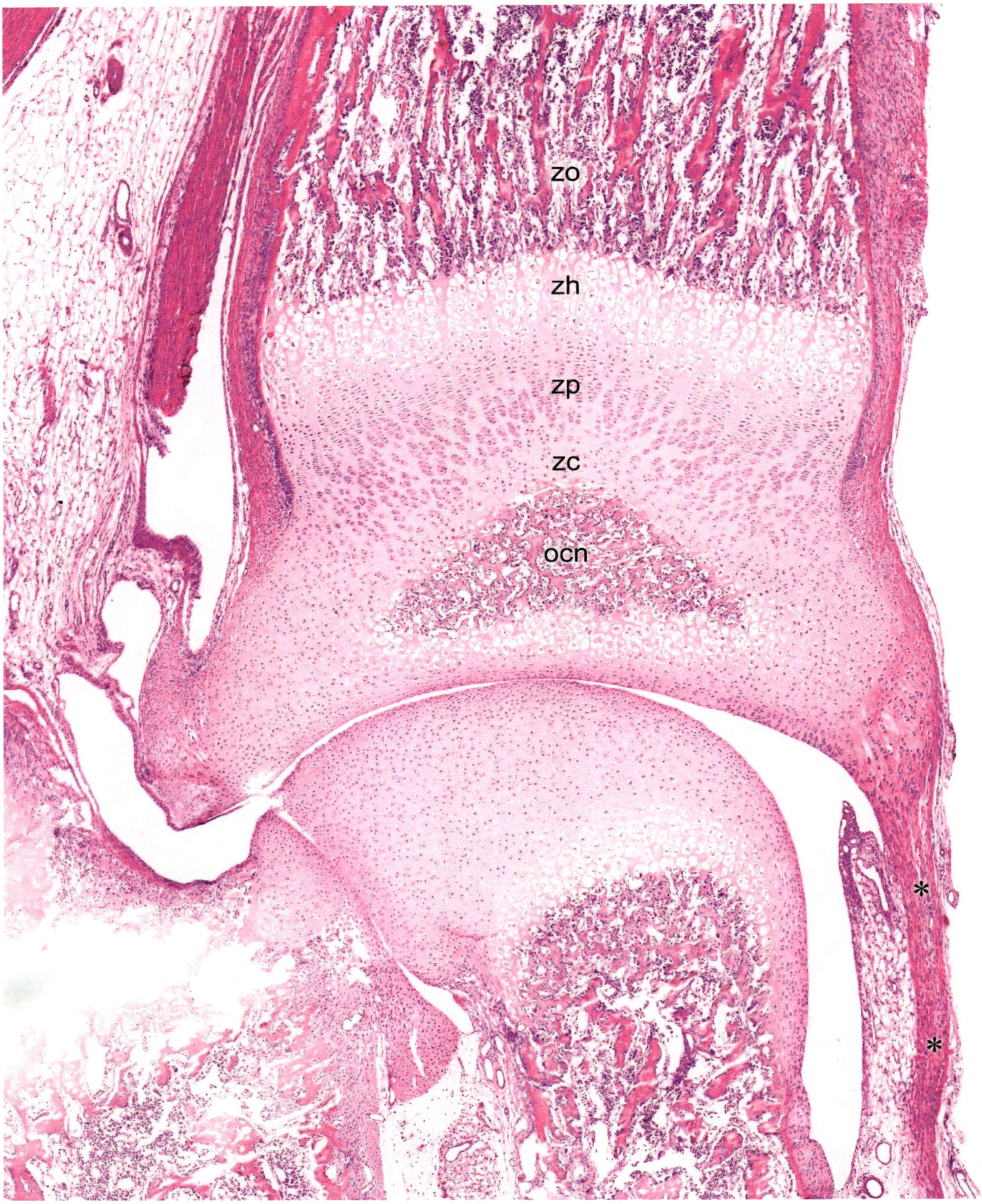

Figure 25. Endochondral ossification. Vessels break into the model of the developing bone that is formed by hyaline cartilage and ossification centers develop. Between the diaphyseal and epiphyseal ossification centers (ocn) cells of the epiphyseal plate drive the longitudinal growth of the bone. Chondrocytes in the zone of reserve cartilage (zc) proliferate forming cell columns (zone of proliferation, zp). These cells undergo degeneration (zone of hypertrophy, zh) and they are eventually replaced by bone (zone of ossification, zo). Dense connective tissue joint capsule (asterisks) is lined by synovial membrane from inside producing the articular (synovial) fluid.

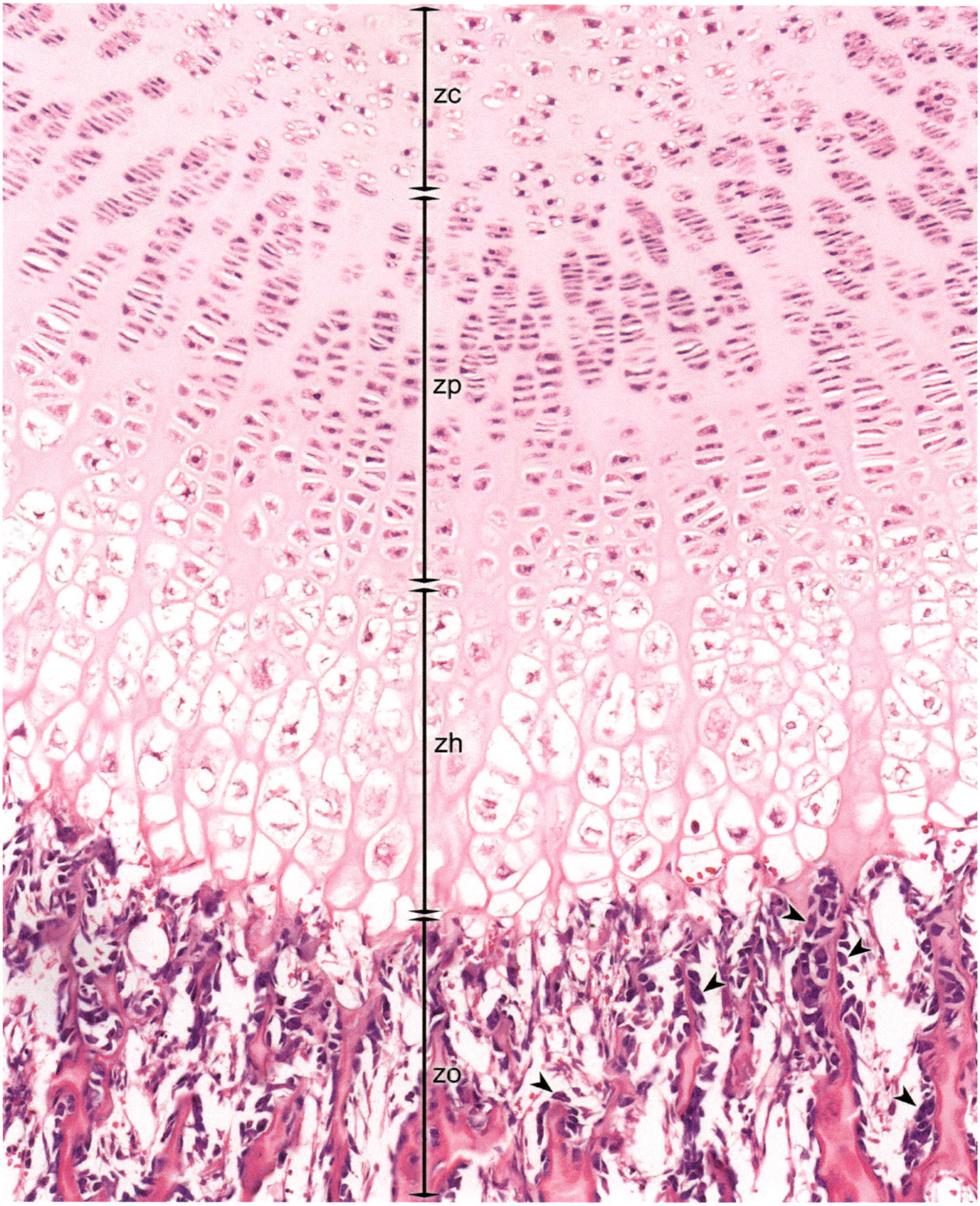

Figure 26. Endochondral ossification, layers of the epiphyseal plate. Chondrocytes in the zone of reserve cartilage (zc) proliferate forming cell columns (zone of proliferation, zp). These cells undergo degeneration (zone of hypertrophy, zh) and eventually die. Cavities of the dead chondrocytes are populated by osteoblasts (arrowheads) that form the bone trabeculae (zone of ossification, zo).

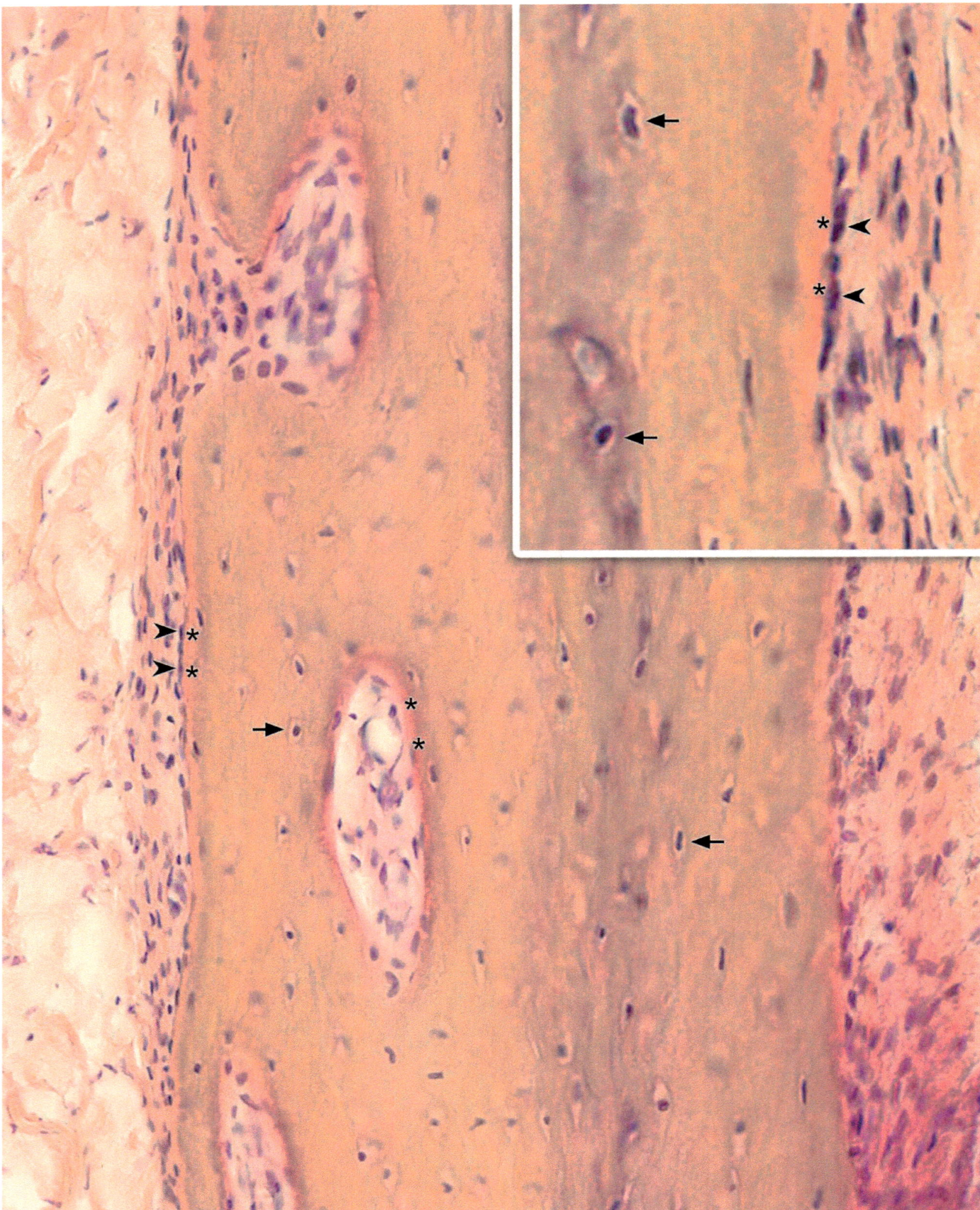

Figure 27. Intramembranous ossification. Flat skull bones develop with intramembranous ossification. Mesenchymal cells in the connective tissue membrane directly turn into low osteoblasts (arrowheads) that produce the organic component of the bone matrix. Osteoblasts form a single layer lining on the eosinophilic layer of osteoid that is not fully mineralized (asterisks) and they eventually enclose themselves into the matrix turning into osteocytes (arrows).

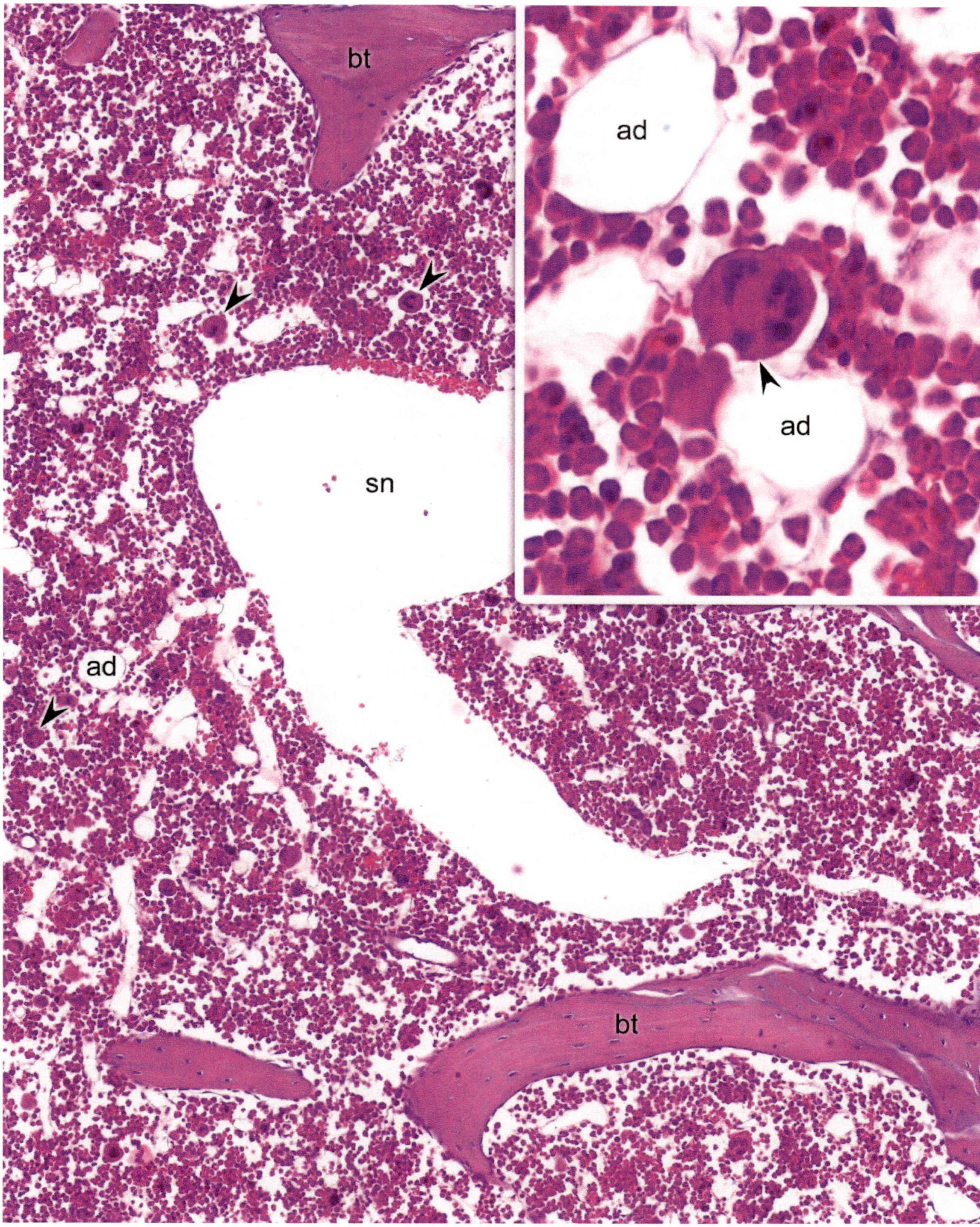

Figure 28. Bone marrow. Developing blood cells at various stages of development populate the bone marrow between the trabeculae of the spongy bone (bt). Bone marrow contains adipocytes as well as sinusoids (sn) lined with discontinuous endothelium. Platelets (thrombocytes) are cytoplasmatic fragments of large spherical bone marrow cells, the megakaryocytes (arrowheads, inset).

Blood

Blood is a special tissue where the intercellular matrix is liquid; consequently, cellular elements are surrounded by blood plasma. The vast majority (99%) of the cells are erythrocytes (red blood cells, RBCs) that are biconcave eosinophilic discs without nuclei. Reticulocytes are immature form of red blood cells with clumped ribosomal complexes in their cytoplasm;they are slightly larger than erythrocytes and exhibit basophilia. White blood cells form approximately 1% of the cellular elements of the blood; they can be categorized as granulocytes (neutrophils, eosinophils, basophils) and agranulocytes (monocytes and lymphocytes). The most common white blood cells are the neutrophils (60-70%) and lymphocytes (25-30%). Neutrophil granulocytes (neutrophils) have characteristic multilobed nucleus and small granules that give the characteristic neutral (light pink) staining of the cytoplasm with H&E. The cytoplasm of eosinophils is filled with large eosinophilic granules that surround the bilobed nucleus. Eosinophils can often be observed in the lamina propria of the intestines and their number increases in helminthic and protozoan infections. Basophils are generally smaller cells; the large basophilic granules typically obscure the bilobed nucleus, hence they can be easily distinguished from mastocytes that possess observable round-shaped nuclei. Occasionally, immature forms of granulocytes with band-shaped nucleus (band granulocyte, stab cells or "stab form") can also be observed in the blood.

Monocytes are larger cells with basophilic cytoplasm and embryo-shaped nucleus that migrate into the tissues and turn into macrophages and antigen presenting cells. Lymphocytes have a prominent round-shaped nucleus that is surrounded by a pale basophilic cytoplasmatic ring. B and T lymphocytes are morphologically undistinguishable, but NK (natural killer) lymphocytes, as well as activated B and T cells are generally larger with an indentation on their nucleus. Platelets (thrombocytes) are cytoplasmatic fragments of larger cells named megakaryocytes that are located in the bone marrow.

Nervous tissue

The nervous system is composed of neurons and glial cells and it can be divided into central nervous system (CNS) and peripheral nervous system (PNS). Neurons are large cells with pale nucleus, prominent nucleolus and well-developed rER that can be observed as basophilic clumps in the cytoplasm (Nissl substance). Neurons also exhibit processes (neurites) emanating from the cell body (perikaryon); these processes can be divided to a single, long axon and shorter, usually multiple dendrites that often branch. Cell organelles, including rER are missing from the region of the perikaryon that is located at the root of the axon (axon hillock). Accumulations of the neurons in the CNS are termed as nuclei while ganglia are groups of neurons in the PNS. According to the shape, neurons can be multipolar, bipolar, and unipolar; a special type of the neurons are the pseudounipolar cells with rounded cell body and a single axon that divides into central and peripheral processes; these cells are located in the sensory ganglia. Large multipolar neurons can be observed in the ventral horn of the spinal cord, in autonomic ganglia and in the cerebral (pyramidal cell) and cerebellar cortices (Purkinje cells).

Axons can be myelinated or unmyelinated; myelin that is composed of multiple layers of cell membrane, covers the axons in multiple sections with gaps between them where the axon also have a minor swelling (nodes of Ranvier). Myelin sheath is formed by Schwann cells in the PNS and oligodendrocytes (oligodendroglia) in the CNS. Schwann cells not only form the myelin sheath in the PNS, but they also surround it with their cytoplasm (Schwann sheath, neurilemma) while unmyelinated axons in the PNS possess only Schwann sheath. In the CNS myelin covers the axons without cytoplasmatic sheath that is also missing in unmyelinated axons.

In addition to the Schwann cells and oligodendrocytes that are responsible for the formation of the myelin sheath, there are other types of glial cells in the nervous system. Satellite cells are located in the PNS surrounding ganglion cells in the sensory and autonomic ganglia. Located in the CNS, astrocytes are star-shaped cells with multiple processes, and they can be subdivided into fibrous and protoplasmic types, the former located predominantly in the white matter and the latter prevalent in the grey matter. Astrocytic processes insulate the walls of vessels in the CNS forming the blood-brain-barrier that separates brain parenchyma from the blood stream. Microglial cells derive from the mesoderm, and they function as macrophages in the CNS; they are extremely plastic cells possessing long, branching processes and a small cellular body when resting and amoeboid shape when activated. Finally, the fourth major type of glial cells in the CNS is the ependymal cell that lines cerebrospinal fluid-filled spaces including ventricles of the brain. These cuboidal or columnar neuroepithelial cells are ciliated or possess microvilli and form a single layer on their basement membrane. Tanycytes are specialized ependymal cells with long basal processes that extend radially into the parenchyma over long distances and connect with vessels.

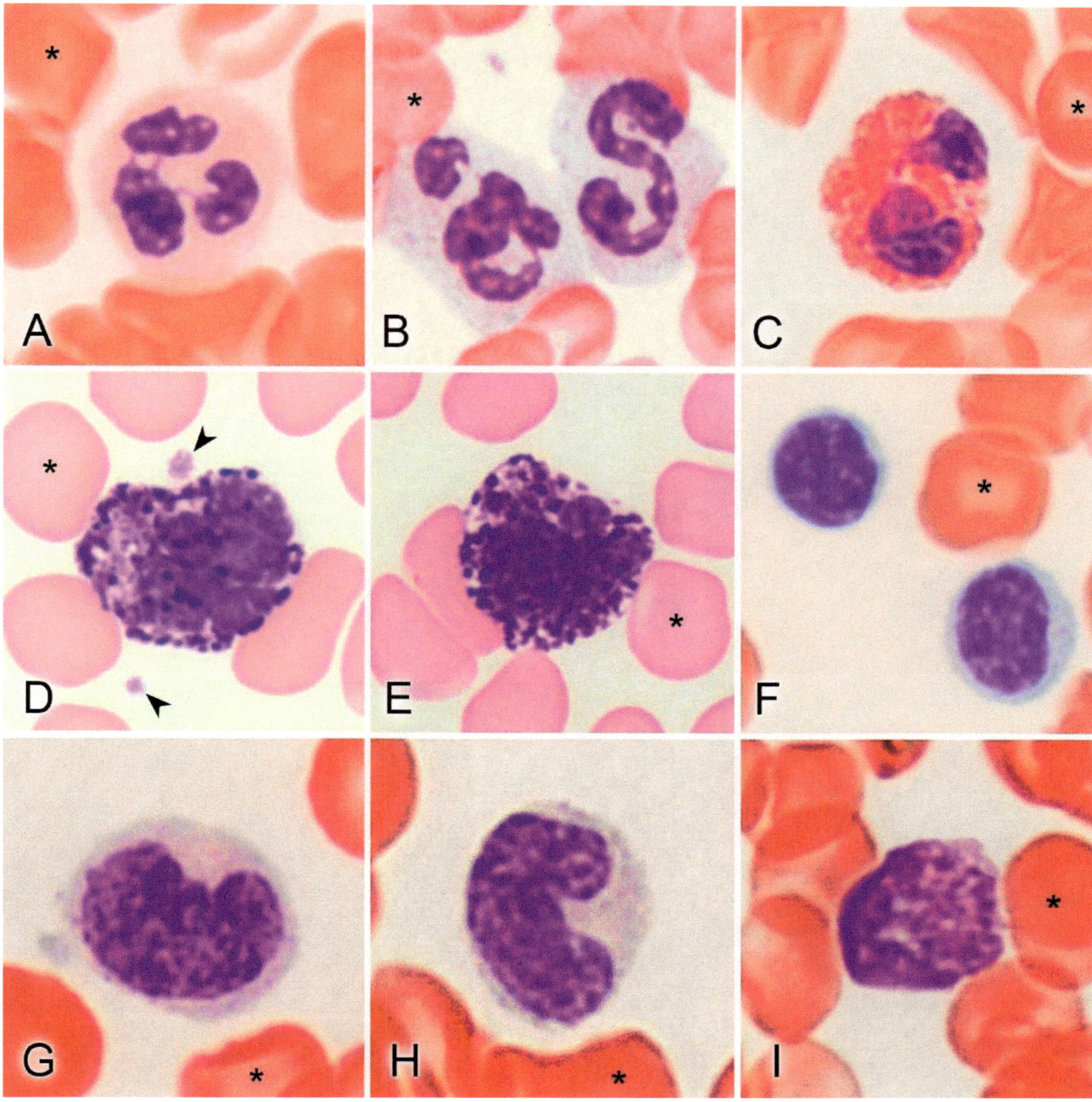

Figure 29. Cellular elements of the blood. (A) Mature neutrophil granulocyte with multilobed nucleus. (B) Neutrophil band granulocyte is an immature neutrophil with a ribbon-shaped nucleus. (C) Eosinophil granulocyte with a bilobed nucleus and large eosinophil granules in the cytoplasm. (D, E) Basophil granulocyte. Basophilic cytoplasmatic granules obscure the typically bilobed nucleus. Thrombocytes (platelets) are denoted with arrowheads; they are cytoplasmatic fragments of the megakaryocytes that are located in the bone marrow. (F) Small lymphocyte (upper) and large lymphocyte with an indentation of the nucleus (lower). Lymphocytes have an intensively stained, spherical nucleus surrounded by a thin shell of basophilic cytoplasm. Large lymphocytes are either activated cells or natural killer (NK) cells. (G, H) Monocytes are cells with embryo (bean)-shaped nucleus. Monocytes are antigen presenting cells that are capable of phagocytosis. (I) Reticulocytes are immature erythrocytes with clumps of ribosomes in their cytoplasm. Mature erythrocytes are denoted with asterisks.

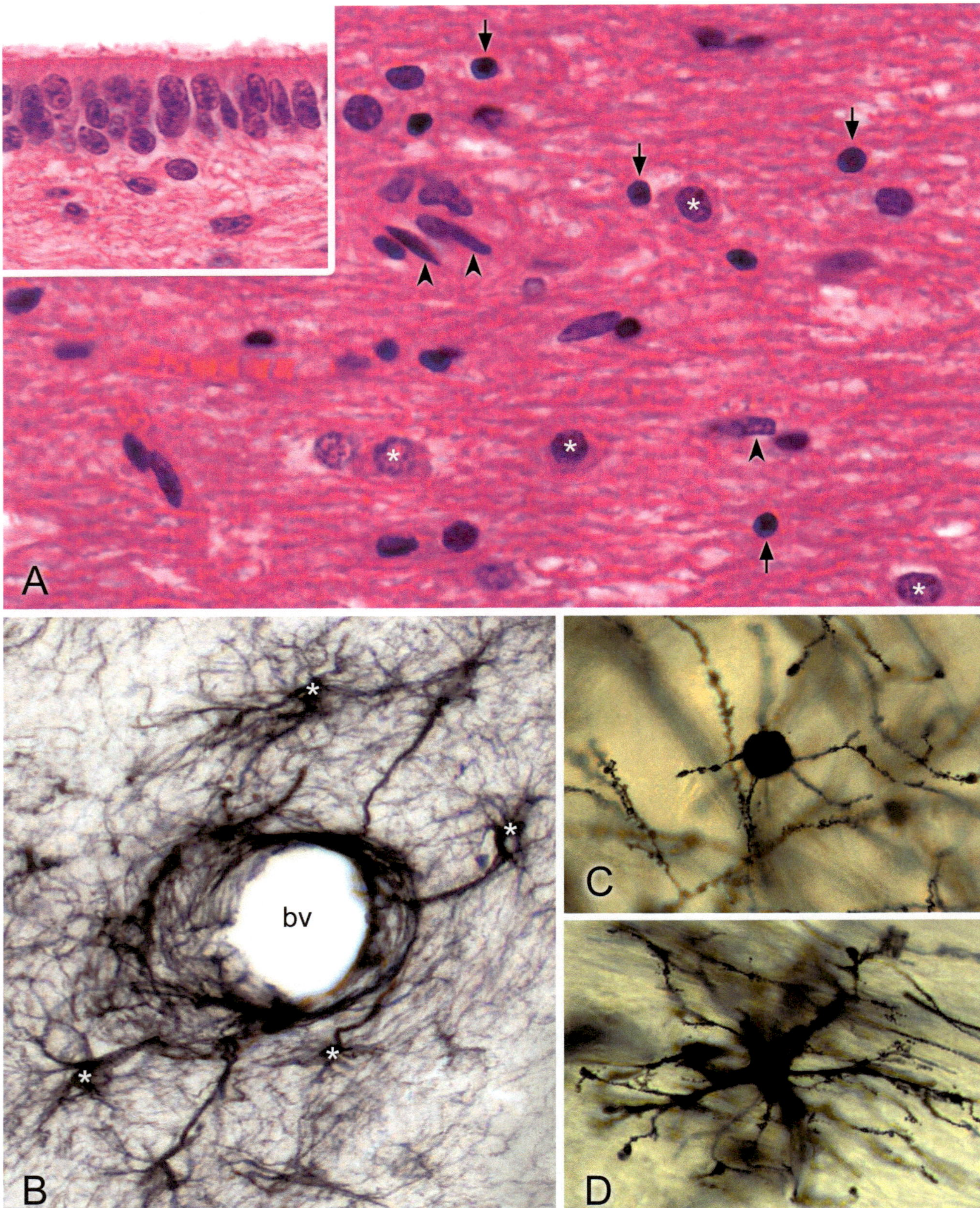

Figure 30. Glial cells in the central nervous system. (A) H&E staining of the brain reveals oligodendrocytes with spherical, darkly stained nucleus (arrows) while astrocytic nuclei are larger and less intensely stained (asterisks). The nuclei of the microglial cells are elongated (arrowheads). Ciliated, cuboidal, or columnar ependymal cells line cerebrospinal fluid-filled cavities in the CNS (inset). (B) Astrocytes (asterisks) form a major component of the blood-brain barrier by surrounding and isolating vessels (bv) with their processes (GFAP immunohistochemistry). (C) Oligodendrocytes (silver impregnation) are small, rounded cells with short processes. (D) Microglia (silver impregnation) are phagocytic cells in the CNS.

Chapter 2

Lymphatic System

Lymphatic follicles

Lymphatic follicles (lymphatic nodules) are spherical or ovoid structures composed of aggregated lymphocytes and a meshwork of reticular cells. Lymphatic follicles lack capsule, and they typically populate the lamina propria of the mucous membranes in the alimentary (gut-associated lymphoid tissue, GALT) and respiratory tracts (bronchus-associated lymphoid tissue, BALT). Primary follicles are mostly composed of small lymphocytes, while secondary follicles contain a central light zone, the germinal center that is formed by activated, large lymphocytes. Mucous membrane may contain solitary lymphatic follicles or aggregated follicles that often elevate the mucous membrane forming tonsils. These aggregated follicles also appear in the vermiform appendix and the Peyer's patches in the ileum.

Tonsils

Tonsils are accumulation of lymphocytes, particularly lymph follicles in the lamina propria of the mucous membranes. The abundance of lymphoid elements elevates the mucous membrane and lymphocytes often infiltrate the epithelium. Tonsils are located at the transition between the oral/nasal cavities and the oropharynx, and they are covered either with respiratory epithelium or the epithelium of the oropharyngeal isthmus. Palatine tonsils are characterized by stratified squamous nonkeratinized epithelium lining that penetrates the lamina propria forming narrow and deep crypts. Pharyngeal tonsils have similar structure, but they are covered with ciliated pseudostratified columnar epithelium. Palatine and pharyngeal tonsils along with the tubal tonsils and the lingual tonsil at the root of the tongue form a lymphatic ring at the oropharyngeal isthmus (Waldeyer's ring).

Lymph node

Lymph nodes are biological filters of the lymphatic circulation, and they are often located along major vessels and in the subcutaneous adipose tissue. Lymph nodes are bean-shaped or elliptical structures that can be several millimeters long and they are covered with a dense connective tissue capsule that forms trabeculae penetrating into the parenchyma. Under the capsule, the cortex of the lymph node contains lymph follicles, in contrast to the medulla at the center of the node that is composed of sinuses separated by trabeculae of lymphatic tissue (medullary cords). Between the cortex and the medulla is the zone of paracortex or deep cortex that is free of lymphatic nodules and rich in T lymphocytes; it also contains high endothelial venules (HEV) with unusually tall endothelium. Reticular fibers produced by reticular cells provide a framework for the lymphatic elements. Afferent lymph vessels enter at the convexity of the lymph node and lymph drains into the subcapsular sinus right under the capsule. Subcapsular sinus continues into the trabecular sinuses located along the penetrating trabeculae; trabecular sinuses eventually lead into the medullary sinuses in the medulla. Lymph from the medullary sinuses leaves through the efferent lymph vessels at the convexity (hilum) of the node. The hilum also receives blood vessels; arteries support capillary networks in the cortex while the postcapillary venules located in the paracortex are typically lined with tall endothelial cells that appear cuboidal in cross section (high endothelial venules, HEV). Lymphocytes leave the blood through these high endothelial venules; the T cells typically remain in the paracortex, while B cells populate the cortex and the medulla, and they eventually filter back to the efferent lymph vessels. Veins leave the lymph node at the hilum.

http://doi.org/10.1016/B978-0-323-91891-6.50002-7,

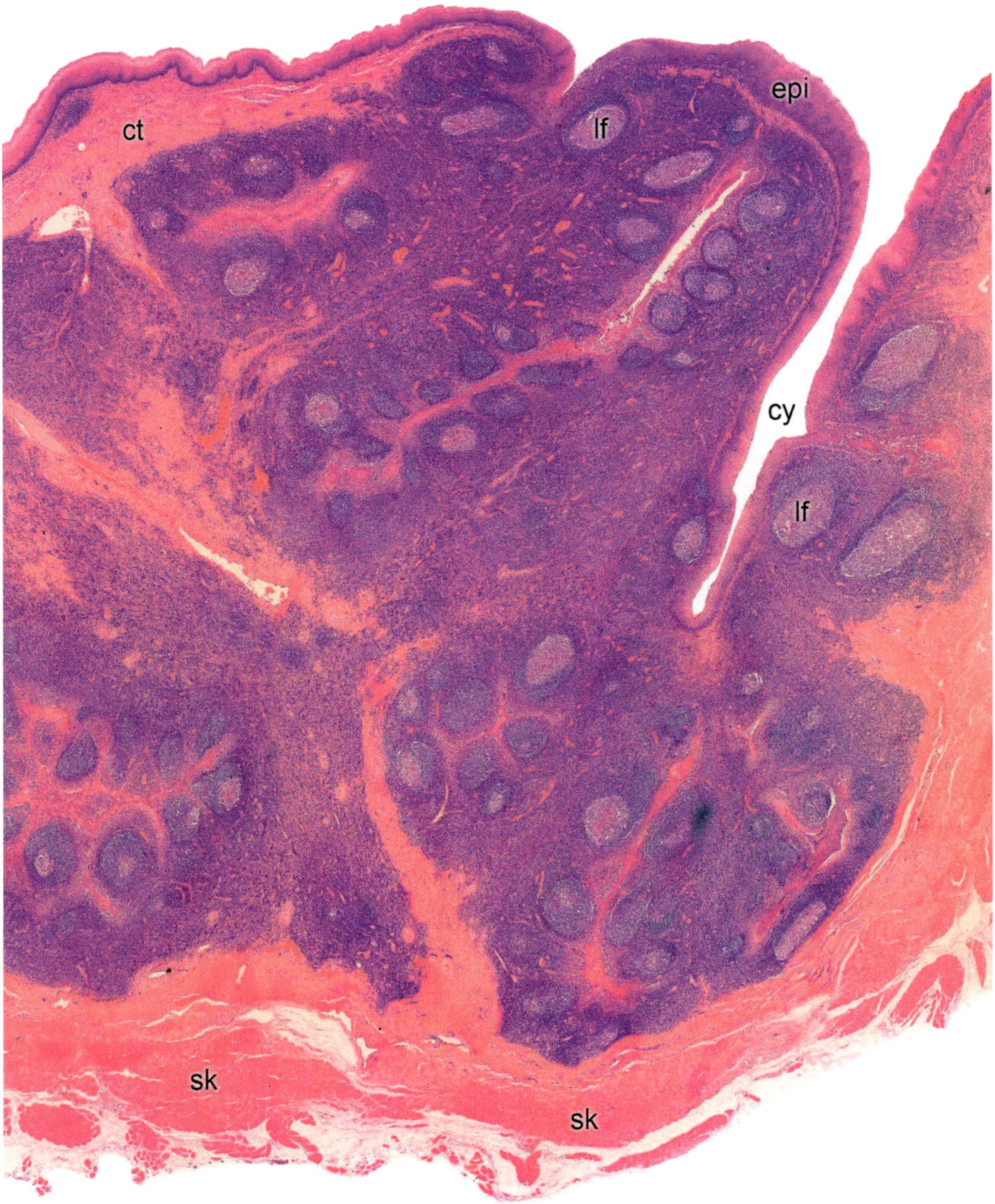

Figure 31. Palatine tonsil. Aggregation of lymphatic follicles (lf) in the lamina propria of the oropharyngeal isthmus elevates the surface of the mucous membrane forming the palatine tonsil. Nonkeratinized stratified squamous epithelium lining the surface (epi) forms deep invaginations, the crypts (cy). Under the epithelium, follicles are embedded into loose connective tissue with large number of collagen fibers (ct) and lie on a skeletal muscle layer formed by the pharyngeal constrictors.

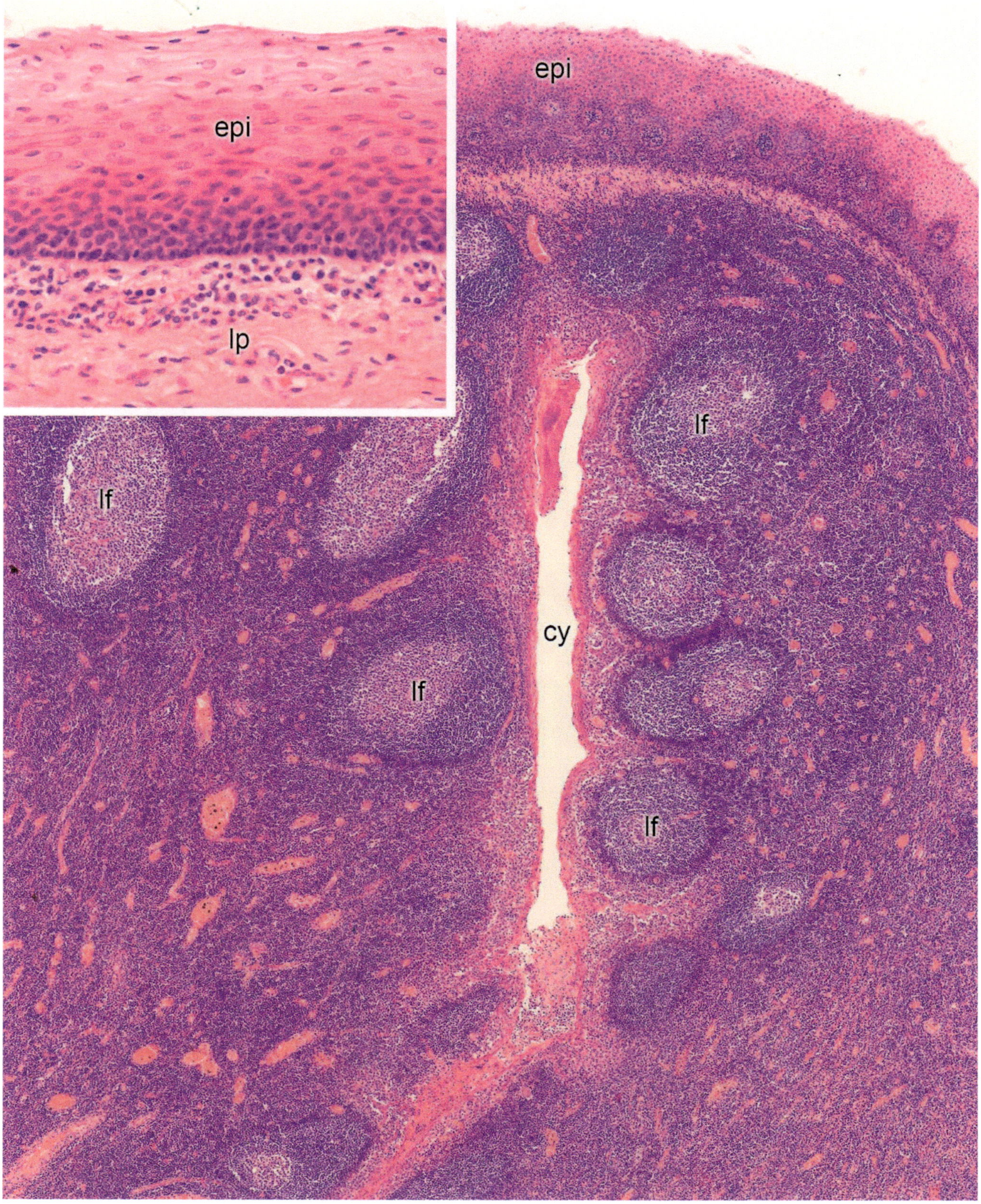

Figure 32. Palatine tonsil. Aggregation of lymphatic follicles (lf) in the lamina propria of oropharyngeal the isthmus elevates the surface of the mucous membrane forming the palatine tonsil. Nonkeratinized stratified squamous epithelium lining the surface (epi) forms deep invaginations called crypts (cy) that often retain antigens and cellular debris. Loose connective tissue rich in collagen fibers form the lamina propria mucosae (lp) under the surface epithelium (inset).

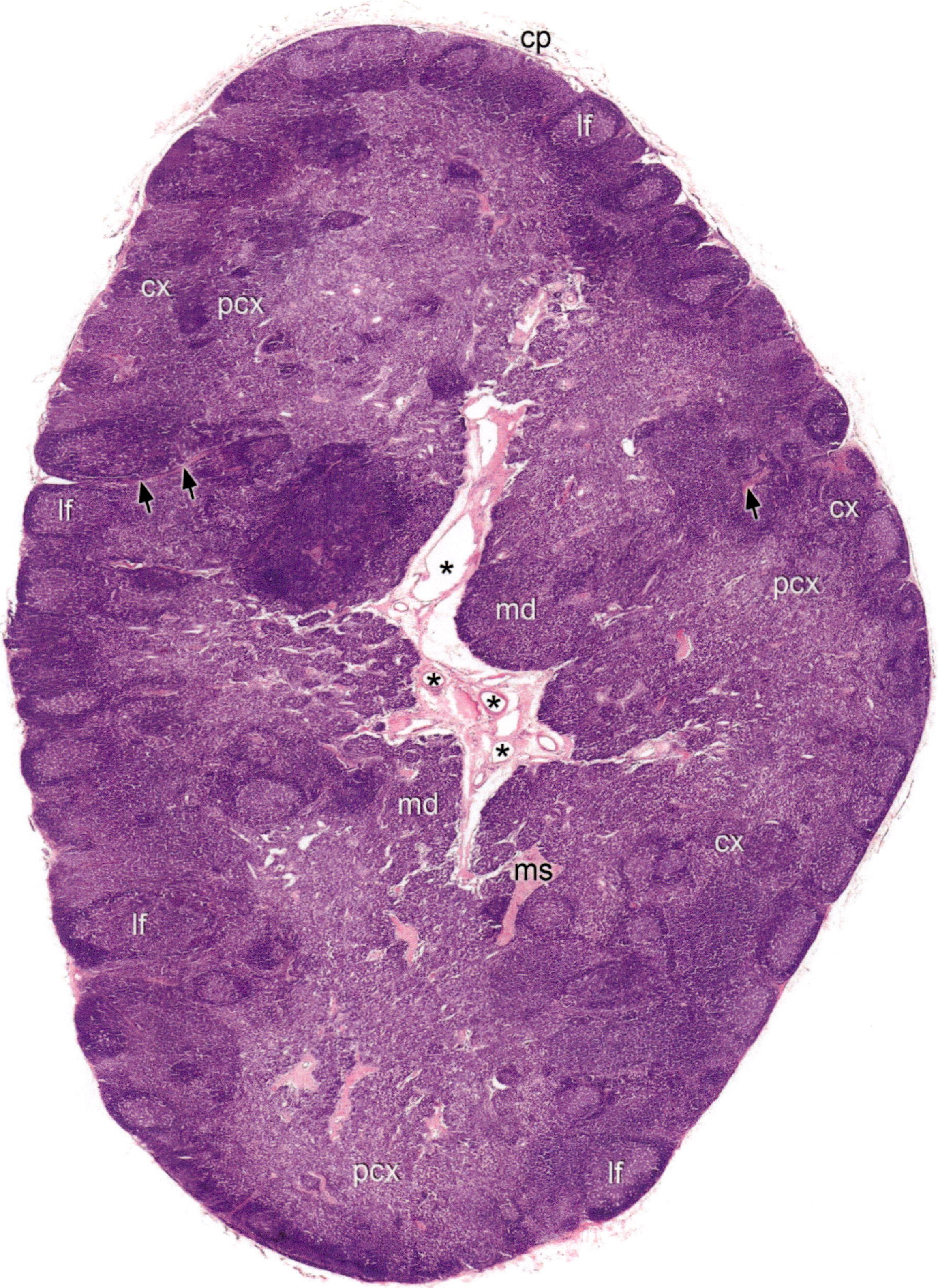

Figure 33. Lymph node. Lymph node is covered by connective tissue capsule (cp) that forms trabeculae penetrating the parenchyma (arrows). Outer cortex (cx) is formed by lymphatic follicles (lf) that are absent in the medulla (md). Paracortex (pcx) between the cortex and the medulla lacks follicles. Medulla contains medullary sinuses (ms) as well as blood and lymph vessels (asterisks).

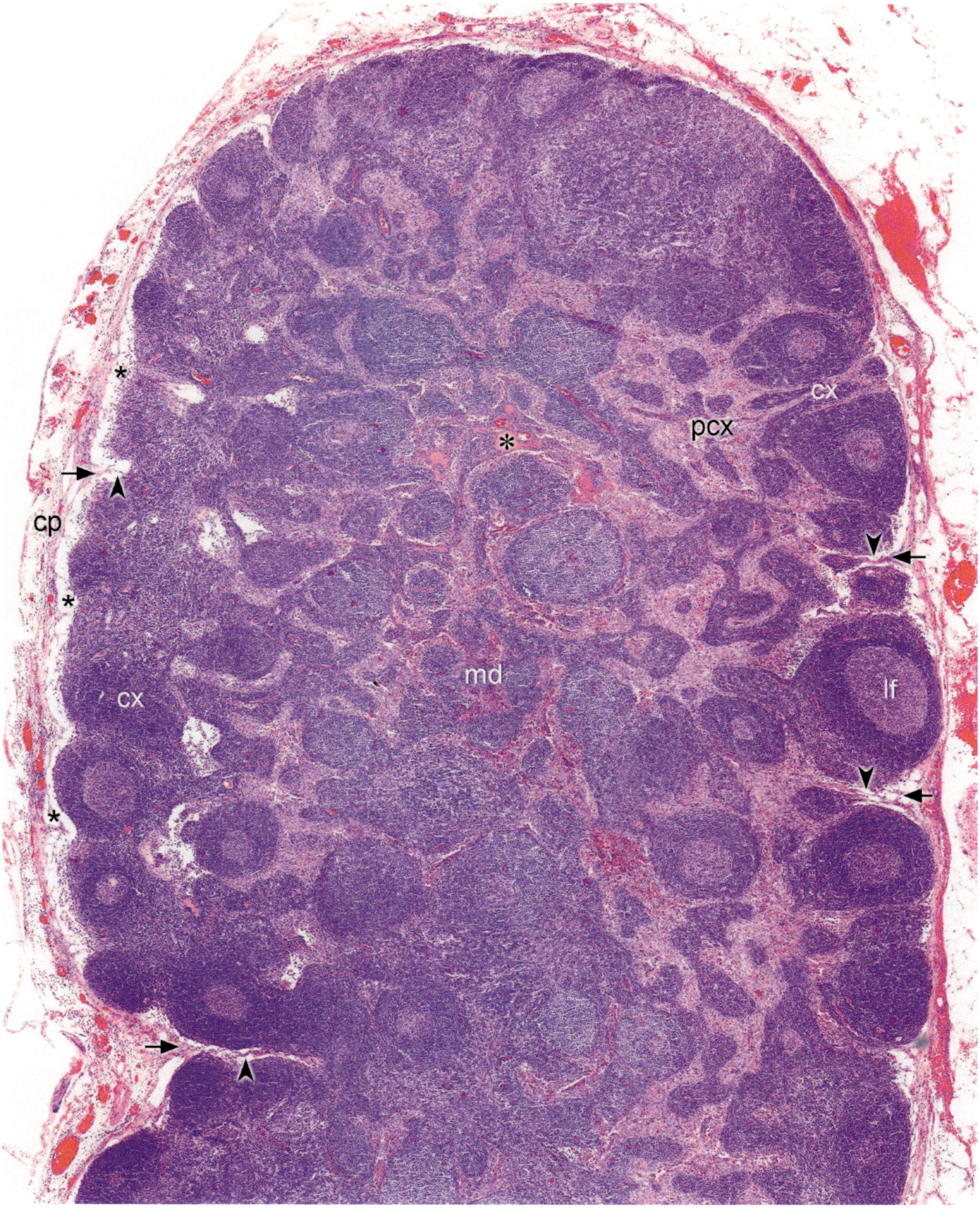

Figure 34. Lymph node. Lymph node is covered by connective tissue capsule (cp) that forms trabeculae penetrating the parenchyma (arrows). Outer cortex (cx) is formed by lymphatic follicles (lf) that are absent in the medulla (md). Paracortex (pcx) between the cortex and the medulla lacks follicles. Afferent lymph vessels enter the subcapsular sinus (five-pointed asterisks) that drains into the trabecular sinuses (arrowheads) located along the trabeculae (arrows). Trabecular sinuses open into medullary sinuses (six-pointed asterisks).

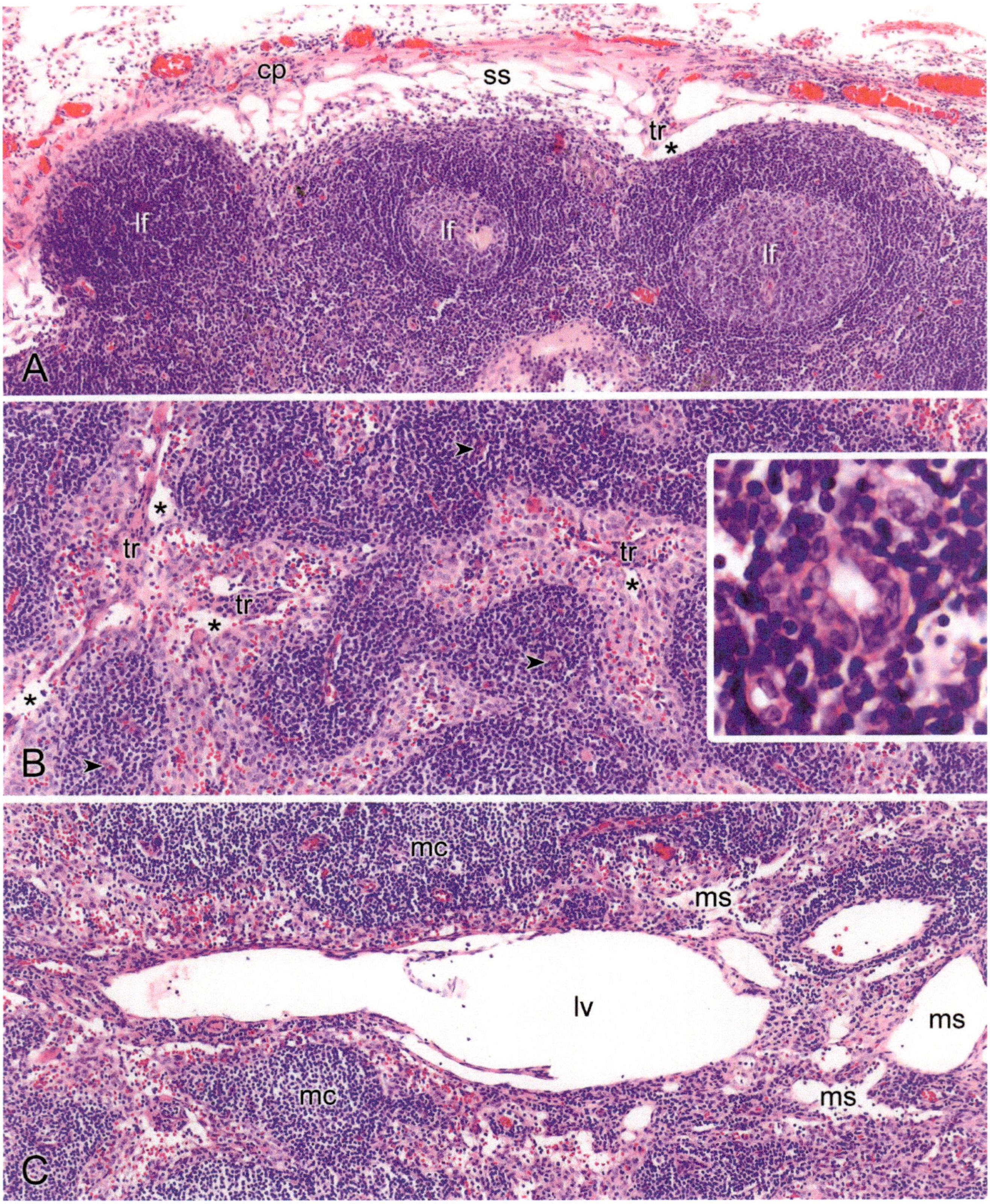

Figure 35. Lymph node. (A) Outer cortex is populated by lymphatic follicles (lf) and it is covered by connective tissue capsule (cp) that forms trabeculae (tr) penetrating the parenchyma. Afferent lymph vessels enter the subcapsular sinus (ss) that drains into the trabecular sinuses (asterisks) along the trabeculae. (B) Paracortex (deep cortex) is located between the outer cortex and the medulla. Paracortexcontains trabecular sinuses (asterisks) running along the connective tissue trabeculae (tr) as well as high endothelial venules (arrowheads, inset). (C) Medulla lacks follicles but contains medullary sinuses (ms) between the medullary cords (mc) and lymph vessels (lv).

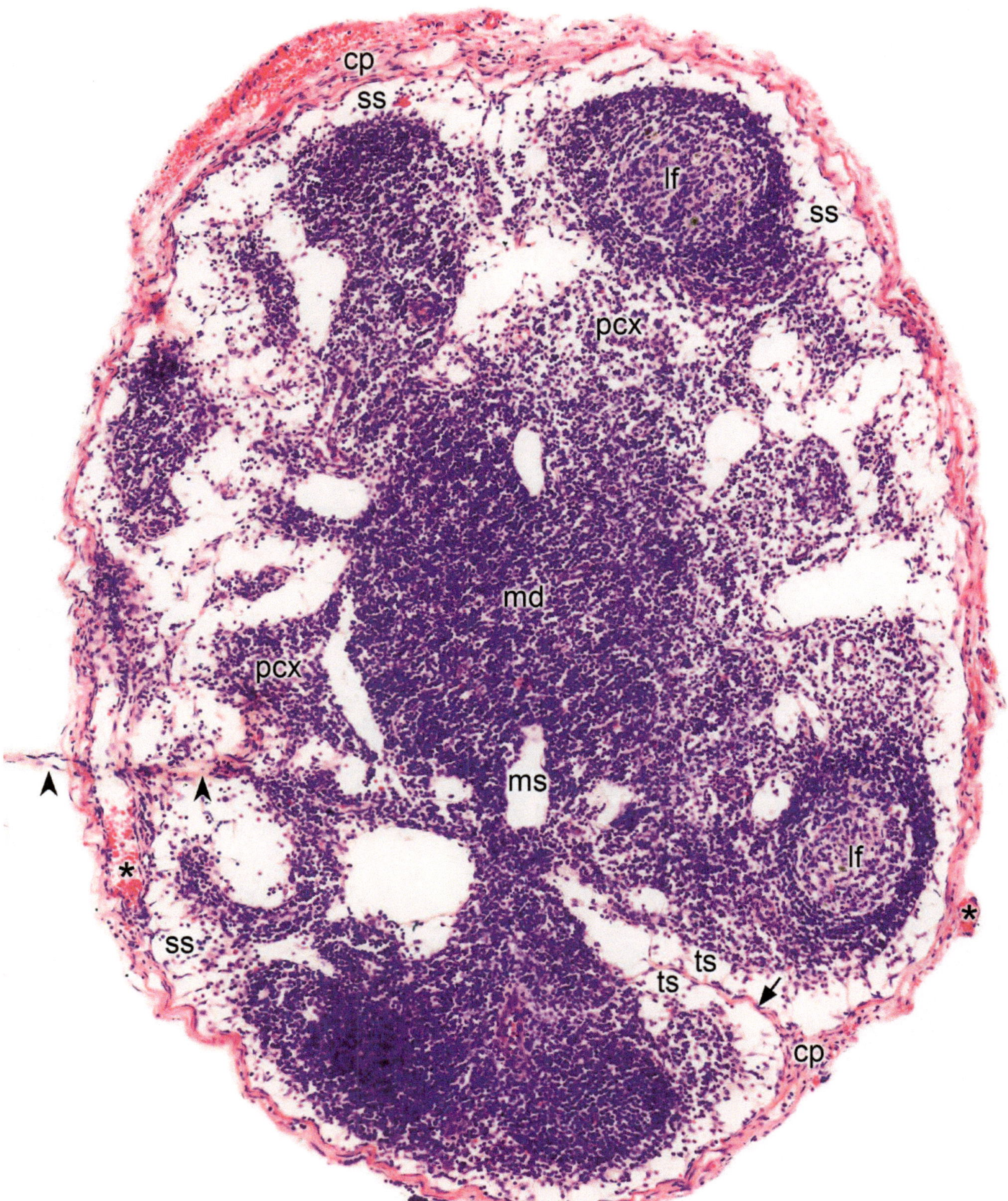

Figure 36. Lymph node (from the tunica adventitia of the caecum). Lymph node is covered by connective tissue capsule (cp) that contains blood vessels (asterisks) and forms trabeculae penetrating the parenchyma (arrow). Lymphatic follicles (lf) form the outer cortex while the paracortex (pcx) and the medulla (md) lacks follicles. Afferent lymph vessels (arrowheads) enter the subcapsular sinus (ss) that drains into the trabecular sinuses (ts) located along the trabeculae (arrows). Trabecular sinuses open into medullary sinuses (ms).

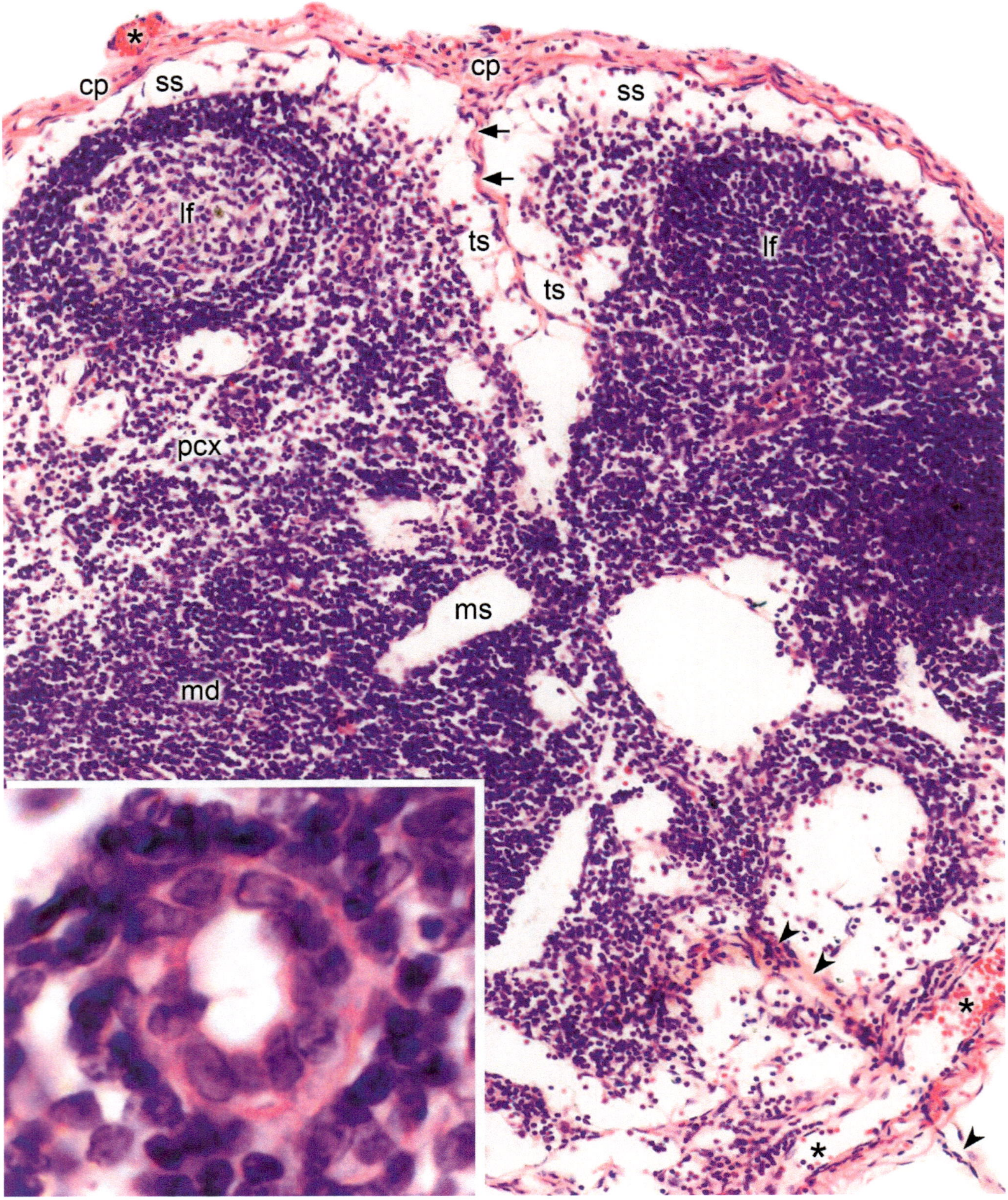

Figure 37. Lymph node (from the tunica adventitia of the caecum). Lymph node is covered by connective tissue capsule (cp) that contains blood vessels and sends trabeculae into the parenchyma (arrows). Lymphatic follicles (lf) form the outer cortex while the paracortex (pcx) and the medulla (md) lacks follicles. Afferent lymph vessels (arrowhead) enter the subcapsular sinus (ss) that drains into the trabecular sinuses (ts) located along the trabeculae (arrows). Trabecular sinuses open into medullary sinuses (ms). Paracortex contains high endothelial venules (inset).

Thymus

Thymus is a lymphatic organ located posterior to the sternum and it is gradually replaced by adipose tissue after puberty during normal development. The connective tissue capsule of the thymus forms trabeculae that divide the thymus into lobules. Each lobule has an outer darker cortex and an inner lighter medulla. Thymus is the primary site of the maturation of T lymphocytes that undergo T cell education while moving from the cortex to the medulla. Epithelioreticular cells that form the framework of the thymus are crucial for this process; from the six types, types I, II and III are located in the cortex while types IV, V and VI populate the medulla. Type II and type V cells are easily detectable as large cells with pale oval nuclei and light cytoplasm in the cortex and medulla, respectively. Concentrically arranged type VI epithelioreticular cells form thymic (Hassall's) corpuscles that are characteristic for the medulla. Type I, III and IV epithelioreticular cells are flattened; type I is located under the capsule while types III and IV delineate the corticomedullary border. Thymus is normally replaced by adipose tissue after puberty; if thymus does not regress (thymus persistens), islands of thymic parenchyma remain embedded into adipose tissue. The cortex and the medulla are ill defined in thymus persistens when compared to the prepubertal thymus.

Spleen

Spleen is surrounded by dense connective tissue capsule that sends trabeculae, containing trabecular vessels, into the parenchyma. When cut, the surface of the unfixed spleen reveals white globular structures (white pulp) surrounded by red substance (red pulp). White pulp is lymphatic tissue, while red pulp is mostly composed of blood. In histological sections stained with H&E, white pulp is represented by basophilic lymphatic aggregations, primarily lymph follicles. In spleen, these follicles are special, since they are perforated by a central artery/arteriole; these follicles are termed as splenic or Malpighian corpuscles and they are primarily composed of B lymphocytes. Between the corpuscles, the red pulp contains splenic sinusoids separated by splenic (Billroth's) chords.

The circulation of the spleen is complex. Splenic arteries open into the trabecular arteries that form the central arteries that eventually become central arterioles. Central arteries/arterioles are surrounded by the periarterial lymphatic sheath (PALS) that is primarily composed of T lymphocytes. After leaving the PALS, central arterioles form penicillar arterioles that are surrounded by macrophage sheaths that filter out cellular debris. Penicillar arterioles may directly open into the splenic sinusoids that are covered by discontinuous endothelium where endothelial cells are arranged like staves of a barrel with gaps between them (closed circulation). In some cases, penicillar arterioles terminate before entering the sinusoids and then the blood filters into the sinusoids through the red pulp (open circulation). Sinusoids drain into the trabecular veins that are tributaries of the splenic veins.

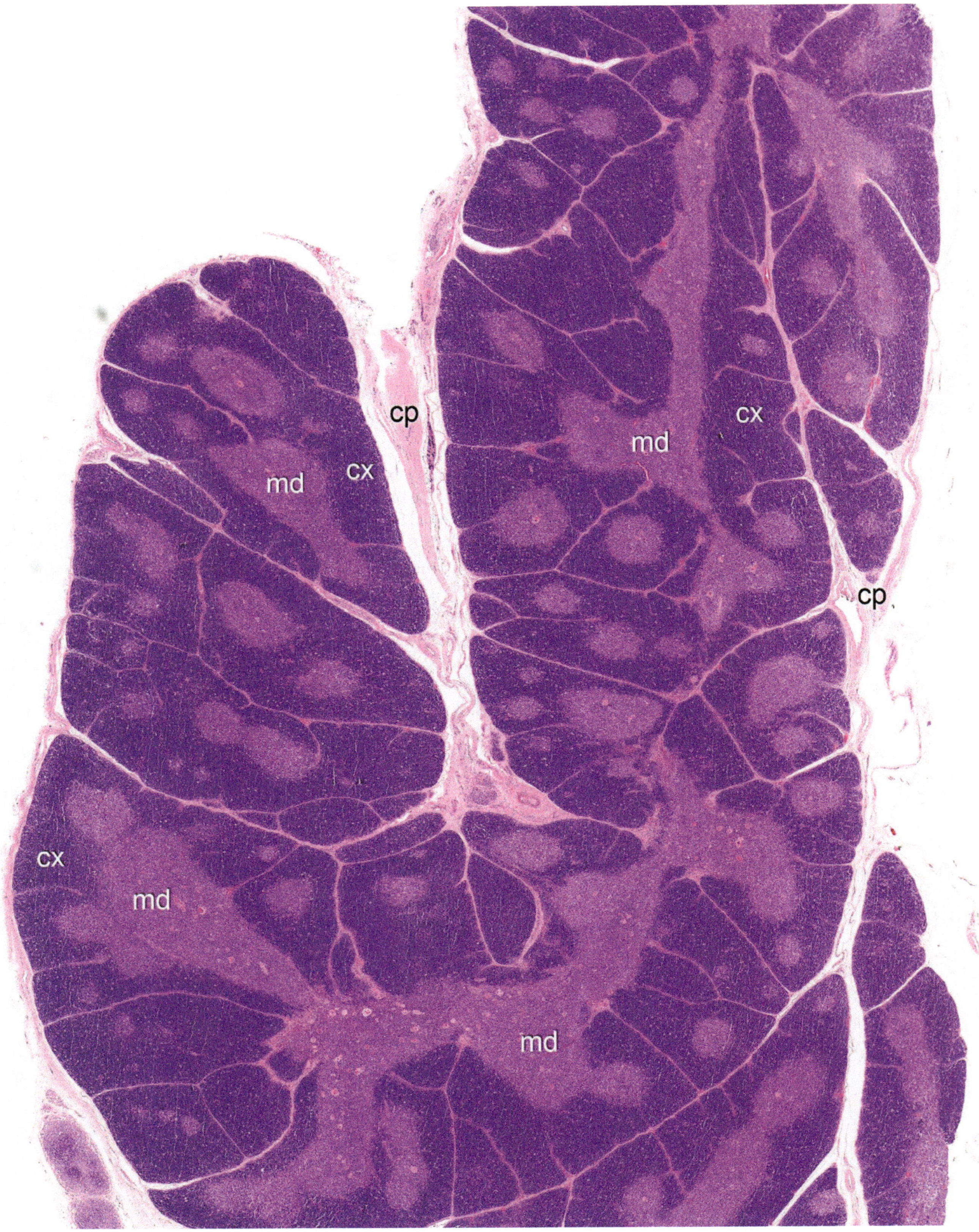

Figure 38. Thymus. Thymus is covered by connective tissue capsule (cp) that sends septa into the parenchyma and divides it into lobules. Each lobule has an outer darker cortex (cx) and inner lighter medulla (md). Small T lymphocytes form the cortex while the medulla is composed of larger T cells. Lymphocytes are supported by a meshwork of epithelioreticular cells. Thymus lacks reticular fibers.

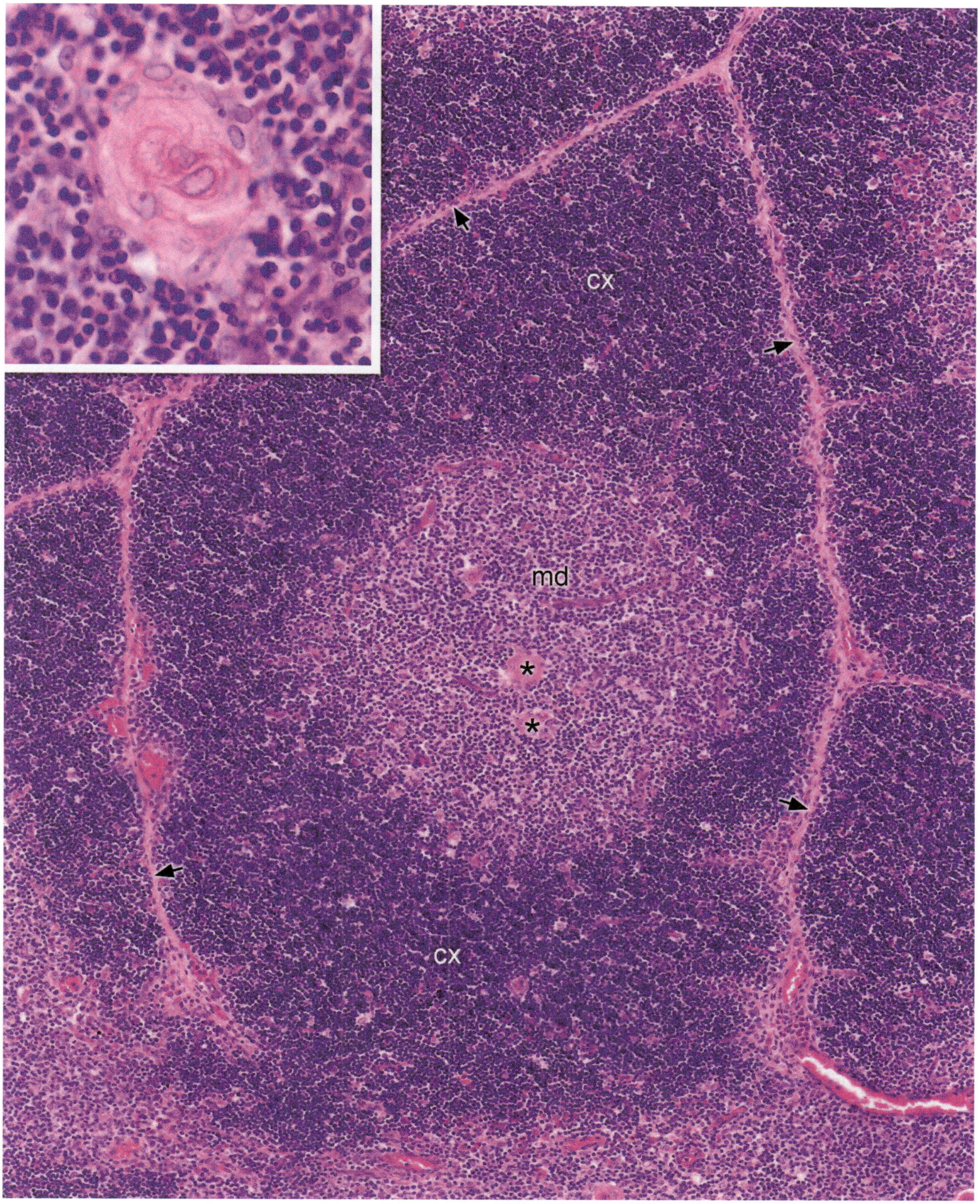

Figure 39. Thymic lobule. Connective tissue septa (arrows) divide the thymus into lobules. Each lobule has an outer darker cortex (cx) and inner lighter medulla (md). Small T lymphocytes form the cortex while the medulla is composed of larger T cells. Thymus lacks reticular fibers; lymphocytes are supported by a meshwork of epithelioreticular cells. Concentrically arranged type VI epithelioreticular cells form Hassal's corpuscles (asterisks) in the medulla.

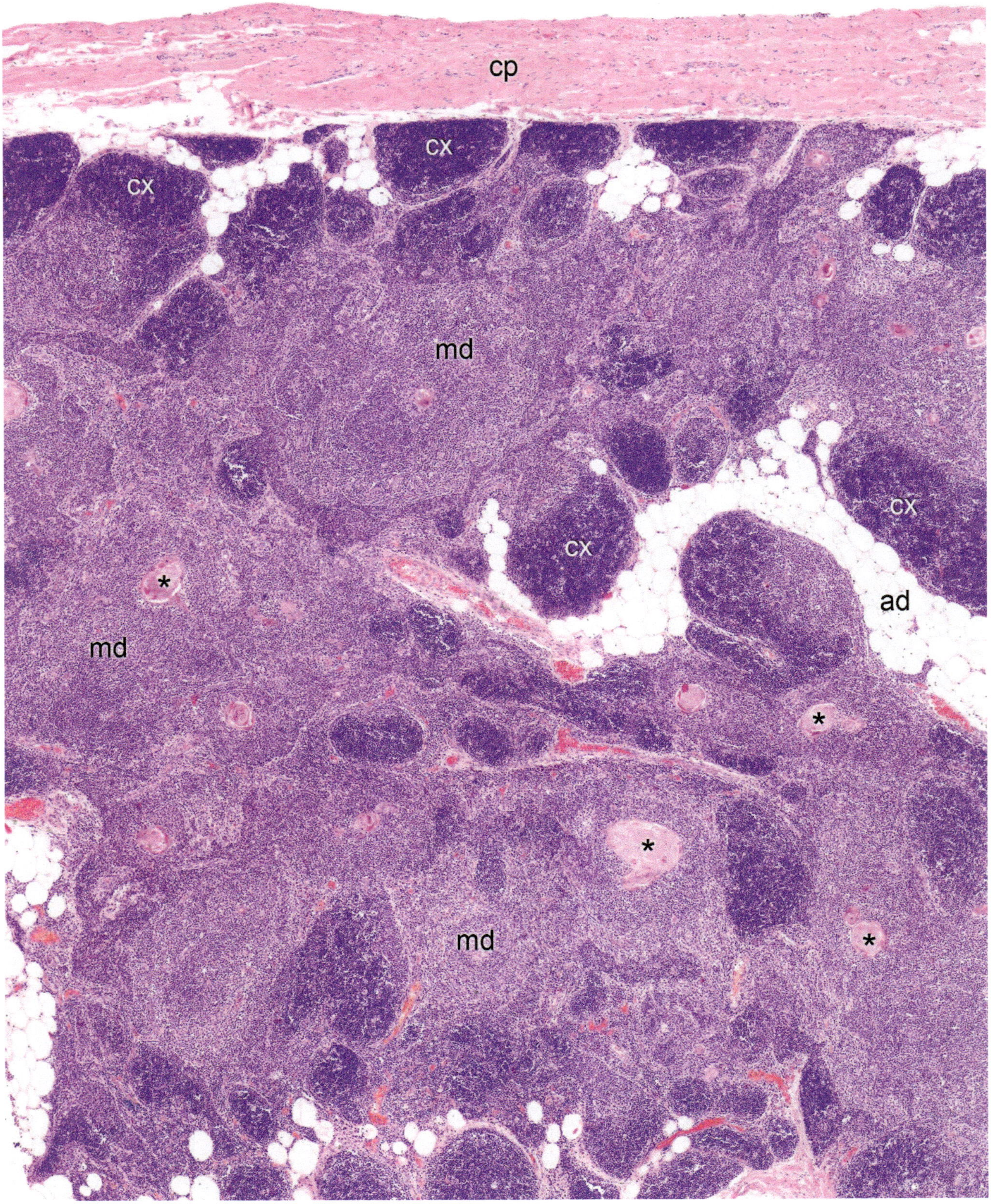

Figure 40. Thymus. Thymus is covered by connective tissue capsule (cp) that sends septa into the parenchyma and divides it into lobules. Each lobule has an outer darker cortex (cx) and inner lighter medulla (md). Thymus lacks reticular fibers; lymphocytes are supported by a meshwork of epithelioreticular cells. In the medulla, concentrically arranged type VI epithelioreticular cells form Hassal's corpuscles (asterisks). Adipocytes (ad) often penetrate the parenchyma.

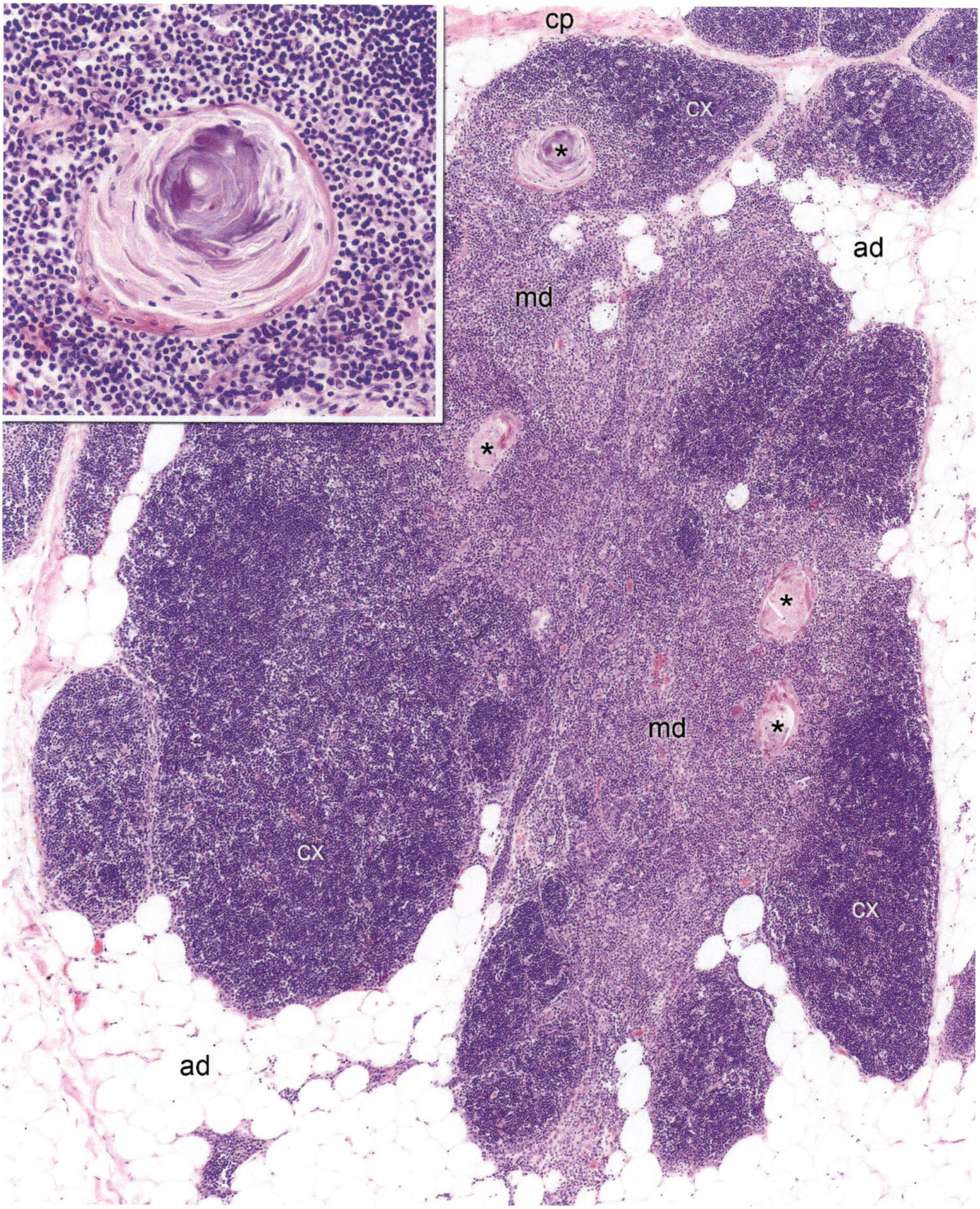

Figure 41. Thymus. Thymus is covered by connective tissue capsule (cp) that sends septa into the parenchyma and divides it into lobules. Each lobule has an outer darker cortex (cx) and inner lighter medulla (md). Thymus lacks reticular fibers; lymphocytes are supported by a meshwork of epithelioreticular cells. In the medulla, concentrically arranged type VI epithelioreticular cells form Hassal's corpuscles (asterisks, inset). Adipocytes (ad) often penetrate the parenchyma.

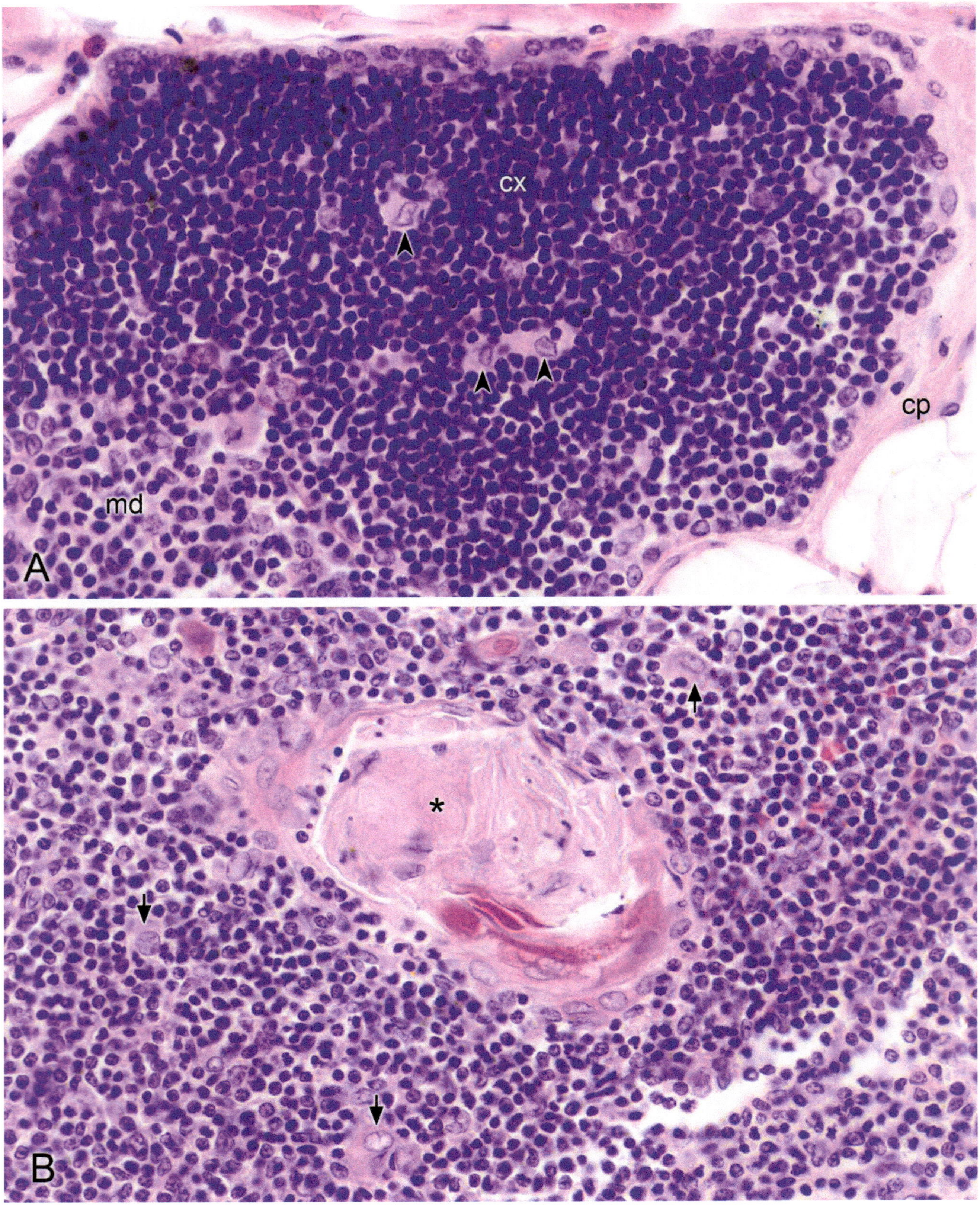

Figure 42. Thymus. (A) Thymus is covered by connective tissue capsule (cp) that forms septa dividing the parenchyma into lobules. Each lobule has an outer darker cortex (cx) and inner lighter medulla (md). Cortex contains type I, II and III epithelioreticular cells; type II cells are large, and they have a light cytoplasm and lightly stained nucleus (arrowheads). (B) Medulla contains type IV, V and VI epithelioreticular cells; type V cell (arrows) has a similar morphology to type II cell while type VI cells form Hassal's corpuscles (asterisk).

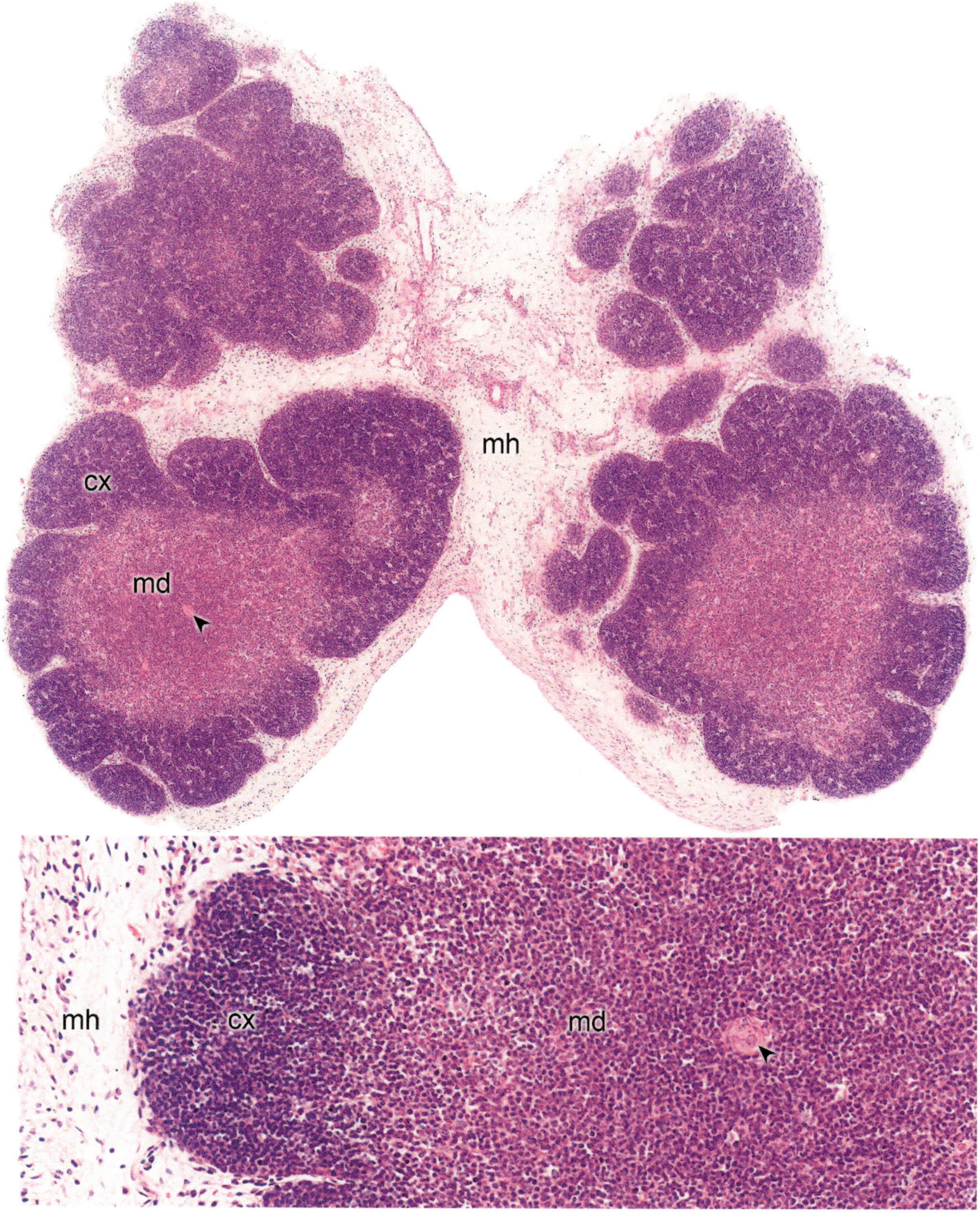

Figure 43. Developing thymus (3-month-old fetus). Fetal thymus exhibits lobular structure with distinct cortex (cx) and medulla (md). Thymic parenchyma is embedded into mesenchyme (mh) that is composed of mesenchymal cells with thin and long processes. In the medulla, concentrically arranged type VI epithelioreticular cells form Hassal's corpuscles (arrow) that are detectable at early age of development.

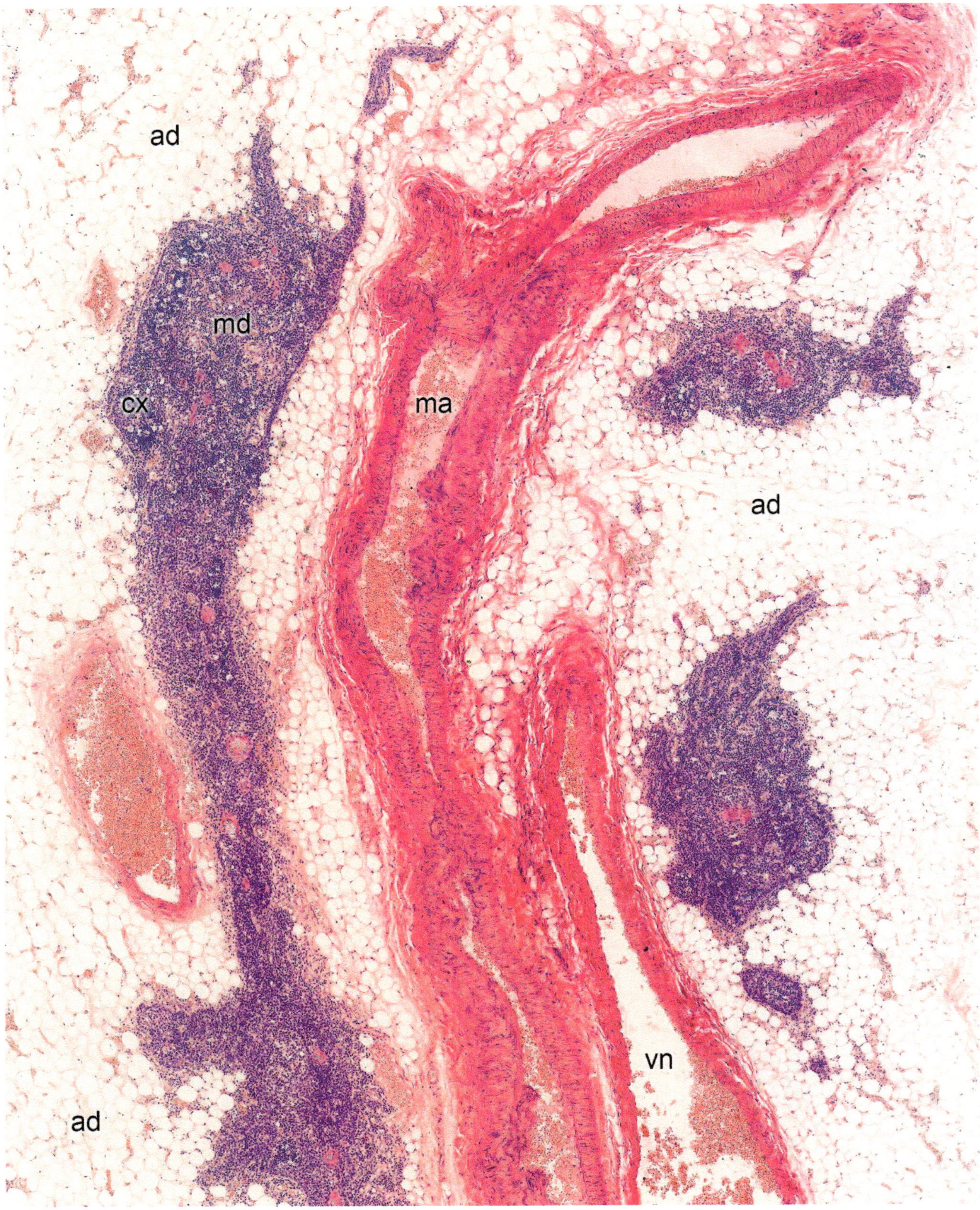

Figure 44. Thymus persistens. Thymus is normally replaced by adipose tissue after puberty; in case of thymus persistens, thymic parenchyma is retained as islands embedded into adipose tissue (ad) that also contains vessels (ma, muscular artery; vn, vein). In thymus persistens, the separation of cortex (cx) and medulla (md) is rather obscure.

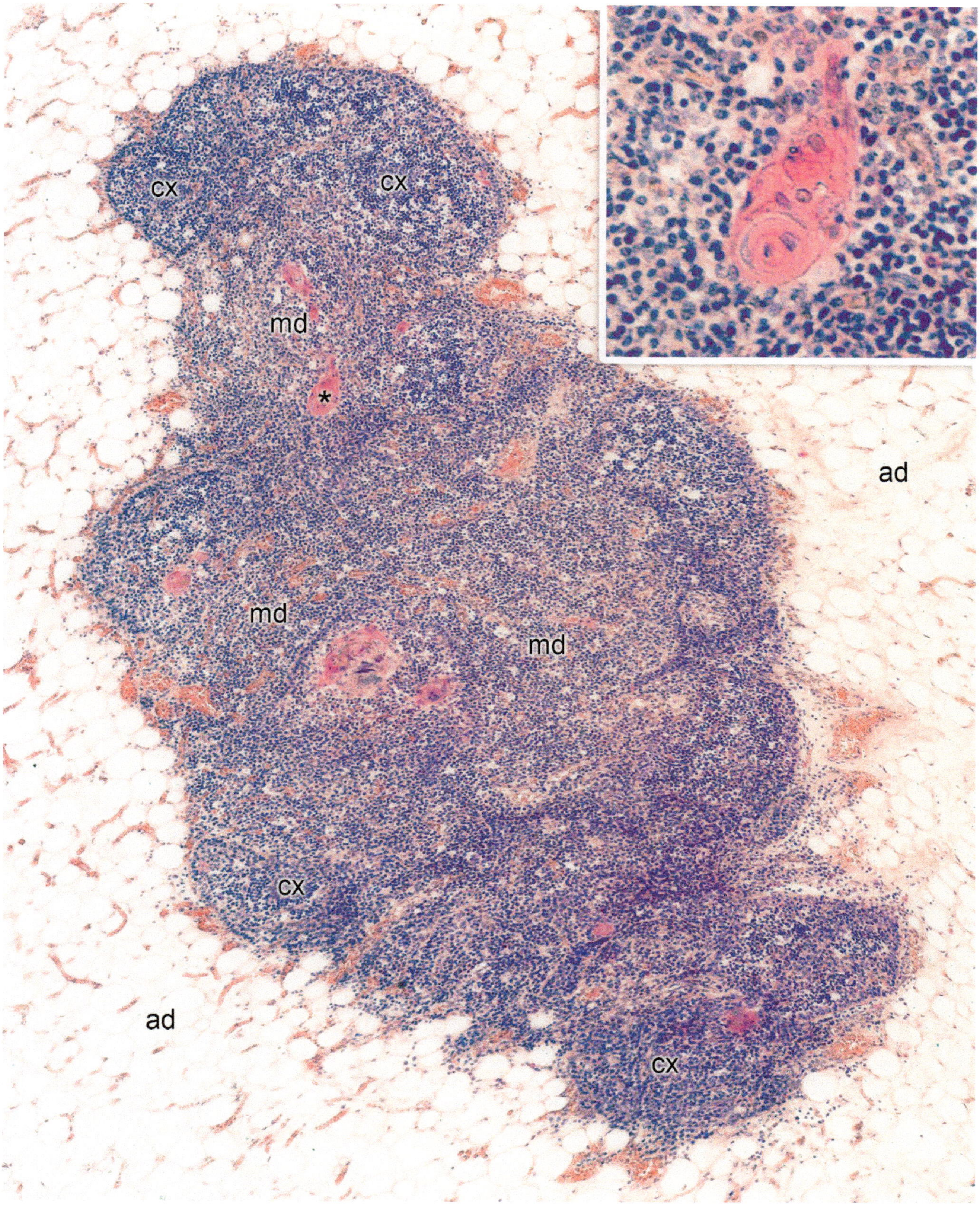

Figure 45. Thymus persistens. Thymus is normally replaced by adipose tissue after puberty; in case of thymus persistens, thymic parenchyma is retained as islands embedded into adipose tissue (ad). In thymus persistens, the separation of cortex (cx) and medulla (md) is rather obscure, but Hassal's corpuscles are clearly recognizable (asterisk, inset).

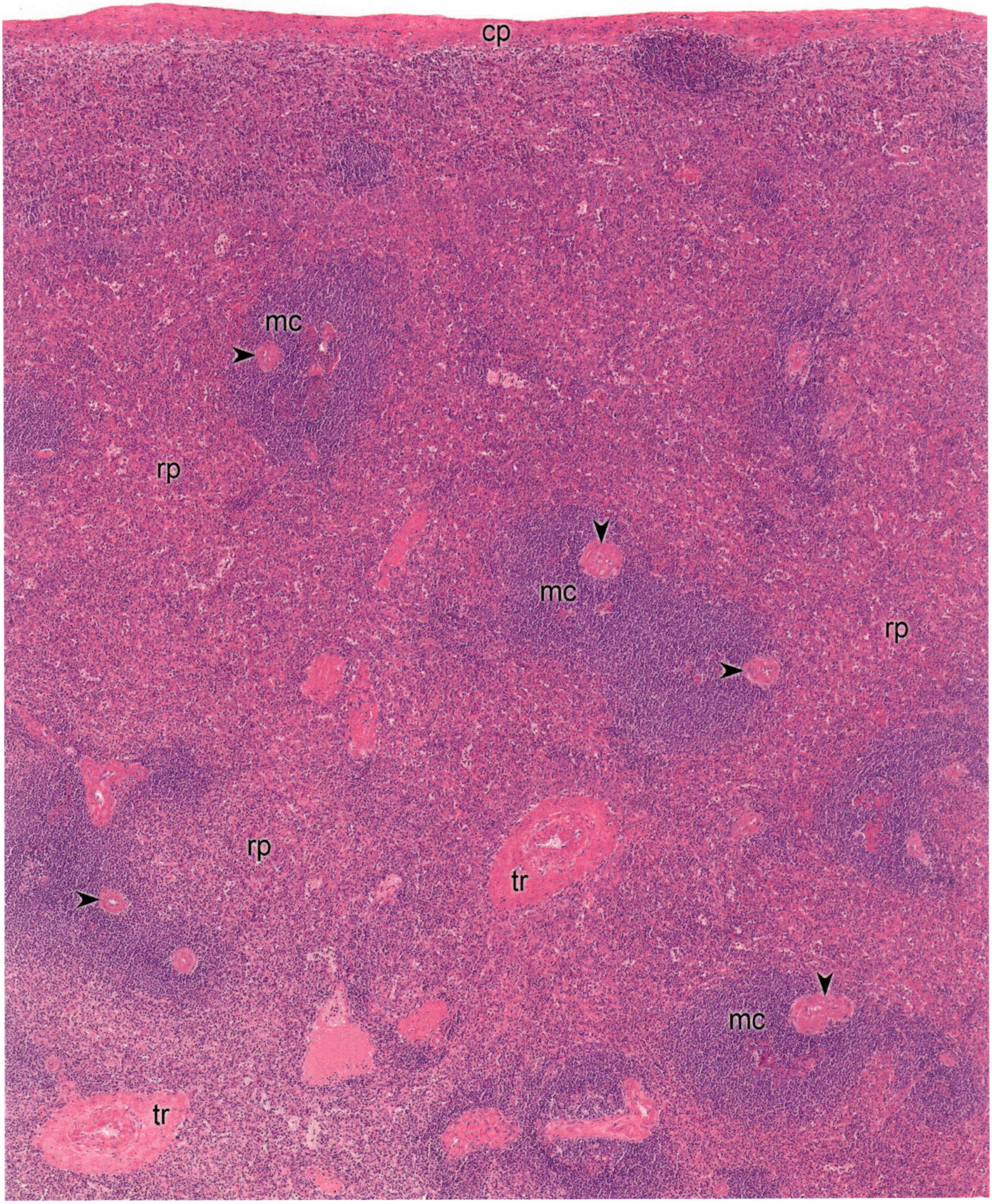

Figure 46. Spleen. Spleen is covered with a dense connective tissue capsule (cp) that sends trabeculae (tr) into the parenchyma. Trabeculae contain trabecular vessels (arteries and veins). The splenic parenchyma is composed of white pulp and red pulp (rp); the former is composed of lymphatic tissue while the latter is formed primarily of blood. White pulp is mostly represented by Malpighian (splenic) corpuscles (mc) that are lymphatic follicles perforated by a central artery or arteriole (arrowheads).

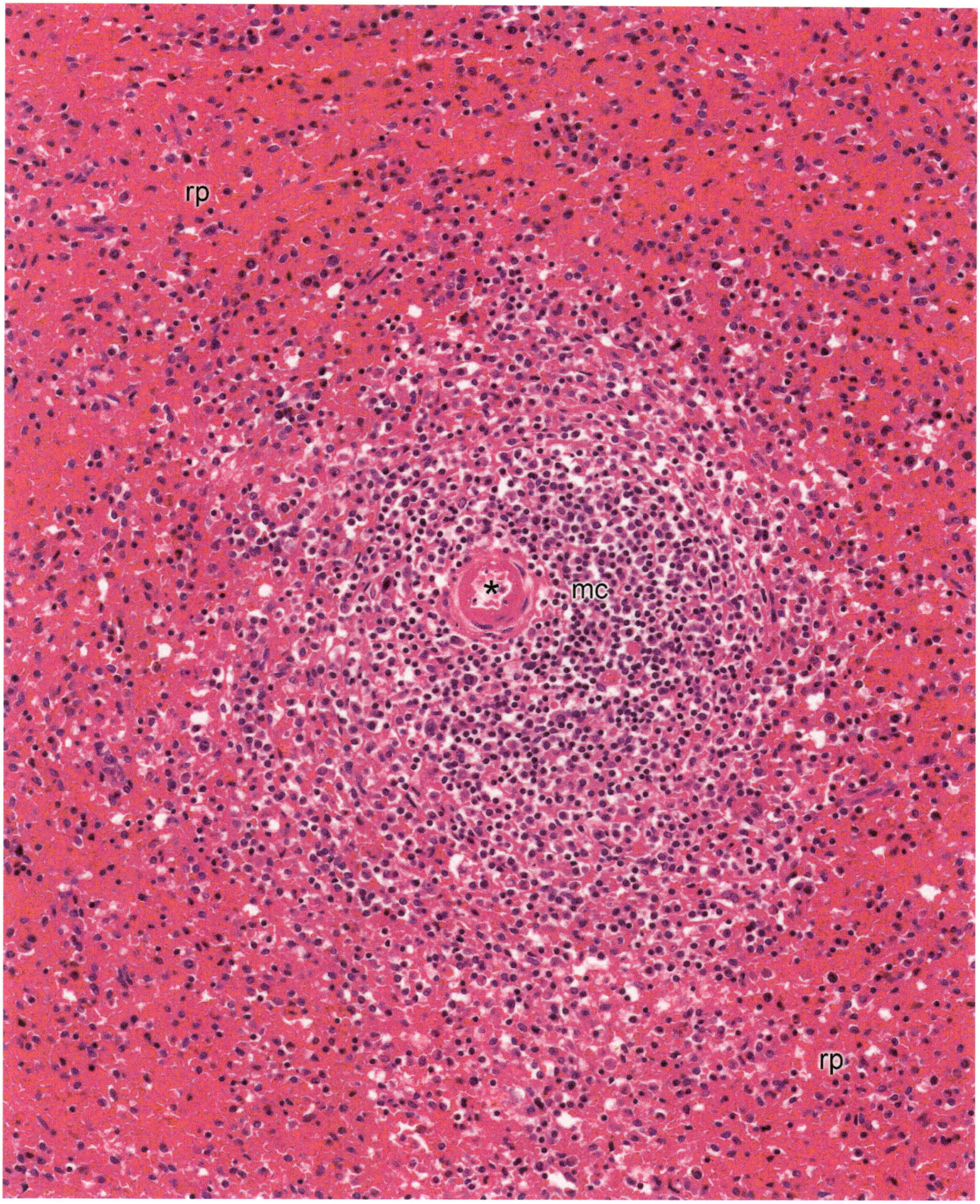

Figure 47. Malpighian (splenic) corpuscle. White pulp is mostly represented by Malpighian (splenic) corpuscles (mc) that are lymphatic follicles perforated by a central artery or arteriole (asterisk). Malpighian corpuscles are surrounded by the red pulp (rp) that is mostly composed of blood.

Chapter 3

Cardiovascular System

Heart

The primary component of the heart is the myocardium composed of cardiac muscle that has been described previously with the basic tissues. Briefly, cardiac muscle cells are elongated branched cells with centrally located, round or oval nuclei, many cells are binucleated. Cytoplasm is eosinophilic and features cross striation; cell boundaries are prominent lines identified as intercalated discs. The innermost surface of the cardiac wall is covered by the endocardium that covers not only the myocardium but the heart valves as well; it is thick in atria and much thinner in ventricles. Endocardium is formed by endothelial cell lining that is continuous with the endothelium of the large vessels and a loose connective tissue subendothelium. This layer is connected to the myocardium by the more fibrous subendocardium that contains the conducting system of the heart. Myocardium is covered by the epicardium on the outer surface of the heart, and it is composed of a single layer of mesothelial cells resting on loose connective tissue. Epicardium can be closely attached to the myocardium, but certain areas, particularly near vessels, can contain large amount of adipose tissue. Heart valves are generally formed by a dense connective tissue core that is covered by endocardium at the atrial and ventricular sides.

The conducting apparatus of the heart is composed of modified cardiac muscle cells. In the nodes (sinoatrial and atrioventricular nodes) these cells are small, without cross striation and visible intercalated discs, and they are embedded into a stroma rich in collagen fibers and capillary networks. The bundle of His and Purkinje fibers are formed by modified cardiac muscle cells that are substantially enlarged, cytoplasm is pale due to the glycogen content and intercalated discs and cross striation are clearly observable. Nuclei of these cells are larger than the ones in the in the normal cardiac muscle, however, they often appear to be small due to the size of the enlarged conducting cells.

General structure of vessels

Vessels are composed of three layers, the tunica intima, media and adventitia. Tunica intima is covered with endothelium and the underlying basal lamina that rests on a loose connective tissue layer with scattered smooth muscle cells (subendothelium). In arteries, a compact layer of elastic fibers, the internal elastic membrane, separates the tunica intima from the tunica media. Tunica media contains circularly arranged smooth muscle cells ensheathed by the basal lamina as well as by elastic fibers. Tunica media is encircled by the external elastic membrane that has a similar structure to the internal one. The outermost layer of the vessels is the tunica adventitia that is composed of loose connective tissue often containing nerves and smaller vessels that supply the vessel wall (nervi vasorum, vasa vasorum, respectively). This layer is interlaced with longitudinally arranged smooth muscle bundles in veins. In arteries the tunica media, while in veins the tunica adventitia is the most prominent.

http://doi.org/10.1016/B978-0-323-91891-6.50003-9,

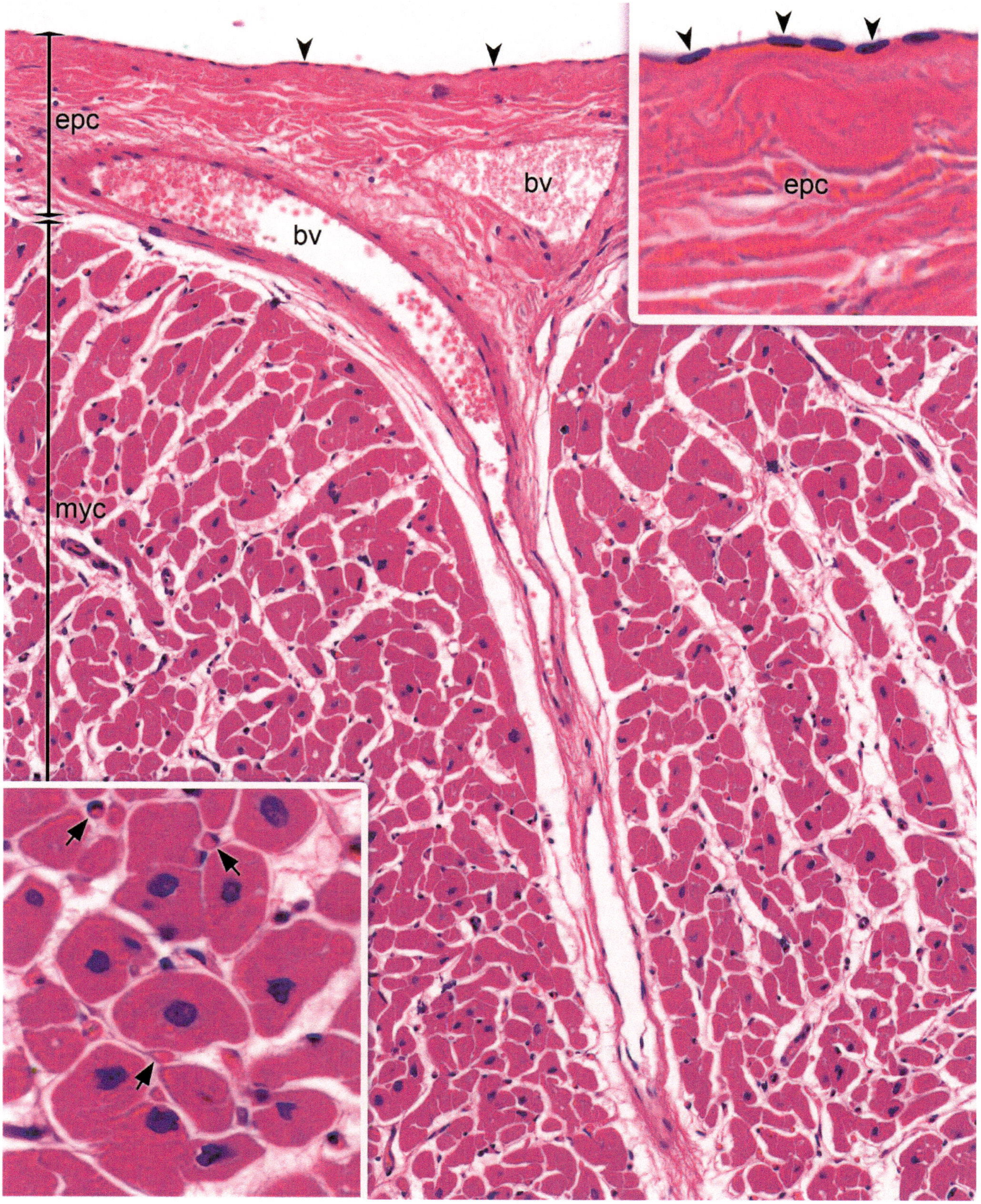

Figure 48. Myocardium and epicardium. Myocardium (myc) is formed by cardiac muscle cells. In cross section, cells exhibit centrally located spherical nuclei (lower inset). Cells are ensheathed by delicate connective tissue envelope with rich capillary network (lower inset, arrows). Myocardium is covered outside by the epicardium (upper inset) that is a connective tissue layer lined with mesothelial cells (arrowheads). Epicardial blood vessels (bv) penetrate the myocardium.

Arteries

The thickest layer of the arteries is the tunica media. Elastic arteries are the largest arteries comprising the aorta and its major branches. The most striking feature of the elastic arteries is the large number of circularly arranged fenestrated elastic lamellae in the tunica media. These lamellae form wavy sheaths between the concentric smooth muscle layers. In elastic arteries, tunica media lacks fibroblasts; the connective tissue fibers are formed by the smooth muscle cells. Elastic fibers in the tunica media help in maintaining diastolic blood pressure. The subendothelium of tunica intima is rich in smooth muscle cells and the internal and external elastic laminae blend with the elastic layers of tunica media. Tunica adventitia is much thinner than the media and contains vessels and nerves embedded into loose connective tissue.

Muscular arteries are medium sized and small vessels that are dominated by the tunica media composed of almost exclusively circularly arranged smooth muscle cells. Tunica intima is thin, and the internal elastic membrane is prominent while the external elastic membrane is usually less defined. Tunica adventitia is composed of loose connective tissue with small vessels and nerves in the larger muscular arteries (vasa and nervi vasorum).

As muscular arteries are getting smaller, they continue in arterioles that are that primary resistance vessels and flow regulators. The transition between the small arteries and arterioles is elusive; generally, arterioles are identified as vessels with only few circular layers of smooth muscle cells. Tunica intima and adventitia are thin and lack well-defined internal and external elastic membranes.

Capillaries

Capillaries convey blood between the arterioles and venules, and they are the smallest vessels of the circulation with the diameter of 5-10 μm. They are composed of a single layer with endothelial cells resting on basal lamina. Pericytes with large, flattened nuclei are commonly present under the endothelium and they are ensheathed by the basal lamina. Fenestrated capillaries have openings on the endothelial cells that are typically covered with a thin diaphragm, and they are characteristic to the capillaries of the endocrine organs and intestines. Continuous capillaries are the most common type located throughout the body including muscle, adipose tissue, and nervous system, and they lack these fenestrations. Finally, discontinuous capillaries have spaces between the endothelial cells, and they commonly form sinusoids in liver, spleen, and bone marrow.

Arteries and veins can also be directly connected by arteriovenous shunts that are vessels with substantial tunica media. AV shunts are located at the acral parts of the body, i.e., fingertips, nose and lips, and they play a key role in thermoregulation and protecting the area against low temperature.

Veins

Capillaries drain into postcapillary venules that have similar wall structure to the capillaries containing endothelial cells, basal lamina and pericytes; however, their diameter is much larger. A specialized type of the postcapillary venules are the high endothelial venules (HEV) of the lymph node that possess tall, almost cuboidal looking epithelium. Postcapillary venules are succeeded by muscular venules with few layers of smooth muscle cells in the tunica media and thin tunica adventitia; they are recognizably different from the arterioles since they have a much larger lumen.

Venules eventually open into small and medium sized veins. These vessels have delicate tunica intima and thin tunica media with few layers of circularly arranged smooth muscle cells; loose connective tissue tunica adventitia is the thickest layer. Medium sized veins commonly contain venous valves that are folds of tunica intima pointing toward the heart. These valves prevent the backflow of the blood. Large veins have similar wall structure, but they lack valves, and usually contain longitudinally arranged smooth muscle bundles in the tunica adventitia.

Lymphatic vessels

Lymphatic circulation starts with blind-ended lymphatic capillaries that converge into larger lymphatic vessels typically containing valves similar to those in the medium sized veins. Lymphatic capillaries have similar wall structure to the blood capillaries, but their diameter is larger. Lymphatic vessels can be identified in the tissue by the absence of erythrocytes in the fluid filling their lumen. Lymph eventually enters the venous circulation by the thoracic duct and the right lymphatic trunk; therefore, lymphatic circulation works in tandem with the general circulation.

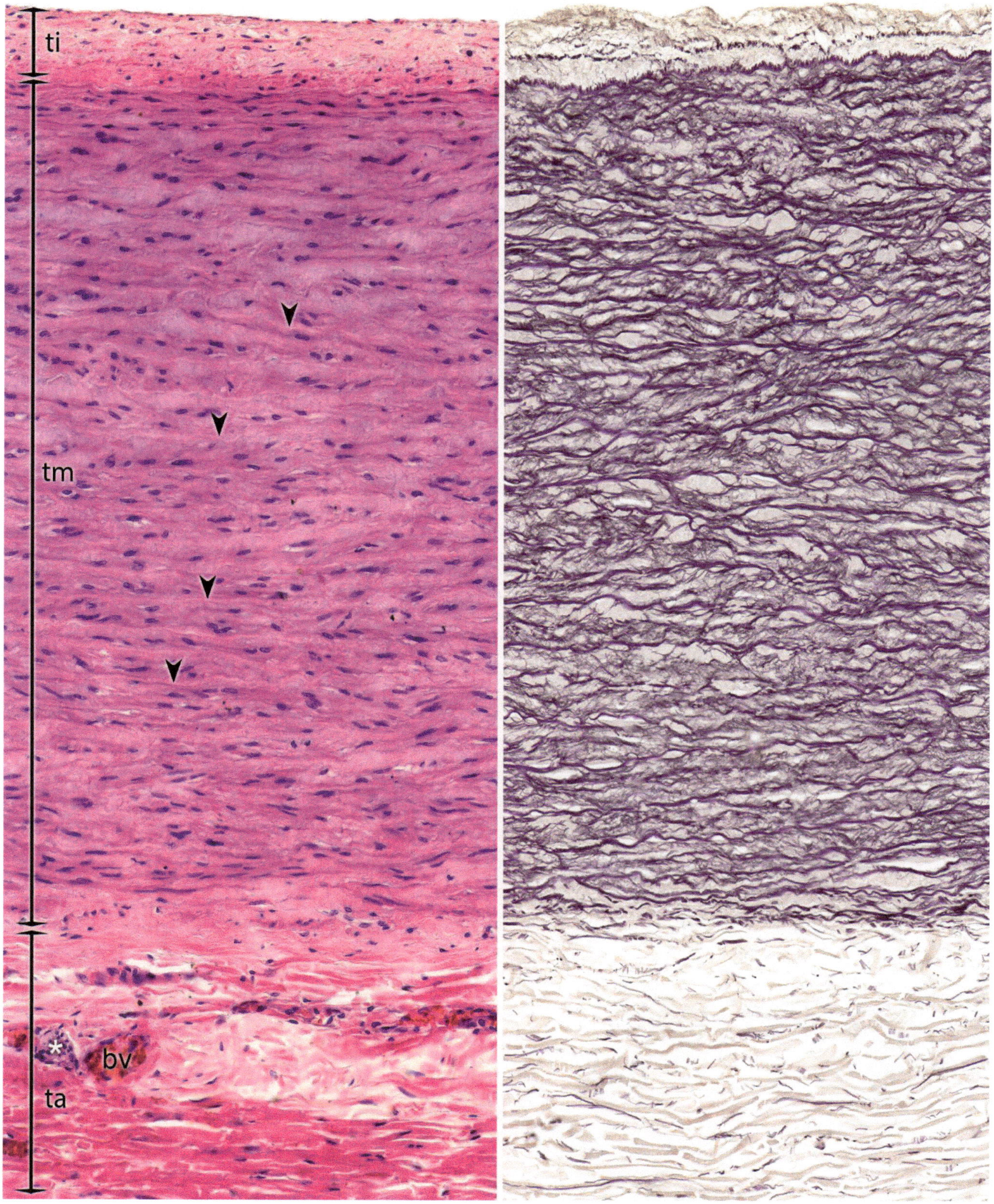

Figure 49. Elastic artery. H&E staining (left) reveals the cellular elements while Weigert's resorcin-fuchsin method stains the elastic fibers (right). Tunica intima (ti) is composed of connective tissue with smooth muscle cells. Tunica media (tm) contains circularly arranged fenestrated elastic lamellae (arrowheads) and smooth muscle cells between them. Tunica adventitia (ta) is formed by connective tissue with blood vessels (bv) and nerves (nv).

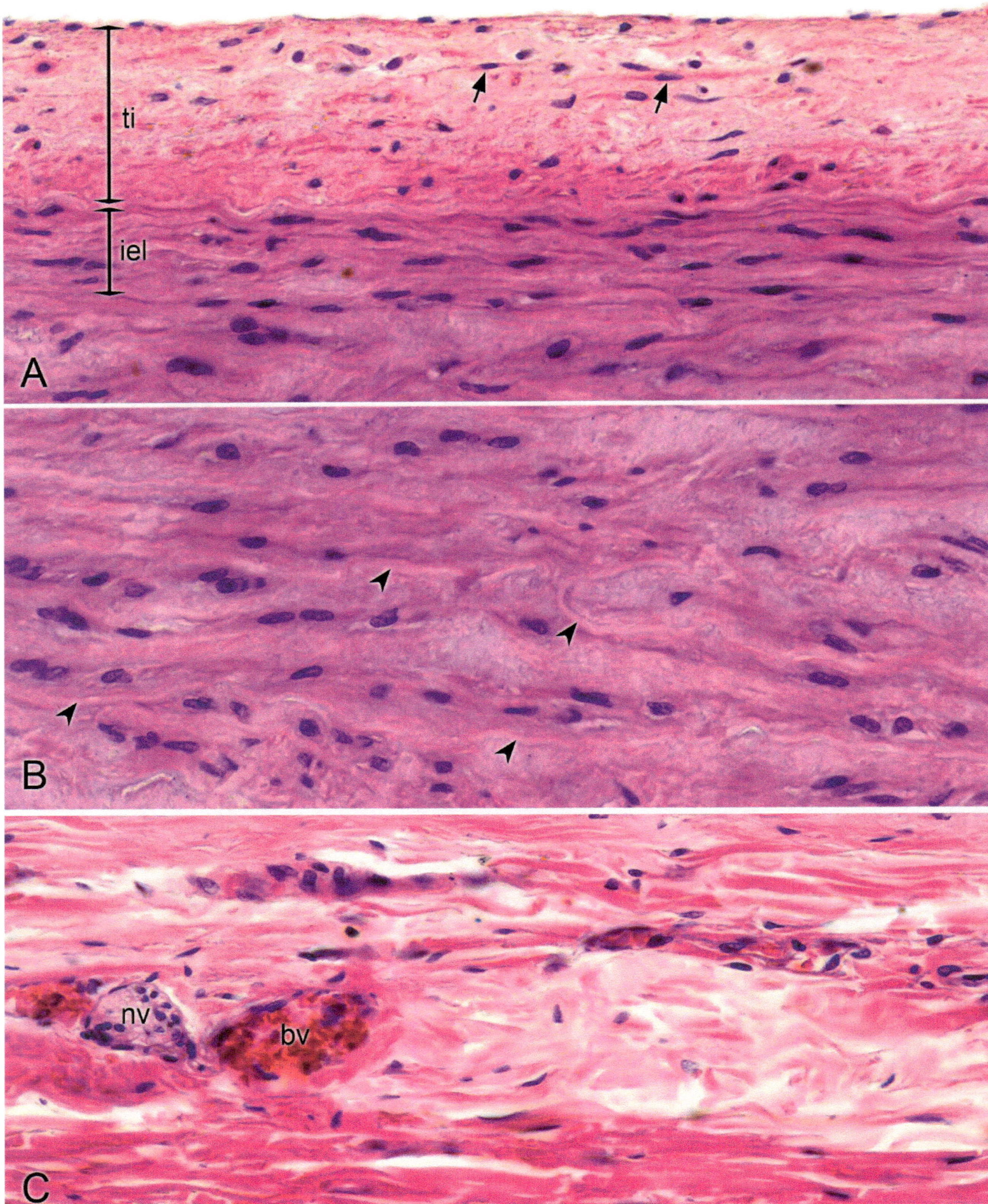

Figure 50. Layers of the elastic artery. (A) Tunica intima (ti) is composed of connective tissue with smooth muscle cells (arrows). Under the intima, the internal elastic lamina (iel) is barely distinguishable from the rest of the tunica media. (B) Tunica media contains circularly arranged fenestrated elastic lamellae (arrowheads) and smooth muscle cells between them. (C) Tunica adventitia is formed by connective tissue with blood vessels (bv) and nerves (nv).

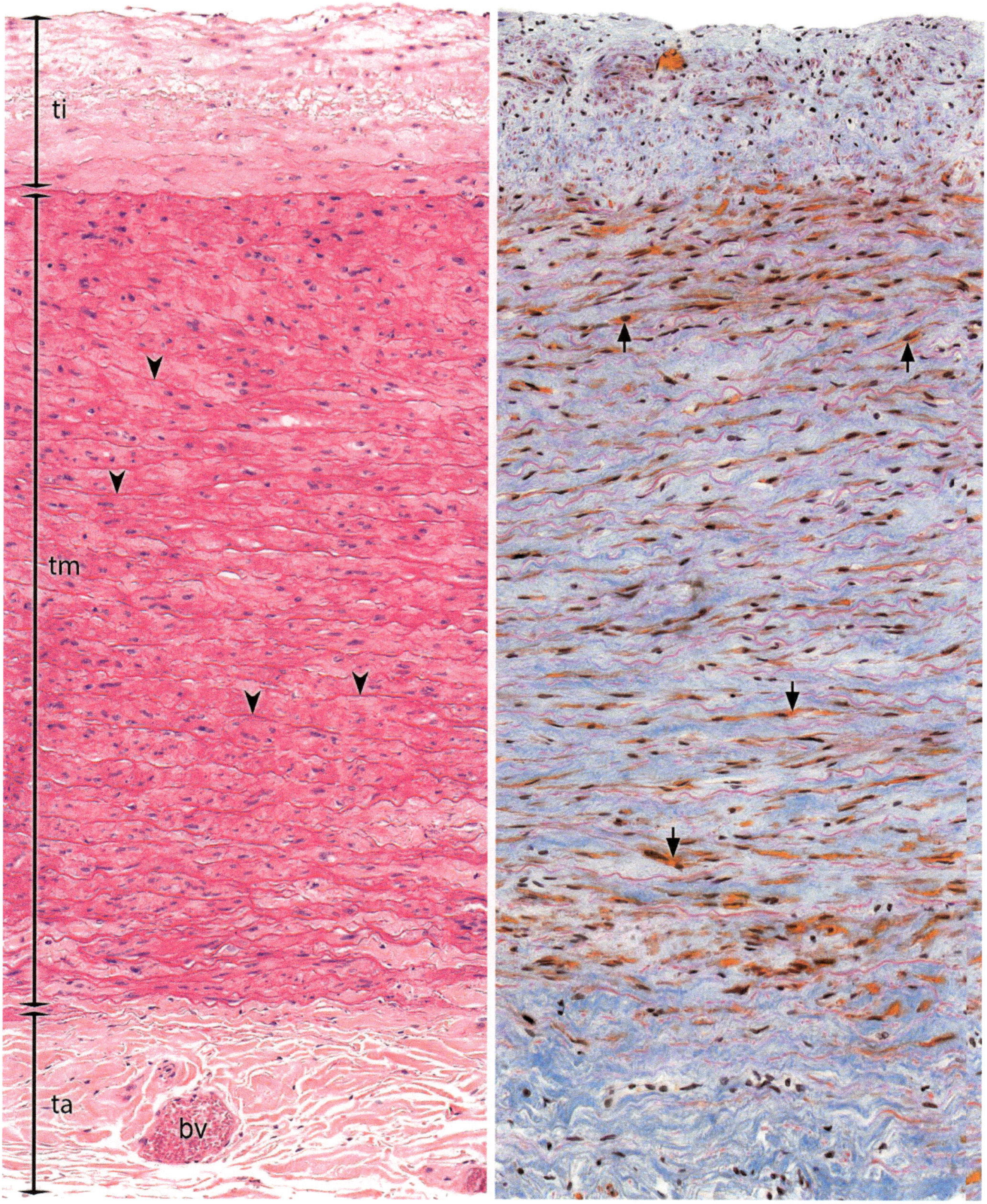

Figure 51. Elastic artery. H&E staining (left) reveals the cellular elements and Masson's trichrome method stains the elastic fibers pink and the smooth muscle cells orange (right). Tunica intima (ti) is thickened, and it is composed of connective tissue with smooth muscle cells. Tunica media (tm) contains circularly arranged fenestrated elastic lamellae (arrowheads) and smooth muscle cells between them. Tunica adventitia (ta) is formed by connective tissue with blood vessels (bv) and nerves.

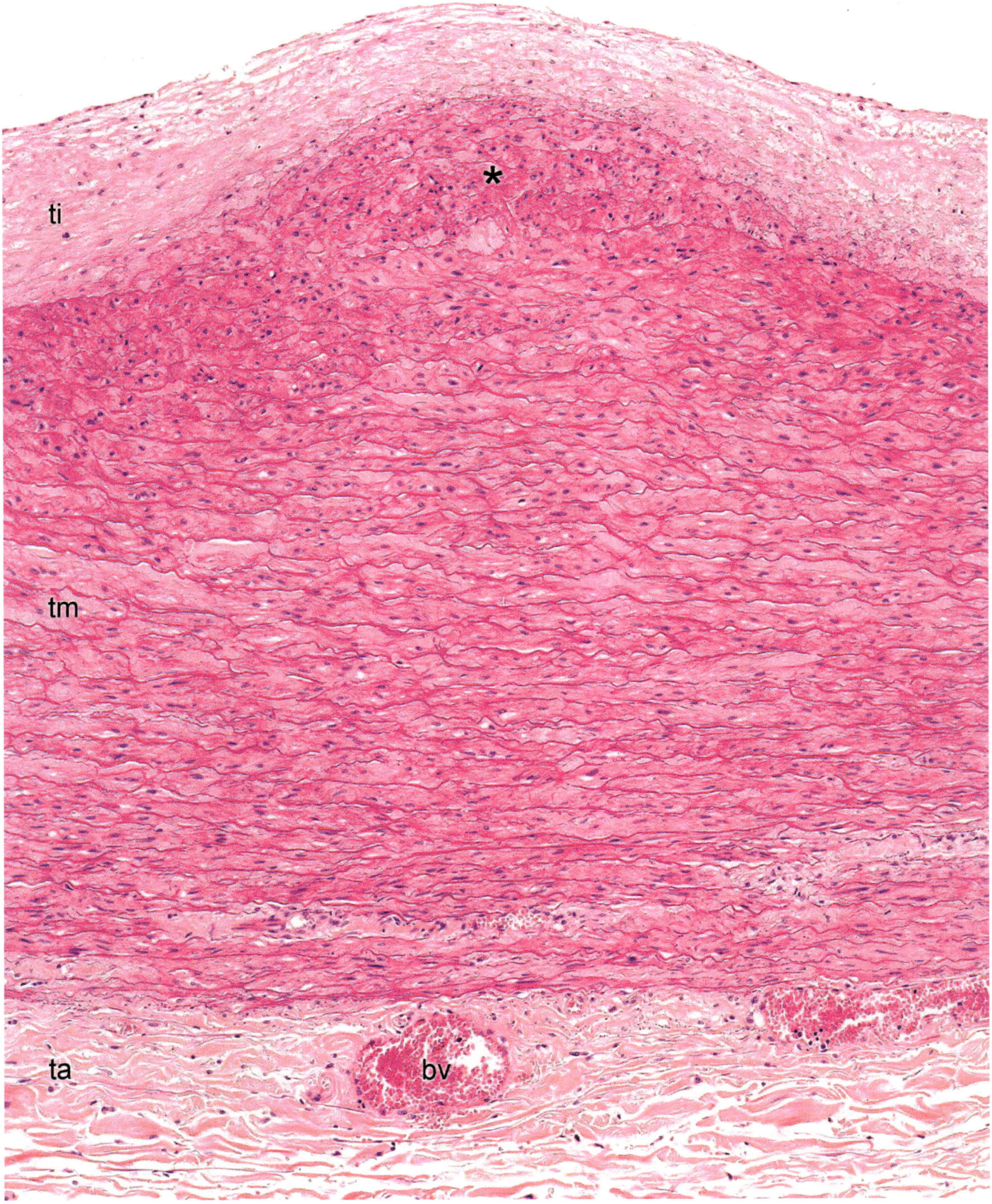

Figure 52. Elastic artery. Tunica intima (ti) is thickened and contains connective tissue fibers, smooth muscle cells and macrophages that accumulate oxidized lipids (foam cells), eventually forming an atherosclerotic plaque (asterisk). Tunica media (tm) is composed of circularly arranged fenestrated elastic lamellae and smooth muscle cells. Tunica adventitia (ta) is formed by connective tissue with blood vessels (bv) and nerves.

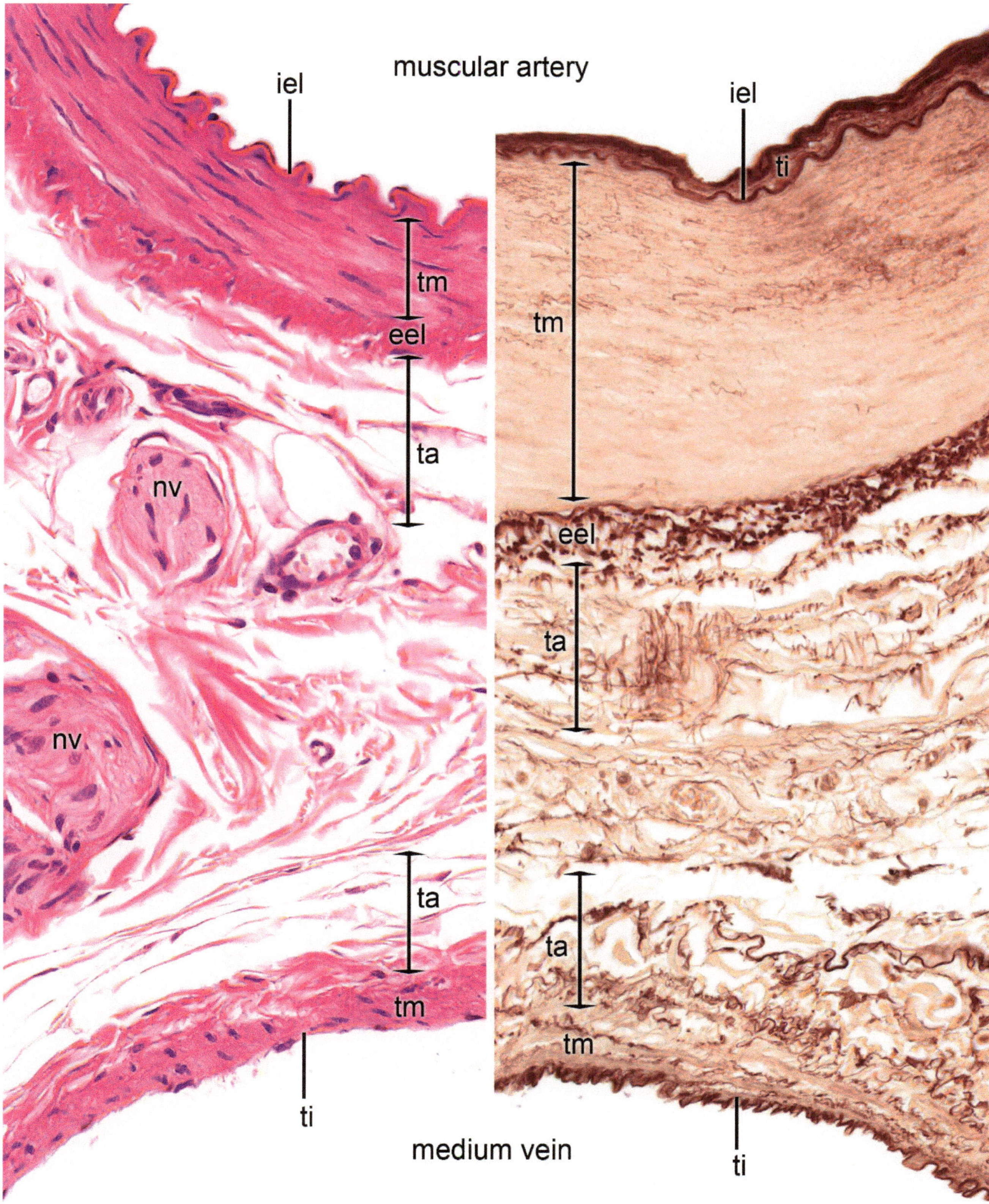

Figure 53. Muscular artery with a medium sized vein. H&E staining on the left reveals the cellular elements while elastic fibers are stained with orcein on the right. In the muscular artery (top) the internal elastic lamina (iel) is between the tunica intima (ti) and media (tm), while external elastic lamina (eel) is between the tunica media and adventitia (ta). Tunica media is thin in veins (bottom) and the dominating layer is the tunica adventitia). Nerves (nv) are often in the close vicinity of vessels.

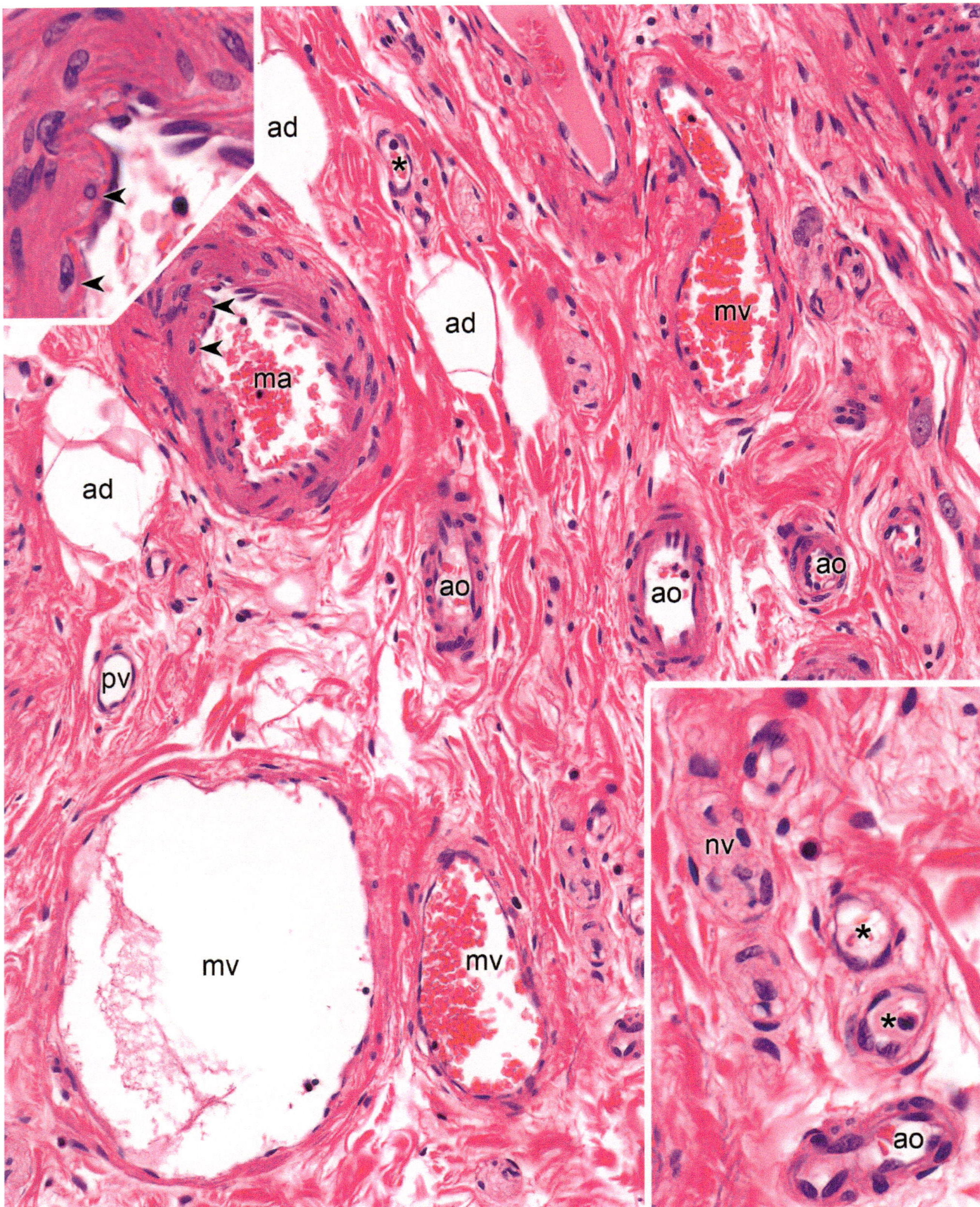

Figure 54. Microvasculature. Small muscular artery (ma) contains internal elastic lamina (arrowheads, upper inset) that is missing from the arterioles (ao) that contain few concentric layers of smooth muscle cells. Asterisk denote capillaries transitioning into postcapillary veins (pv); both vessels are formed by endothelial cells and adjacent pericytes. Smooth muscle appears in the wall of muscular venules (mv). Adipocytes (ad) and small nerves (nv) often surround the microvasculature.

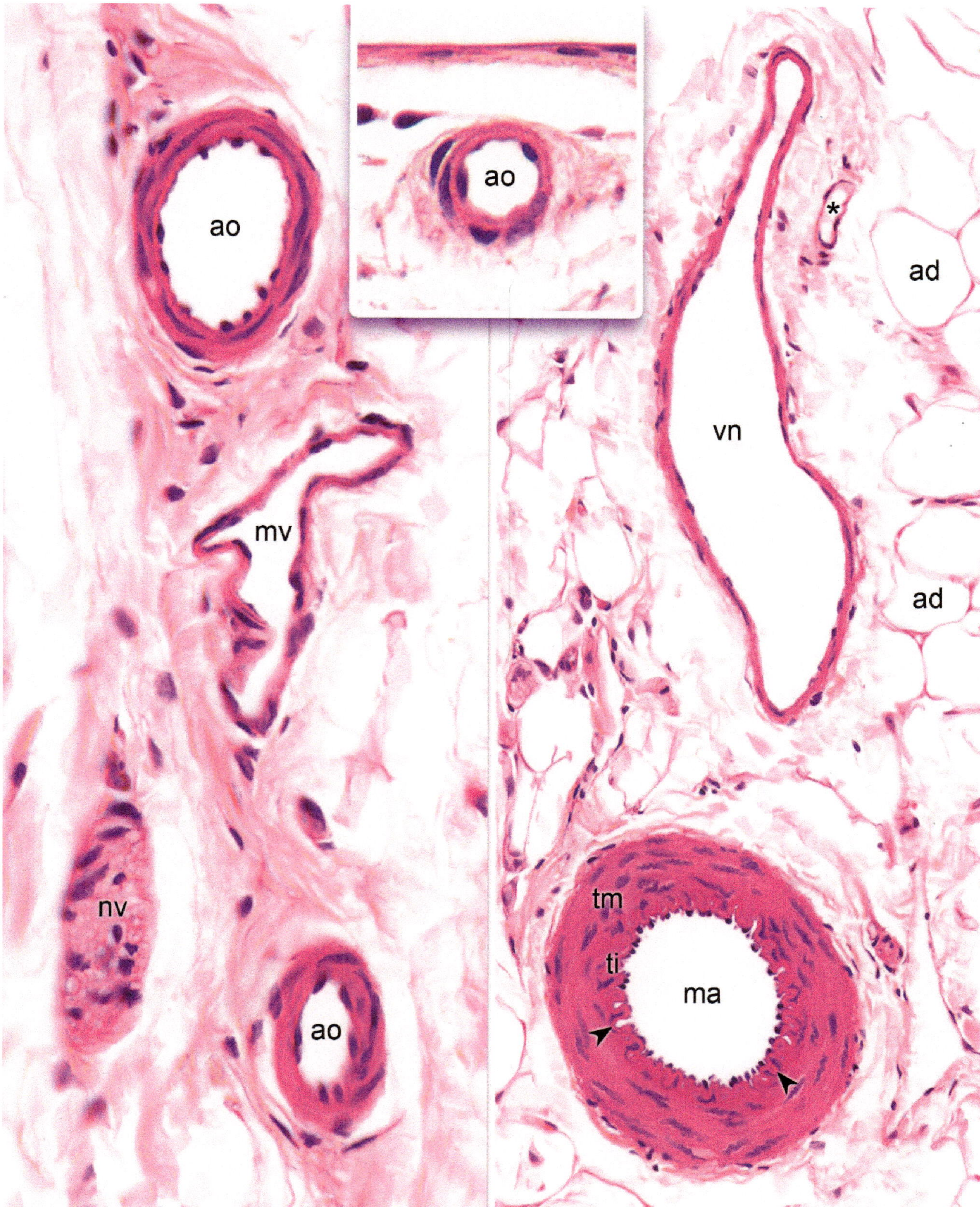

Figure 55. Microvasculature. Between the tunica intima and media (tm), small muscular artery (ma) contains internal elastic lamina (arrowheads) that is missing from the arterioles (ao) that contain few concentric layers of smooth muscle cells. Asterisk denotes a capillary formed by endothelial cells and adjacent pericytes. Smooth muscle appears in the wall of muscular venules (mv), and it forms several layers in small veins. Adipocytes (ad) and small nerves (nv) often surround the microvasculature.

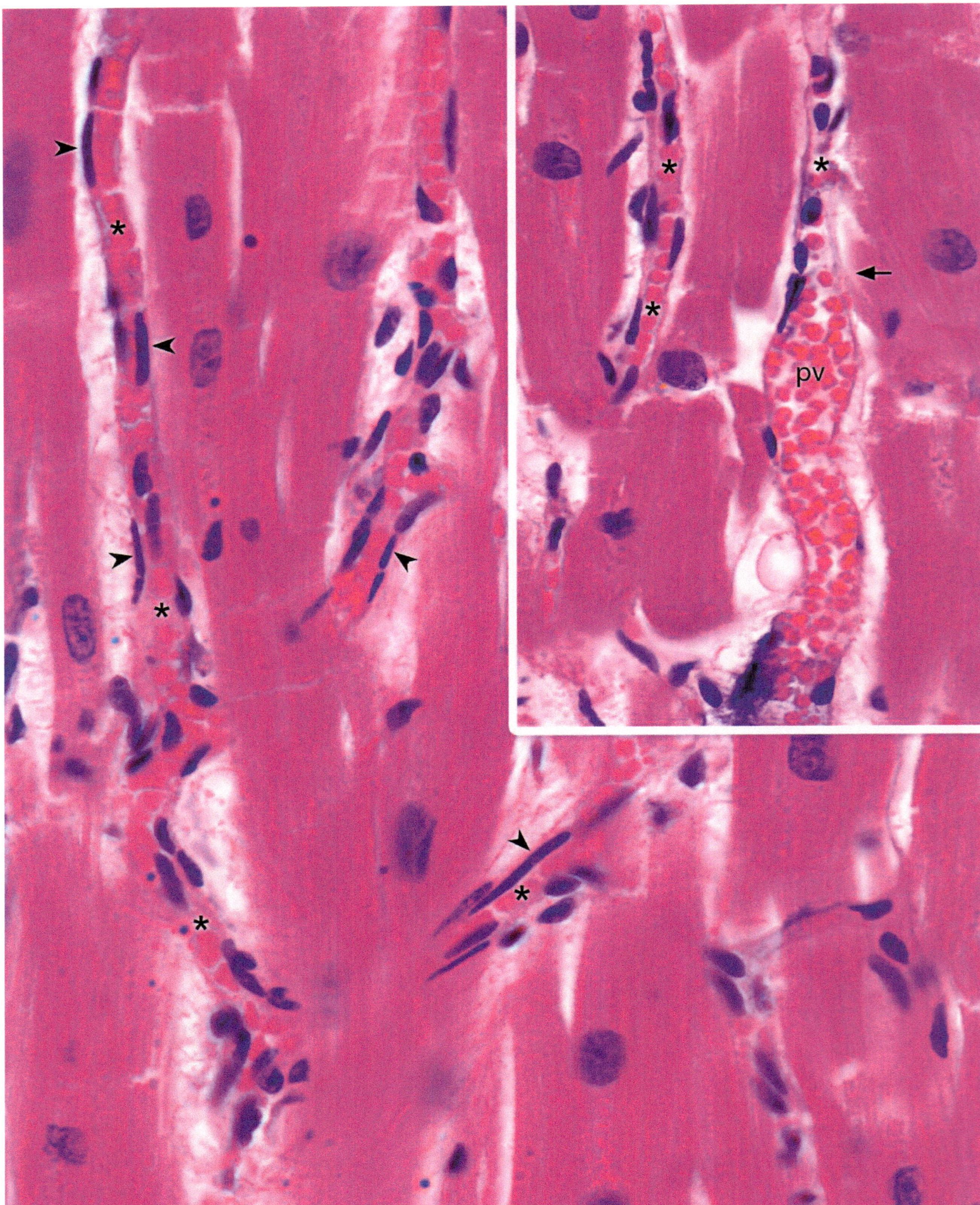

Figure 56. Microvasculature in cardiac muscle. The wall of the capillaries (asterisks) is formed by endothelial cells (arrowheads) and adjacent pericytes. Capillaries eventually transition (arrow) to postcapillary venules (pv) that have a larger diameter, but a similar wall structure.

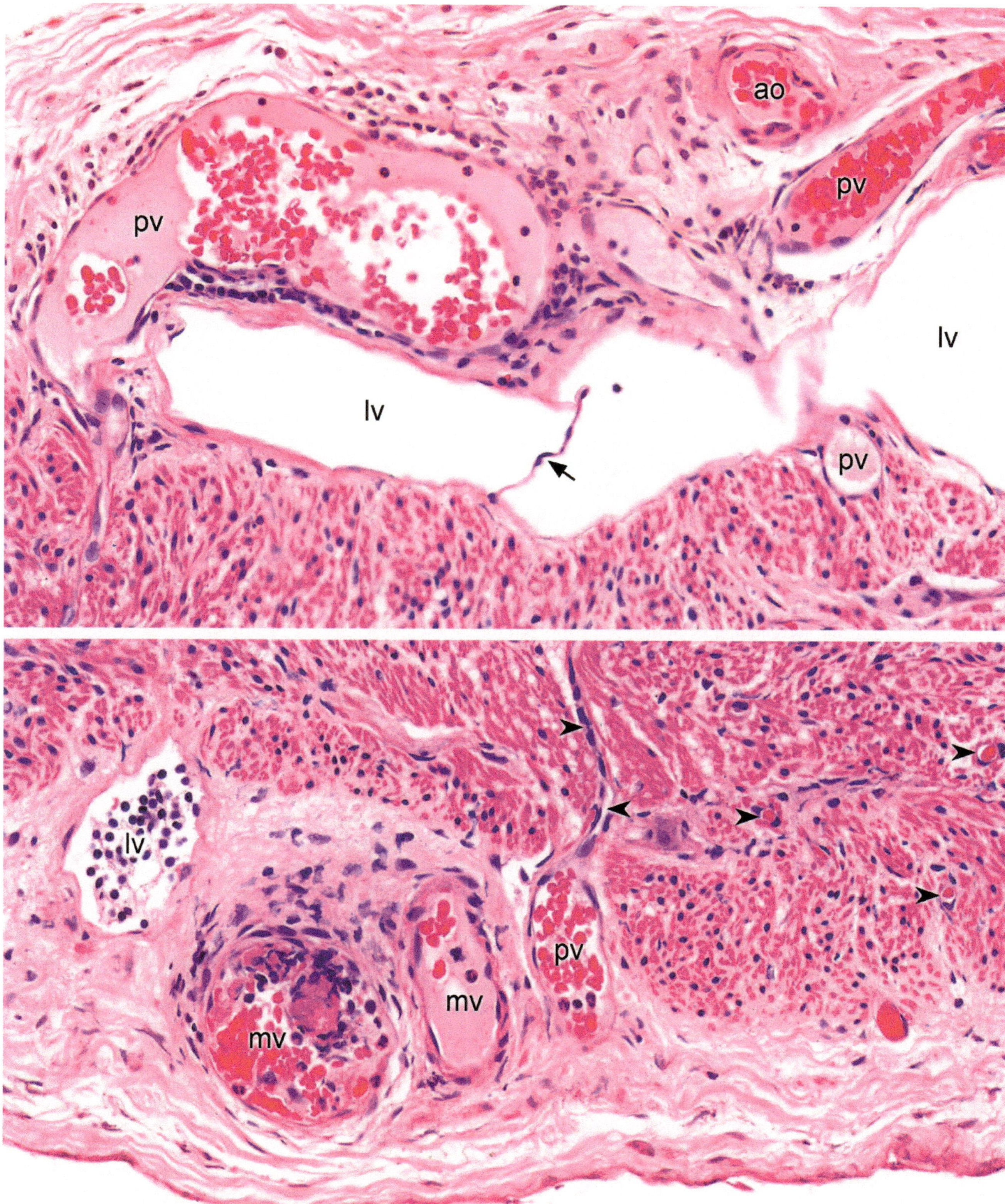

Figure 57. Microvasculature. Arterioles (ao) drain into the capillary bed. Capillaries (arrowheads) transition to postcapillary venules (pv) that have a larger diameter, but a similar wall structure. Smooth muscle appears in the wall of the muscular venules (mv) that drain the postcapillary venules. Lymph vessels (lv) often appear to be empty or contain lymphocytes; their wall is formed by discontinuous endothelium. Lymph vessels often contain valves (arrow).

Chapter 4

Integumentary System

Skin

Skin is covered with stratified squamous keratinized epithelium, the epidermis that is mostly composed of keratinocytes. The basal layer (stratum basale or stratum germinativum) of the epidermis is formed by a single layer of low columnar keratinocytes that are closely packed while resting on the basement membrane. These cells are continually undergoing mitosis forming the more superior layers of the epidermis that gradually move towards the surface and eventually will be sloughed off and replaced by the underlying cell layers. Above the stratum basale, the stratum spinosum is composed of several layers of cuboidal cells with round-shaped nuclei. As these cells mature, they become flattened and move towards the epithelial surface forming the stratum granulosum that is composed of 1-3 layers of flattened keratinocytes with dark basophilic cytoplasm with keratohyalin granules. Stratum granulosum is followed by a layer of similar thickness but with bright eosinophilic staining, the stratum lucidum, which is often missing in thin epidermis. Keratinocytes in the stratum lucidum lack nuclei as they start to keratinize; the cells are eventually filled up with keratin and form the most superficial layer of the skin, the stratum corneum. The thickness of the stratum corneum varies; it is the thickest on the soles and the palm and thinner at most of the body surface.

Apart from the keratinocytes, epidermis also contains melanocytes as well as Langerhans cells that are dendritic antigen presenting cells. Since the contours of these cells are hard to observe with general histological techniques, they have similar histological appearance with H&E staining; both cells possess light cytoplasm that can be easily distinguished from the surrounding keratinocytes. However, the location of these cells is different, as melanocytes can be usually observed in the stratum basale while dendritic cells are located primarily in the stratum spinosum. Melanocytes possess long processes penetrating between the keratinocytes that phagocytose the tip of the processes with the melanin-containing melanosomes inside (pigment donation). Stratum basale also contains disk-shaped, sensory Merkel's cells, but they are difficult to identify with H&E staining.

The layer dermis lies under the epidermis, and it is composed of dense irregular connective tissue. The papillary layer of the dermis forms dermal papillae that penetrate the epidermis and often contain capillary networks and encapsulated sensory nerve endings called Meissner's corpuscles. Under the papillary layer, the reticular layer of the dermis is much thicker and less cellular and may contain large amount of smooth muscle at certain regions of the body including perineum, scrotum, and areola (tunica dartos). The deepest layer of the skin is the hypodermis, an adipose tissue blanket containing most of the epithelial appendages that derive from the epidermis: hair follicles, sebaceous and sweat glands as well as Pacinian corpuscles that are encapsulated sensory nerve endings. Hypodermis may also contain striated muscle (platysma, facial muscles).

In addition to the barrier function, skin is also a sensory organ containing several cutaneous receptors. Merkel's cells (disks) are unicellular non-capsulated mechanoreceptors located in the basal layers of the epidermis as well as in hair follicles and mucous membranes close to the cutaneous areas (oral and anal mucosa). Merkel's cells are disk-shaped cells that are associated with a nerve terminal forming a serotoninergic synapse. Nerve fibers can innervate numerous Merkel's cells while passing by.

The most common encapsulated nerve endings in the skin are the Pacinian corpuscles and the Meissner's corpuscles. Meissner's corpuscles are encapsulated mechanoreceptors located in the dermal papillae. The receptor contains 1-2 unmyelinated nerve fibers spiraling in the center of the corpuscle among plate-like lamellae that derive from the Schwann cells. Pacinian corpuscles are large ovoid or spherical structures located in the subcutis, commonly embedded into adipose tissue and often accompanied by nerve bundles; they can also be found in joints and capsules of internal organs as well as in the periosteum. The corpuscle can be over a millimeter in diameter, and it is surrounded by a dense connective tissue capsule. The axis of the corpuscle is occupied by an unmyelinated nerve fiber that loses its myelin sheath after it enters the corpuscle. This axial nerve fiber is covered by several layers of Schwann cells that form the inner core. The inner core is surrounded by concentric lamellae composed of flattened cells that derive from the cells of endoneurium. The space between the lamellae is filled by fluid similar to lymph. Pacinian corpuscles are rapidly adapting mechanoreceptors detecting pressure changes and vibratory stimuli.

http://doi.org/10.1016/B978-0-323-91891-6.50004-0,

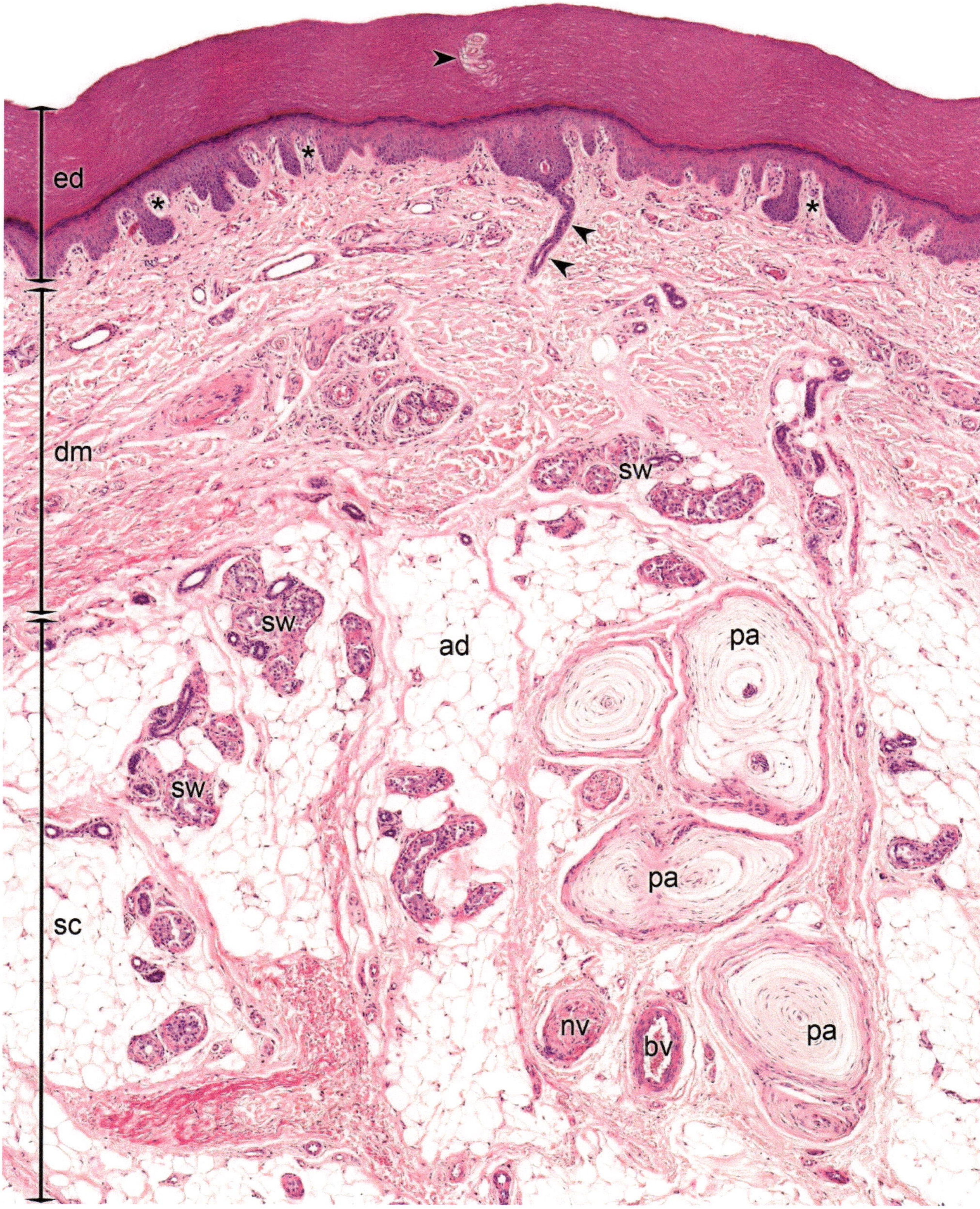

Figure 58. Skin (fingertip). Hairless (glabrous) skin of the fingertip is covered by the epidermis (ed) that is keratinized stratified squamous epithelium. The underlying dermis (ds) is formed by dense irregular connective tissue with dermal papillae (asterisks) penetrating the epidermis. The deepest layer is the subcutis (sc) that is mostly composed of adipose tissue (ad) that contains Pacinian corpuscles (pa), nerves (nv), blood vessels (bv) and sweat glands (sw) whose ducts perforate the dermis and the epidermis (arrowheads).

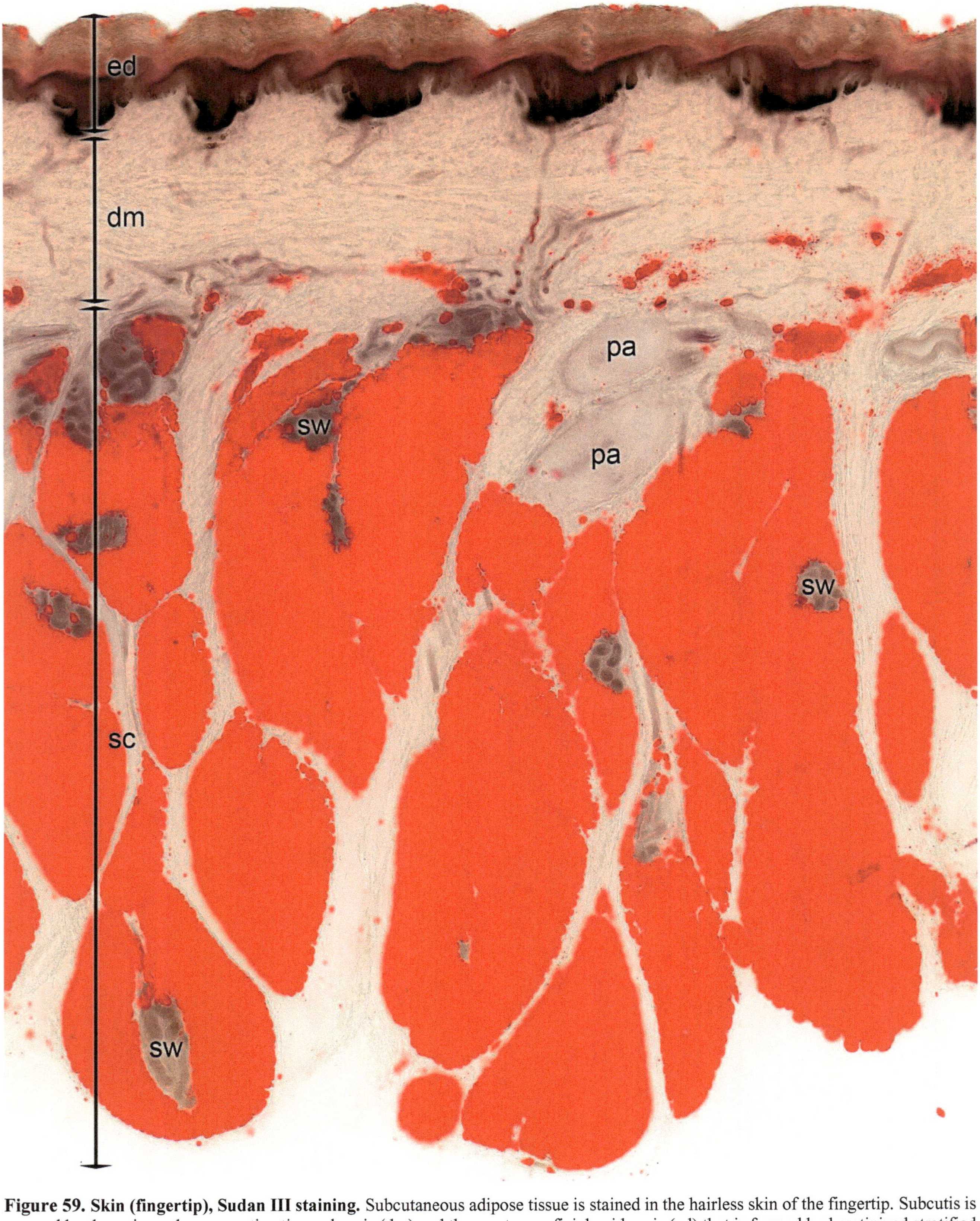

Figure 59. Skin (fingertip), Sudan III staining. Subcutaneous adipose tissue is stained in the hairless skin of the fingertip. Subcutis is covered by dense irregular connective tissue dermis (dm) and the most superficial epidermis (ed) that is formed by keratinized stratified squamous epithelium. Subcutis (sc) is mostly composed of adipose tissue that is stained orange with Sudan III and surrounds Pacinian corpuscles (pa) and sweat glands (sw).

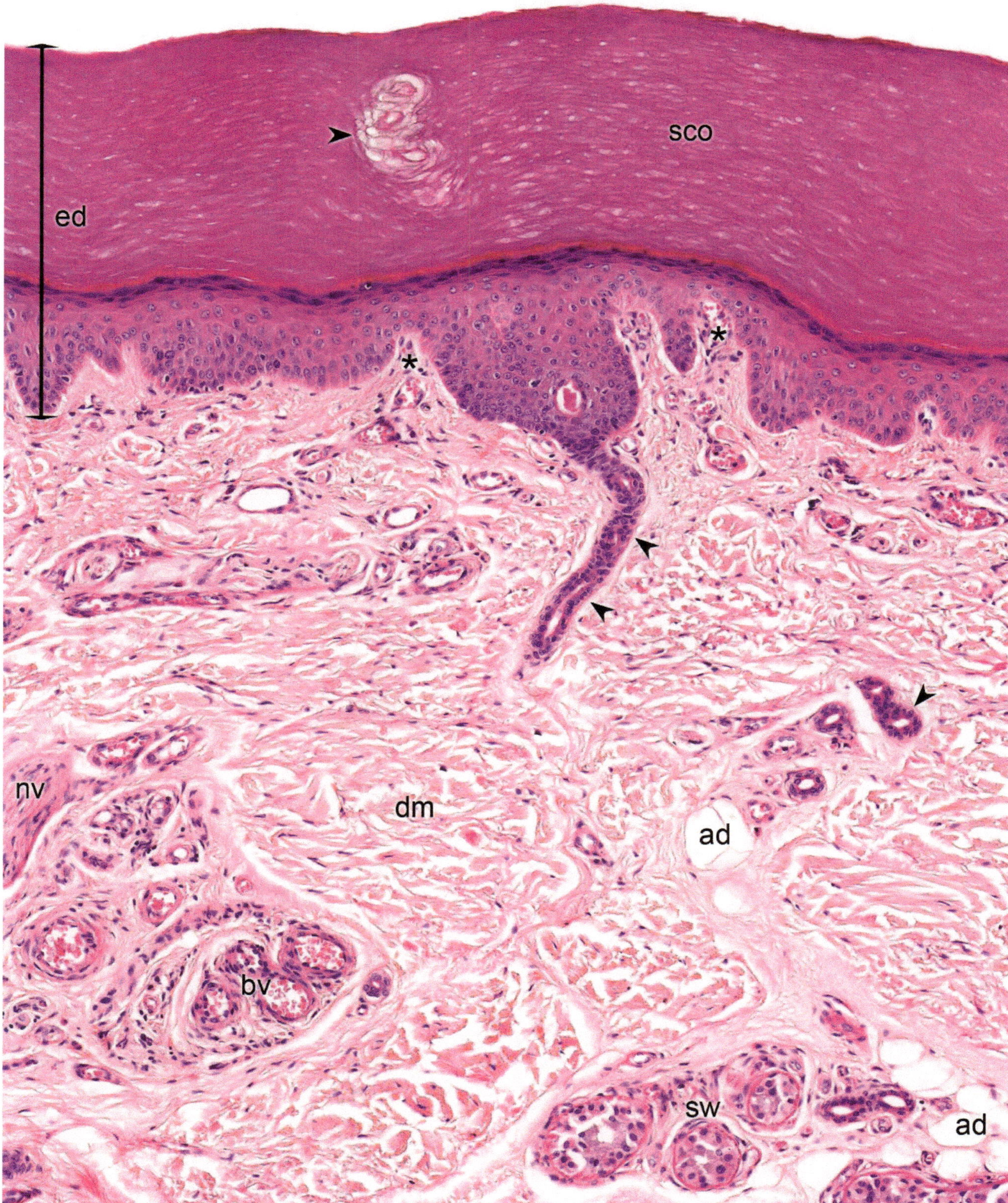

Figure 60. Skin (fingertip). Hairless (glabrous) skin of the fingertip is covered by the epidermis (ed) that is keratinized stratified squamous epithelium with the superficial stratum corneum (sco). The underlying dermis is formed by dense irregular connective tissue with dermal papillae (asterisks) penetrating the epidermis. Subcutaneous adipose tissue (ad) containing the base of the sweat glands (sw) can protrude into the dermis that also contains vessels (bv) and nerves (nv). Ducts of the sweat glands pierce the dermis and the epidermis (arrowheads).

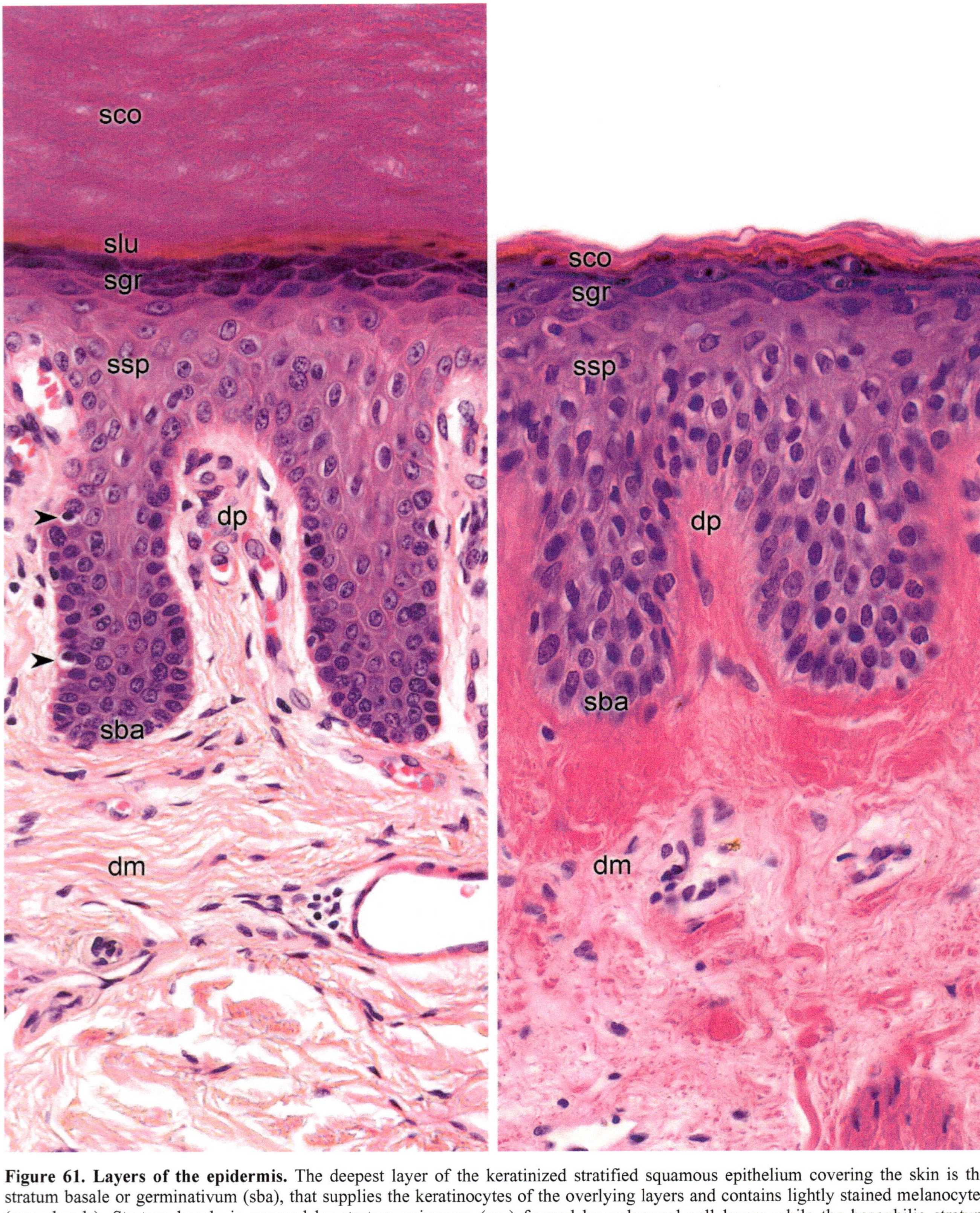

Figure 61. Layers of the epidermis. The deepest layer of the keratinized stratified squamous epithelium covering the skin is the stratum basale or germinativum (sba), that supplies the keratinocytes of the overlying layers and contains lightly stained melanocytes (arrowheads). Stratum basale is covered by stratum spinosum (ssp) formed by polygonal cell layers while the basophilic stratum granulosum, the eosinophilic stratum lucidum (slu) and the keratinized stratum corneum (sco) are composed of flattened cells. Stratum lucidum and stratum corneum normally lack cell nuclei. Hairless (glabrous) skin covering the sole and the palm is lined with epidermis with thick keratinized layer (left) while thin skin has missing stratum lucidum and has a thin stratum corneum (right). Dermis (dm) is formed by dense irregular connective tissue with dermal papillae (dp) penetrating the epidermis.

Hair follicles

Hair covers the majority of the body except the glabrous (hairless) skin areas that include the palms, fingers, soles and the lip. The hair follicle itself is formed by the nonkeratinized layers of the epidermis while most of the hair is the derivative of the keratinized stratum corneum.

Longitudinally, the hair follicle can be divided into the infundibulum, which is between the surface opening and the opening of the pilosebaceous canal, the isthmus that extends from the infundibulum to the arrector pili muscle that is responsible for the bristling of the hair, and the inferior segment distally to the isthmus. The inferior segment ends in the bulb, a thickened part at the base of the follicle.

Hair follicle is enveloped by the dermal sheath that is a connective tissue sheath formed by the dermis under the basement membrane of the epidermis (glassy membrane or vitreous layer). Dermis protrudes into the distal part of the bulb forming the dermal papilla and here it is surrounded by a layer of matrix cells that derive from the stratum germinativum of the epidermis, and contain melanocytes. Under the dermal sheath, the external and internal root sheaths surround the growing hair; they are the derivatives of the nonkeratinized layers of the epidermis. The iternal root sheath is composed of the layers of Henle and Huxley, both formed by cuboidal cells as well as the innermost cuticle that is formed by squamous cells. The cuticle of the follicle is adjacent to the cuticle of the hair that has the similar morphology; flattened cells of these corresponding layers are interlocked. Finally, the hair itself is composed of the outer cortex, formed by cuboidal cells and the inner medulla that consists of large vacuolated cells with light staining. Moving distally, the epithelial cells of the hair contain increasing amount of keratin as it grows out of the follicle.

Epithelial glands

Eccrine (merocrine) sweat (sudoriferous) glands are simple tubular structures where the base of the gland is coiled and located in the hypodermis or deeper parts of the dermis. The duct of the glands perforates the dermis and the epidermis and opens on the surface of the stratum corneum. The base of the glands is lined with cuboidal cells, some of them darker, secreting the protein component of the sweat and some of them lighter producing the watery component with less proteins. Myoepithelial cells with flattened nuclei surround the coiled base of the glands and facilitate the emptying of the tubules. The duct segment of the sweat glands is covered by stratified cuboidal epithelium (generally two layers of cuboidal cells). Similar to the eccrine sweat glands, apocrine sweat glands are simple (sometimes branched) coiled tubular glands, but their lumen is distended, and the duct segment usually opens into a hair follicle above the duct of the sebaceous glands. The apical portion of the glandular epithelial cells of the secretory segment appears to be sloughing off forming a bleb-like extension; however, ultrastructural studies indicate that the secretory mechanism is merocrine. The secretory product of the gland is complex, and it may function as pheromone. While eccrine sweat glands are distributed throughout the body, apocrine sweat glands are characteristic for certain body areas including the genitalia, nipple, and armpit and become fully functional at puberty. Modified apocrine sweat glands are also located in the eyelid (gland of Moll) and in the external ear (ceruminous glands).

Sebaceous glands of the skin produce the oily substance sebum by holocrine secretion. These glands are surrounded by layers equivalent to stratum basale and spinosum, and the keratinocytes gradually degrade by accumulating lipid droplets in their cytoplasm and developing pyknotic, shrunken nuclei as they move towards the center of the gland. Sebaceous glands commonly open into the hair follicles by the pilosebaceous canal but they can also exist independently in the eyelids (Meibomian glands), nipples (Montgomery's glands), corner of the mouth (Fordyce spots) and around the genitalia.

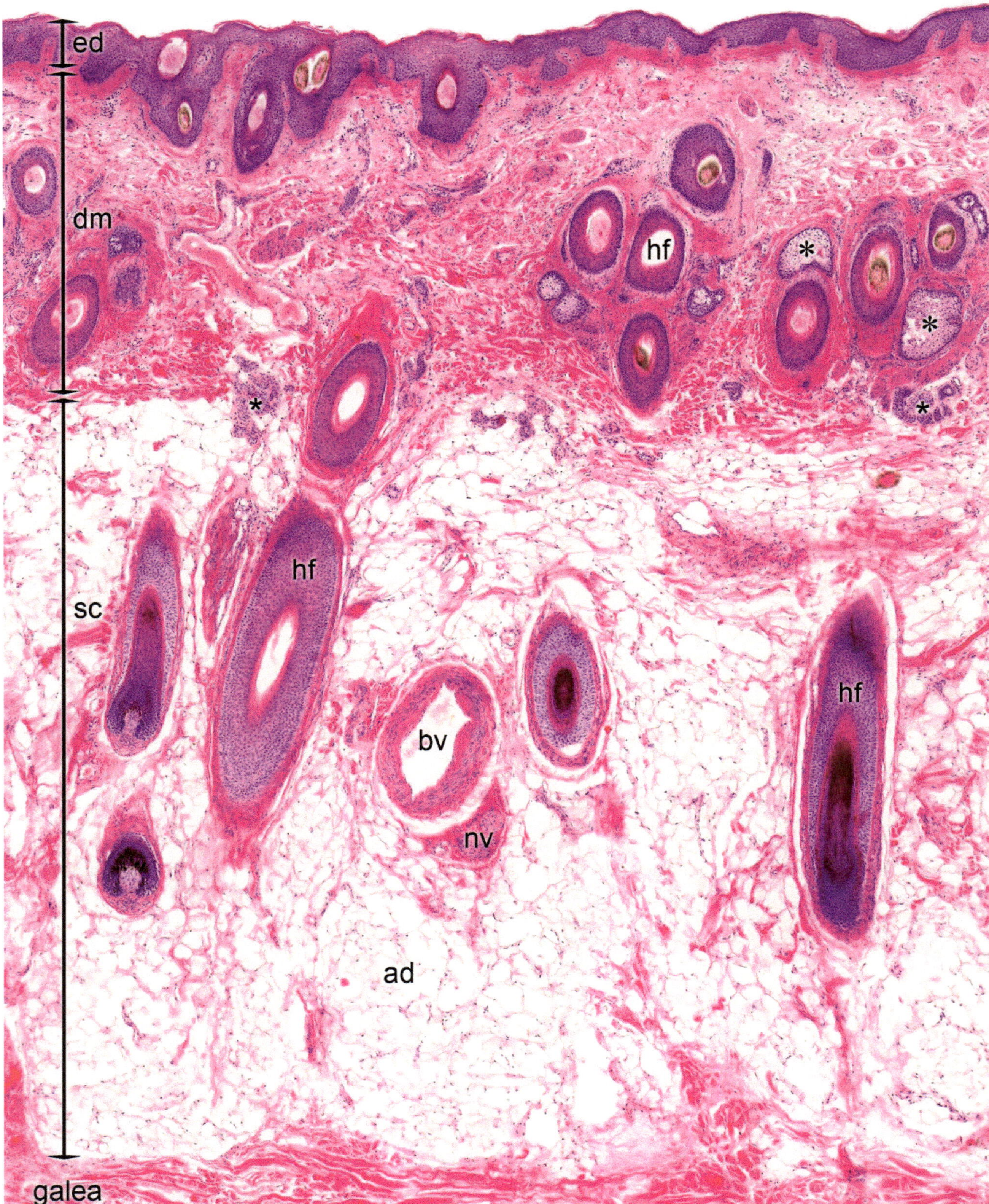

Figure 62. Skin (scalp). Hairy skin is characterized with a relatively thin keratinized layer in the epidermis (ed) and the presence of hair follicles (hf) that receive holocrine sebaceous glands (six-pointed asterisks). Under the dermis (dm), subcutis contains adipose tissue (ad) surrounding the base of the hair follicles, as well as sweat glands (five-pointed asterisks), blood vessels (bv) and nerves (nv). Subcutis rests on a dense connective tissue layer, the galea aponeurotica.

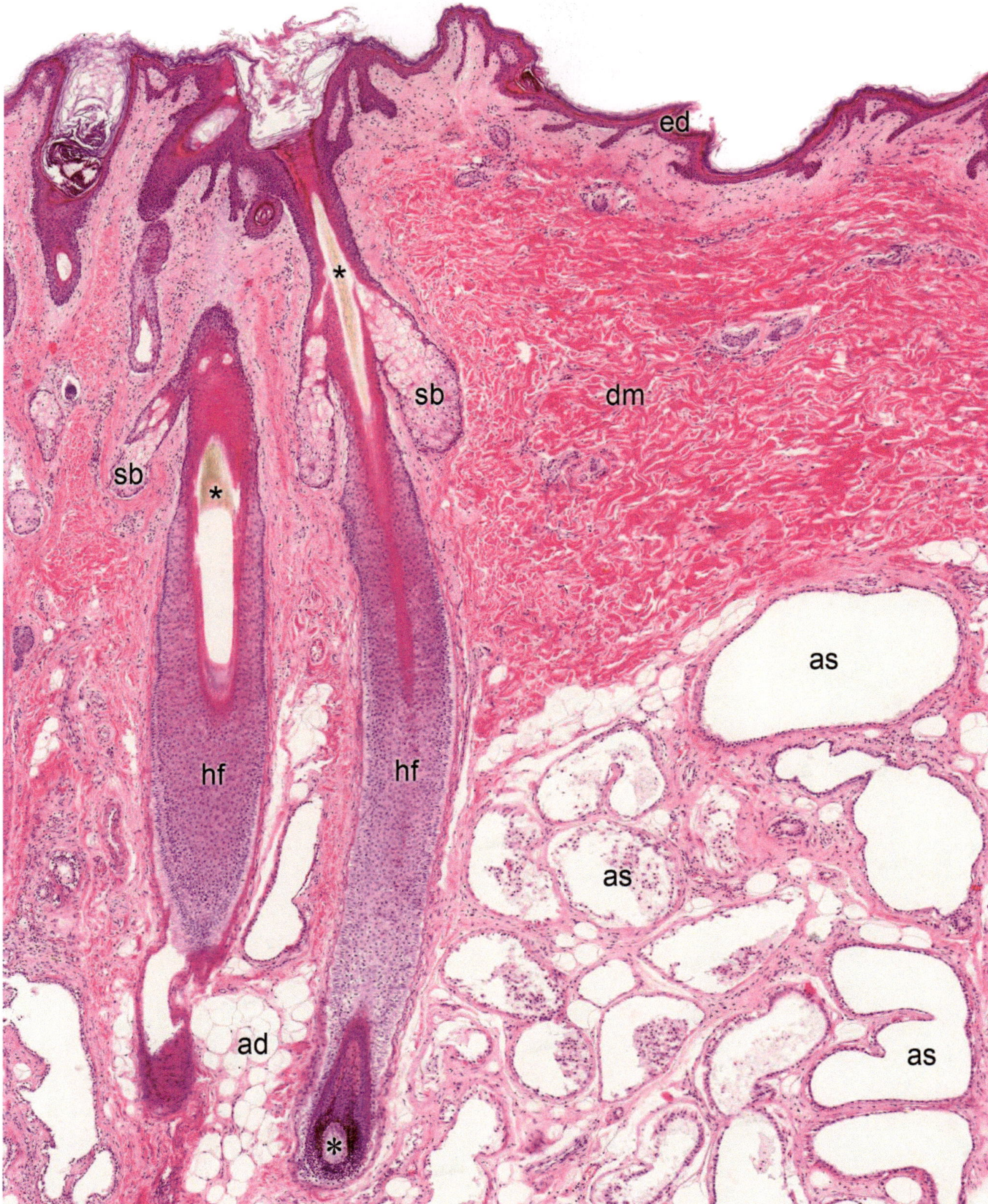

Figure 63. Skin (armpit). Thin (hairy) skin is characterized with thin epidermis (ed) and the presence of hair follicles (hf) containing hair (five-pointed asterisks). The basal segments of the hair follicles are penetrated by connective tissue papillae (six-pointed asterisks). Holocrine sebaceous glands (sb) open into the hair follicles that are surrounded by adipocytes (ad). Under the dermis (dm), subcutis contains apocrine sweat glands (as) characterized with large lumen. These glands are abundant in the skin of the armpit and genitalia.

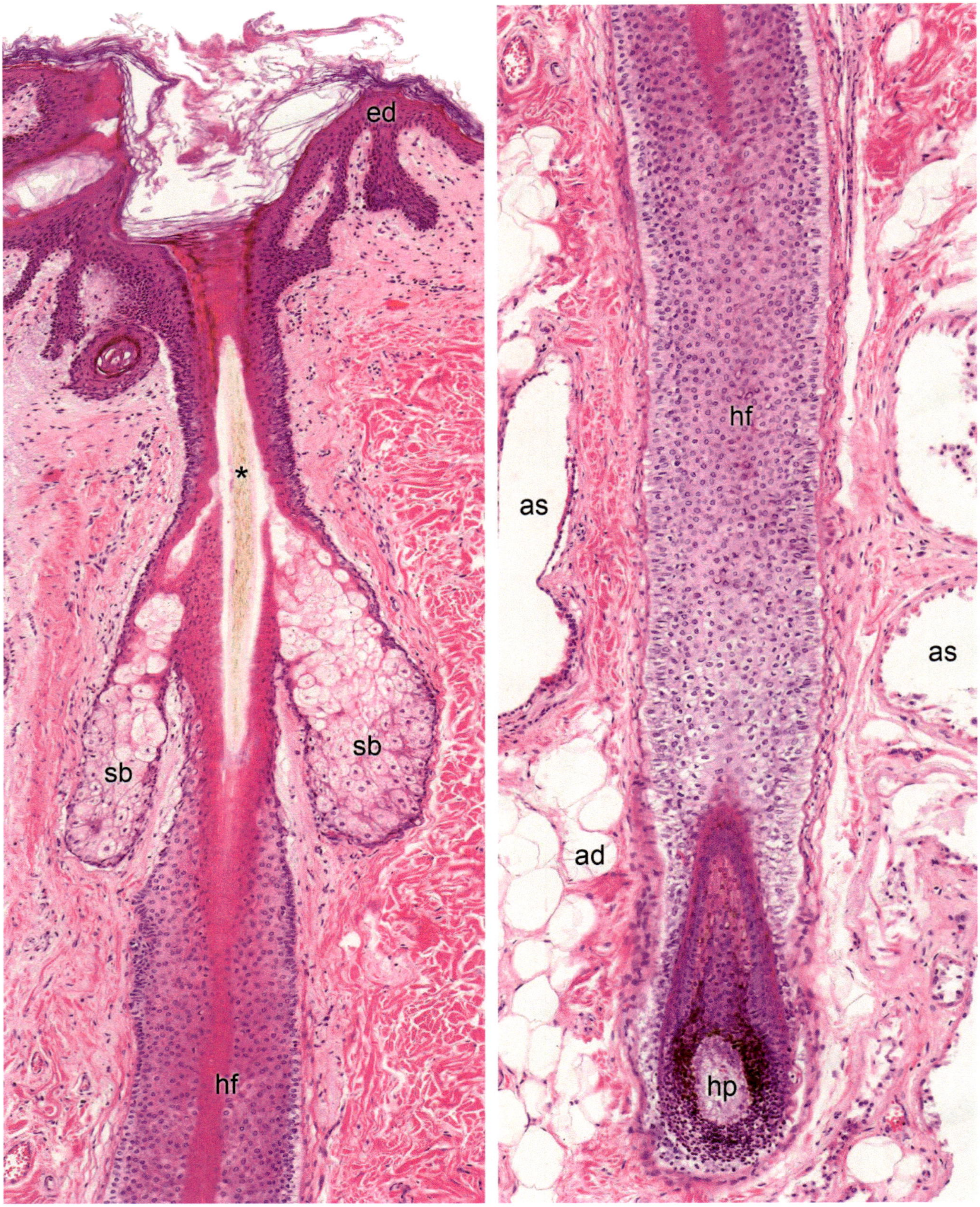

Figure 64. Hair follicle. Hair follicles (hf) containing the hair (asterisks) are the derivatives of the epidermis (ed). The basal segment of the hair follicle is invaginated by a connective tissue papilla (hp). Holocrine sebaceous glands (sb) open into the hair follicles that are surrounded by adipocytes (ad).

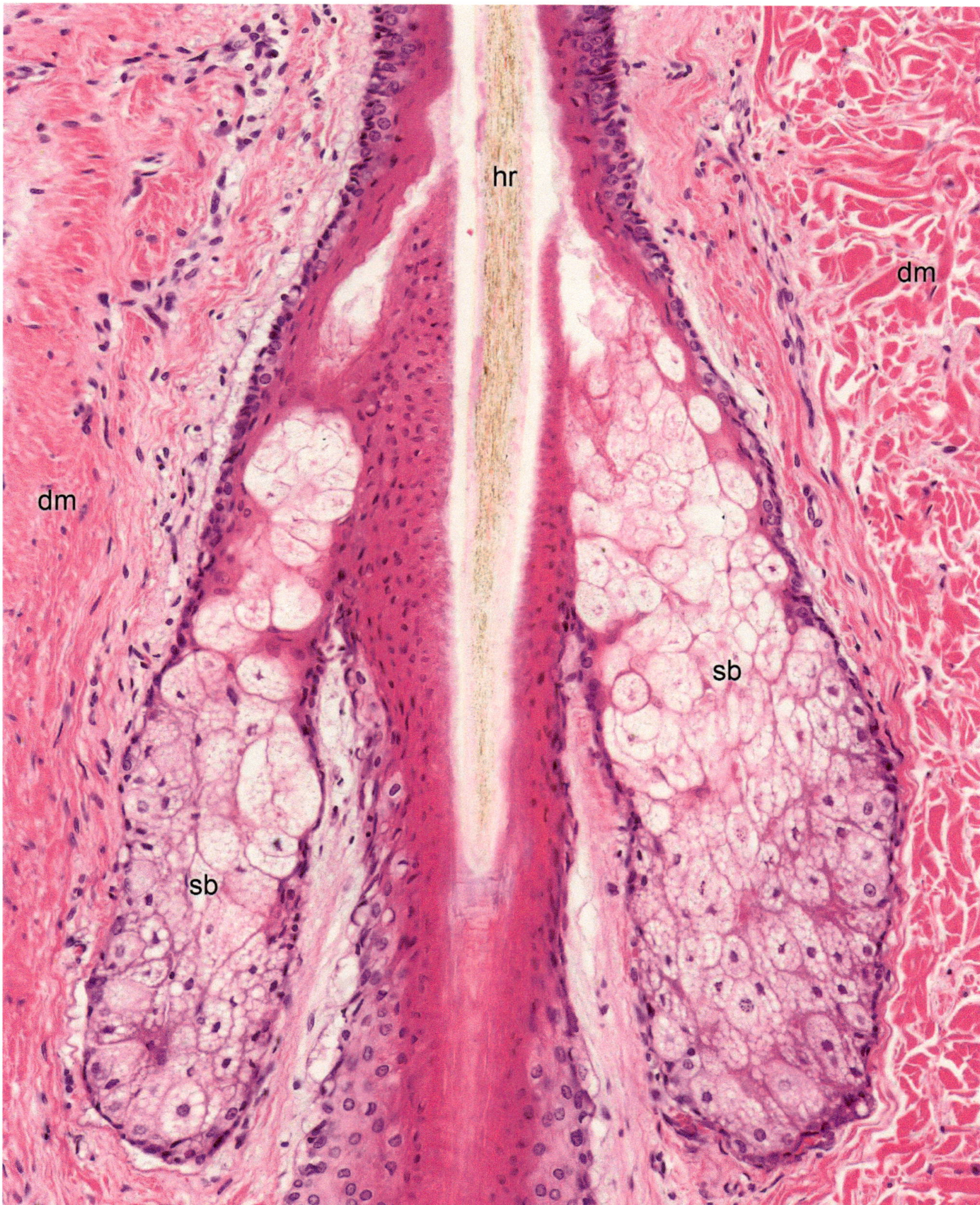

Figure 65. Sebaceous glands. Sebaceous glands (sb) develop from the epidermis, and with few exceptions, they commonly open into hair follicles. Glandular cells accumulate lipid droplets in the cytoplasm as they move toward the center. Pycnotic, shrunken nuclei that eventually disappear from the cells indicate that these cells are dying as they form the secretory product, the sebum. Sebaceous glands are located in the dermis (dm).

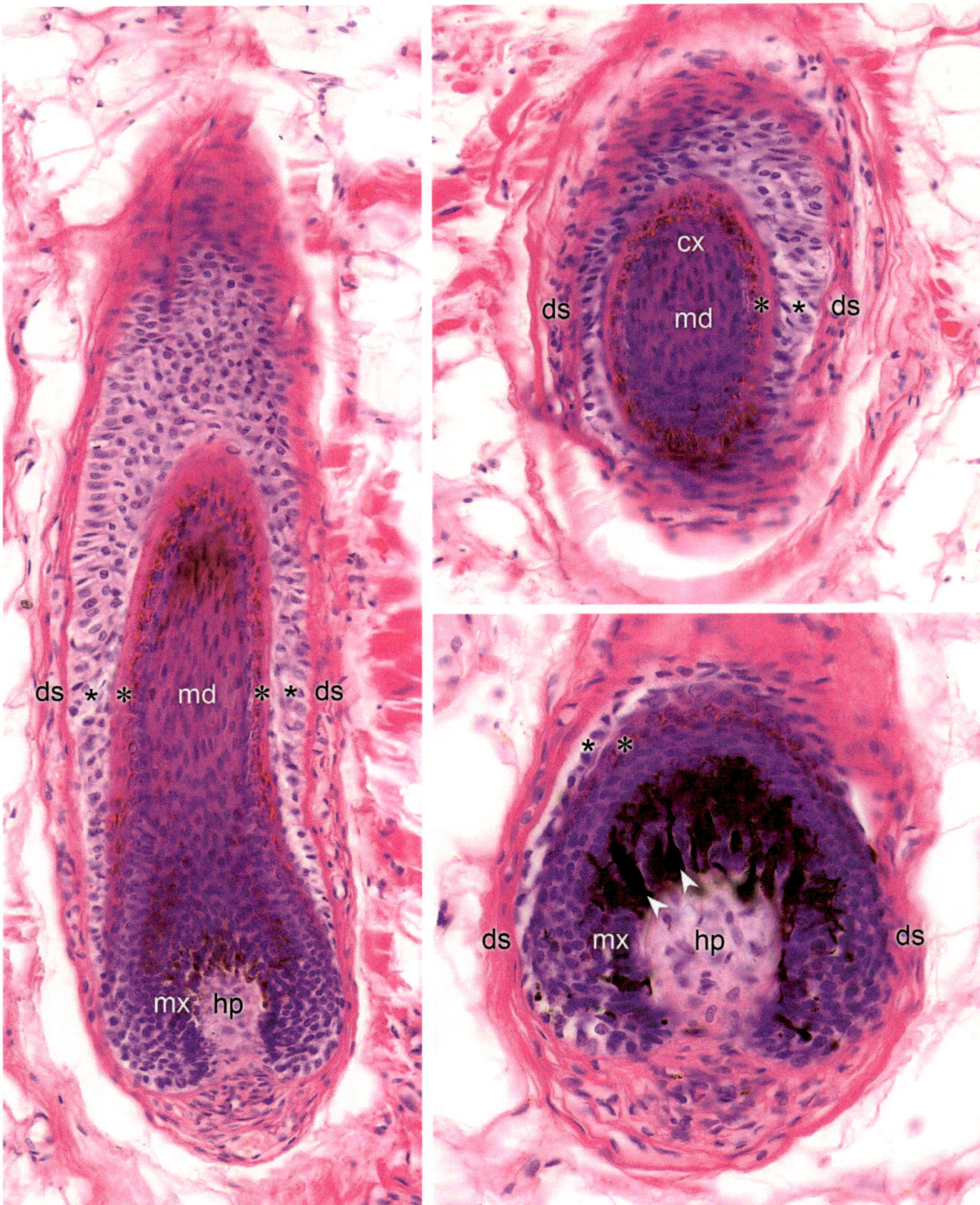

Figure 66. Basal segment of the hair follicle. The bulb of the hair follicle is invaginated by a connective tissue papilla (hp) that is surrounded by matrix cells (mx) forming the growing hair. Hair follicle is surrounded by a dermal sheath (ds) composed of connective tissue. Epithelial external root sheath (five-pointed asterisks) and internal root sheath (six-pointed asterisks) cover the hair itself that is composed of the outer cortex (cx) and the inner medulla (md). Melanocytes around the dermal papilla are denoted by arrowheads.

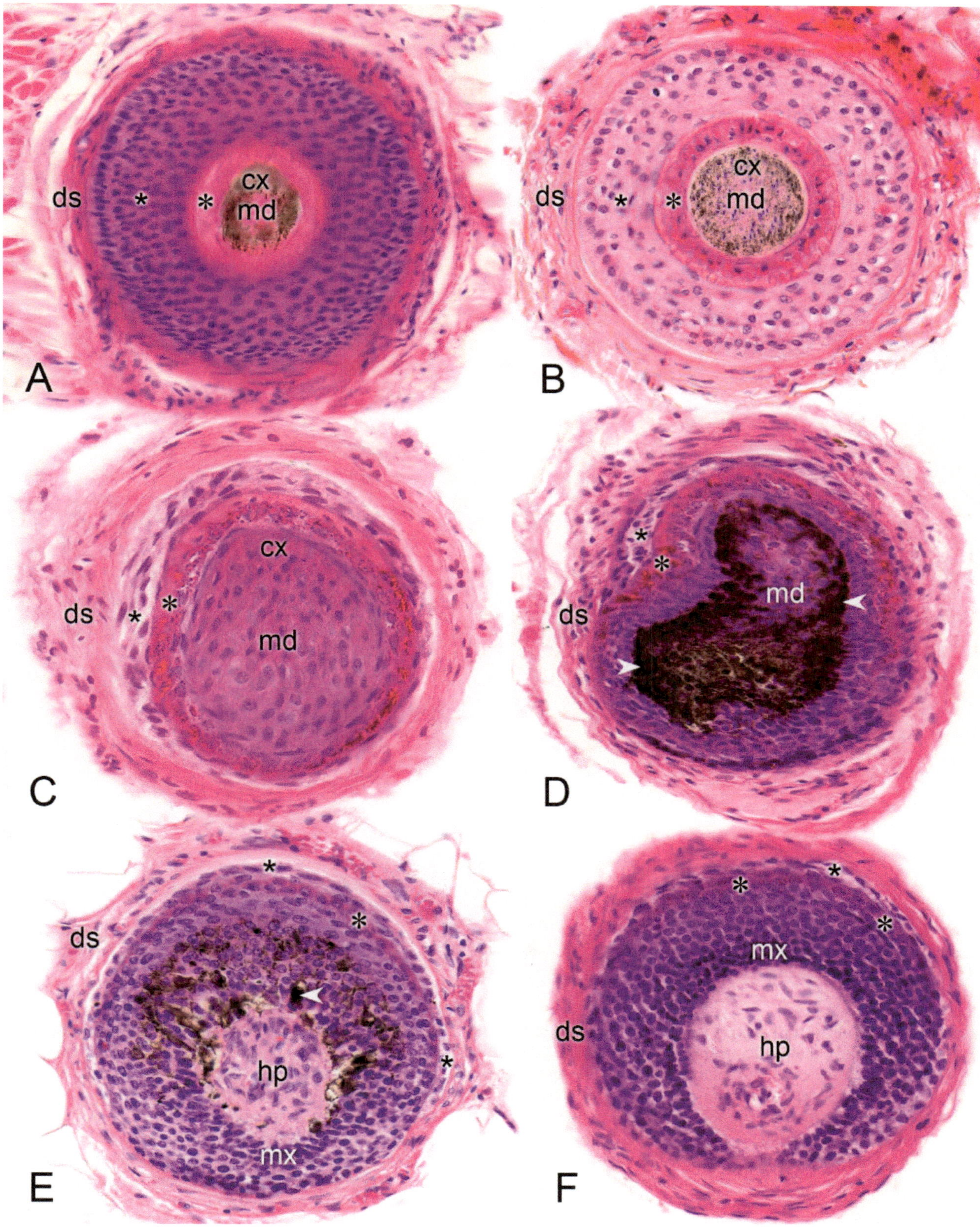

Figure 67. Cross sections of the hair follicle from superficial (A) to deep (F). (A) Superficial level of the hair shaft, (B) level slightly below, (C) section right above the papilla, and (D) at the tip of the papilla, followed by sections at the (E) upper and the (F) lower papillary level. The hair follicle is surrounded by the connective tissue dermal sheath (ds). Epithelial external root sheath (five-pointed asterisks) and internal root sheath (six-pointed asterisks) cover the hair that is composed of the outer cortex (cx) and the inner medulla (md). Matrix cells (mx) surround the papilla that is also covered with melanocytes (arrowheads).

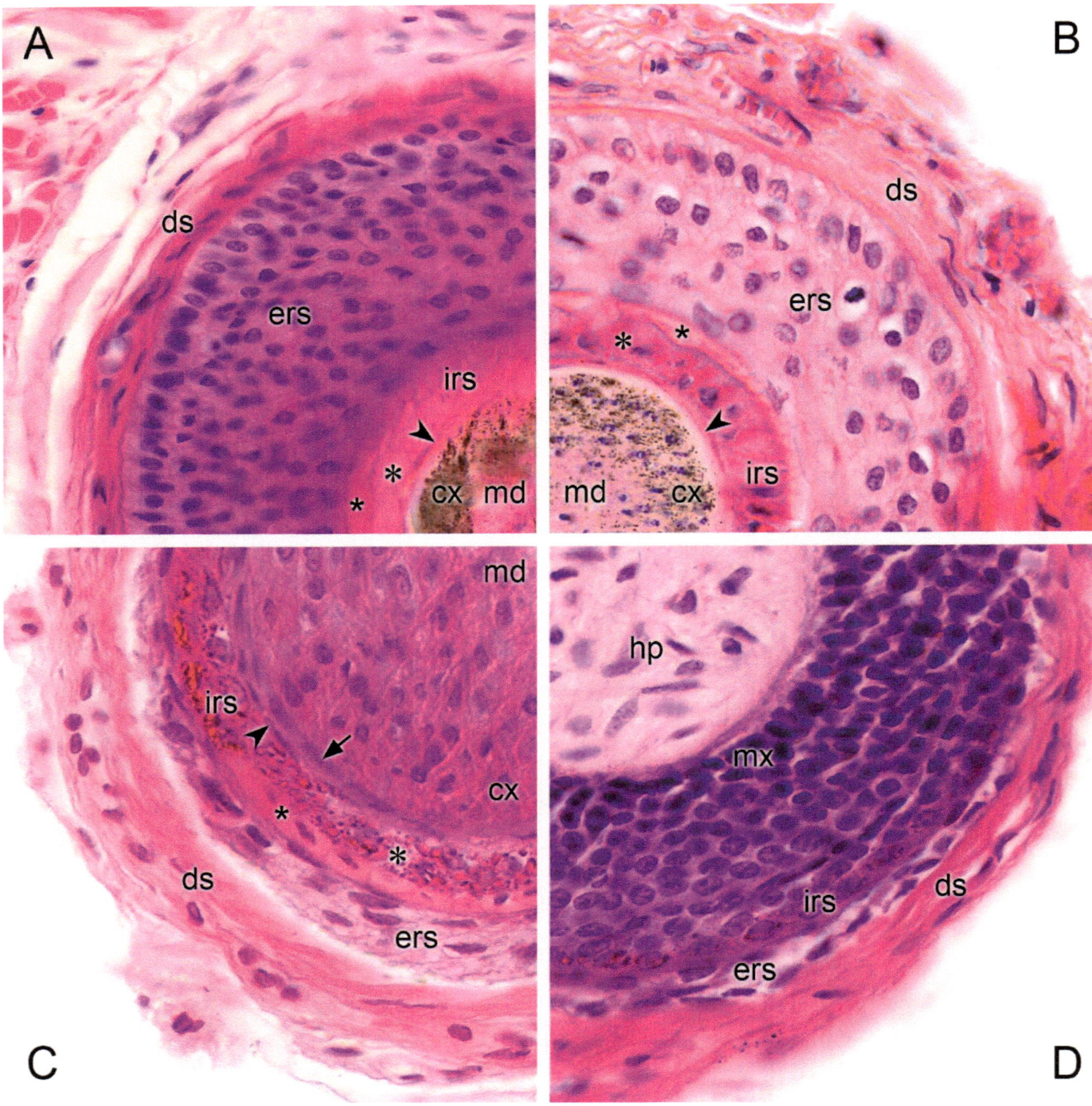

Figure 68. Layers of the hair follicle from superficial (A) to deep (D). (A) Superficial level of the hair shaft, (B) level slightly below, (C) section right above the papilla, and (D) at the papillary level. The hair follicle is surrounded by the connective tissue dermal sheath (ds), and it is composed of the external root sheath (ers) and internal root sheath (irs) covering the hair shaft. The internal root sheath can be separated into three layers. The outermost layer of Henle (five-pointed asterisk) is composed of flattened or low cuboidal cells while the layer of Huxley (six-pointed asterisk) is formed by cuboidal cells. The innermost layer of the internal root sheath is the cuticle (arrowheads) formed by a single layer of squamous cells and it is adjacent to the similarly structured cuticle of the hair (arrows). Under the cuticle, hair is composed of the outer cortex (cx) and the inner medulla (md). The external root sheath corresponds to the stratum basale and the stratum spinosum of the epidermis that is invaginated by the connective tissue papilla (hp) at the base of the hair follicle. Stratum basale surrounding the papilla yields the mitotically active matrix cells (mx) that are responsible for the longitudinal growth of the hair.

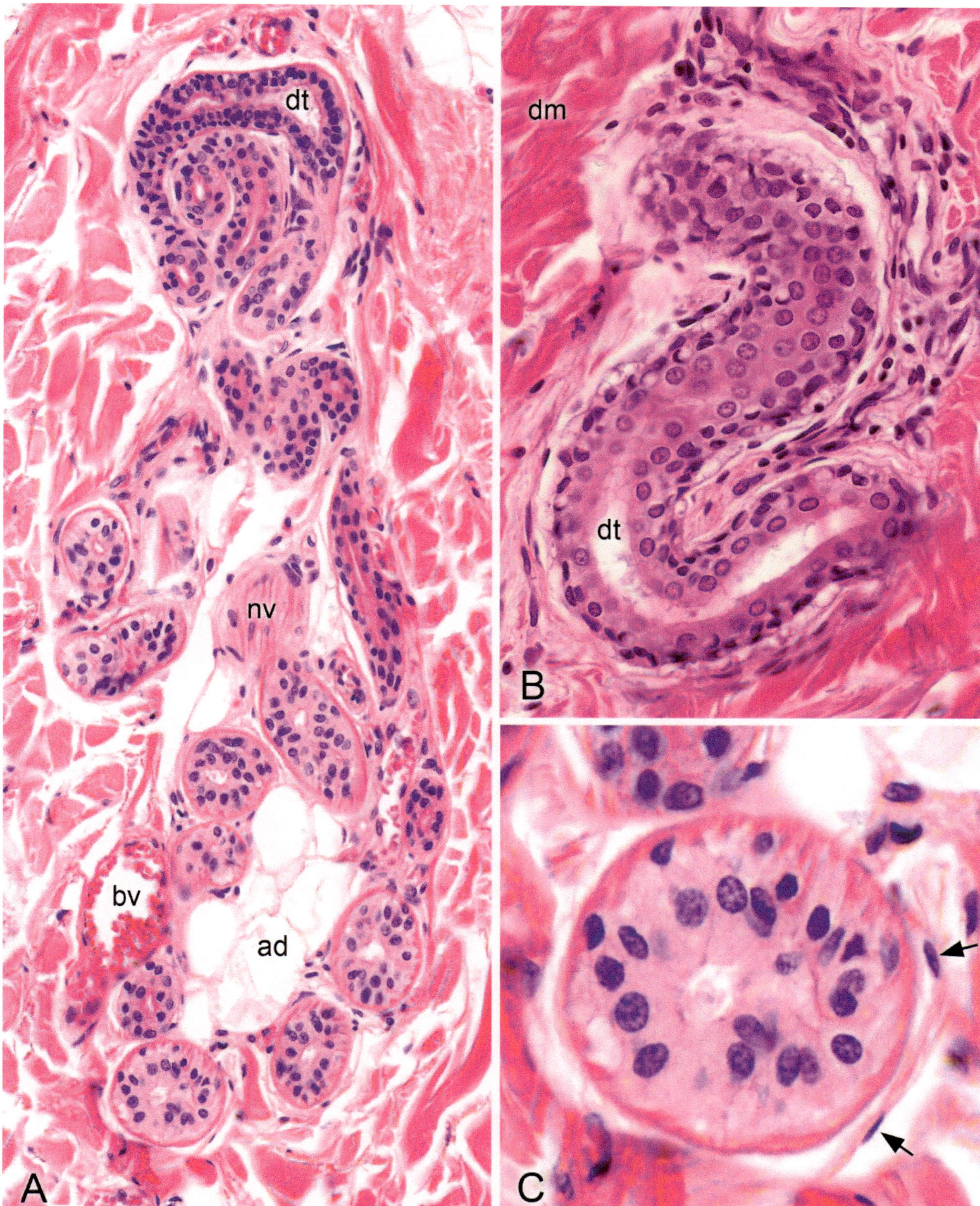

Figure 69. Merocrine (eccrine) sweat gland. (A) Sweat glands are simple tubular glands with a coiled basal segment that is typically located in the superficial layer of subcutis in the close vicinity of adipocytes (ad), blood vessels (bv) and nerves (nv). The duct segment (dt) is lined with darker cells. (B) Sweat gland ducts perforate the dermis (dm) and they are covered with stratified cuboidal epithelium. (C) The base of the sweat glands is surrounded with myoepithelial cells (arrows) that facilitate the discharge of the sweat.

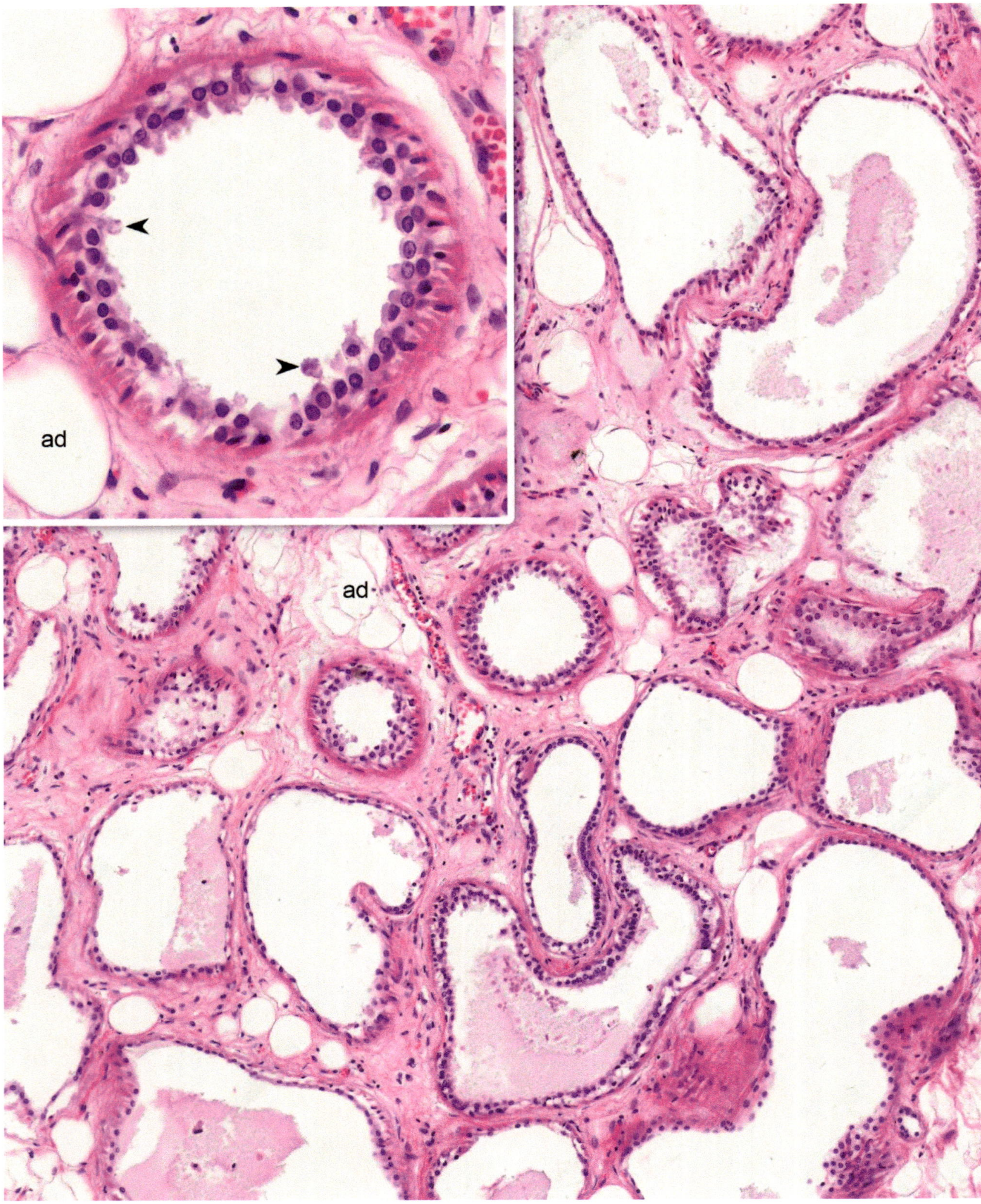

Figure 70. Apocrine sweat gland. Apocrine sweat glands are simple (sometimes branched) coiled tubular glands typically located in the subcutis of the armpit and genitalia where they are surrounded by adipocytes (ad). These glands are characterized with large, distended lumen and they open into hair follicles usually above the duct of the sebaceous glands. The apical part of the secretory cells lining the basal segment appears to be sloughing off forming a bleb-like extension (inset, arrowheads); however, ultrastructural evidence suggests merocrine secretion. The secretory product of the glands is complex and may function as a pheromone.

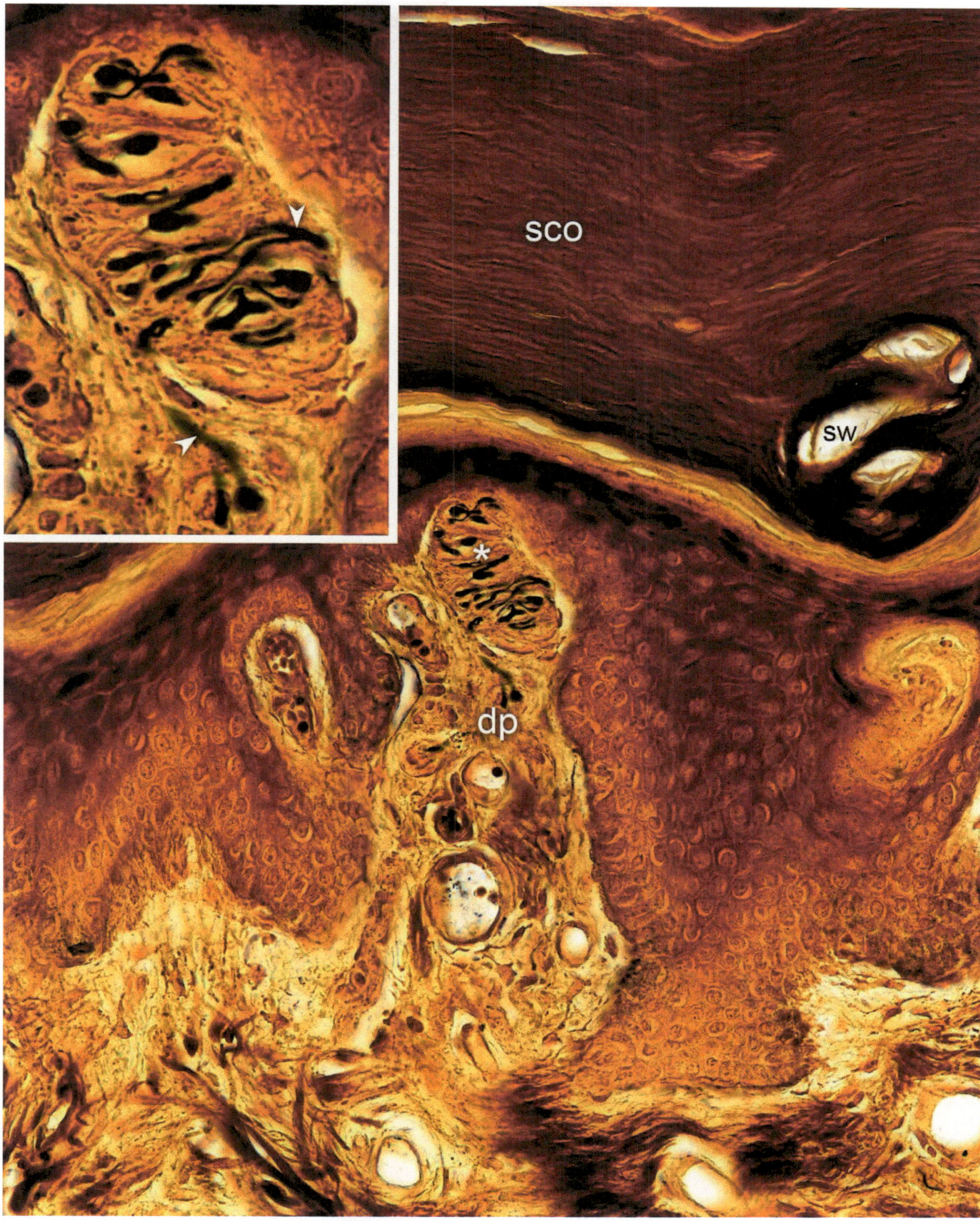

Figure 71. Meissner's corpuscle, silver impregnation. Meissner's corpuscles (asterisk) are encapsulated sensory nerve endings that are located at the tip of the dermal papillae (dp). They contain one or two unmyelinated axons (arrowheads) spiraling between flattened cells that look like stacked plates. Meissner's corpuscles are touch receptors sensitive to mechanical stimuli. Adjacent to the corpuscle, stratum corneum (sco) of the epidermis is often perforated by ducts of sweat glands (sw).

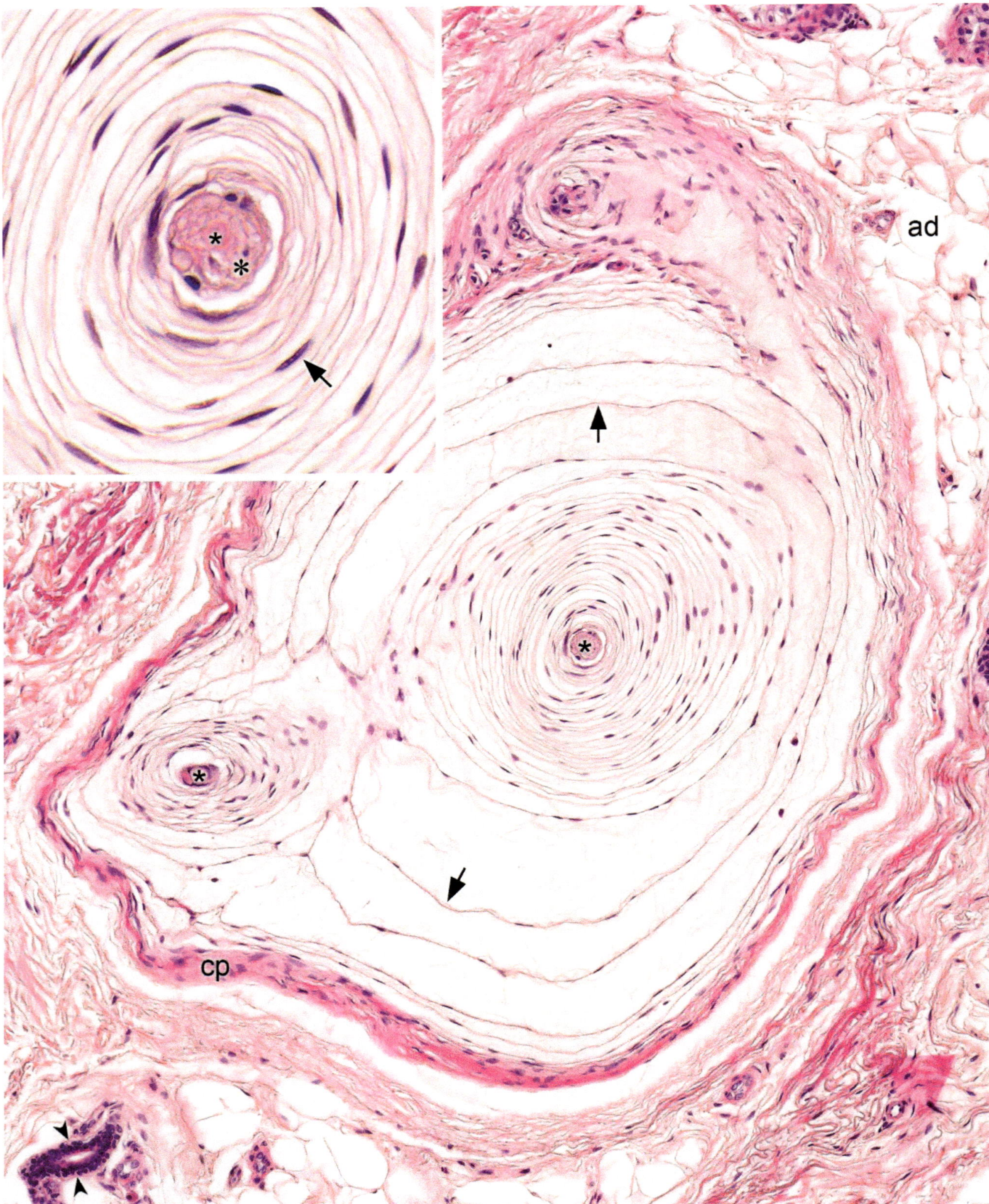

Figure 72. Pacinian corpuscle. Pacinian corpuscles are large, encapsulated sensory nerve endings located in the subcutis where they are surrounded with adipocytes (ad) and occasional sweat glands (arrowheads). Pacinian corpuscles are composed of concentrically arranged lamellae of flattened cells (arrows) covered with a connective tissue capsule (cp). A myelinated axon (five-pointed asterisk) enters the corpuscle at the axis, loses its myelin sheath after a short run and it is covered with layers of Schwann cells (six-pointed asterisks). The Schwann sheath of the axon is referred as inner core while concentric lamellae represent the outer core of the corpuscle.

Chapter 5

Digestive System I: Upper Alimentary Tract

Layers of the gastrointestinal tract

The wall structure of the tubular (hollow) organs follows a general organization that is uniform along the gastrointestinal tract. The lumen of the organs is covered with *tunica mucosa* (mucous membrane) that is composed of the superficial *epithelial lining*, the underlying *lamina propria mucosae* and the *tunica muscularis mucosae*. Lamina propria is composed of loose connective tissue that often contains glands and lymphocytic aggregations while the thin muscularis mucosae is usually formed by inner circular and outer longitudinal layers of smooth muscle cells. The mucous membrane is attached to the tunica muscularis externa by the *tunica submucosa* that is composed of dense irregular connective tissue with occasional glands embedded into the layer. Submucosa also contains autonomic ganglion cells and branches of nerves forming the submucosal or Meissner's plexus that innervates the muscularis mucosae. *Tunica muscularis externa* is a thick layer of smooth muscle cells under the submucosa that is organized most commonly into an inner layer of circularly and outer layer of longitudinally arranged cells. Ganglion cells and nerve fibers located between the circularand longitudinal layers form the myenteric or Auerbach's plexus that is responsible for the innervation of the smooth muscle cells of the layers. If the organ is covered by peritoneum, the muscularis externa is covered by tunica serosa that is composed of a loose connective tissue layer (lamina propria serosae) lined outside with a single layer of squamous mesothelial cells. In the absence of the peritoneal coverage, muscularis externa is ensheathed by a simple loose connective tissue coat, the tunica adventitia. Both lamina propria serosae and tunica adventitia contain nerves and vessels that supply the wall of the organ. Organs that are not tubular (e.g., glands associated with the digestive system), commonly contain a functional part, the *parenchyma* that is surrounded by a connective tissue *stroma* that provides the structural support and carries vessels and nerves.

Lip

Oral cavity can be subdivided into the vestibule and the oral cavity proper; the former is located between the lips, cheeks, and the teeth while the latter represents the rest of the oral cavity. Vestibule is bordered anteriorly by the lips that contain a core of striated muscle mostly composed of the orbicularis oris. The muscle core is covered by three distinct layers. Pars cutanea represents the outermost (anterior) zone covered by skin, and it contains hair follicles, sebaceous glands, and eccrine sweat glands. The interior (posterior) surface of the lip is covered by lining mucosa that is composed of stratified squamous nonkeratinized epithelium with occasional parakeratinized areas and the underlying lamina propria that sends few short papillae into the epithelium (pars mucosa). The mucous membrane is attached to the core muscle layer by a dense irregular connective tissue submucosa that contains seromucous glands. Sebaceous glands without hair follicles may also appear in the submucosa around the lip (Fordyce spots). Pars cutanea and pars mucosa is connected by the thinly keratinized vermilion zone (rubor labii) that is characterized with high connective tissue papillae and extensive capillary networks that are responsible for the red coloration of this region. The cheeks (bucca) have a similar structural organization to the lips with the exception of the missing vermillion zone.

http://doi.org/10.1016/B978-0-323-91891-6.50005-2,

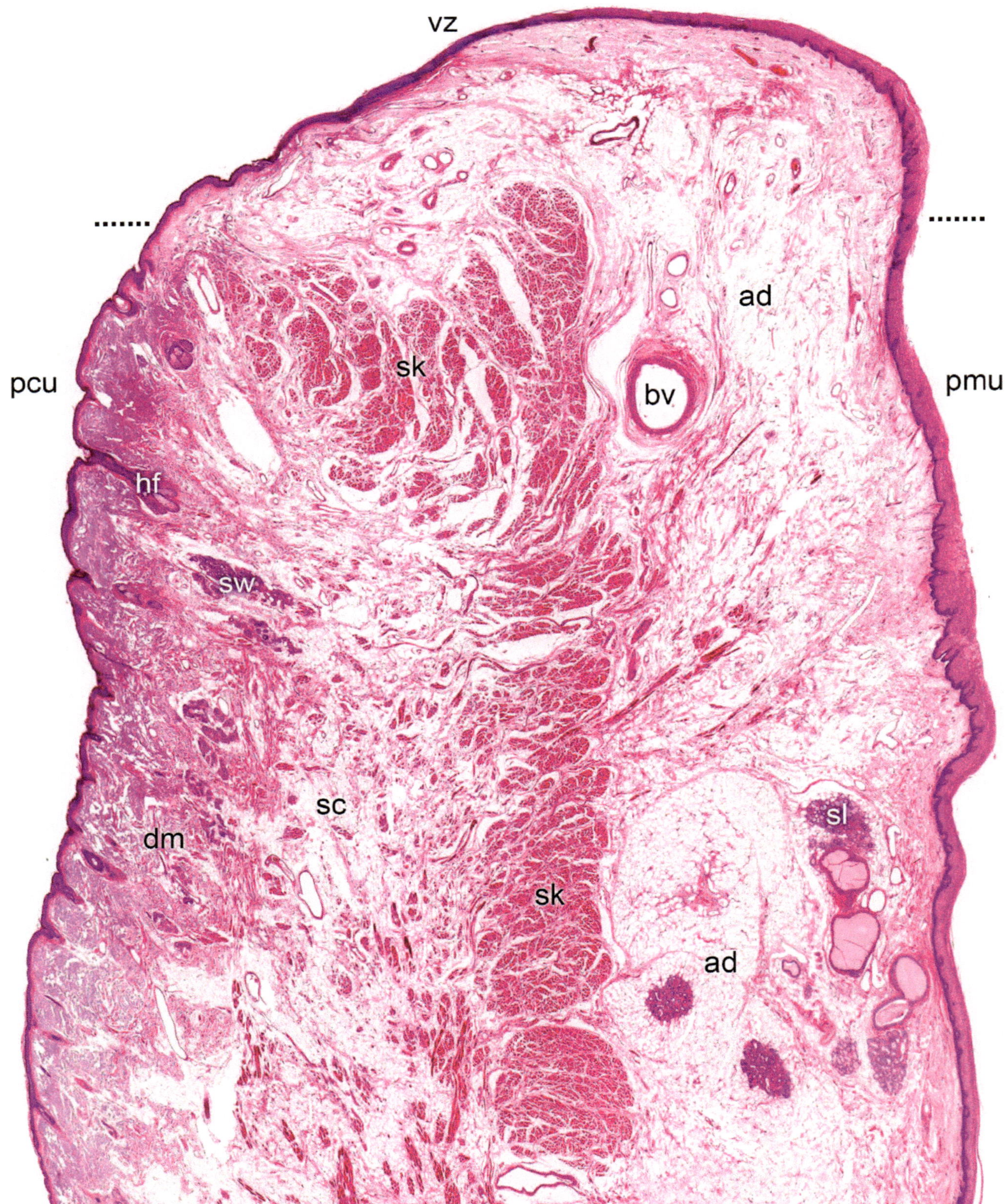

Figure 73. Lip (labium). The outer surface of the lip, the pars cutanea (pcu), is covered with skin. Dermis (dm) and subcutis (sc) contains hair follicles (hf) and sweat glands (sw). Pars mucosa (pmu) represents the inner surface of the lip covered with mucous membrane lined with nonkeratinized stratified squamous epithelium. Lamina propria contains small seromucous salivary glands (sl) and adipose tissue (ad). Pars cutanea and the mucosa form borders (dotted lines) with the vermillion zone (vz) lined with slightly keratinized or parakeratinized epithelium. Skeletal muscle (sk) of the orbicularis oris and branches of the labial artery and vein (bv) occupy the core of the lip.

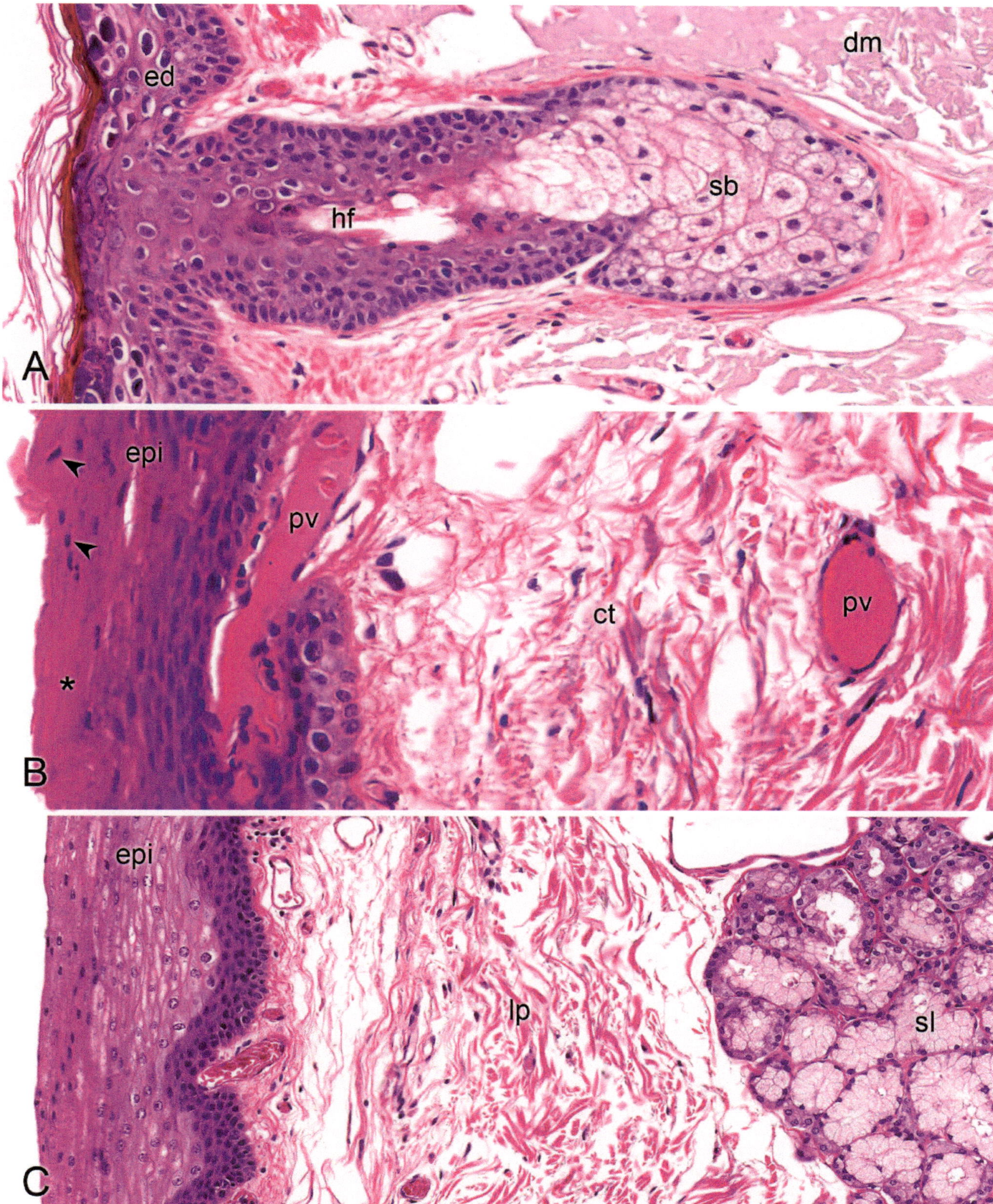

Figure 74. Zones of the lip. (A) Pars cutanea is covered with skin. Under the epidermis, the dermis (dm) contains sebaceous glands (sb) opening into hair follicles (hf). (B) The vermillion zone (vz) is lined with stratified squamous epithelium (epi) with lightly keratinized (asterisk) or parakeratinized superficial layer, where keratinized cells retain their nuclei (arrowheads). Under the epithelium, collagen rich connective tissue contains vessels, including postcapillary venules. (C) Pars mucosa in lined with nonkeratinized stratified squamous epithelium (epi); lamina propria (lp) is rich in collagen fibers and contains small seromucous salivary glands (sl).

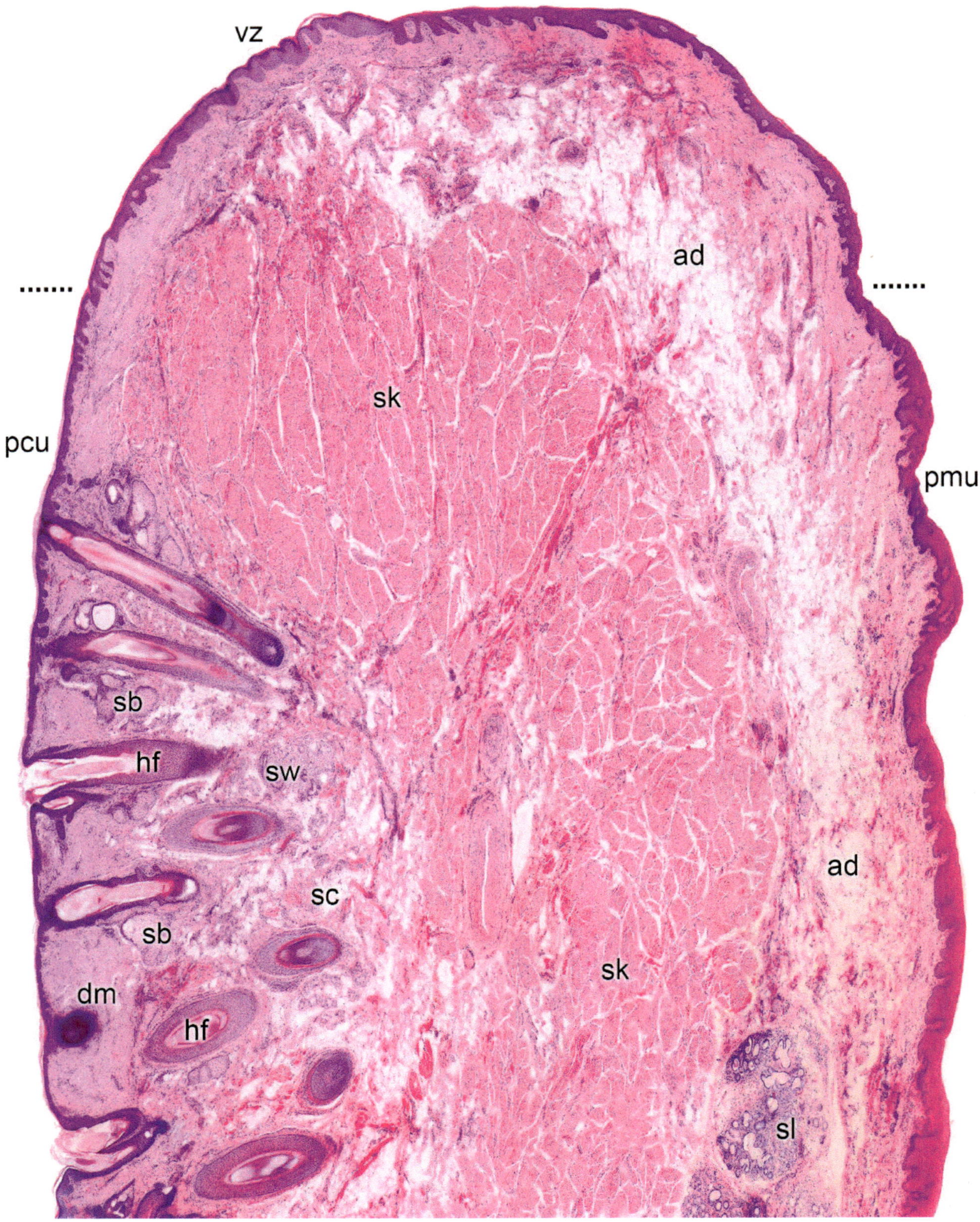

Figure 75. Lip (labium). Pars cutanea (pcu) represents the outer surface of the lip, covered with skin. Dermis (dm) and subcutis (sc) contain hair follicles (hf), sebaceous (sb) and sweat glands (sw). Pars mucosa (pmu) represents the inner surface of the lip covered with mucous membrane lined with nonkeratinized stratified squamous epithelium. Lamina propria contains small seromucous salivary glands (sl) and adipose tissue (ad). Between the pars cutanea and the mucosa (borders marked with dotted line), the vermillion zone (vz) is lined with slightly keratinized or parakeratinized epithelium. Skeletal muscle (sk) of the orbicularis oris occupies the core of the lip.

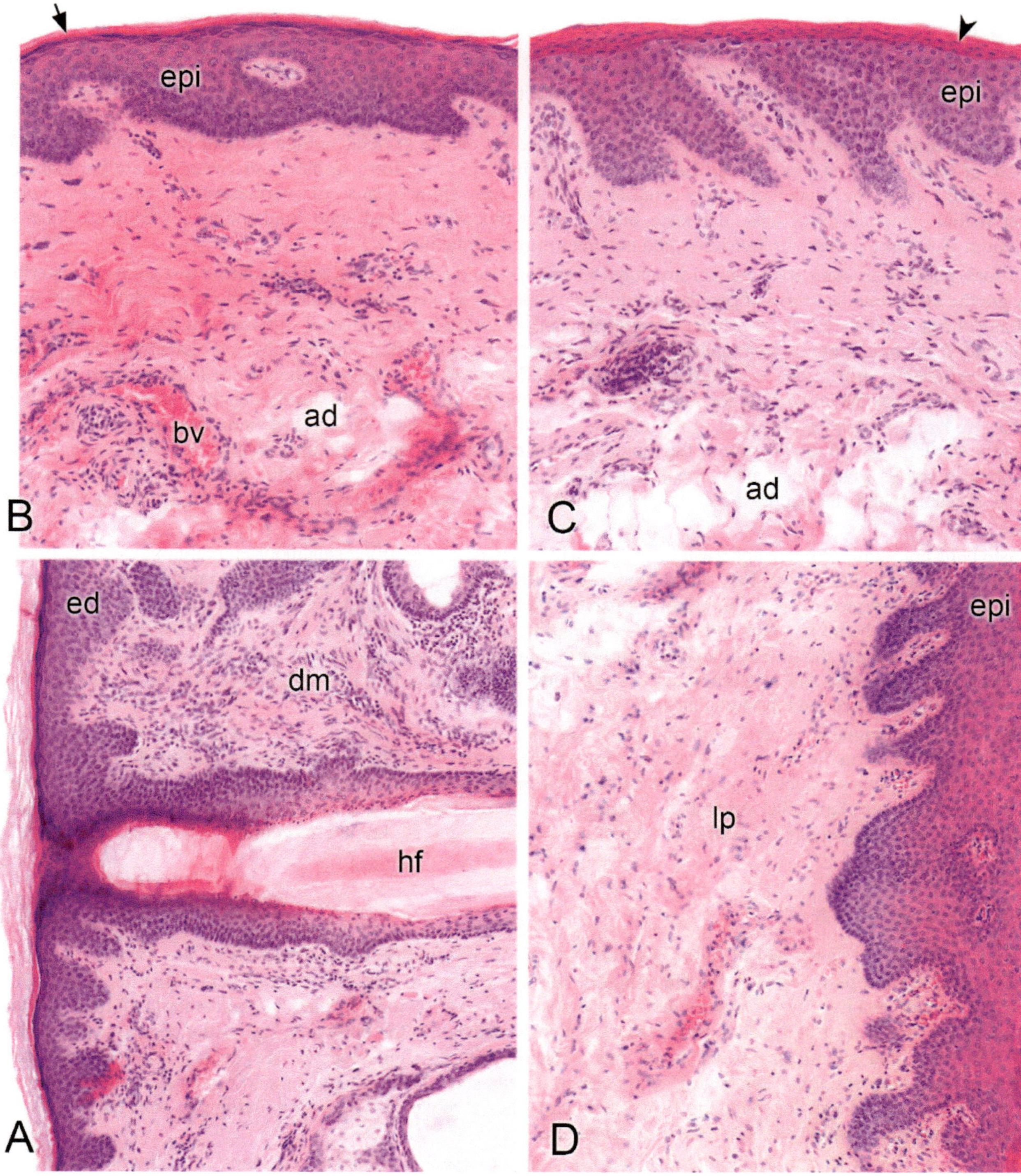

Figure 76. Lip (labium). (A) Pars cutanea is covered with skin. Under the epidermis (ed), dermis (dm) contains hair follicles (hf). (B) The vermillion zone (vz) is lined with stratified squamous epithelium (epi) with lightly keratinized stratum corneum (arrow) that transitions to (C) epithelium with parakeratinized superficial layer (arrowhead), where keratinized cells retain their nuclei. Under the epithelium, collagen rich connective tissue contains blood vessels (bv) and adipose tissue (ad). (D) Pars mucosa in lined with nonkeratinized stratified squamous epithelium (epi); lamina propria (lp) is rich in collagen fibers and contains small seromucous salivary glands.

Tongue

Tongue is primarily a skeletal muscle organ covered by lining mucosa at the ventral (inferior) surface while on the dorsal surface, the specialized mucosa is thrown into projections called papillae. The skeletal muscle core contains large amount of adipocytes dispersed between the muscle fibers, as well as vessels, nerves and the bases of minor salivary glands. The lining mucosa at the inferior surface rests directly on the skeletal muscle without tunica submucosa. The papillae of the specialized mucosa can be categorized as filiform, fungiform, foliate and circumvallate papillae, they are covered with stratified squamous epithelium that can show signs of keratinization and may contain intraepithelial spherical structures, the taste buds. Lining mucosa is lined with nonkeratinized stratified squamous epithelium that can occasionally transition into parakeratinized epithelium.

Filiform papillae are long, thin, conical structures resembling candle flames with the keratinized tips pointing posteriorly. They are the most abundant form of papillae in humans causing the velvety appearance of the dorsum and they have purely mechanical role since they lack taste buds. Fungiform papillae are mushroom-shaped structures on the back and the tip of the tongue; they are commonly surrounded by filiform papillae and associated with taste buds. Foliate papillae are closely packed, leaf-shaped projections located mainly at the lateral borders of the tongue; their epithelium contains taste buds, and they may be associated with minor serous glands opening into the deep clefts bordering the papillae. Circumvallate papillae are large, rounded protrusions surrounded by a circular sulcus. They are located along the lamina terminalis that marks the border between the dorsum of the tongue and the root of the tongue. Circumvallate papillae are associated with von Ebner's serous glands that are embedded between the muscle fibers of the tongue and empty into the moat-like circular sulcus around the papilla. Circumvallate papillae contain numerous taste buds, commonly facing the circular sulcus.

Taste buds are lightly stained, spherical or oval structures inside the stratified squamous epithelium extending from the stratum basale to the luminal surface where they open with the taste pore. Taste buds are composed of long, spindle-shaped, gustatory receptor cells and supporting (sustentacular) cells with oval nuclei that are arranged like the staves of a barrel. The neuroepithelial receptor cells tend to occupy the center of the taste bud, and at their base they are associated with a nerve fiber, while the microvillated apex projects into the taste pore. Basal cells with spherical nucleus function as stem cells for the receptor and supporting cells. Taste buds are also present in the epithelium covering the oropharyngeal isthmus as well as the oro- and laryngopharynx.

The root of the tongue is covered with lining mucosa with occasional epithelial crypts and large amount of aggregated lymph follicles that form the lingual tonsils. These follicles are associated with mucous glands embedded into the muscular layer. In addition to these mucous glands and the serous von Ebner's glands, the tip of the tongue also contains a pair of pea-sized seromucous glands that are embedded among the muscle fibers. These are the anterior lingual or Nuhn-Blandin glands and their ducts open on the ventral surface of the tongue close to the lingual frenulum.

Teeth and gingiva

Teeth are primarily composed of hydroxyapatite that gives the rigidity to the tissue. Teeth are formed by a dentin core that contains the pulp chamber. Starting slightly below the gingival line, the superior part of the dentin is covered with an acellular enamel layer (crown) that is composed of enamel rods formed by ameloblasts during the tooth development. The inferior part of the dentin (root), which is attached to the bony alveolus of the jaw, is covered by the layer cementum that is similar to the bone, but avascular tissue. Cementocytes are enclosed in the lacunae of cementum, and they resemble osteocytes, while cementoblasts show close resemblance to the osteoblasts and they line the outer surface. The gap between the cementum and the alveolus is filled by bridging collagen fibers, the periodontal ligament, which extends into the cementum.

Dentin is produced by the odontoblasts that form a single layer of cuboidal cells lining the interior of the pulp chamber. Odontoblasts rest on the bright eosinophilic layer of newly produced predentin that is not mineralized yet. During the formation of the dentin, part of the odontoblasts remain in the dentinal tubules as thin processes that reach the outer surface of the dentin and participate in the transmission of nociceptive stimuli to the nerve fibers of the pulp. The opposite surfaces of the odontoblasts lay on the dental pulp that fills the pulp chamber and it is composed of loose connective tissue rich in mesenchymal cells as well as vessels and nerves.

Teeth are surrounded by masticatory mucosa that forms the gingiva (gum). Masticatory mucosa also covers the hard palate, and it is lined with keratinized or parakeratinized epithelium with a thin stratum corneum and absent stratum lucidum. The base of the loose connective tissue lamina propria is rich in collagen fibers and it merges with the periosteum of the underlying bone. The epithelium of the masticatory mucosa is firmly attached to the enamel at the base of the periodontal sulcus. This junctional epithelium rests on an external basal lamina and secretes an internal basal lamina at to top of the epithelium that adheres to the tooth.

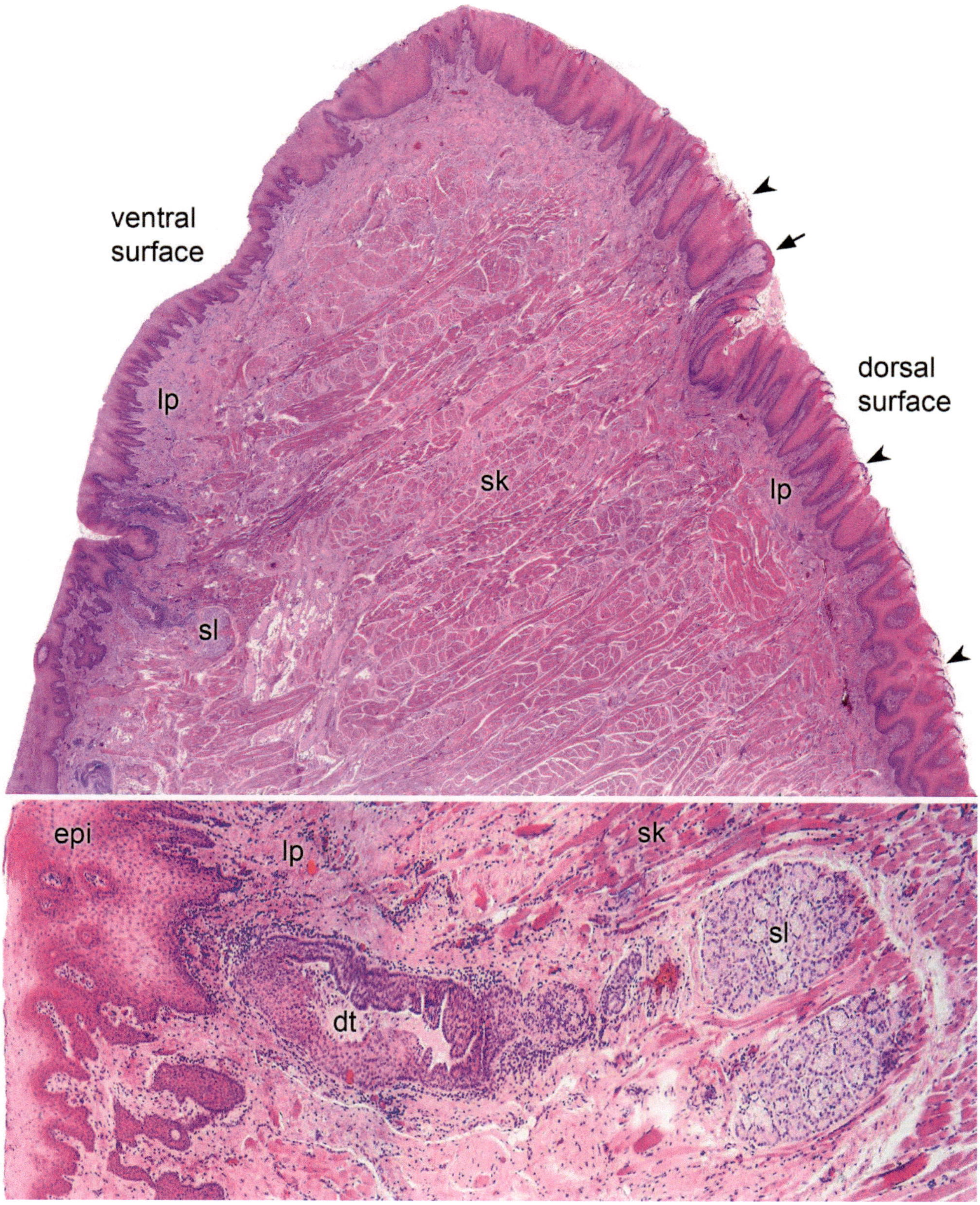

Figure 77. Tip of the tongue. The dorsal surface of the tip is lined with specialized epithelium that forms filiform (arrowheads) and fungiform (arrows) papillae with keratinized surface epithelium. Lamina propria (lp) lies on a skeletal muscle core that contains adipocytes between the muscle fibers. Ventral surface (bottom inset) lacks papillae and it has a smooth mucous membrane with stratified squamous nonkeratinized epithelium (epi); lamina propria (lp) contains ducts (dt) of seromucous salivary glands (sl) embedded into the skeletal muscle core (sk). These glands commonly form a pair of pea-sized masses called anterior lingual glands or Nuhn-Blandin glands; they open on the ventral surface.

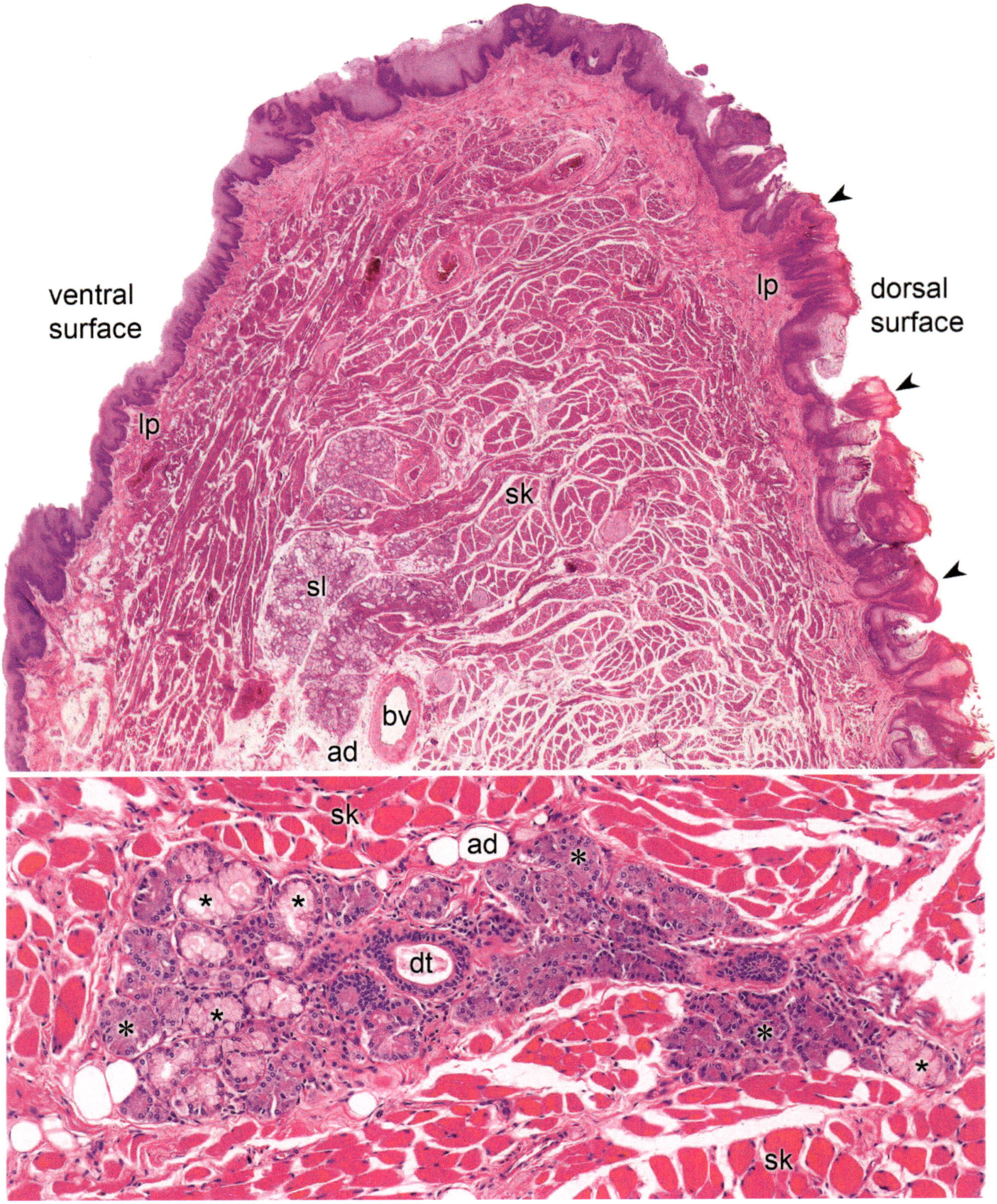

Figure 78. Tip of the tongue, lateral margin. The dorsal surface of the tip is lined with specialized epithelium that forms foliate papillae close to the lateral margins (arrowheads). Papillae have keratinized surface epithelium. Lamina propria (lp) lies on a skeletal muscle core that contains adipocytes (ad) and blood vessels (bv) between the muscle fibers. Ventral surface lacks papillae and it has a smooth mucous membrane with stratified squamous nonkeratinized epithelium; lamina propria (lp) contains seromucous salivary glands (sl, Nuhn-Blandin glands, bottom inset) with serous (six-pointed asterisks) and mucous acini (five-pointed asterisks) and ducts (dt) embedded into the skeletal muscle core (sk).

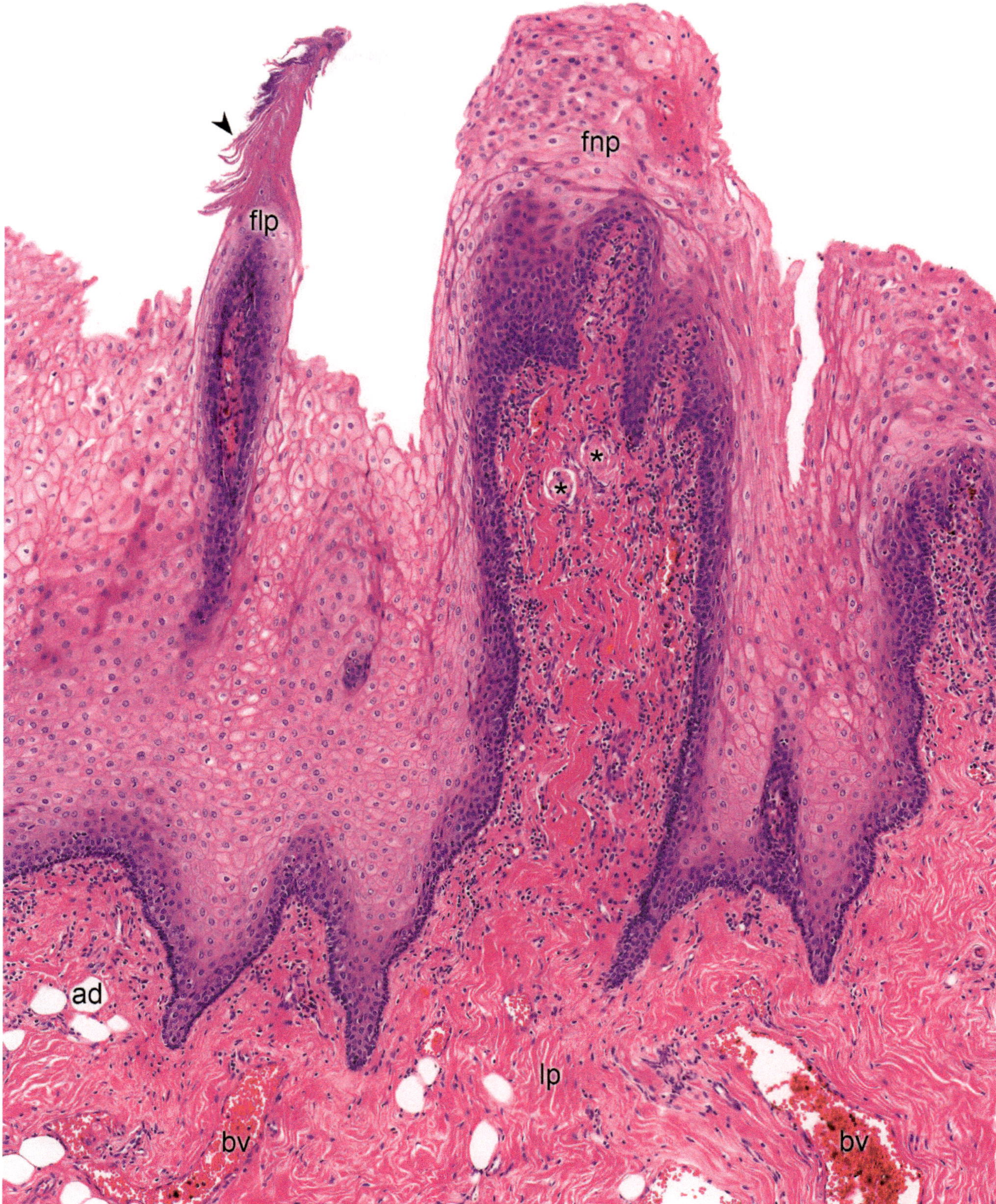

Figure 79. Dorsum of the tongue. The dorsal surface of the tongue is lined with specialized epithelium that forms filiform (flp) and fungiform papillae (fnp). The luminal surfaces of the papillae are keratinized (arrowhead). Filiform papillae have purely mechanical function, while fungiform papillae contain taste buds that are supplied by nerve fibers (asterisks). Lamina propria (lp) is rich in collagen fibers and contains blood vessels (bv) and adipocytes (ad).

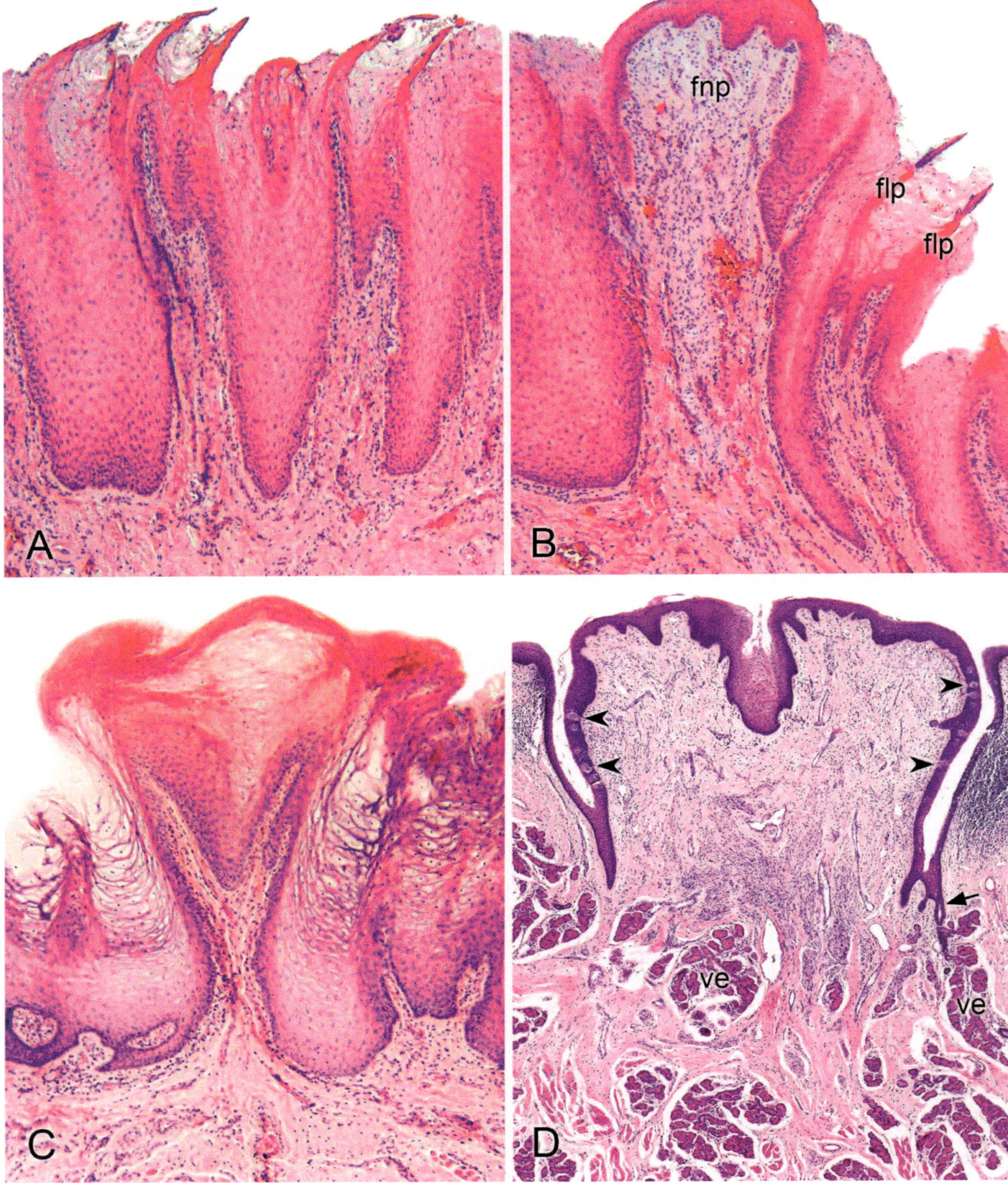

Figure 80. Papillae of the tongue. (A) Filiform papillae cover the majority of the dorsum; they lack taste buds and have purely mechanical function. (B) Fungiform papillae (fnp) are surrounded by filiform papillae (flp) and they contain taste buds. (C) Foliate papillae also contain taste buds and populate the lateral margins of the tongue. (D) Circumvallate papillae are located at the border between the dorsum and the root of the tongue; they are surrounded by a moat-like circular sulcus that drains the ducts (arrow) of the serous von Ebner's glands (ve). Taste buds populate the epithelium facing the circular sulcus (arrowheads).

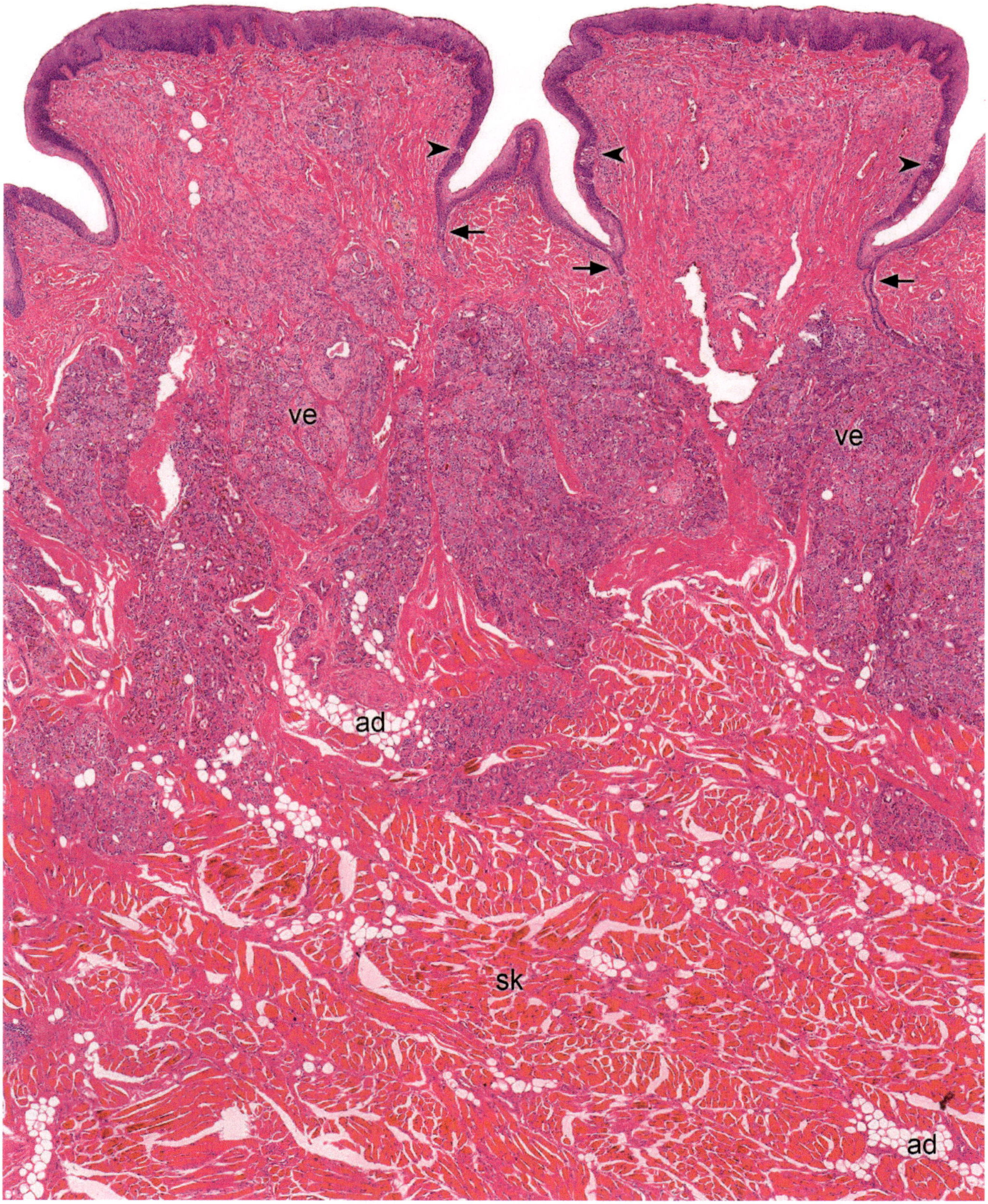

Figure 81. Circumvallate papillae of the tongue. Circumvallate papillae are surrounded by a moat-like circular sulcus that drains the ducts (arrow) of the serous von Ebner's glands (ve). They are located at the border between the dorsum and the root of the tongue and they contain taste buds populating the epithelium facing the circular sulcus (arrowheads). Von Ebner's glands are located in the lamina propria, and they protrude into the skeletal muscle core of the tongue (sk) that contains dispersed adipocytes (ad).

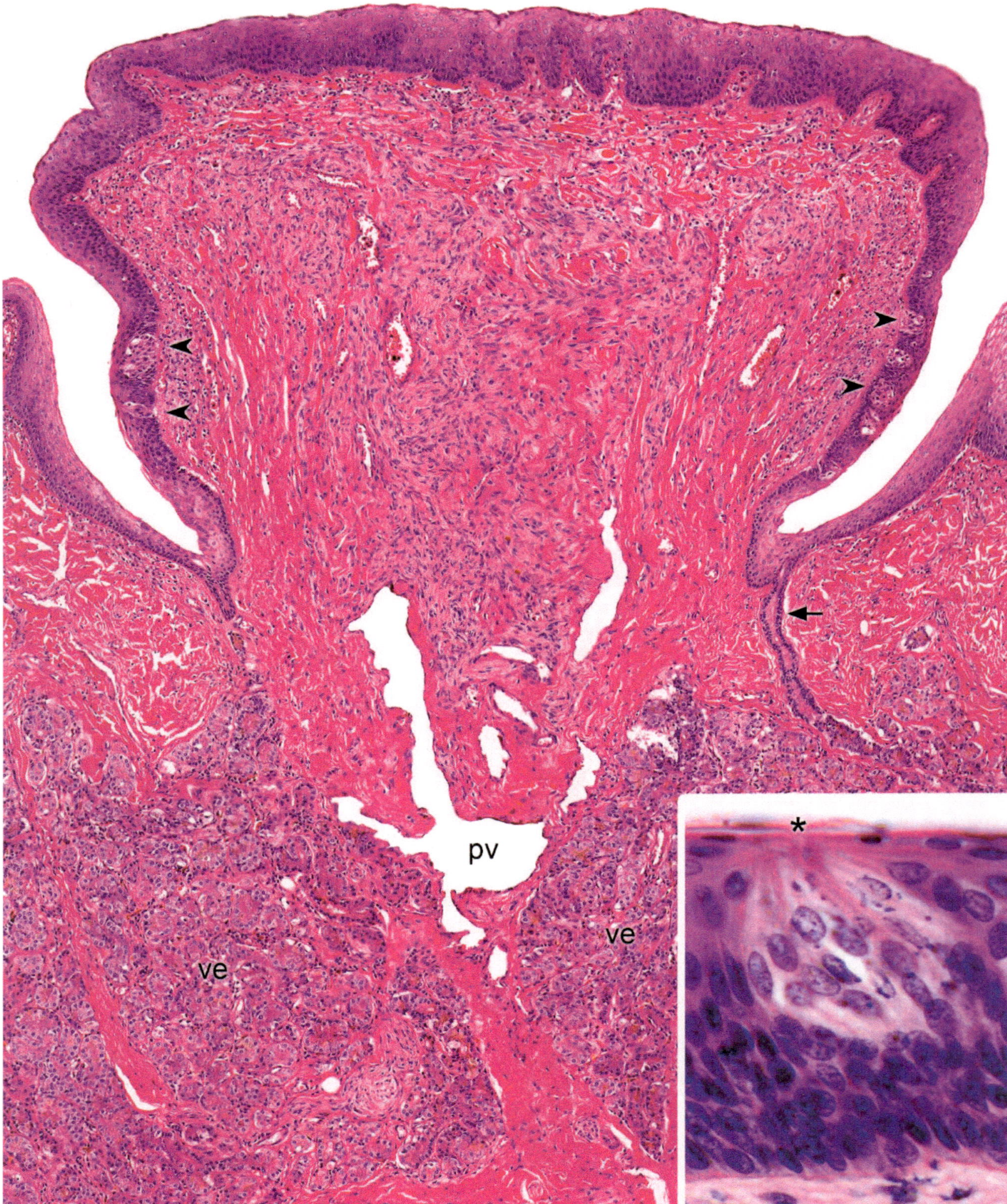

Figure 82. Circumvallate papillae of the tongue. Circumvallate papillae are located at the border between the dorsum and the root of the tongue and they are surrounded by a moat-like circular sulcus that drains the ducts (arrow) of the serous von Ebner's glands (ve). They contain taste buds populating the epithelium facing the circular sulcus (arrowheads). Lamina propria under the epithelium contains a large number of vessels including postcapillary venules (pv). Taste buds (inset) are composed of gustatory receptor cells and supporting cells that are arranged like staves of a barrel; they extend from the stratum basale and open into the taste pore (asterisk).

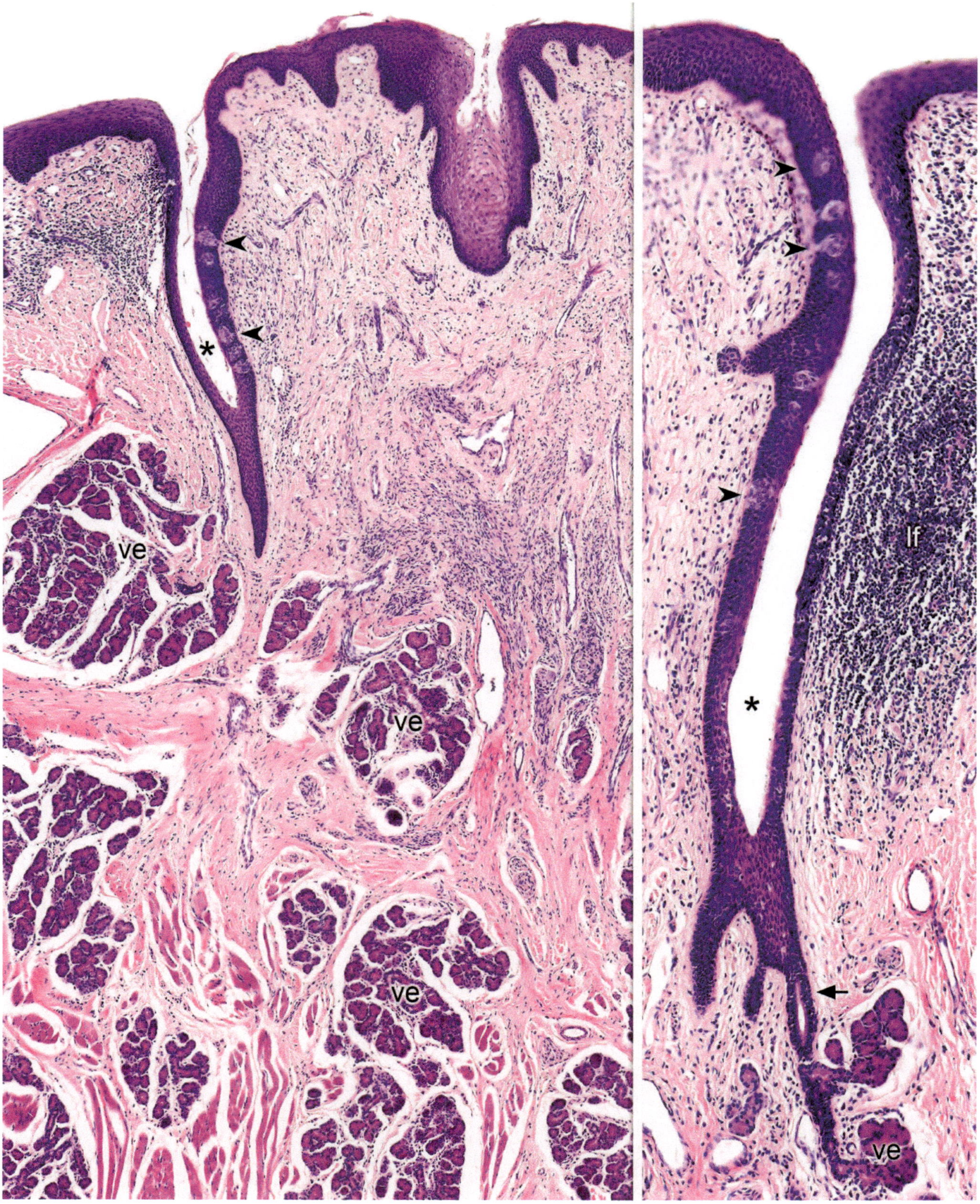

Figure 83. Circumvallate papillae of the tongue. The ducts (arrow) of the serous von Ebner's glands (ve) drain into a moat-like circular sulcus (asterisk) surrounding the circumvallate papillae. These papillae are located at the border between the dorsum and the root of the tongue and they contain taste buds (arrowheads) populating the epithelium facing the circular sulcus. Inset on the right shows the sulcus with high magnification; lymphocytes commonly infiltrate the lamina propria forming lymphatic follicles (lf).

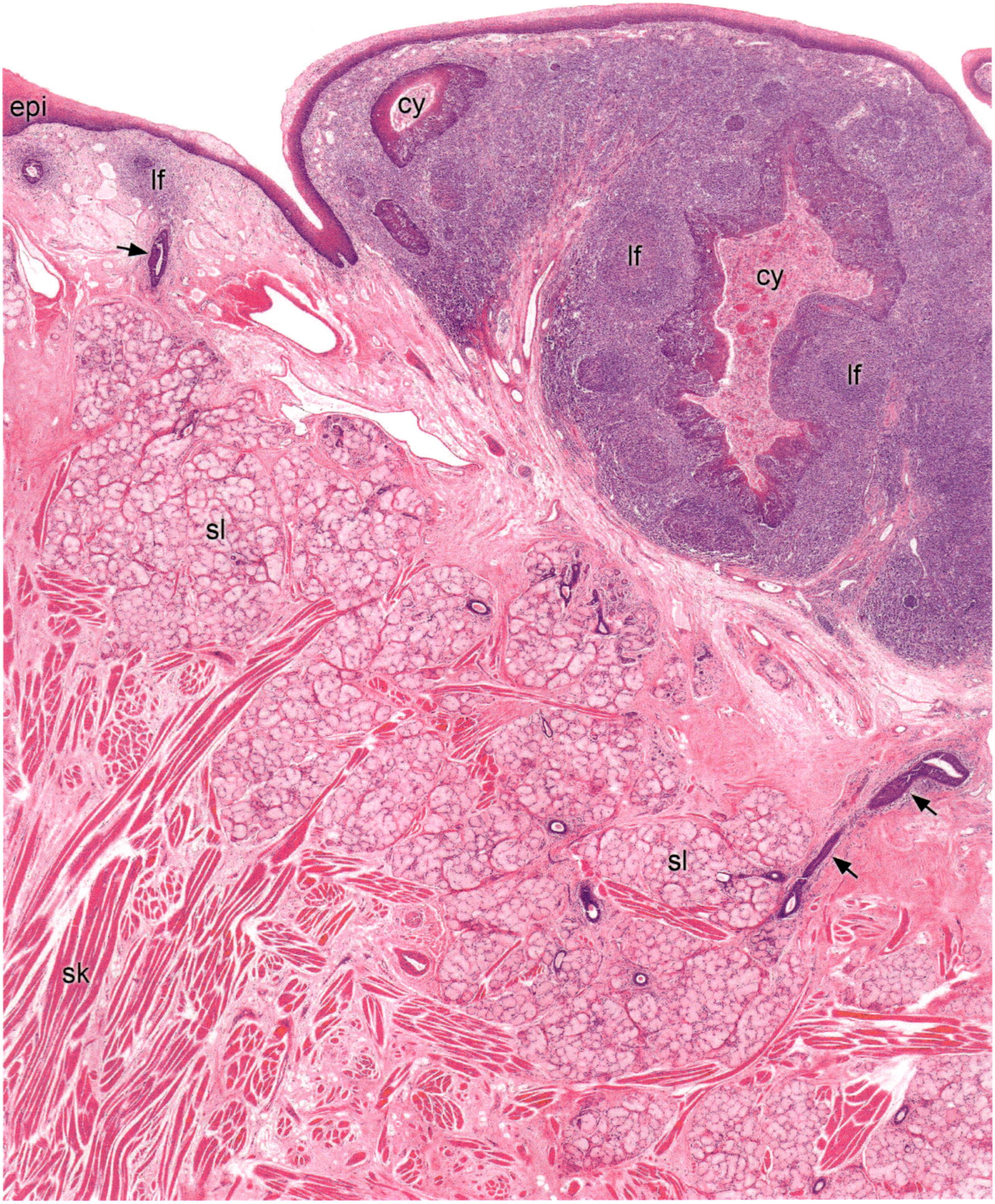

Figure 84. Root of the tongue. Aggregated lymphatic follicles (lf) elevate the mucous membrane of the root of the tongue forming the lingual tonsil. The surface is covered with stratified squamous nonkeratinized epithelium (epi) that may show signs of keratinization. Mucous salivary glands (sl) are embedded into the skeletal muscle (sk) under the lymphatic follicles; their ducts (arrows) perforate the lamina propria. Deep invaginations of the surface epithelium form crypts (cy).

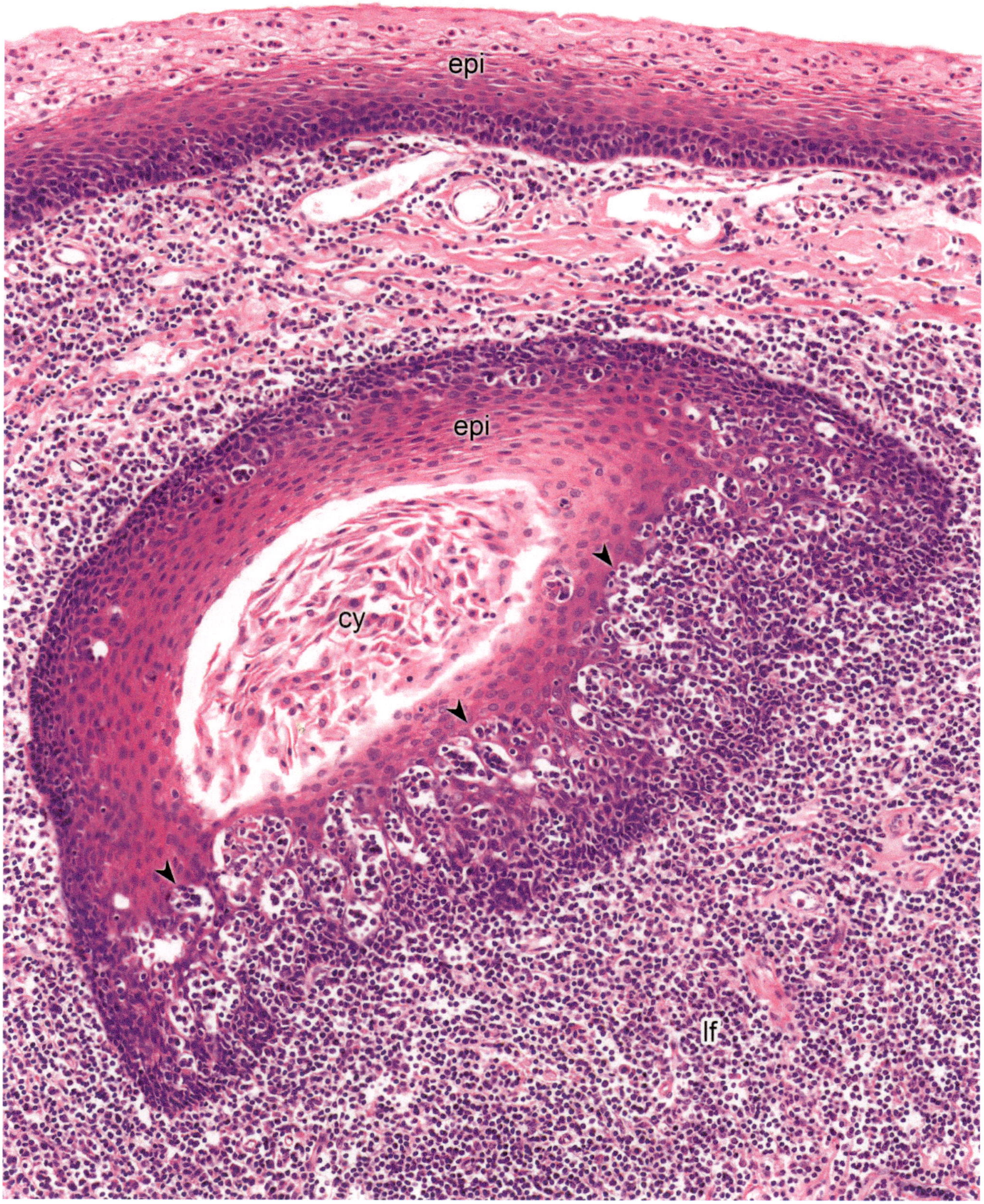

Figure 85. Root of the tongue, cross section of a crypt. Stratified squamous nonkeratinized epithelium (epi) covering the surface forms deep and narrow invaginations called crypts (cy) that penetrate the lamina propria. Lymphocytes of the adjacent lymphatic follicles (lf) commonly invade the epithelium (arrowheads).

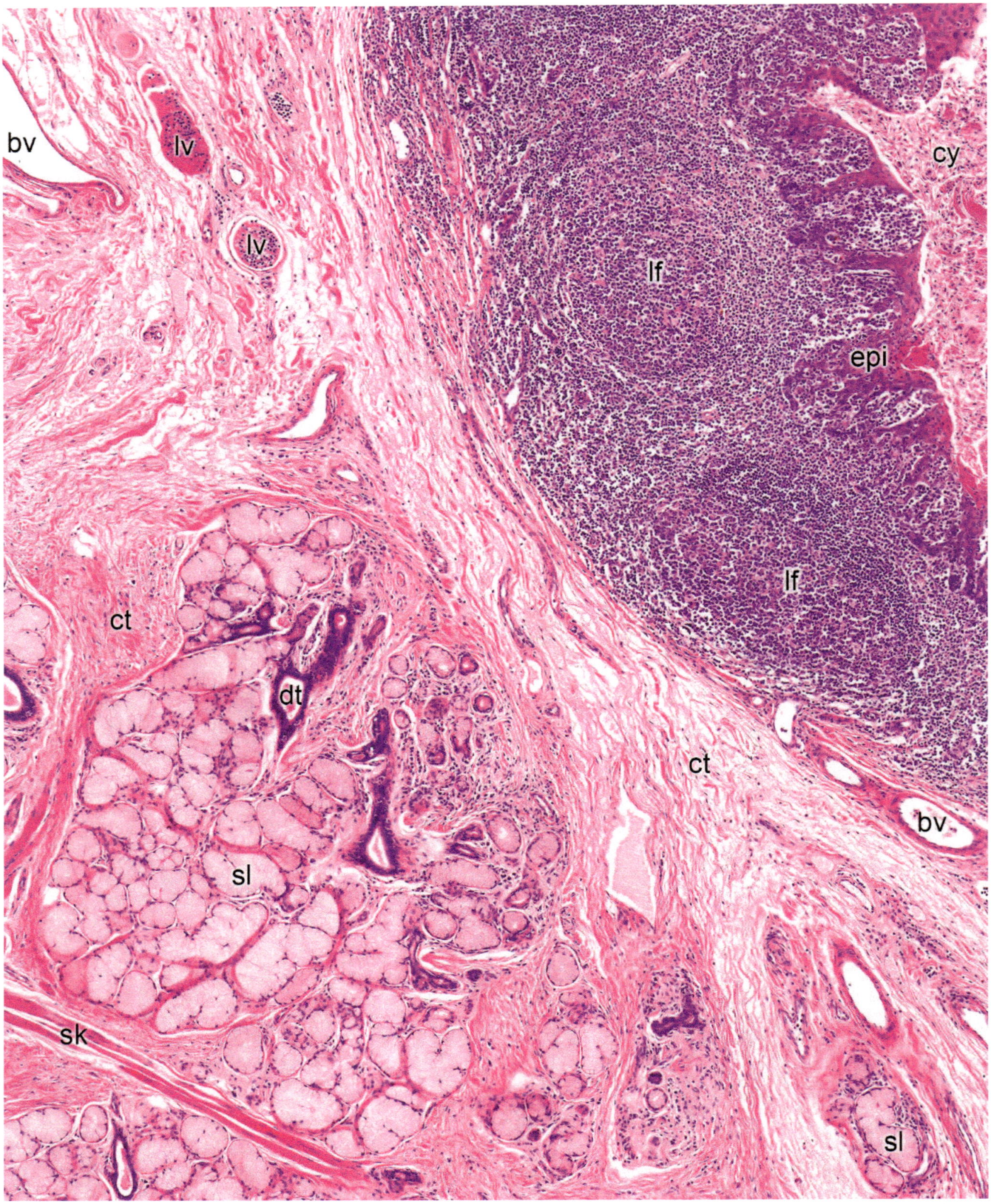

Figure 86. Root of the tongue. Aggregated lymphatic follicles (lf) elevate the mucous membrane of the root of the tongue forming the lingual tonsil. The surface is covered with stratified squamous nonkeratinized epithelium (epi) that is often invaded by lymphocytes. Mucous salivary glands (sl) are embedded into skeletal muscle (sk) and connective tissue (ct) under the lymphatic follicles; their ducts (dt) perforate the lamina propria that contains blood vessels (bv) and lymph vessels (lv). Deep invaginations of the surface epithelium form crypts (cy).

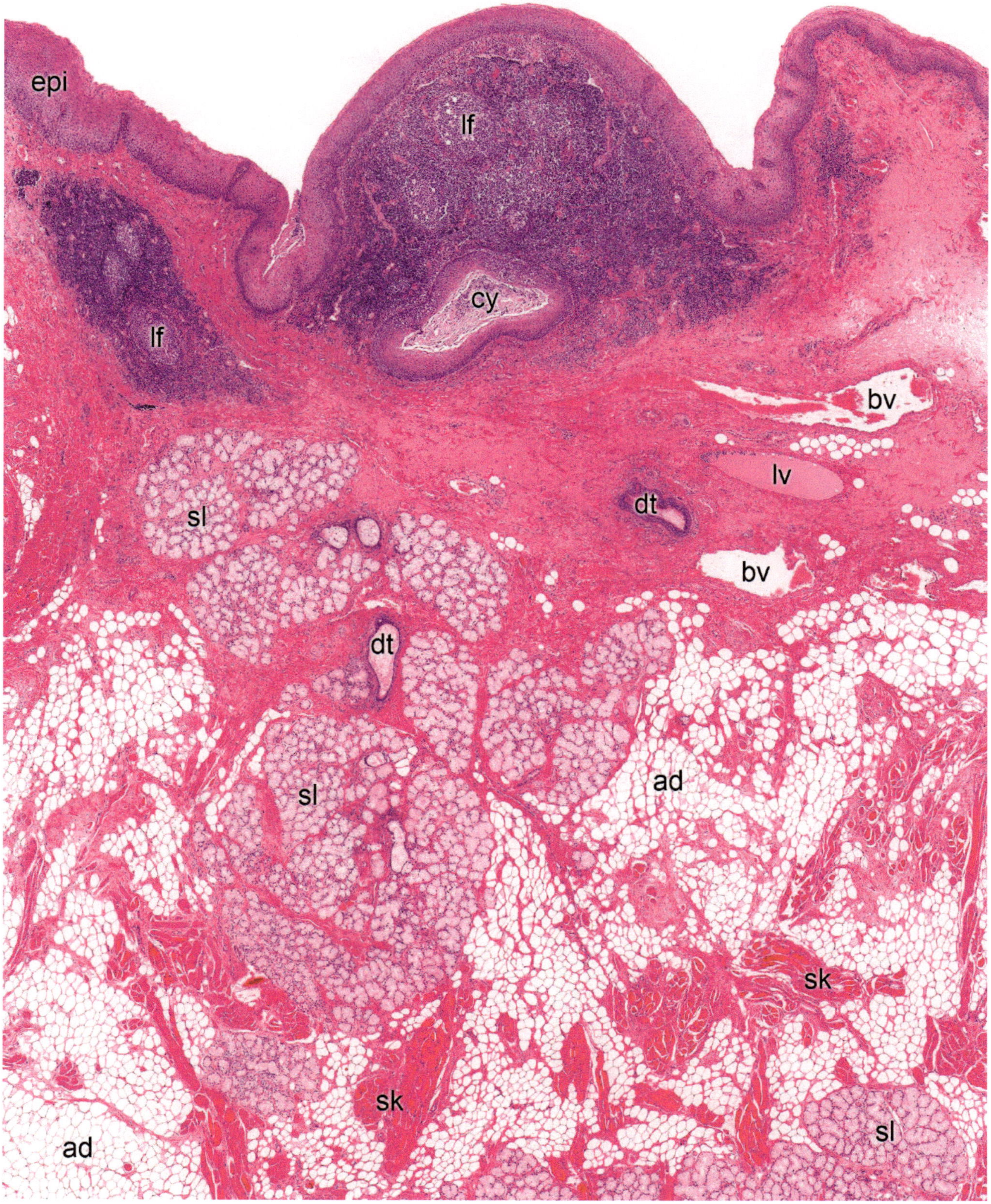

Figure 87. Root of the tongue. Aggregated lymphatic follicles (lf) elevate the mucous membrane of the root of the tongue forming the lingual tonsil. The surface is covered with nonkeratinized stratified squamous epithelium (epi). Mucous salivary glands (sl) are embedded into the skeletal muscle (sk) and adipose tissue (ad) under the lymphatic follicles; their ducts (dt) perforate the lamina propria that contains blood vessels (bv) and lymph vessels (lv). Deep invaginations of the surface epithelium form crypts (cy).

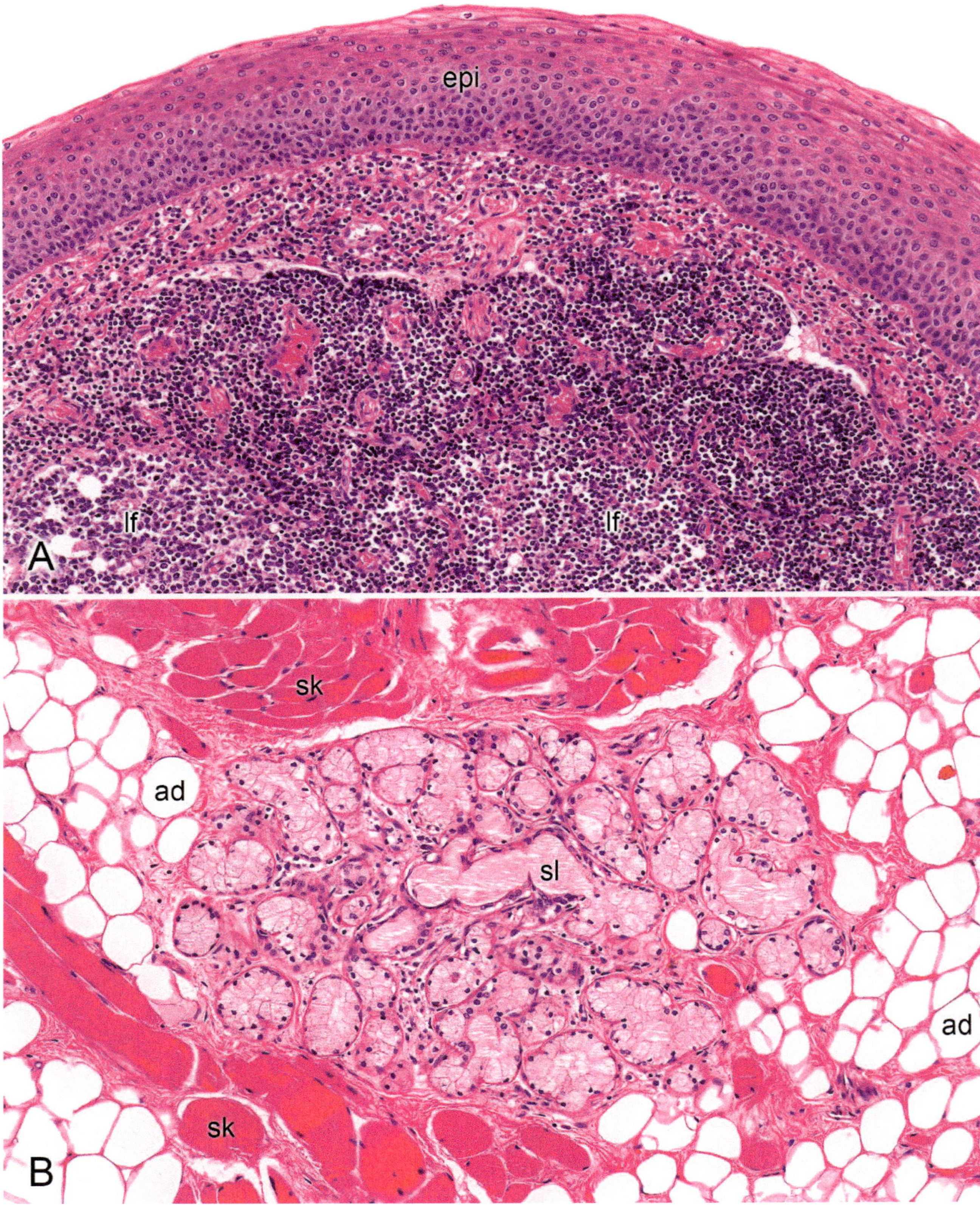

Figure 88. Lingual tonsil at the root of the tongue. (A) Aggregated lymphatic follicles (lf) elevate the mucous membrane of the root of the tongue forming the lingual tonsil. The surface is covered with stratified squamous nonkeratinized epithelium (epi). (B) Mucous salivary glands (sl) are embedded into the skeletal muscle (sk) and adipose tissue (ad) under the lymphatic follicles.

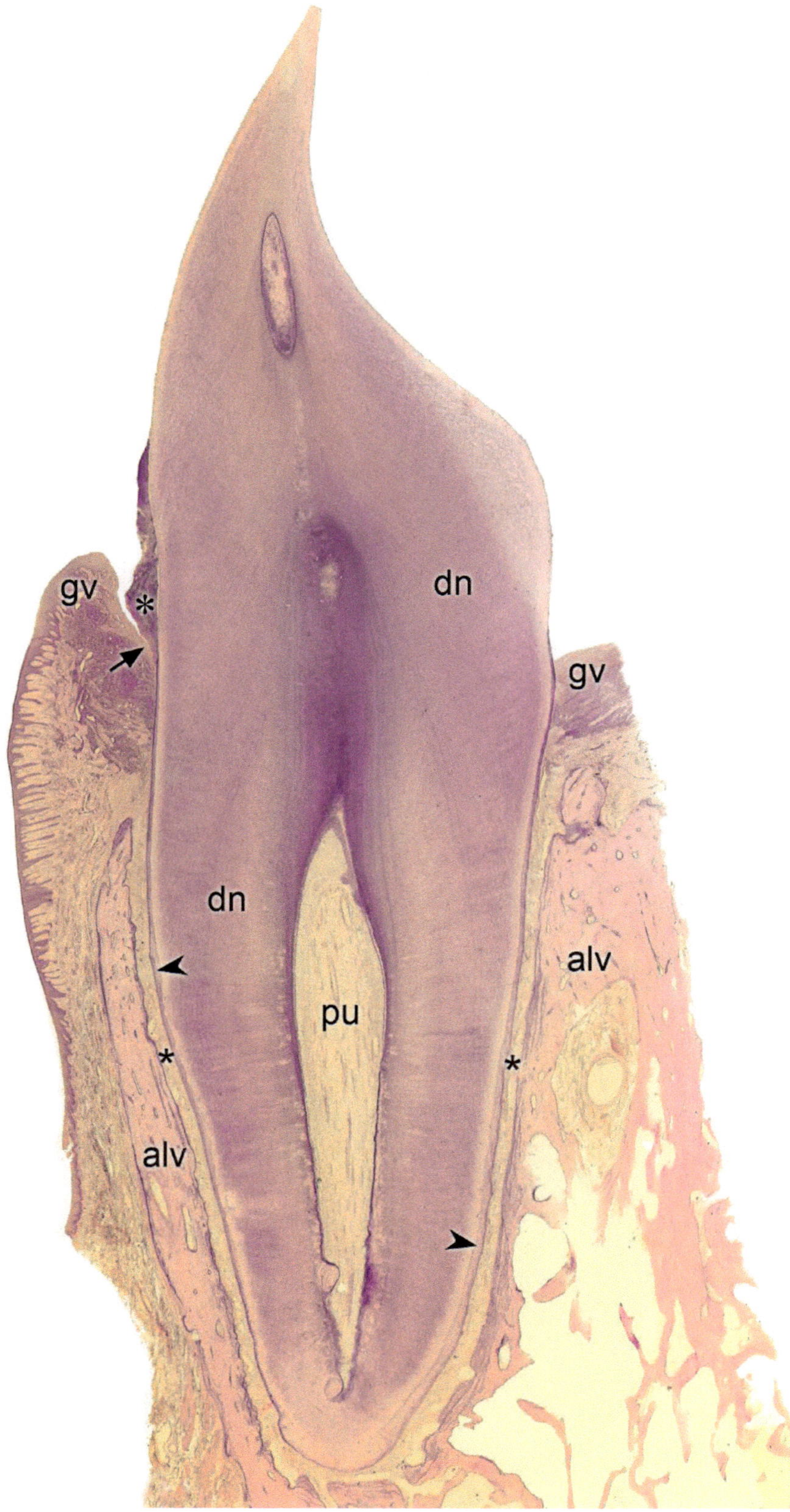

Figure 89. Tooth. The root of the tooth is connected to the alveolar bone (alv) by the periodontal ligament (five-pointed asterisks) while the crown is covered by enamel that has been dissolved during the decalcification process with only a small part remaining (six-pointed asterisk). The majority of tooth is formed by the avascular dentin (dn) enclosing the pulp chamber that contains the loose connective tissue dental pulp (pu) rich in mesenchymal cells. At the root of the tooth, dentin is lined with the cementum (arrowheads). Gingiva (gv) is attached to the neck of the tooth by a basal lamina-like material that is formed by the junctional epithelium (arrow).

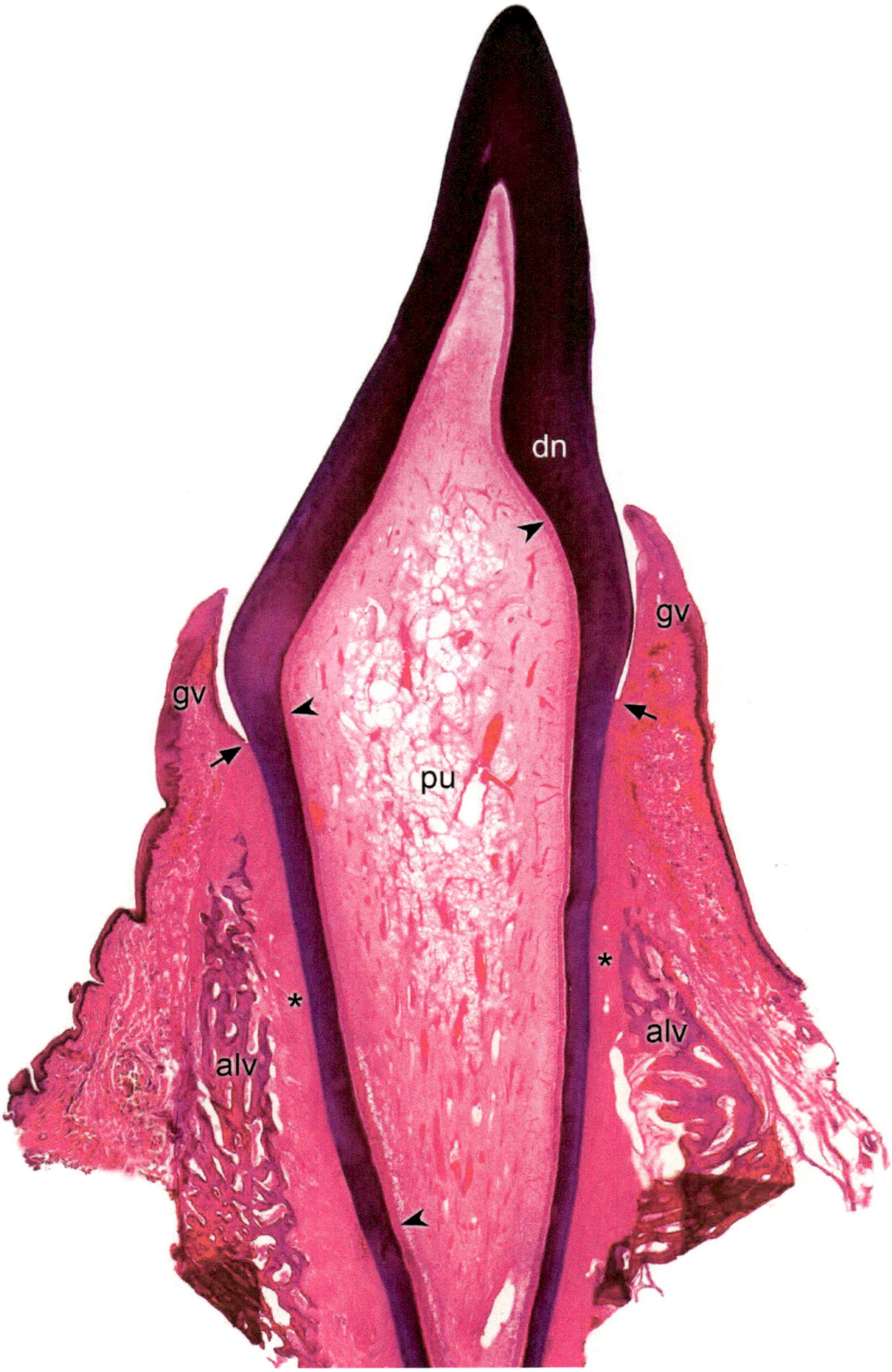

Figure 90. Tooth. The root of the tooth is connected to the alveolar bone (alv) by the periodontal ligament (five-pointed asterisks) while the crown is covered by enamel that has been dissolved during the decalcification process. The majority of tooth is formed by the avascular dentin (dn) enclosing the pulp chamber that contains the loose connective tissue dental pulp (pu) rich in mesenchymal cells. Inside the pulp chamber, dentin is lined with a single layer of columnar odontoblast cells (arrowheads) whose processes perforate the dentin. Gingiva (gv) is attached to the neck of the tooth by a basal lamina-like material that is formed by the junctional epithelium (arrows).

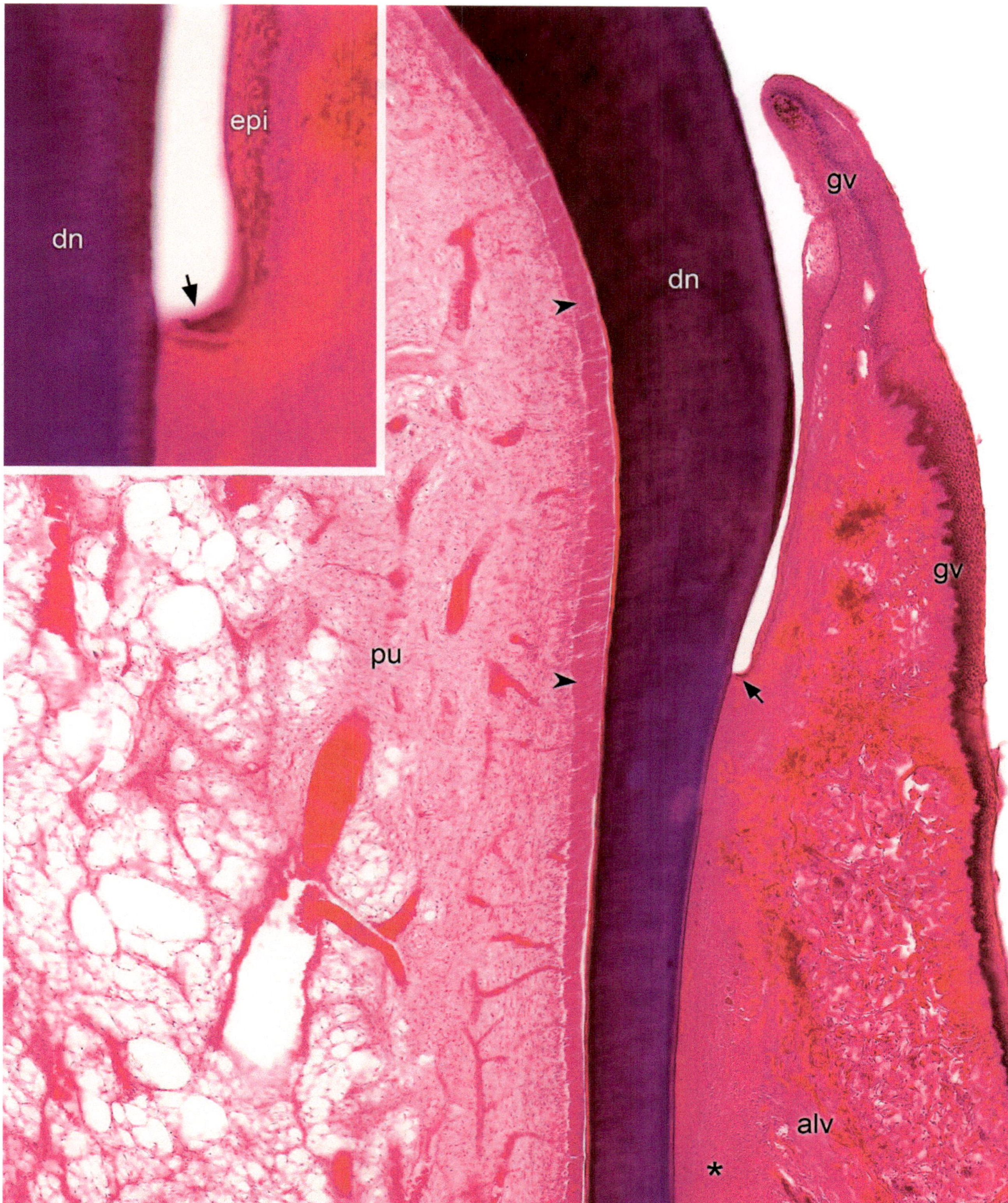

Figure 91. Tooth. The majority of tooth is formed by the avascular dentin (dn) enclosing the pulp chamber that contains the loose connective tissue dental pulp (pu) rich in mesenchymal cells. The root of the tooth is connected to the alveolar bone (alv) by the periodontal ligament (five-pointed asterisks) while the crown is covered by enamel that has been dissolved during the decalcification process. Inside the pulp chamber, dentin is lined with a single layer of columnar odontoblast cells (arrowheads) whose processes perforate the dentin and reach the external surface through dentinal tubules. Gingiva (gv) is lined with parakeratinized or keratinized stratified squamous epithelium (epi) that is attached to the neck of the tooth by a basal lamina-like material (junctional epithelium, arrows).

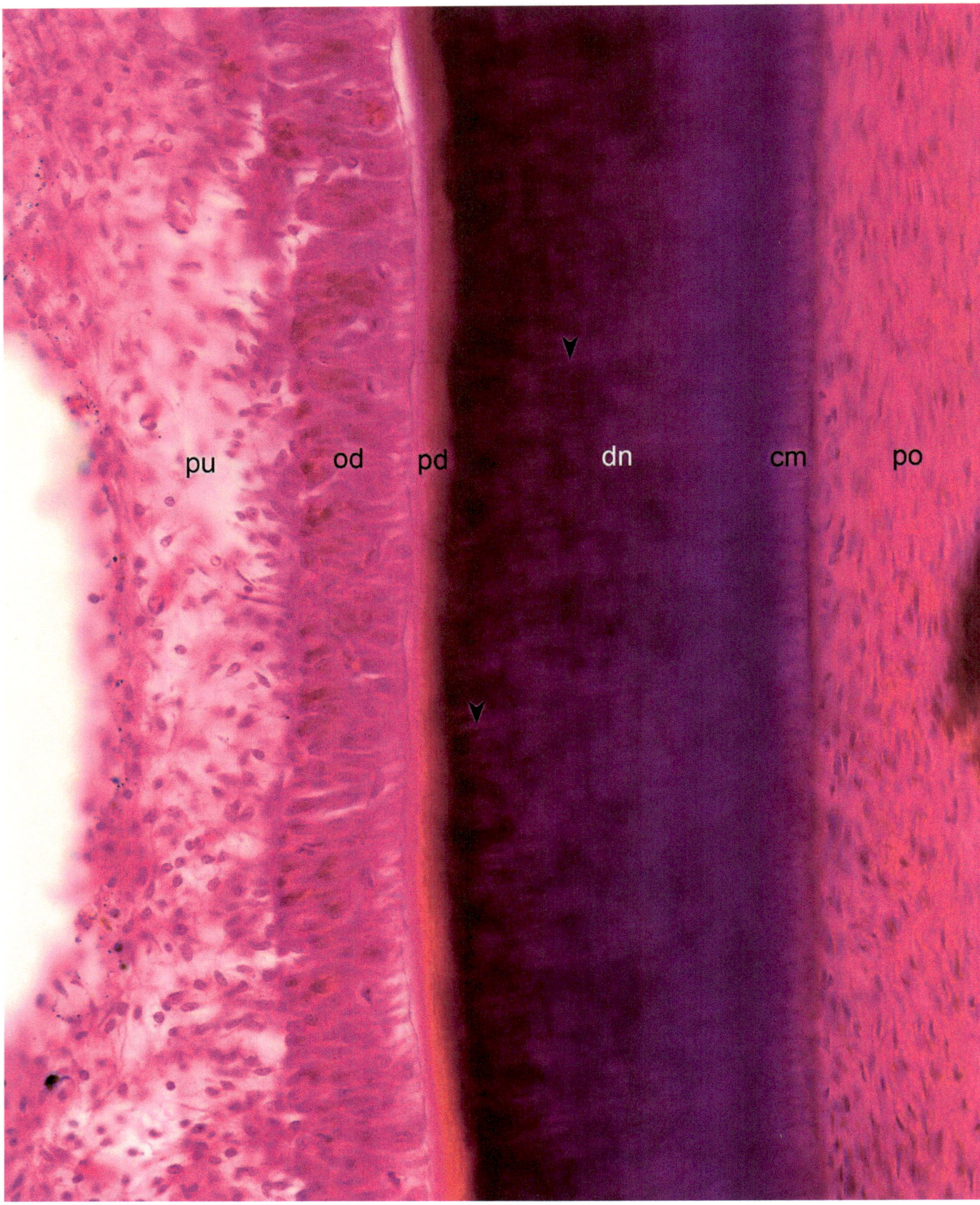

Figure 92. Root of the tooth. The avascular dentin (dn) forms the majority of the tooth enclosing the pulp chamber that contains the loose connective tissue dental pulp (pu) rich in mesenchymal cells. The root of the tooth is connected to the alveolar bone by the periodontal ligament (po). Inside the pulp chamber, dentin is lined with a single layer of columnar odontoblast cells (od) that send processes to the external surface of the dentin through dentinal tubules (arrowheads). Odontoblast cells form the dentin, and they rest on an eosinophilic predentin layer (pd) that is not fully calcified. Dentin is covered by the avascular cementum (cm) on the external surface.

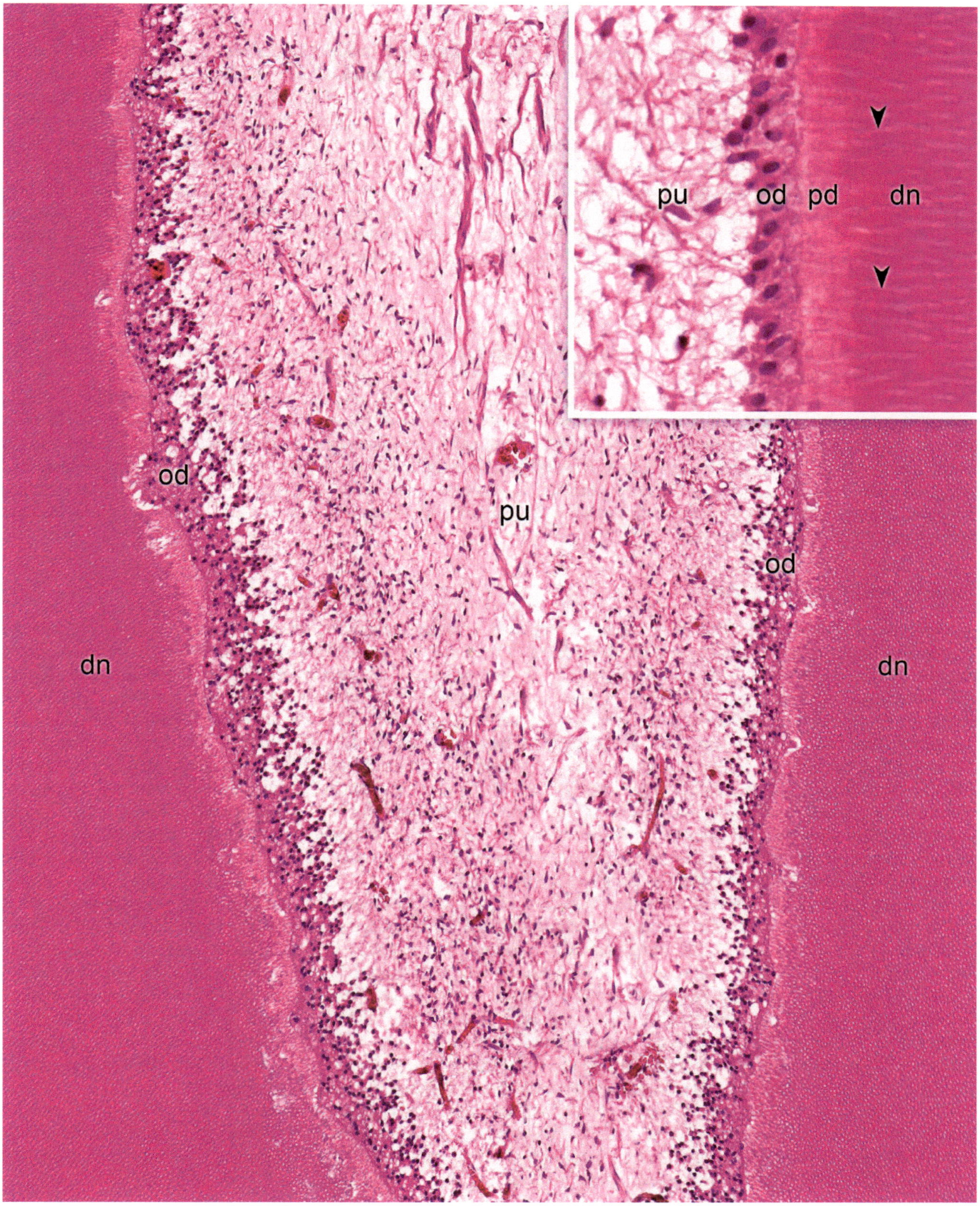

Figure 93. Root of the tooth. The avascular dentin (dn) forms the majority of the tooth enclosing the pulp chamber that contains the loose connective tissue dental pulp (pu) rich in mesenchymal cells. Inside the pulp chamber, dentin is lined with a single layer of columnar odontoblast cells (od) that send processes to the external surface of the dentin through dentinal tubules (arrowheads). Odontoblast cells form the dentin, and they rest on an eosinophilic predentin layer (pd) that is not fully calcified.

Soft palate

Soft palate (velum palatinum) is the posterior continuation of the hard palate, and it marks the transition between the naso- and the oropharynx. Consequently, the anterior part of the upper surface of the soft palate is lined with respiratory epithelium (ciliated pseudostratified columnar epithelium with goblet cells) while the inferior surface is covered with stratified squamous nonkeratinized epithelium of the lining mucosa of the oral cavity; the nonkeratinized stratified squamous epithelium rests on the lamina propria that is attached to the muscle core by a dense irregular connective tissue submucosa. The lamina propria of the superior and inferior mucosal coverage contains seromucous glands and adipocytes. The core of the soft palate is occupied by skeletal muscle fibers surrounded with dense connective tissue fasciae and mucous salivary glands. Uvula is a cylindrical process attached to the midline of the posterior border of the soft palate; it is covered by stratified squamous nonkeratinized epithelium and contains axial skeletal muscle fibers (musculus uvulae) as well as lymphatic follicles, adipose tissue and seromucous glands.

Pharynx

Nasopharynx is lined with ciliated pseudostratified columnar epithelium with goblet cells (respiratory epithelium), while oropharynx and hypopharynx (laryngopharynx) are covered with nonkeratinized stratified squamous epithelium. The lamina propria mucosae of the pharynx is composed of fibroelastic connective tissue particularly rich in elastic fibers. Oropharynx and laryngopharynx have muscularis mucosae and submucosa that attaches to the underlying skeletal muscle. This muscle layer is composed of the inner longitudinal stylopharyngeus and palatopharyngeus muscles and the outer, mostly circular pharyngeal constrictors. Between the muscle layers, an autonomic plexus with nerve fibers and ganglion cells can be observed. Outside, pharynx is covered with loose connective tissue forming the tunica adventitia. Pharyngeal mucosa contains lymphatic aggregations (pharyngeal and tubal tonsils) as well as occasional taste buds in the oro- and laryngopharynx.

Major salivary glands

Oral cavity is associated with large, paired, compound alveolar (acinar) or tubuloalveolar (tubuloacinar) salivary glands that drain into the cavity through their ducts. These salivary glands are surrounded by a connective tissue capsule that sends septa into the gland subdividing it into lobules. The parenchyma of the glands contains salivons that are salivary secretory units composed of basal secretory acini and the duct system that transports and modifies the saliva by absorption and secretion. The acini of the salivary glands are composed of either serous or mucous cells; seromucous (mixed) acini contain both. Serous cells are basophilic due to their rER content and their secretion is rich in enzymes (proteins). In contrast, mucous cells have light cytoplasm because mucin has been dissolved from the tissue during the general histological procedure. In basic histological sections, a mixed acinus is represented by a serous demilune on the surface of a mucous acinus (serous demilune of Gianuzzi), that is considered to be an artifact where serous cells are squeezed out from between the enlarged mucous cells. Flattened myoepithelial cells surround the base of the gland and facilitate the release of the secretory product. The connective tissue stroma of the salivary glands often contains lymphatic aggregations that may penetrate the parenchyma.

Acini open into the intercalated ducts that have generally smaller diameter than that of the acinus. Intercalated ducts are covered by low cuboidal cells, and they are short and indistinct in mucous glands. Intercalated ducts from different acini join in compound glands, including the salivary gland, and continue in striated ducts that have usually larger diameter than that of the acini. These ducts are covered initially with simple cuboidal epithelium that gradually becomes columnar. Epithelial cells with centrally located spherical nuclei exhibit basal infoldings with mitochondria packed between the folds that results in eosinophilic cytoplasm. Both striated and intercalated ducts are intralobular ducts. Interlobular, excretory ducts succeed striated ducts, and they are lined with simple cuboidal or columnar epithelium that transitions into pseudostratified and then stratified cuboidal or columnar epithelia. The terminal zone of the main ducts before they open into the oral cavity is lined with stratified squamous epithelium.

Parotid gland is a compound alveolar gland that is exclusively composed of serous acini. The gland contains substantial amounts of adipose tissue and nerve cross sections since branches of the facial nerve perforate the gland. Submandibular and sublingual glands are seromucous compound tubuloalveolar glands; the former is primarily serous while the latter is mostly composed of mucous acini.

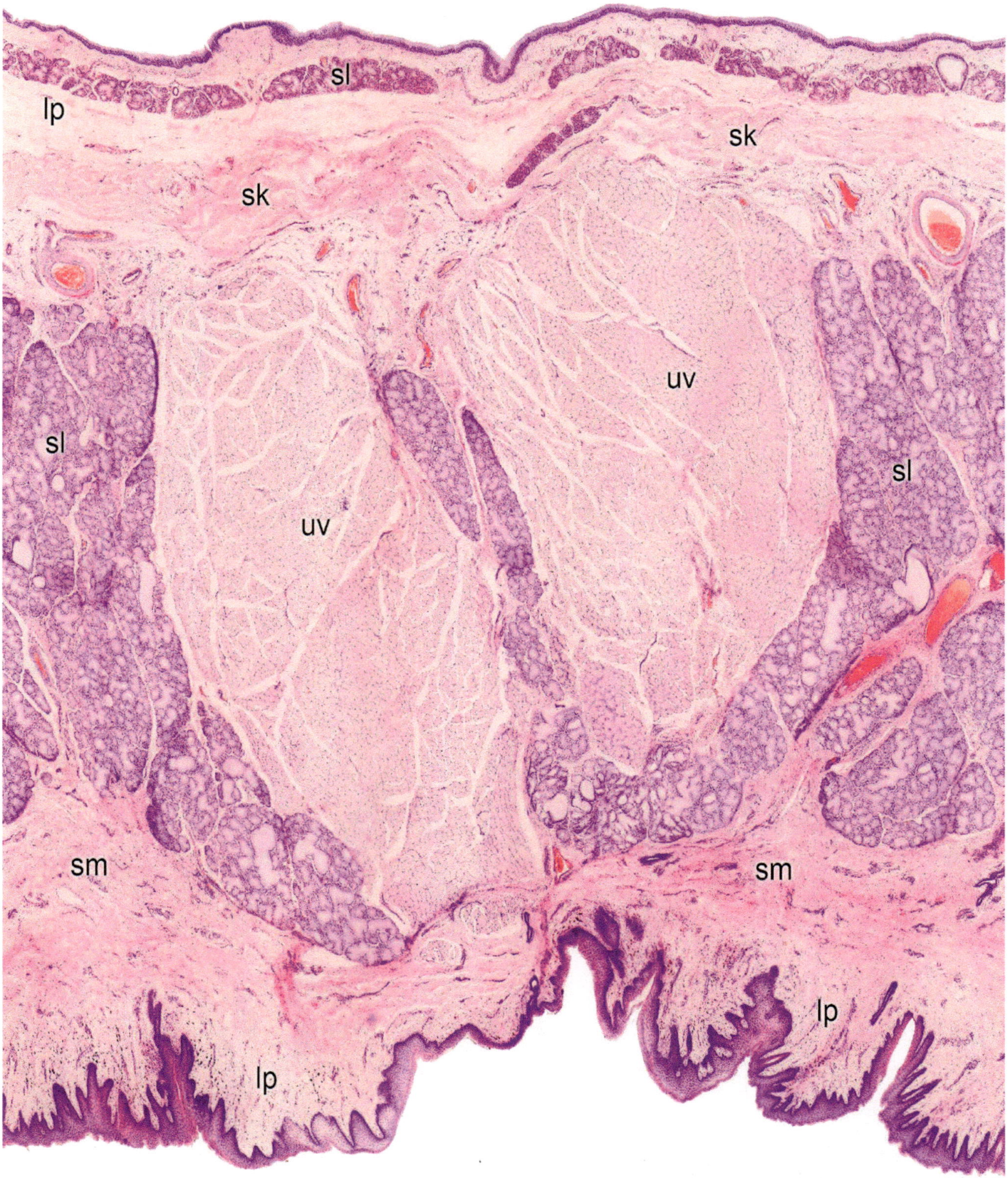

Figure 94. Soft palate (palatum molle, velum palatinum), cross section. The central part of the soft palate contains cross sections of a paired skeletal muscle, the musculus uvulae (uv). The muscle is surrounded with mucous salivary glands (sl). On the top of the musculus uvulae, skeletal muscle fibers (sk) radiate from lateral to medial into the soft palate. The upper part of the soft palate is lined with ciliated pseudostratified columnar epithelium. The underlying loose connective tissue lamina propria (lp) contains mixed salivary glands (sl). The inferior surface of the soft palate is covered with lining mucosa of the oral cavity; nonkeratinized stratified squamous epithelium rests on the lamina propria (lp) that is attached to the muscle core by a dense irregular connective tissue submucosa.

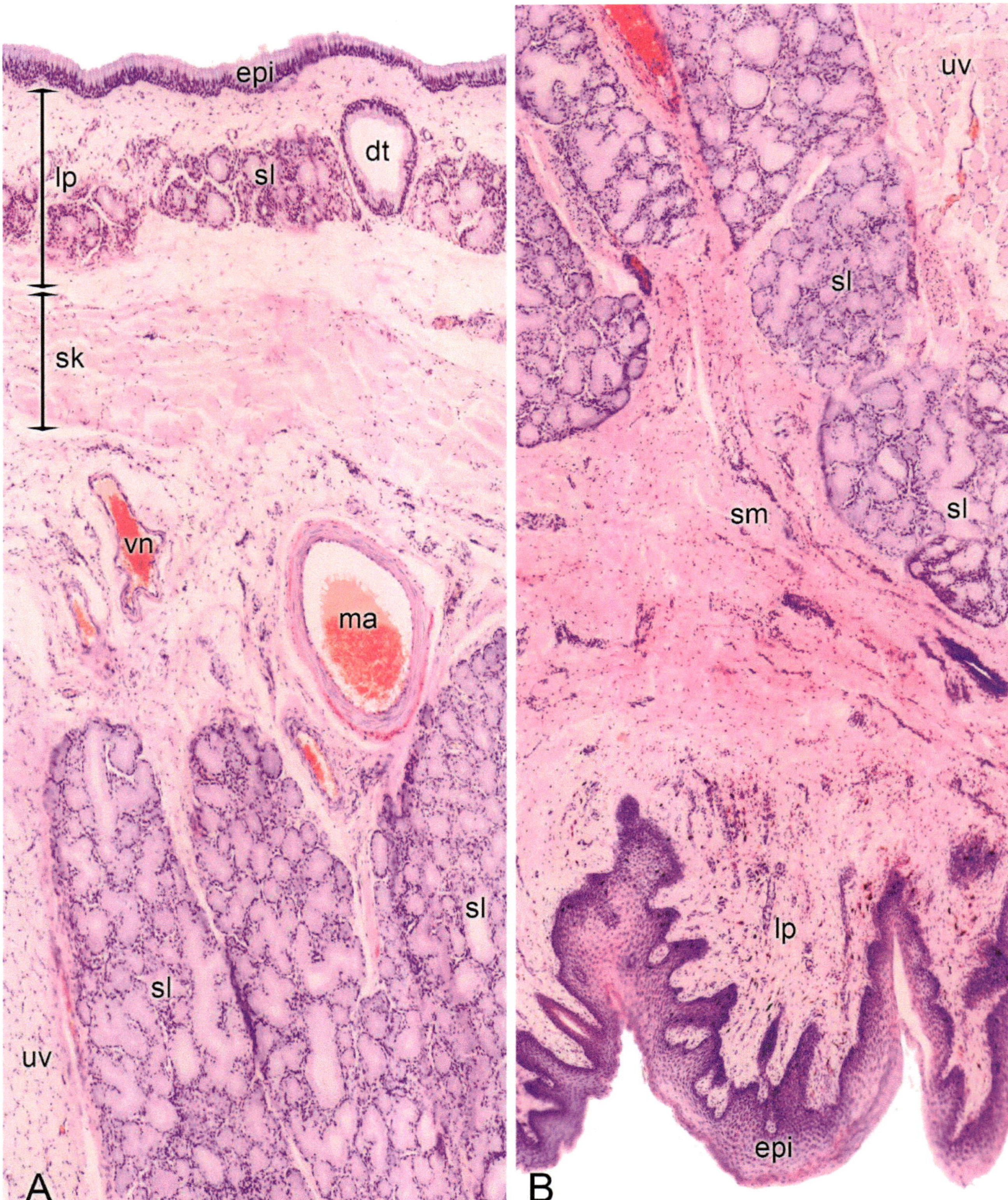

Figure 95. Soft palate (palatum molle, velum palatinum), (A) superior, (B) inferior layers. The central part of the soft palate contains cross sections of a paired skeletal muscle, the musculus uvulae, mucous salivary glands (sl), arteries (ma) and veins (vn). On the top of the musculus uvulae, skeletal muscle fibers (sk) radiate from lateral to medial into the soft palate. (A) The upper part of the soft palate is lined with ciliated pseudostratified columnar epithelium (epi). The underlying loose connective tissue lamina propria (lp) contains mixed salivary glands (sl). (B) The inferior surface of the soft palate is covered with lining mucosa of the oral cavity; the nonkeratinized stratified squamous epithelium rests on the lamina propria (lp) that is attached to the muscle core by a dense irregular connective tissue submucosa (sm).

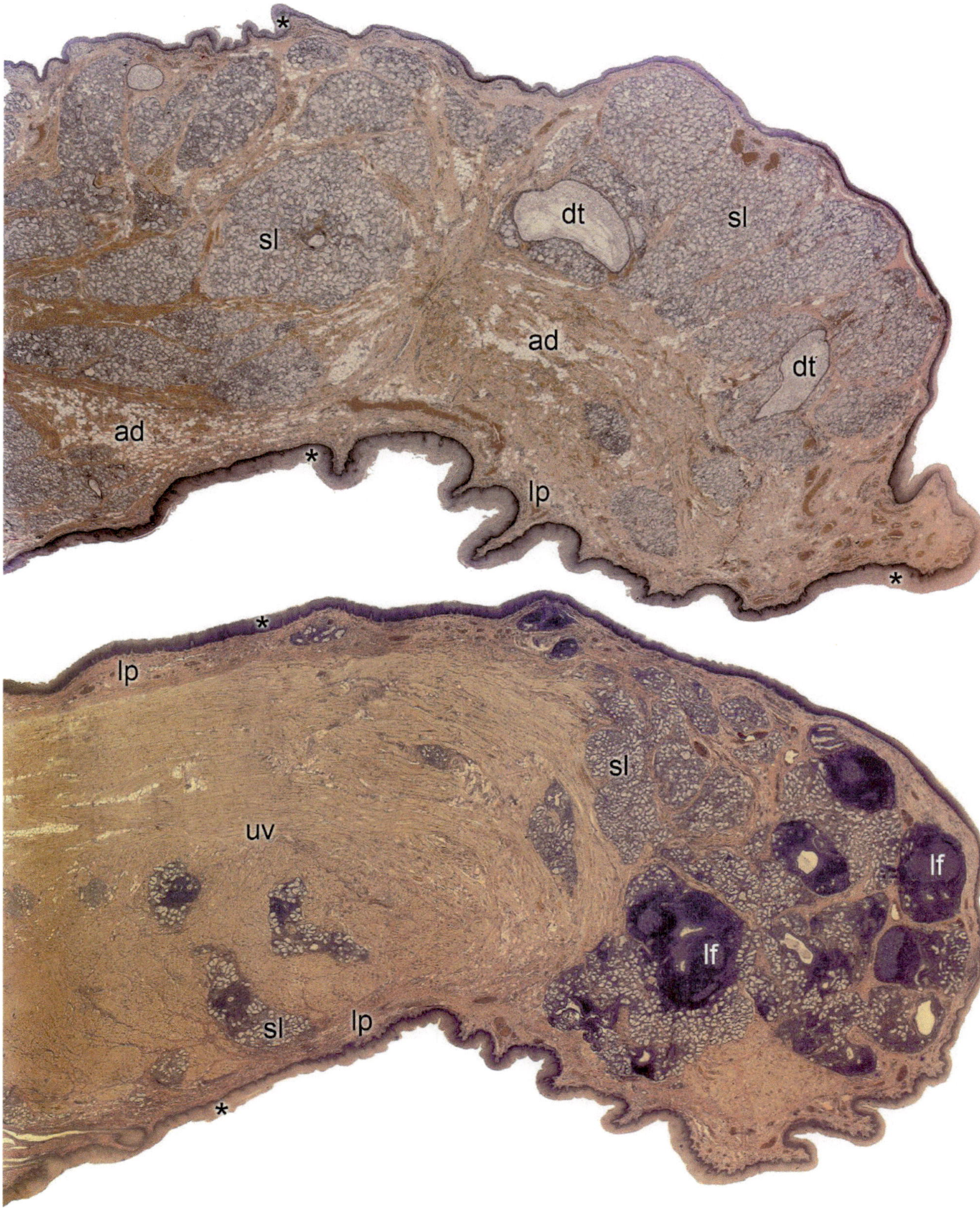

Figure 96. Uvula, longitudinal sections, Van Gieson staining. Upper inset depicts a lateral paramedian sagittal section of the uvula while the lower inset is at a more medial sagittal plane. The skeletal muscle core of the uvula is formed by the musculus uvulae (uv) that is surrounded by seromucous salivary glands (sl) and their ducts (dt) as well as adipose tissue (ad). Uvula is covered with nonkeratinized stratified squamous epithelium (asterisks) and the underlying loose connective tissue lamina propria (lp) that often contains lymphatic follicles (lf).

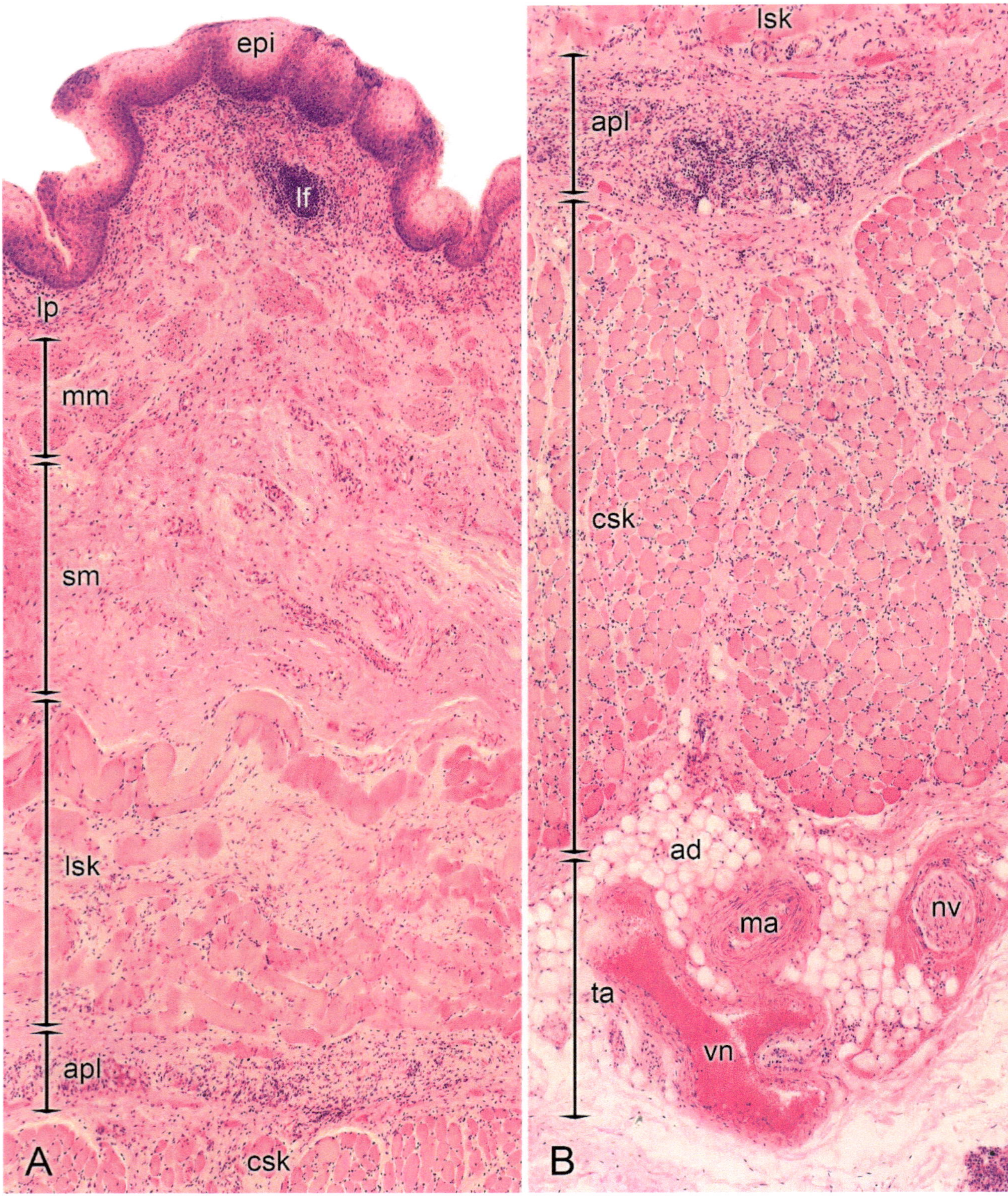

Figure 97. Hypopharynx, (A) internal, (B) external layers. Hypopharynx is covered with nonkeratinized stratified squamous epithelium (epi). Lamina propria mucosae (lp) is composed of loose connective with occasional lymph follicles (lf) and it lies on the tunica muscularis mucosae (mm), a circular smooth muscle layer. Submucosa (sm) is rich in elastic fibers and it is attached to the skeletal muscle layer. The inner longitudinal skeletal muscle fibers (lsk) are formed by the palato- and stylopharyngeus muscles while pharyngeal constrictors form the outer circular fibers (csk). Autonomic plexus (apl) with nerve fibers and ganglion cells can be observed between the muscle layers. The outermost tunica adventitia (ta) contains adipocytes (ad), nerves (nv) as well as vessels, including muscular arteries (ma) and veins (vn), embedded into loose connective tissue.

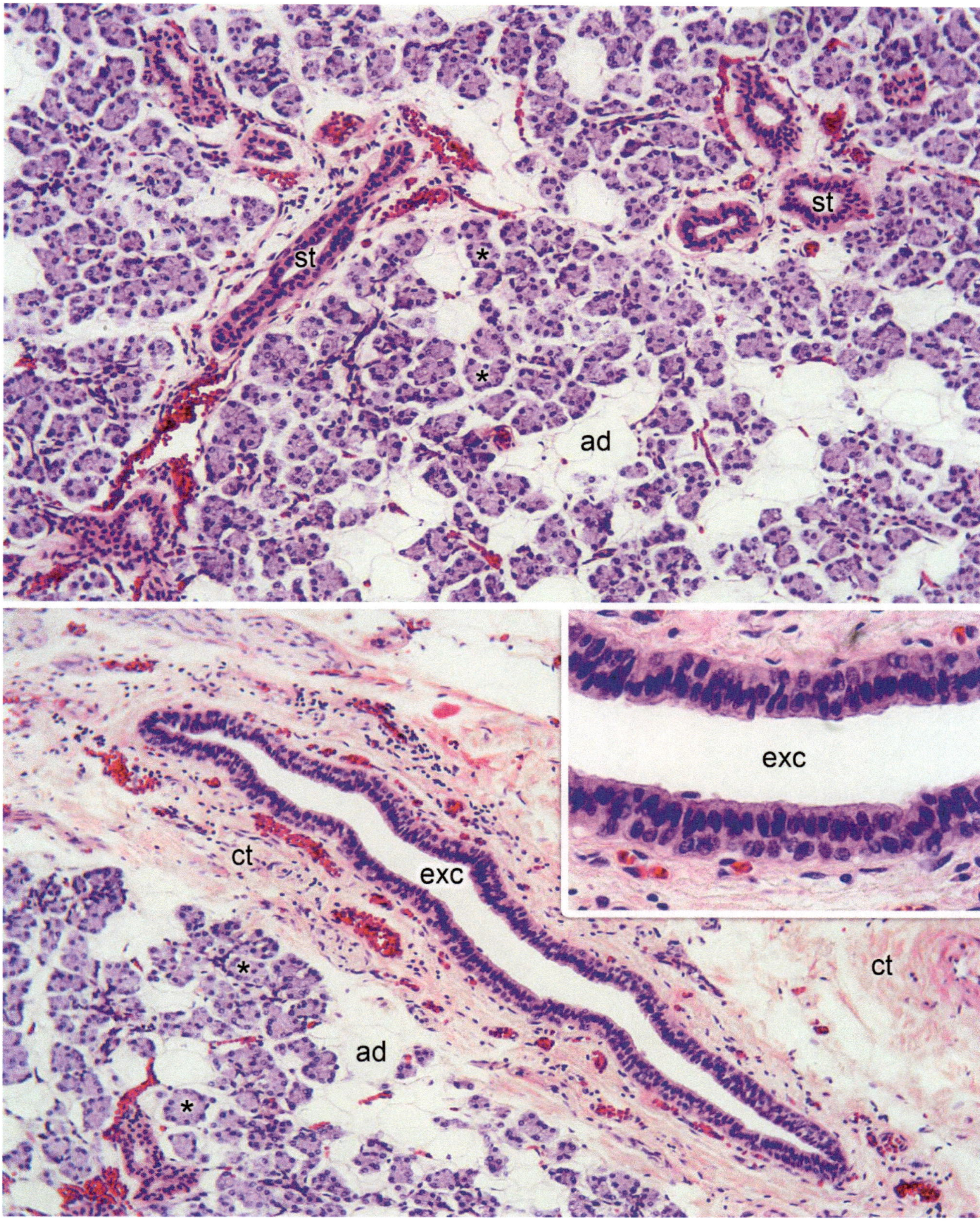

Figure 98. Parotid gland. The largest salivary gland is the parotid that is composed of serous acini (asterisks) with dispersed adipocytes (ad). Acini drain into intercalated ducts that empty into larger striated ducts (st) lined with cuboidal or columnar cells with spherical or oval nuclei. The cytoplasm is eosinophilic and exhibits basal striation due to the numerous mitochondria occupying the infoldings of the basal plasma membrane. Excretory ducts (exc) are lined with stratified columnar epithelium (inset).

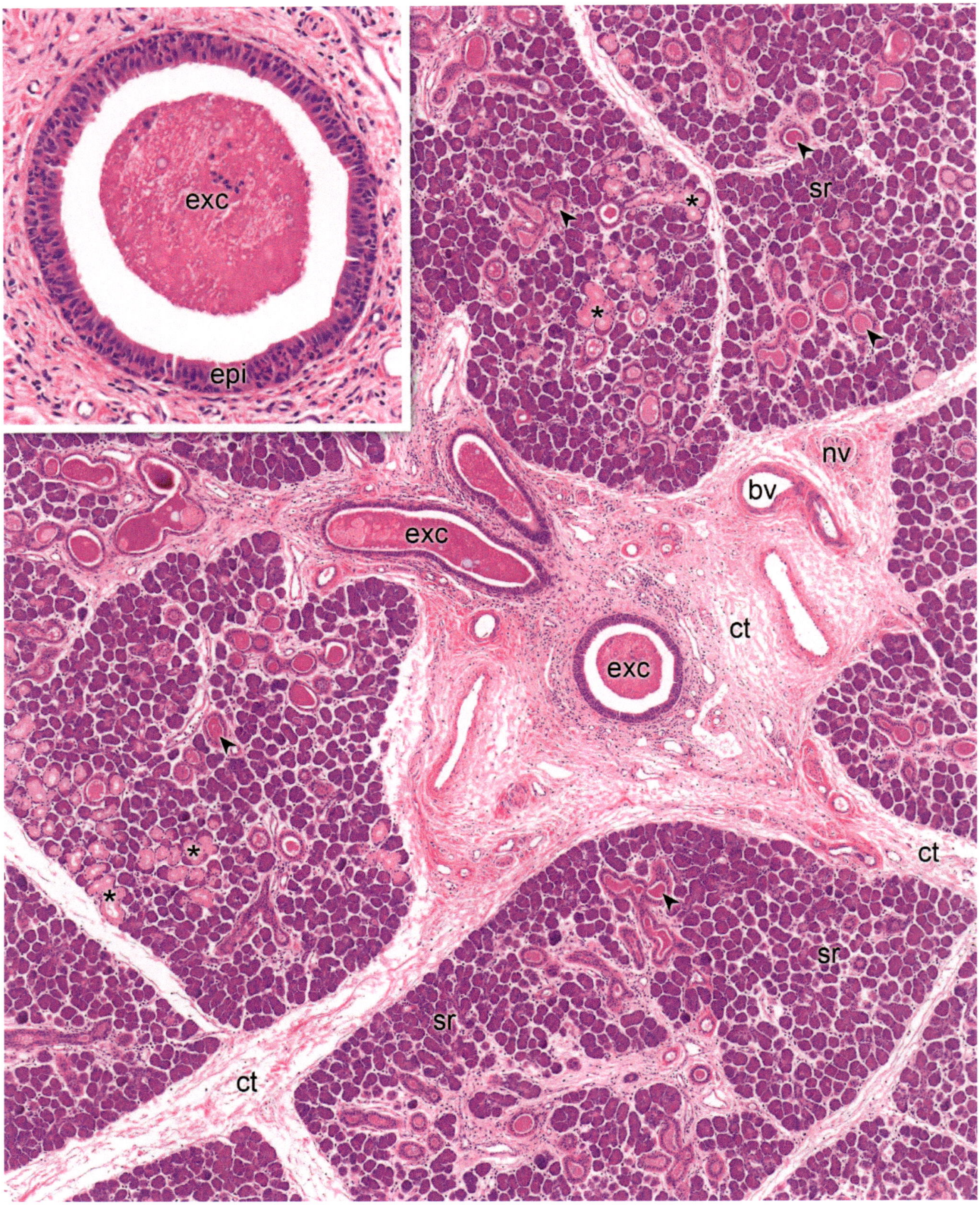

Figure 99. Submandibular gland. Connective tissue septa (ct) separate the parenchyma of the submandibular gland into lobules that are formed primarily by serous acini (sr) and a smaller number of mucous acini (asterisk). Connective tissue surrounds the smaller striated ducts (arrowheads) that are lined with cuboidal or columnar cells with spherical or oval nuclei. Larger excretory ducts (exc) are lined with stratified columnar epithelium (inset). Septa also contain blood vessels (bv) and nerves (nv).

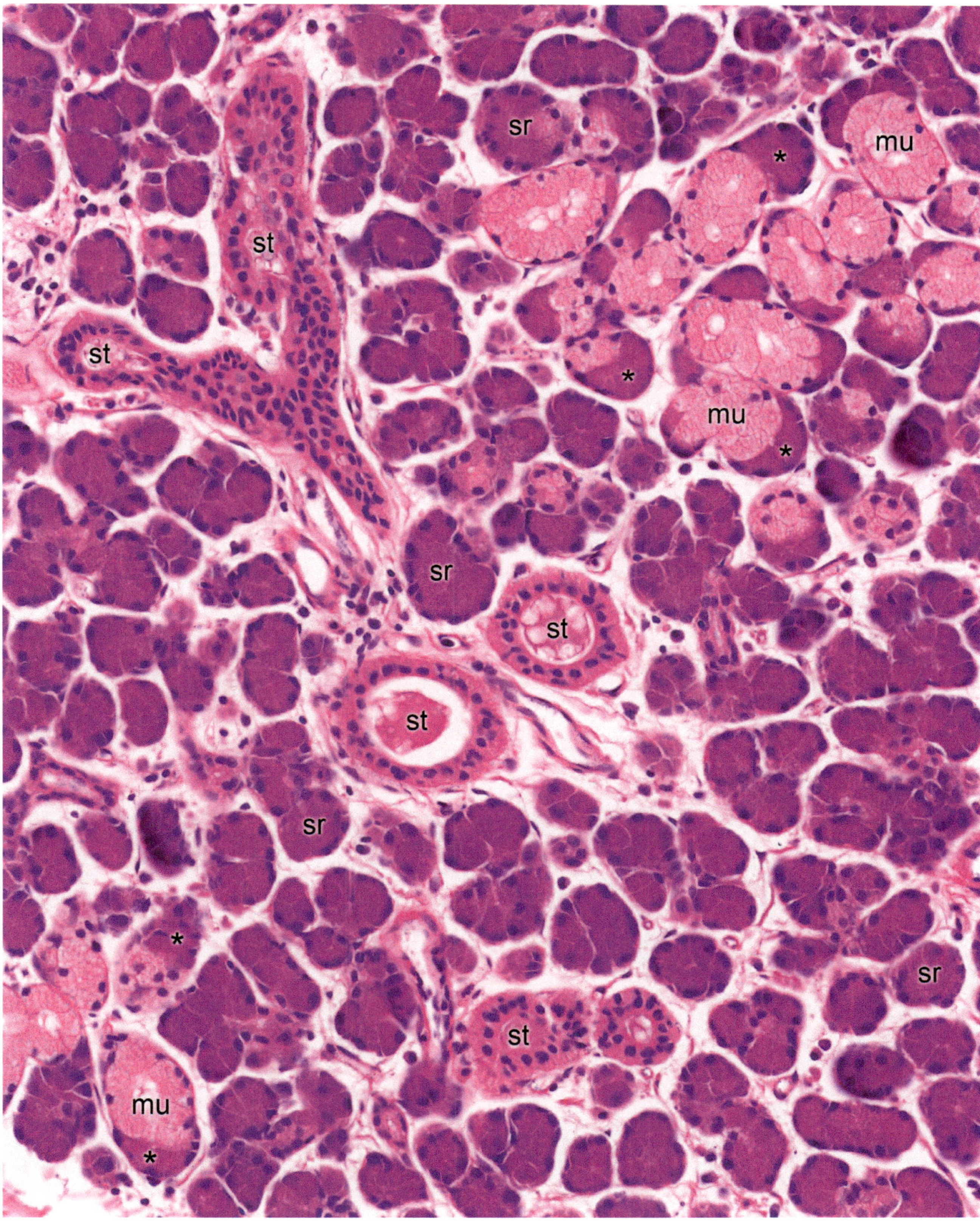

Figure 100. Submandibular gland. Submandibular gland is mostly composed of by serous acini (sr) and a smaller number of mucous acini (mu). Seromucous acini are represented by serous demilunes (asterisks) that are serous caps on the surface of mucous acini. Connective tissue surrounds the smaller striated ducts (st) lined with cuboidal or columnar cells with spherical or oval nuclei. The cytoplasm is eosinophilic and exhibits basal striation due to the numerous mitochondria at the base of the cell.

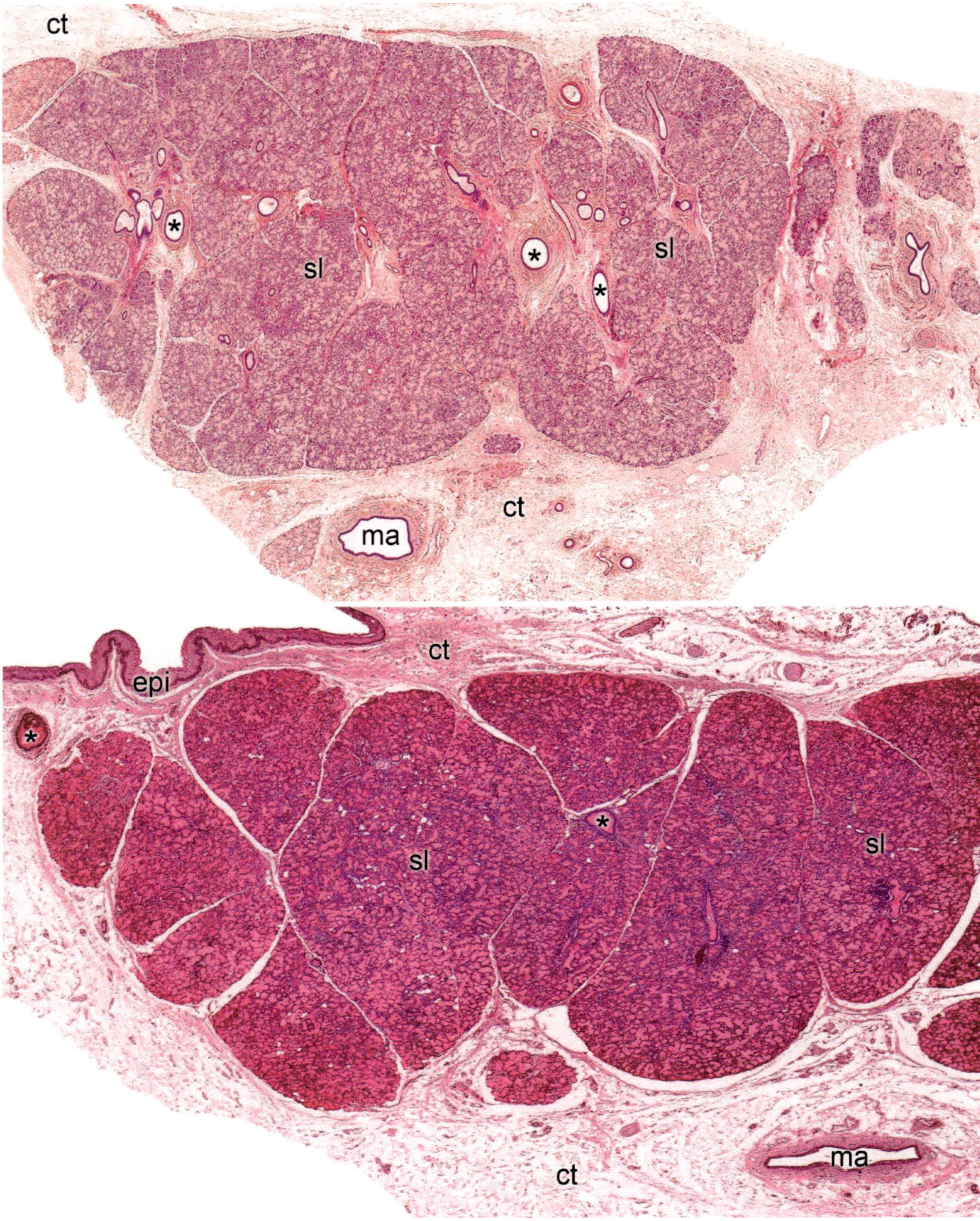

Figure 101. Sublingual gland, H&E (top) and mucicarmin stainings (bottom). Sublingual gland is located under the tongue, on the oral diaphragm, where it is lined with oral lining mucosa with nonkeratinized stratified squamous epithelium (epi). The parenchyma of this salivary gland (sl) is mostly composedof mucous acini with occasional larger excretory ducts (asterisks). The connective tissue capsule (ct) contains muscular arteries (ma) and veins.

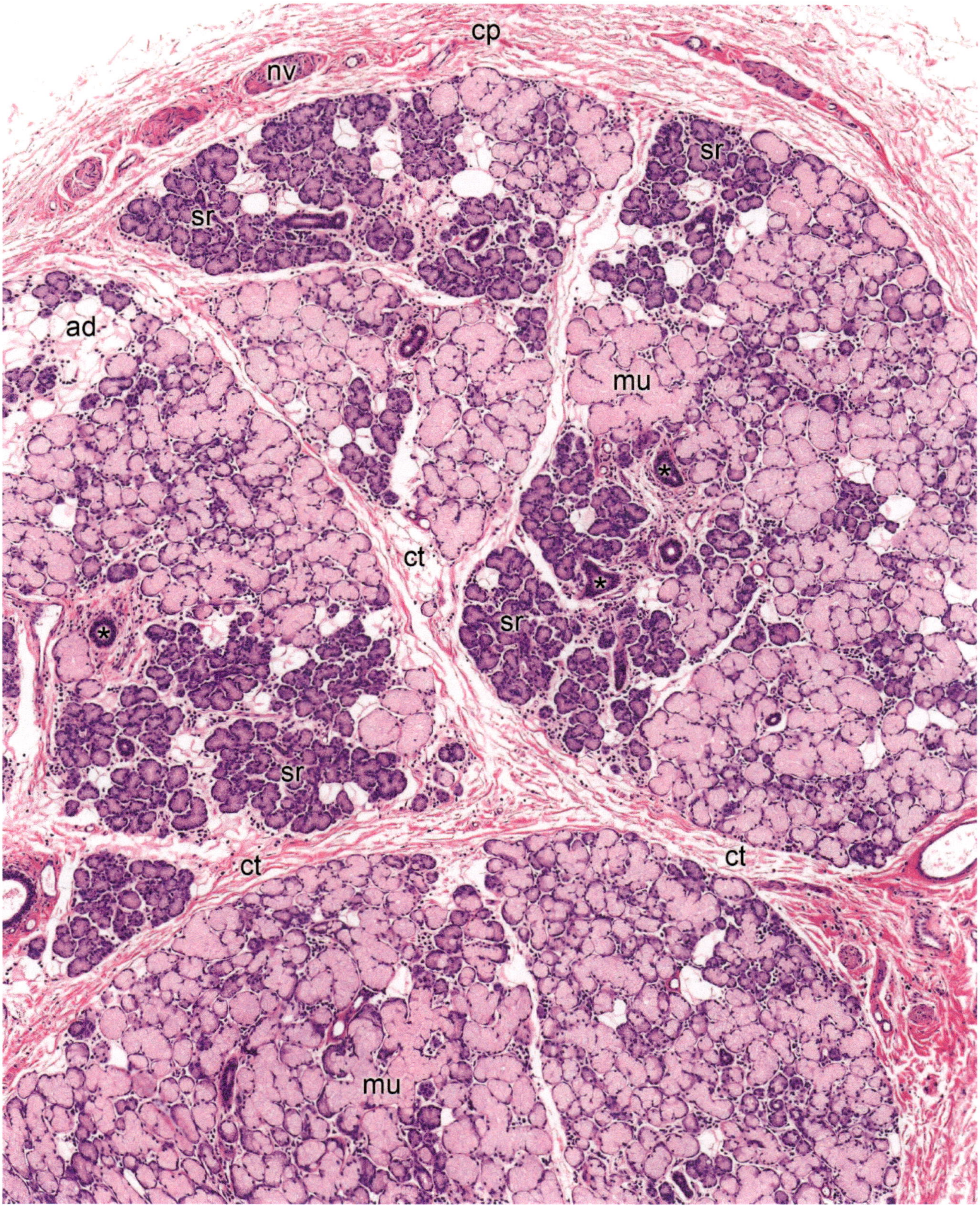

Figure 102. Sublingual gland. Sublingual gland is surrounded by a capsule (cp) that penetrates the parenchyma forming connective tissue septa (ct). The lobules formed by the septa are mostly composed of mucous acini (mu) and a smaller number of serous acini (sr). Connective tissue surrounds the smaller striated ducts (asterisks) lined with cuboidal or columnar cells with spherical or oval nuclei and eosinophilic cytoplasm due to the numerous mitochondria.

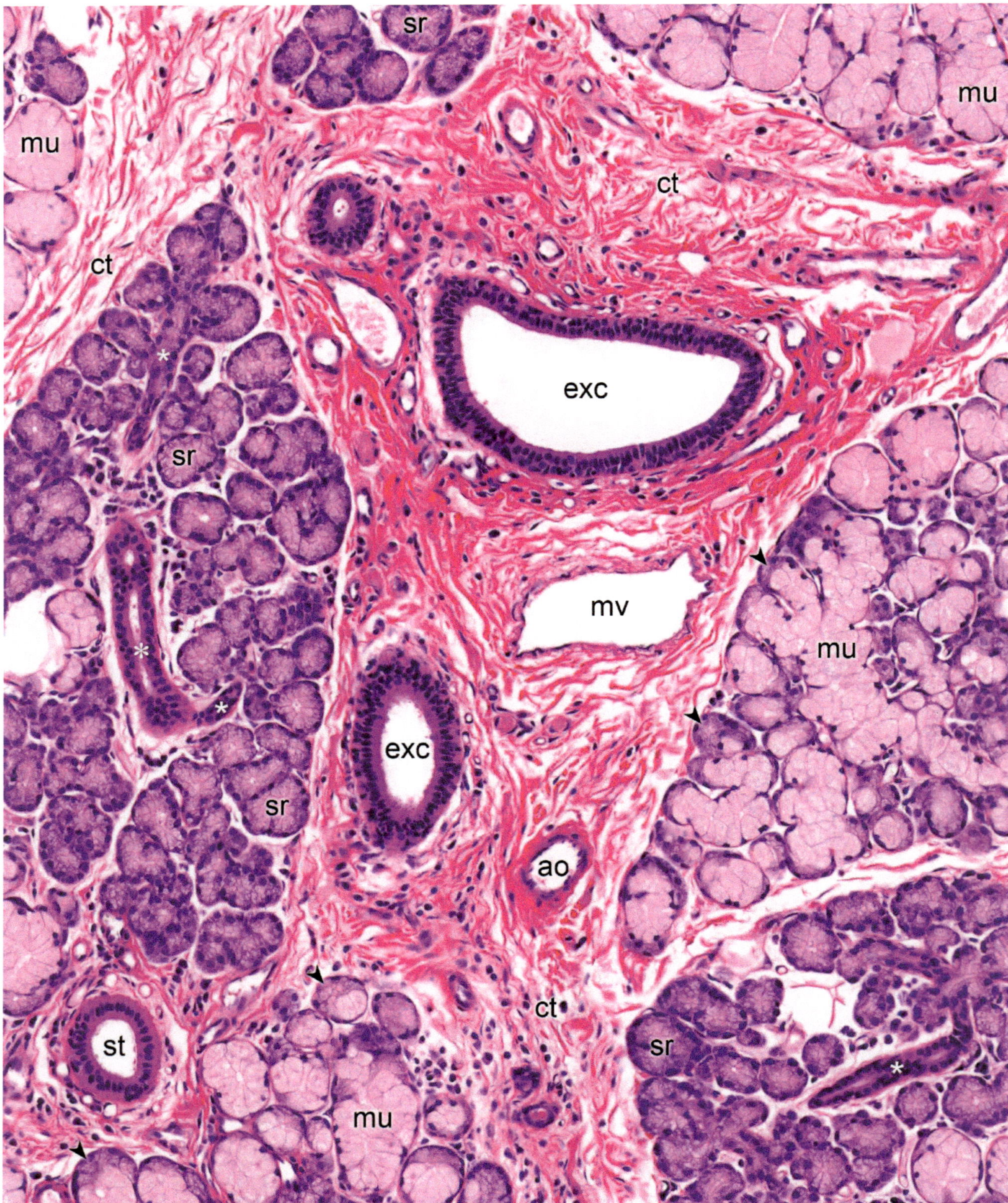

Figure 103. Sublingual gland. Sublingual gland is subdivided into lobules by connective tissue septa (ct). The lobules are mostly composed of mucous acini (mu) and a smaller number of serous acini (sr). Seromucous acini are represented by serous demilunes (arrowheads) that are serous caps on the surface of mucous acini. Acini drain into intercalated ducts (five-pointed asterisks) that empty into larger striated ducts that are initially small in diameter (six-pointed asterisks) and then become larger (st). Striated ducts are lined with eosinophilic cuboidal or columnar cells with spherical or oval nuclei, while larger excretory ducts (exc) are covered with stratified columnar epithelium. Septa also contain nerves and blood vessels including arterioles (ao) and muscular venules (mv).

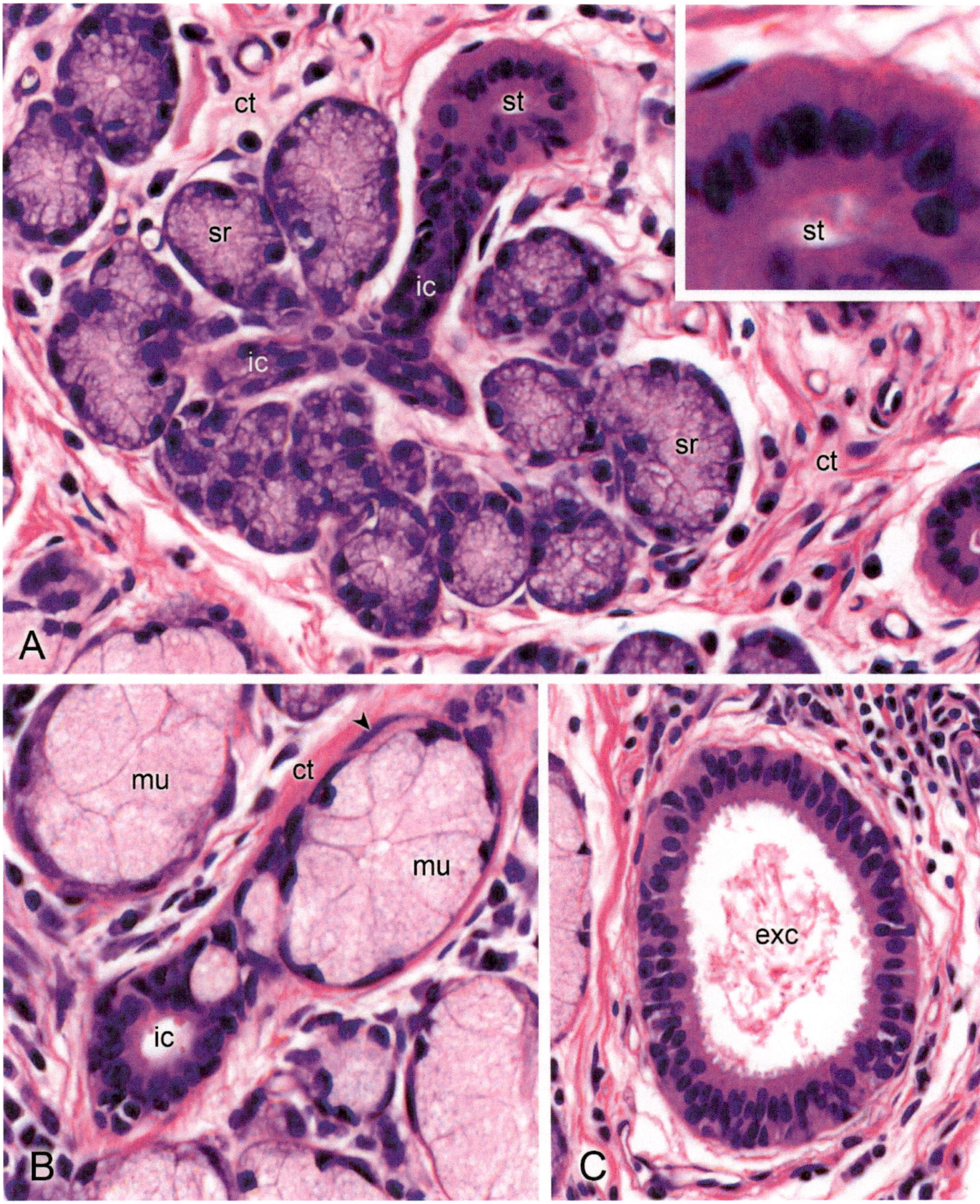

Figure 104. Sublingual gland. Sublingual gland is subdivided into lobules by connective tissue septa (ct). (A) Serous acini open into intercalated ducts (ic) covered with low cuboidal cells. Intercalated ducts drain into larger striated ducts that are lined with eosinophilic cuboidal or columnar cells with spherical or oval nuclei (inset). (B) Intercalated ducts (ic) are short and indistinct in mucous glands (mu). Acini are surrounded by myoepithelial cells (arrowheads) that facilitates the discharge of the saliva from the acini enveloped in connective tissue (ct). (C) Larger excretory ducts (exc) are covered with stratified columnar epithelium.

Chapter 6

Digestive System II: Lower Alimentary Tract

Esophagus

The mucous membrane of the esophagus (oesophagus) is covered by stratified squamous nonkeratinized epithelium. The lamina propria is composed of loose connective tissue with lymphatic elements that form occasional lymphatic follicles. The muscularis mucosae is remarkably thick and it is formed by longitudinally arranged smooth muscle bundles. The underlying submucosa is composed of dense irregular connective tissue with occasional ganglion cells and nerve fibers (Meissner's plexus). Tunica muscularis externa is composed of inner circular and outer longitudinal layers and it is quite unusual since it contains skeletal muscle in the upper third of the esophagus that is mixed with smooth muscle in the middle third while the distal part of the muscularis externa is composed of smooth muscle only. Elements of myenteric (Auerbach's) plexus are abundant between the muscle layers. Outside, the esophagus is covered by loose connective tissue tunica adventitia, only the lowermost abdominal part has serosal coverage. Esophageal glands are mucous tubuloalveolar glands. Proper esophageal glands are scattered throughout the entire length of the submucosa (submucosal glands), while glands in the lamina propria (mucosal glands) are restricted to the cardiac region (esophageal cardiac glands) and to the proximal part of the esophagus.

Stomach

Histologically, stomach can be divided into three distinct regions that differ from each other primarily by the gastric glands of the mucosa. The cardia of the stomach represents the most proximal part surrounding the esophageal junction, while the pylorus is the most distal part that transitions into the duodenum. Between the cardia and the pylorus, the corpus and the fundus are histologically identical. Gastric mucosa is covered by simple columnar epithelium that is composed of mucous cells accumulating mucus at their apex. This mucus protects the mucosal surface forming a viscous protective layer. Gastric glands are branched tubular glands occupying the lamina propria mucosae and they open into the gastric pits covered by the surface epithelium. Since gastric glands are tightly packed in the lamina propria, there is a relatively small amount of stroma between them, formed mainly by fibroblasts and reticular fibers with occasional smooth muscle cells, eosinophils, plasma cells and lymphocytes that often form lymphatic follicles. The base of the gastric glands extends to the thin muscularis mucosae that is composed of inner circular and outer longitudinal layers of smooth muscle.

Fundic gastric glands are simple, branched tubular glands that contain four different cell types. Mucous neck cells populate the neck of the glands below the gastric pits; they are shorter columnar cells without distinct mucous cap and spherical or slightly oval nuclei in contrast to the surface epithelium. These cells produce soluble mucus. Chief cells occupy primarily the basal segment of the glands above the muscularis mucosae; due to the high rER content, the cytoplasm is basophilic. Chief cells have spherical nucleus, and they are responsible for the production of the enzyme pepsinogen, the inactive form of pepsin. Parietal cells populate primarily the middle zone of the gland including the neck and they are large bright eosinophilic cells with spherical nucleus and extensive cytoplasm containing intracellular canaliculi that are crucial for their hydrogen chloride (HCl) production. The produced HCl activates the pepsinogen produced by the chief cells and converts it to the active pepsin. In addition to the HCl, parietal cells also produce intrinsic factor that plays a key role in vitamin B_{12} absorption by forming a complex with the vitamin. Finally, enteroendocrine cells are scattered throughout the entire gastric gland. These cells are much less characteristic with traditional staining; they are smaller triangular cells with light cytoplasm and spherical nucleus; they rest on the basement membrane often not extending to the luminal surface. Enteroendocrine cells secrete gastrointestinal peptides, including gastrin that is the primary stimulus for the HCl production by the parietal cells. The neck region of the gastric glands also contains precursor cells that provide a supply for the four main cell types.

Cardiac and pyloric glands are coiled, branched tubular glands populating the cardiac and pyloric regions, respectively. They contain mucous secreting cells and enteroendocrine cells only, although occasionally parietal and chief cells can also be observed in these glands close to the corpus and fundus. The gastric pits in the pylorus penetrate deeply into the lamina propria and they are more shallow in the cardiac region.

A dense connective tissue submucosa connects the mucous membrane of the stomach to the muscularis externa. Apart from vessels and adipose tissue, submucosa contains the Meissner's plexus, a network of ganglion cells and autonomic nerve fibers. Tunica muscularis externa is composed of inner oblique, middle circular, and outer longitudinal layers,

http://doi.org/10.1016/B978-0-323-91891-6.50006-4,

although the boundaries between these strata are sometimes elusive and parts of the layers may be absent at certain gastric regions. Ganglion cells with associated nerve fibers are abundant between the layers, forming the Auerbach's (myenteric) plexus. Muscularis externa is enveloped by the gastric serosa that is composed of an outer mesothelial coverage and the underlying loose connective tissue lamina propria.

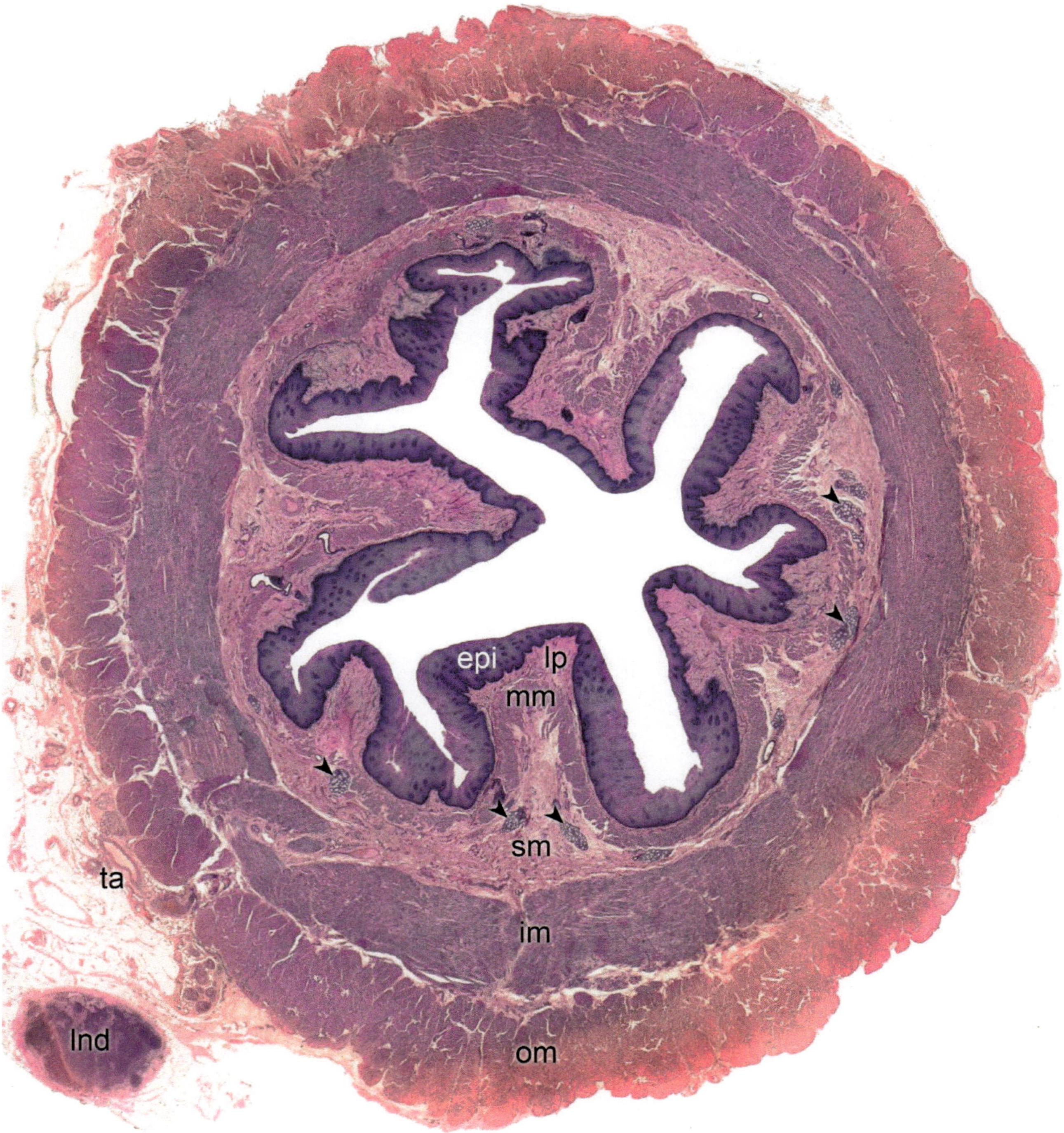

Figure 105. Esophagus. Tunica mucosa of the esophagus is composed of nonkeratinized stratified squamous epithelium (epi), and the underlying loose connective tissue lamina propria mucosae (lp) and tunica muscularis mucosae (mm). Mucous membrane is attached to the tunica muscularis externa by the dense irregular connective tissue tunica submucosa (sm) that contains mucous glands (submucosal glands, arrowheads). Muscularis externa has an inner circular (im) and outer longitudinal (om) smooth muscle layer that is covered by the loose connective tissue tunica adventitia (ta) containing vessels, nerves, and occasional small lymph nodes (lnd).

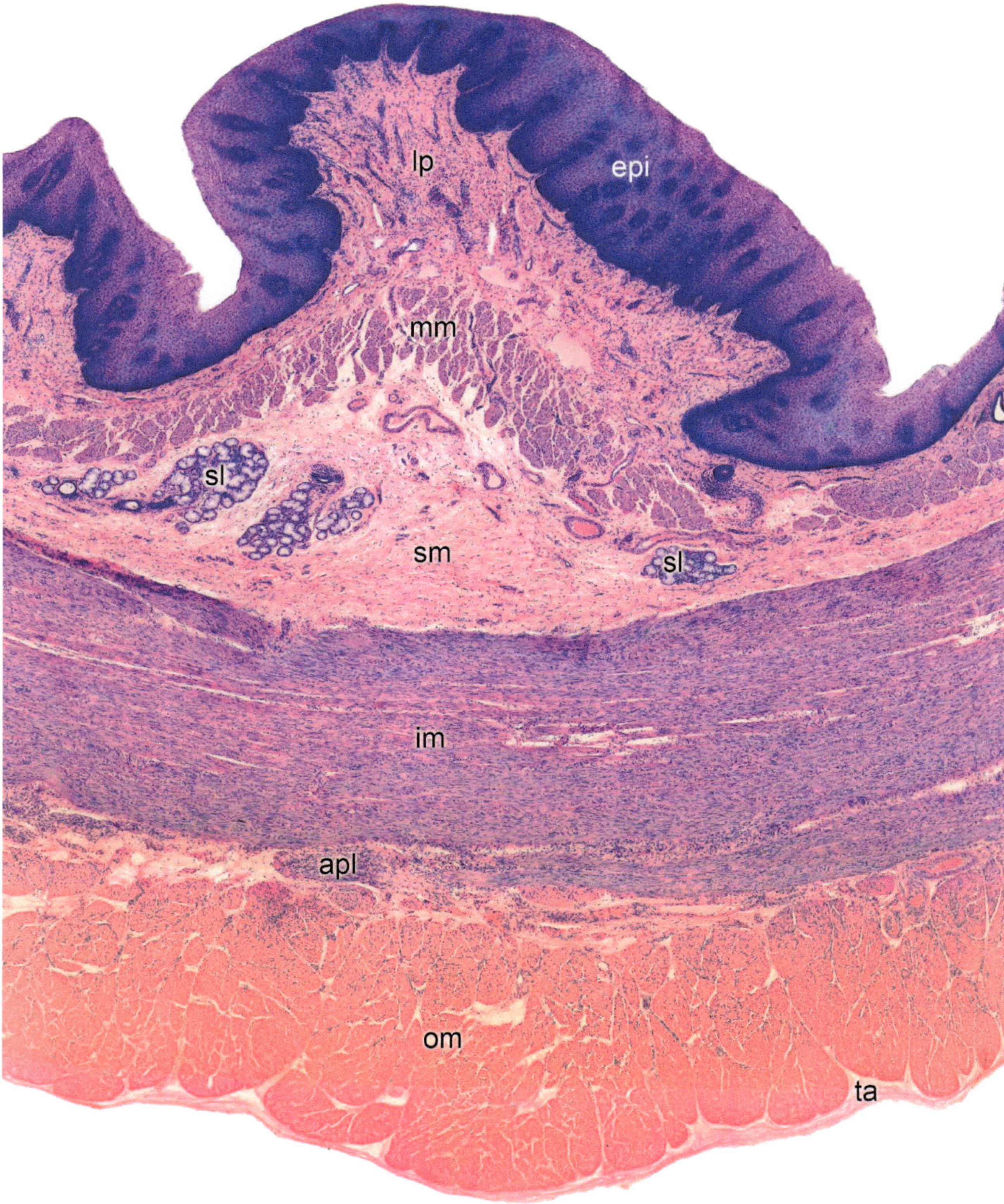

Figure 106. Esophagus. Tunica mucosa of the esophagus is composed of nonkeratinized stratified squamous epithelium (epi), and the underlying loose connective tissue lamina propria mucosae (lp) and tunica muscularis mucosae (mm). Mucous membrane is attached to the tunica muscularis externa by the dense irregular connective tissue submucosa (sm) that contains mucous salivary glands, the submucosal glands (sl). Muscularis externa has an inner circular (im) and outer longitudinal (om) smooth muscle layer that is covered by tunica adventitia (ta).

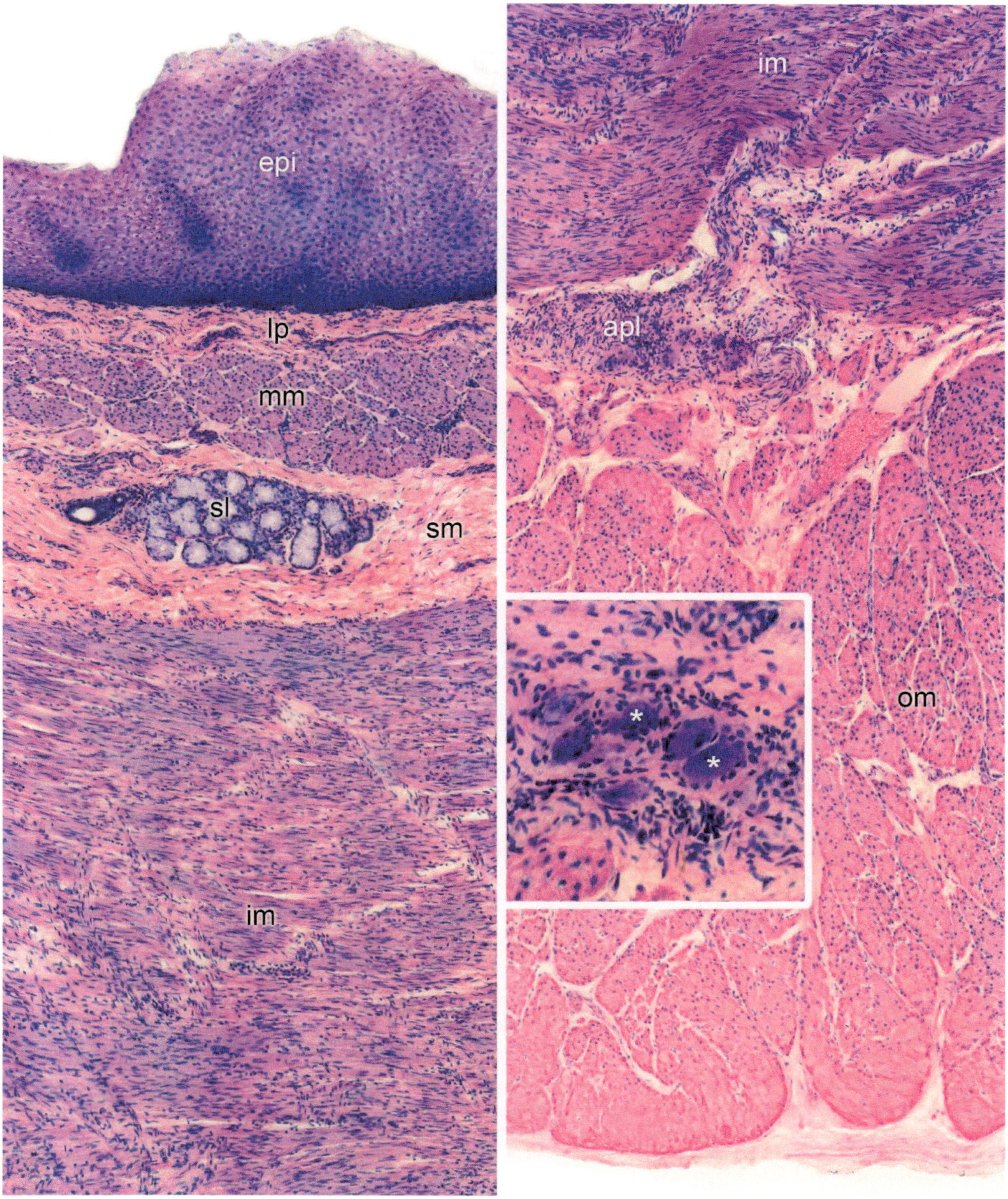

Figure 107. Esophagus, internal (left) and external (right) layers. Tunica mucosa of the esophagus is composed of nonkeratinized stratified squamous epithelium (epi), and the underlying loose connective tissue lamina propria mucosae (lp) and tunica muscularis mucosae (mm). Mucous membrane is attached to the tunica muscularis externa by the dense irregular connective tissue submucosa (sm) that contains mucous salivary glands, the submucosal glands (sl). Between the inner circular (im) and outer longitudinal (om) smooth muscle layers of the tunica muscularis externa, the Auerbach's autonomic plexus contains ganglion cells (inset, asterisks) and nerve fibers. Esophagus is covered by a loose connective tissue tunica adventitia (ta).

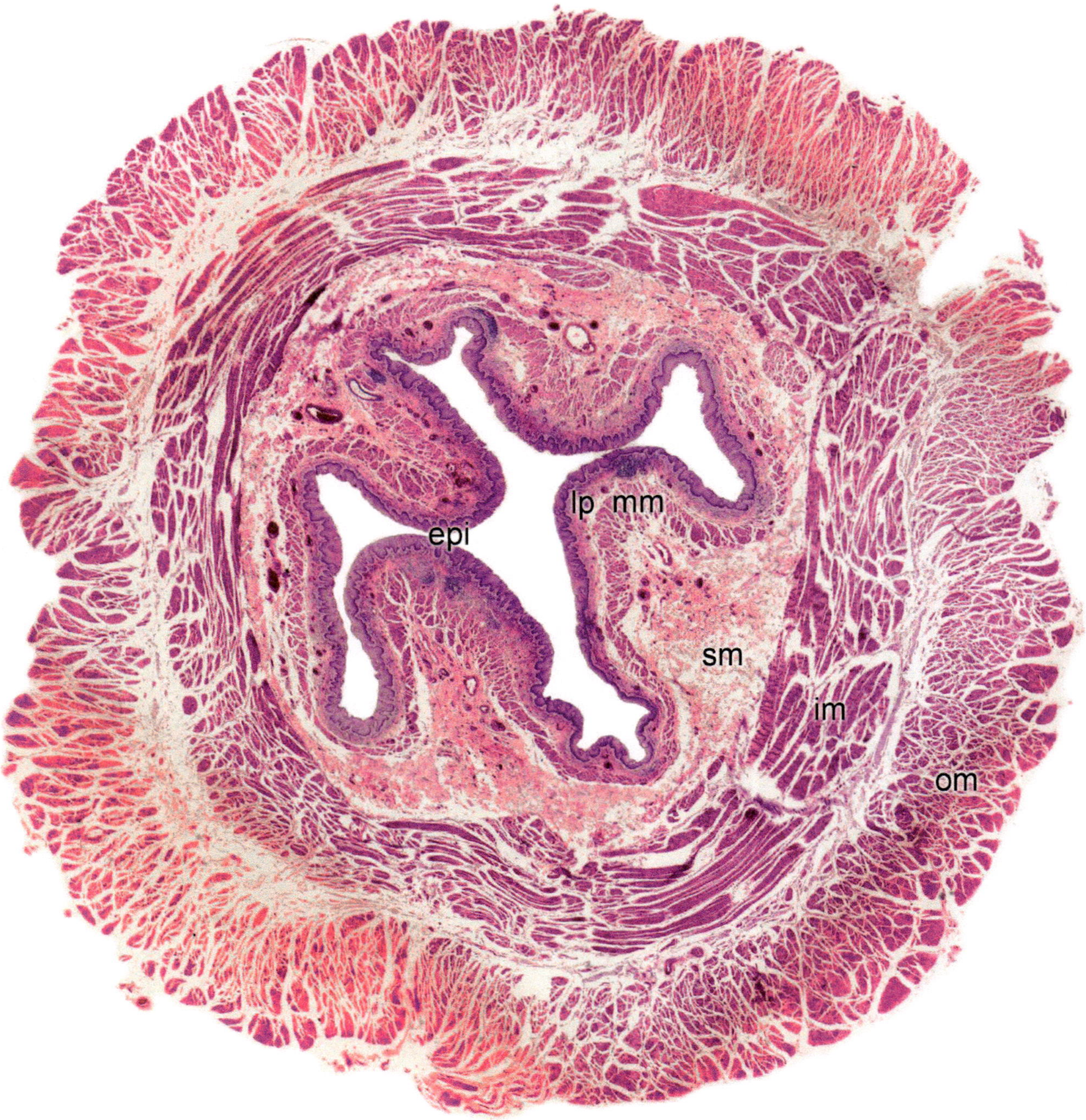

Figure 108. Esophagus. Tunica mucosa of the esophagus is composed of nonkeratinized stratified squamous epithelium (epi), and the underlying loose connective tissue lamina propria mucosae (lp) and tunica muscularis mucosae (mm). The dense connective tissue tunica submucosa (sm) contains the submucosal glands that are mucous glands (arrowheads). Muscularis externa has an inner circular (im) and outer longitudinal (om) smooth muscle layer that is covered by tunica adventitia (ta).

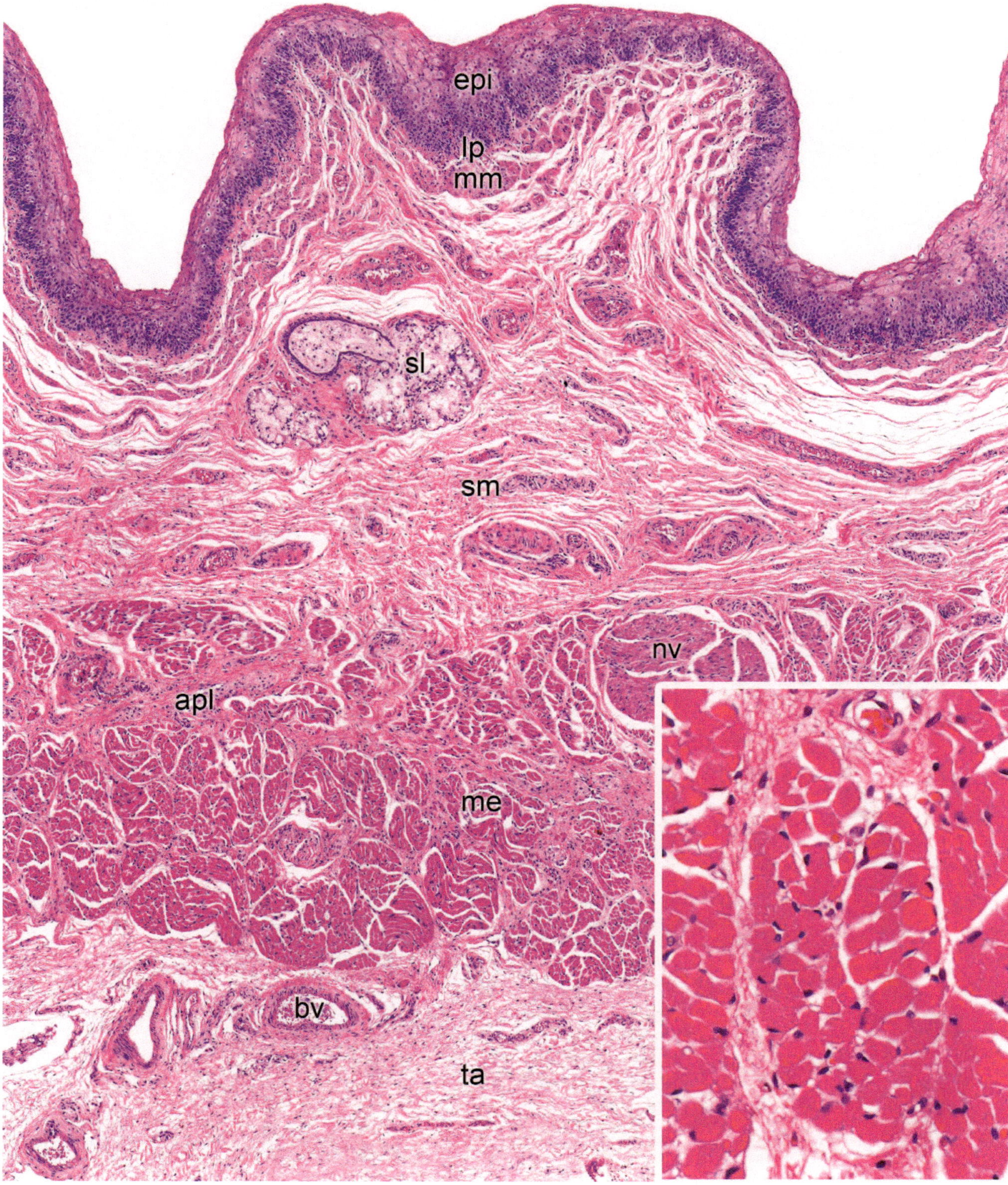

Figure 109. Esophagus. Tunica mucosa of the esophagus is composed of nonkeratinized stratified squamous epithelium (epi), and the underlying loose connective tissue lamina propria mucosae (lp) and tunica muscularis mucosae (mm). The dense irregular connective tissue submucosa (sm) contains submucosal glands that are mucous salivary glands (sl). Tunica muscularis externa (me) is composed of skeletal muscle at the upper one third of the esophagus (inset) with autonomic plexus (apl) and nerves (nv) between the muscle layers. Loose connective tissue tunica adventitia (ta) contains blood vessels (bv) and nerves.

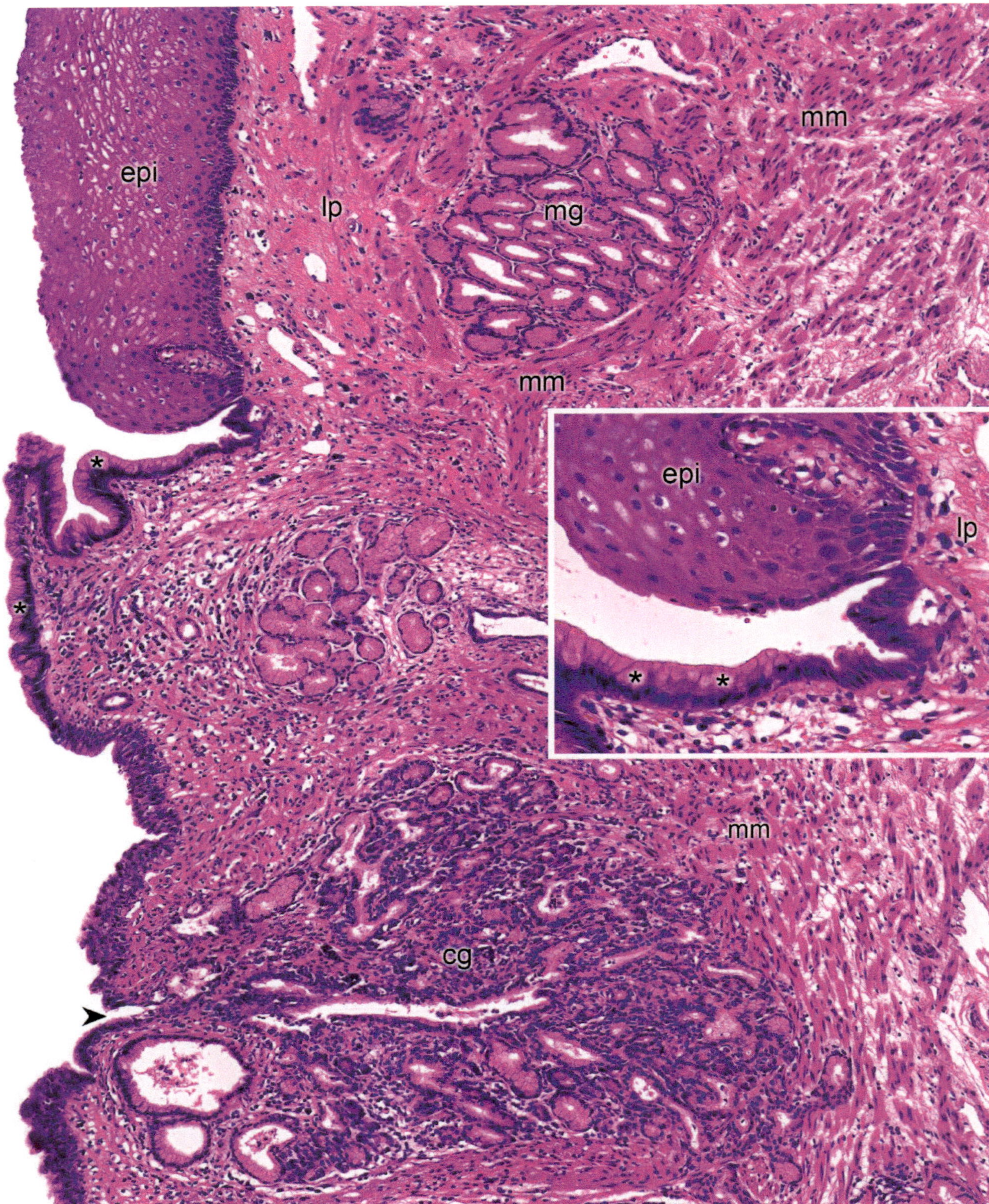

Figure 110. Esophageal-cardiac junction (mucicarmine and H&E staining). The stratified squamous nonkeratinized epithelium of the esophagus (epi) transitions into the simple columnar surface epithelium of the stomach at the cardia (asterisks; inset). The surface columnar cells accumulate mucus at their apex. Esophageal mucosal glands (mg) are located in the lamina propria mucosae (la) of the esophageal segment, and they are composed of mucous acini, while the coiled, branched tubular gastric cardiac glands (cg), containing mucous secreting and enteroendocrine cells, occupy the lamina propria of the gastric segment. Gastric cardiac glands open into gastric pits (arrowhead) lined with gastric surface epithelium. Tunica muscularis mucosae (mm) is well developed.

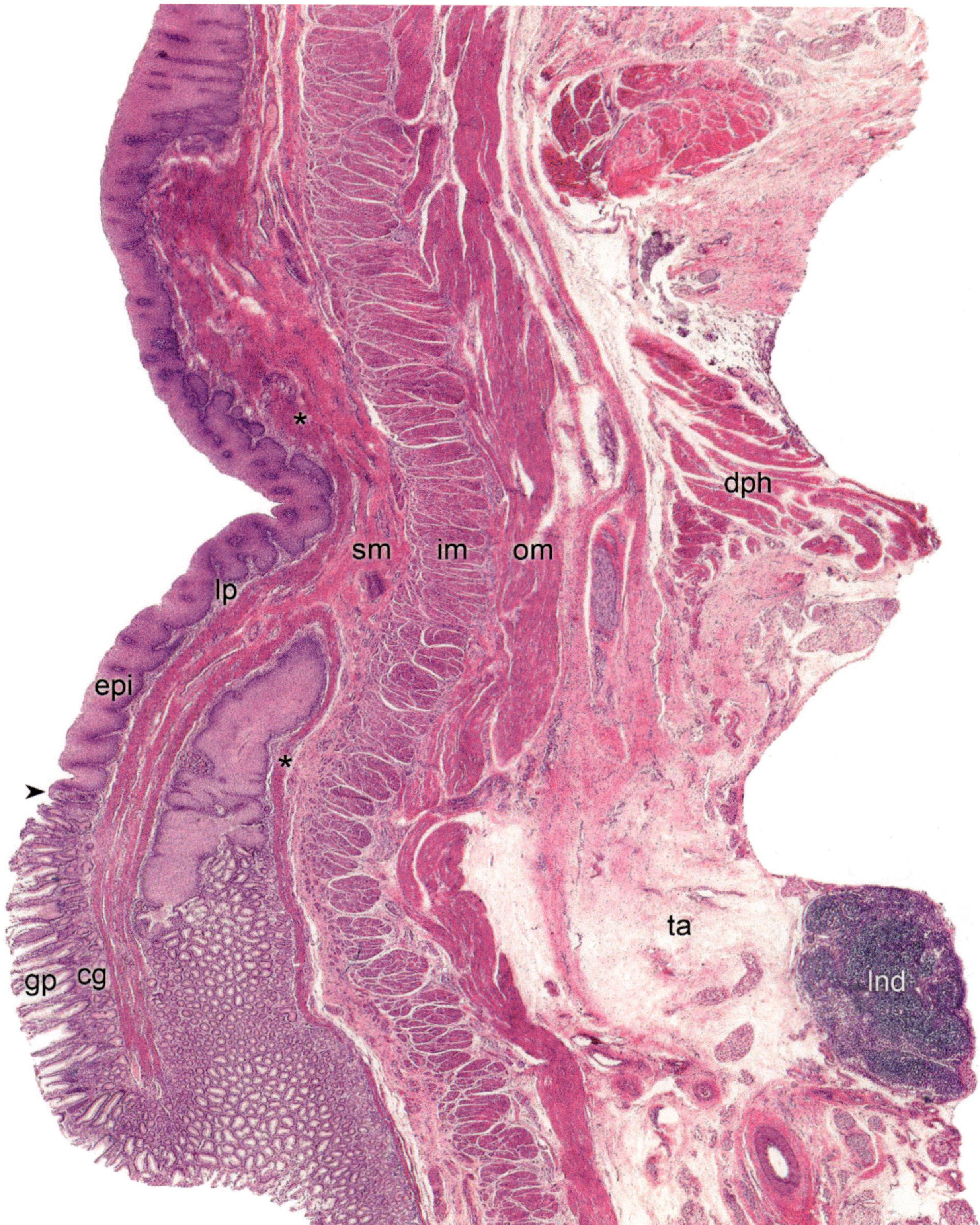

Figure 111. Esophageal-cardiac junction. The stratified squamous nonkeratinized epithelium of the esophagus (epi) abruptly transitions (arrowhead) into the simple columnar surface epithelium of the stomach at the cardia. Coiled branched tubular gastric cardiac glands (cg) containing mucous secreting and enteroendocrine cells occupy the lamina propria (lp) of the gastric segment and they open into gastric pits (gp). Tunica muscularis mucosae (asterisk) is well developed. Dense connective tissue submucosa connects the mucous membrane to the inner circular (im) and outer longitudinal smooth muscle layers of the tunica muscularis externa. Tunica adventitia (ta) contains occasional small lymph nodes (lnd) and merges with the fascia of the thoracic diaphragm that is composed of skeletal muscle (dph).

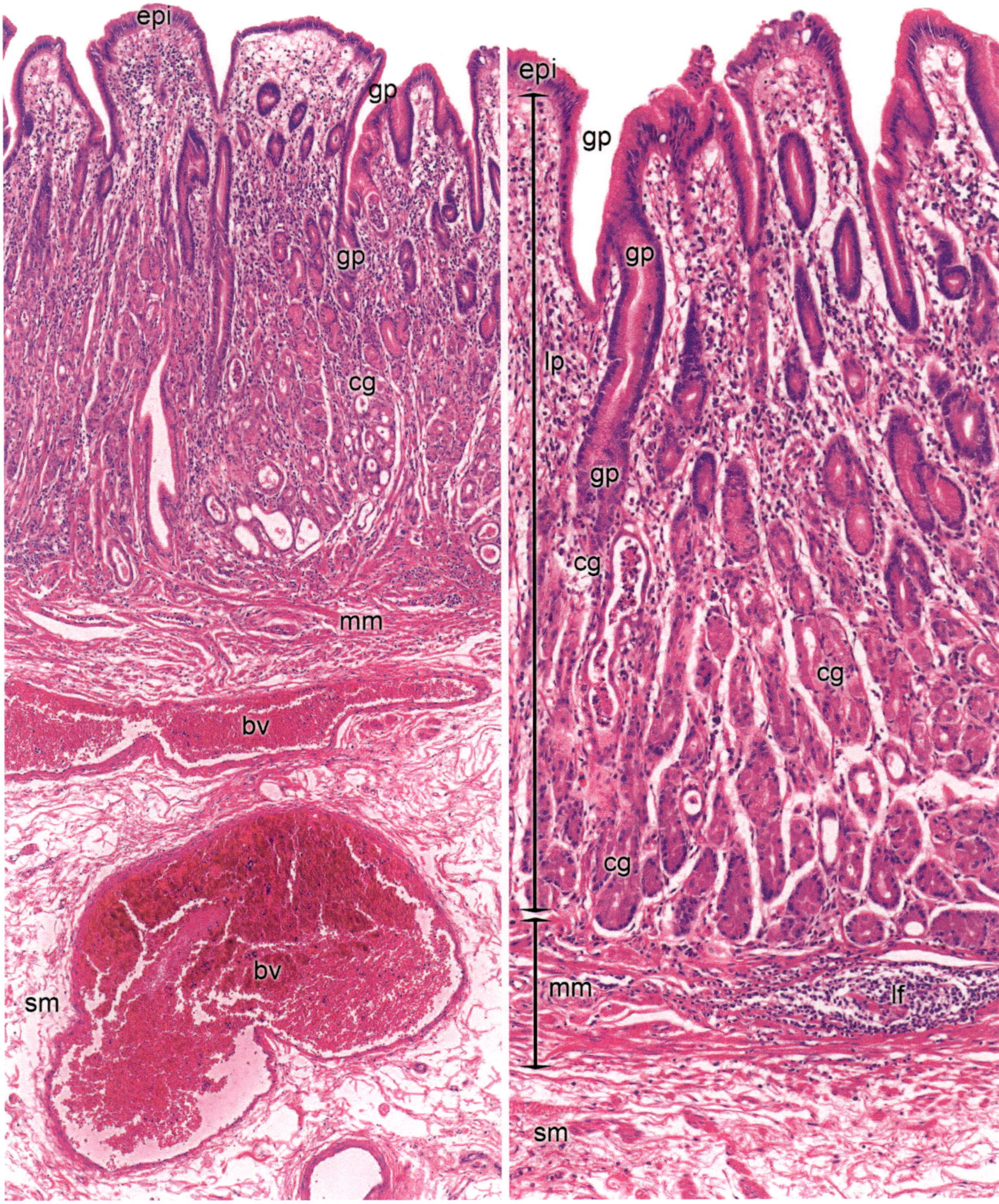

Figure 112. Gastric cardiac glands. Stomach is lined with a single layer of columnar cells that accumulate mucus at their apex. Coiled branched tubular gastric cardiac glands (cg) containing mucous secreting and enteroendocrine cells occupy the lamina propria (lp) and they open into gastric pits (gp) lined with gastric surface epithelium. Submucosa contains blood vessels (bv) as well as occasional lymph follicles (lf) that may penetrate the muscularis mucosae (mm).

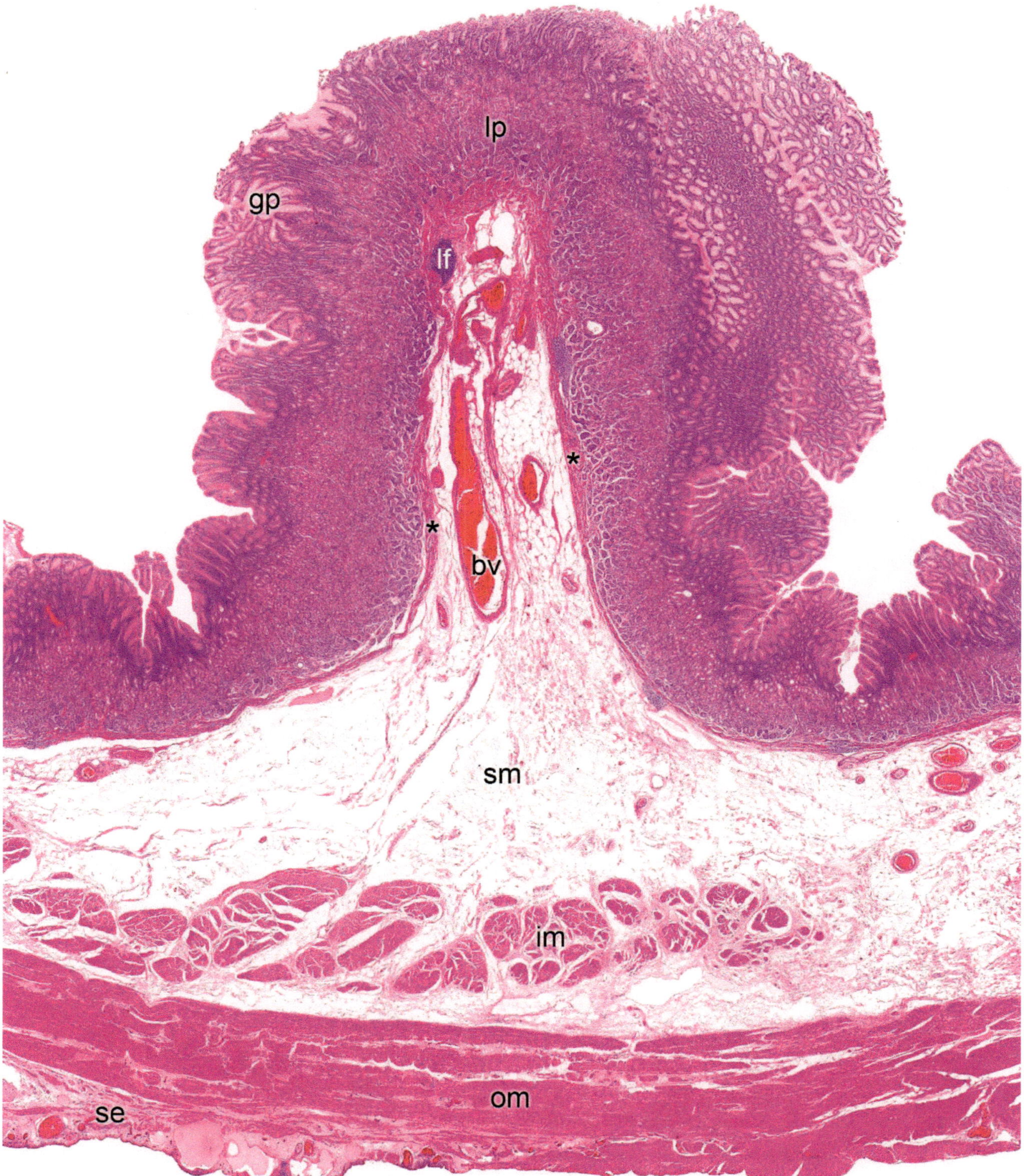

Figure 113. Corpus of the stomach. Stomach is lined with a single layer of columnar cells that are accumulate mucus at their apex. Simple branched tubular fundic glands containing parietal and chief cells as well as mucous secreting and enteroendocrine cells occupy the lamina propria (lp) and they open into gastric pits (gp) lined with gastric surface epithelium. Occasional lymph follicles (lf) can be also observed in the lamina propria and the submucosa. The submucosa connects the tunica muscularis mucosae (asterisk) to the tunica muscularis externa that is composed of inner oblique and circular layers (im) and an additional outer longitudinal layer (om) of smooth muscle cells. Stomach is covered by the tunica serosa (se) with mesothelial coverage.

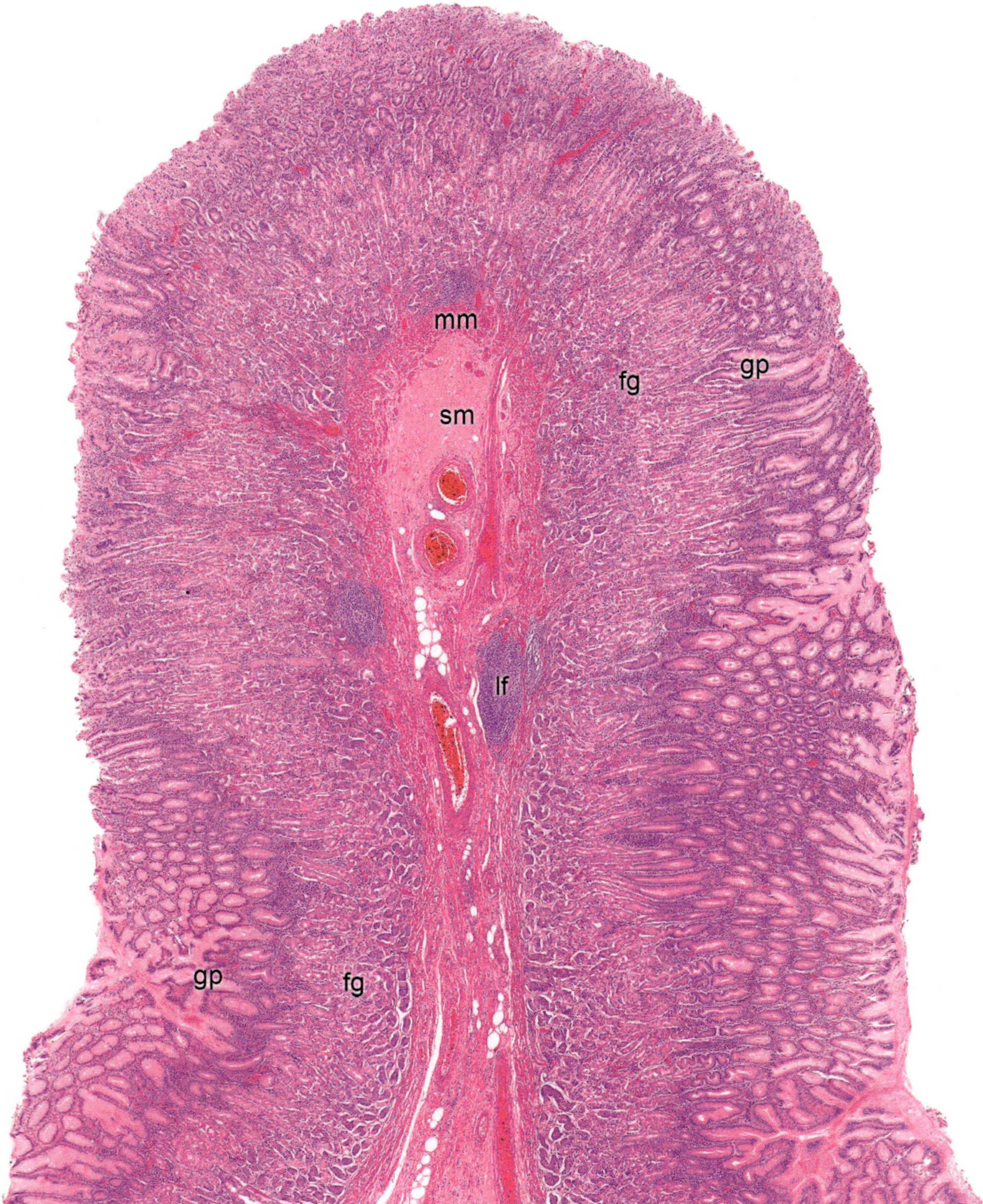

Figure 114. Corpus of the stomach. Cross section of a gastric fold (ruga) reveals a submucosal core (sm). The luminal surface of the stomach is lined with a single layer of lightly stained, mucous columnar cells. Lamina propria is tightly packed with simple branched tubular fundic glands (fg) that open into gastric pits (gp) lined with gastric surface epithelium. The glands contain parietal and chief cells as well as mucous secreting and enteroendocrine cells. Lymph follicles (lf) can be also observed in the lamina propria and the submucosa.

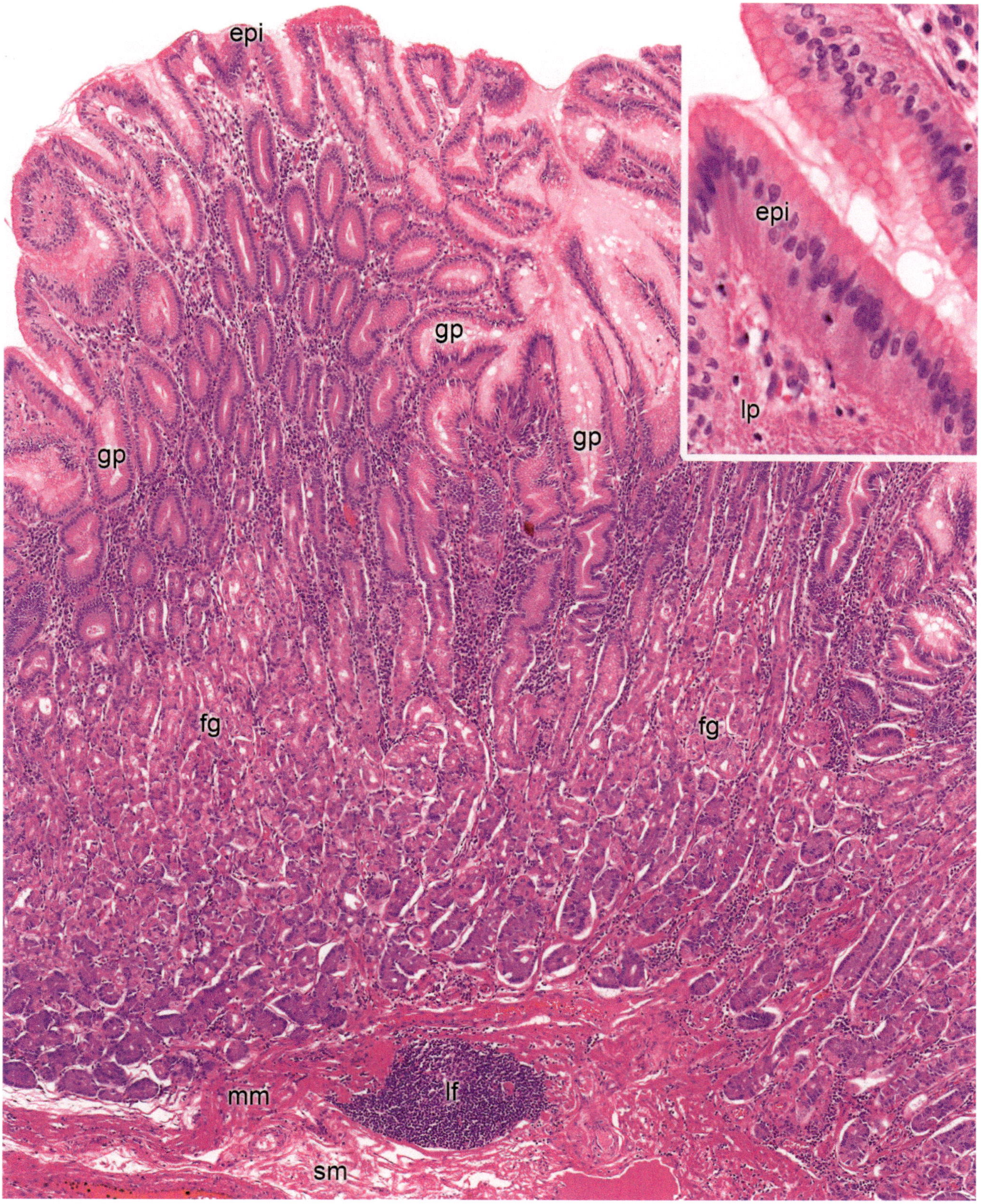

Figure 115. Corpus of the stomach, tunica mucosa. Stomach is lined with a single layer of columnar cells (epi) that accumulate mucus at their apex. Simple branched tubular fundic glands (fg) containing parietal and chief cells as well as mucous secreting and enteroendocrine cells occupy the lamina propria (lp) and they open into gastric pits (gp) lined with gastric surface epithelium. Occasional lymph follicles (lf) can be also observed in the lamina propria, muscularis mucosae (mm) and in the submucosa.

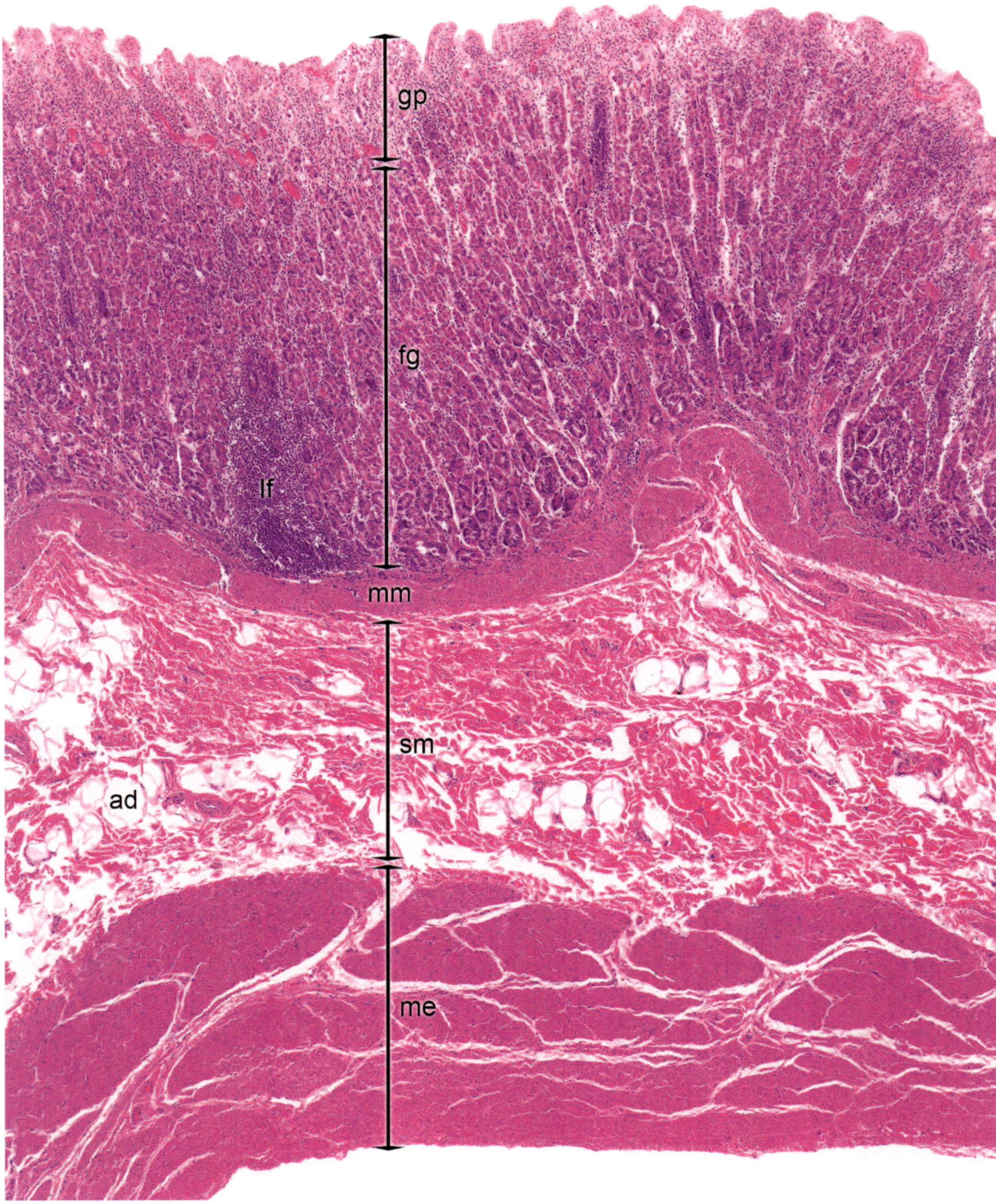

Figure 116. Fundus of the stomach. Stomach is lined with a single layer of columnar cells that accumulate mucus at their apex. Simple branched tubular fundic glands (fg) containing parietal and chief cells as well as mucous secreting and enteroendocrine cells occupy the lamina propria and they open into gastric pits (gp) that are invaginations of the gastric surface epithelium. Occasional lymph follicles (lf) can also be observed in the lamina propria. Tunica submucosa (sm) contains dense irregular connective tissue with dispersed adipocytes (ad), and it connects the muscularis mucosae (mm) to the tunica muscularis externa (me).

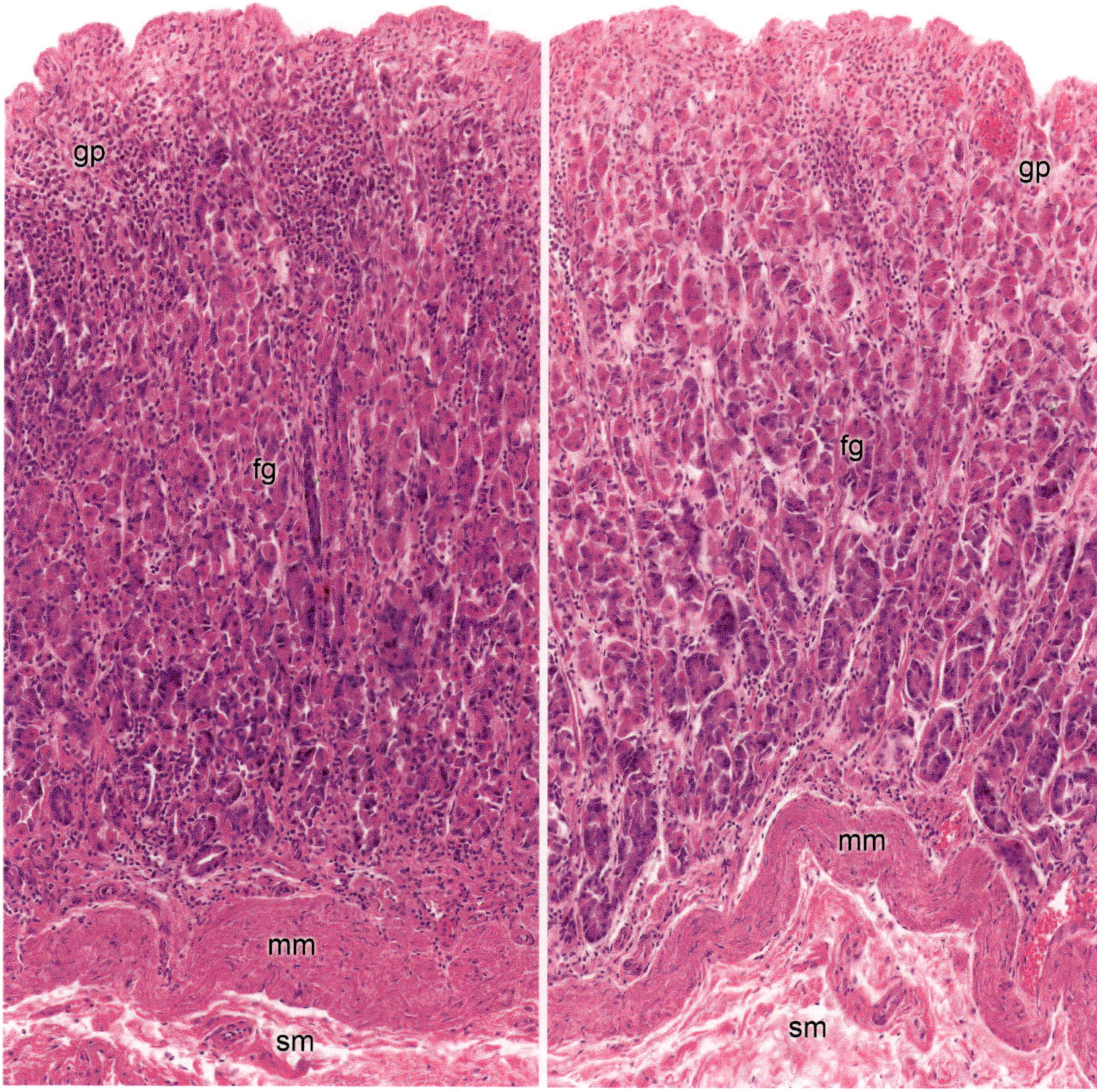

Figure 117. Fundic mucosa of the stomach. The mucosa of the stomach is lined with a single layer of columnar cells that accumulate mucus at their apex. Simple branched tubular fundic glands (fg) containing parietal and chief cells as well as mucous secreting and enteroendocrine cells are located in the lamina propria and they open into gastric pits (gp) that are invaginations of the gastric surface epithelium. Parietal cells are large and acidophilic, and they occupy mainly the middle region of the glands while basophilic chief cells are more abundant at the base of the glands. Tunica muscularis mucosae (mm) separates the lamina propria from the underlying tunica submucosa (sm) that is composed of dense irregular connective tissue.

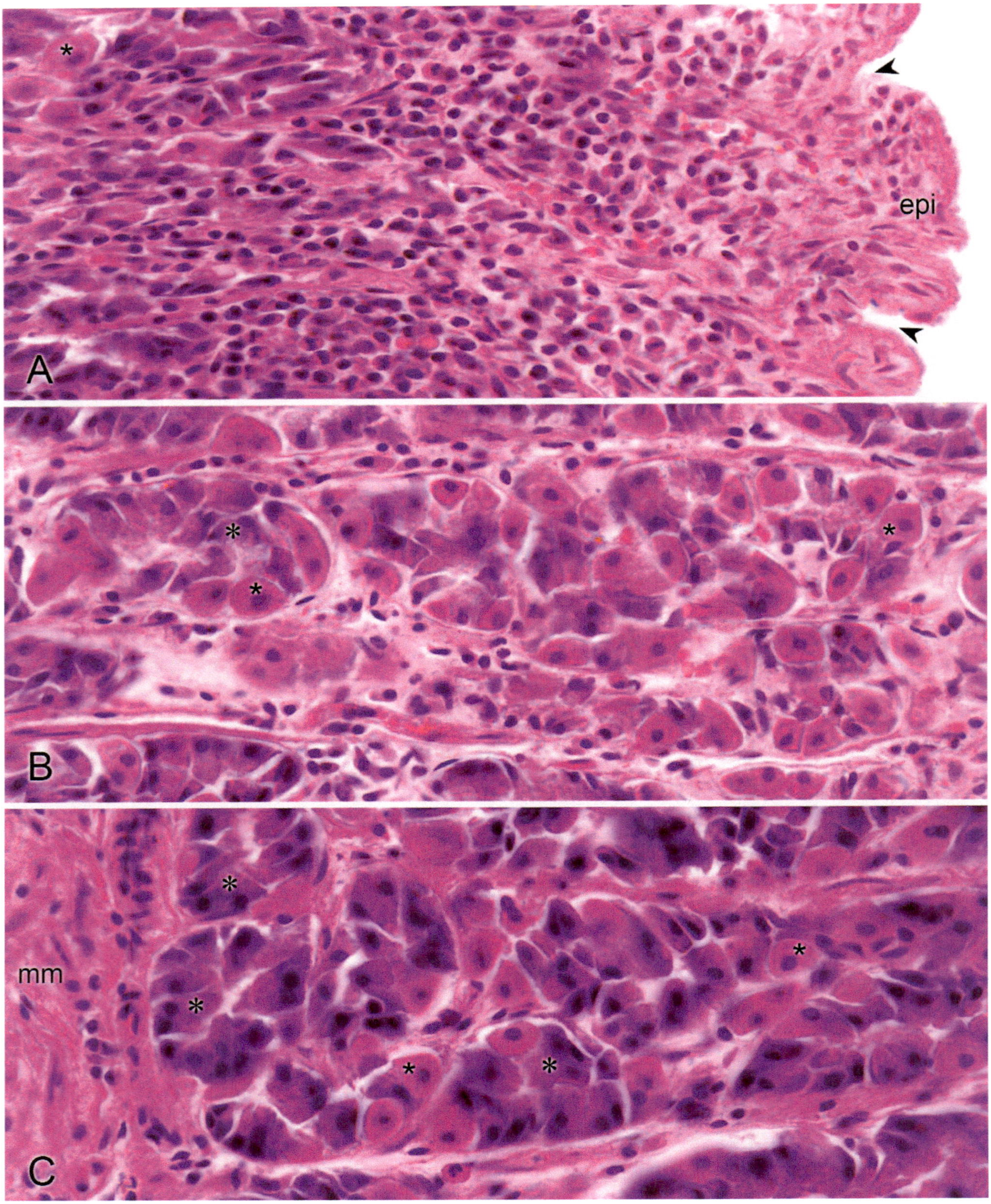

Figure 118. Fundic glands. (A) Luminal part, (B) middle segment, (C) base of the fundic gland, resting on the muscularis mucosae (mm). The surface is lined with a single layer of columnar cells (epi), which also line the gastric pits that are surface invaginations (arrowheads). The simple branched tubular fundic glands contain parietal and chief cells as well as mucous secreting and enteroendocrine cells. Parietal cells are large and acidophilic (five-pointed asterisks), and they occupy mostly the middle region of the glands while basophilic chief cells (six-pointed asterisks) are more characteristic to the base of the glands.

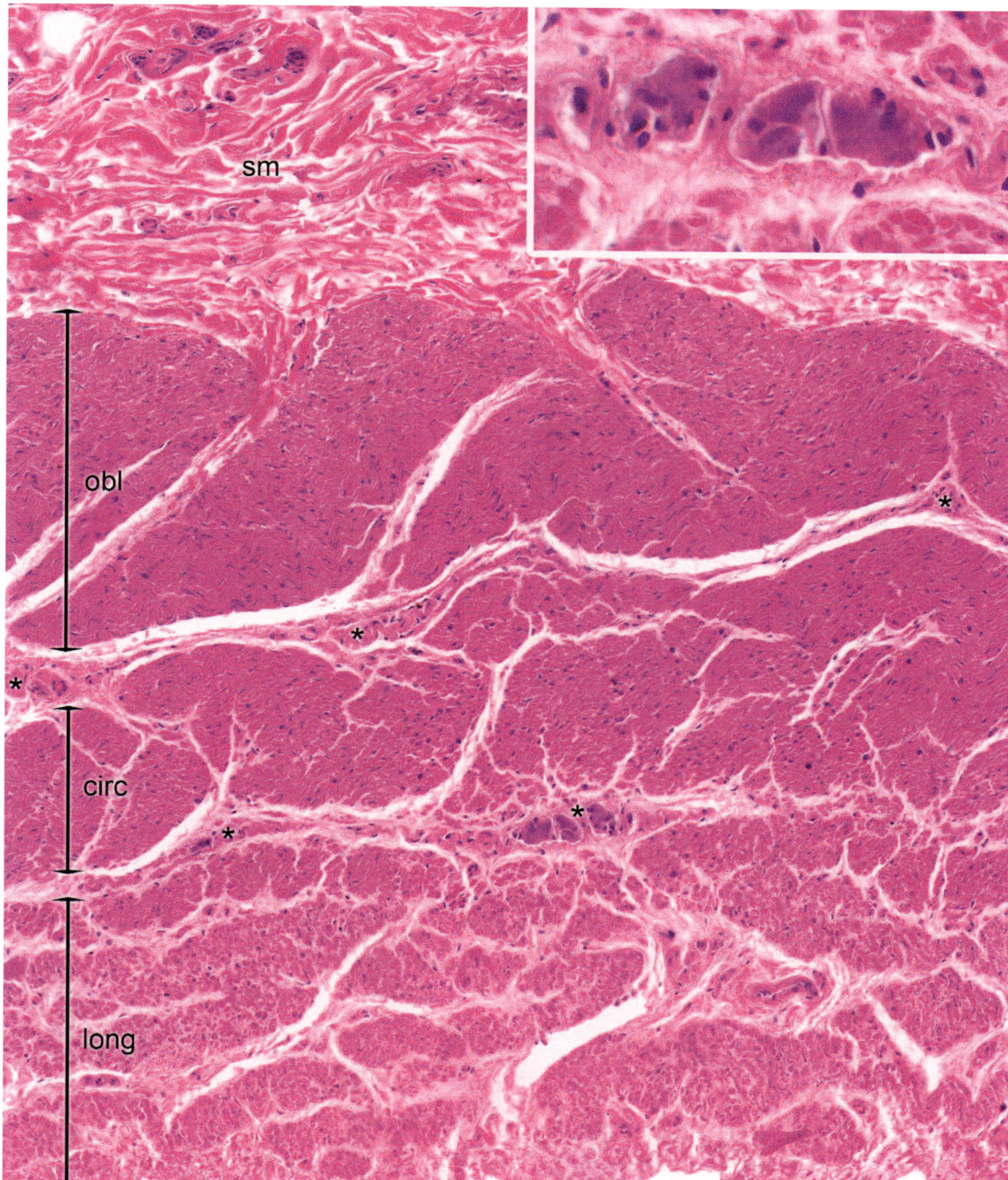

Figure 119. Muscle layers of the gastric wall. Tunica muscularis externa of the stomach is composed of three layers, the innermost oblique (obl), the middle circular (circ) and the outer longitudinal (long) layers. Muscle cells are innervated by ganglion cells that are located between the muscle layers (asterisks), forming the autonomic myenteric plexus (inset). At the luminal surface, tunica muscularis externa is covered by tunica submucosa (sm).

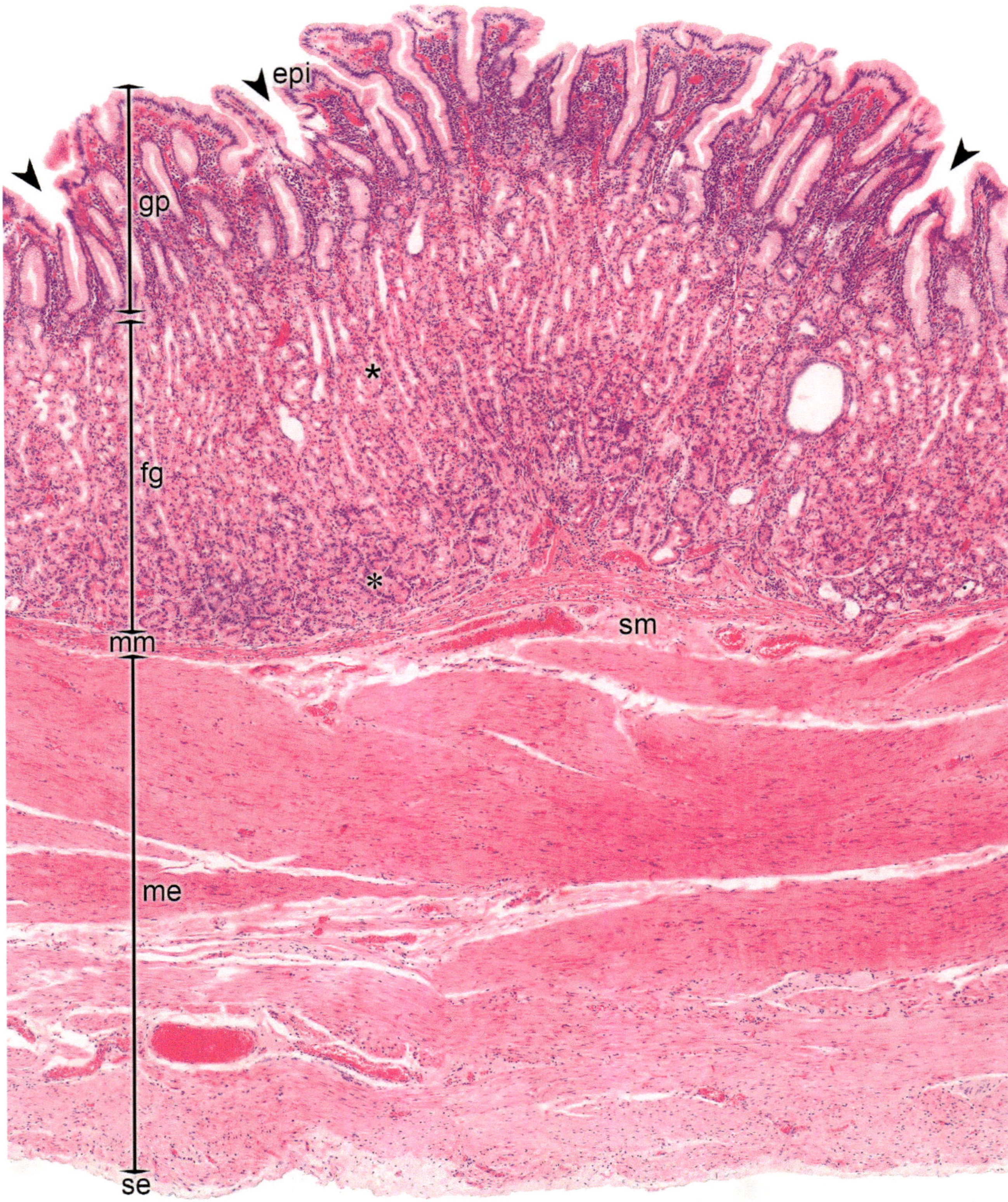

Figure 120. Fundus of the stomach. Stomach is lined with a single layer of lightly stained columnar cells that accumulate mucus at their apex (epi). Simple branched tubular fundic glands (fg) containing parietal and chief cells as well as mucous secreting and enteroendocrine cells occupy the lamina propria and they open into gastric pits (gp) that are invaginations of the gastric surface epithelium (arrowheads). Dense irregular connective tissue tunica submucosa (sm) connects the muscularis mucosae (mm) to the tunica muscularis externa (me). Tunica serosa (se) with mesothelial lining covers the external surface of the stomach.

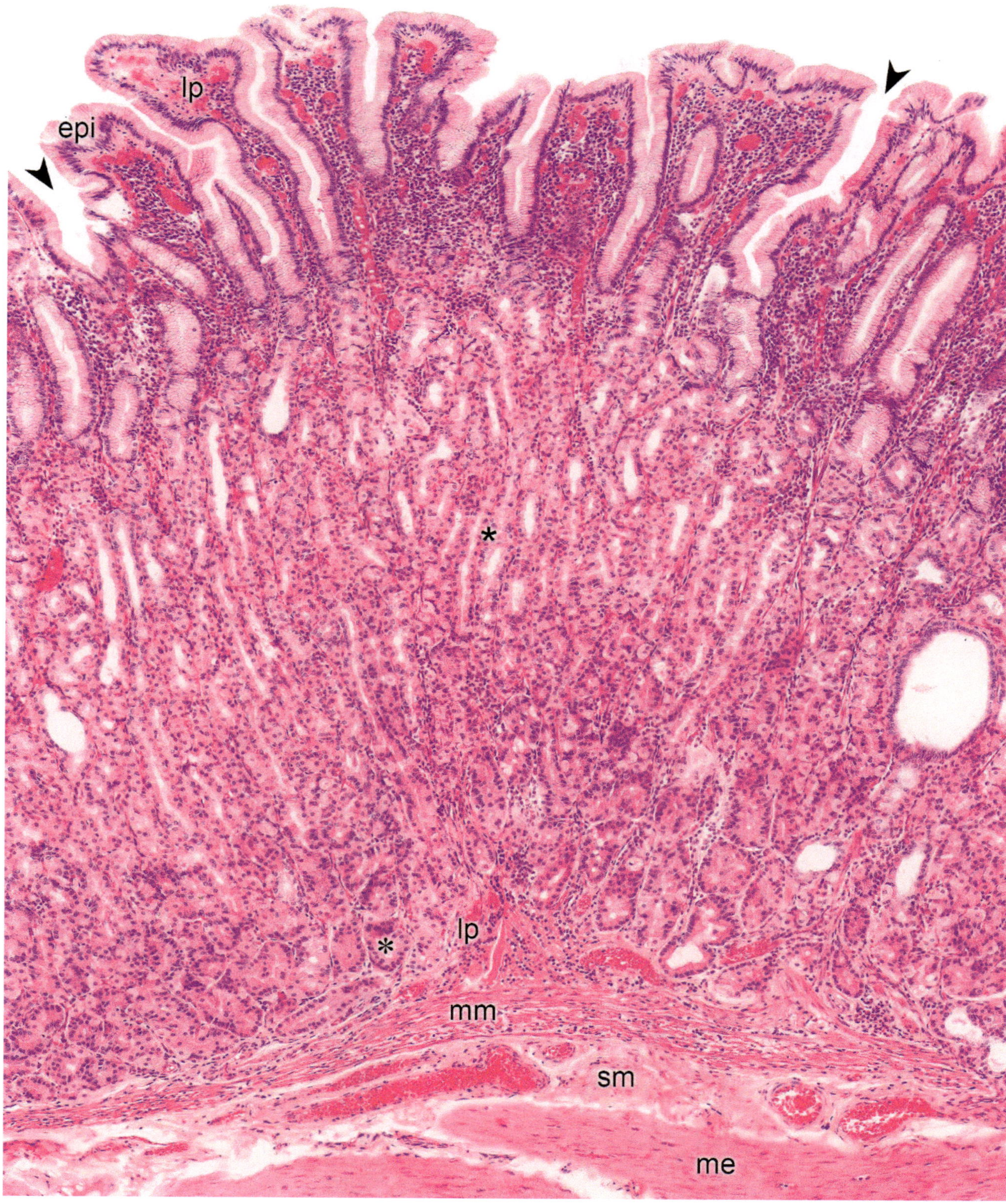

Figure 121. Fundic mucosa of the stomach. The mucosa of the stomach is lined with a single layer of lightly stained columnar cells that accumulate mucus at their apex (epi). Simple branched tubular fundic glands containing parietal and chief cells as well as mucous secreting and enteroendocrine cells are located in the lamina propria (lp) and they open into gastric pits (arrowheads) that are invaginations of the gastric surface epithelium. Parietal cells are large and acidophilic, and they occupy mainly the middle region of the glands (five-pointed asterisks) while basophilic chief cells are more characteristic to the base of the glands (six-pointed asterisks). Dense irregular connective tissue tunica submucosa (sm) connects the tunica muscularis mucosae (mm) to the tunica muscularis externa (me).

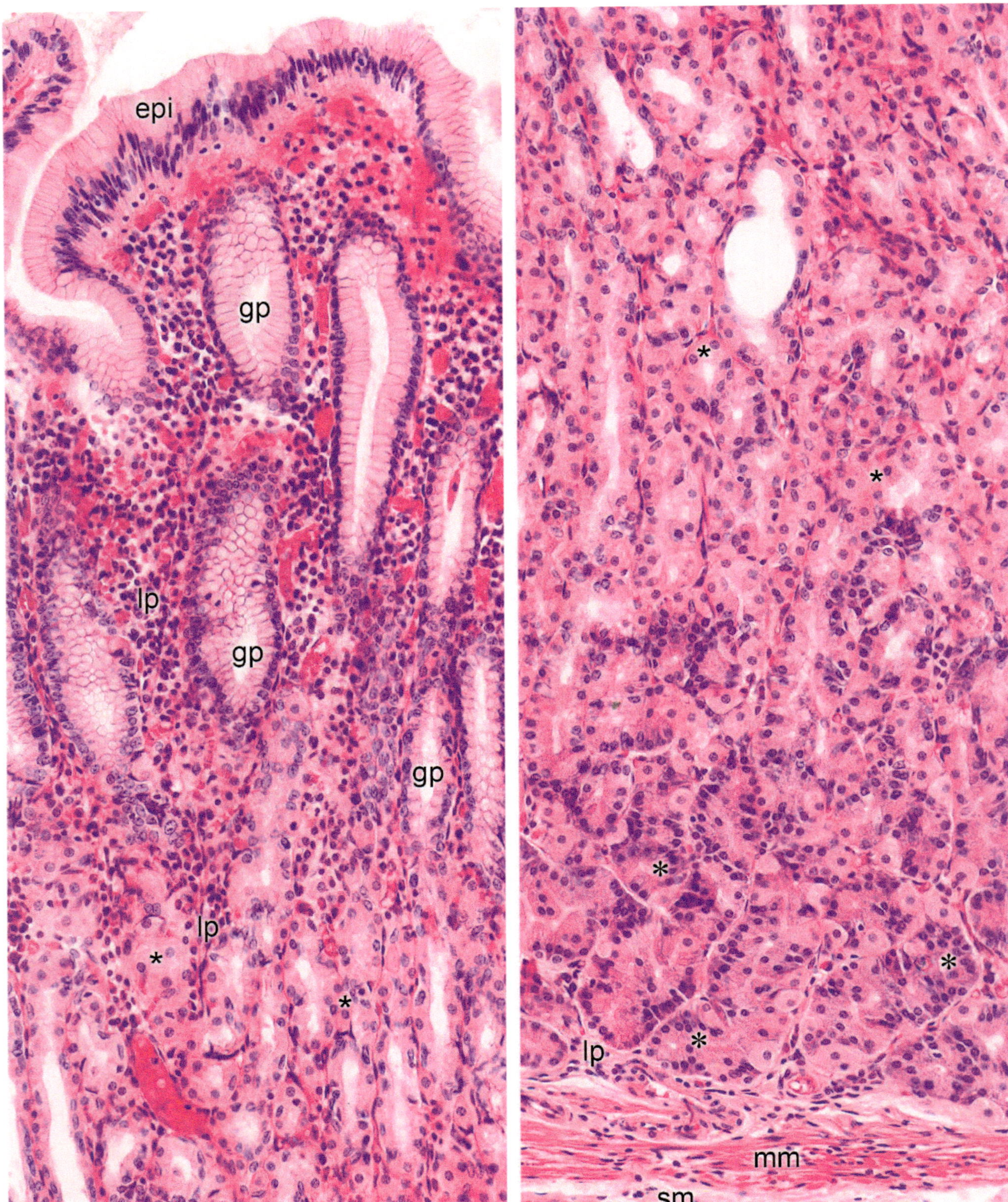

Figure 122. Fundic glands. Luminal (left) and basal segments of the fundic glands, resting on the muscularis mucosae (mm). Surface epithelium (epi) is formed by a single layer of lightly stained columnar cells. These cells also line the gastric pits (gp) that drain the simple branched tubular fundic glands occupying the loose connective tissue lamina propria (lp) under the surface epithelium. Fundic glands contain parietal and chief cells as well as mucous secreting and enteroendocrine cells. Parietal cells are large and acidophilic (five-pointed asterisks), and they occupy mostly the middle region of the glands while basophilic chief cells (six-pointed asterisks) are more characteristic to the base of the glands. The base of the glands rests on tunica muscularis mucosae (mm) and submucosa (sm).

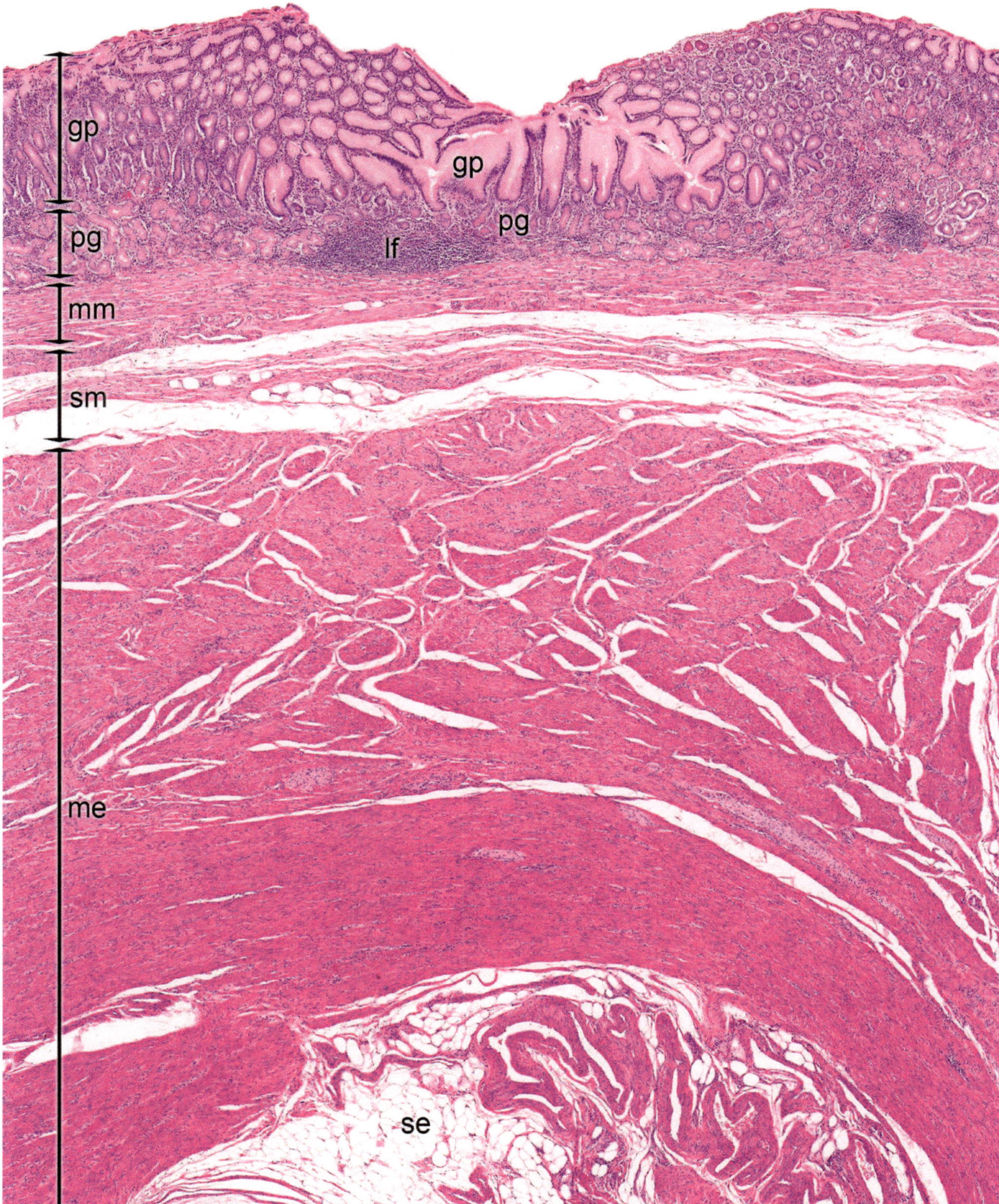

Figure 123. Pylorus of the stomach. Coiled branched tubular pyloric glands (pg) containing only mucous secreting and enteroendocrine cells occupy the lamina propria and they open into deep gastric pits (gp) that are invaginations of the gastric surface epithelium. Dense irregular connective tissue tunica submucosa (sm) connects the muscularis mucosae (mm) to the tunica muscularis externa (me). Loose connective tissue tunica serosa (se) with adipocytes and mesothelial lining covers the wall of the stomach otside. Lamina propria often contains lymphatic follicles (lf).

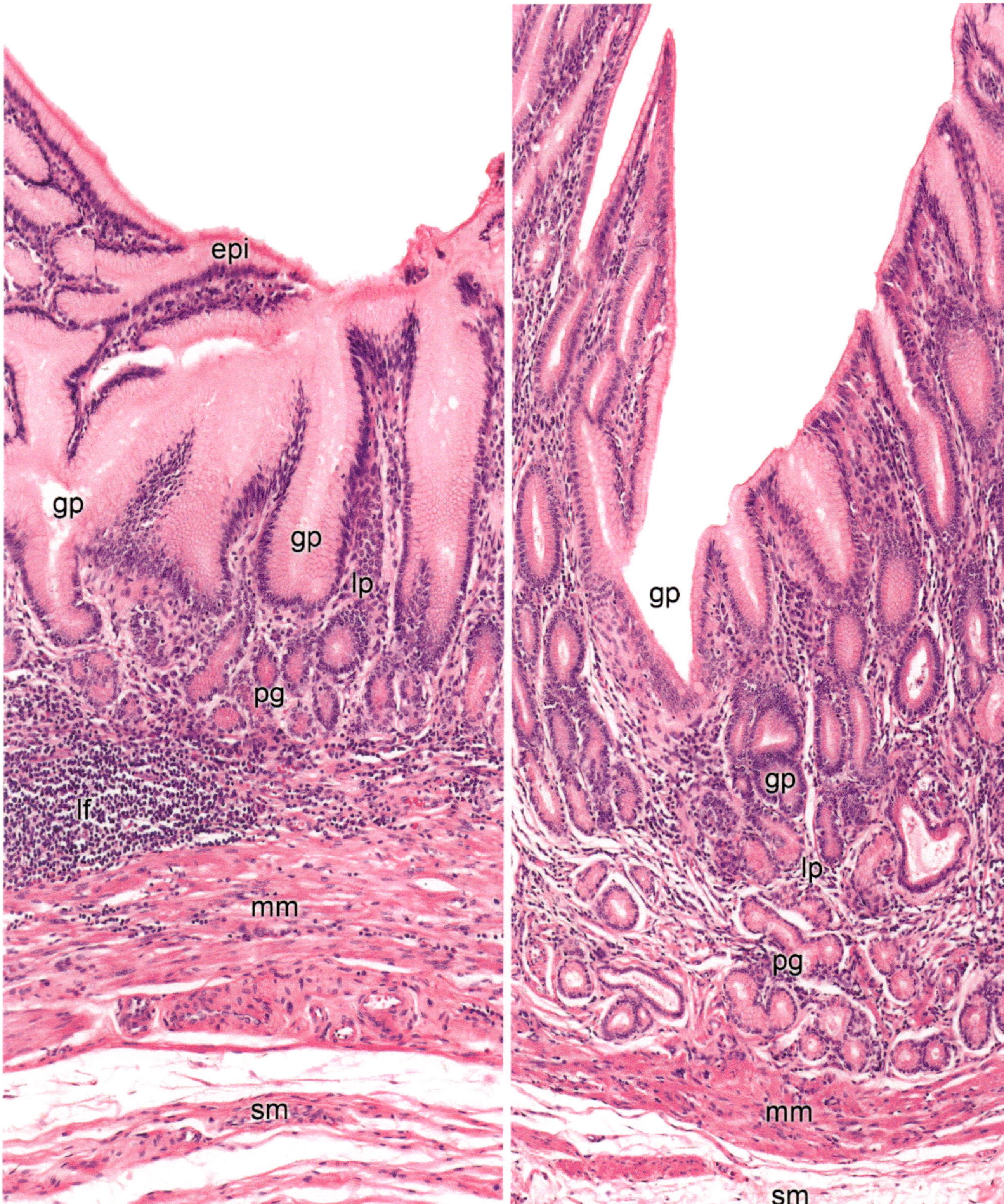

Figure 124. Pyloric glands. The surface of the pylorus is lined with a single layer of lightly stained columnar cells (epi) that line the large pyloric gastric pits (gp). Coiled branched tubular pyloric glands (pg) containing only mucous secreting and enteroendocrine cells occupy the lamina propria (lp) and they open into deep gastric pits (gp) that are invaginations of the gastric surface epithelium. Lamina propria often contains lymphatic follicles (lf). Dense irregular connective tissue tunica submucosa (sm) connects the muscularis mucosae (mm) to the tunica muscularis externa.

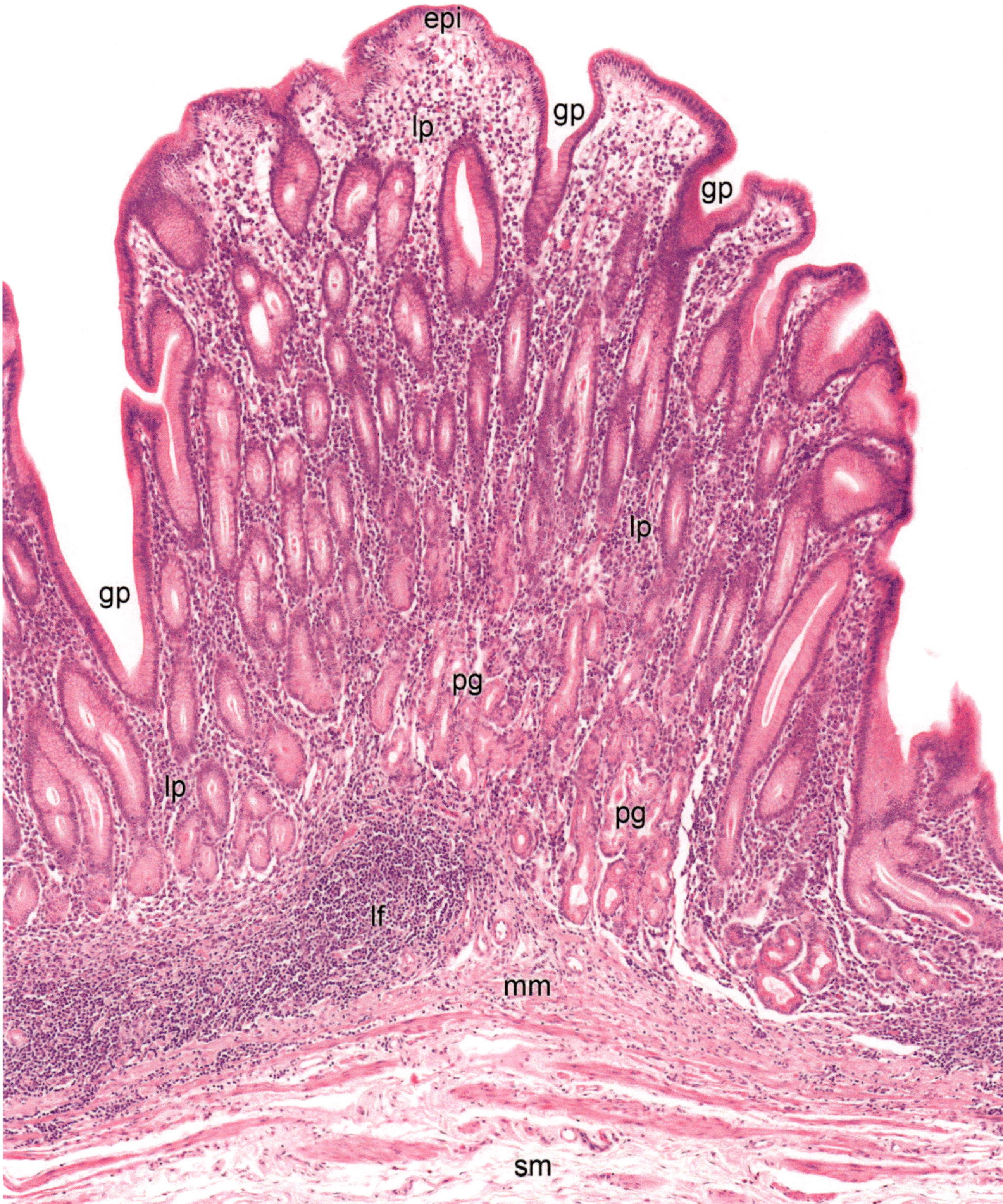

Figure 125. Pyloric glands. The surface epithelium (epi) of the pylorus is formed by a single layer of lighly stained columnar cells that also line the large pyloric gastric pits (gp). Coiled branched tubular pyloric glands (pg) containing only mucous secreting and enteroendocrine cells occupy the lamina propria (lp) and they open into deep gastric pits (gp) that are invaginations of the gastric surface epithelium. Lamina propria often contains lymphatic follicles (lf). Dense irregular connective tissue tunica submucosa (sm) connects the muscularis mucosae (mm) to the tunica muscularis externa.

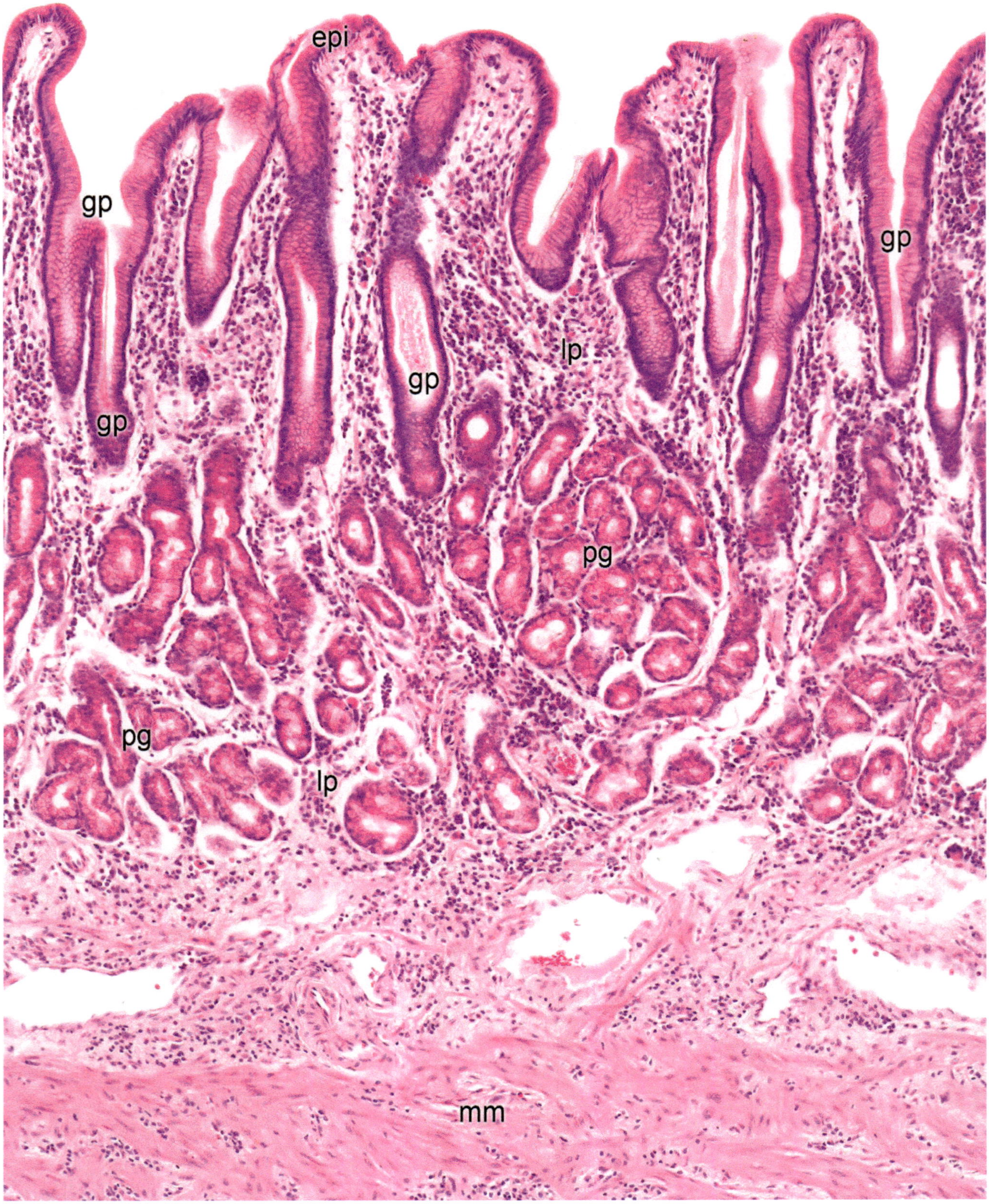

Figure 126. Pyloric glands. The surface of the pylorus is lined with a single layer of lightly stained columnar cells (epi) that also line the large pyloric gastric pits (gp). Coiled branched tubular pyloric glands (pg) containing only mucous secreting and enteroendocrine cells occupy the lamina propria (lp) and they open into deep gastric pits (gp) that are invaginations of the gastric surface epithelium. Lamina propria mucosae rests on the tunica muscularis mucosae (mm).

Small intestine

The luminal surface of the small intestine is enlarged by circular folds (plicae circulares) and villi. Plicae circulares contain a submucosal core while villi are finger- or leaf-shaped projections formed only by the mucous membrane; they commonly contain a centrally located lymphatic capillary (central lacteal). Mucosal surface is covered by enterocytes that are tall columnar cells with basal nucleus and apical microvilli (brush border). The primary function of these cells is absorption, but they also secrete enzymes. Goblet cells are scattered among the enterocytes. Apart from these two main cell types of the surface epithelium, intestine also contains enteroendocrine and microfold cells (M cells); the latter cell type lines the luminal surface of the occasional aggregated lymphatic follicles occupying the mucous membrane, particularly in the ileum (Peyer's patches). Microfold cells are large, often flattened cells with small folds on the apical surface; they are believed to transport macromolecules from the lumen to the underlying lymphatic elements, thus initiating immune response. Enteroendocrine cells are hormone-producing cells.

Intestinal glands (crypts of Lieberkühn) are simple tubular glands opening at the base of the intestinal villi. Intestinal glands occupy the entire thickness of the lamina propria mucosae spanning from the muscularis mucosae, which is composed of inner circular and outer longitudinal layers of smooth muscle, to the surface epithelium. In addition to the cell types that build up the surface epithelium, the bases of the glands also contain Paneth cells with large, eosinophilic cytoplasmatic granules that contain lysozyme and probably regulate the intestinal flora. Intestinal glands are surrounded with lamina propria mucosae rich in immune cells (lymphocytes, plasma cells, eosinophils, macrophages) that often form lymphatic follicles, especially in the ileum, where these lymphatic aggregations elevate the mucous membrane (Peyer's patches). These lymphatic follicles often penetrate the muscularis mucosae and the submucosa that is composed of dense connective tissue with adipocytes. In the proximal part of the duodenum, submucosa contains the compound tubular Brunner's glands that are histologically similar to the glands located in the distal gastric mucosa (pyloric glands). Brunner's glands have alkaline secretion containing mucin to protect mucous membrane from the acidic flow of the stomach; they perforate the muscularis mucosae and open into the base of the Lieberkühn's crypts.

The tunica muscularis externa of the small intestine is composed of inner circular and outer longitudinal smooth muscle layers, and the myenteric (Auerbach's) plexus between them with ganglion cells and nerve fibers. The outer tunica serosa (intraperitoneal segments) or tunica adventitia (retroperitoneal segments of duodenum) completes the wall structure of the small intestine.

Large intestine

Small intestine abruptly transitions to large intestine with the ileocaecal (Bauhin) valve. The mucosa of the large intestine lacks circular folds or villi. Epithelial lining is formed by the same absorptive columnar, and secretory goblet cells that cover the small intestine; however, there is no enzyme secretion by the surface epithelium. Lieberkühn's crypts occupy the lamina propria mucosae and they contain essentially the surface epithelial cells, but apart from the proximal part of the large intestine, including the cecum and the appendix, Paneth cells are missing from the glands. Lamina propria is a cell rich layer similar to that of the small intestine with frequent aggregated lymph follicles that often penetrate the submucosa. This feature is particularly characteristic to the vermiform appendix where extensive mucosal and submucosal lymphatic aggregations protrude into the narrow lumen. In contrast to the small intestine, large intestine usually contains solitary lymph follicles. Tunica muscularis mucosae is unique in the large intestine since most of the outer longitudinal layer of smooth muscle is organized into three thickened bands that correspond to the taeniae coli. This feature is missing from the vermiform appendix and rectum, where taeniae cannot be observed; in the vermiform appendix, the inner circular layer is also particularly thick. As in the small intestine, elements of the myenteric plexus can be observed between the inner circular and outer longitudinal smooth muscle layers.

Large intestine transitions into the rectum that also forms the upper part of the anal canal, the colorectal zone that is covered with the mucosa of the large intestine. Mucous membrane is thrown into folds here that form the anal columns, with the anal sinuses between, therefore the region is also termed as zona columnaris. Distally, the columnar epithelium of the zona columnaris is abruptly replaced by stratified cuboidal epithelium (linea anorectalis) and then by stratified squamous nonkeratinized and parakeratinized epithelia representing the anal transitional zone or zona intermedia with multiple epithelia. This zone transitions into the zona cutanea (cutaneous or squamous zone) that is covered with skin; the border between them is delineated by the linea anocutanea (Hilton's line, linea alba). The internal hemorrhoidal venous plexus is located in the submucosa, particularly at the level of zona columnaris and intermedia. Glands of the anal canal include tubular anal glands in the anal transitional zone that secrete mucus. In the zona cutanea, sebaceous glands associated with hair follicles as well as apocrine circumanal glands can be observed. The internal anal sphincter is formed by the enlarged circular layer of tunica muscularis externa that is surrounded by longitudinally arranged smooth muscle layers originating from the outer longitudinal muscle layers of the intestine. These longitudinal fibromuscular bundles form an internal layer under the hemorrhoidal plexus (anal submucosal muscle, Treitz's muscle) and an additional external layer (conjoined longitudinal muscle) as well as bundles between, penetrating the internal sphincter (mucosal suspensory ligament, Park's ligament).

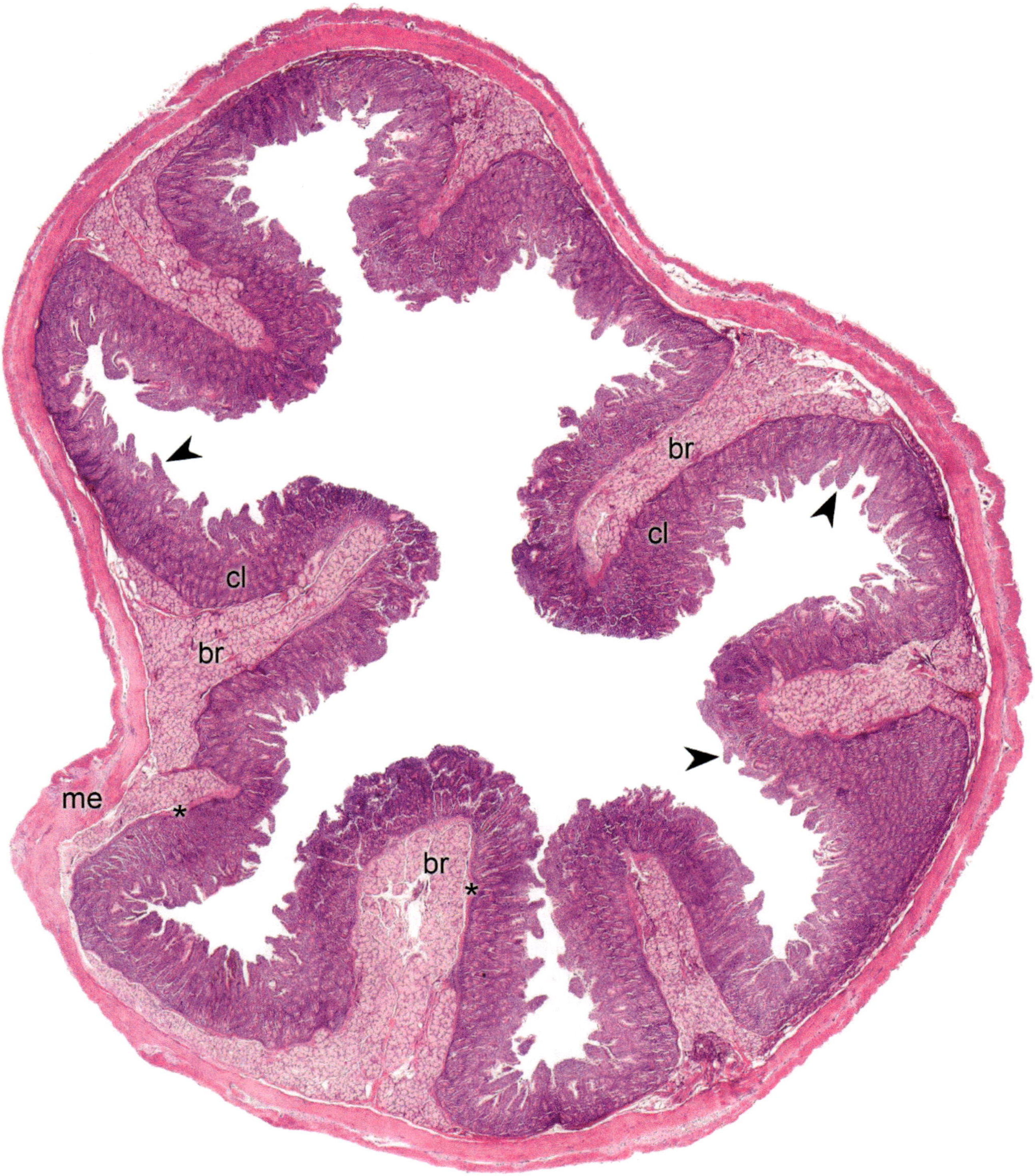

Figure 127. Duodenum. Duodenal mucosa is lined with simple columnar absorptive epithelium with microvilli. Mucous membrane forms fingerlike projections called villi that are relatively small in duodenum (arrowheads). Lamina propria contains intestinal glands, the crypts of Lieberkühn (cl). Under the thin tunica muscularis mucosae (asterisks), submucosa is filled with mucous Brunner's glands (br) that are compound tubular glands characteristic to the proximal part of the duodenum. Brunner's glands open into the base of the crypts of Lieberkühn and produce alkaline secretion containing mucin that protects the mucosa from gastric acid. Tunica muscularis externa (me) is composed of inner circular and outer longitudinal smooth muscle layers and it is covered outside with tunica adventitia.

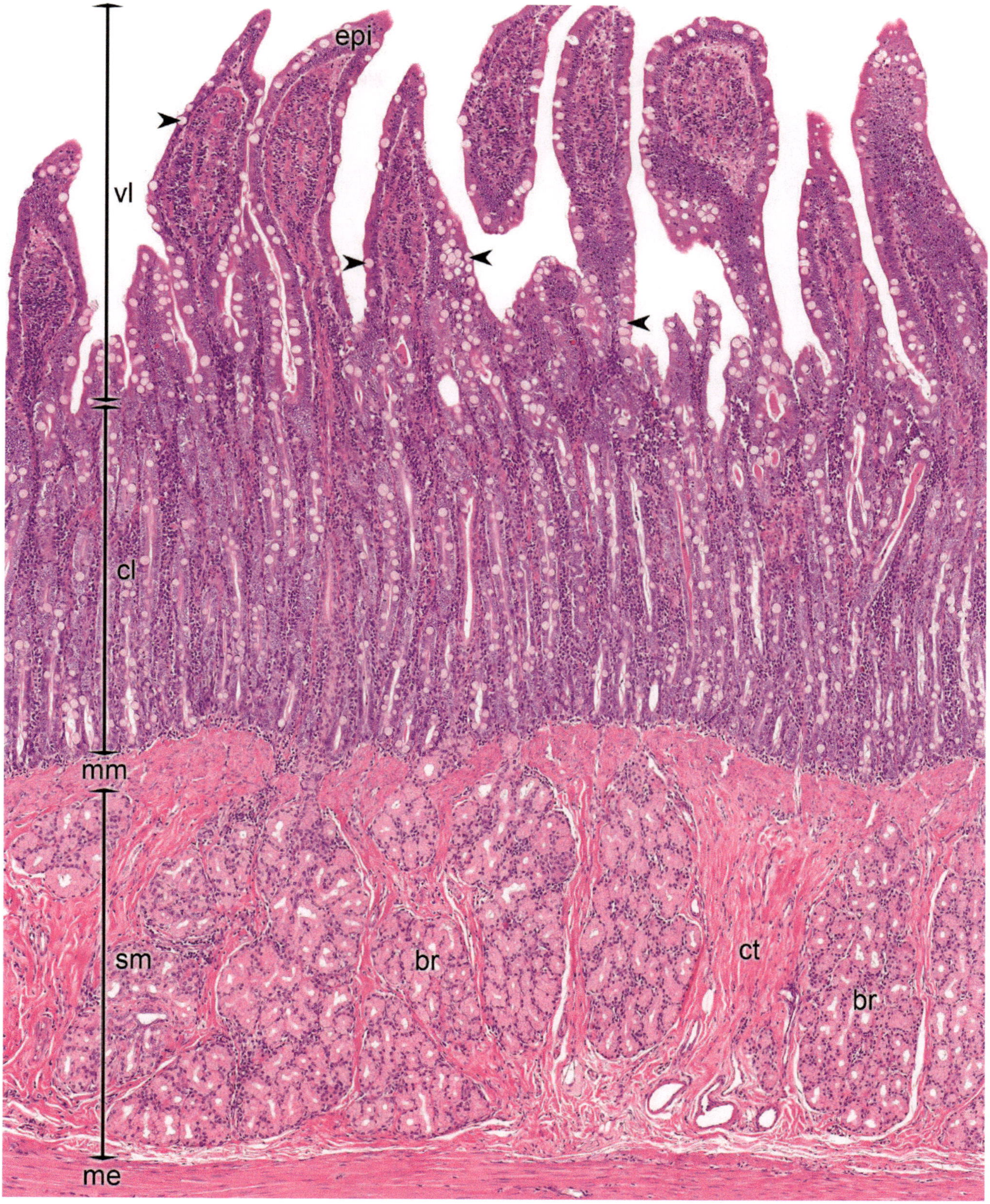

Figure 128. Duodenum. Duodenal mucosa is lined with microvillated simple columnar epithelium (epi) with goblet cells (arrowheads). Mucous membrane forms fingerlike projections called villi (vl). Lamina propria contains intestinal glands, the crypts of Lieberkühn (cl). Under the thin tunica muscularis mucosae (mm), submucosa (sm) contains compound tubular Brunner's glands (br) embedded into dense irregular connective tissue (ct). Brunner's glands open ino the base of the crypts of Lieberkühn and produce alkaline mucin that protects the mucosa from gastric acid. Tunica muscularis externa (me) is composed of inner circular and outer longitudinal smooth muscle layers.

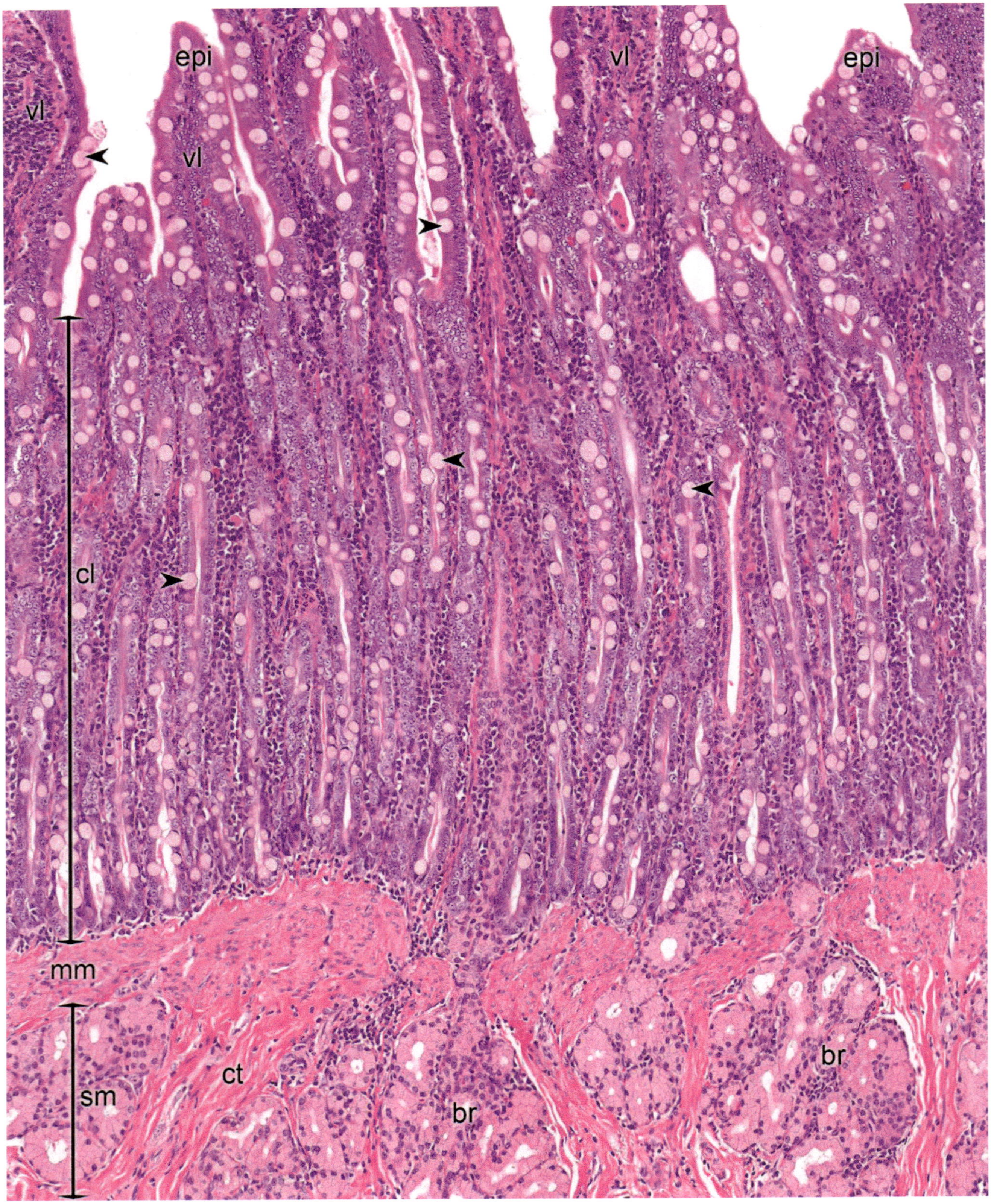

Figure 129. Duodenum. Duodenal mucosa forms fingerlike projections called villi (vl) lined with microvillated simple columnar epithelium (epi) with goblet cells (arrowheads). This epithelium also extends into the crypts of Lieberkühn (cl). Under the thin tunica muscularis mucosae (mm), submucosa (sm) contains compound tubular Brunner's glands (br) embedded into dense irregular connective tissue (ct). Brunner's glands open into the base of the crypts of Lieberkühn and produce alkaline mucin that protects the mucosa from gastric acid.

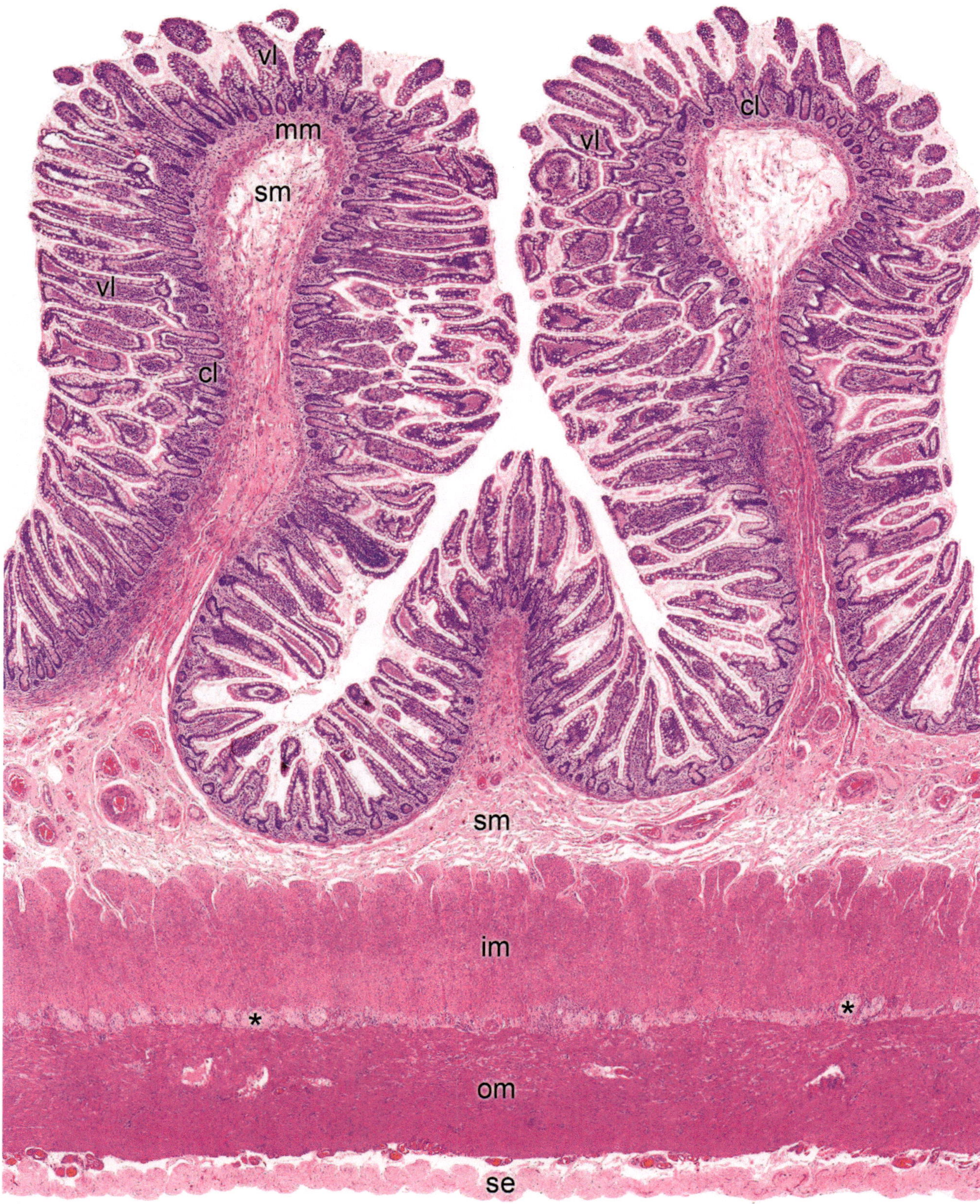

Figure 130. Jejunum. The tunica mucosa of the jejunum forms long villi (vl) lined with microvillated simple columnar epithelium with goblet cells (arrowheads). This epithelium also lines the crypts of Lieberkühn (cl). Under the tunica muscularis mucosae (mm), the tunica submucosa (sm) forms the core of the circular folds. Tunica muscularis externa is composed of inner circular (im) and outer longitudinal (om) smooth muscle layers. Jejunum is covered with tunica serosa (se) that has an outer mesothelial lining.

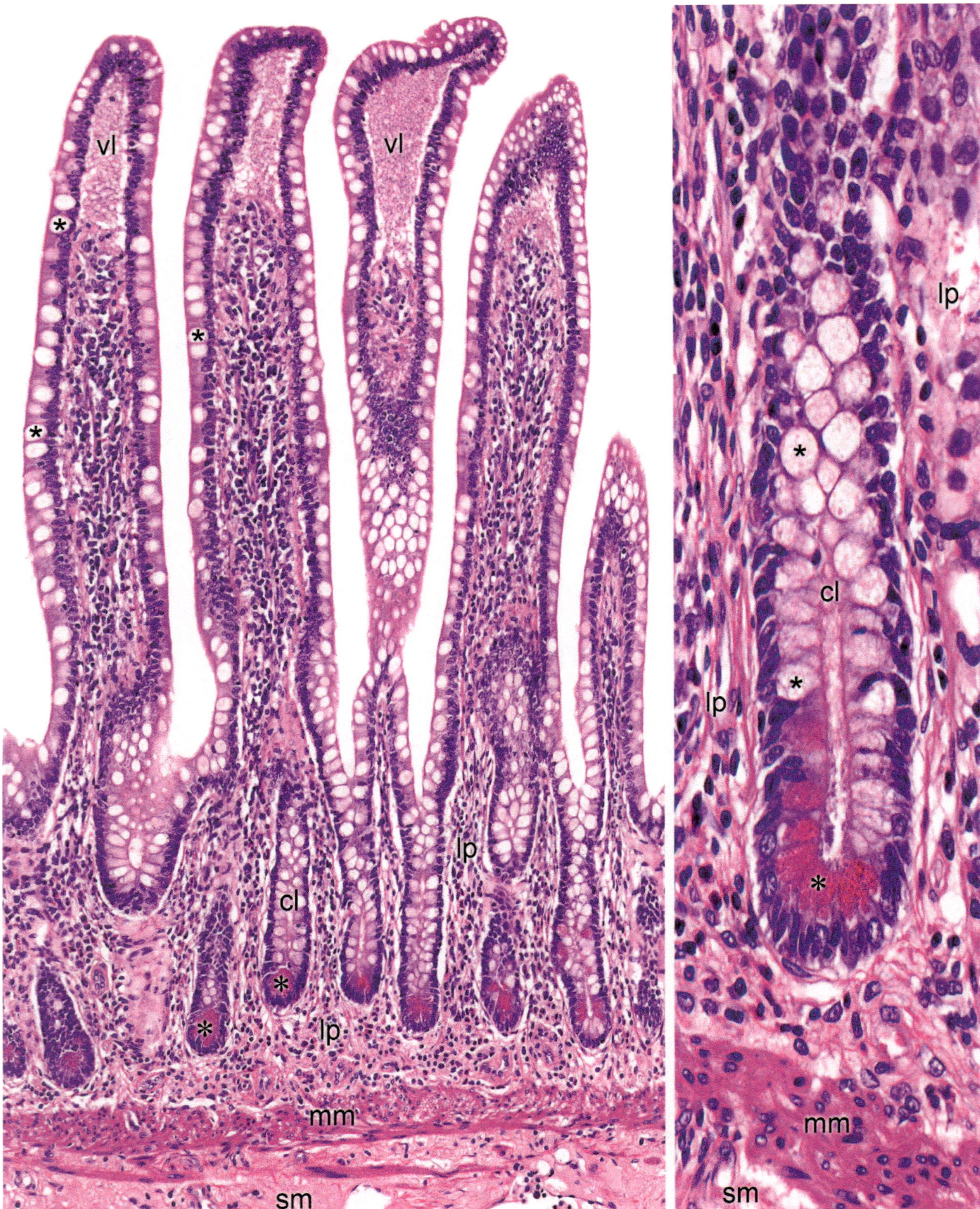

Figure 131. Jejunum. Tunica mucosa of the jejunum forms long villi (vl) lined with microvillated simple columnar epithelium with goblet cells (five-pointed asterisks). This epithelium also lines the crypts of Lieberkühn (cl) located in the lamina propria (lp) that also forms the core of the villi. Paneth cells with eosinophilic granules (six-pointed asterisks) populate the base of the glands. Under the tunica muscularis mucosae (mm), submucosa (sm) is composed of dense irregular connective tissue.

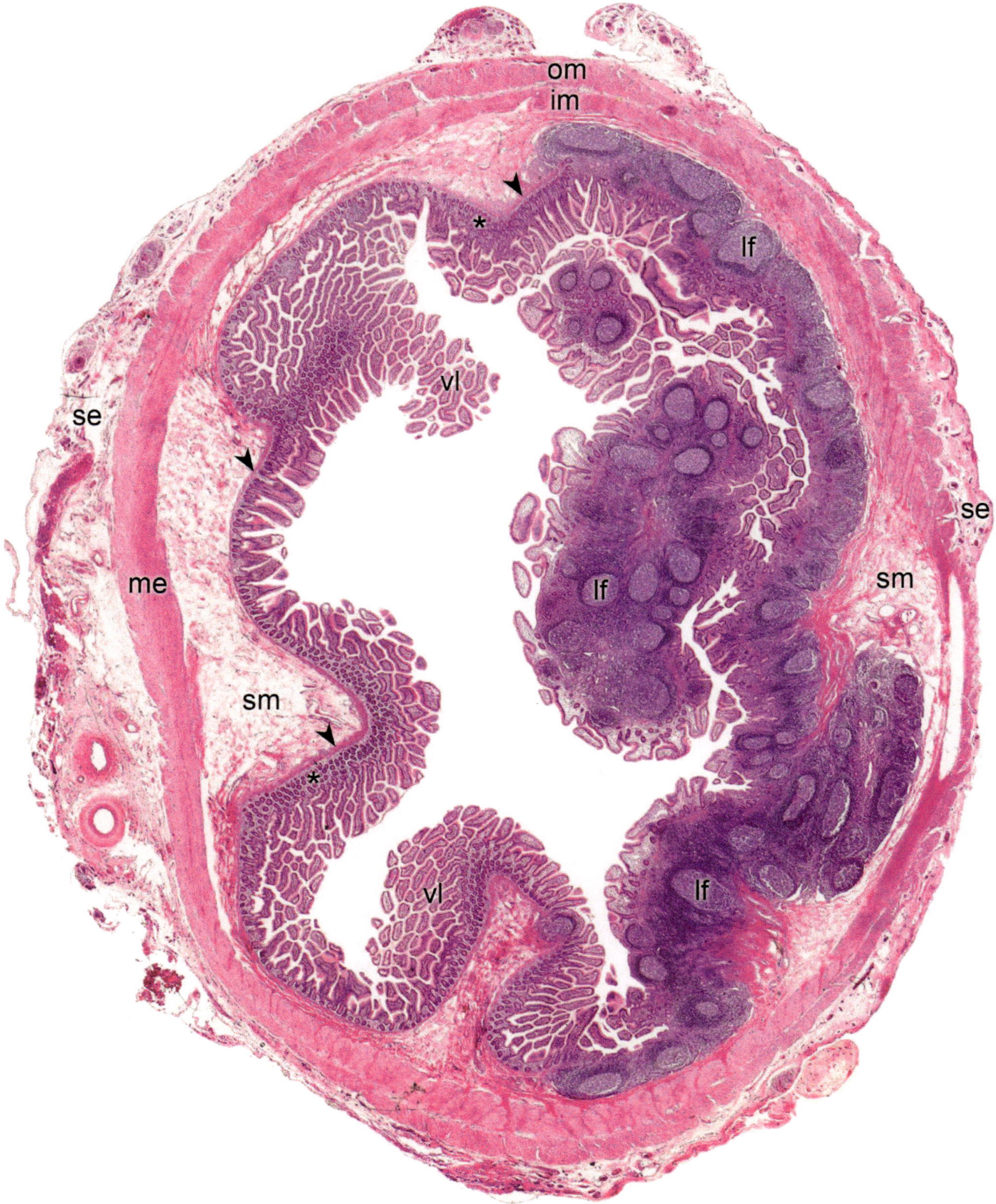

Figure 132. Ileum. The tunica mucosa of the ileum forms long villi (vl) lined with microvillated simple columnar epithelium with goblet cells. This epithelium also lines the crypts of Lieberkühn (asterisks). Lamina propria of the ileum contains aggregated lymph follicles (lf) that are called Peyer's patches; they elevate the mucosal surface and often invade the tunica submucosa (sm). Tunica muscularis externa (me) is composed of inner circular (im) and outer longitudinal (om) smooth muscle layers. Ileum is covered with tunica serosa (se) that has an outer mesothelial lining.

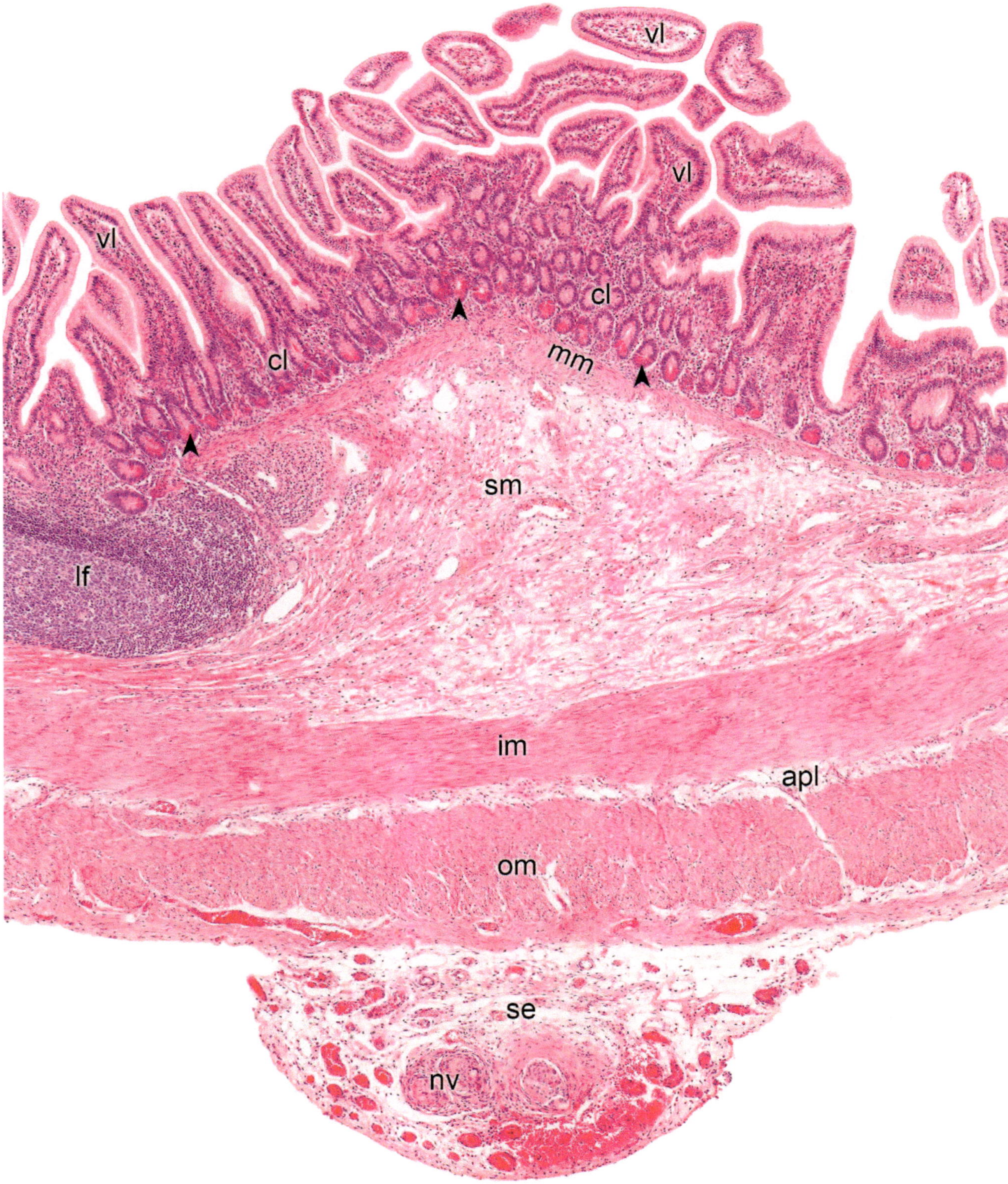

Figure 133. Ileum. Tunica mucosa of the ileum forms long villi (vl) lined with microvillated simple columnar epithelium with goblet cells. This epithelium also lines the crypts of Lieberkühn (cl) that contain eosinophilic Paneth cells at the base (arrowheads). Lamina propria of the ileum contains lymph follicles (lf) that often aggregate and elevate the mucosal surface (Peyer's patches). These lymphatic follicles often invade the tunica submucosa (sm). Tunica muscularis externa (me) is composed of inner circular (im) and outer longitudinal (om) smooth muscle layers with rich autonomic plexus between them. Ileum is covered with tunica serosa (se) that has an outer mesothelial lining and contains nerves (ne) and vessels.

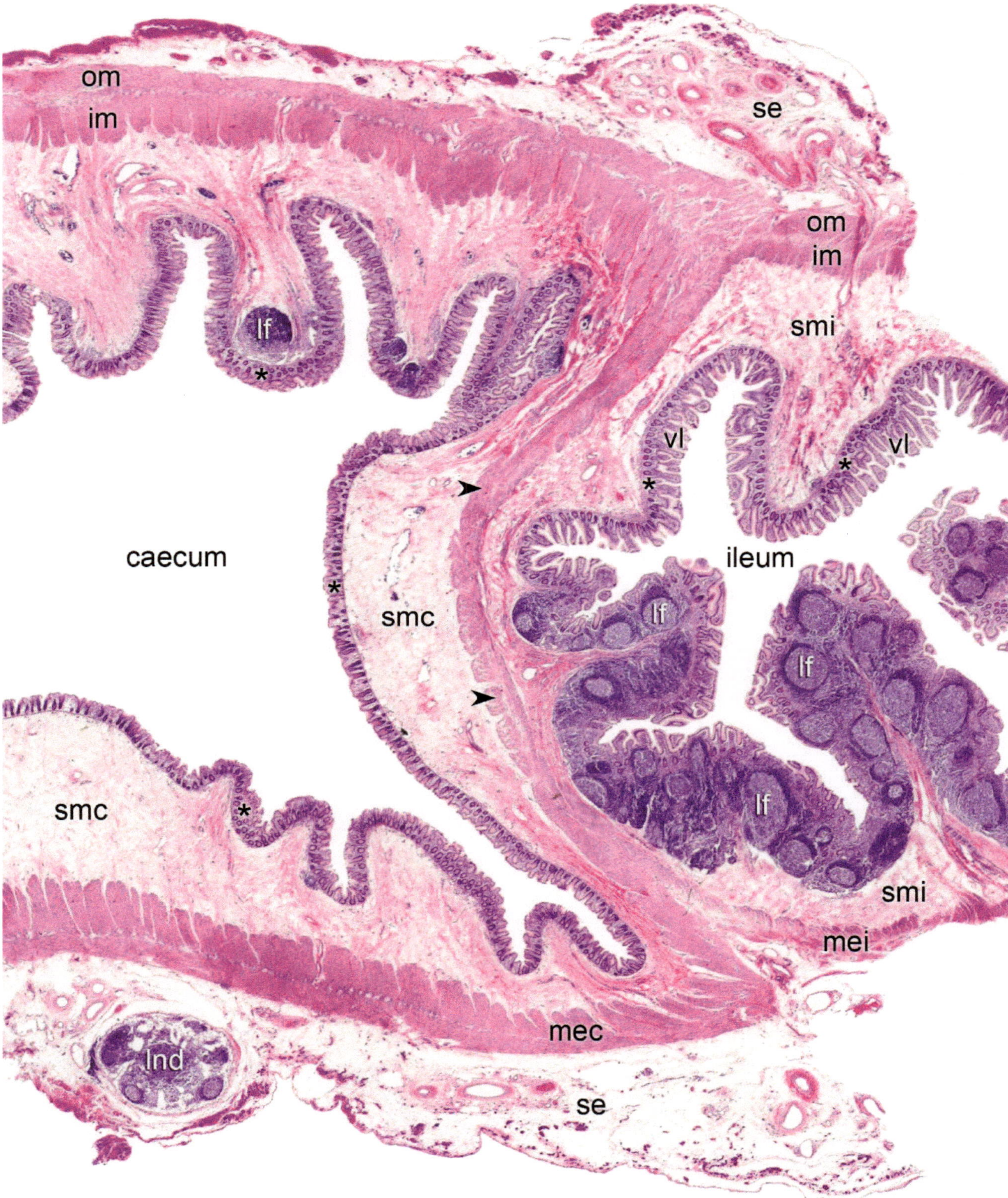

Figure 134. Ileocaecal junction. Intestinal mucosa contains crypts of Lieberkühn (asterisks) and it forms villi in the ileum (vl) that are missing from the large intestine. Ileal mucosa and submucosa contains aggregated lymph follicles (lf) forming Peyer's patches; solitary follicles also appear in large intestine. The submucosa of the caecum (smc) is continuous with the ileal submucosa (smi) at the ileocaecal valve. Similarly, the caecal (mec) and ileal tunica muscularis externa (mei), which are both composed of inner circular (im) and outer longitudinal smooth muscle layers (om), merge at the ileocaecal junction (arrowheads). The junction is covered by tunica serosa (se) that occasionally contains small lymph nodes (lnd).

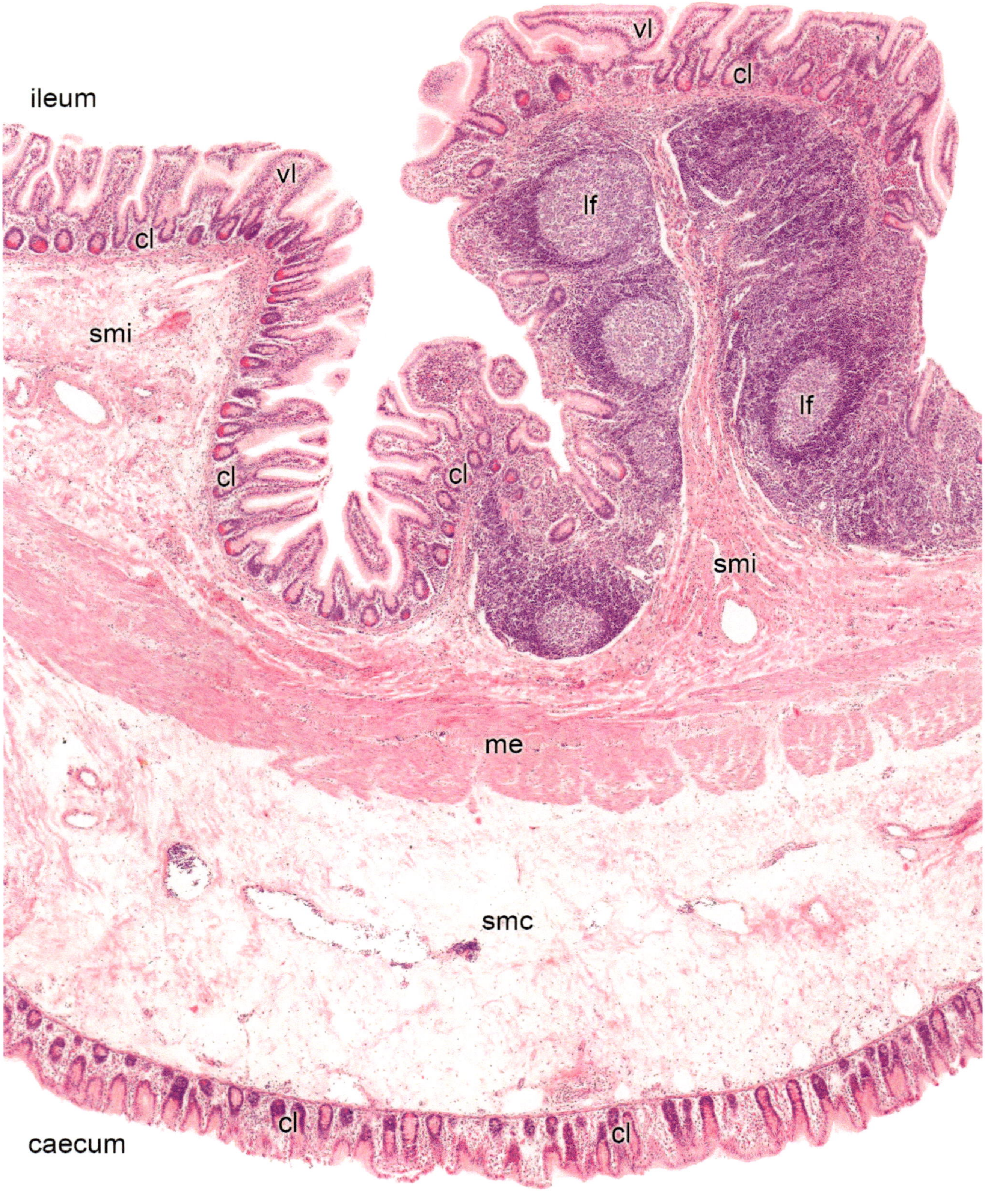

Figure 135. Ileocaecal valve (ileal and caecal surfaces). Intestinal mucosa contains crypts of Lieberkühn (cl) and it forms villi in the ileum (vl) that are missing from the large intestine. Ileal mucosa and submucosa contains aggregated lymph follicles (lf) forming Peyer's patches. The submucosa of the caecum (smc) is continuous with the ileal submucosa (smi) at the ileocaecal valve; between the two submucosal layers, the caecum and the ileum share a common tunica muscularis externa (me).

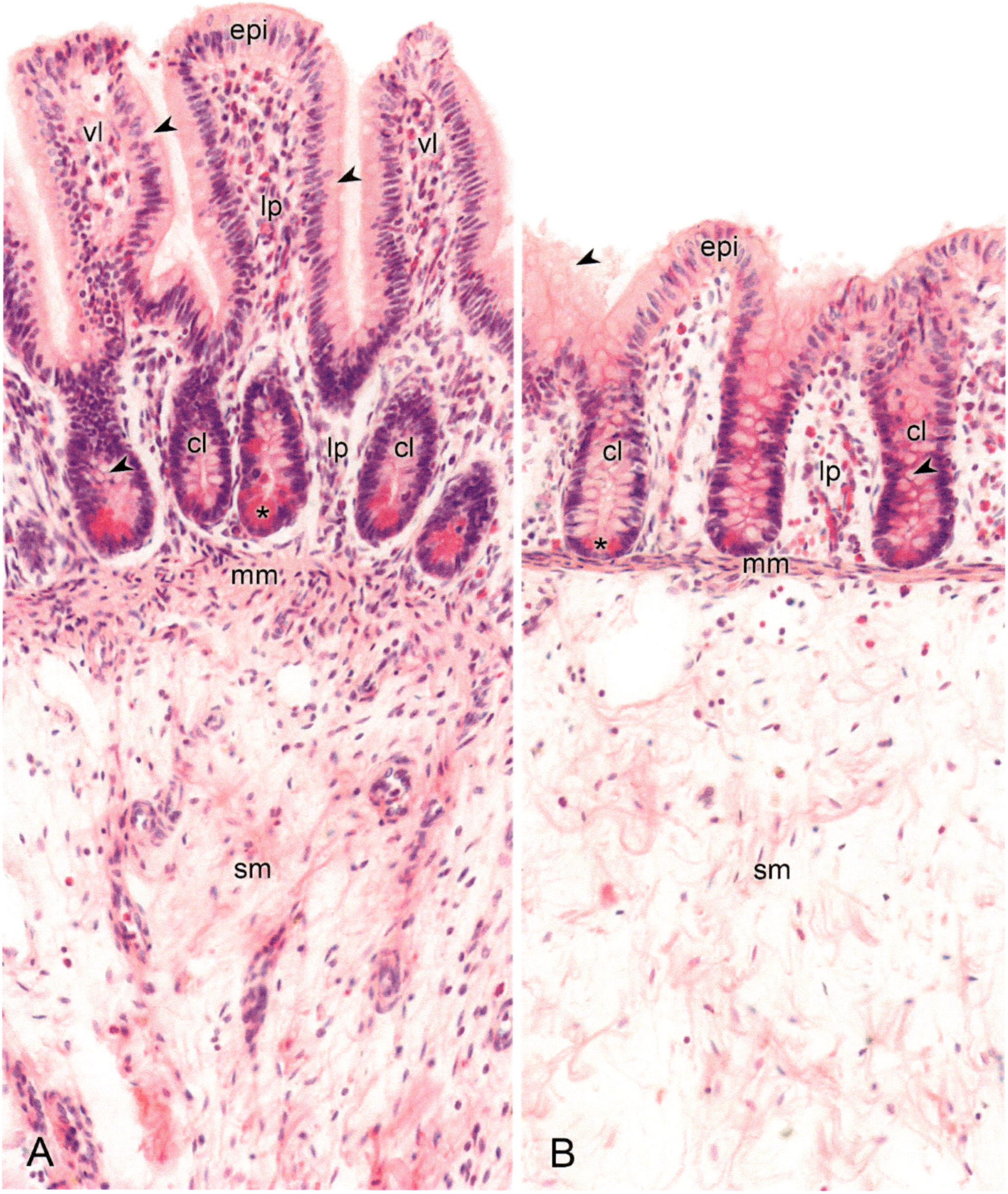

Figure 136. Ileocaecal valve. (A) Tunica mucosa covering the ileal side and (B) the caecal side of the ileocaecal valve. The epithelial lining of the intestines (epi) is formed by absorptive columnar cells as well as goblet cells (arrowheads). These cells cover the villi of the ileum (vl) that are missing from the large intestine, and they extend into the crypts of Lieberkühn (cl). Paneth cells with eosinophilic granules (asterisks) populate the base of the intestinal glands in the lamina propria (lp) of the small intestine; occasionally they can also be found in the proximal segment of the large intestine. Intestinal submucosa (sm) is located under the tunica muscularis mucosae (mm).

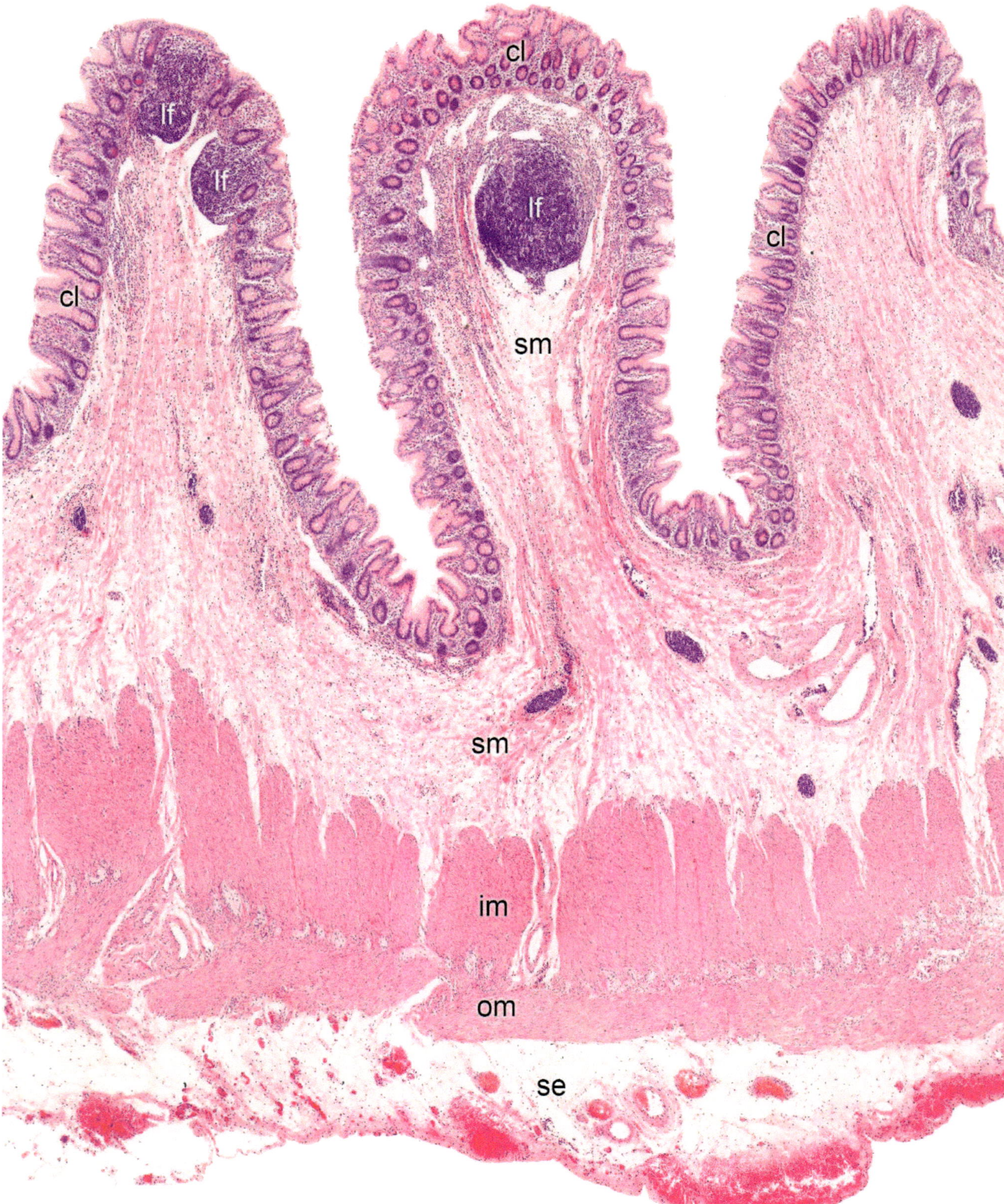

Figure 137. Caecum. Caecum is the most proximal part of the large intestine. The epithelial lining is formed by absorptive columnar as well as goblet cells that extend into the crypts of Lieberkühn (cl). Paneth cells with eosinophilic granules (asterisks) can be occasionally observed in the caecal glands. Caecal mucosa and submucosa often contains lymph follicles (lf). Under the submucosa (sm), tunica muscularis externa is formed by inner circular (im) and outer longitudinal layers (om). Caecum is covered by tunica serosa (se).

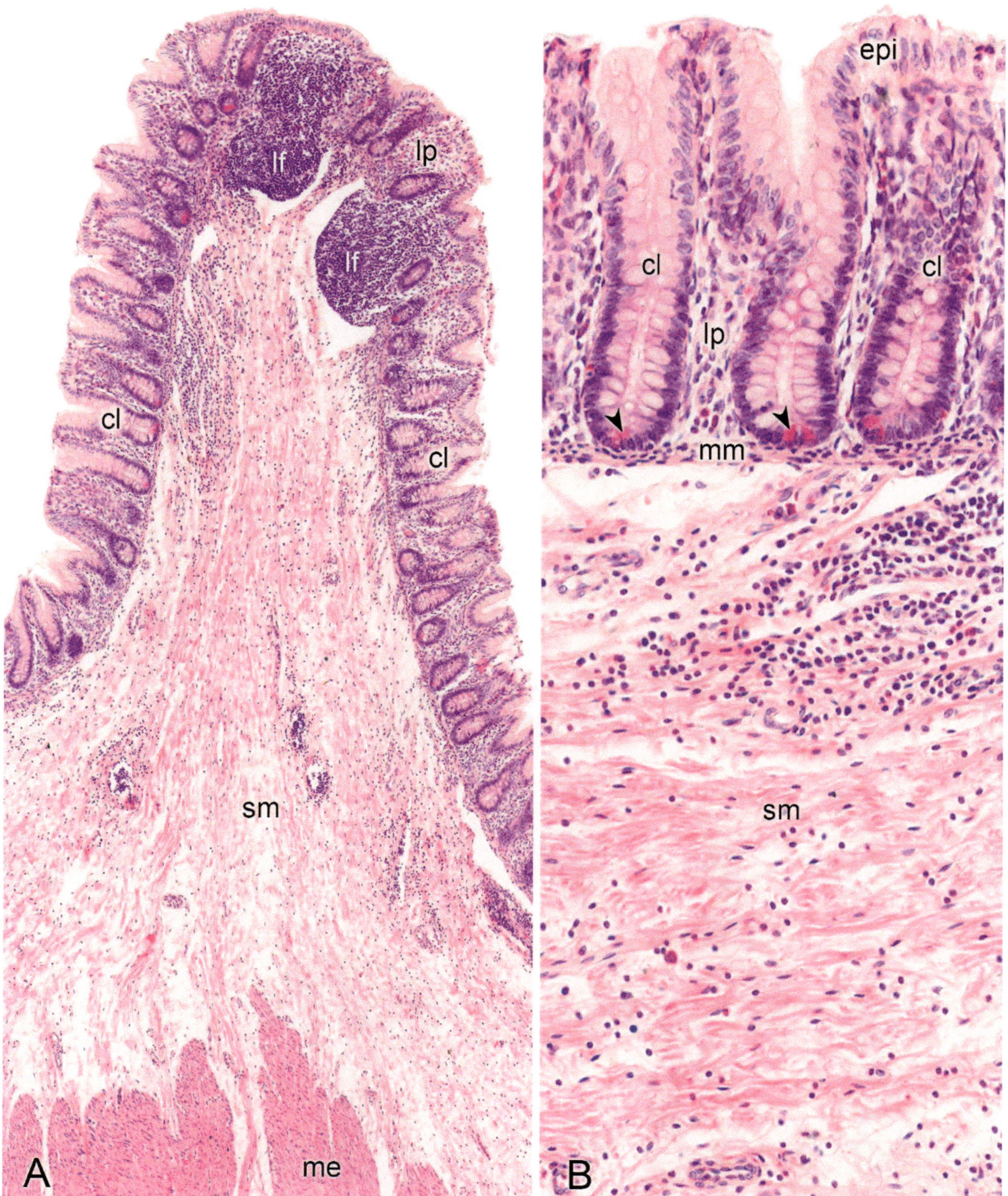

Figure 138. Caecum. Caecum is the most proximal part of the large intestine. (A) Caecal mucosa and submucosa often contains lymph follicles (lf). Tunica muscularis externa (me) is located under the submucosa (sm). The epithelial lining of the mucosa is formed by absorptive columnar as well as goblet cells that extend into the crypts of Lieberkühn (cl) located in the lamina propria (lp). (B) Paneth cells with eosinophilic granules (arrowheads) can be observed in the base of the caecal glands resting on the muscularis mucosae (mm).

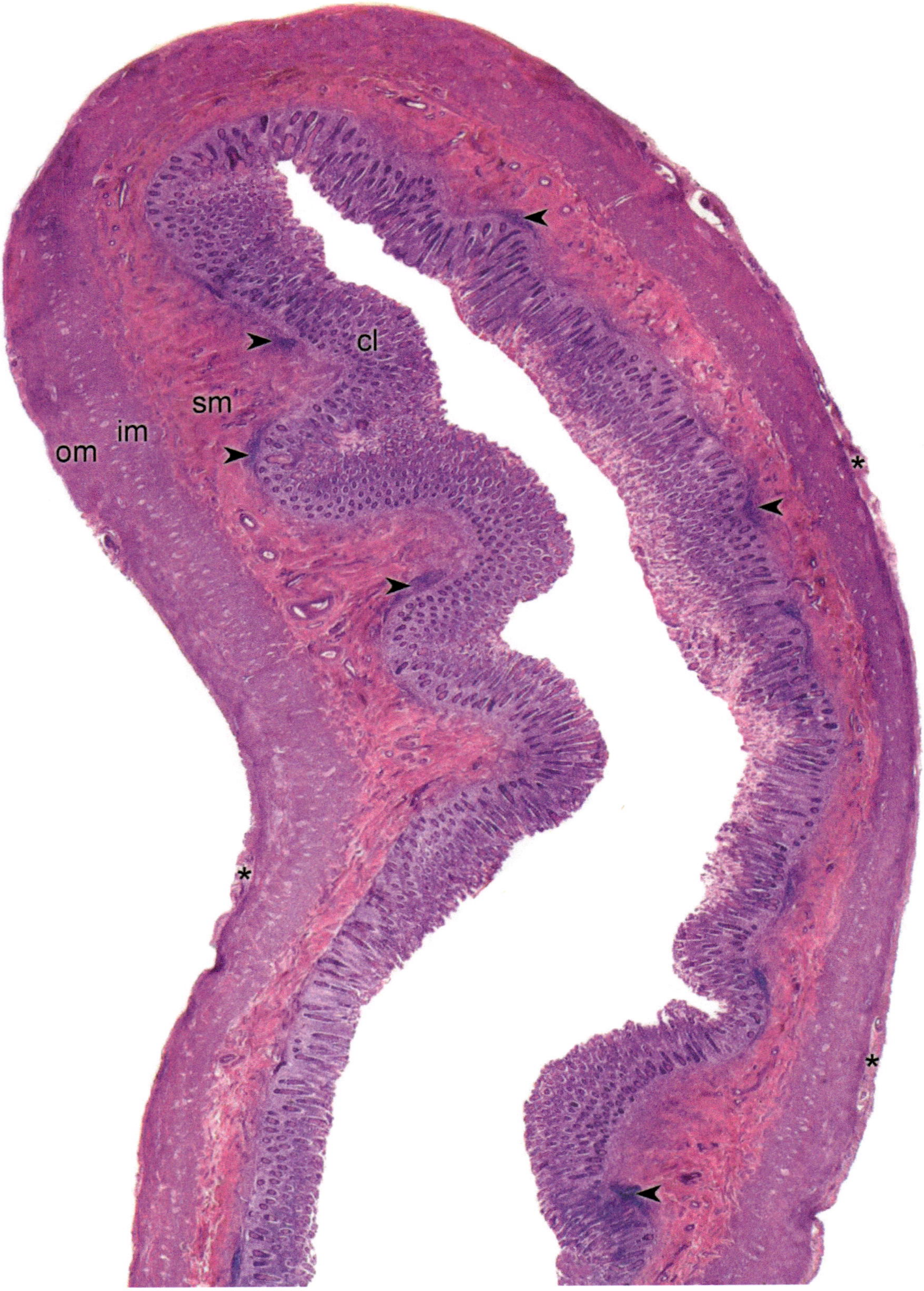

Figure 139. Vermiform appendix, longitudinal section. Vermiform appendix is covered with large intestinal mucosa and it opens into the caecum. The tunica mucosa contains crypts of Lieberkühn (cl) and lymph follicles (arrowheads) that often invade the tunica submucosa (sm). The epithelial lining of the mucosa is formed by absorptive columnar as well as goblet cells that extend into the crypts of Lieberkühn. Tunica muscularis externa is composed of inner circular (im) and outer longitudinal layers (om). Appendix is covered by tunica serosa (asterisks).

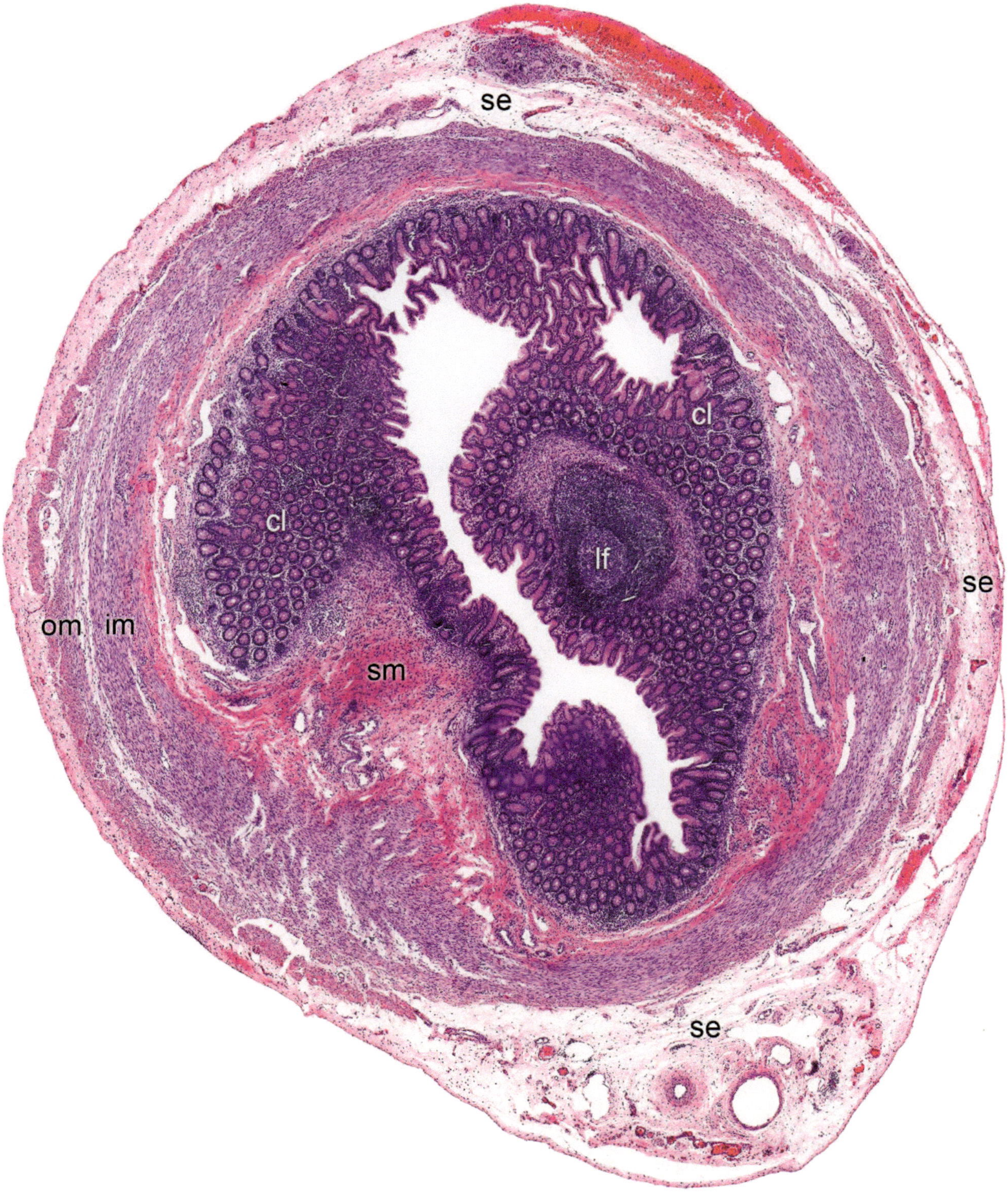

Figure 140. Vermiform appendix. Tunica mucosa contains crypts of Lieberkühn (cl) and lymph follicles (lf). The epithelial lining of the mucosa is formed by absorptive columnar as well as goblet cells that extend into the crypts of Lieberkühn. Tunica submucosa (sm) connects the mucosa to the tunica muscularis externa that is composed of inner circular (im) and outer longitudinal layers (om). Appendix is covered by tunica serosa (se).

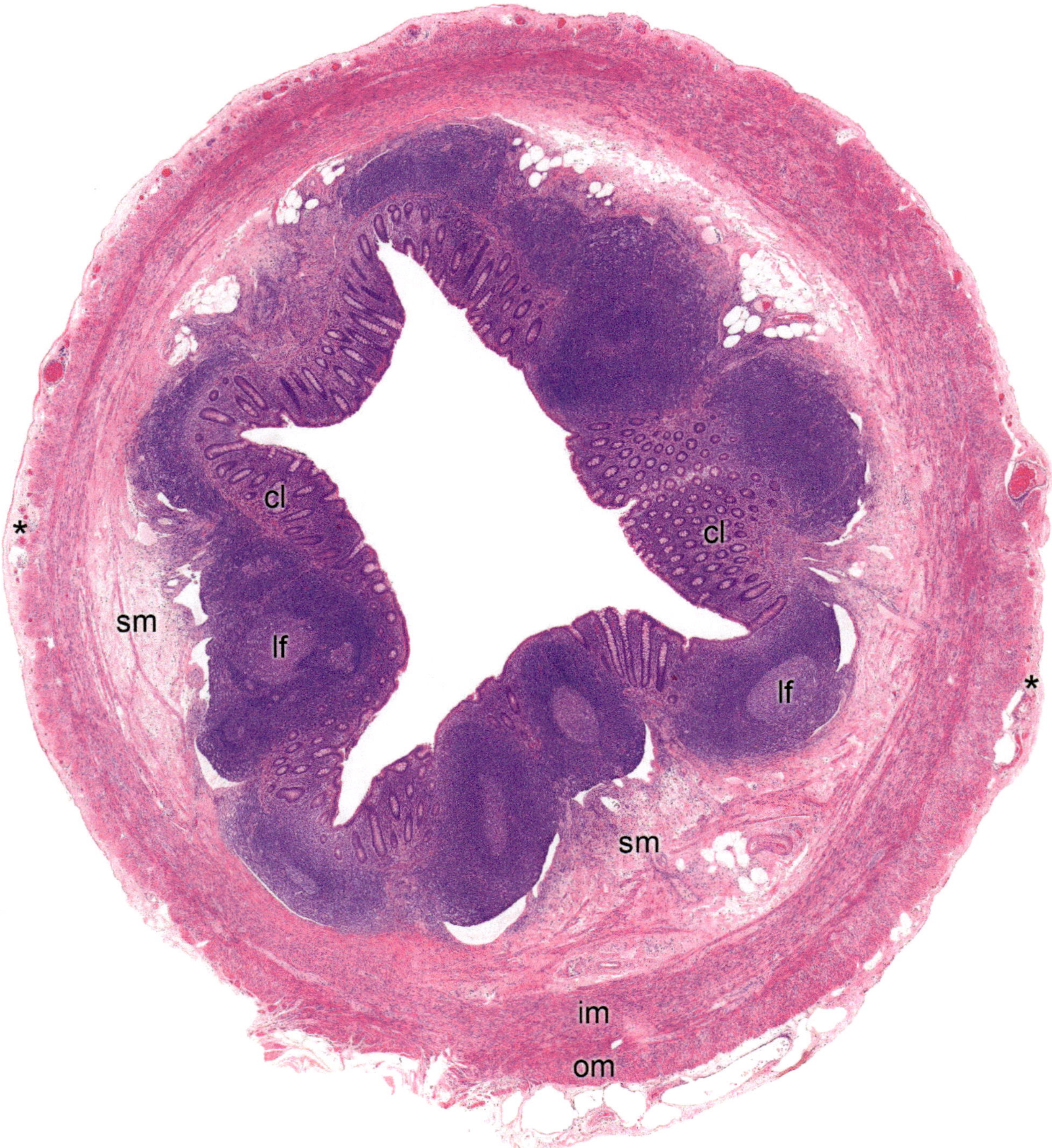

Figure 141. Vermiform appendix. Tunica mucosa contains crypts of Lieberkühn (cl) and lymph follicles (lf) that often invade the tunica submucosa (sm). The epithelial lining of the mucosa is formed by absorptive columnar cells as well as goblet cells that extend into the crypts of Lieberkühn. Tunica submucosa connects the mucosa to the tunica muscularis externa that is composed of inner circular (im) and outer longitudinal layers (om). Appendix is covered by tunica serosa (asterisks).

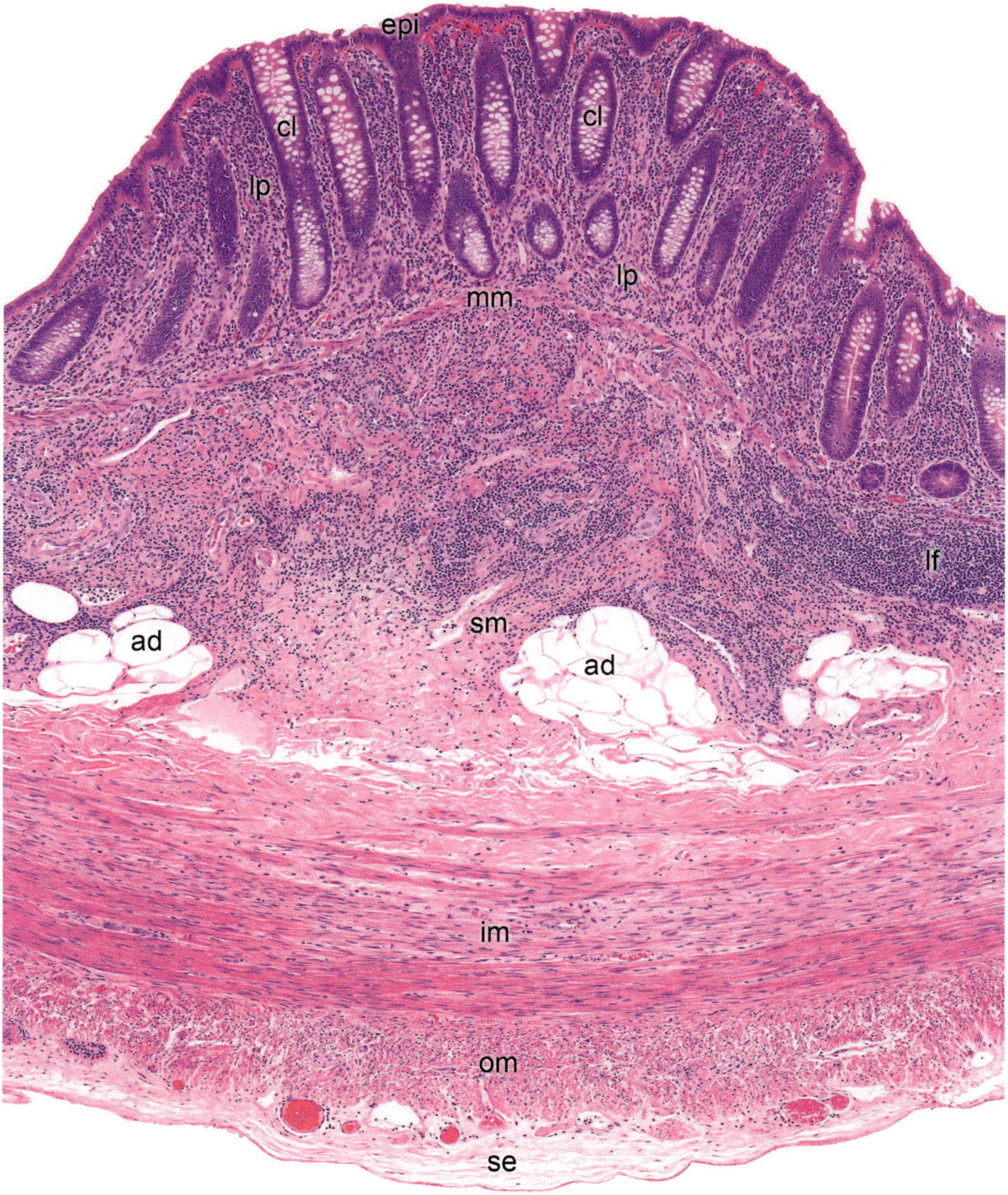

Figure 142. Vermiform appendix. Lamina propria mucosae (lp) contains crypts of Lieberkühn (cl) and lymph follicles (lf) that often invade the tunica submucosa (sm). The epithelial lining of the mucosa is formed by absorptive columnar as well as goblet cells that extend into the crypts of Lieberkühn where goblet cells become abundant. Tunica submucosa (sm) is composed of dense irregular connective tissue with adipocytes (ad) and connects the tunica muscularis mucosae (mm) to the tunica muscularis externa that is composed of inner circular (im) and outer longitudinal layers (om). Appendix is covered by tunica serosa (se).

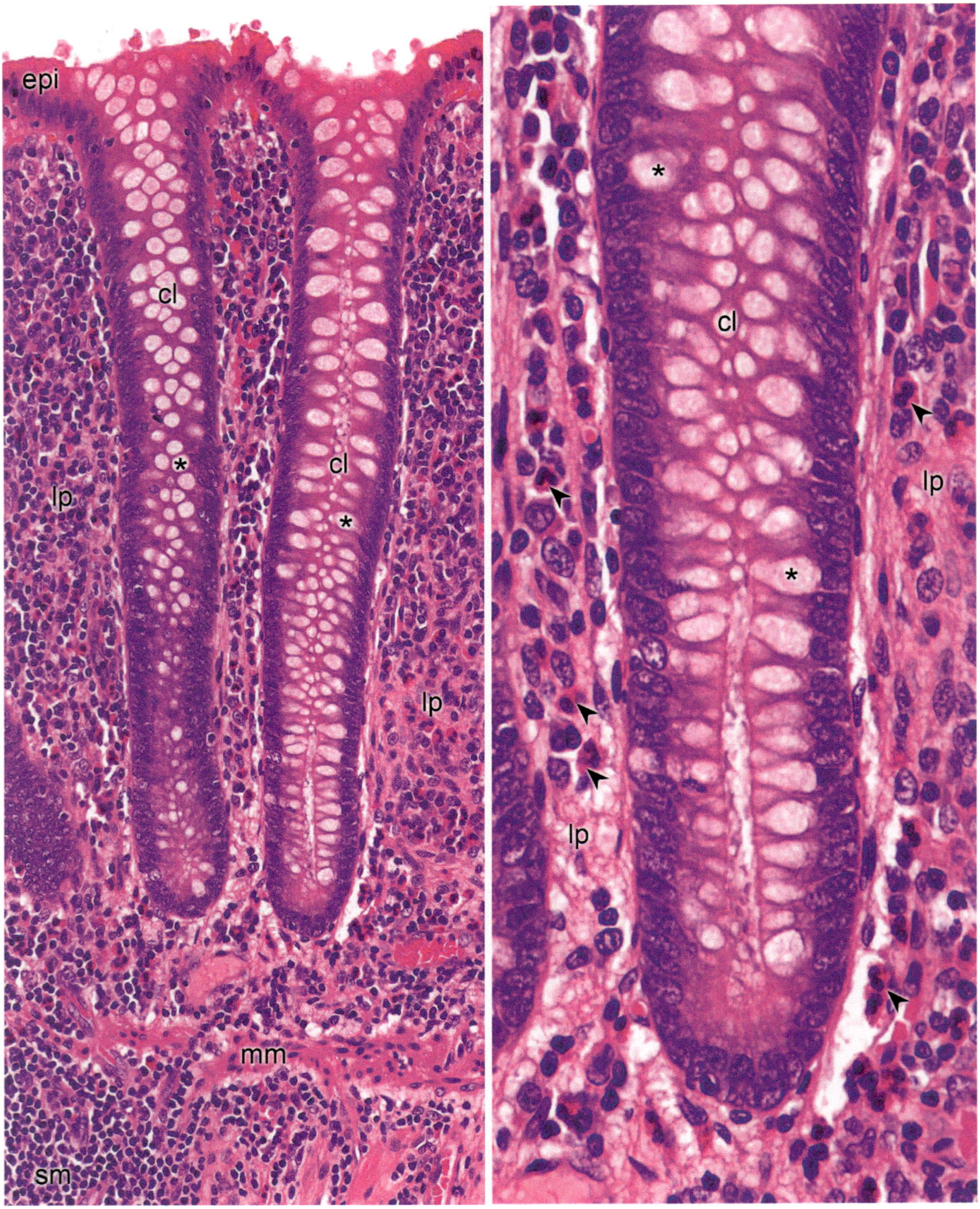

Figure 143. Vermiform appendix, glands. Lamina propria mucosae (lp) contains crypts of Lieberkühn (cl) and large number of eosinophil granulocytes between the gland (arrowheads). The epithelial lining of the mucosa (epi) is formed by absorptive columnar as well as goblet cells that extend into the crypts of Lieberkühn where goblet cells become abundant (asterisks). Tunica submucosa (sm) is located under the tunica muscularis mucosae (mm).

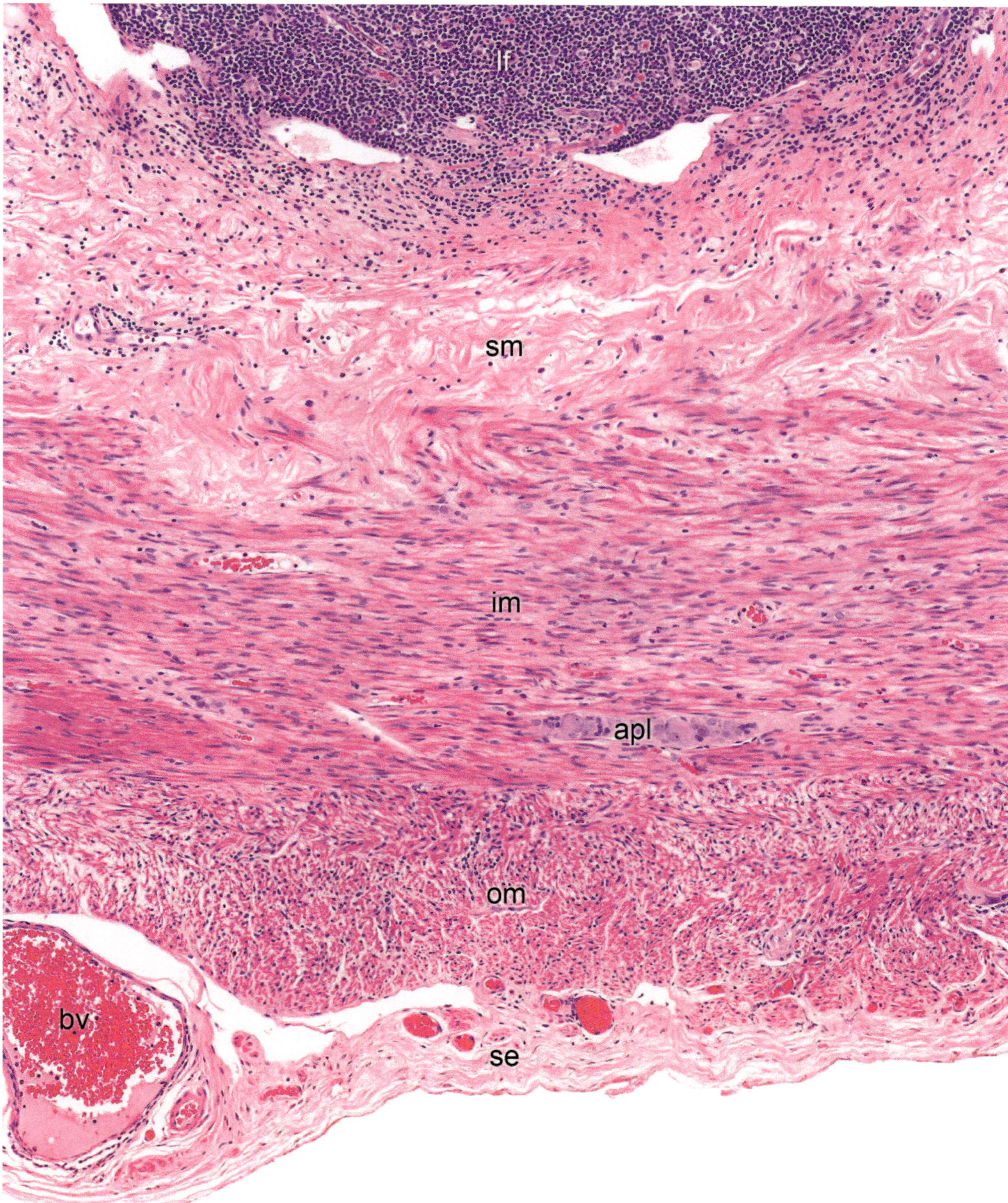

Figure 144. Vermiform appendix, external layers. Tunica mucosa of the appendix contains aggregated lymph follicles (lf) that often invade the tunica submucosa (sm). Tunica submucosa (sm) is composed of dense irregular connective tissue with adipocytes (ad) and connects the tunica muscularis mucosae (mm) to the tunica muscularis externa that is composed of inner circular (im) and outer longitudinal layers (om), with the myenteric autonomic plexus (apl) between the layers. Appendix is covered by tunica serosa (se) that contains blood vessels (bv) and nerves and has a mesothelial lining.

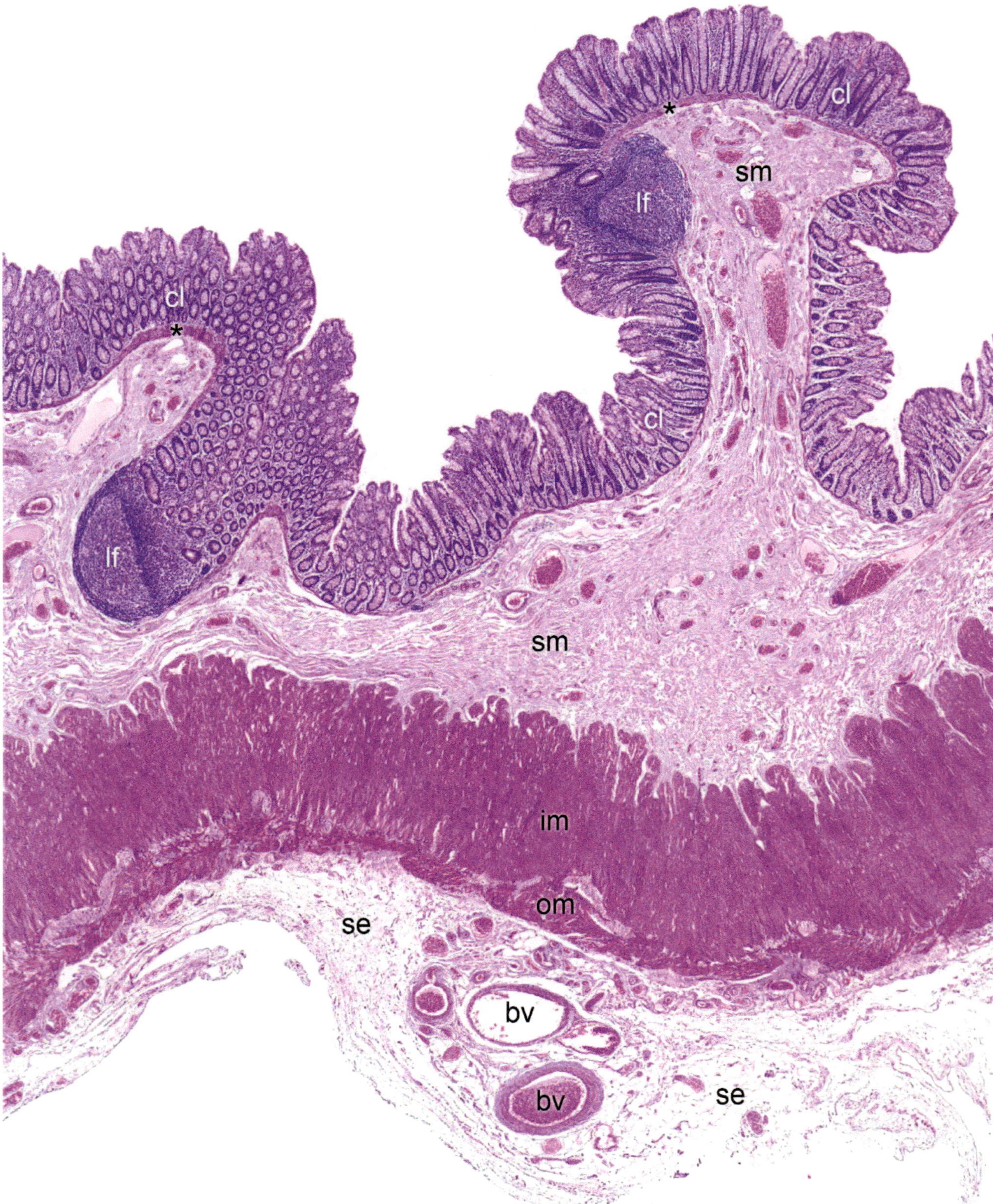

Figure 145. Transverse colon. The surface of the tunica mucosa as well as the crypts of Lieberkühn (cl) are lined with microvillated simple columnar epithelium with goblet cells. Lamina propria contains solitary lymph follicles (lf) that may penetrate the muscularis mucosae (asterisks) and invade the tunica submucosa (sm). Tunica muscularis externa is composed of inner circular (im) and outer longitudinal layers (om). Transverse colon is covered by tunica serosa (se) that contains blood vessels (bv) and nerves.

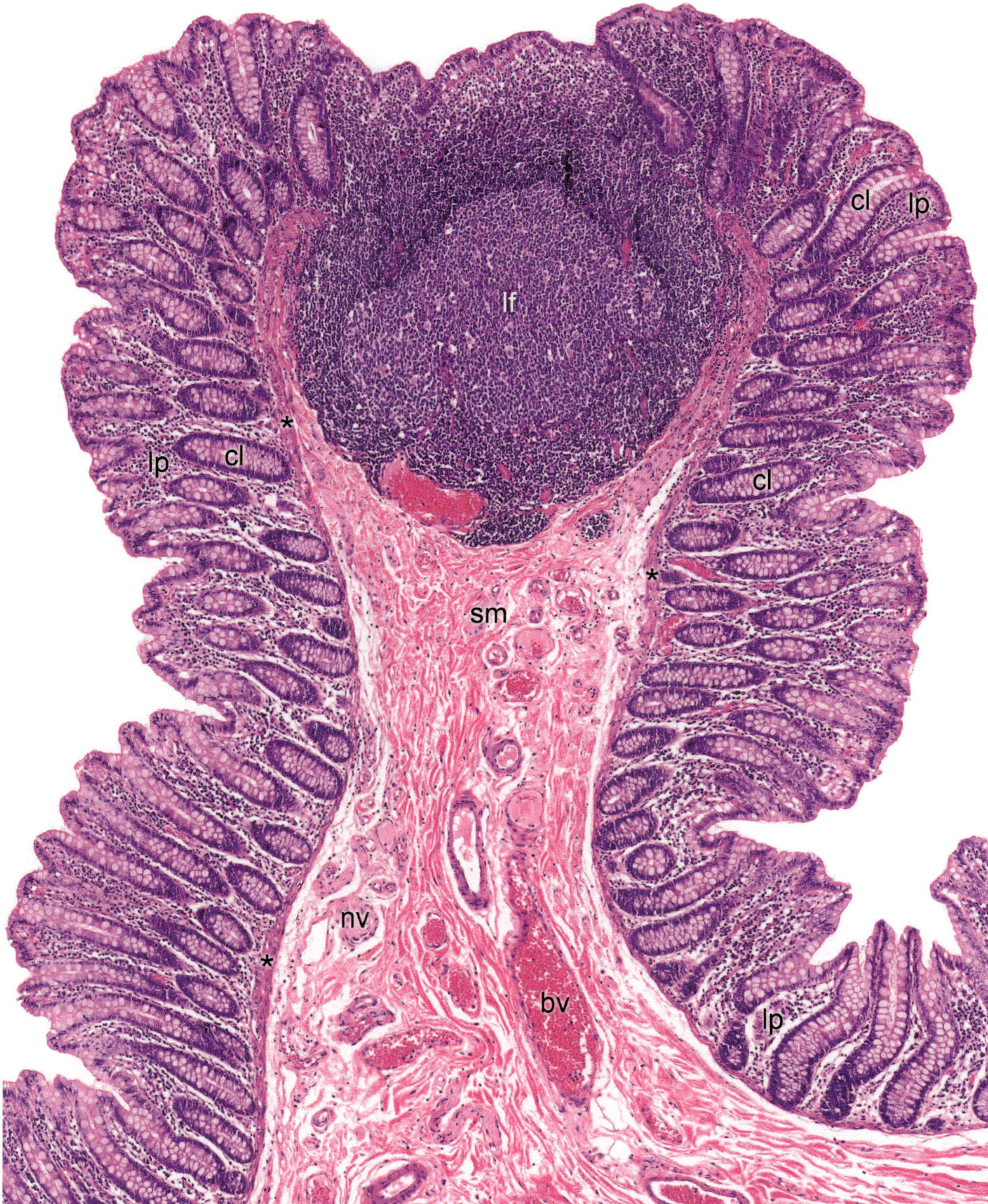

Figure 146. Transverse colon. Cross section of a circular fold with a submucosal core. Tunica mucosa of the colon lacks villi and it is lined with microvillated simple columnar epithelium with goblet cells. The epithelium extends into the crypts of Lieberkühn (cl) where goblet cells dominate. Lamina propria (lp) contains lymph follicles (lf) that often penetrate the muscularis mucosae (asterisks) and invade the tunica submucosa (sm) that is composed of dense irregular connective tissue with blood vessels (bv) and nerves (nv).

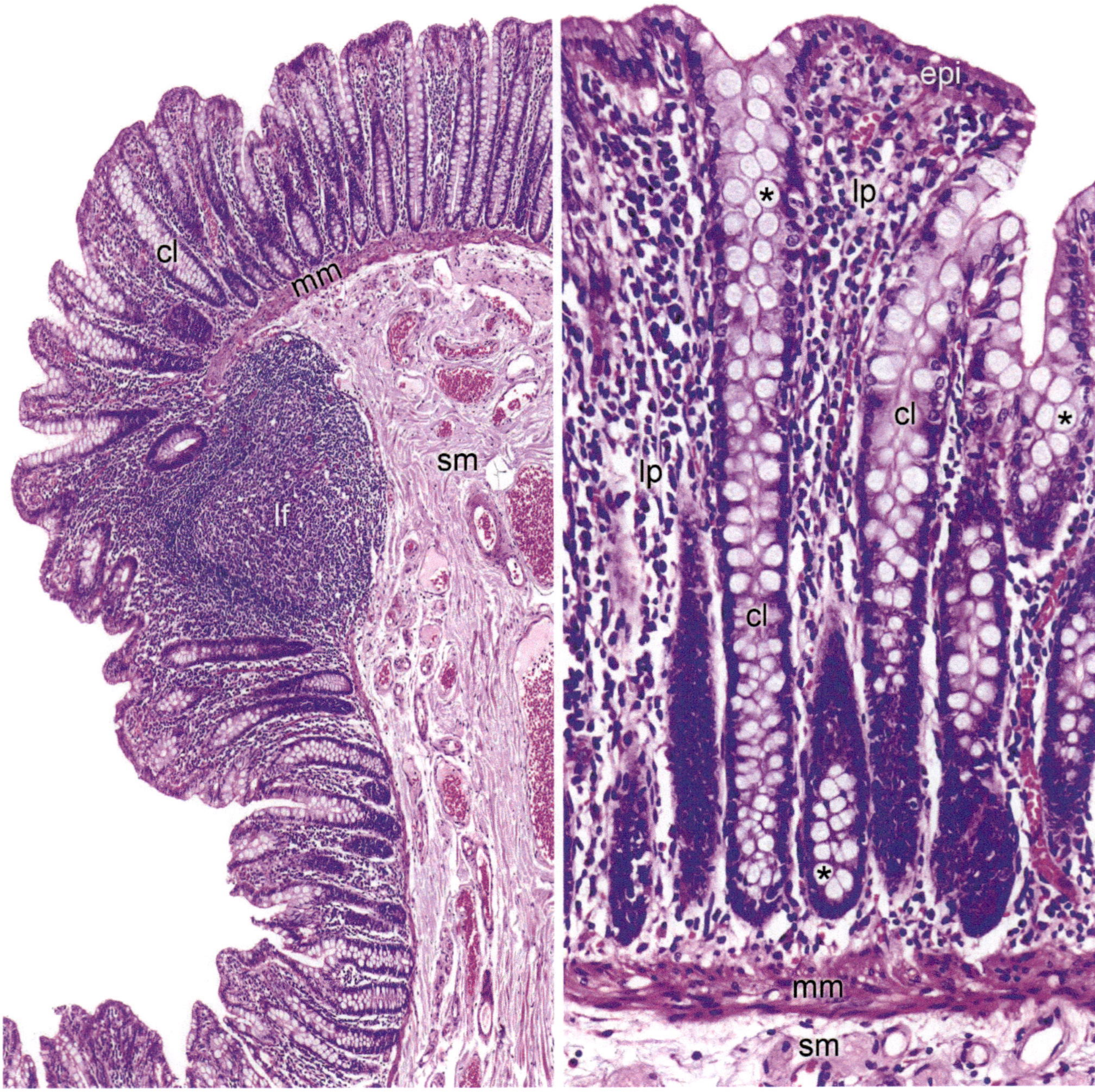

Figure 147. Transverse colon. Cross section of a circular fold (left) and the covering tunica mucosa with the intestinal glands (right). Tunica mucosa lacks villi in the large intestine and it is lined with microvillated simple columnar epithelium (epi) with goblet cells (asterisks). The epithelium extends into the crypts of Lieberkühn (cl) where goblet cells dominate. Lamina propria (lp) contains solitary lymph follicles (lf) that may penetrate the muscularis mucosae (mm) and invade the tunica submucosa (sm) that is composed of dense irregular connective tissue.

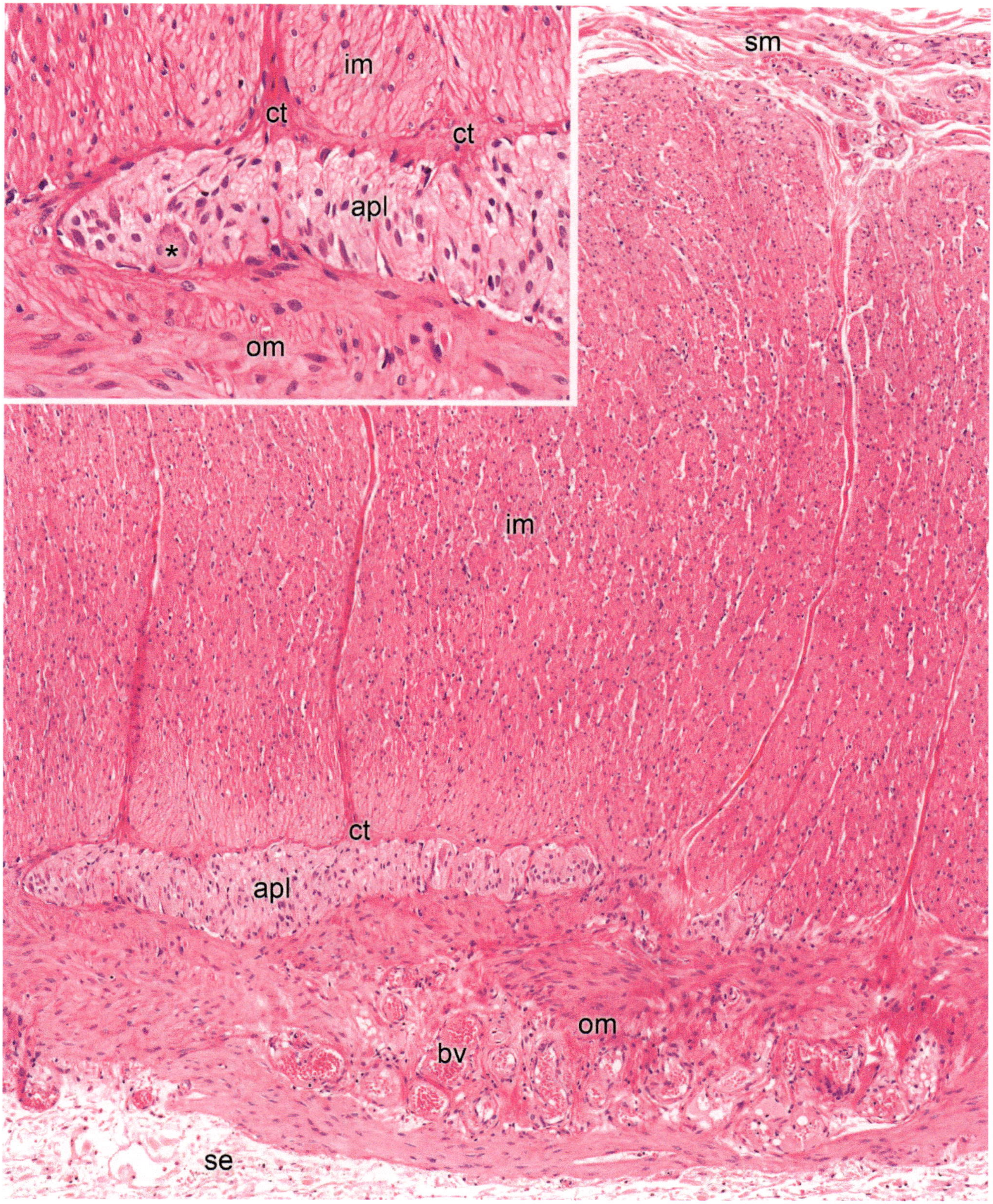

Figure 148. Transverse colon, tunica muscularis externa. Under the submucosa (sm), tunica muscularis externa is composed of inner circular (im) and outer longitudinal layers (om) of smooth mucle that contain islands of blood vessels (bv). Between the layers, the myenteric autonomic (Auerbach's) plexus (apl) contains nerve fibers and large ganglion cells (inset, asterisk) embedded into dense connective tissue envelopes that penetrate into the muscle layers. Transverse colon is covered by tunica serosa (se) while the ascending and descending colons are only partially covered by serous membrane on the anterior surface.

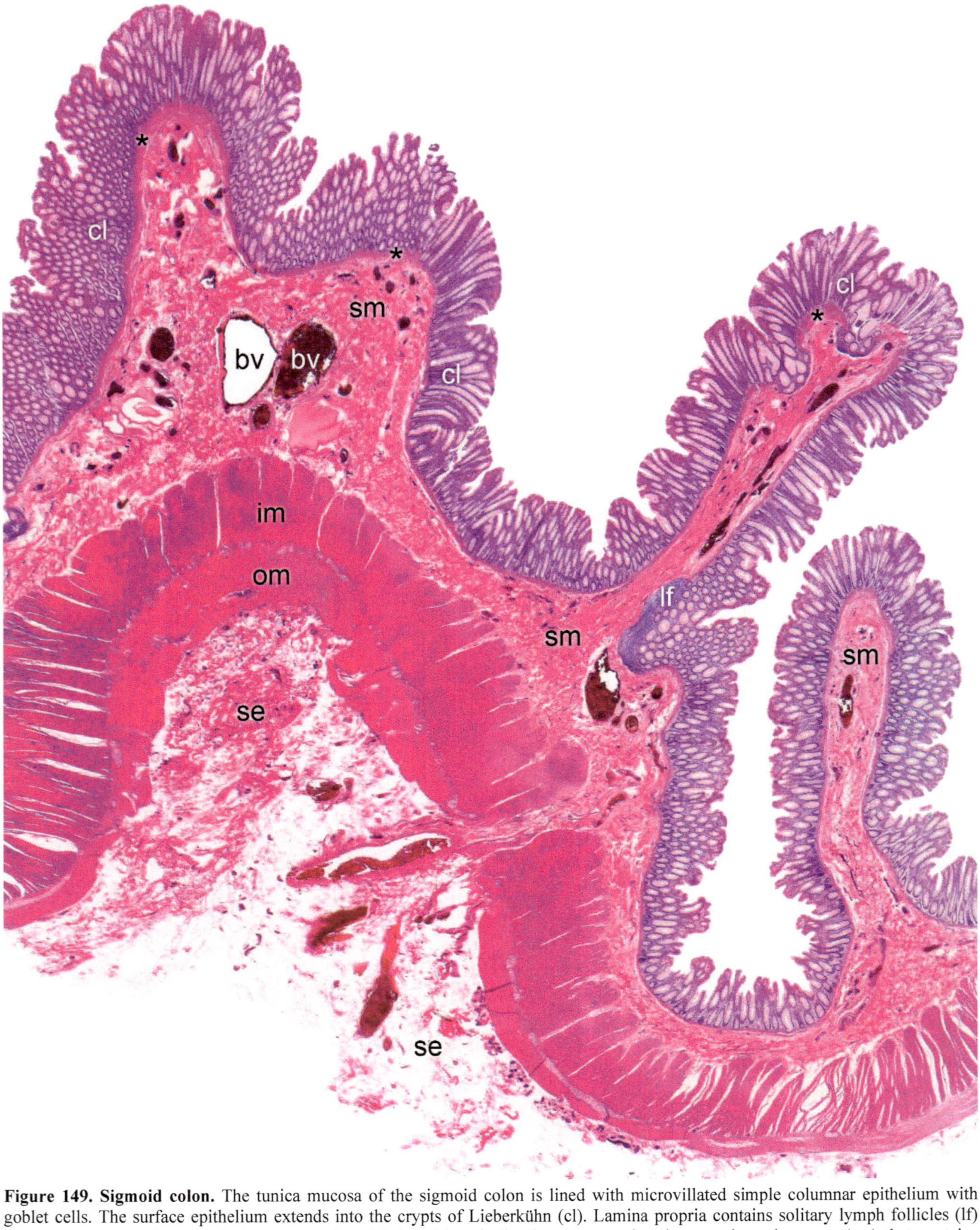

Figure 149. Sigmoid colon. The tunica mucosa of the sigmoid colon is lined with microvillated simple columnar epithelium with goblet cells. The surface epithelium extends into the crypts of Lieberkühn (cl). Lamina propria contains solitary lymph follicles (lf) that may penetrate the muscularis mucosae (asterisks) and invade the dense connective tissue tunica submucosa (sm) that contains large vessels (bv). Tunica muscularis externa is composed of inner circular (im) and outer longitudinal layers (om). Transverse colon is intraperitoneal, therefore, it is covered by tunica serosa.

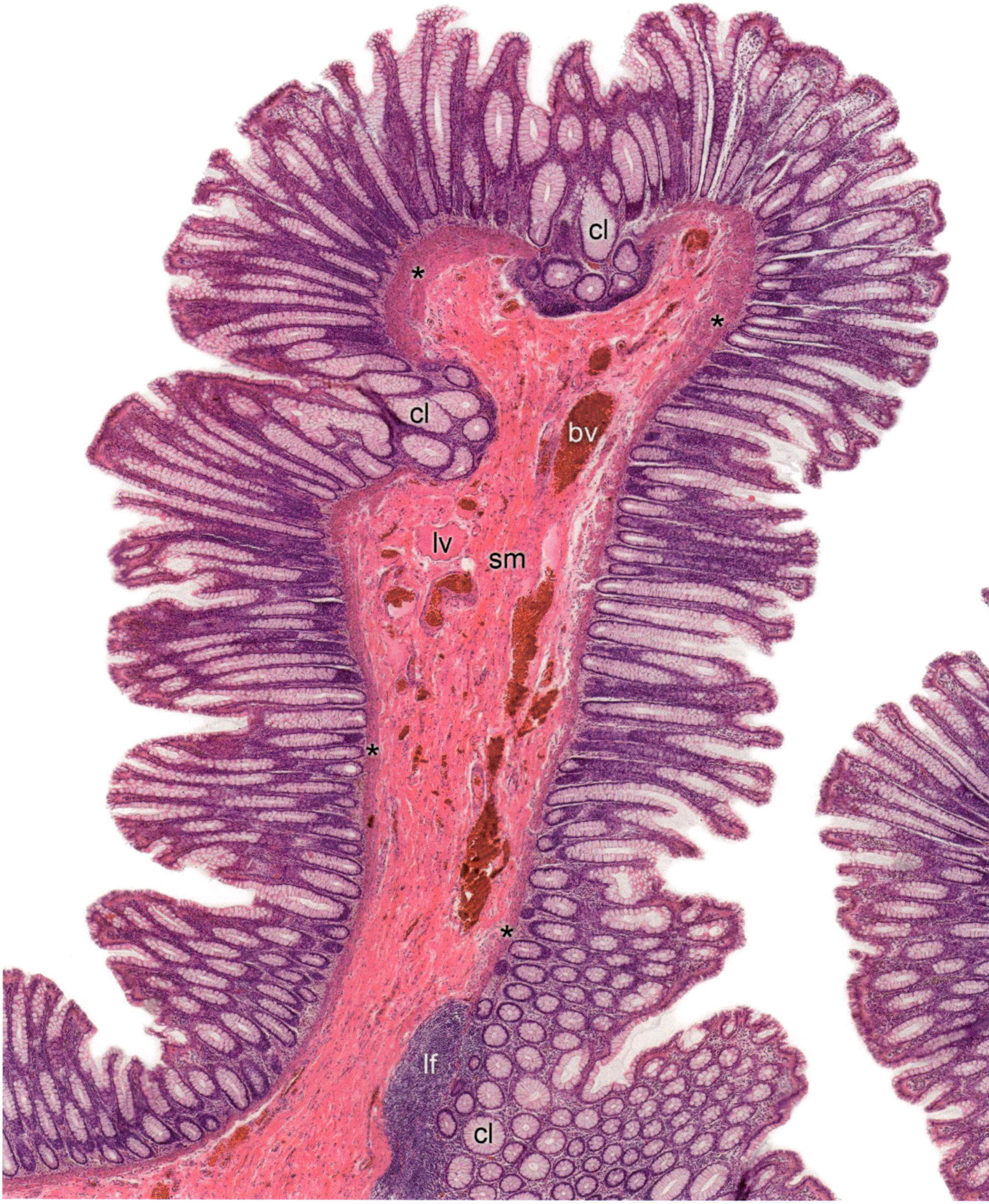

Figure 150. Sigmoid colon. Cross section of a circular fold (plica circularis) with a submucosal core. Tunica mucosa lacks villi in the large intestine including the sigmoid colon and it is lined with microvillated simple columnar epithelium with goblet cells. The epithelium extends into the crypts of Lieberkühn (cl) where goblet cells dominate. In addition to the intestinal glands, lamina propria contains lymph follicles (lf) that often penetrate the muscularis mucosae (asterisks) and invade the tunica submucosa (sm) that is composed of dense irregular connective tissue with large blood (bv) as well as lymph vessels (lv).

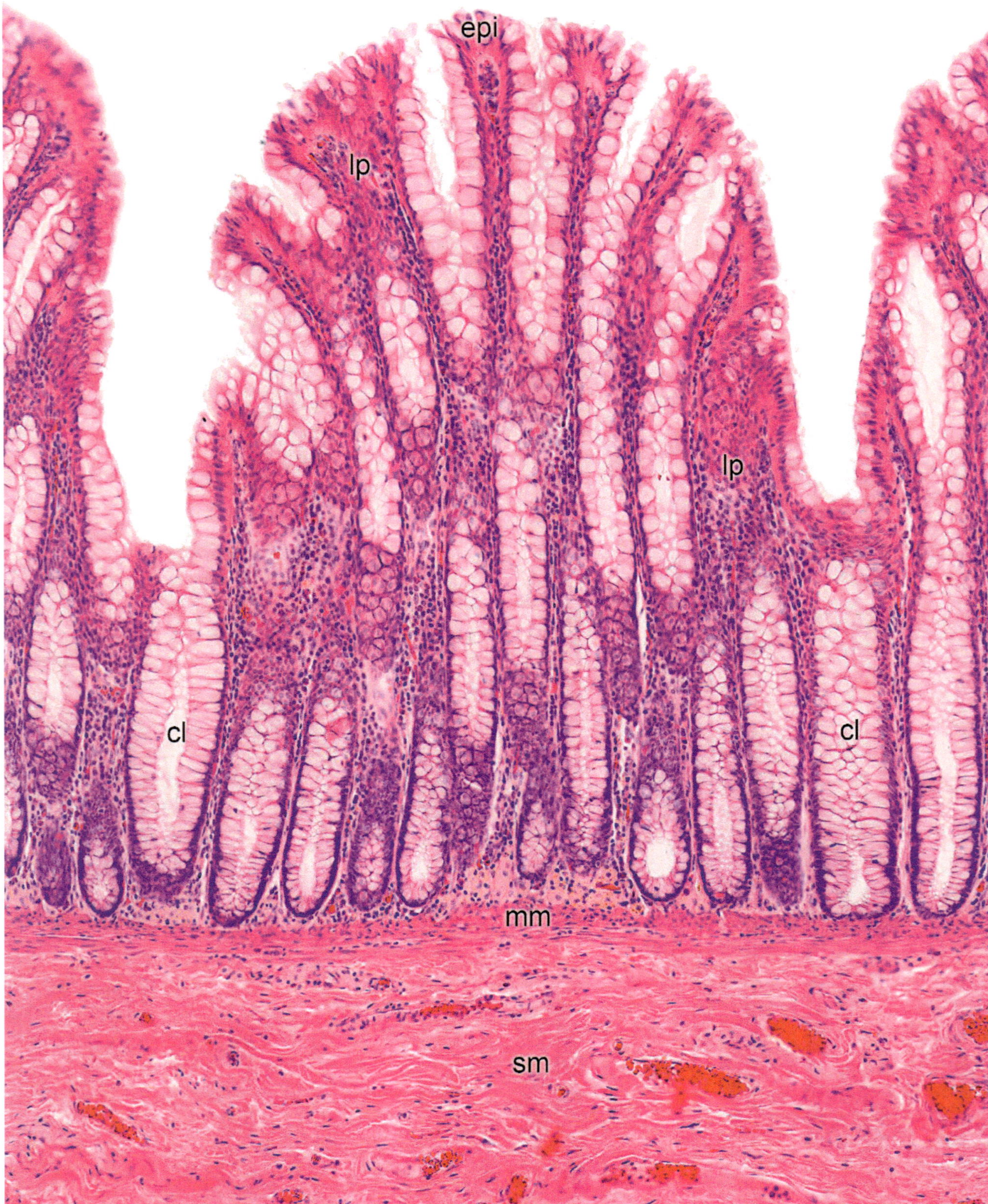

Figure 151. Sigmoid colon, mucosal glands. Tunica mucosa lacks villi in the large intestine and it is lined with microvillated simple columnar epithelium (epi) with abundant goblet cells that extend into the crypts of Lieberkühn (cl) where goblet cells dominate. These intestinal glands are surrounded by the loose connective tissue lamina propria (lp). Tunica muscularis mucosae (mm) is formed by circularly arranged smooth muscle cells. Tunica submucosa (sm) is composed of dense irregular connective tissue.

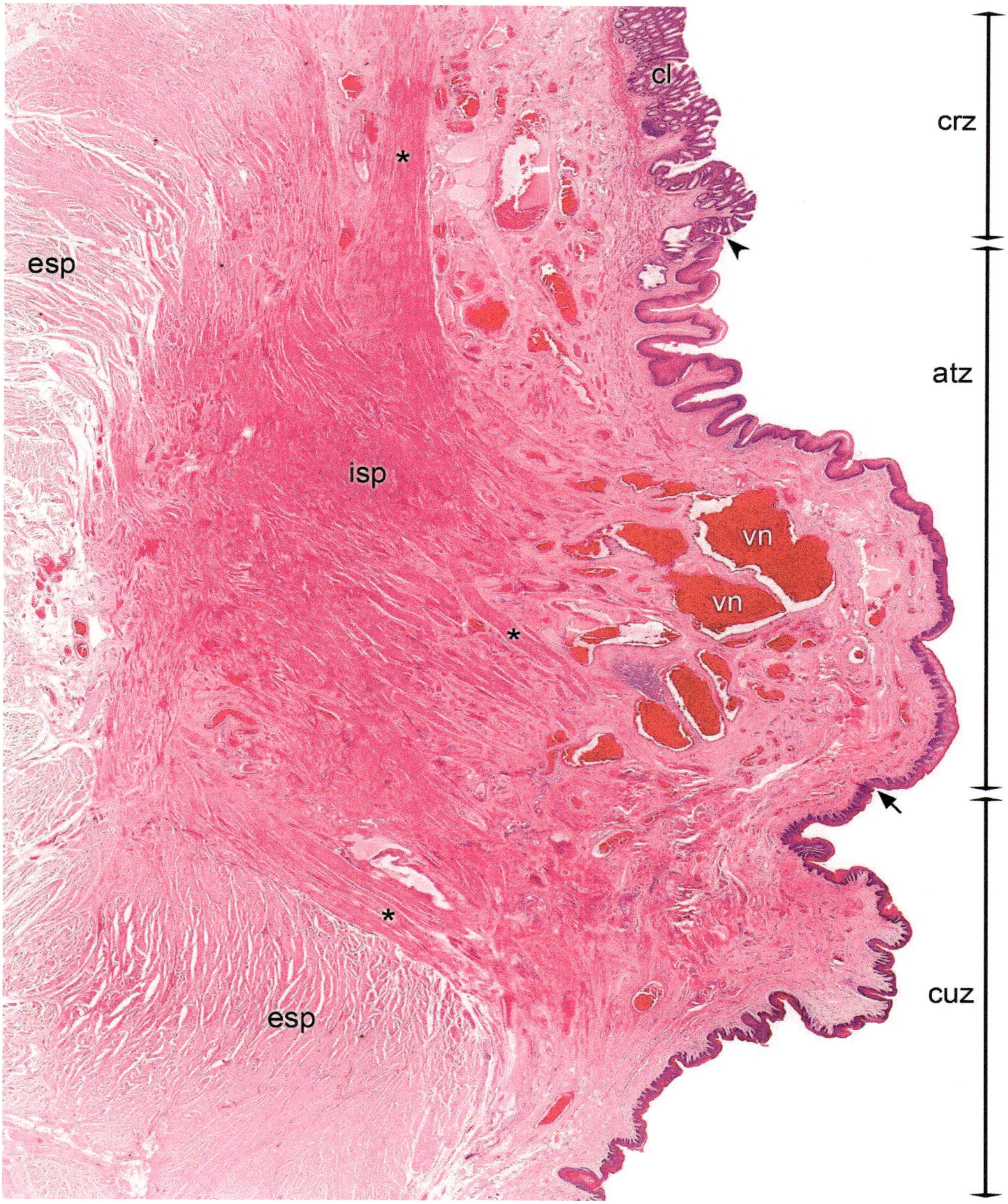

Figure 152. Anal canal. The proximal colorectal zone (crz) is covered with the mucosa of the large intestine with crypts of Lieberkühn (cl). The columnar epithelium of the colorectal zone is abruptly replaced by stratified cuboidal epithelium at the linea anorectalis (arrowhead) and then by stratified squamous nonkeratinized or parakeratinized epithelium, lining the anal transitional zone (atz). This zone transitions into the cutaneous zone (cuz) that is covered with skin; the border between them is delineated by the linea anocutanea (arrow). Anal transitional zone contains veins of the hemorrhoidal plexus (vn). The internal anal sphincter (isp) is surrounded and perforated by longitudinal smooth muscle layers (asterisks) and bordered by the skeletal muscle of the external anal sphincter (esp).

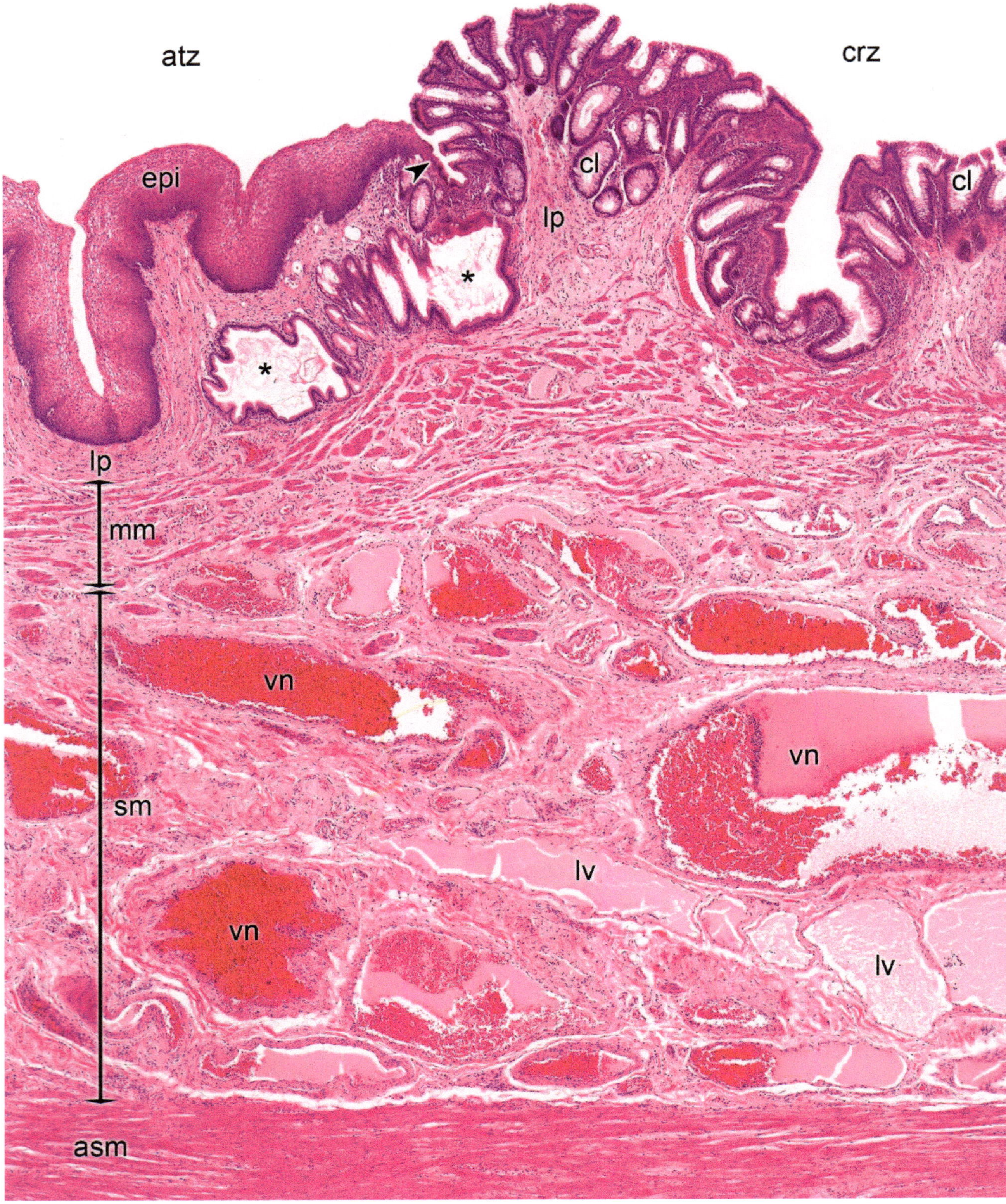

Figure 153. Anal canal, transition between the colorectal and anal transitional zones. The proximal colorectal zone (crz) is covered with the mucosa of the large intestine with crypts of Lieberkühn (cl). The columnar epithelium of the colorectal zone is abruptly replaced by stratified cuboidal epithelium at the linea anorectalis (arrowhead) and then by stratified squamous nonkeratinized epithelium (epi) lining the anal transitional zone (atz). Lamina propria (lp) contains invaginations lined with the surface epithelium (asterisks). Under the tunica muscularis mucosae (mm), tunica submucosa (sm) contains lymph vessels (lv) as well as large veins of the hemorrhoidal plexus (vn) that rest on longitudinally arranged smooth muscle layers (asterisks) of the anal submucosal muscle of Treitz (asm).

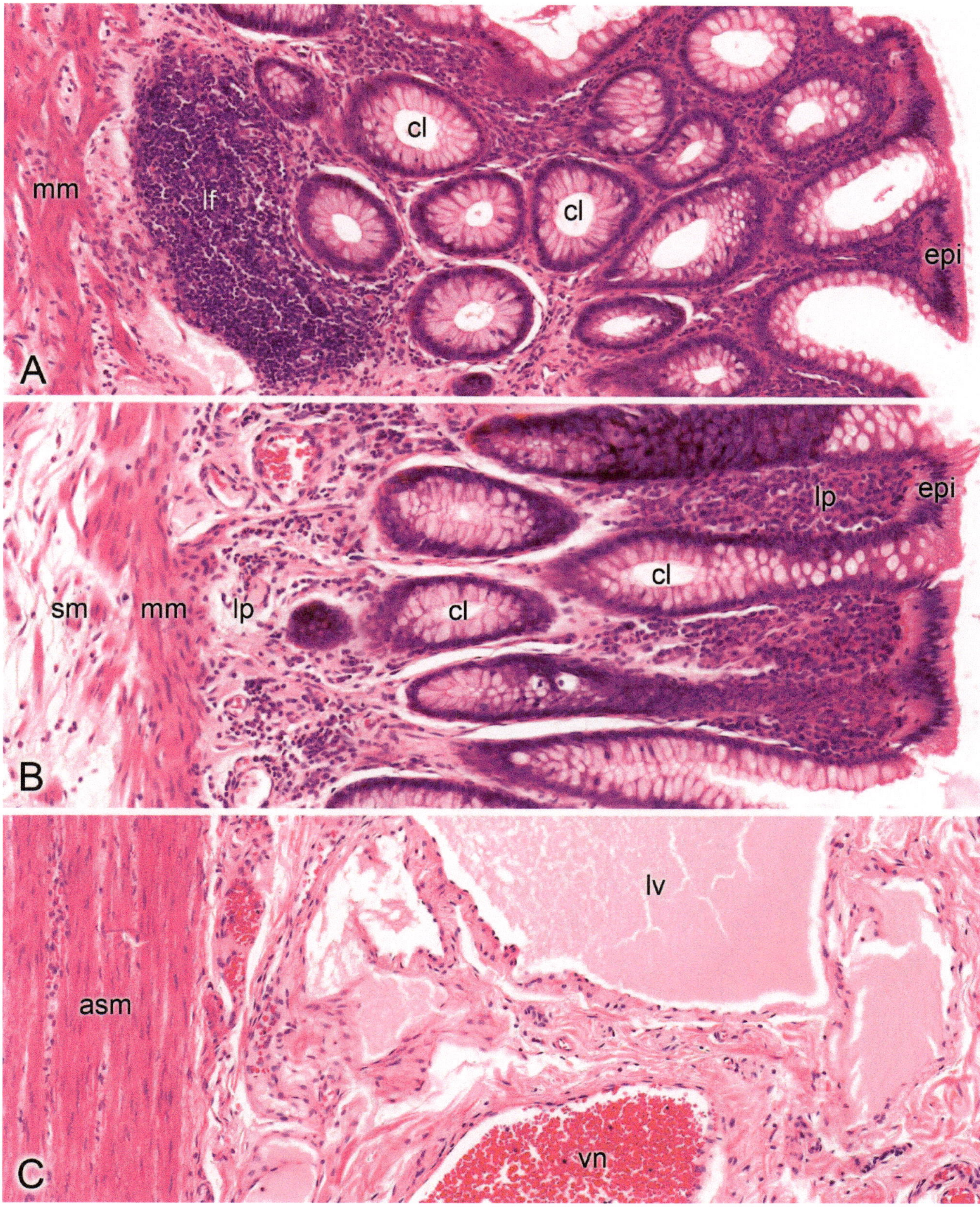

Figure 154. Anal canal, colorectal zone. (A,B) The colorectal zone is covered with the mucosa of the large intestine with crypts of Lieberkühn (cl) and simple columnar epithelium with goblet cells (epi). Lamina propria often contains lymphatic follicles that may penetrate the muscularis mucosae (mm) and the submucosa (sm). (C) The lateral part of the colorectal zone contains lymph vessels (lv) as well as large veins of the hemorrhoidal plexus (vn) that rest on longitudinally arranged smooth muscle layers of the anal submucosal muscle of Treitz (asm).

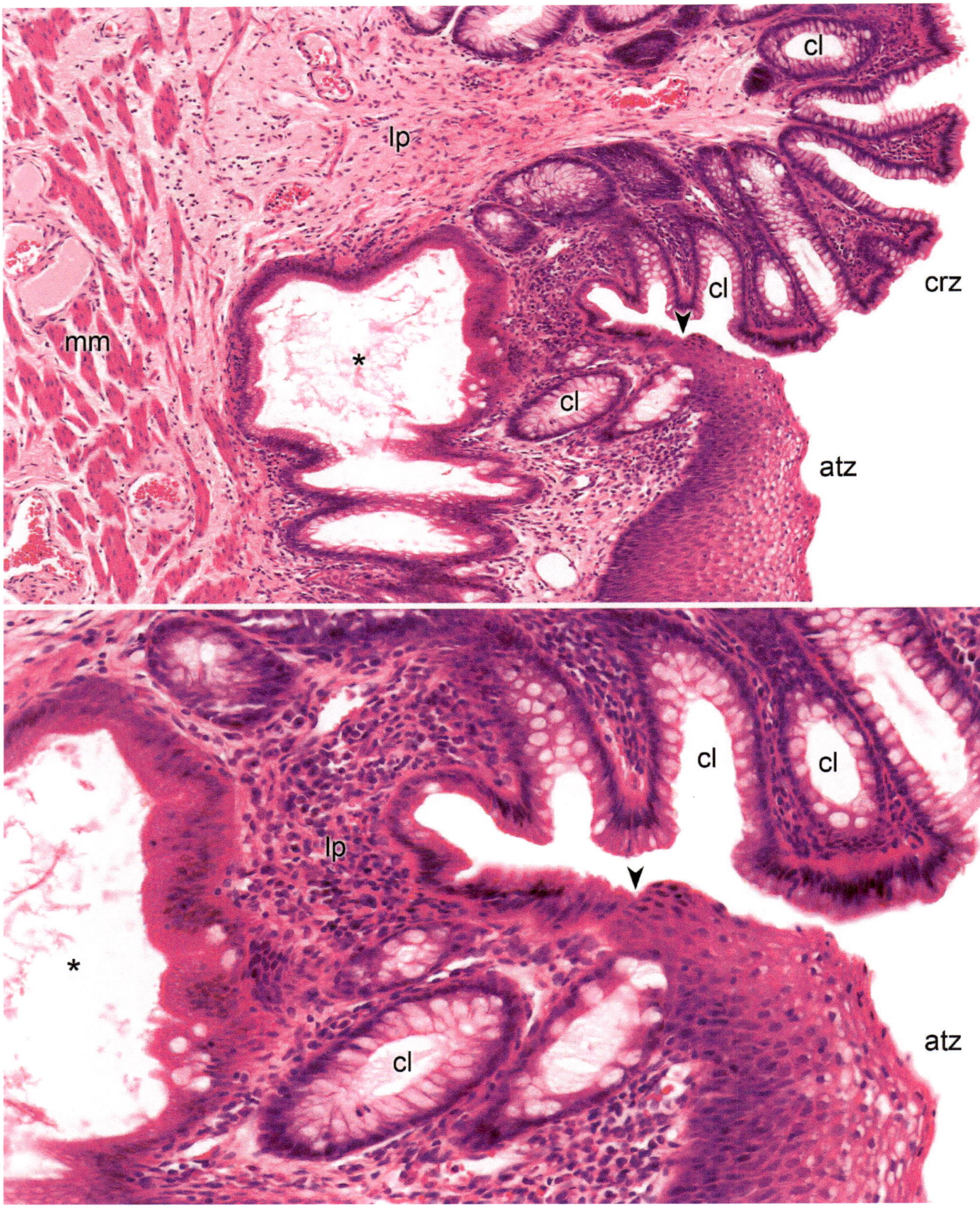

Figure 155. Anal canal, linea anorectalis. Low magnification (top) and high magnification (bottom) of the epithelial transition. The columnar epithelium of the colorectal zone is abruptly replaced by stratified cuboidal epithelium at the linea anorectalis (arrowhead) that transitions into nonkeratinized stratified squamous epithelium, lining the anal transitional zone (atz). The proximal colorectal zone (crz) is covered with the mucosa of the large intestine with crypts of Lieberkühn (cl). Above the tunica muscularis mucosae (mm), lamina propria (lp) contains invaginations lined with the surface epithelium (asterisks).

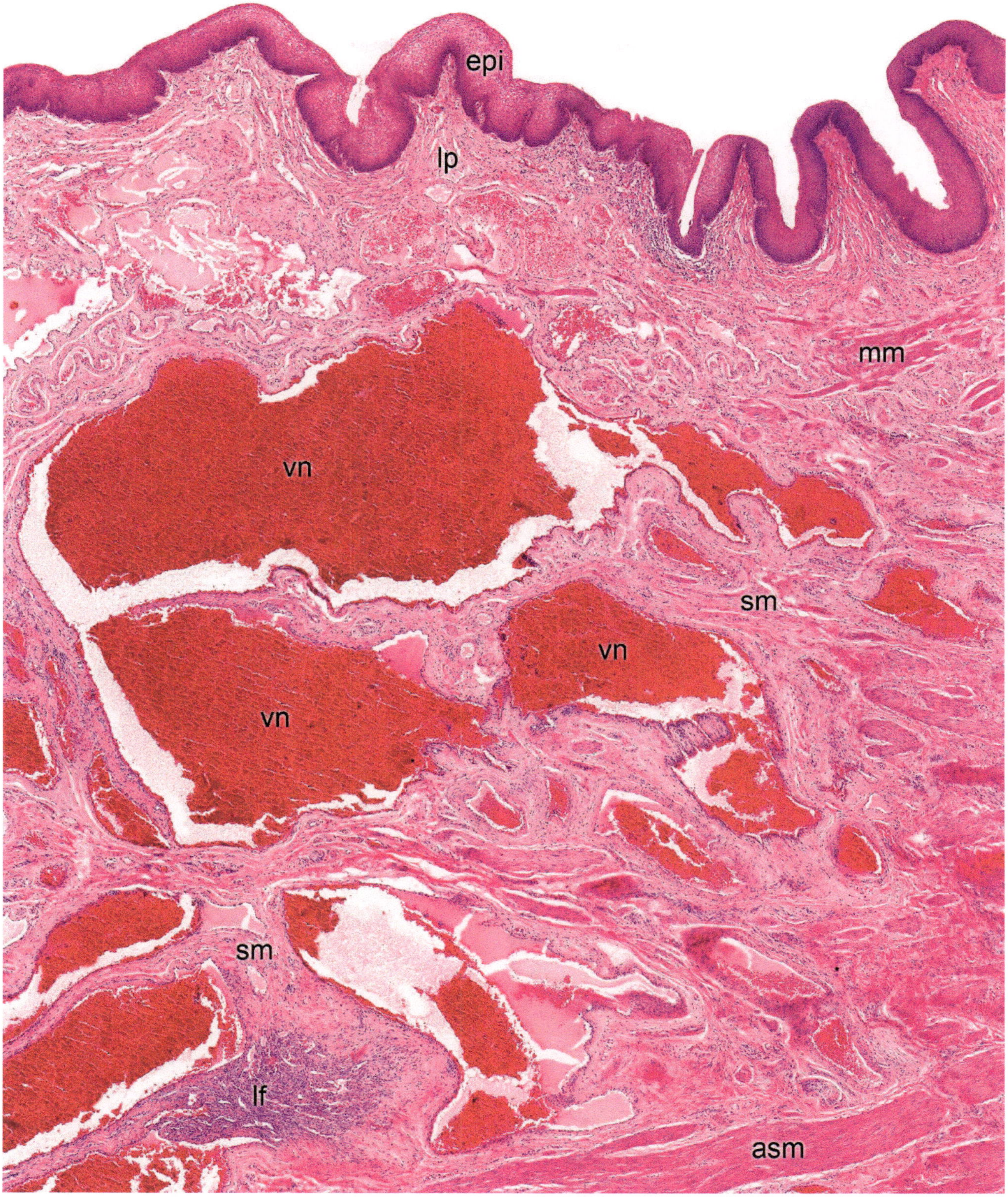

Figure 156. Anal canal, anal transitional zone. Anal transitional zone is located between the colorectal and cutaneous zones and it is mostly covered by stratified squamous nonkeratinized epithelium (epi) that shows signs of keratinization distally. Under the lamina propria (lp), tunica muscularis mucosae (mm) is often discontinuous due to the large protruding veins (vn) of the hemorrhoidal plexus occupying the submucosa (sm). Lymph follicles (lf) are often present between the vessels. Submucosa rests on the longitudinally arranged smooth muscle layers of the anal submucosal muscle of Treitz (asm).

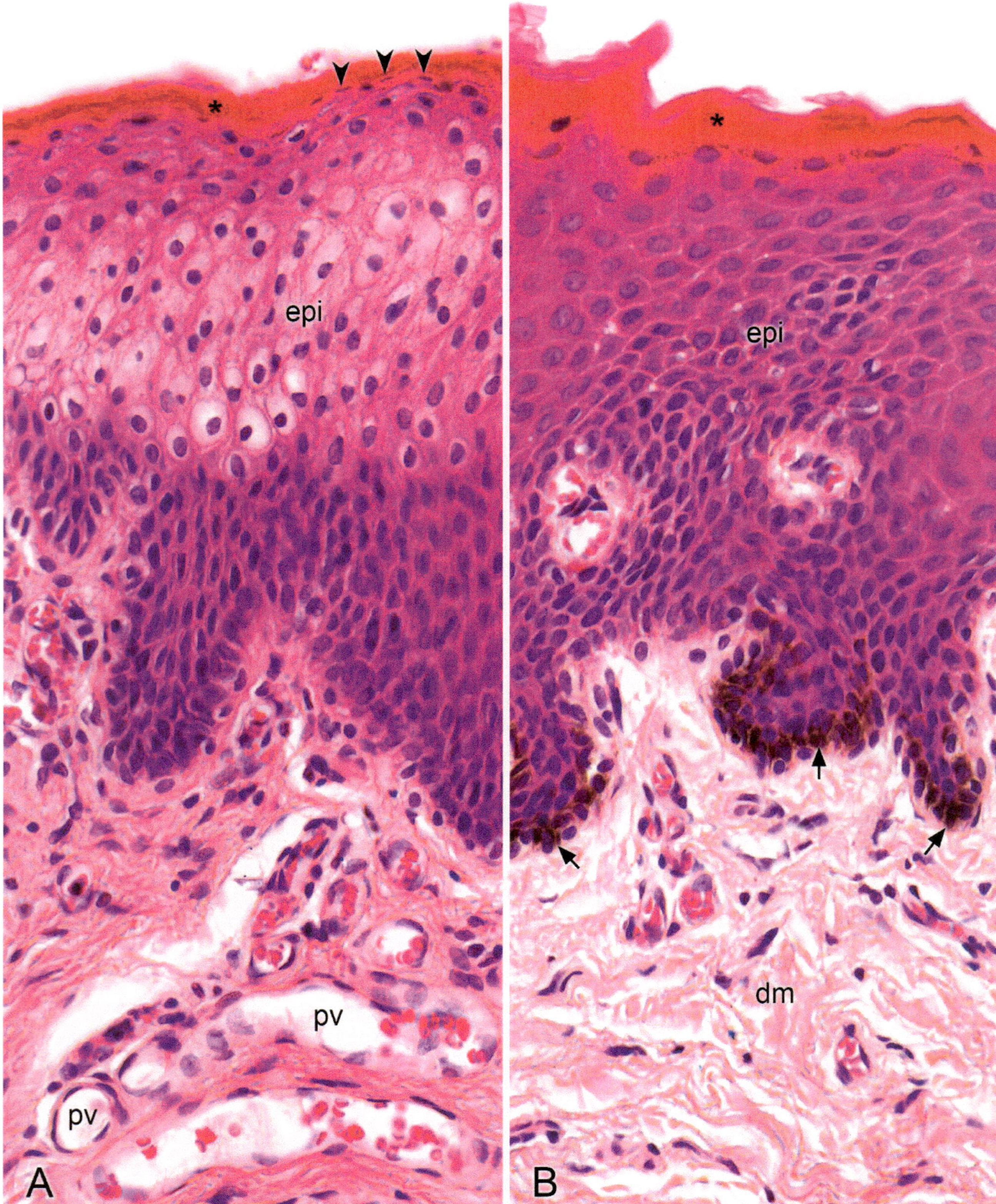

Figure 157. Anal canal. (A) Epithelium (epi) of the distal anal transitional zone shows signs of keratinization (parakeratinized epithelium) with occasional nuclei (arrowheads) remaining in the keratinized layer (asterisk). The connective tissue lamina propria under the epithelium is rich in collagen fibers and contains numerous postcapillary venules (pv). (B) The cutaneous zone is covered with keratinized stratified squamous epithelium (epi) with an eosinophilic stratum corneum (asterisk). Stratum basale contains numerous melanocytes with brown melanin pigmentation (arrows). The epithelium rests on the dense irregular connective tissue dermis (dm).

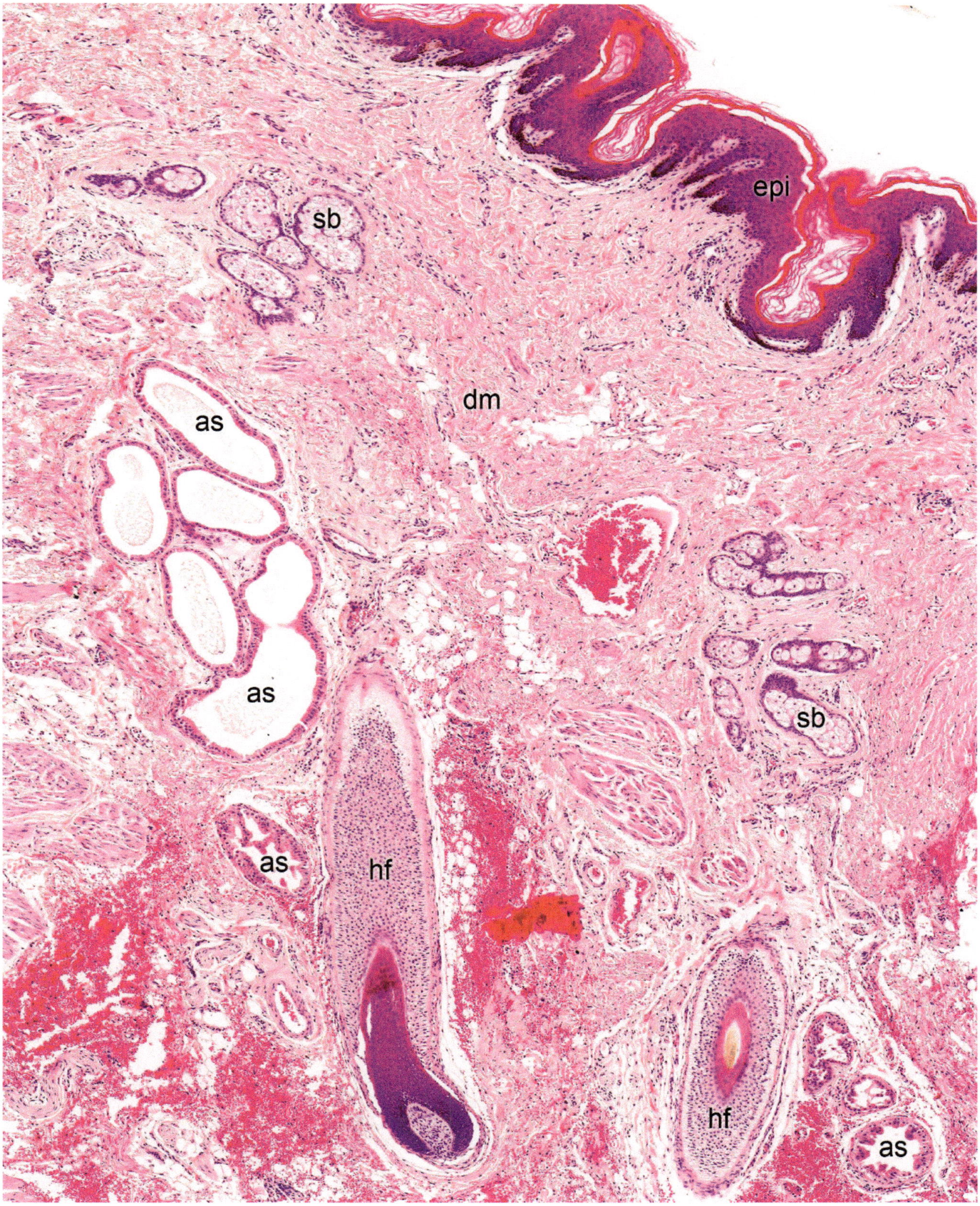

Figure 158. Anal canal, cutaneous zone. The skin of the **c**utaneous zone is covered with keratinized stratified squamous epithelium (epi). The dense irregular connective tissue dermis (dm) contains sebaceous glands that are associated with hair follicles. Apocrine sweat glands with distended lumen are abundant in the zona cutanea (circumanal glands).

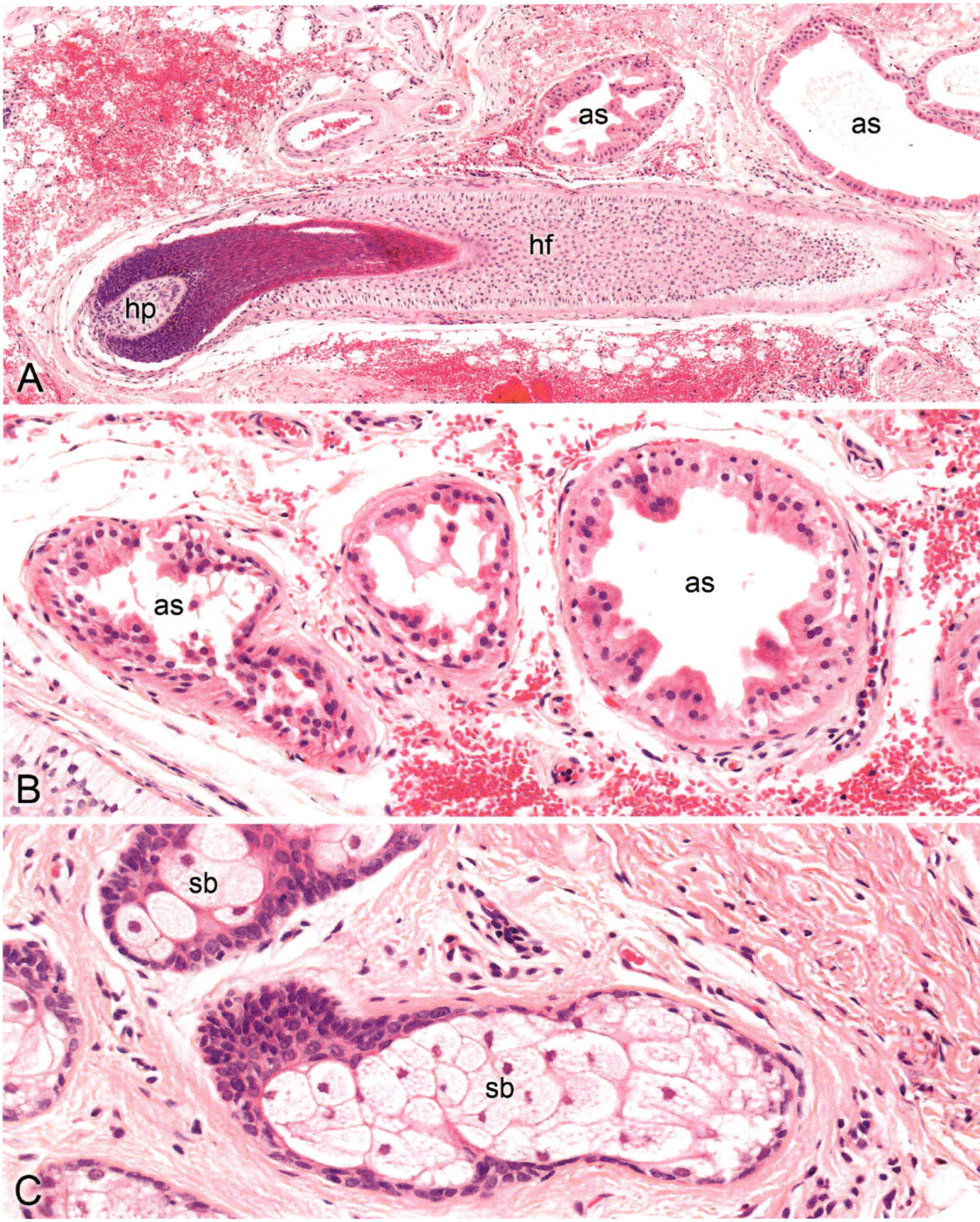

Figure 159. Cutaneous zone, dermis. (A) Dermis of the cutaneous zone contains hair follicles (hf) with basal connective tissue papillae (hp), and apocrine sweat glands (as). (B) Circumanal apocrine sweat glands (as) have distended lumen; although they exhibit the morphology of apocrine secretion, there is an indication that they secrete with merocrine mechanism. (C) Sebaceous glands (sb) are associated with hair follicles in the cutaneous zone of the anal canal.

Chapter 7

Digestive System III: Liver and Pancreas

Liver

The parenchyma of the liver is primarily composed of hepatocytes that are large polygonal cells containing centrally located spherical nucleus with prominent nucleolus (nucleoli) and clumps of heterochromatin. Hepatocytes have eosinophilic cytoplasm due to the large number of mitochondria and may be binucleated. In the parenchyma, hepatocytes form a single cell thick plates that are the structural building blocks of the *hepatic lobules*, the morphological units of the liver. Lobules are formed by anastomosing plates of hepatocytes radially arranged around a centrally located vessel, the central vein. Between the plates, hepatic sinusoids contain blood and scattered sinusoidal macrophages specific to the liver, the Kupffer cells. Sinusoids are covered by discontinuous endothelium and basal lamina as well as occasional Kupffer cells. Endothelial cells do not rest directly on the surface of the hepatocytes; instead, there is a perisinusoidal space between them (space of Disse) containing short microvilli extending from the luminal surface of the hepatocytes. Hepatic stellate cells (Ito cells) are wedged between the hepatocytes; these cells store vitamin A and they are activated during liver injury. When the liver is damaged, Ito cells lose their vitamin A content, and they play key a role in the production of extracellular matrix and collagen resulting in scar tissue formation.

In the human liver, lobules are not separated by connective tissue, and the sinusoids of the adjacent lobules communicate. Instead, the periphery of the hepatic lobules is demarcated by 3-5 sets of vessels and biliary ducts embedded into a loose connective tissue envelope. These structures are the portal canals (portal triad) and they contain branches of hepatic artery, portal vein, lymph vessels, autonomic nerve fibers and a bile duct that is covered with simple cuboidal epithelium surrounded by a basal lamina. Between the stroma of the portal canal and the adjacent hepatocytes a small space can be observed that is connected to the perisinusoidal space; this space of Mall is considered to be the initial site of lymph collection of the liver and eventually drains into the lymph vessels of the portal canal. Arterial blood from the hepatic artery, and venous blood from the portal vein is mixed in the hepatic sinusoids where it is modified by the hepatocytes and this mixed blood leaves through the central vein that eventually drains into the hepatic vein. Consequently, the periphery of the hepatic lobule is most exposed to toxins, while the center of the lobule receives the least oxygenated blood; therefore, it is most sensitive to hypoxia. Thus, ischemia due to various conditions may result in centrilobular necrosis, when hepatocytes around the central vein degenerate by accumulating lipid in the cytoplasm. These pathologic conditions outline the functional unit of the liver, the *liver acinus* that is described as zones of the adjacent hepatic lobules between their central veins. The center of the acinus receives blood from vessels of the portal canals and this blood is transported toward the center of the hepatic lobule that corresponds to the periphery of the liver acinus, the site of the centrilobular necrosis.

Apart from the key role in the metabolism, liver is also responsible for the production of bile. Corresponding grooves on the surfaces of adjacent hepatocytes form the bile canaliculi that start from the center of the hepatic lobule and eventually drain into the intrahepatic ductules (canals of Hering) located at the periphery of the lobules. Intrahepatic ductules lead into the interlobular bile ducts that are components of the portal canals; both are covered by simple cuboidal epithelium. Consequently, bile flow is centrifugal in contrast to the centripetal flow of the blood in the hepatic lobule. Larger bile ducts that drain the interlobular bile ducts are lined with simple columnar epithelium with microvilli and eventually lead to the hepatic ducts leaving the liver at the hilum. The secretory function of liver is represented by the *portal lobule* that contains the draining field of a single interlobular bile duct and therefore it involves areas of several adjacent hepatic lobules around a single portal canal.

http://doi.org/10.1016/B978-0-323-91891-6.50007-6,

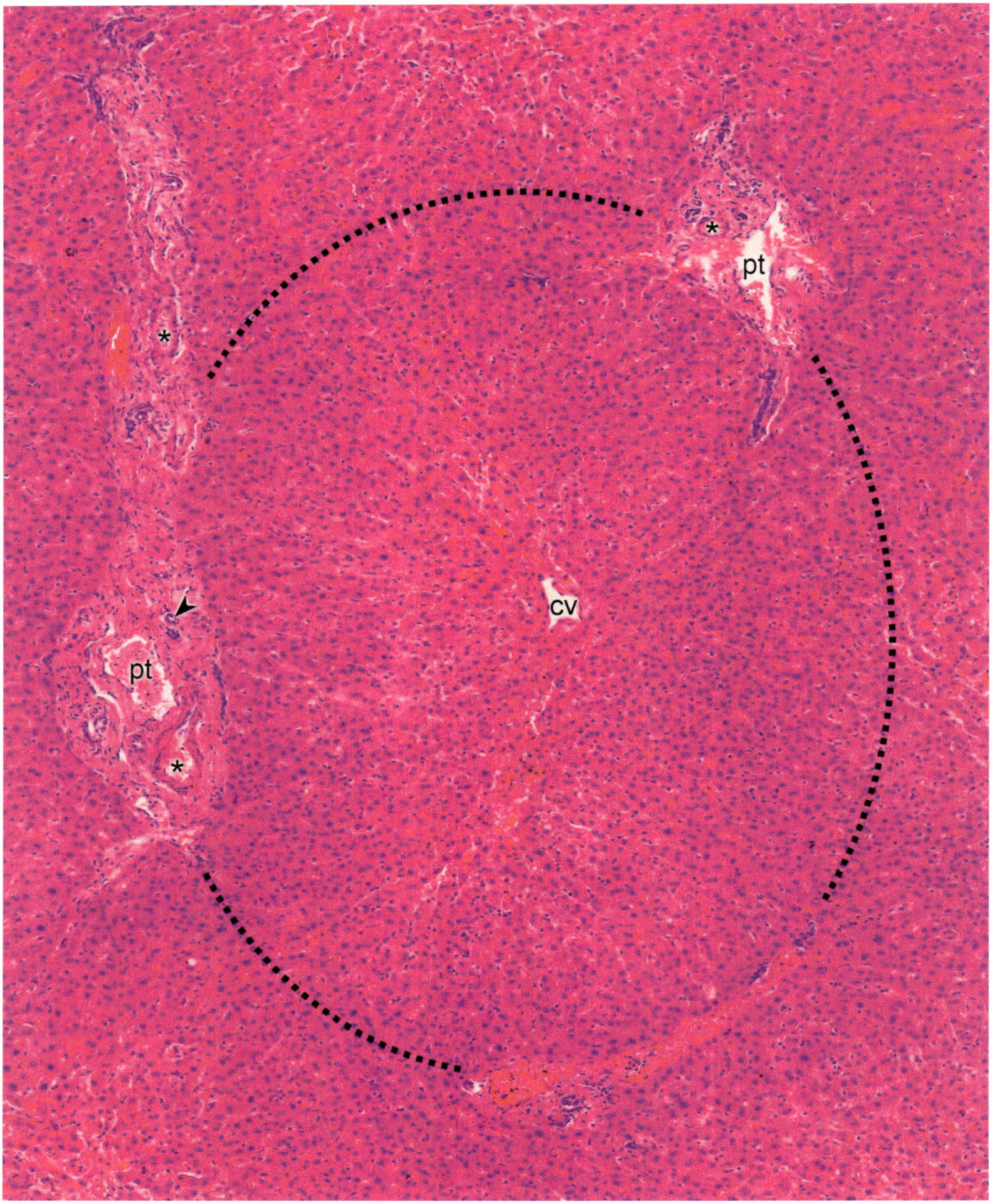

Figure 160. Liver, hepatic lobule (dotted line). Hepatic lobule is formed by anastomosing plates of hepatocytes arranged radially around the central vein (cv) that is a tributary of the hepatic vein. The space between the adjacent plates of hepatocytes is occupied by the hepatic sinusoids and the surrounding perisinusoidal space of Disse. At the periphery, hepatic lobules are bordered by the portal canals that are composed of the branches of the portal vein (pt), hepatic artery (asterisks), bile duct with simple cuboidal lining (arrowhead) as well as autonomic nerves and lymph vessels embedded into connective tissue.

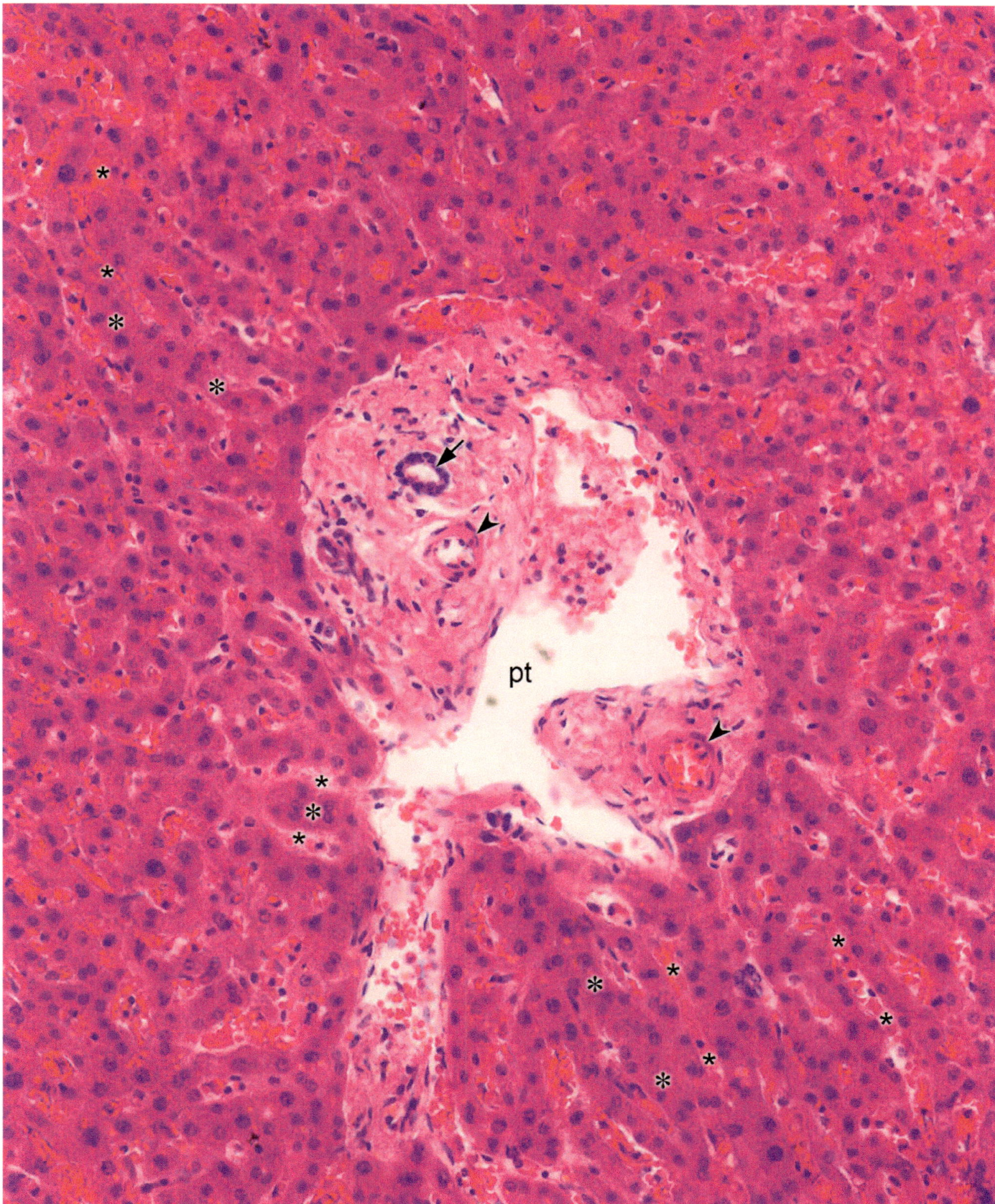

Figure 161. Liver, portal canal. Hepatic lobule is formed by anastomosing plates of hepatocytes (six-pointed asterisks) arranged radially around the central vein. The space between the adjacent plates of hepatocytes is occupied by the hepatic sinusoids (five-pointed asterisks) and the surrounding perisinusoidal space of Disse. At the periphery, hepatic lobules are bordered by the portal canals that are composed of the branches of the portal vein (pt), hepatic artery (arrowheads), bile duct with simple cuboidal lining (arrow) as well as autonomic nerves and lymph vessels embedded into connective tissue.

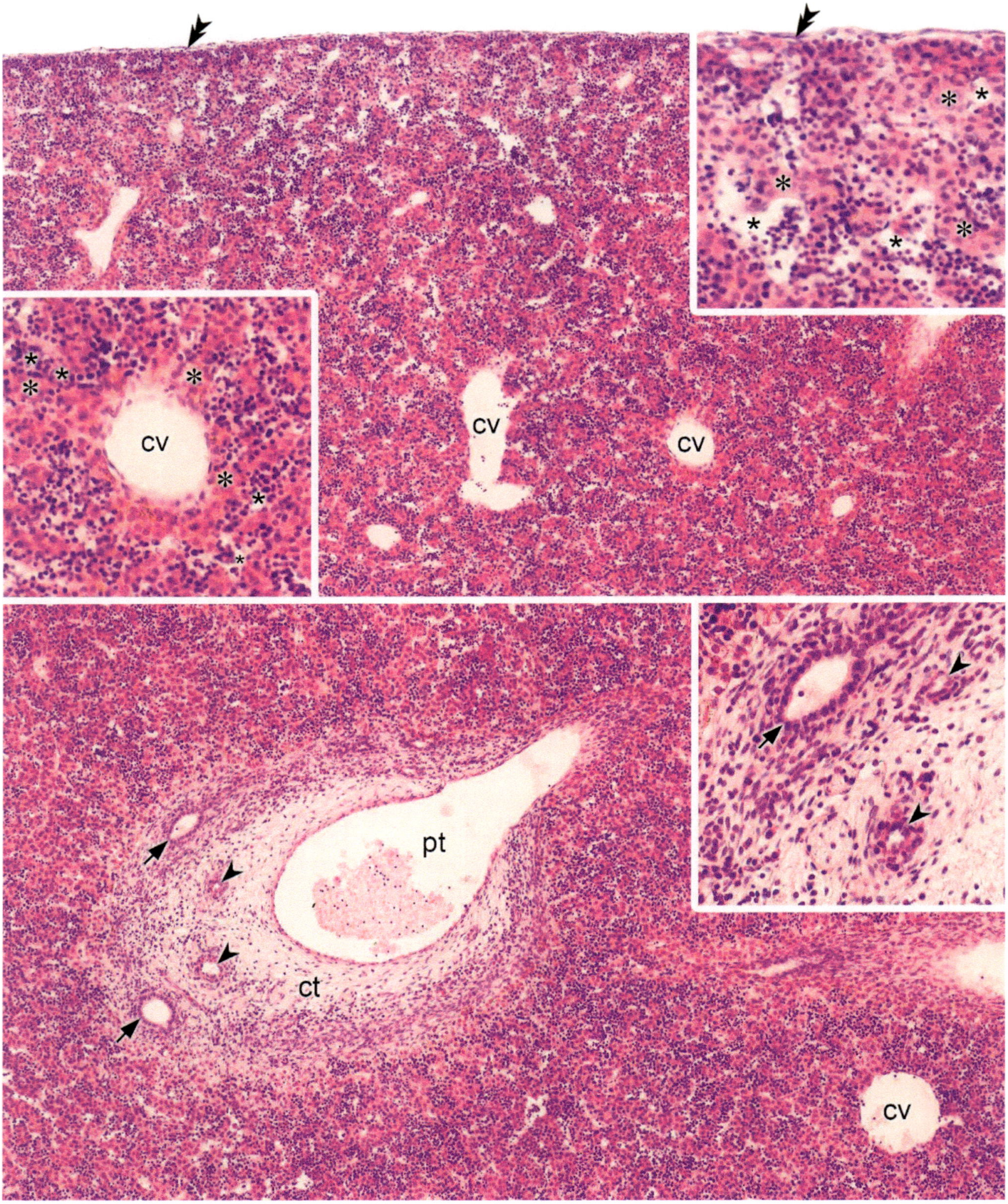

Figure 162. Liver, 3 month-old-fetus. Hepatic lobule is formed by anastomosing plates of hepatocytes (six-pointed asterisks) arranged radially around the central vein (cv) and hepatic sinusoids between the plates (five-pointed asterisks). Developmental forms of blood cells occupy the sinusoids in the fetus, since hemopoiesis takes place in the liver in most of the fetal life. At the periphery, hepatic lobules are bordered by the portal canals that are composed of the branches of the portal vein (pt), hepatic artery (arrowheads), bile duct (arrow) as well as autonomic nerves and lymph vessels embedded into connective tissue. Liver is has a connective tissue capsule with mesothelial coverage (double arrowheads).

Gallbladder

The mucous membrane of the gallbladder is lined with tall simple columnar epithelium with apical microvilli. Epithelium rests on the loose connective tissue lamina propria that is directly attached to the muscularis externa, since gallbladder lacks tunica muscularis mucosae as well as submucosa. Occasional mucous glands can be observed in the lamina propria but they are rare; epithelium, however, forms deep invaginations that often penetrate the muscularis externa (Rokitansky-Aschoff sinuses). These diverticula are covered with the surface epithelium, and they are not glands. Lamina propria also contains macrophages (histiocytes) that may accumulate cholesteryl esters and triglycerides in their cytoplasm (cholesterolosis). Muscularis externa contains randomly oriented smooth muscle bundles without distinct organization. Externally, the gallbladder is covered with connective tissue rich in vessels and autonomic nerve fibers; at the superior surface, where the bladder is attached to the liver this layer represents the tunica adventitia, while inferiorly the loose connective tissue has a mesothelial cell lining and corresponds to the peritoneal coverage.

Exocrine pancreas

The exocrine pancreas is a serous tubuloalveolar gland into which spherical structures, the Langerhans islets are embedded that represent the endocrine part. Pancreas is covered by a loose connective tissue capsule that sends septa into the parenchyma subdividing it into lobules. Septa contain occasional mucous glands that open into the ducts. The acini of the pancreas are composed of triangular cells; the base of the cells is basophilic due to the rER content while the luminal surface contains zymogen granules and therefore, it is acidophilic. Zymogen granules contain precursors of numerous enzymes that participate in the process of digestion.

The duct system of the pancreas is unique since the thin intercalated ducts lined with flattened cells start within the acinus forming the centroacinar cells with their initial segments. Intercalated ducts drain into the intralobular ducts covered with cuboidal epithelium while larger interlobular ducts have columnar epithelial lining with occasional goblet cells. Similar epithelium lines the main pancreatic and the accessory pancreatic ducts.

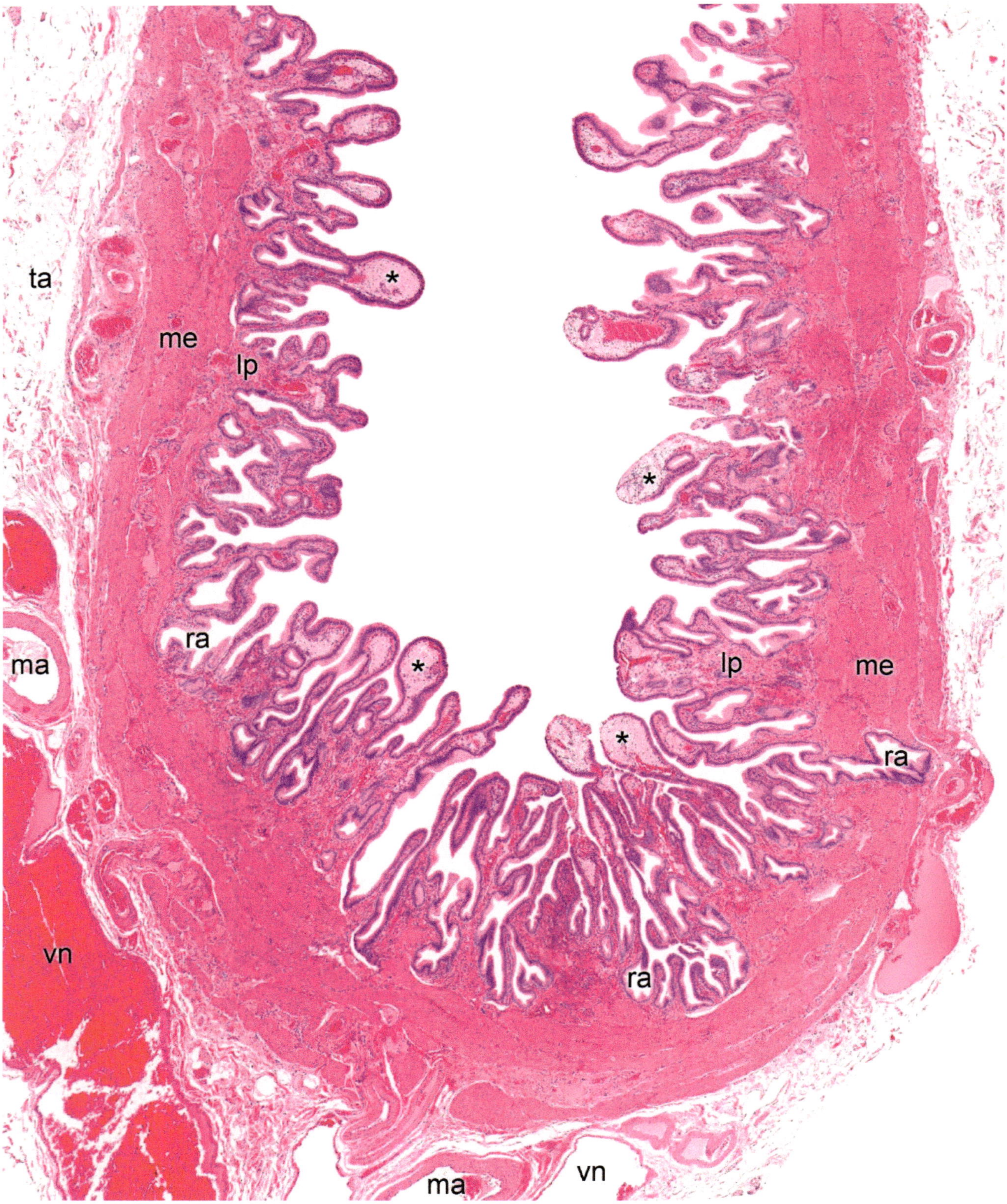

Figure 163. Gallbladder. Tunica mucosa of the gallbladder is composed of microvillated simple columnar epithelium that forms deep mucosal invaginations called Rokitansky-Aschoff sinuses (ra) often penetrating the muscularis externa. Tunica muscularis mucosae and submucosa is missing; lamina propria (lp) is directly attached to the tunica muscularis externa (me) that is composed of randomly oriented smooth muscle bundles. The surface of the gallbladder adjacent to the liver is covered by tunica adventitia (ta) containing muscular arteries (ma) and veins (vn), while the free inferior surface is lined with tunica serosa. Lamina propria contains macrophages (histiocytes) that may accumulate cholesteryl esters and triglycerides in their cytoplasm resulting in cholesterolosis (asterisks).

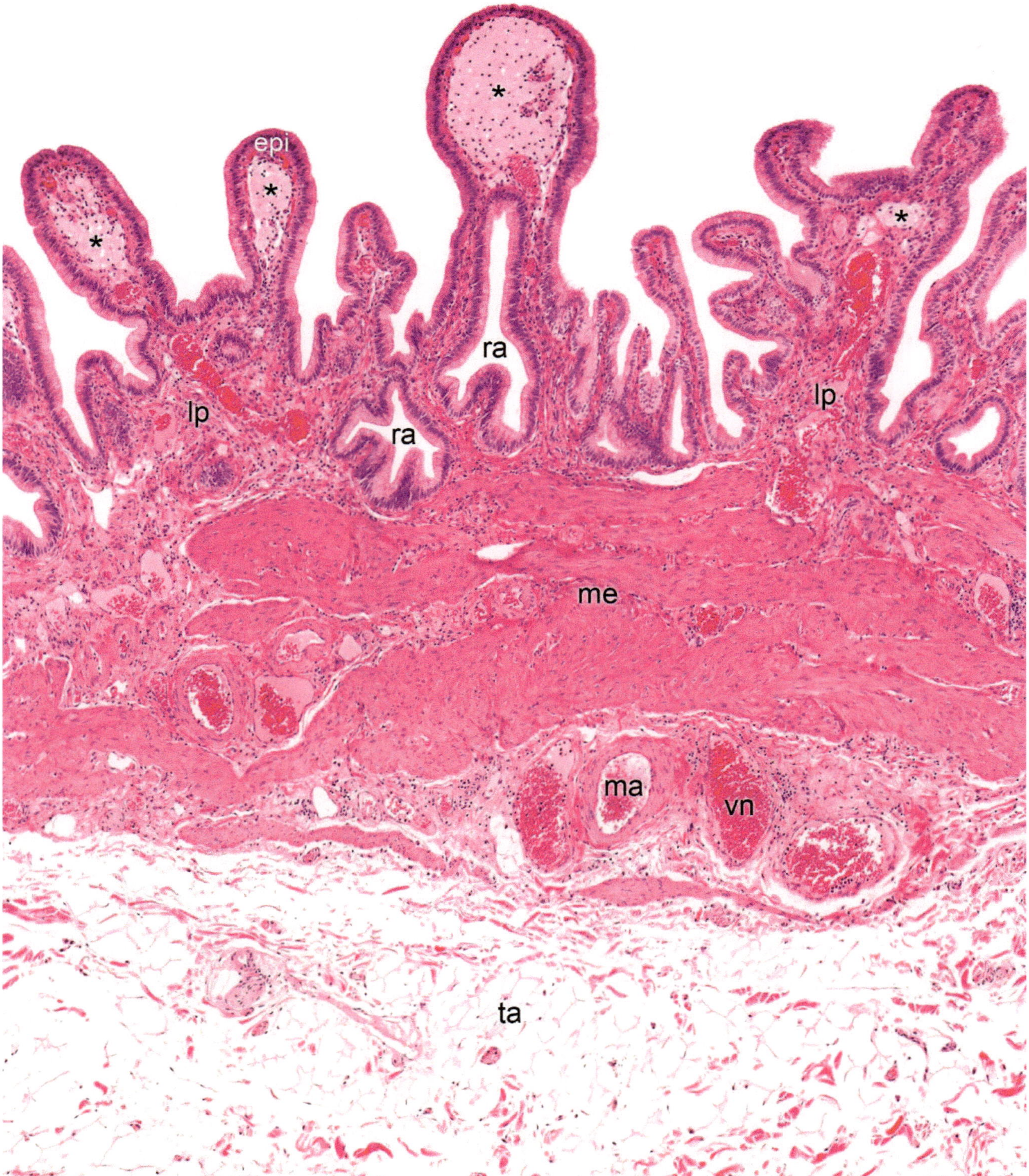

Figure 164. Gallbladder. Tunica mucosa of the gallbladder is composed of tall, microvillated simple columnar epithelium (epi) that also lines deep mucosal invaginations called Rokitansky-Aschoff sinuses (ra) often penetrating the muscularis externa. Tunica muscularis mucosae and tunica submucosa is missing; lamina propria (lp) is directly attached to the tunica muscularis externa (me) that is composed of randomly oriented smooth muscle bundles. The surface of the gallbladder adjacent to the liver is covered by tunica adventitia (ta) containing muscular arteries (ma) and veins (vn), while the inferior surface is lined with serosa. Lamina propria contains macrophages (histiocytes) that may accumulate cholesteryl esters and triglycerides in their cytoplasm (cholesterolosis; asterisks).

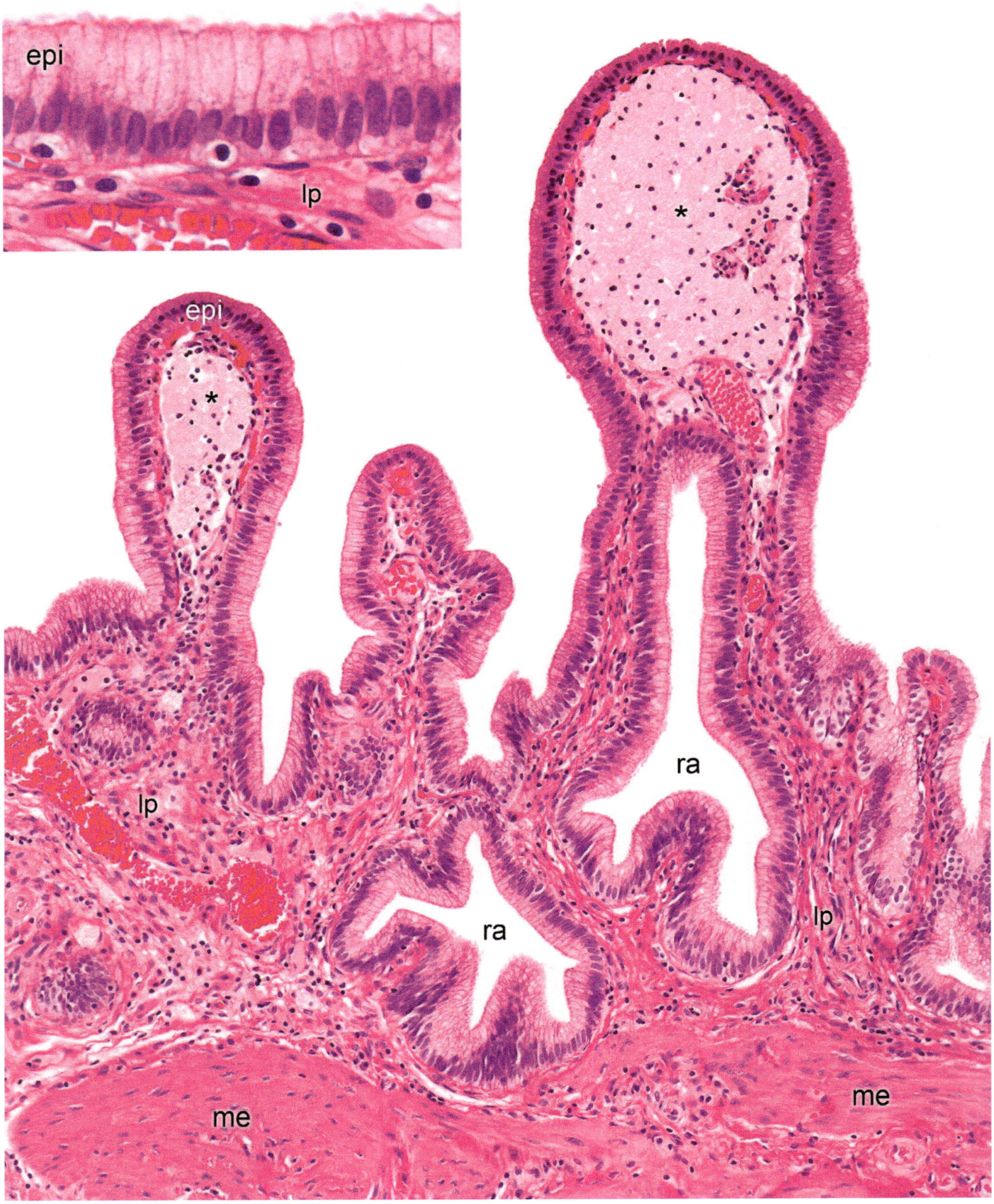

Figure 165. Gallbladder mucous membrane. The epithelium of the gallbladder (epi) is composed of microvillated columnar cells (inset) that also line deep mucosal invaginations called Rokitansky-Aschoff sinuses (ra) often penetrating the muscularis externa. Tunica muscularis mucosae and tunica submucosa is missing; lamina propria (lp) is directly attached to the tunica muscularis externa (me) that is composed of randomly oriented smooth muscle bundles. Macrophages in the lamina propria (histiocytes) may accumulate cholesteryl esters and triglycerides in their cytoplasm resulting in cholesterolosis (asterisks).

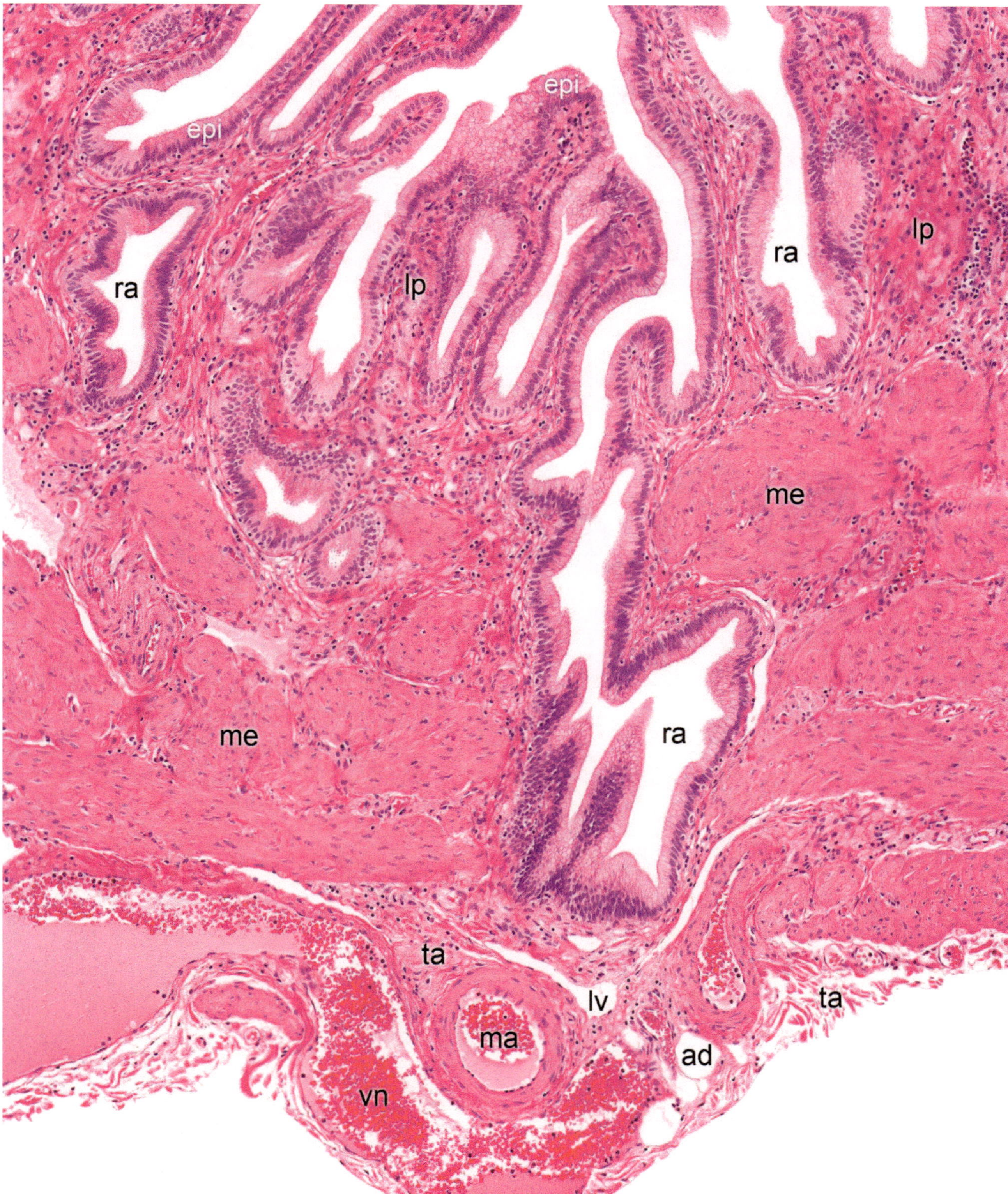

Figure 166. Gallbladder. The epithelium of the gallbladder (epi) is composed of microvillated columnar cells that also line deep mucosal invaginations called Rokitansky-Aschoff sinuses (ra) often penetrating the muscularis externa. Tunica muscularis mucosae and tunica submucosa is missing; lamina propria (lp) is directly attached to the tunica muscularis externa (me) that is composed of randomly oriented smooth muscle bundles. The surface of the gallbladder adjacent to the liver is covered by tunica adventitia (ta) containing muscular arteries (ma) and veins (vn), while the free inferior surface is lined with tunica serosa.

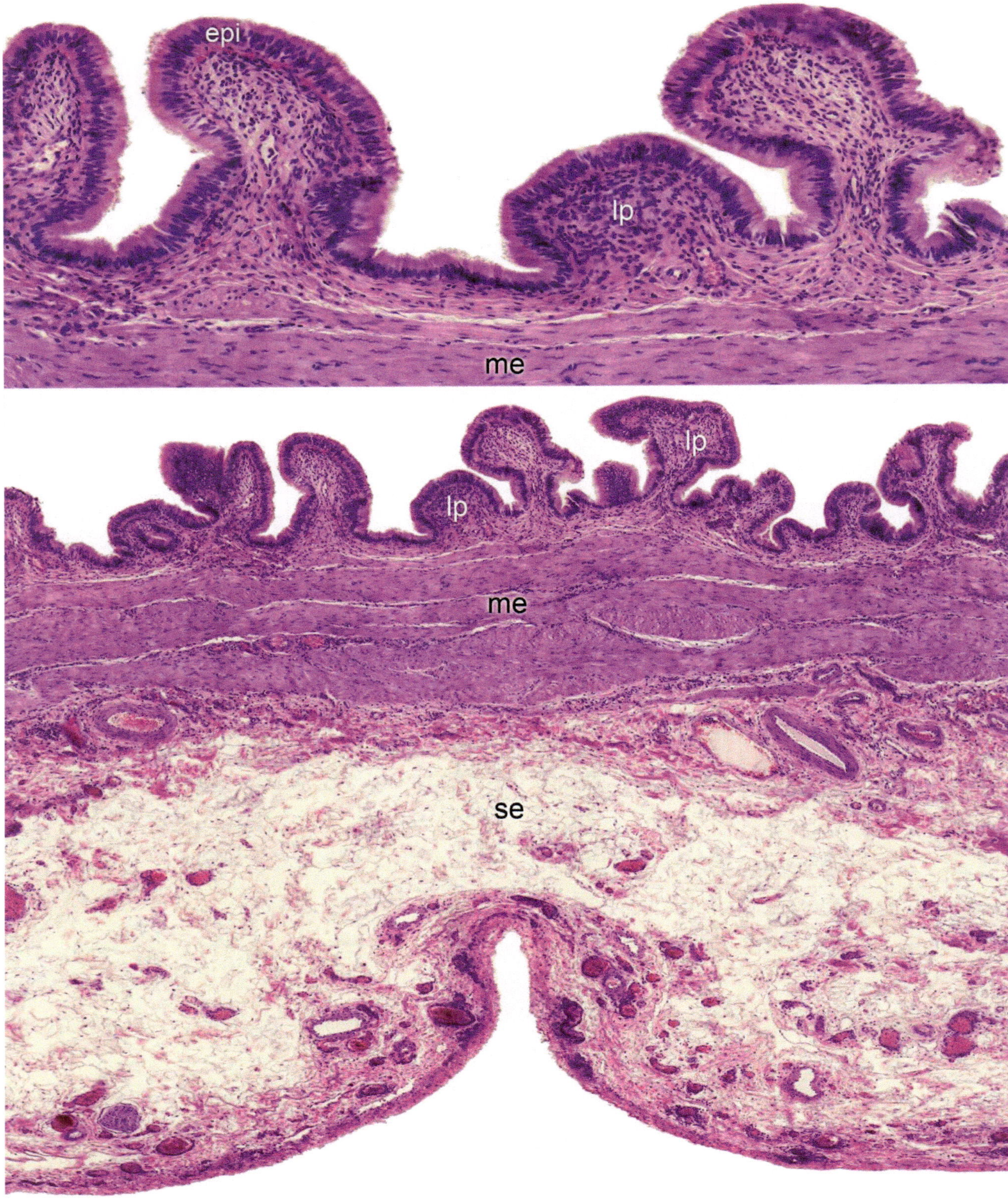

Figure 167. Gallbladder. Tunica mucosa of the gallbladder is composed of tall, microvillated simple columnar epithelium (epi). Tunica muscularis mucosae and tunica submucosa is missing; lamina propria (lp) is directly attached to the tunica muscularis externa (me) that is composed of randomly oriented smooth muscle bundles. The surface of the gallbladder adjacent to the liver is covered by tunica adventitia while the free inferior surface is lined with tunica serosa (se) formed by loose connective tissue with mesothelial coverage.

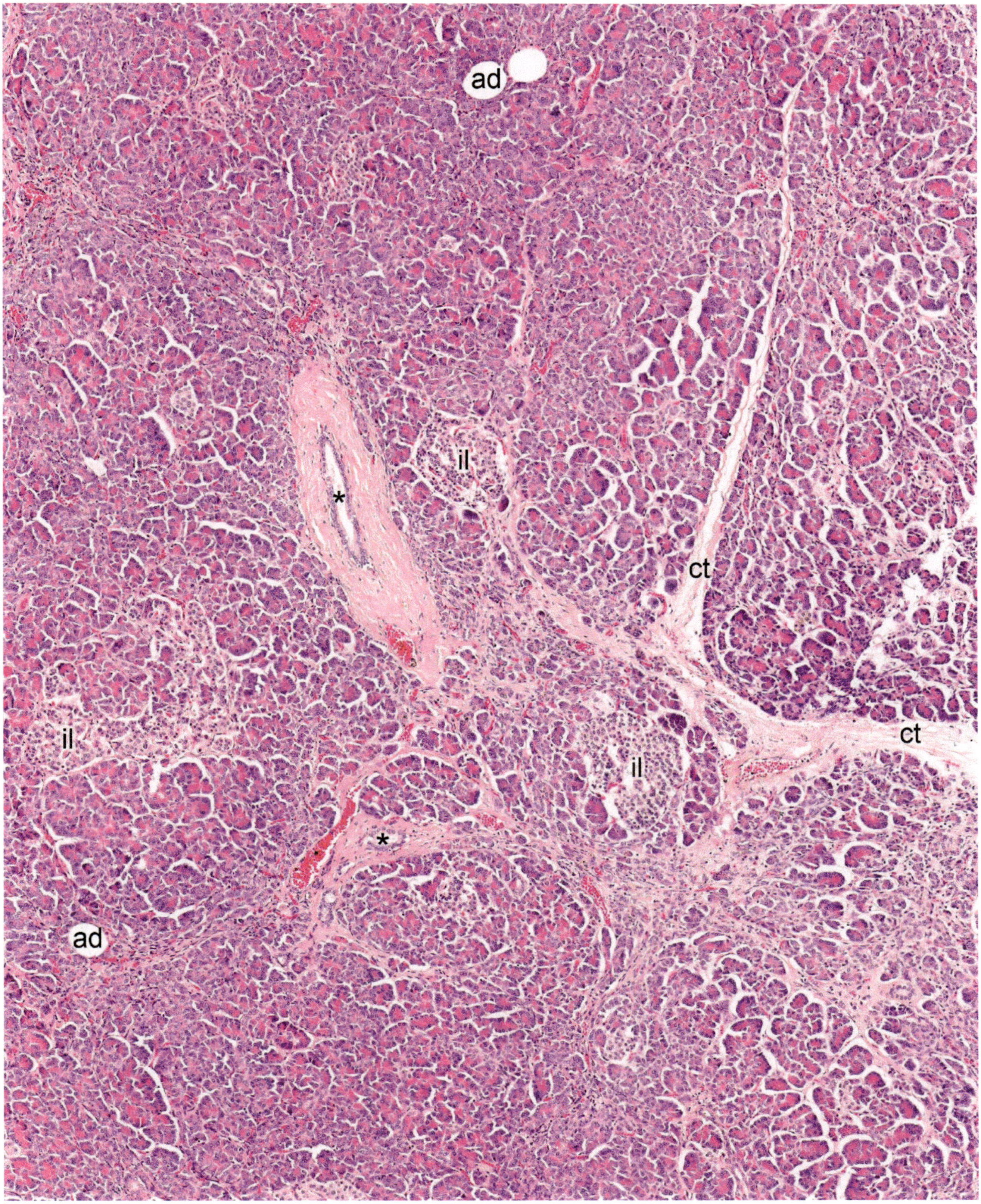

Figure 168. Pancreas. Pancreas is an exocrine serous tubuloalveolar gland that contains spherical endocrine structures, the islets of Langerhans (il). Connective tissue septa (ct) subdivide the parenchyma into lobules and ensheath intralobular ducts covered with simple cuboidal epithelium (asterisk). The acini of the pancreas are composed of triangular cells; the base of the cells is basophilic due to the rER content while the luminal surface contains zymogen granules and therefore, it is acidophilic.

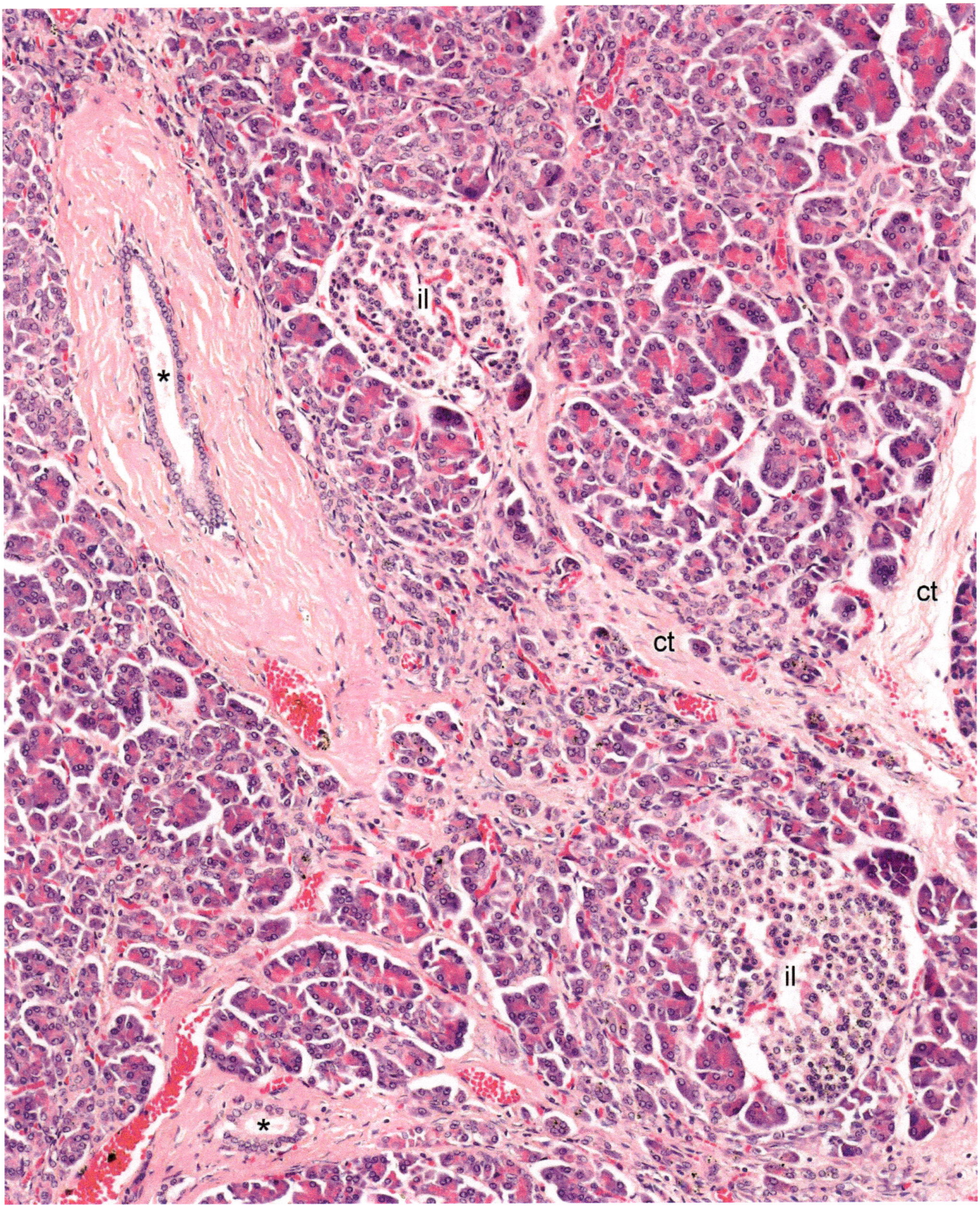

Figure 169. Pancreas. Pancreas is an exocrine serous tubuloalveolar gland that contains spherical endocrine structures, the islets of Langerhans (il). Connective tissue septa (ct) subdivide the parenchyma into lobules and ensheath intralobular ducts covered with simple cuboidal epithelium (asterisk). The acini of the pancreas are composed of triangular cells; the base of the cells is basophilic due to the rER content while the luminal surface contains zymogen granules and therefore, it is acidophilic.

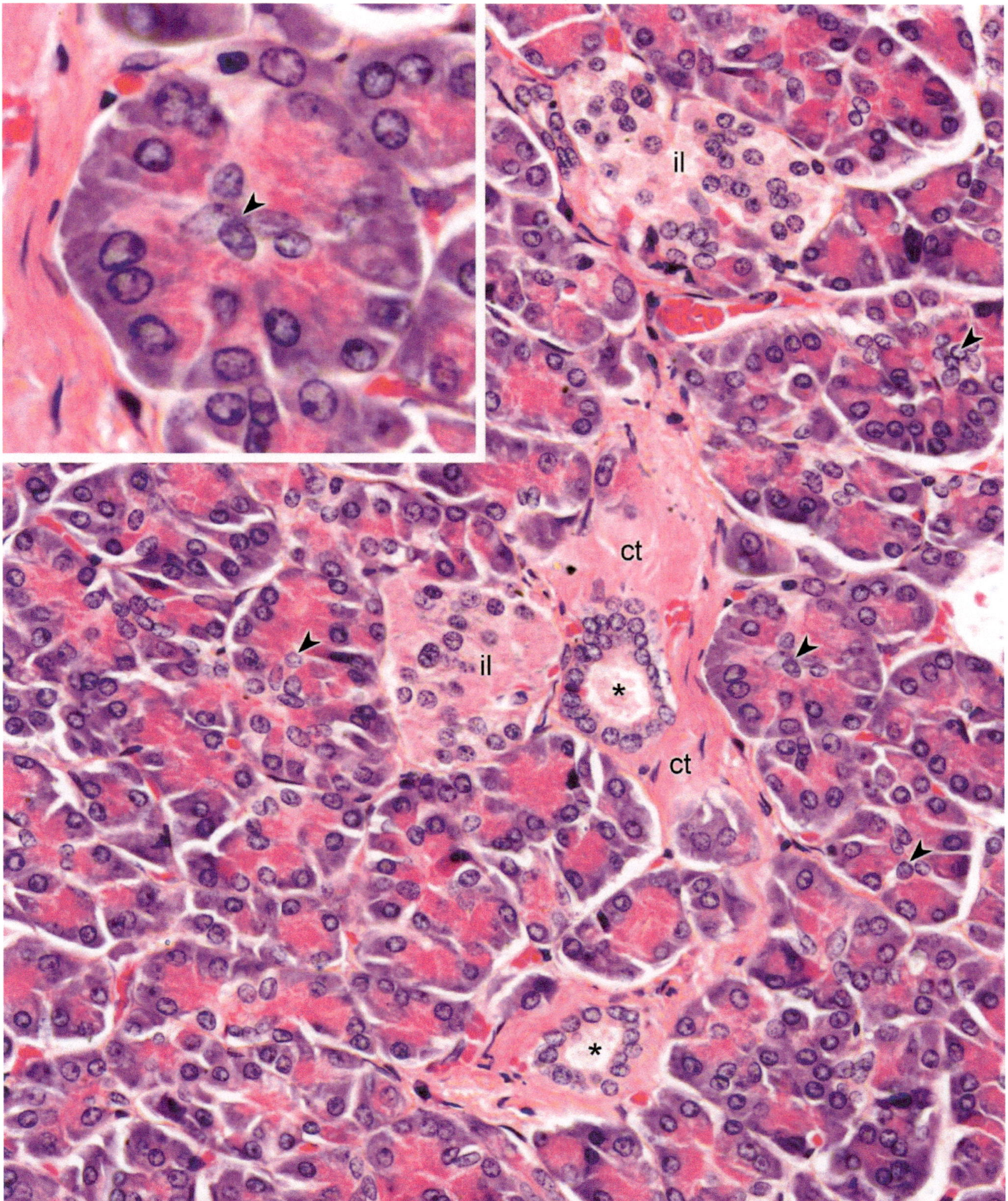

Figure 170. Pancreas. Pancreas is an exocrine serous tubuloalveolar gland that contains spherical endocrine structures, the islets of Langerhans (il). Connective tissue septa (ct) subdivide the parenchyma into lobules and ensheath intralobular ducts covered with simple cuboidal epithelium (asterisk). The acini of the pancreas are composed of triangular cells; the base of the cells is basophilic due to the rER content while the luminal surface contains zymogen granules and therefore, it is acidophilic (inset). Intercalated ducts lined with flattened cells start within the acinus forming the centroacinar cells (arrowheads).

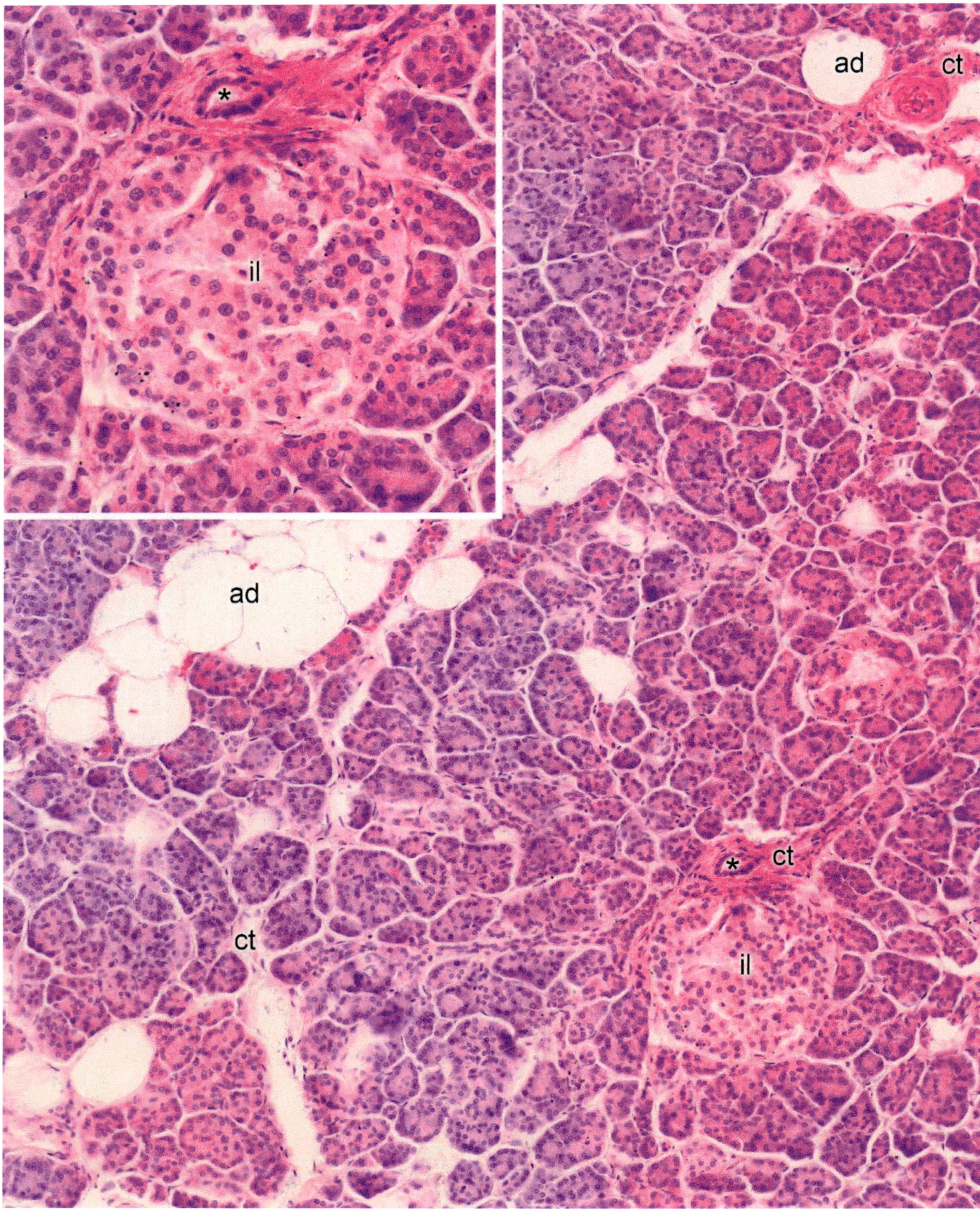

Figure 171. Pancreas. Pancreas is an exocrine serous tubuloalveolar gland that contains spherical endocrine structures, the islets of Langerhans (il; inset). Connective tissue septa (ct) subdivide the parenchyma into lobules with dispersed adipocytes (ad) and ensheath intralobular ducts covered with simple cuboidal epithelium (asterisk). The acini of the pancreas are composed of triangular cells; the base of the cells is basophilic due to the rER content while the luminal surface contains zymogen granules and therefore, it is acidophilic.

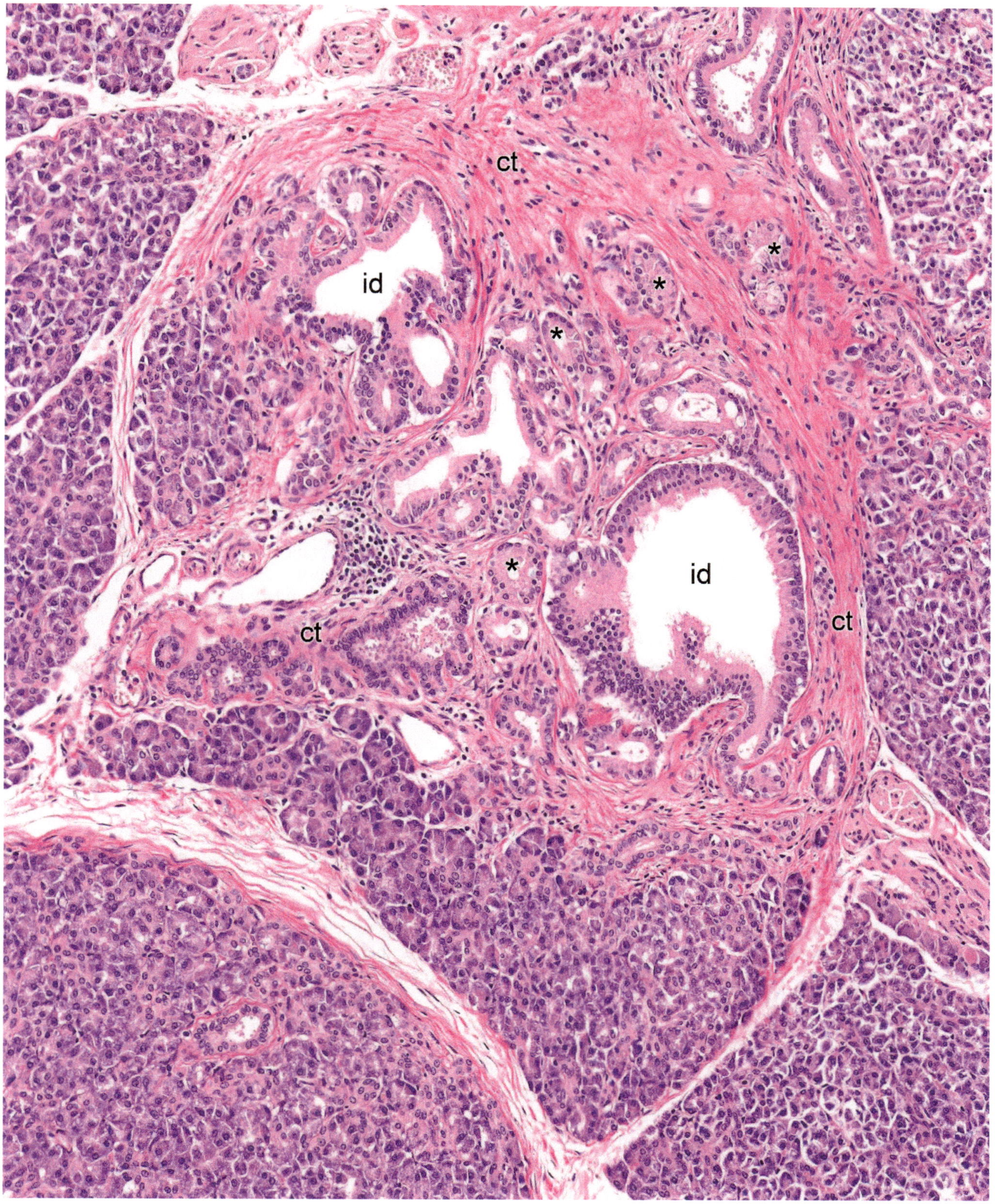

Figure 172. Pancreas. Pancreas is an exocrine serous tubuloalveolar gland that contains spherical endocrine structures, the islets of Langerhans. Connective tissue septa (ct) contain occasional mucous glands (asterisks) and ensheath larger interlobular ducts covered with columnar epithelial lining with few goblet cells (id). The acini of the pancreas are composed of triangular cells; the base of the cells is basophilic due to the rER content while the luminal surface contains zymogen granules and therefore, it is acidophilic.

Chapter 8

Respiratory System

Nasal cavity

The initial part of the airways are the nasal cavities that – apart from the vestibule, the cavity of the external nose that is lined with stratified squamous keratinized epithelium - are covered with kinociliated pseudostratified columnar epithelium with goblet cells and occasional sensory brush cells. The lamina propria under the epithelium is attached to the periosteum of the underlying bone and contains scattered mucous glands with occasional serous demilunes. Paranasal sinuses are lined with mucous membrane of similar structure.

Nasopharynx, soft palate

Nasal cavity opens into the nasopharynx via the choanae. The epithelium of the nasopharynx is pseudostratified columnar epithelium with kinocilia and scattered goblet cells. Lamina propria of the mucous membrane contains a large amount of elastic fibers with occasional seromucous glands as well as lymphatic tissue, especially at the site of the tubal and pharyngeal tonsils where lymphatic follicles occupy the mucous membrane and bulge into the pharyngeal lumen. Lamina propria rests on a layer of skeletal muscle of pharyngeal constrictors. The rest of the pharynx is lined with stratified squamous nonkeratinized epithelium. Soft palate marks the boundary between the nasopharynx and oropharynx; consequently, the upper surface is covered with respiratory epithelium while the lower surface is lined with the epithelium of the oropharynx. The core of the soft palate is formed by skeletal muscles and seromucous glands can be observed in the mucous membrane of both sides. The structure of the soft palate and pharynx is discussed in detail in Chapter 5.

Larynx

Larynx is a rather complicated structure formed by a cartilaginous skeleton covered with skeletal muscle and primarily respiratory mucous membrane. Most of the laryngeal cartilages are made of hyaline cartilage; the epiglottis and some small accessory cartilages are formed by elastic cartilage. Larynx is subdivided into three main parts. The middle part, the glottis, contains the vocal folds, and the space between them, the rima glottidis. Above the glottis, the supraglottis contains the vestibulum; below the glottis, the subglottis encloses the subglottic space. Laryngeal mucosa, lining the vestibulum and the subglottic area also covers the ventricle that opens between the vestibular and vocal folds. The mucous membrane is covered with kinociliated pseudostratified columnar epithelium with goblet cells with the exception of the luminal surface of the vocal folds and most of the epiglottis that are covered with nonkeratinized stratified squamous epithelium. Mucous membrane contains occasional seromucous glands and rests on the perichondrium of the laryngeal cartilages as well as on the membranous skeleton of the larynx formed by the quadrangular and triangular membranes that are essentially parts of the tunica submucosa. Laryngeal muscles are composed of skeletal muscle innervated mostly by the inferior laryngeal nerve.

Human Histology
http://doi.org/10.1016/B978-0-323-91891-6.50008-8,

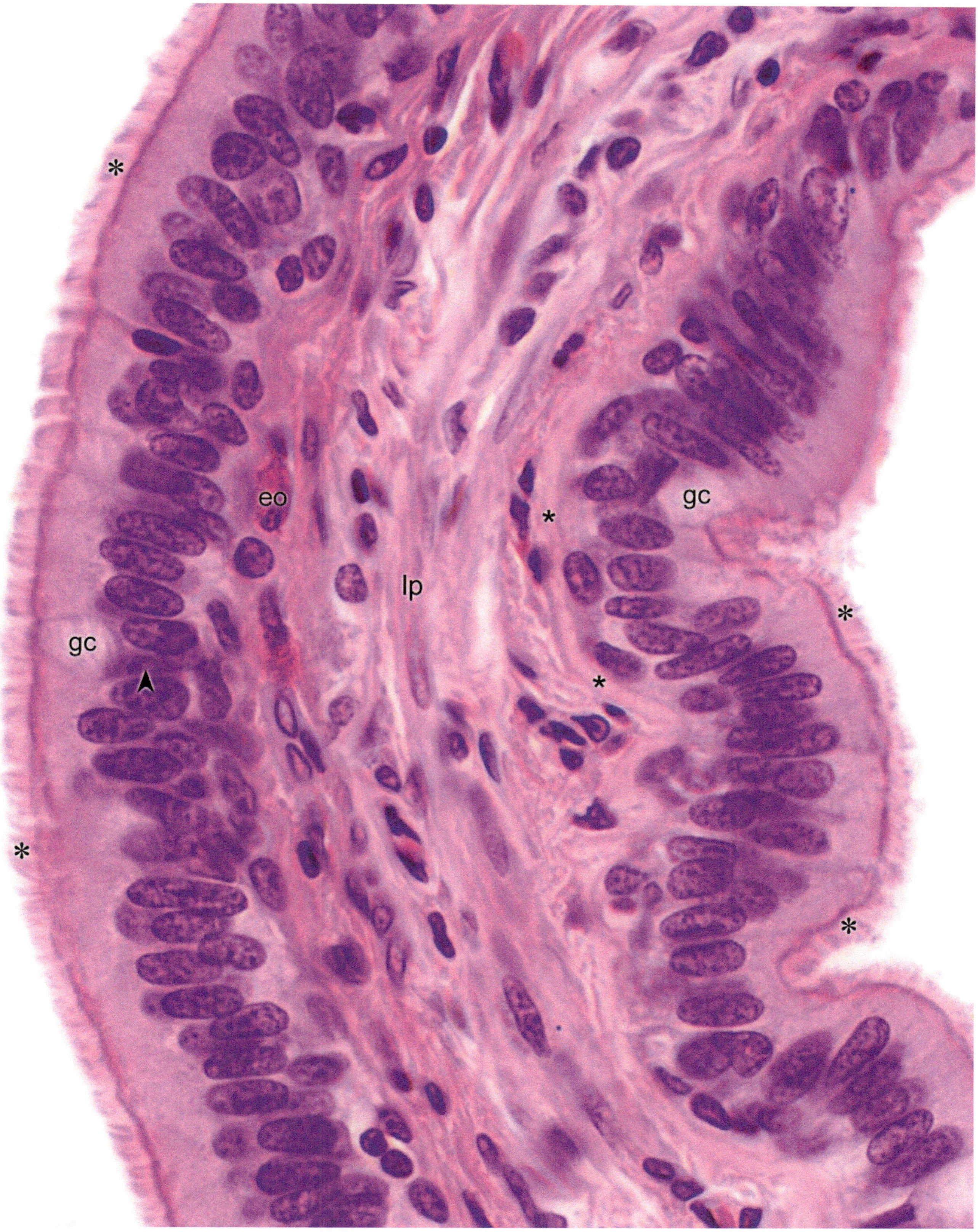

Figure 173. Respiratory epithelium. The majority of the respiratory mucosa is covered with pseudostratified columnar epithelium. Columnar cells have kinocilia on their apical surface (six-pointed asterisks). Goblet cells with mucous apical portion and elongated basal nucleus (arrowhead) can be observed between the columnar ciliated cells. Epithelium rests on a well-defined basement membrane (five-pointed asterisks). The underlying loose connective tissue lamina propria (lp) often contains eosinophils (eo).

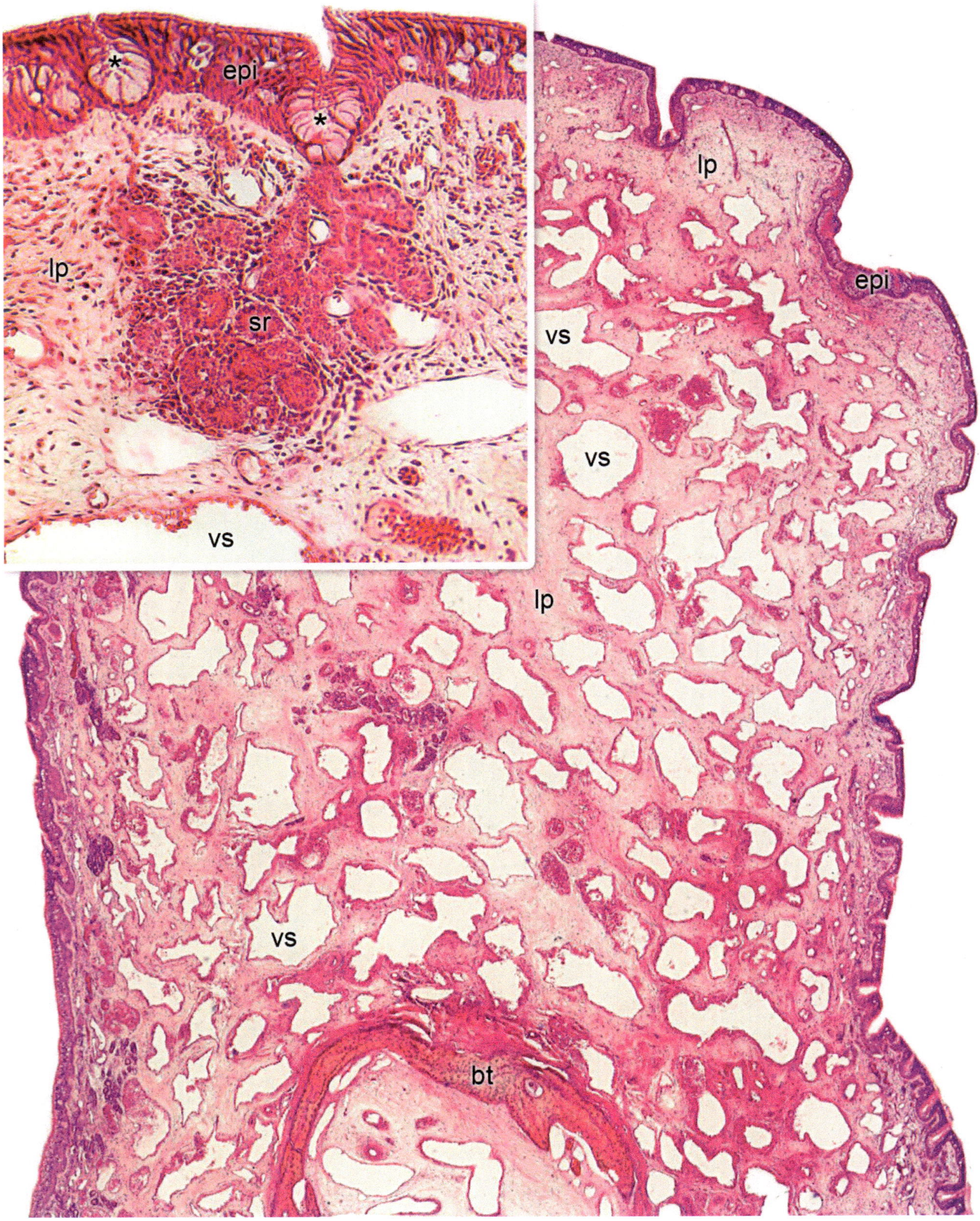

Figure 174. Inferior nasal concha. Nasal cavity is lined with kinociliated pseudostratified columnar epithelium (epi) containing intraepithelial multicellular mucous glands composed of goblet cells (asterisks). Serous glands (sr) open into the base of the intraepithelial glands. Lamina propria mucosae is composed of loose connective tissue and it contains numerous venous sinuses (vs). Mucous membrane is attached to the bone of the nasal concha (bt).

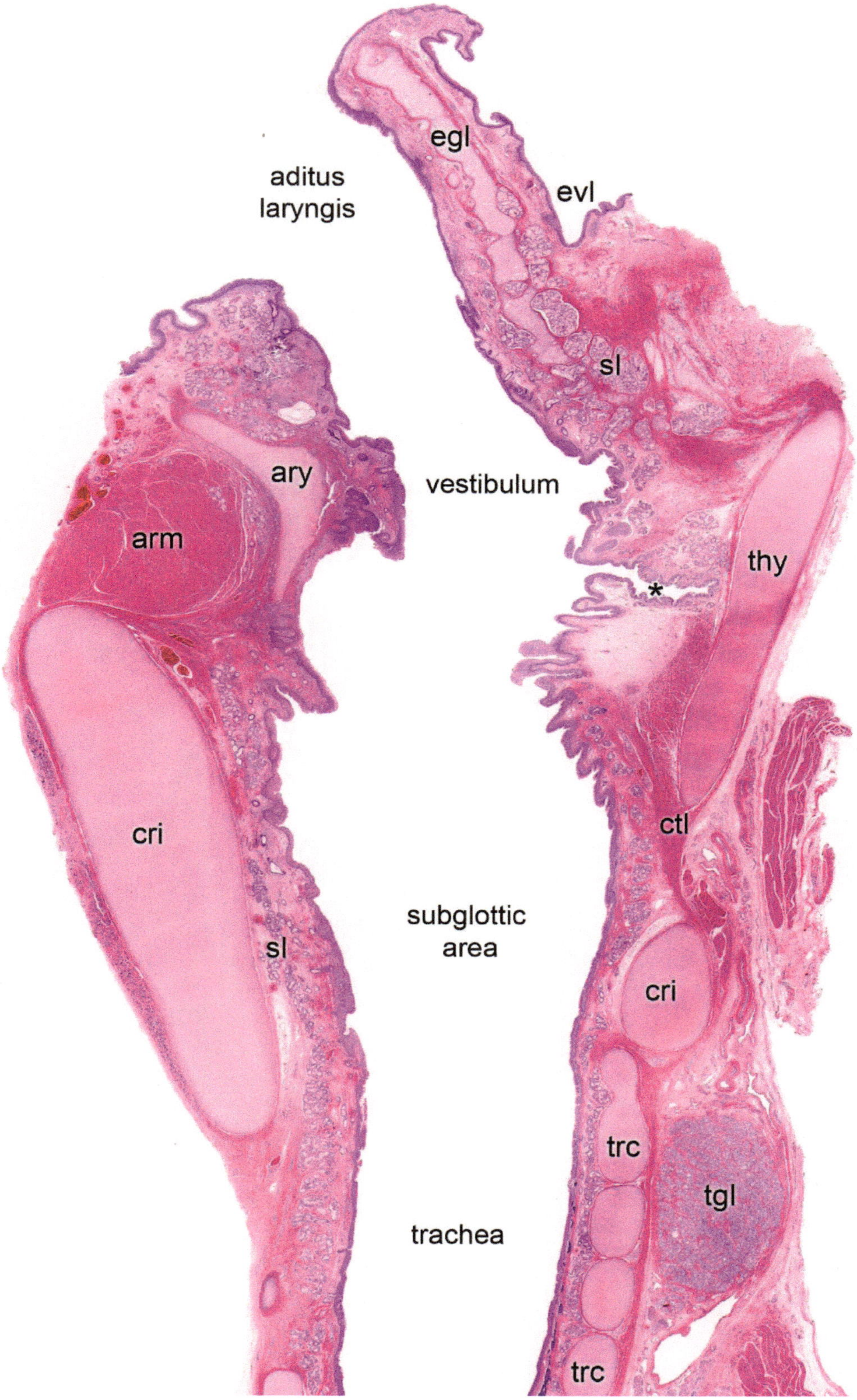

Figure 175. Larynx (newborn), sagittal section. Respiratory mucosa with seromucous salivary glands (sl) and the underlying tunica submucosa covers the cricoid (cri), thyroid (thy), arytenoid (ary) and tracheal cartilages (trc). The mucous membrane is covered with kinociliated pseudostratified columnar epithelium except the luminal surface of the vocal folds and most of the epiglottis (egl) bordered by the epiglottic vallecula (evl) that are covered with stratified squamous nonkeratinized epithelium. Abbreviations: tgl, thyroid gland; arm, arytenoid muscle; ctl, cricothyroid ligament. Asterisk denotes the ventricle.

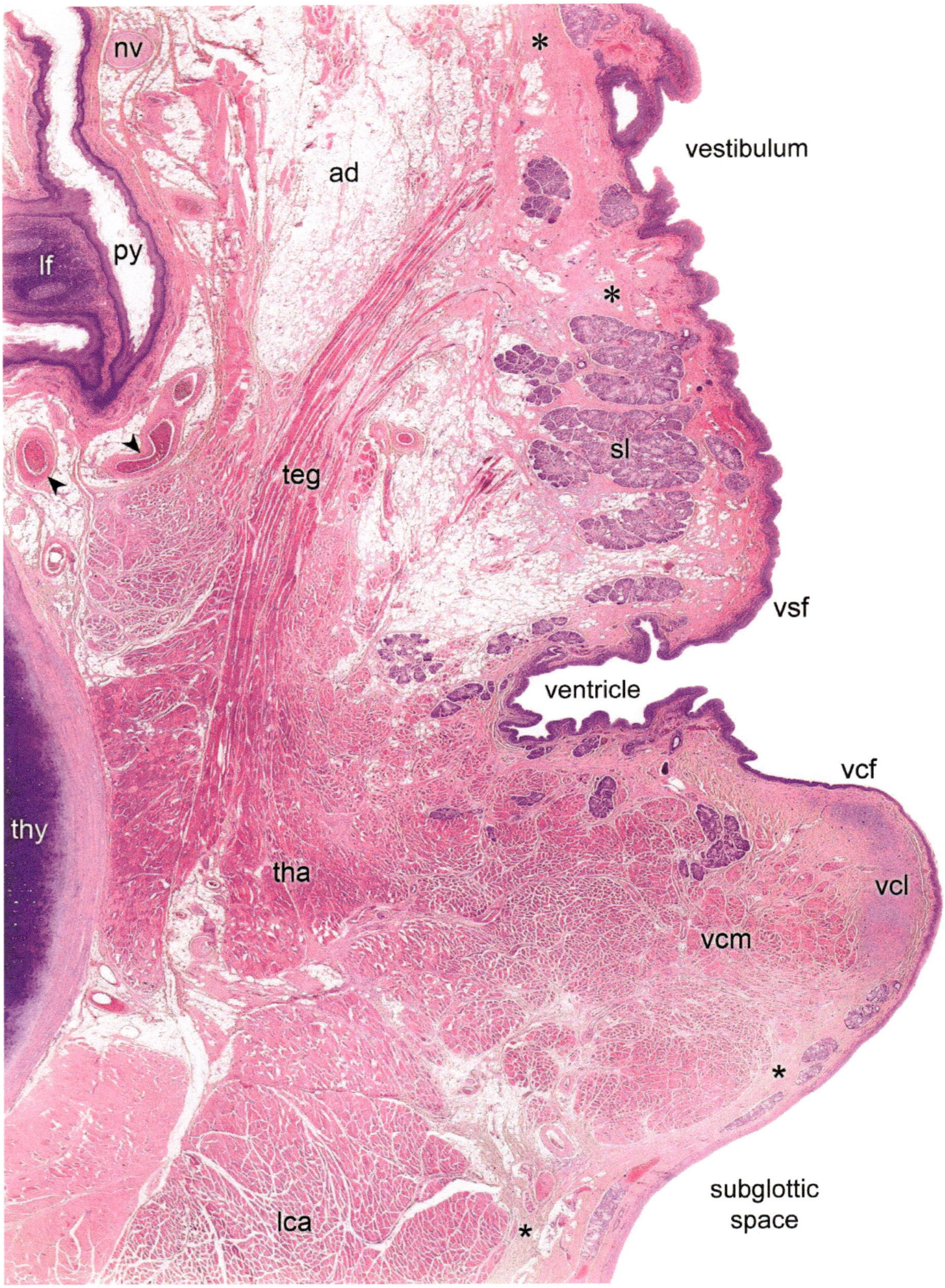

Figure 176. Larynx (newborn), coronal section. Mucous membrane with seromucous salivary glands (sl) forms the vestibular (vsf) and vocal folds (vcf); the latter contains the vocal ligament (vcl) and the vocalis muscle (vcm). Tunica mucosa rests on the submucosa that forms the quadrangular (six-pointed asterisks) and triangular (five-pointed asterisks) membranes. Branches of the superior laryngeal artery (arrowheads) and nerve (nv) surrounded by adipose tissue (ad) can be observed next to the pyriform recess (py). Abbreviations: lf, lymphatic follicle; lca, lateral cricoarytenoid muscle; teg, thyroepiglottic muscle; tha, thyroarytenoid muscle; thy, thyroid cartilage.

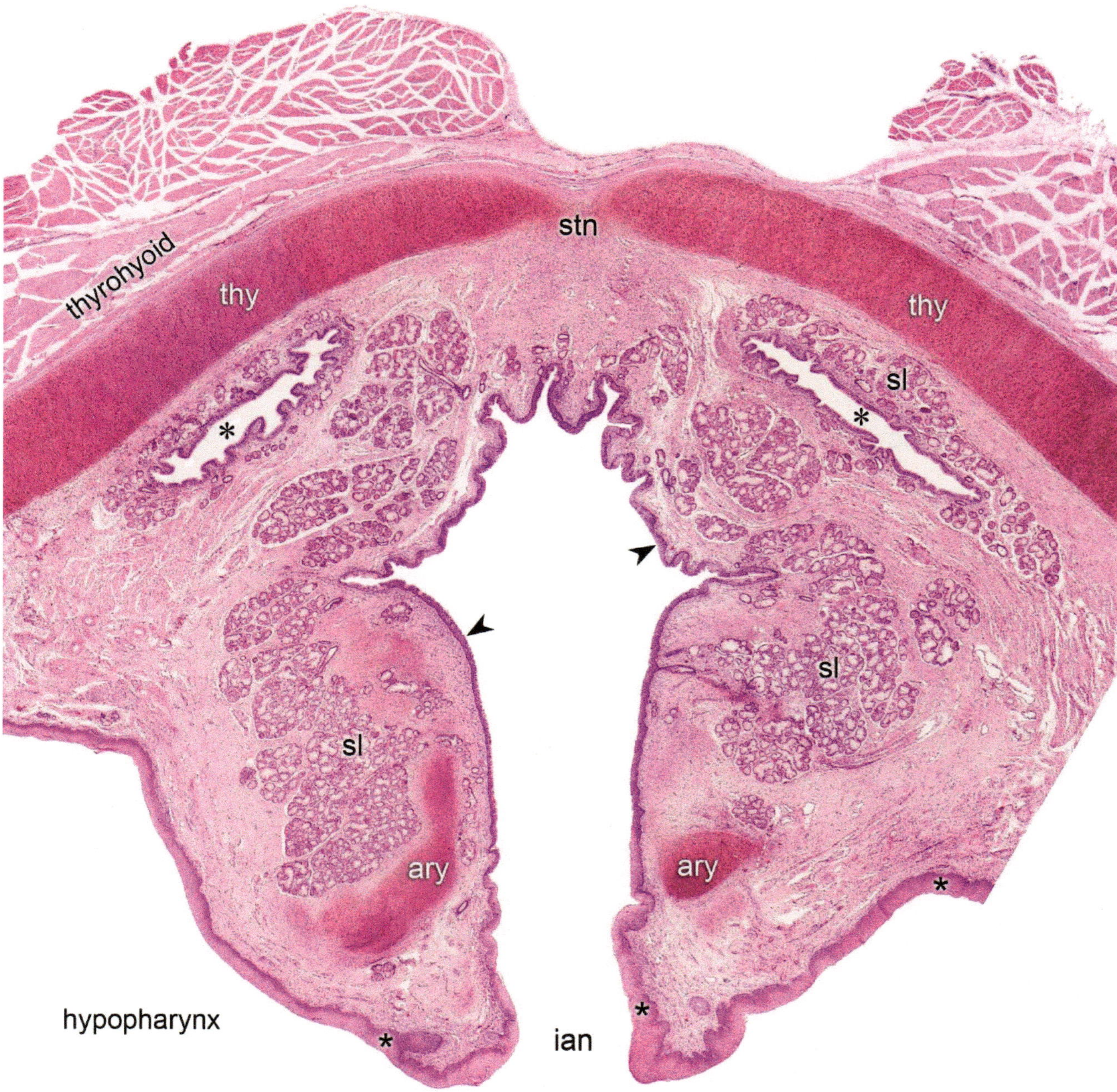

Figure 177. Larynx (newborn). Horizontal section at the level of the superior thyroid (stn) and the interarytenoid notches (ian). The tunica mucosa lining the vestibulum (arrowheads) and the ventricles (six-pointed asterisks) are covered with kinociliated pseudostratified columnar epithelium with goblet cells, while the hypopharynx and part of the laryngeal aditus is lined with stratified squamous nonkeratinized epithelium (five-pointed asterisks). Seromucous salivary glands (sl) are abundant in the mucosa. The lamina of the thyroid cartilage is covered by thyrohyoid muscles externally. Arytenoid cartilages (ary) are composed of hyaline cartilage.

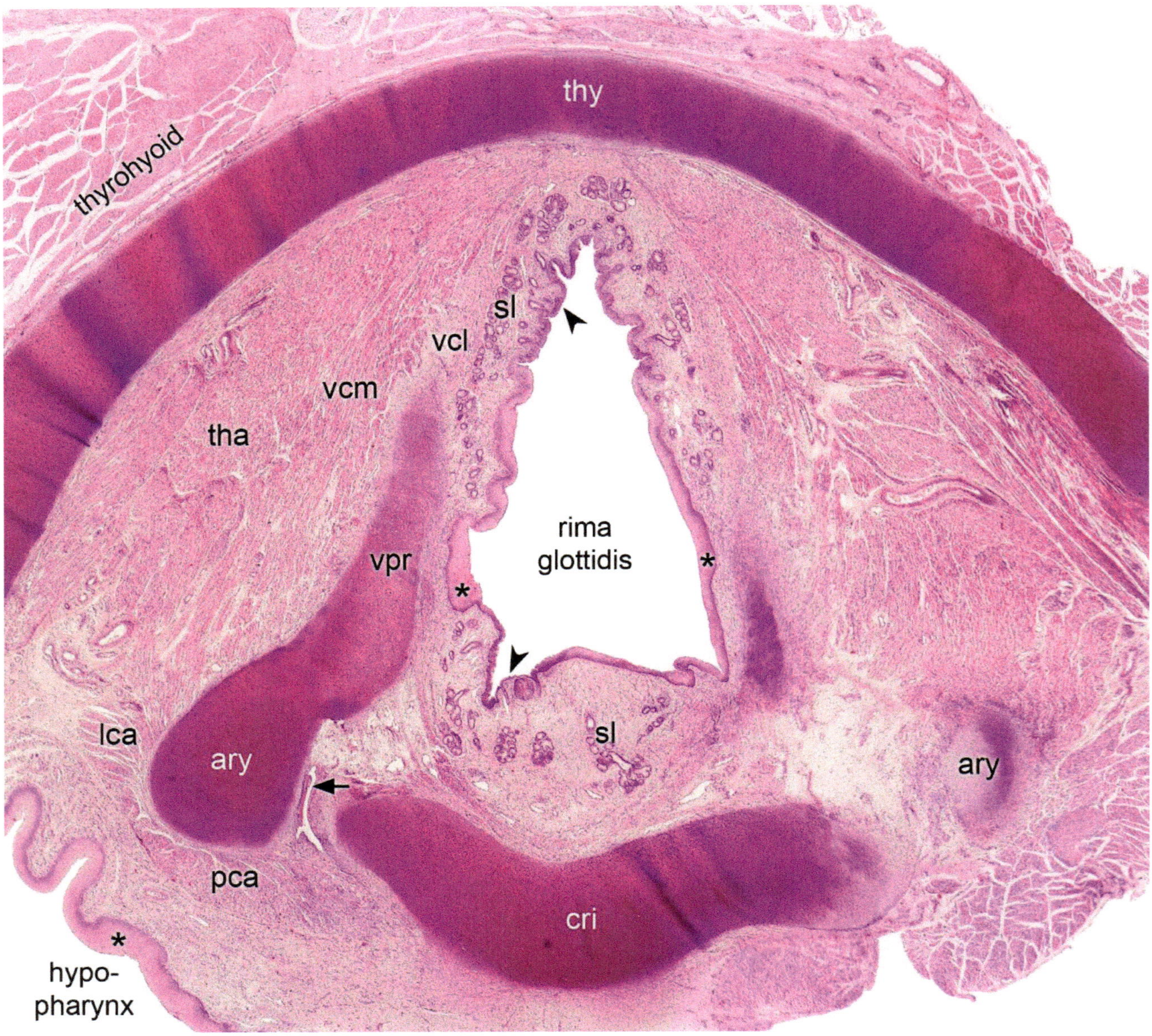

Figure 178. Larynx (newborn). Horizontal section at the level of the vocal fold. Vocal fold is lined mostly with stratified squamous nonkeratinized epithelium (five-pointed asterisks) with occasional kinociliated pseudostratified columnar epithelium coverage at the anterior and posterior segments of the rima glottidis (arrowheads). Stratified squamous nonkeratinized epithelium also lines the hypopharynx. Laryngeal mucosa contains seromucous salivary glands (sl), and it is attached to the tunica submucosa that is composed of dense irregular connective tissue. The lamina of the thyroid cartilage is covered by the thyrohyoid muscles externally. Arytenoid cartilages (ary) form synovial joints (arrow) with the cricoid cartilage (cri); they are moved by the posterior (pca) and the lateral cricoarytenoid muscles (lca). Vocal ligaments (vcl) contain collagen and elastic fibers; they are located within the vocal folds of the mucosa and they are attached to the vocal process of the arytenoid cartilages. Laterally, vocal ligaments are bordered by the vocalis muscle (vcm) and the thyroarytenoid muscles (tha). Laryngeal muscles are composed of skeletal muscle innervated mostly by the inferior laryngeal nerve. Cricoid, thyroid and arytenoid cartilages consist of hyaline cartilage.

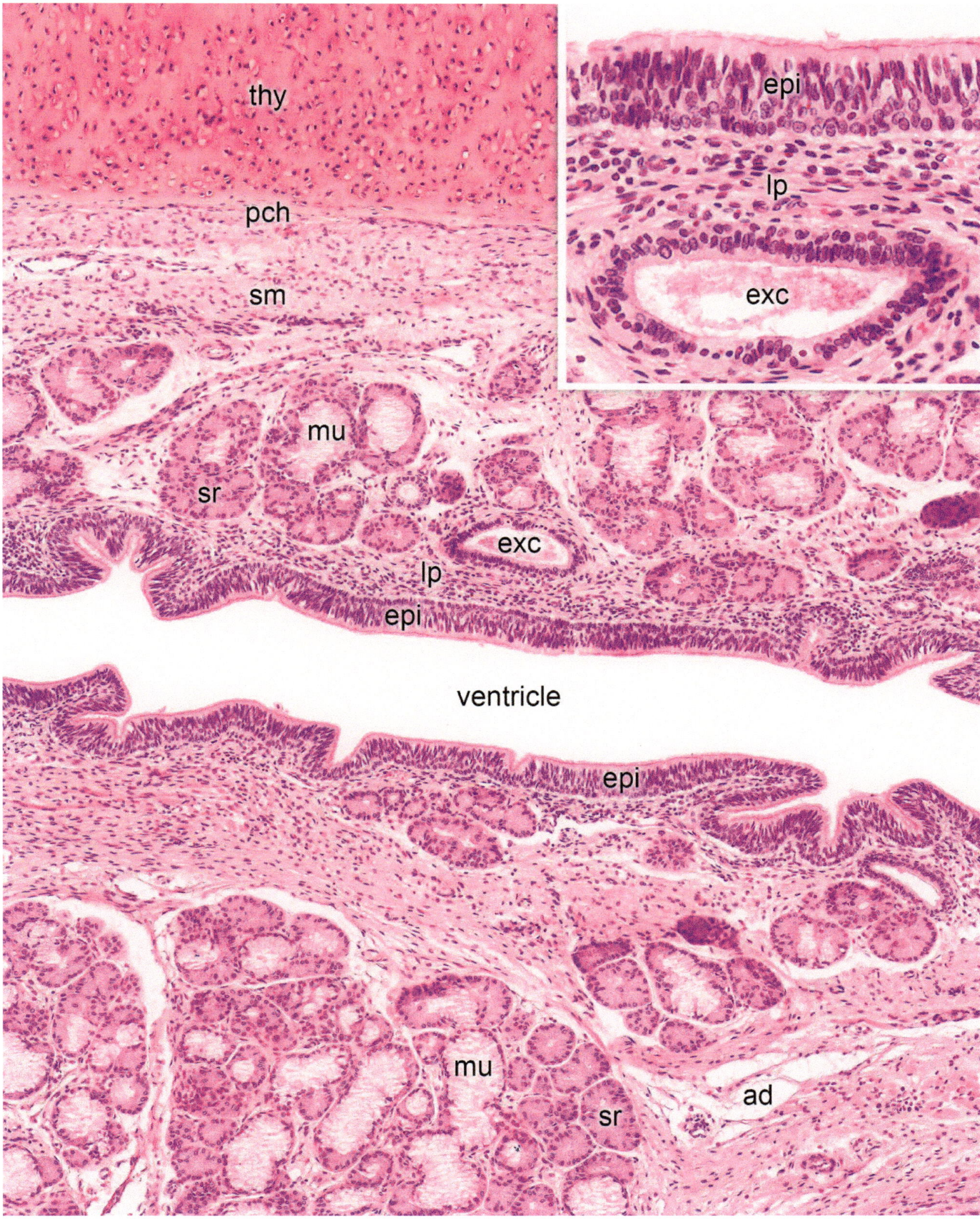

Figure 179. Larynx (newborn), horizontal section. The tunica mucosa lining the ventricles is covered with kinociliated pseudostratified columnar epithelium with goblet cells (epi; inset). Mixed salivary glands composed of mucous (mu) and serous acini (sr) as well as their excretory ducts (exc) covered with stratified columnar epithelium (inset) are abundant in the lamina propria (lp) that contains occasional adipocytes (ad). Tunica submucosa (sm) is attached to the perichondrium (pch) of the thyroid cartilage (thy).

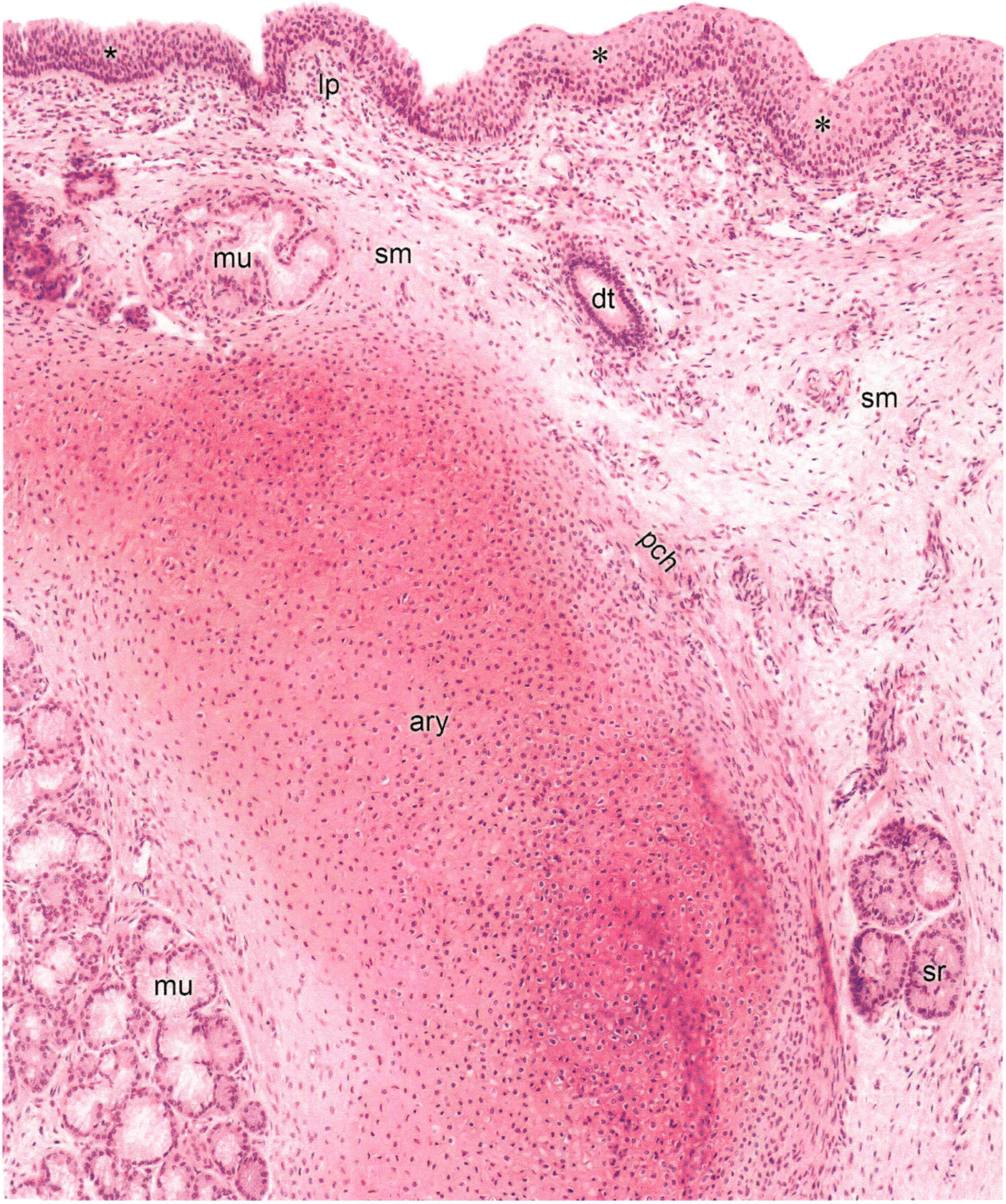

Figure 180. Larynx (newborn), horizontal section. Laryngeal mucosa lining the vestibulum is covered with kinociliated pseudostratified columnar epithelium with occasional goblet cells (five-pointed asterisk) that transitions to the nonkeratinized stratified squamous epithelium of the pharyngeal mucosa (six-pointed asterisk) at the laryngeal inlet (aditus laryngis). Mixed salivary glands composed of mucous (mu) and serous acini (sr) as well as their excretory ducts (exc) covered with columnar epithelium are abundant in the lamina propria (lp). Tunica submucosa (sm) is attached to the perichondrium (pch) of the arytenoid cartilage (ary).

Trachea

The mucous membrane of the trachea is covered by pseudostratified columnar epithelium with kinocilia and mucous goblet cells, as well as occasional cuboidal basal reserve cells resting on the thick basement membrane; additional cell types of the epithelium, including the sensory brush cells with blunt microvilli, and the respiratory endocrine cells, are difficult to identify with H&E staining. Under the thick basement membrane of the epithelium, loose connective tissue lamina propria covers the denser submucosa that is attached to 16-20 incomplete, C-shaped hyaline cartilage rings that often calcify in the elderly. The free edges of the rings are connected by a smooth muscle sheet, the trachealis muscle, located at the posterior membranous part of the trachea facing the esophagus. This smooth muscle layer is composed of internal, circularly arranged cells and occasional external longitudinal bundles embedded into the tunica submucosa that also contains large amount of mixed salivary glands dominated by mucous acini. Similar to the membranous part, the submucosa of the cartilaginous part is populated by tracheal glands that are primarily mucous with occasional serous demilunes and acini. Aggregation of lymphatic tissue can be often observed in the submucosa and the lamina propria. Trachea is covered with loose connective tissue tunica adventitia outside and it eventually transitions to the left and right primary bronchi whose wall structure is similar to that of the trachea.

Lung

As entering the hilum of the lung, primary bronchi are divided to lobar (secondary) and segmental (tertiary) bronchi. In the lung, the cartilaginous rings of the bronchi became less regular and a smooth muscle layer between the lamina propria and the submucosa can be observed (tunica muscularis mucosae). After several subdivisions of the tertiary bronchi, hyaline cartilage and glands disappear from the wall of the airways that, at this point, turn into bronchioli.

The epithelium of the larger bronchioli is pseudostratified columnar with kinocilia that gradually turns into ciliated simple columnar epithelium followed by ciliated simple cuboidal epithelium with dome-shaped nonciliated Clara cells (club cells) in the smaller terminal bronchioli. Clara cells are responsible for the production of pulmonary surfactant that prevents the collapse of the airways. Goblet cells can be observed in the larger bronchioli; their number gradually decreases with the increasing number of Clara cells, and they are completely missing from the terminal bronchioli under normal conditions. Gas exchange starts in the respiratory bronchioli that are covered predominantly with Clara cells dispersed among ciliated cuboidal cells. Respiratory bronchioli open into alveolar ducts that terminate in blind-ended alveolar sacs. Both alveolar ducts and sacs are covered with simple squamous epithelium formed by type I pneumocytes. Type II pneumocytes are secretory cells bulging into the alveolar lumen; they produce surfactant that covers the alveolar surface. Occasional brush cells are also present, but they are difficult to identify with H&E staining. In the wall of the alveolar septa, numerous capillaries can be observed; their endothelium, together with the basement membrane and the pneumocytes covered with a layer of surfactant, forms the air-blood barrier that is the primary site of gas exchange. Lungs are covered with the visceral layer of the pleura that is a serous membrane with a mesothelial lining and the underlying loose connective tissue lamina propria serosae.

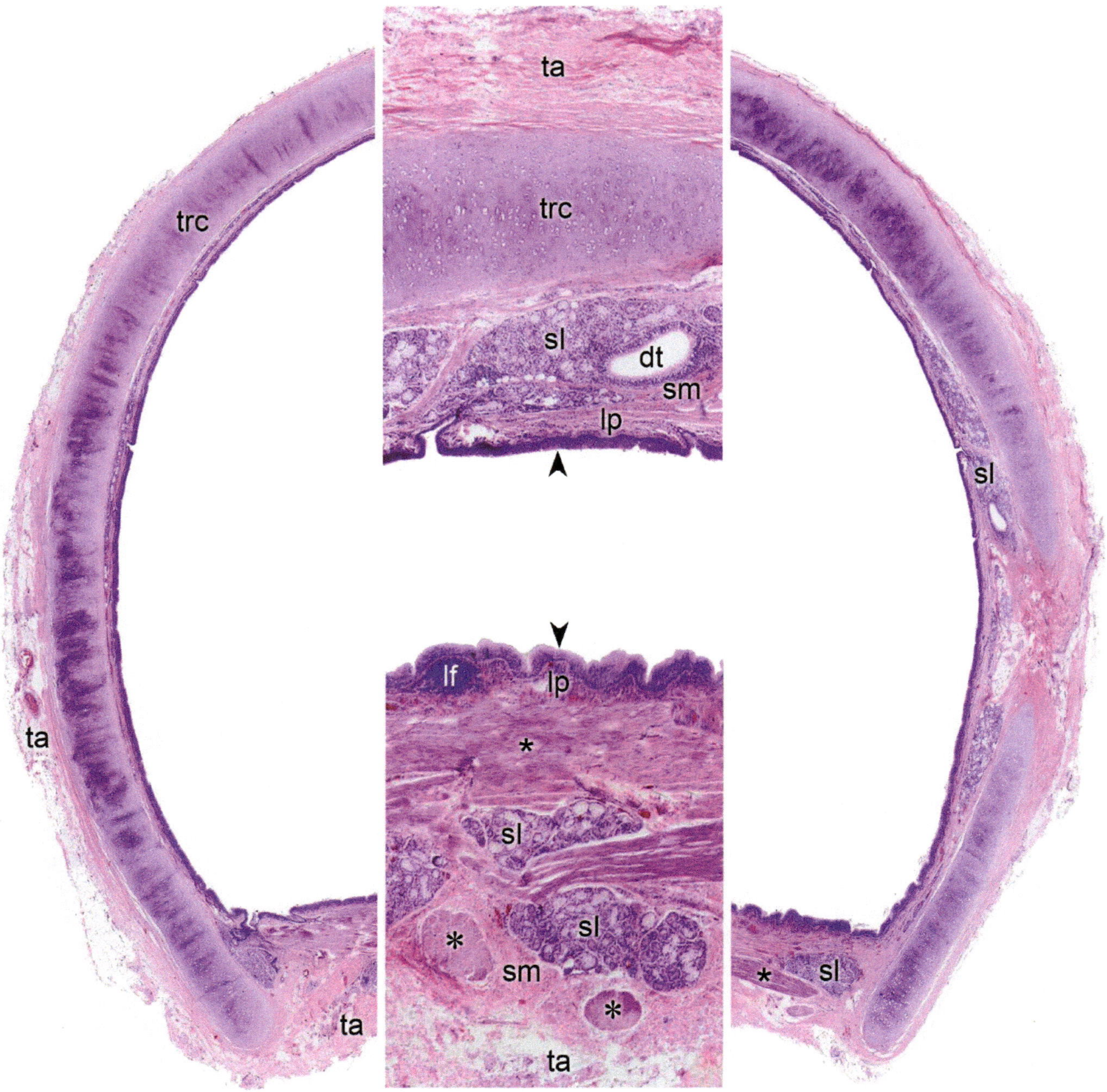

Figure 181. Trachea (adult), horizontal section. Layers of the cartilaginous part are denoted on the upper inset, while the membranous part is illustrated on the lower inset. The mucous membrane of the trachea is lined with kinociliated pseudostratified columnar epithelium (arrowheads, upper and lower insets) with occasional goblet cells. Loose connective tissue lamina propria (lp) transitions to the dense irregular connective tissue submucosa (sm) containing seromucous salivary glands (sl) and their ducts (dt). Submucosa is attached to the perichondrium of the C-shaped tracheal cartilages (trc; upper inset) that are composed of hyaline cartilage, which may undergo ossification in elder age. Posteriorly, facing the esophagus, the tracheal wall lacks cartilage (membranous part, lower inset). Here, the lamina propria, covered by pseudostratified columnar epithelium, rests on the dense connective tissue submucosa that contains circularly (five-pointed asterisks) and longitudinally arranged smooth muscle bundles (six-pointed asterisks). Outside, the trachea is covered with loose connective tissue tunica adventitia (ta) that contains vessels and nerve fibers.

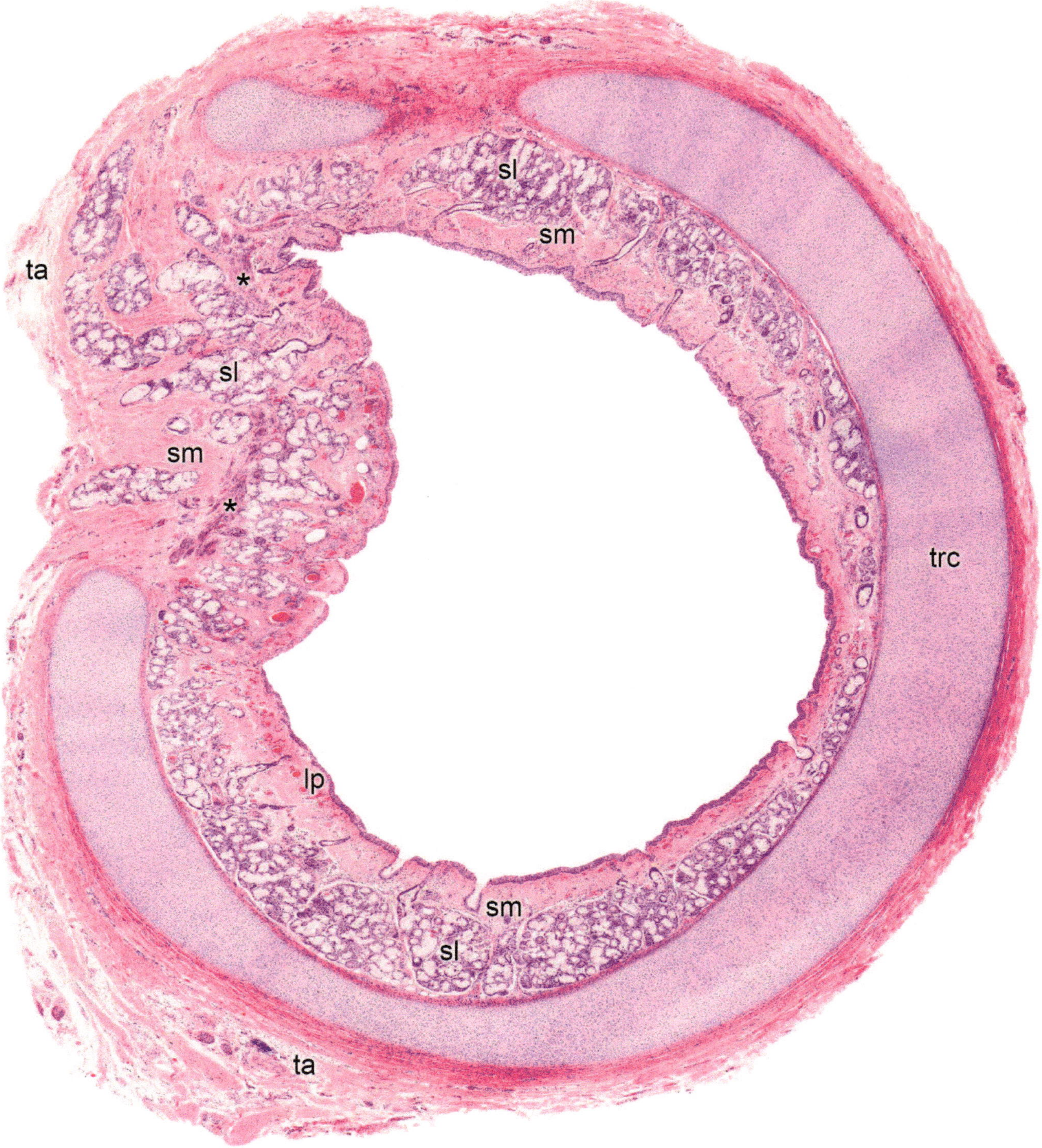

Figure 182. Trachea (newborn), horizontal section. Tracheal mucosa is composed of kinociliated pseudostratified columnar epithelium with occasional goblet cells and the underlying loose connective tissue lamina propria (lp). Dense irregular connective tissue submucosa (sm) contains seromucous salivary glands (sl) and their ducts and it is attached to the perichondrium of the C-shaped tracheal cartilages (trc) that are composed of hyaline cartilage, which may undergo ossification in elder age. Posteriorly, facing the esophagus, the tracheal wall lacks cartilage (membranous part). Here, the lamina propria, covered by pseudostratified columnar epithelium, rests on the dense connective tissue submucosa that contains circularly and longitudinally arranged smooth muscle bundles (asterisks). Outside, the trachea is covered with loose connective tissue tunica adventitia (ta) that contains vessels and nerve fibers.

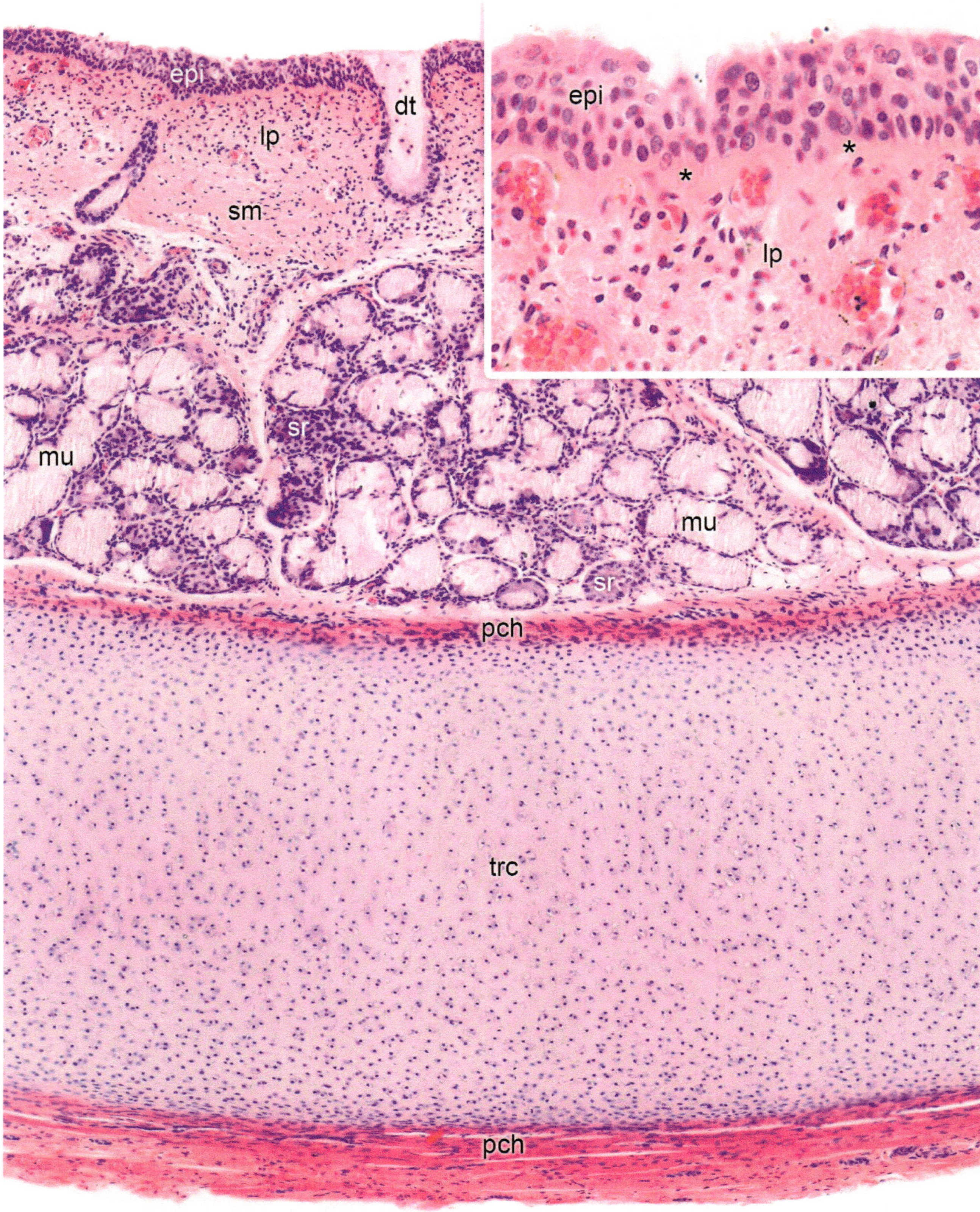

Figure 183. Trachea (newborn). Tracheal mucosa is composed of kinociliated pseudostratified columnar epithelium (epi) with occasional goblet cells and thick basement membrane (asterisks, inset), and the underlying loose connective tissue lamina propria (lp). Dense irregular connective tissue submucosa (sm) contains serous (sr) and mucous acini (mu) and their ducts (dt), and it is attached to the perichondrium (pch) of the C-shaped tracheal cartilages (trc) that are composed of hyaline cartilage.

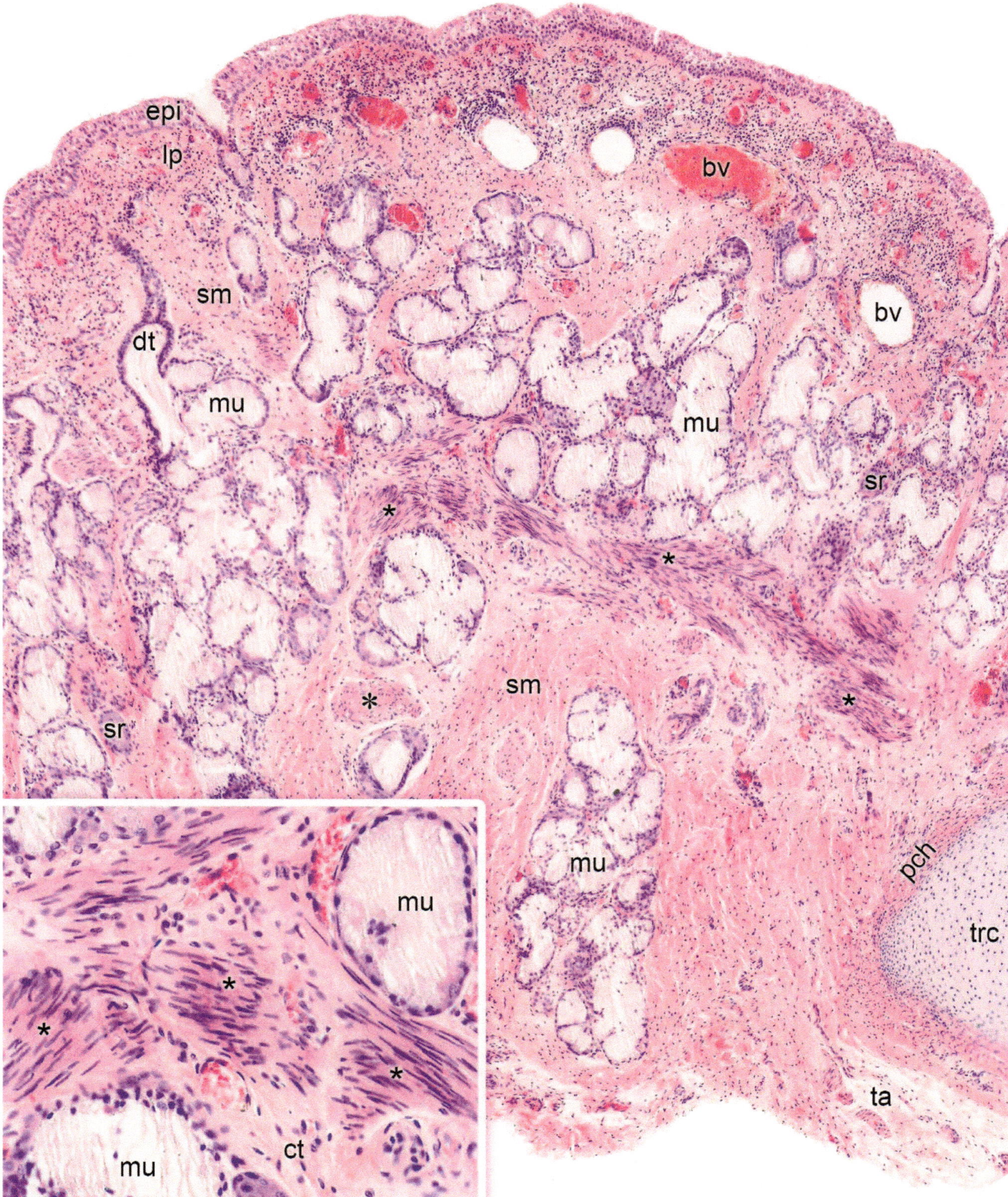

Figure 184. Trachea (newborn), membranous part. Posteriorly, the tracheal wall lacks tracheal cartilage (trc) and the associated perichondrium (pch). Here, the loose connective tissue lamina propria (lp), covered by pseudostratified columnar epithelium (epi), rests on the tunica submucosa (sm) that contains circularly (five-pointed asterisks) and longitudinally arranged smooth muscle bundles (six-pointed asterisks) as well as mucous (mu) and serous acini (sr) with their ducts (dt), surrounded by dense connective tissue (ct). Outside, trachea is covered with loose connective tissue tunica adventitia (ta). Lamina propria contains numerous blood vessels (bv).

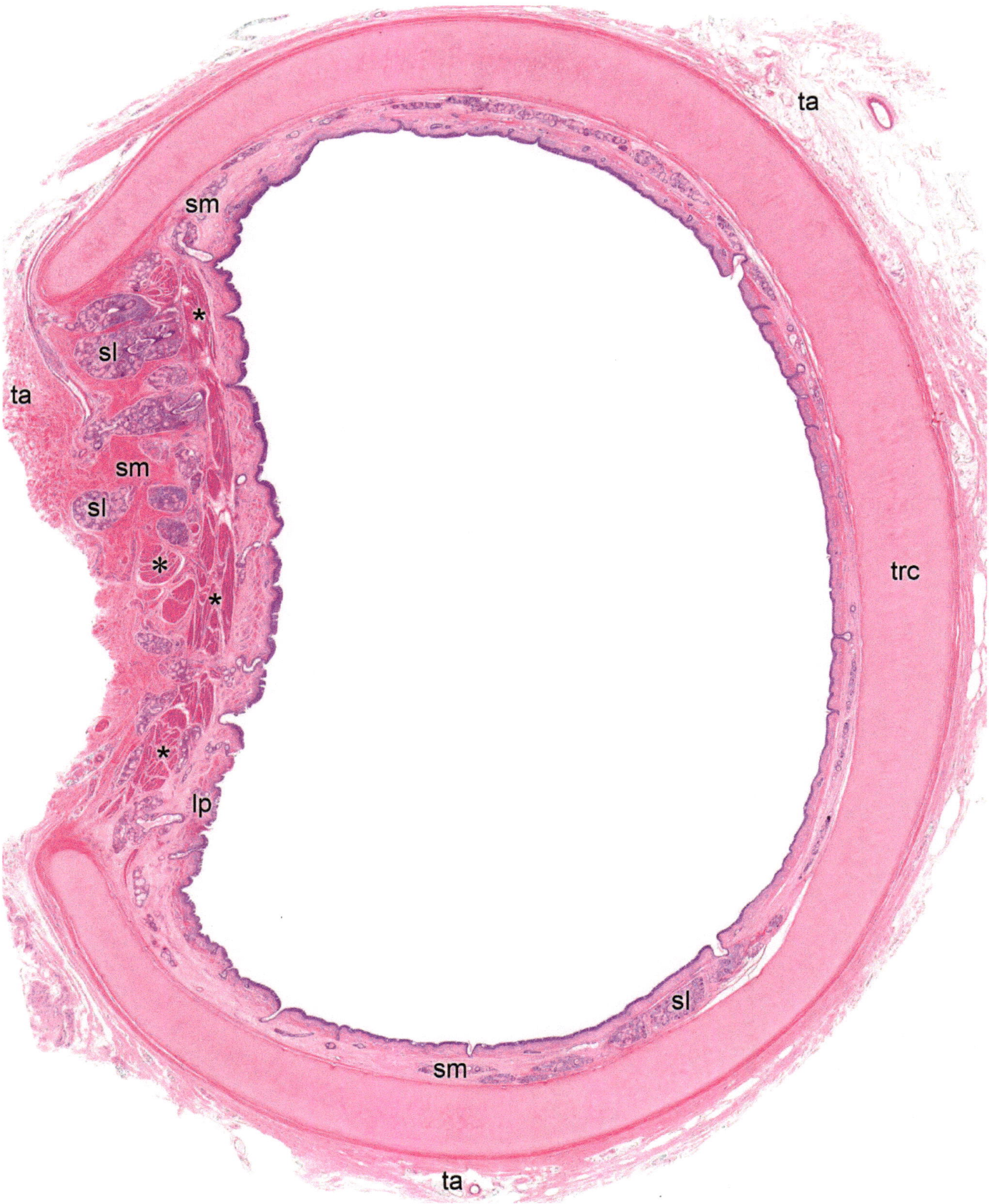

Figure 185. Trachea, horizontal section. The mucous membrane of the trachea is lined with kinociliated pseudostratified columnar epithelium with occasional goblet cells. Loose connective tissue lamina propria (lp) transitions to the dense irregular connective tissue submucosa (sm) containing seromucous salivary glands (sl). Posteriorly, the tracheal wall lacks tracheal cartilage (trc); here, tunica submucosa (sm) contains circularly (five-pointed asterisks) and longitudinally arranged smooth muscle bundles (six-pointed asterisks) as well as seromucous salivary glands (sl). Outside, trachea is covered with loose connective tissue tunica adventitia (ta).

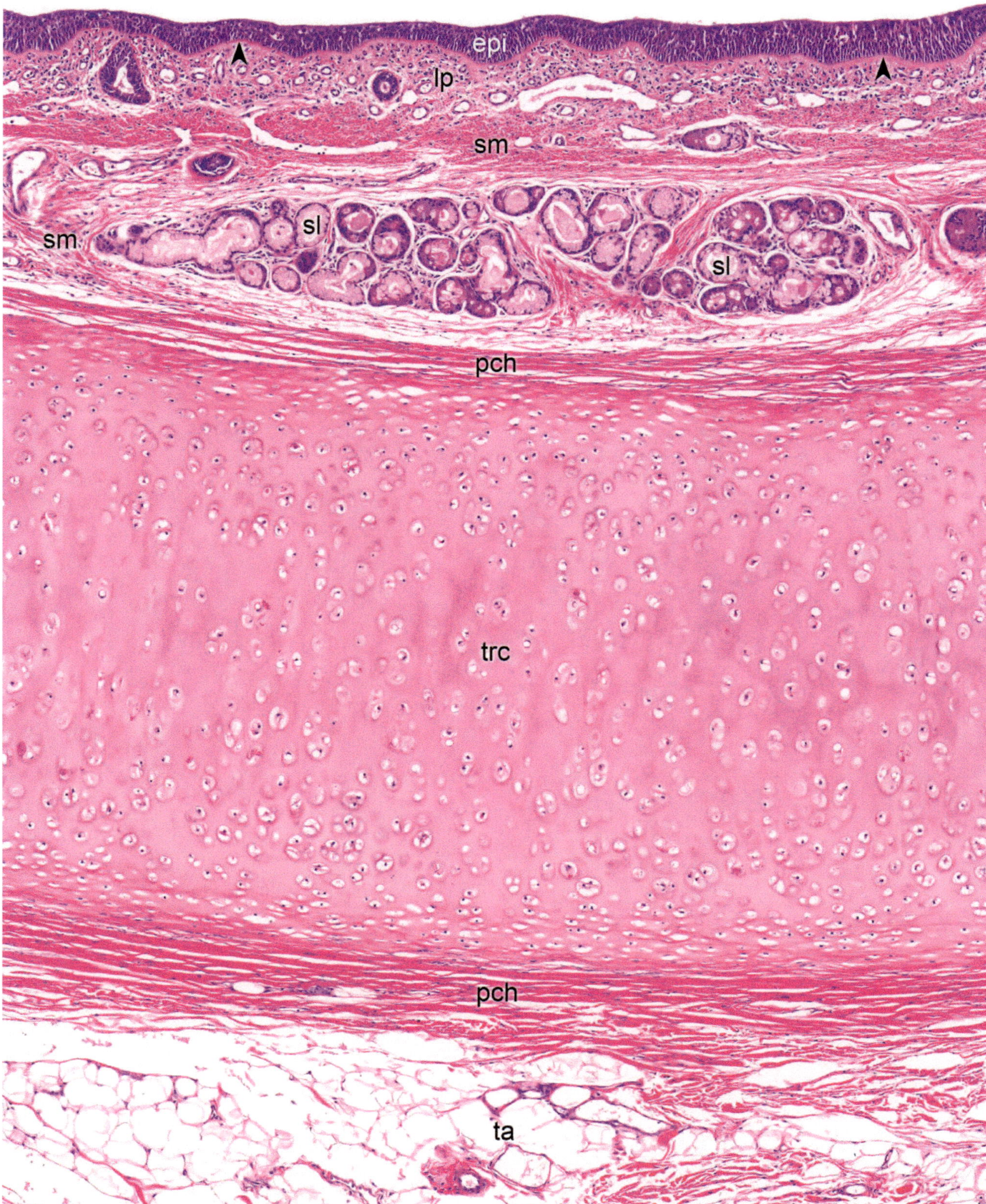

Figure 186. Trachea. Tracheal mucosa is composed of kinociliated pseudostratified columnar epithelium (epi) with occasional goblet cells and thick basement membrane (arrowheads), and the underlying loose connective tissue lamina propria (lp). Dense irregular connective tissue submucosa (sm) contains salivary glands (sl), and it is attached to the perichondrium (pch) of the C-shaped tracheal cartilages (trc) that are composed of hyaline cartilage. Outside, trachea is covered with loose connective tissue tunica adventitia (ta).

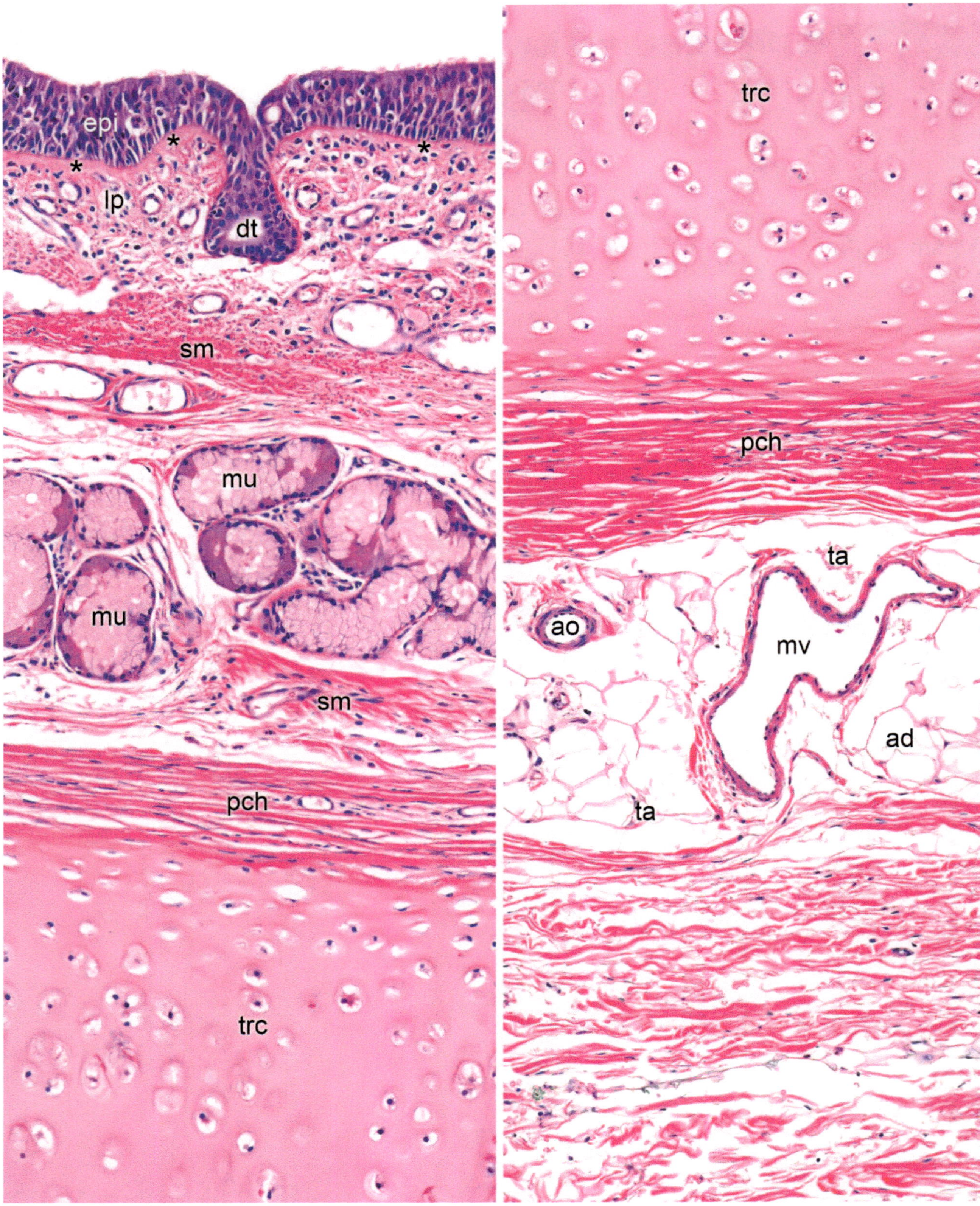

Figure 187. Trachea, internal (left) and external (right) layers. Tracheal mucosa is composed of kinociliated pseudostratified columnar epithelium (epi) with occasional goblet cells and thick basement membrane (asterisks), and the underlying loose connective tissue lamina propria (lp). Dense irregular connective tissue submucosa (sm) contains mostly mucous acini (mu) with serous demilunes and ducts (dt), and it is attached to the perichondrium (pch) of the C-shaped tracheal cartilages (trc) that are composed of hyaline cartilage. Outside, trachea is covered with tunica adventitia (ta) containing adipocytes (ad), arterioles (ao) and muscular venules (mv).

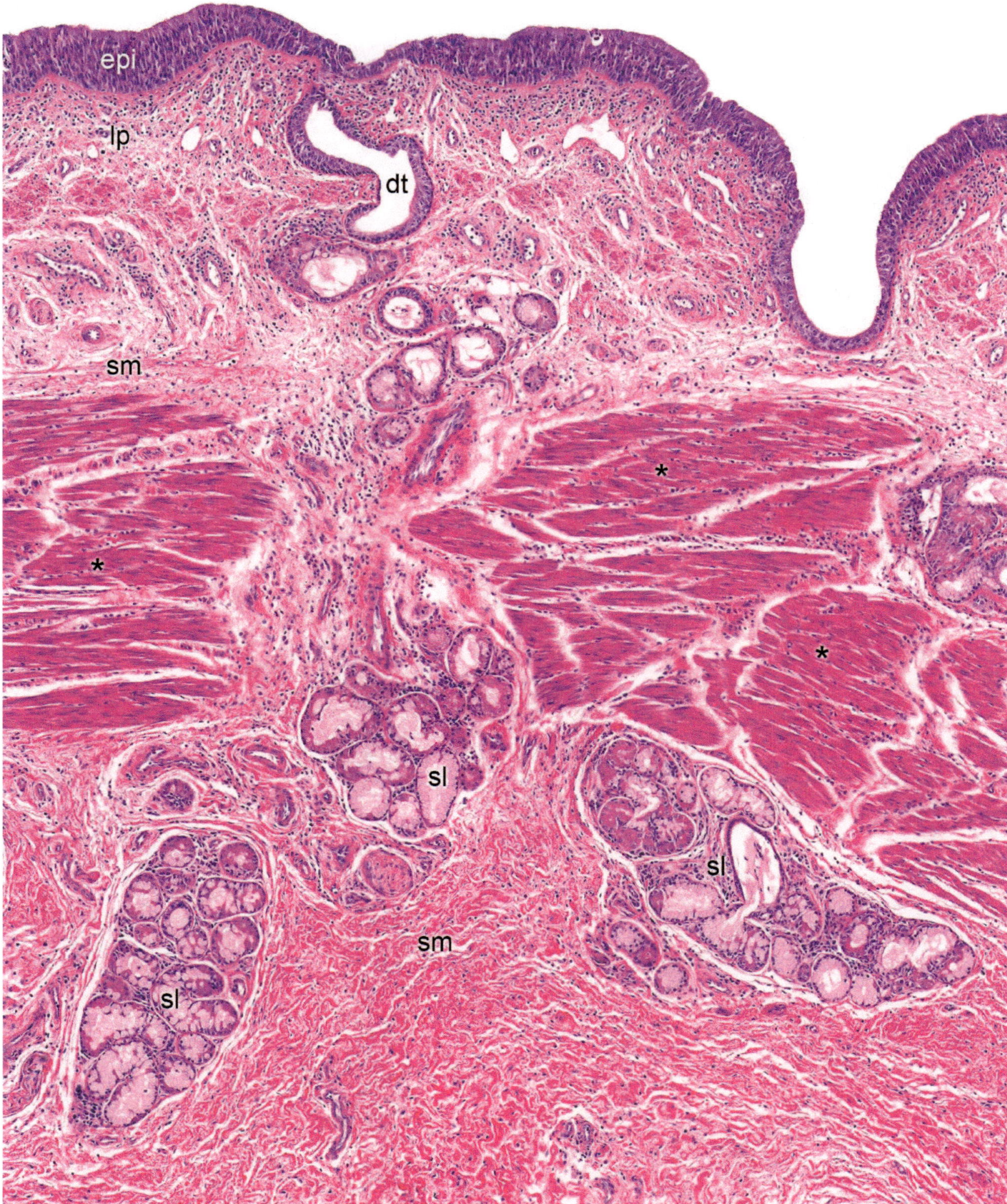

Figure 188. Trachea (newborn), membranous part. Posteriorly, the tracheal wall lacks tracheal cartilage. The loose connective tissue lamina propria (lp), is covered by pseudostratified columnar epithelium (epi) with occasional goblet cells. Dense connective tissue tunica submucosa (sm) contains circularly and longitudinally arranged smooth muscle bundles (asterisks) as well as seromucous salivary glands (sl) with their ducts (dt) opening on the epithelial surface.

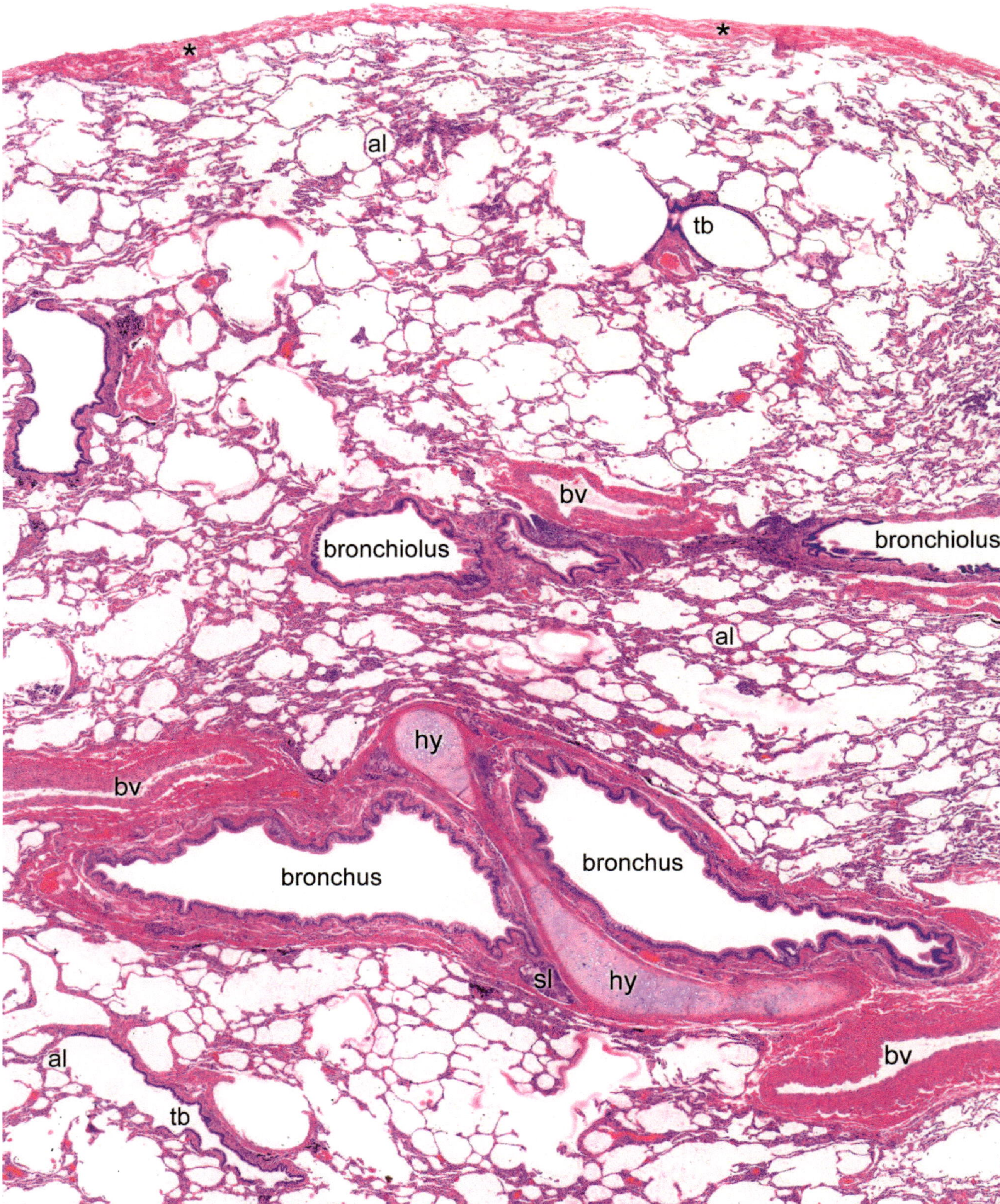

Figure 189. Lung. The surface of the lung is covered with pleura that is a serous membrane (asterisks) with mesothelial lining. Bronchi has a hyalin cartilage (hy) component while bronchioli lack cartilage. Bronchi are covered by ciliated pseudostratified columnar epithelium and contain seromucous salivary glands (sl), while bronchioli are lined by columnar cells that become cuboidal in the terminal bronchioli that also contains Clara cells. Blood vessels (bv) are associated with bronchi and bronchioli. Terminal bronchioli (tb) eventually transition to alveoli (al).

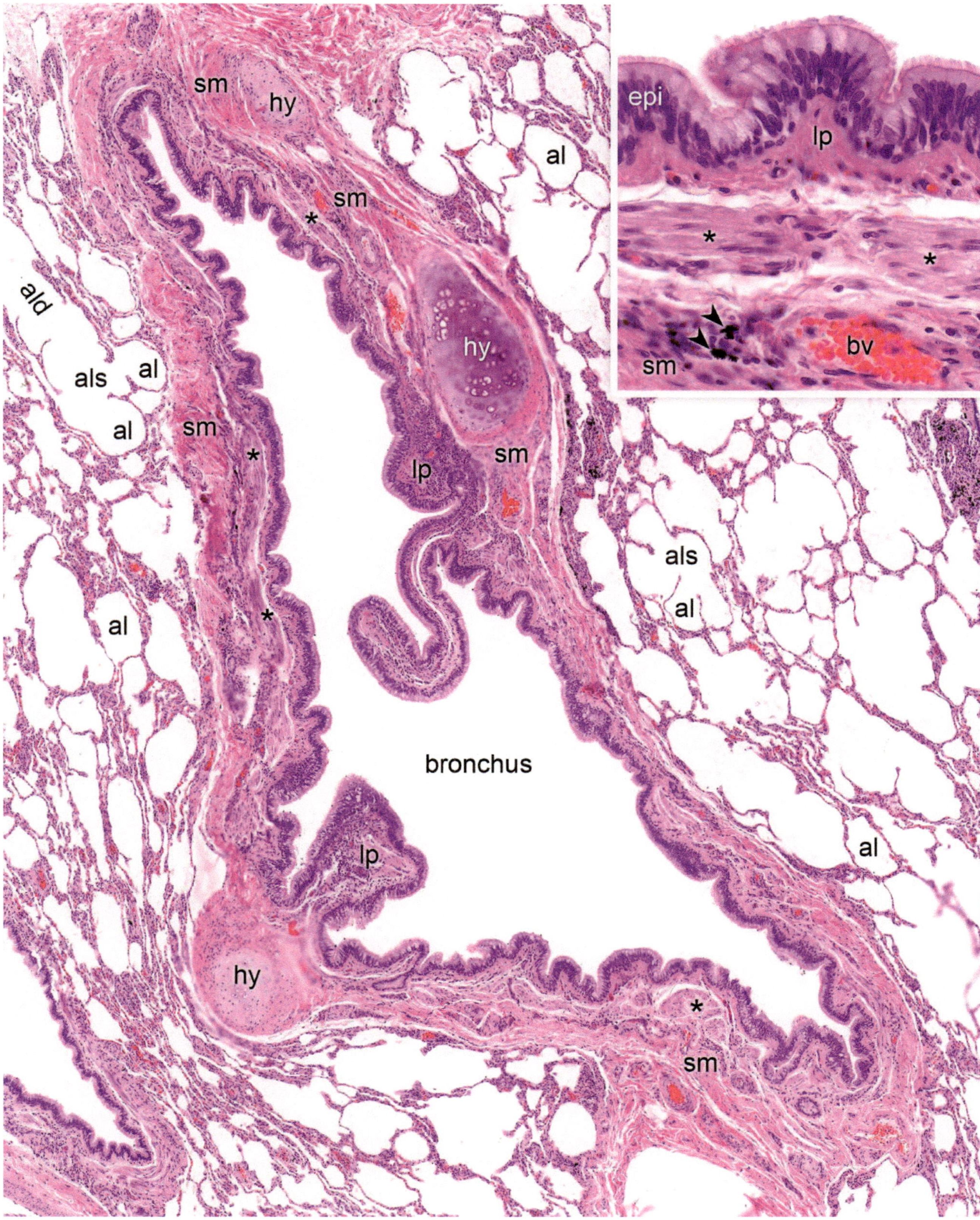

Figure 190. Bronchus. Bronchi has a hyalin cartilage (hy) component while bronchioli lack cartilage. Bronchi are covered by ciliated pseudostratified columnar epithelium (epi, inset). The underlying lamina propria rests on the tunica muscularis mucosae (asterisks) that is substantial in bronchi. Tunica submucosa (sm) is rich in collagen fibers and contains blood vessels (bv, inset). Bronchi and bronchioli eventually lead to alveolar ducts (ad) that terminate in alveolar sacs (als) and alveoli (al).

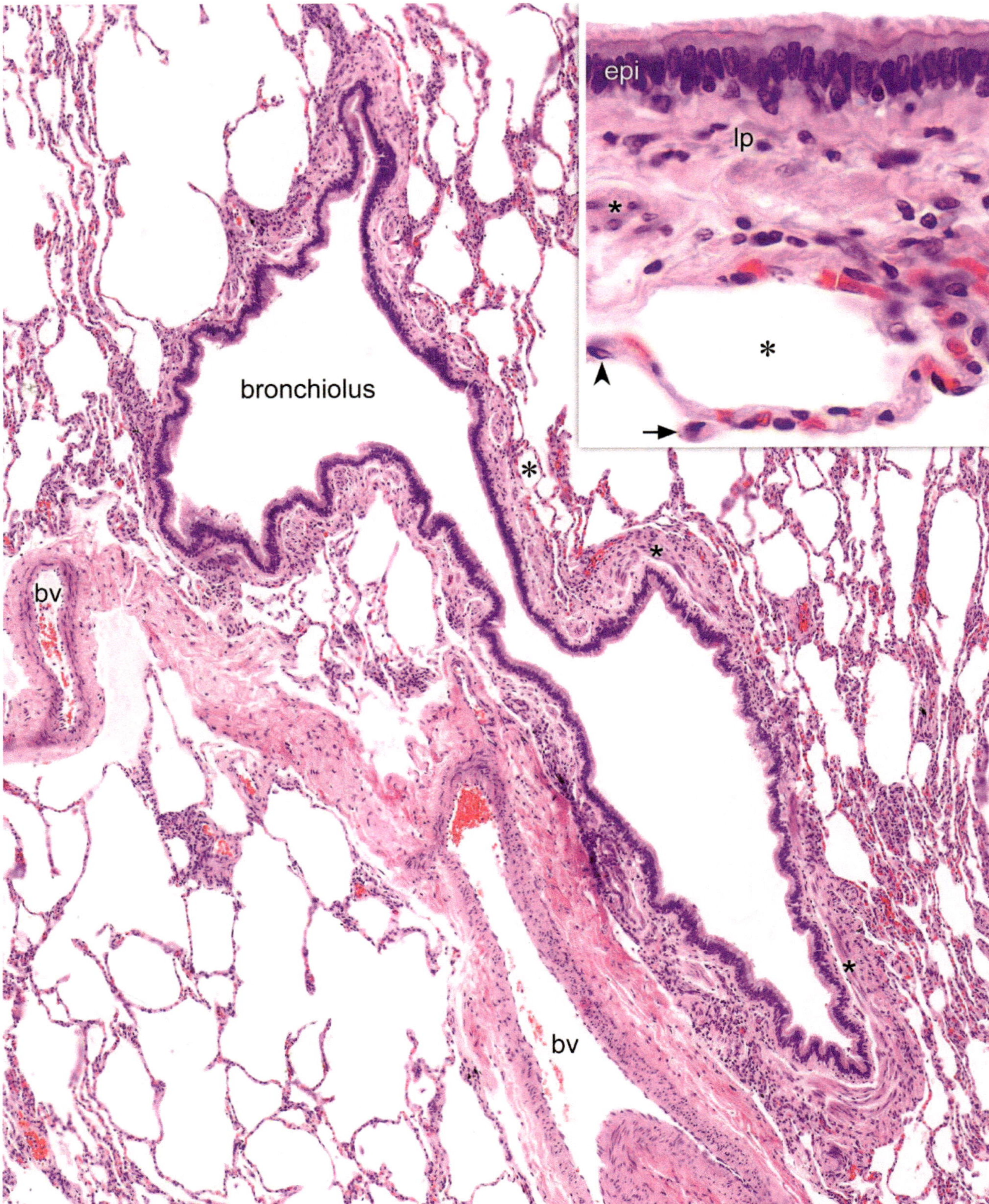

Figure 191. Bronchiolus. Bronchioli lack hyalin cartilage and they are covered initially by ciliated pseudostratified columnar epithelium (epi, inset) that is eventually replaced by cuboidal cells at the terminal bronchioli. Lamina propria (lp) rests on the tunica muscularis mucosae (five-pointed asterisks) that is thin and often incomplete in bronchioli. Tunica submucosa is thin and rich in collagen fibers. Bronchi and bronchioli eventually terminate in alveoli (six-pointed asterisk) that contain squamous type I alveolar cells (inset, arrowhead) and dome-shaped type II pneumocytes (inset, arrow). Blood vessels (bv) are associated with bronchi and bronchioli.

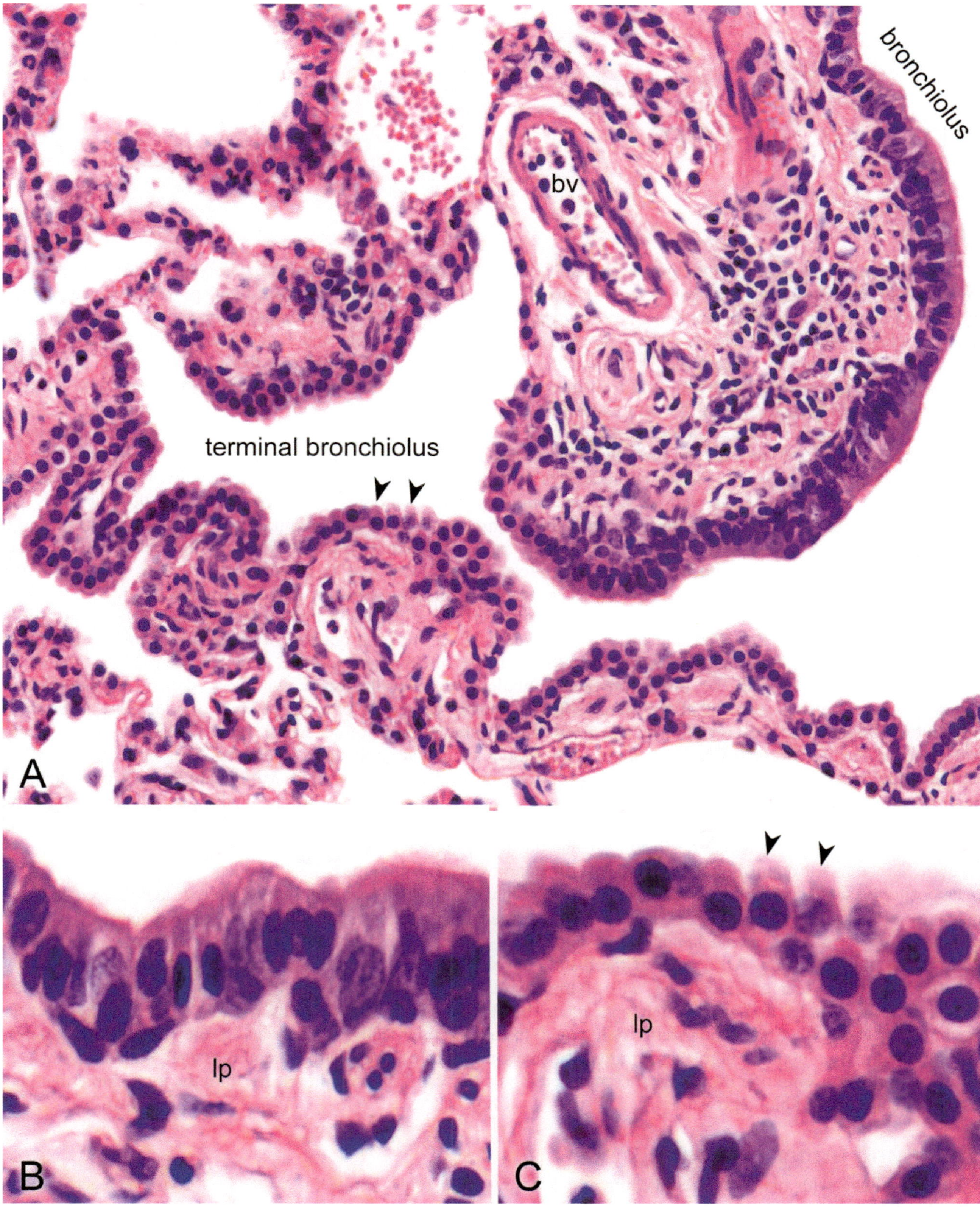

Figure 192. Terminal bronchiolus. (A) The smallest bronchioli are covered by simple columnar epithelium and it is eventually replaced by ciliated cuboidal cells at the terminal bronchioli that also contain Clara cells with tall, dome-shaped apical portion (arrowheads). (B) Simple columnar epithelium of the smallest bronchioli rests on thin lamina propria (lp). (C) Located between the ciliated cuboidal cells of the terminal bronchioli, the taller, dome-shaped Clara cells populate the epithelium (arrowheads). Clara cells produce surfactant lipoproteins and their number increases distally.

Chapter 9

Urinary System

Kidney

Kidney is covered with a fibrous capsule formed by an outer layer of dense connective tissue and an inner layer that is composed of myofibroblasts. The parenchyma under the capsule can be divided into a light, pyramid-shaped medulla surrounded by the darker cortex that penetrates between the pyramids as cortical columns, representing the borders of the adjacent lobes of the kidney clearly visible in embryonic age. The structural and functional unit of the kidney is the nephron, a delicate tubular system that is responsible for the formation of the filtrate from blood plasma; this filtrate eventually turns into urine by secretion and reabsorption of substances by the tubular cells. Nephron starts with the renal corpuscle which is composed of a system of capillary loops, called glomerulus, surrounded by the Bowman's capsule. This tuft of capillaries is located between the afferent and the efferent arterioles, the former having a slightly larger diameter than the latter to provide the pressure gradient that is necessary for the filtration process. The inner, "visceral" wall of the Bowman's capsule is formed by podocytes that surround the capillary loops and together with the fenestrated endothelium of the capillaries as well as the underlying basement membrane they form the blood-filtrate barrier. The parietal wall of the Bowman's capsule is composed of simple squamous epithelium, while the cavity of the capsule opens into the proximal convoluted tubule at the urinary pole of the renal corpuscle. Proximal convoluted tubule leads to the Henle's loop that is composed of a thick descending segment that is histologically similar to the proximal convoluted tubules, a thin segment, as well as a thick ascending segment that is continuous with the distal convoluted tubule and structurally similar to it. Most of the Henle's loop is located in the medulla while the rest of the nephron is in the cortex. Proximal and distal convoluted tubules are both covered with simple cuboidal epithelium; however, the proximal tubules are lined by cells that are darker eosinophilic due to their high content of mitochondria while cells of the distal convoluted tubules are generally lighter with less defined boundaries and bulging nuclei. The thin segment of the Henle's loop is mostly composed of a single layer of squamous cells. Distal convoluted tubules enter the collecting tubule lined with a single layer of cuboidal cells; collecting tubules enter into larger collecting ducts (of Bellini) lined with simple columnar epithelium. Collecting ducts open into the minor calices at the tip of the renal papilla that is the apical portion of the renal pyramid. Minor calices drain into the major calices that lead into the renal pelvis that is covered with transitional epithelium that gradually gets thinner towards the renal papilla. The bundles of the collecting ducts are easily detectable in the cortex forming medullary rays of Ferrein; "medullary" refers to the destination of the ducts and not their location. Medullary rays represent the center of the cortical lobules surrounded by the cortical part of nephrons opening into the collecting ducts.

Opposite to the urinary pole, the vascular pole of the renal corpuscle is defined by the afferent and efferent arteriole and the extraglomerular mesangial (lacis) cells between them. Mesangial cells also extend into the tuft of capillaries, enclosed by the basal lamina; these contractile cells are capable of phagocytosis and they are believed to have smooth muscle origin. A segment of the distal convoluted tubule is located adjacent to the glomerulus, between the afferent and efferent arterioles; the wall of this segment that is facing the glomerulus is composed of closely packed cells and nuclei and therefore it appears to be darker and it is termed as macula densa. Extraglomerular mesangial cells, together with the macula densa and juxtaglomerular cells in the wall of the afferent and efferent arterioles form the juxtaglomerular apparatus that plays a key role in regulating blood pressure. Macula densa cells monitor sodium concentration of the distal convoluted tubule that can trigger renin release from the juxtaglomerular cells resulting in elevation of blood pressure.

Ureter and urinary bladder

Ureter is lined with transitional epithelium that is thrown into folds. Under the epithelium, loose connective tissue forms the lamina propria mucosae. The relatively thick smooth muscle layer under the mucous membrane is composed of inner longitudinal and outer circular layers proximally; distally, a third, outermost longitudinal layer can be observed. Autonomic ganglion cells are dispersed between the muscle layers. Outside, the ureter is covered with loose connective tissue tunica adventitia. The structure of the urinary bladder follows a similar distribution, but the layers of the tunica muscularis (detrusor muscle) are more elusive, and the superior surface of the bladder is covered with serosa.

http://doi.org/10.1016/B978-0-323-91891-6.50009-X,

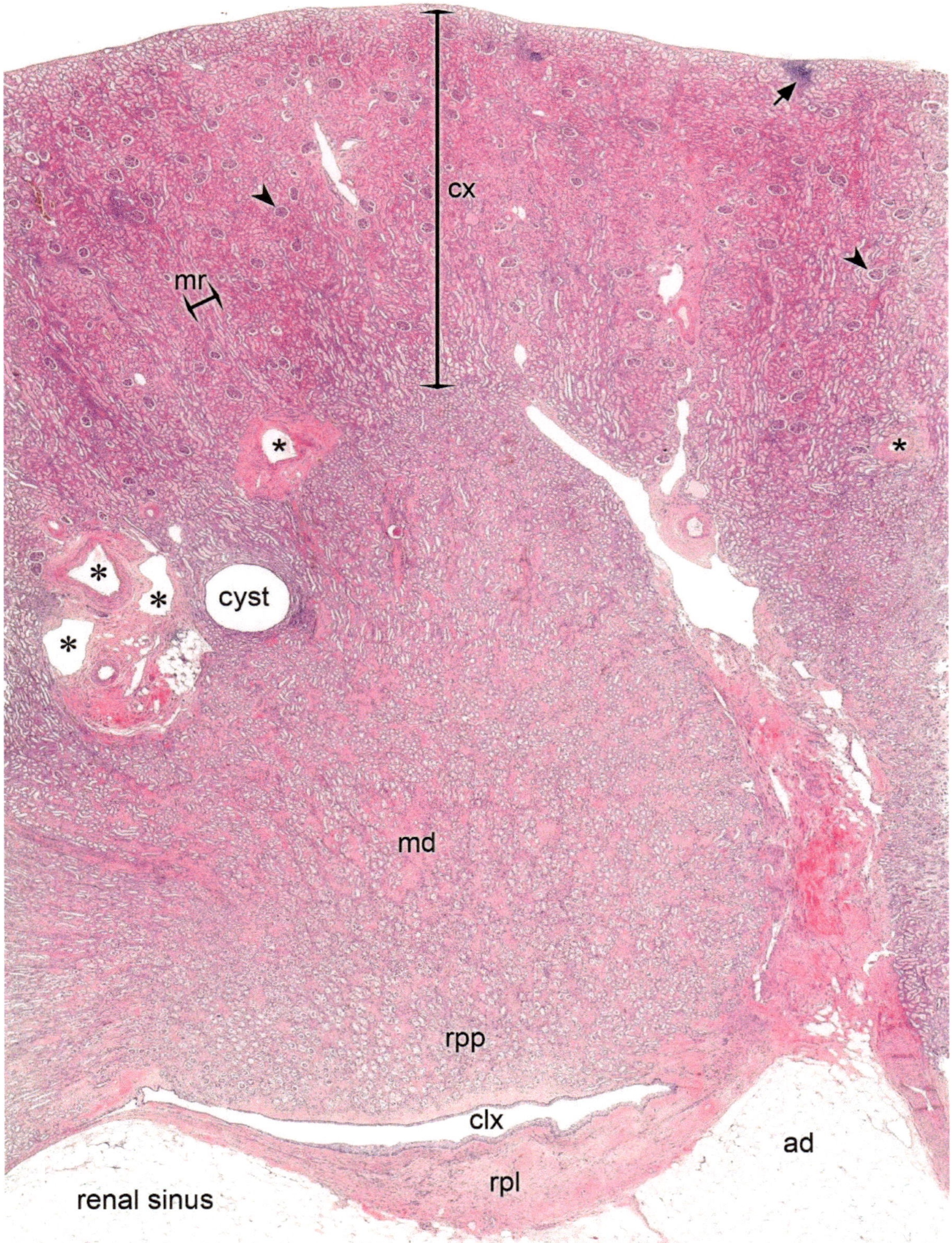

Figure 193. Kidney. Renal cortex (cx) contains nephrons that are composed of spherical renal corpuscles (arrowheads) and uriniferous tubules. Occasional lymphatic aggregations can be observed under the renal capsule (arrow). Nephrons lead into collecting ducts that form bundles in the cortex called medullary rays of Ferrein (mr). Medulla (md) lacks renal corpuscles and forms pyramidal structures (renal pyramids). Collecting ducts running in the medulla open into a minor calyx (clx) at the tip of the pyramid (renal papilla, rpp). Minor and major calices eventually lead into the renal pelvis that is covered with transitional epithelium and it is located in the renal sinus that also contains adipose tissue (ad), as well as vessels and nerves. Between the cortex and the medulla, arcuate arteries and veins can be observed (five-pointed asterisks) that are branches of the interlobar blood vessels (six-pointed asterisks) surrounded by connective tissue. Cysts are relatively common in the kidney at elder age.

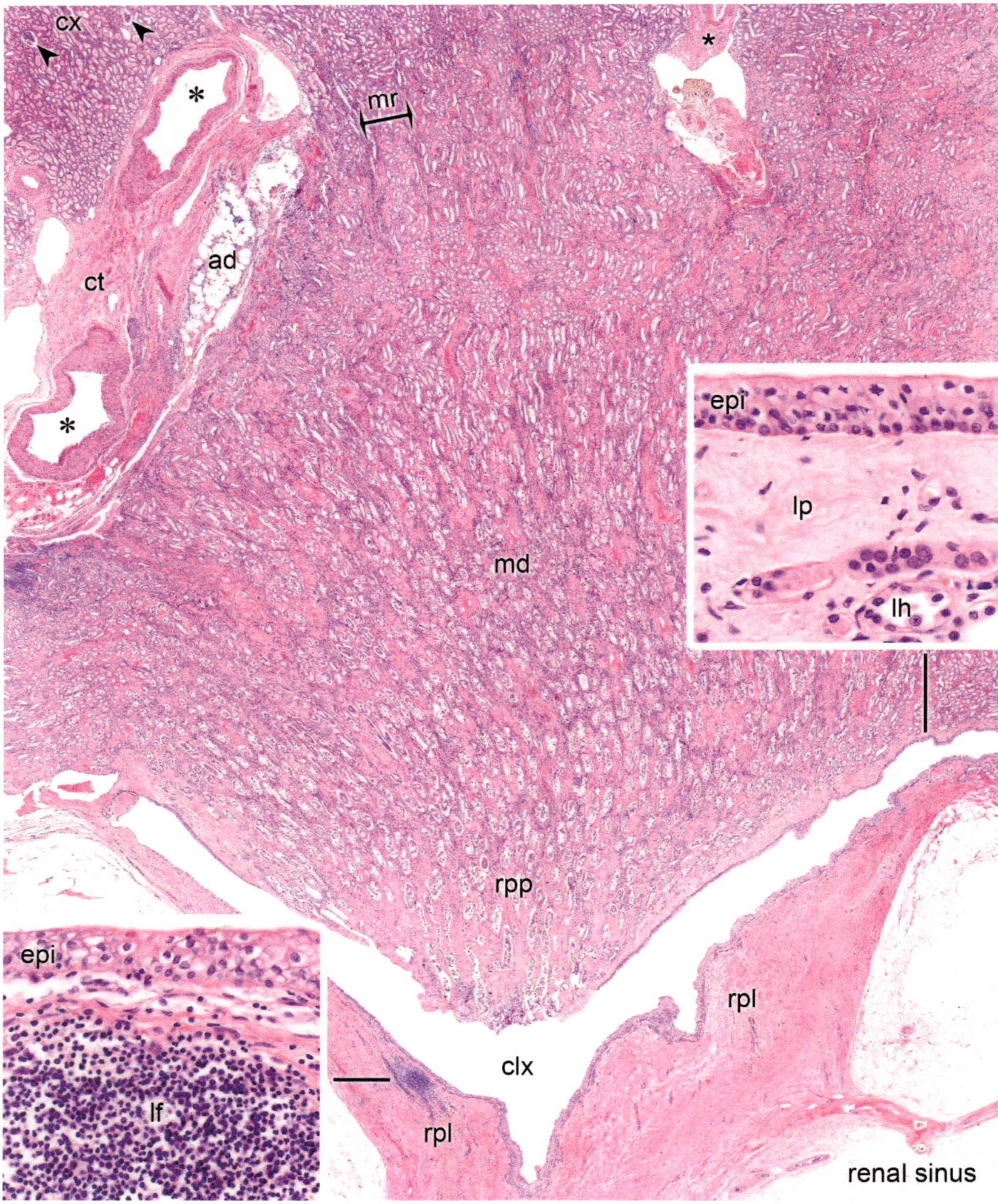

Figure 194. Kidney. Renal cortex (cx) contains spherical renal corpuscles (arrowheads) and bundles of collecting ducts called medullary rays of Ferrein (mr) that penetrate the medulla. Medulla (md) lacks renal corpuscles and it protrudes into the minor calyx (clx) with the renal papilla (rpp) that is covered with thin transitional epithelium (epi) and the underlying lamina propria (lp) (right inset; lh, thick segment of loop of Henle). Renal pelvis (rpl; left inset) is a dense connective tissue sac with occasional lymph follicles (lf) and transitional epithelial lining (epi). Between the cortex and the medulla, arcuate arteries (five-pointed asterisks) derive from the interlobar arteries (six-pointed asterisks) surrounded by connective tissue (ct) with adipocytes (ad).

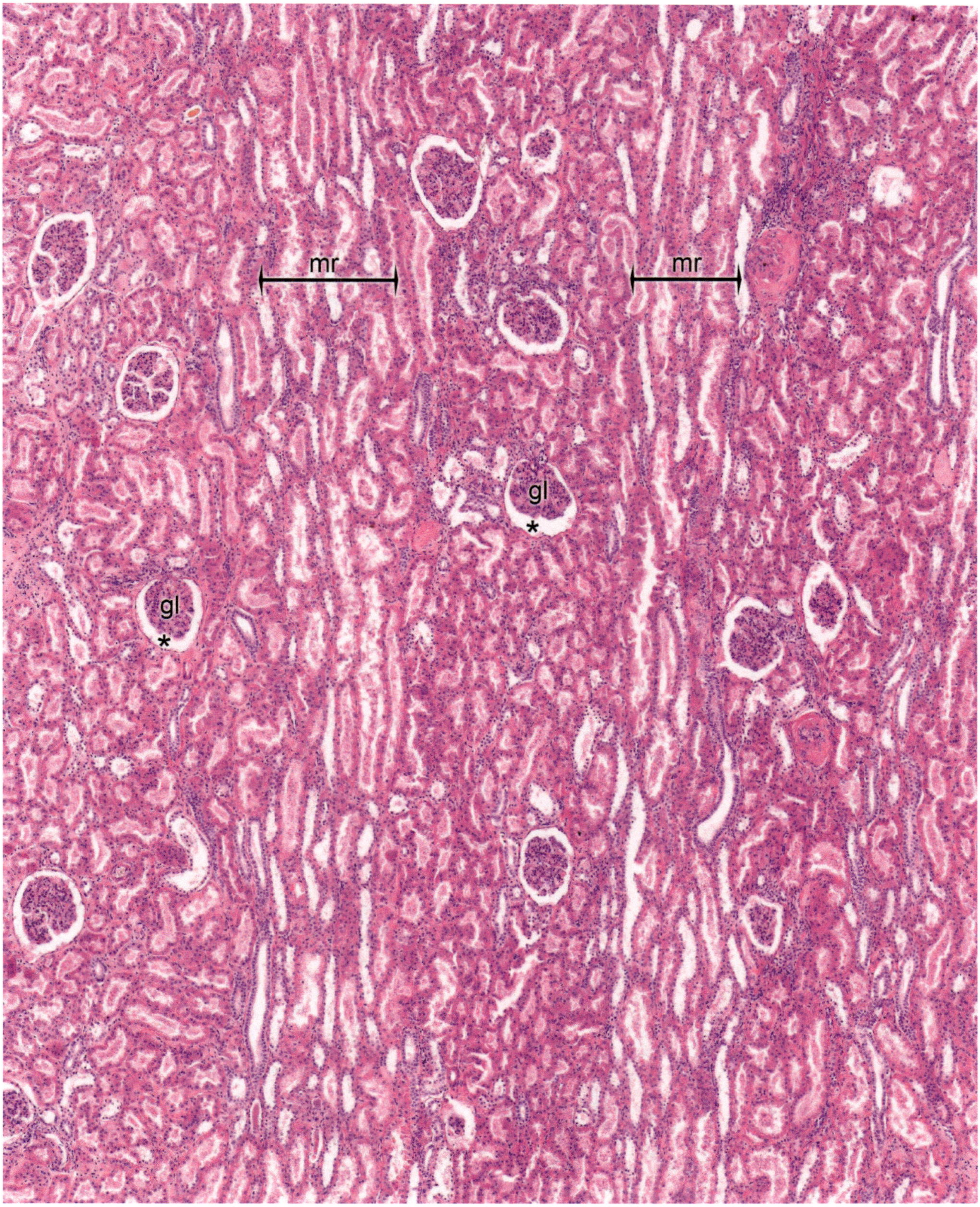

Figure 195. Renal cortex. Renal cortex (cx) contains nephrons that are composed of spherical renal corpuscles and uriniferous tubules. Renal corpuscles are formed by a spherical tuft of capillaries called glomerulus and the surrounding Bowman's capsule (asterisks). Nephrons lead into collecting ducts that form bundles in the cortex called medullary rays of Ferrein (mr) that represent the center of the cortical lobules surrounded by the cortical part of nephrons opening into the collecting ducts.

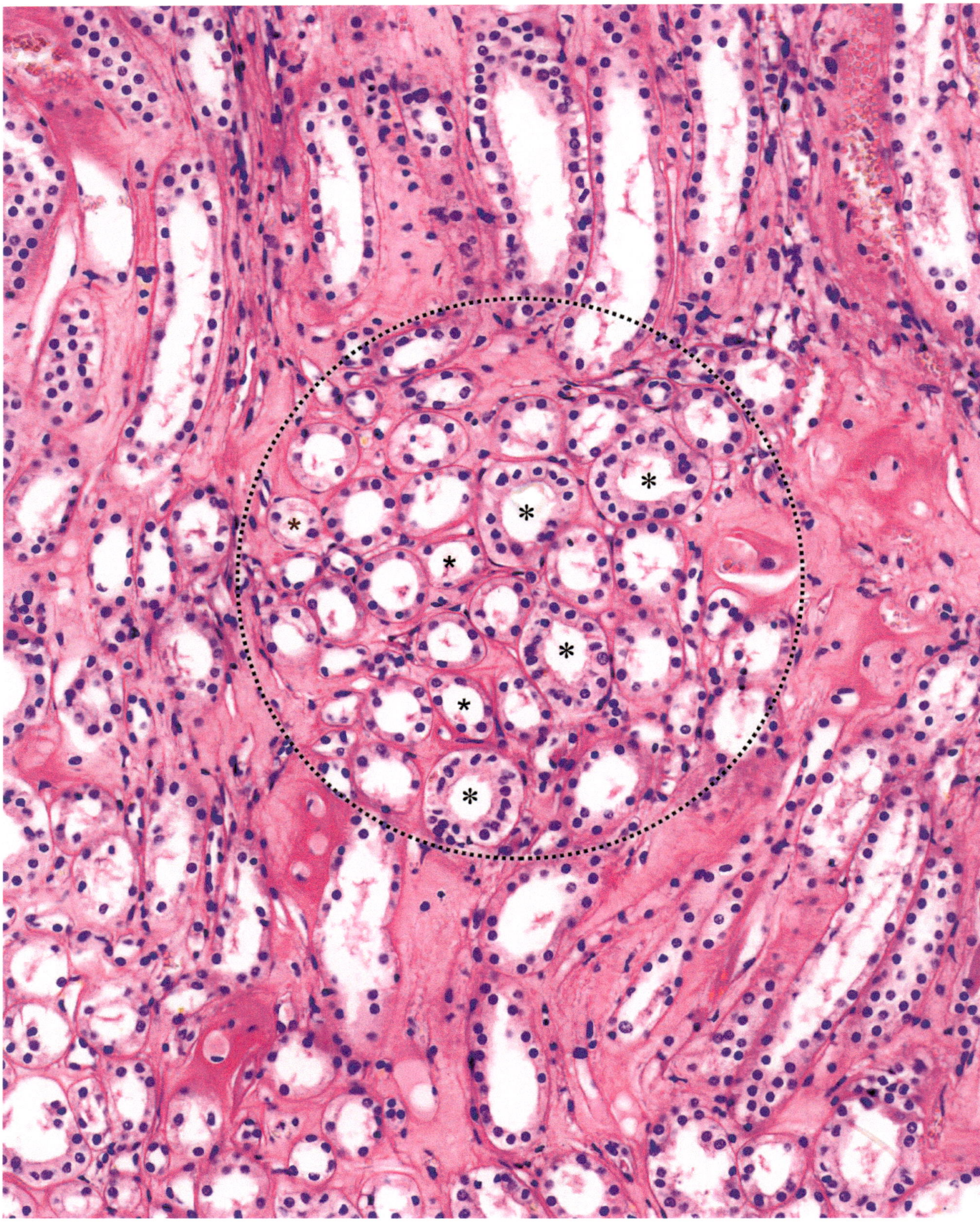

Figure 196. Renal cortex, cross section of the medullary rays. Renal cortex contains nephrons that are composed of spherical renal corpuscles and uriniferous tubules that lead into collecting ducts. Segments of loop of Henle (five-pointed astrerisks) together with the collecting ducts (six-pointed asterisks) form bundles in the cortex called medullary rays of Ferrein, denoted by the dotted line. Medullary rays represent the center of the cortical lobules surrounded by the cortical part of nephrons opening into the collecting ducts.

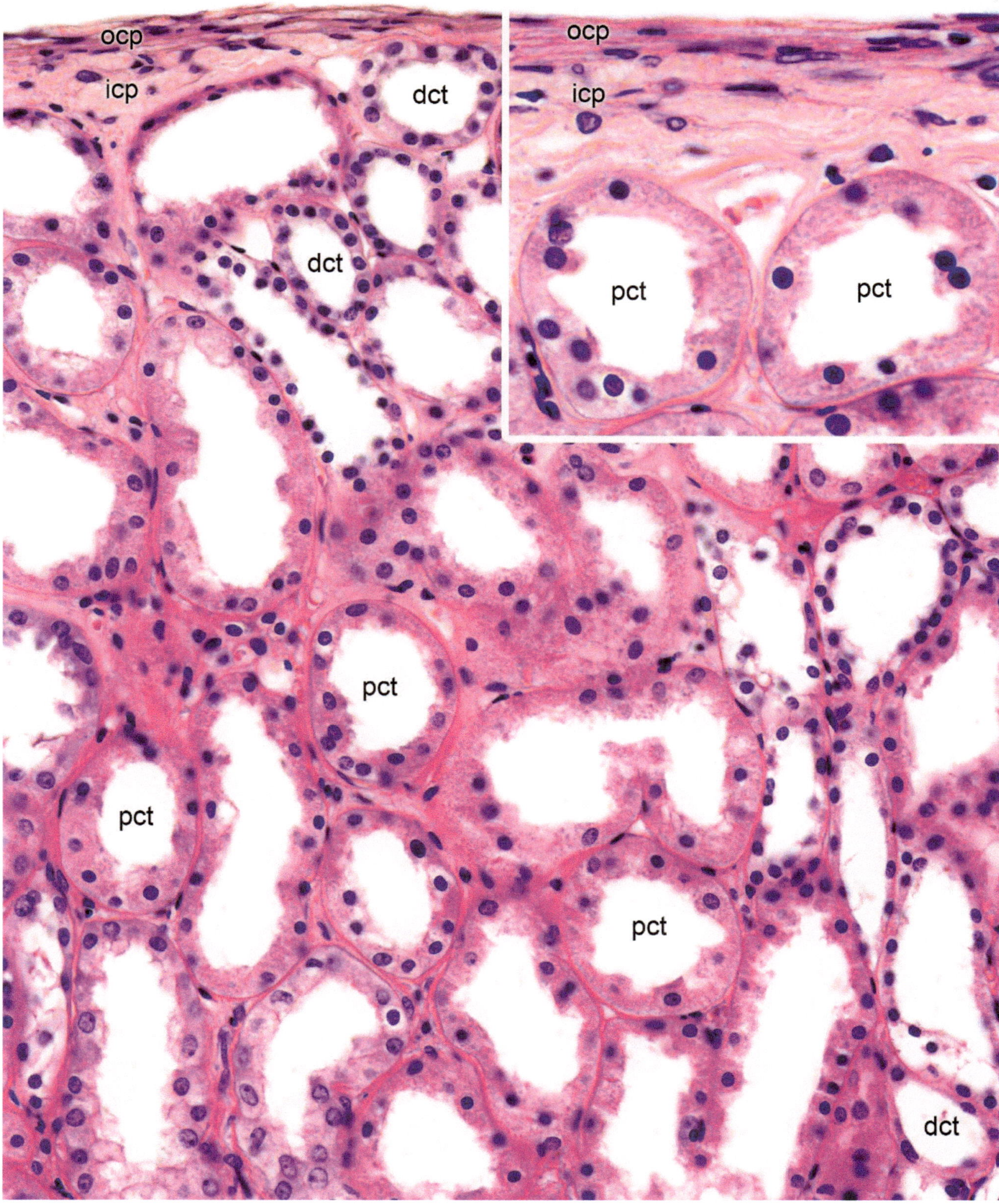

Figure 197. Renal cortex, convoluted tubules. Kidney is covered with a fibrous capsule that is composed of an outer layer of dense connective tissue (ocp) and an inner layer that is composed of myofibroblasts (icp). Under the capsule, proximal (pct) and distal convoluted tubules (dct) are both covered with simple cuboidal epithelium; however, the proximal tubules are lined by cells that are darker eosinophilic due to their higher content of mitochondria, while cells of the distal convoluted tubules are generally lighter with bulging nuclei and less defined boundaries.

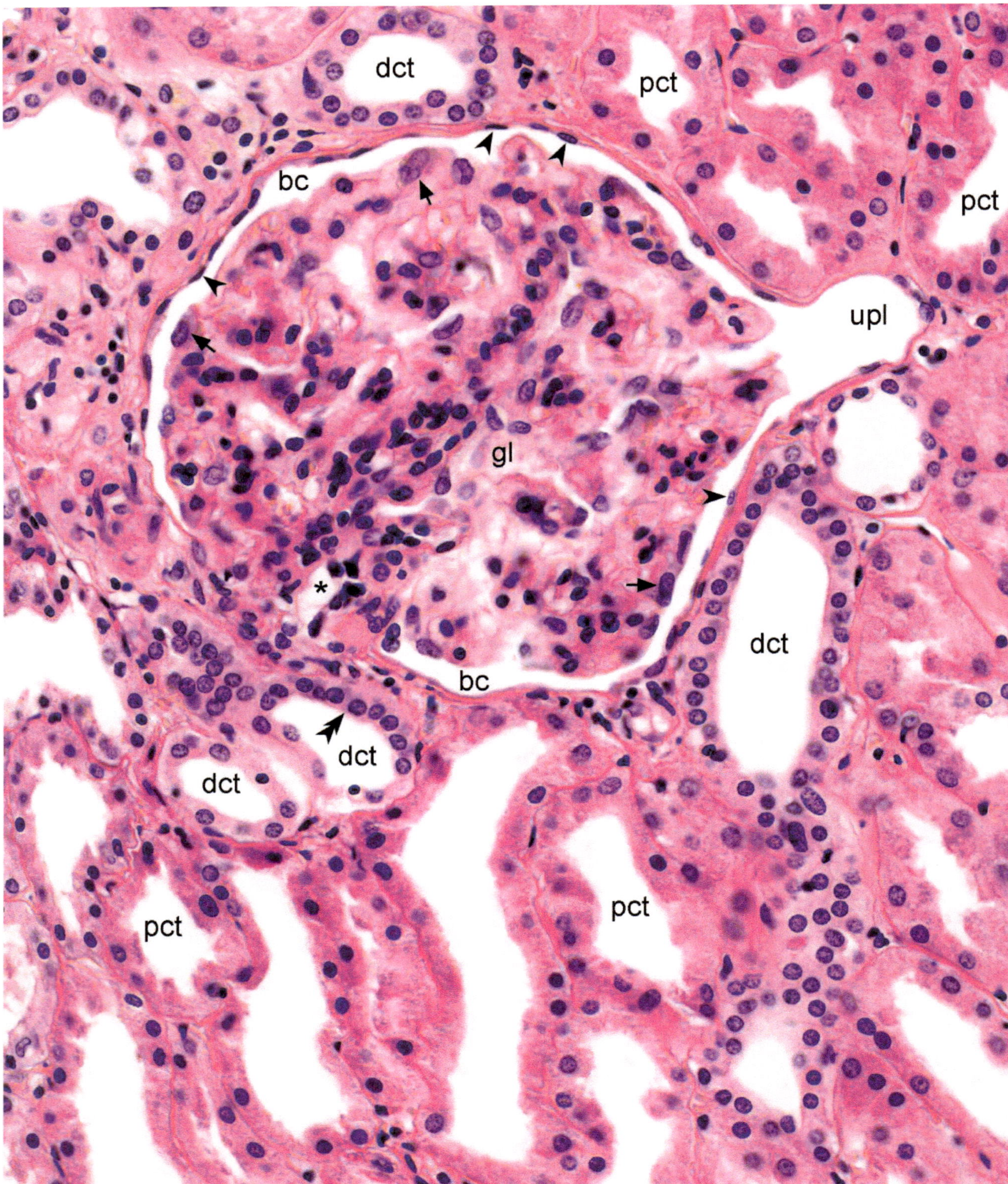

Figure 198. Renal corpuscle. Renal corpuscles are formed by the glomerulus (gl), a spherical tuft of capillaries between the afferent and efferent arterioles (asterisk), surrounded by the Bowman's capsule (bc). The parietal part of the Bowman's capsule is covered with simple squamous epithelium (arrowheads) while the layer covering the glomerulus is composed of podocytes (arrows) with large, lightly stained nuclei. Bowman's capsule drains into the proximal convoluted tubule at the urinary pole (upl). Proximal (pct) and distal convoluted tubules (dct) are both covered with simple cuboidal epithelium; however, the proximal tubules are lined with cells that are more eosinophilic, while cells of the distal convoluted tubules are lighter with more ambiguous borders and bulging nuclei. A segment of a distal convoluted tubule adjacent to the glomerulus contains tightly packed nuclei and it is called macula densa (double arrowheads).

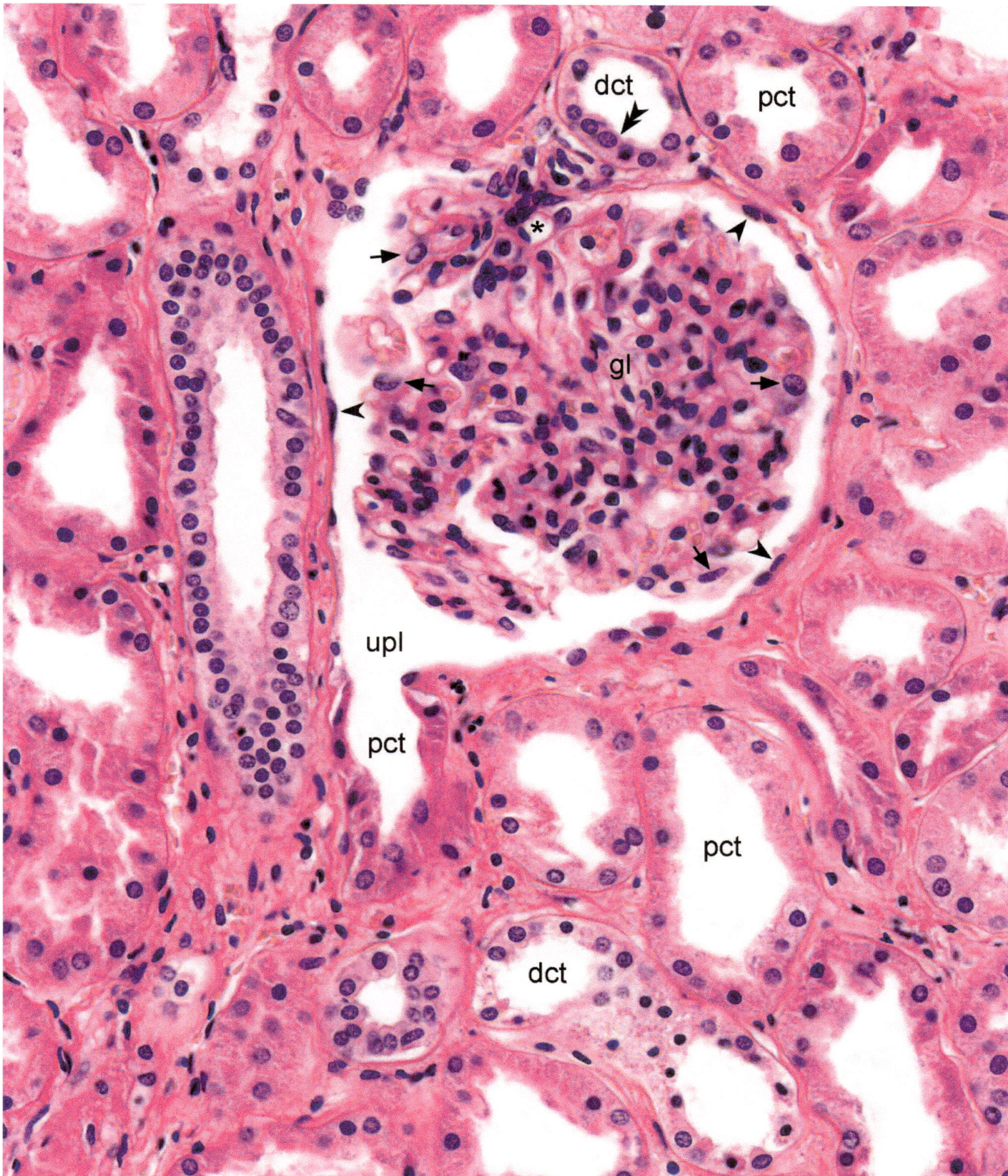

Figure 199. Renal corpuscle. Renal corpuscles are formed by the glomerulus (gl), a spherical tuft of capillaries between the afferent and efferent arterioles (asterisk), surrounded by the Bowman's capsule (bc). The parietal part of the Bowman's capsule is covered with simple squamous epithelium (arrowheads) while the layer covering the glomerulus is composed of podocytes (arrows) with large, lightly stained nuclei. Bowman's capsule drains into the proximal convoluted tubule at the urinary pole (upl). Proximal (pct) and distal convoluted tubules (dct) are both covered with simple cuboidal epithelium; however, the proximal tubules are lined with cells that are more eosinophilic, while cells of the distal convoluted tubules are lighter with more ambiguous borders and bulging nuclei. A segment of a distal convoluted tubule adjacent to the glomerulus contains tightly packed nuclei and it is called macula densa (double arrowheads).

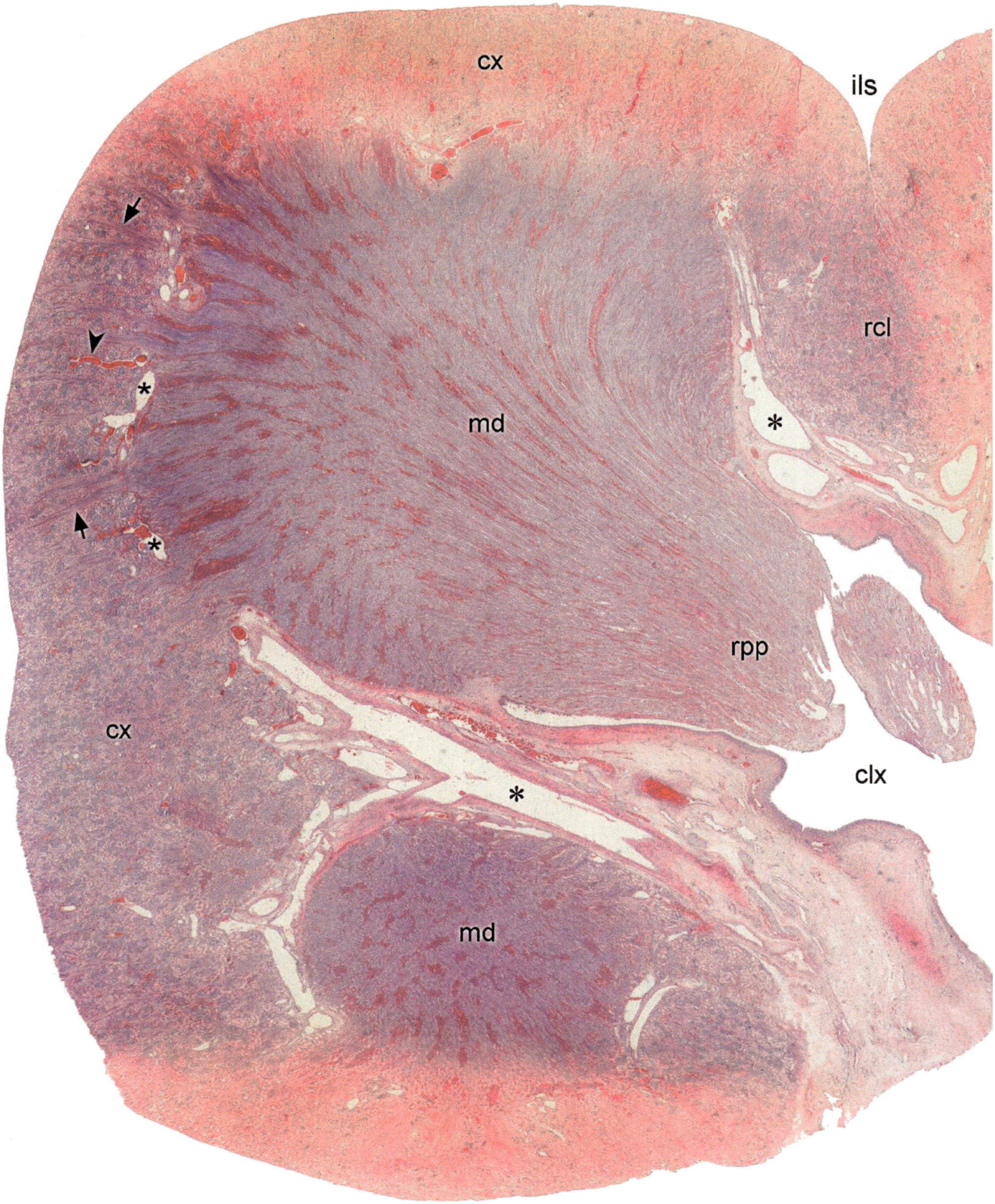

Figure 200. Kidney, newborn. The lobar structure of the kidney at birth is represented by interlobar sulci (ils) and cortical renal columns (rcl) extending between the medullary pyramids. Between the lobes, interlobar arteries and veins (six-pointed asterisks) form arcuate blood vessels (five-pointed asterisks) at the border between the cortex (cx) and medulla (md). Arcuate arteries give interlobular arteries that are located between the cortical lobules (arrowheads) while medullary rays of Ferrein denote the center of the lobules (arrows). Medullary pyramids protrude into the corresponding minor calyx (clx) by the renal papilla (rpp).

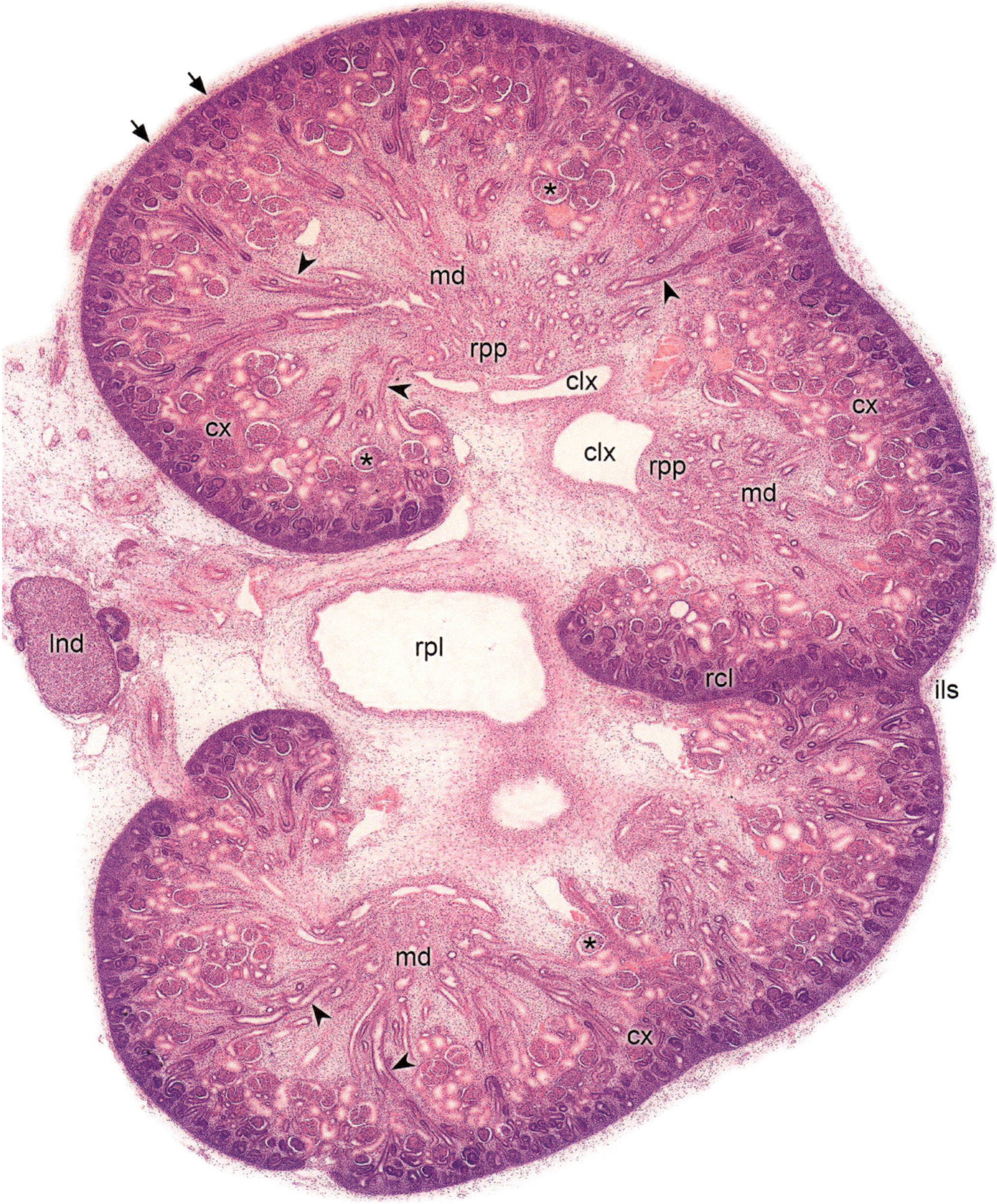

Figure 201. Kidney, 3-month-old fetus. The lobar structure of the fetal kidney is represented by interlobar sulci (ils) and cortical renal columns (rcl) extending between the medullary pyramid (md). Kidney is covered with a fibrous capsule (arrows) that is composed of dense connective tissue and myofibroblasts. Renal cortex (cx) contains spherical renal corpuscles (asterisks) and bundles of collecting ducts (arrowheads) that penetrate the medulla (medullary rays of Ferrein). The internal segment of the medullary pyramids form the renal papillae (rpp) that protrude into the minor calices (clx), which lead into the major calices that open into the renal pelvis (rpl).

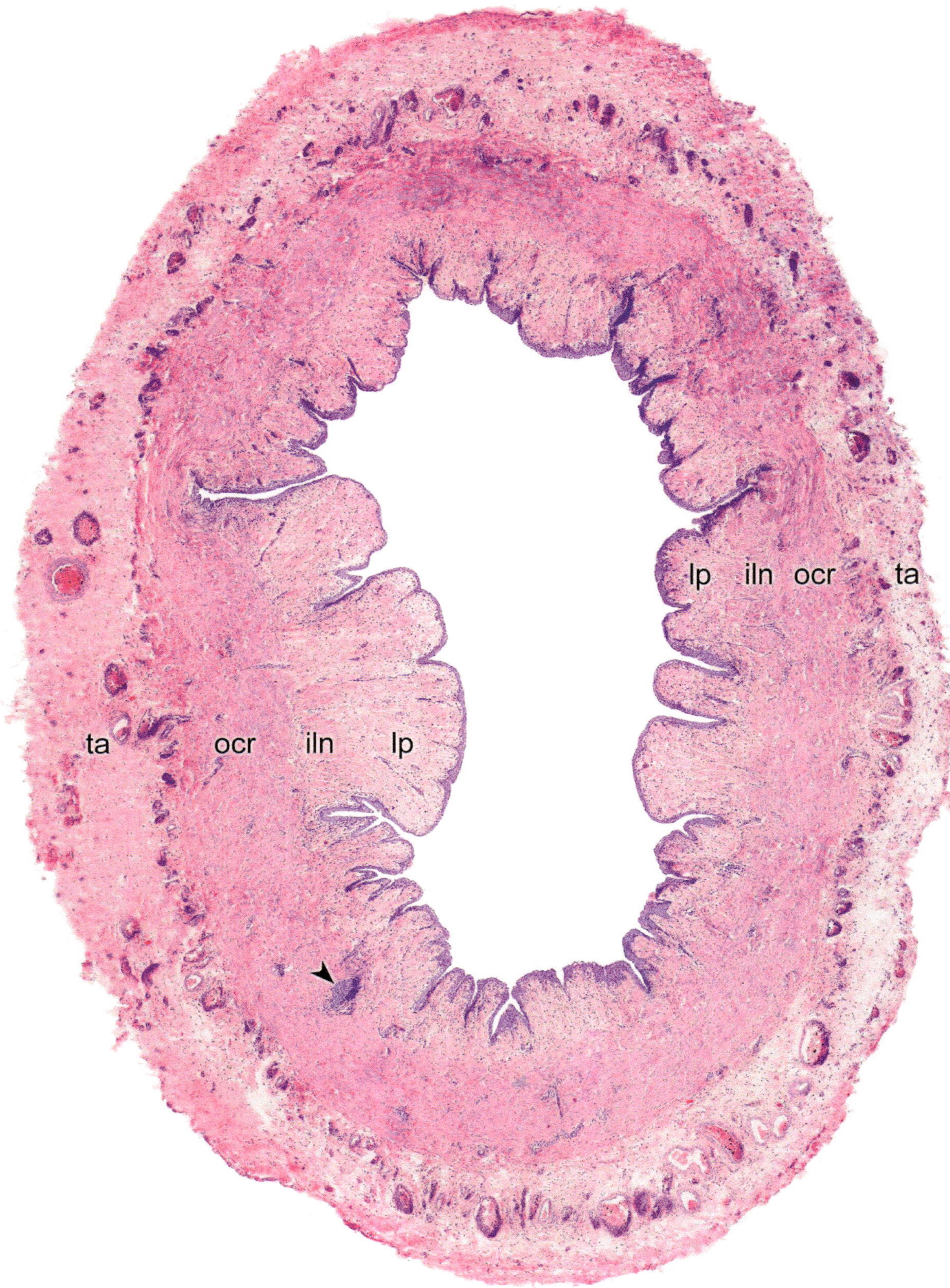

Figure 202. Ureter, proximal part. The tunica mucosa of the ureter is lined with transitional epithelium (urothelium) that is characterized with a luminal layer of umbrella cells. Lamina propria (lp) rests directly on the tunica muscularis; muscularis mucosae and submucosa are missing. Proximally, tunica muscularis is composed of inner longitudinal (iln) and outer circular (ocr) layers, while distally a third, outermost longitudinal layer can be observed. Loose connective tissue tunica adventitia (ta) covers the outer surface.

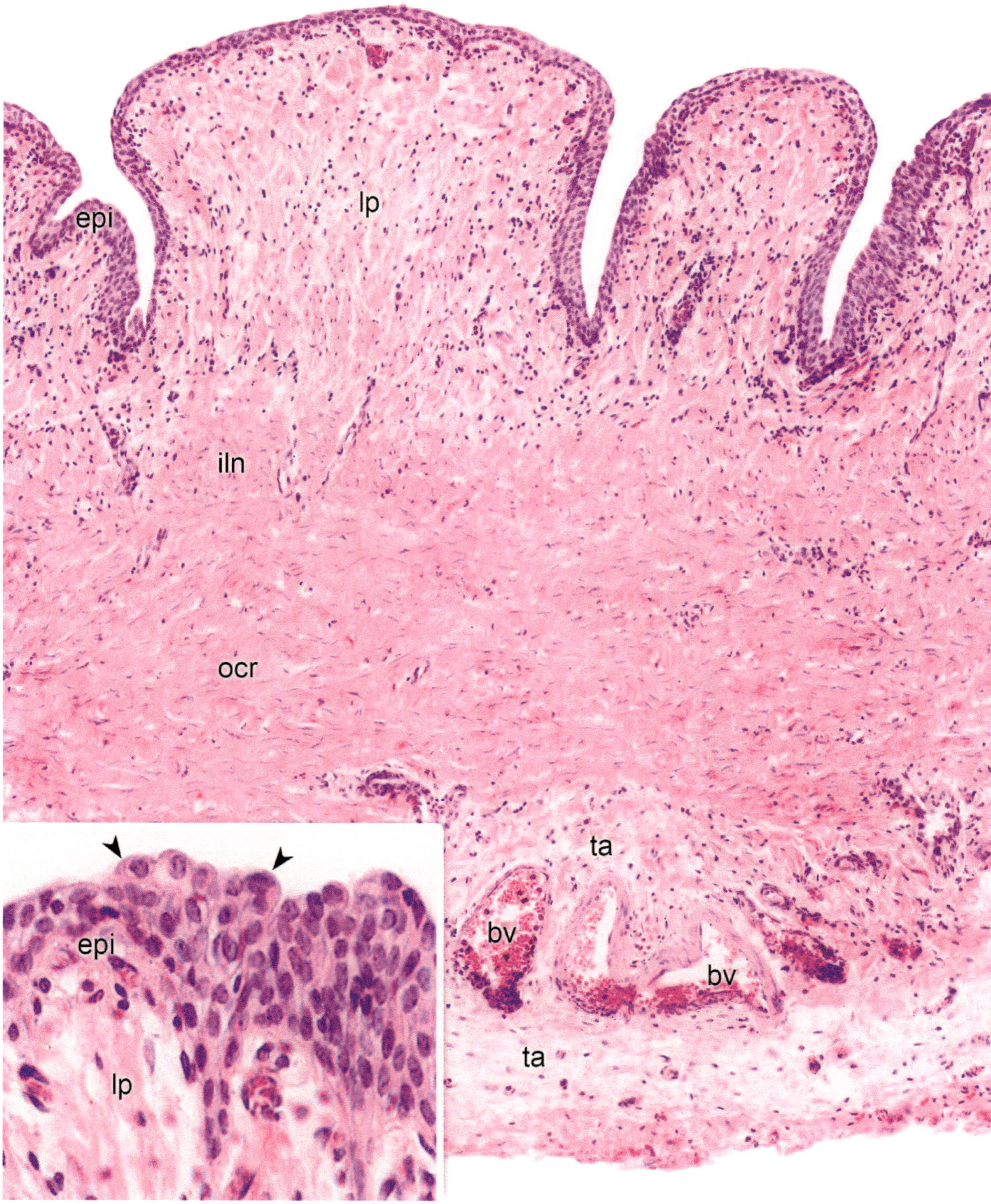

Figure 203. Ureter, proximal part. The tunica mucosa of the ureter is lined with transitional epithelium (epi) that is characterized with a luminal layer of umbrella cells (arrowheads, inset). Lamina propria (lp) rests directly on the tunica muscularis; muscularis mucosae and submucosa are missing. Proximally, tunica muscularis is composed of inner longitudinal (iln) and outer circular (ocr) layers, while distally a third, outermost longitudinal layer can be observed. Loose connective tissue tunica adventitia (ta) containing blood vessels covers the outer surface.

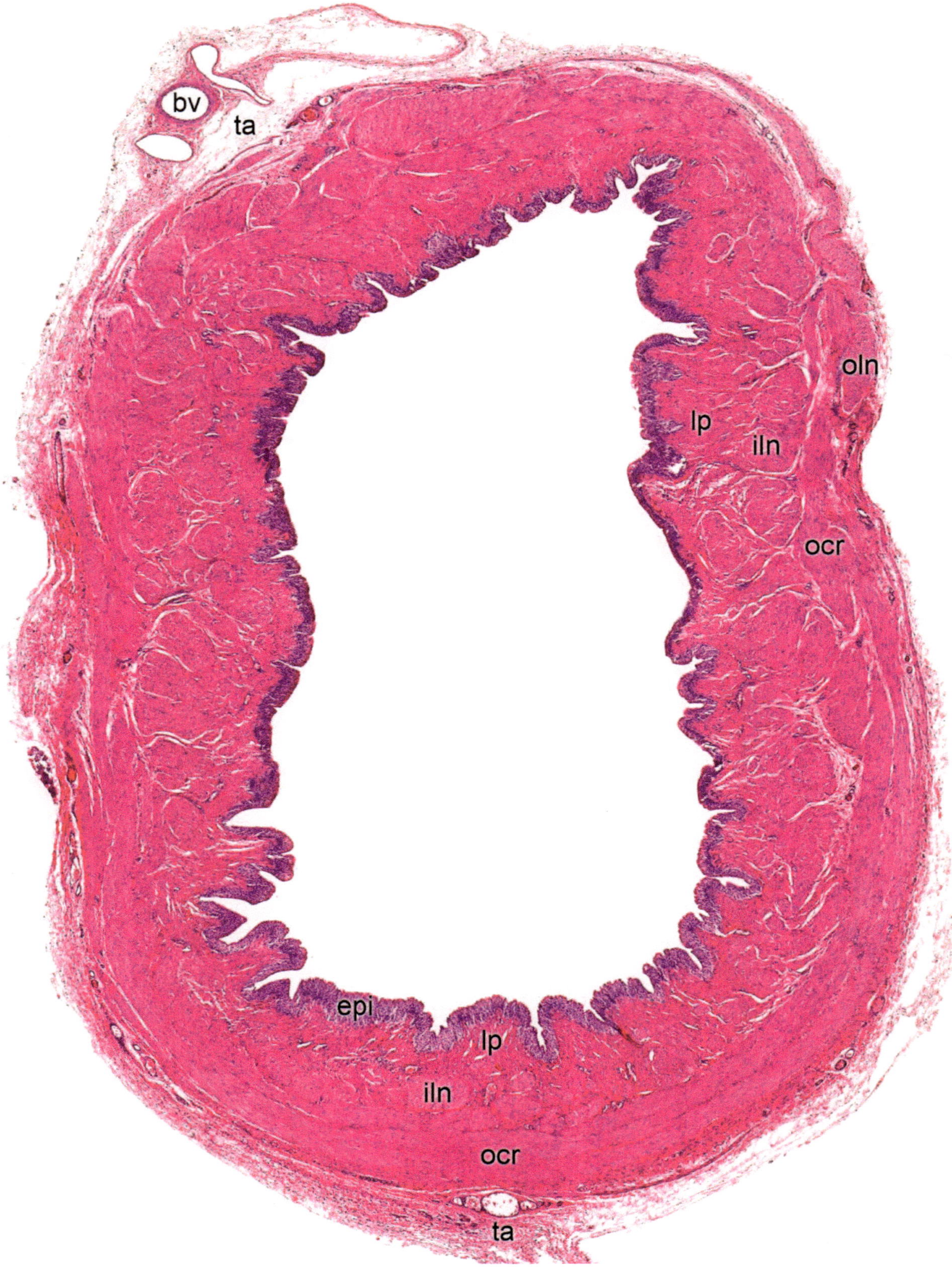

Figure 204. Ureter, distal part. An outermost longitudinal layer (oln) is added distally to the inner longitudinal (iln) and outer circular (ocr) layers of the tunica muscularis. Ureter is lined with transitional epithelium (epi) that is characterized with a luminal layer of umbrella cells. Lamina propria (lp) rests directly on the tunica muscularis; muscularis mucosae and submucosa are missing. Loose connective tissue tunica adventitia (ta), containing blood vessels (bv) covers the outer surface.

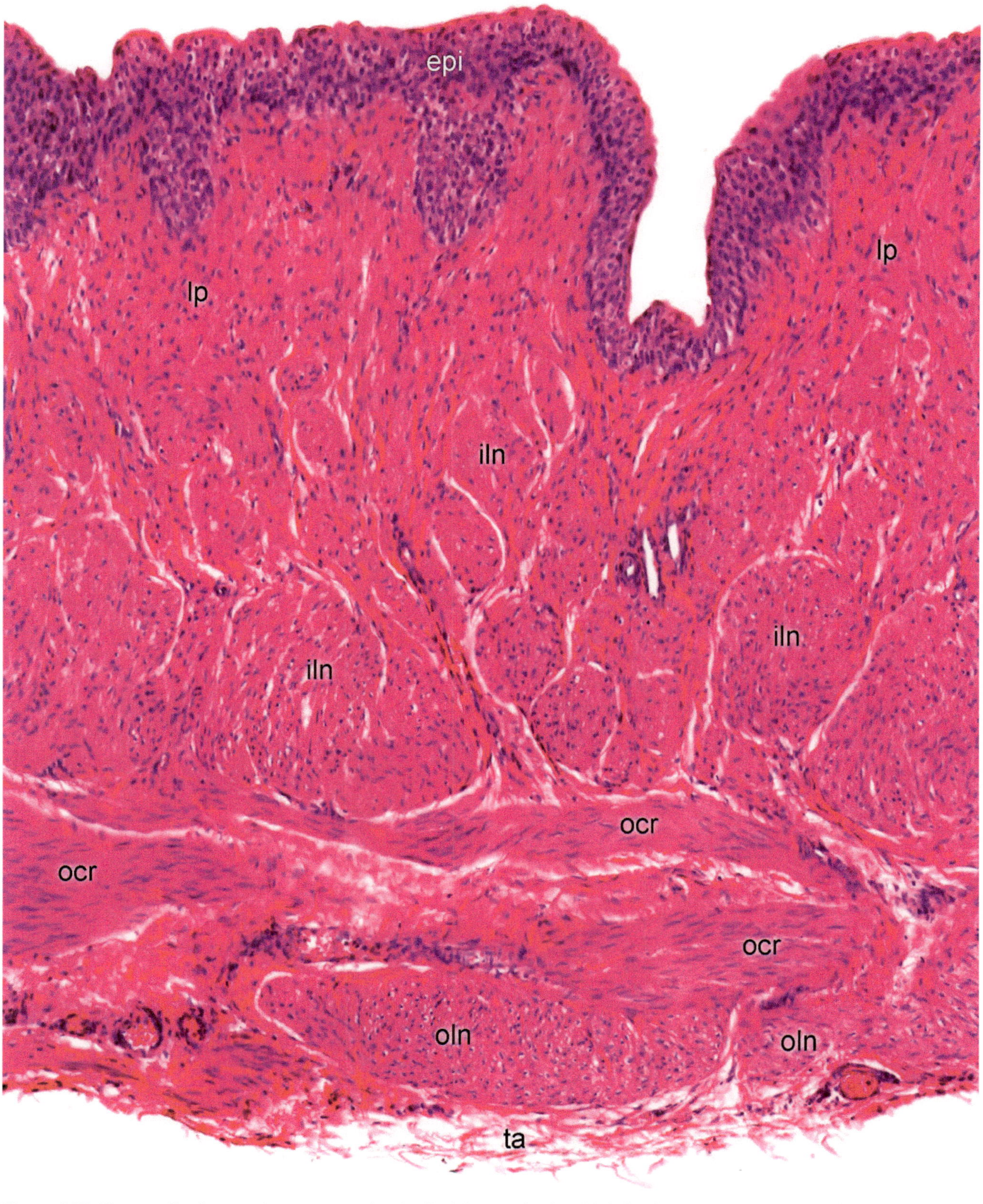

Figure 205. Ureter, distal part. An outermost longitudinal layer (oln) is added distally to the inner longitudinal (iln) and outer circular (ocr) layers of the tunica muscularis. Ureter is lined with transitional epithelium (epi) that is characterized with a luminal layer of umbrella cells. Lamina propria (lp) rests directly on the tunica muscularis; muscularis mucosae and submucosa are missing. The outer surface of the ureter is covered with loose connective tissue tunica adventitia (ta).

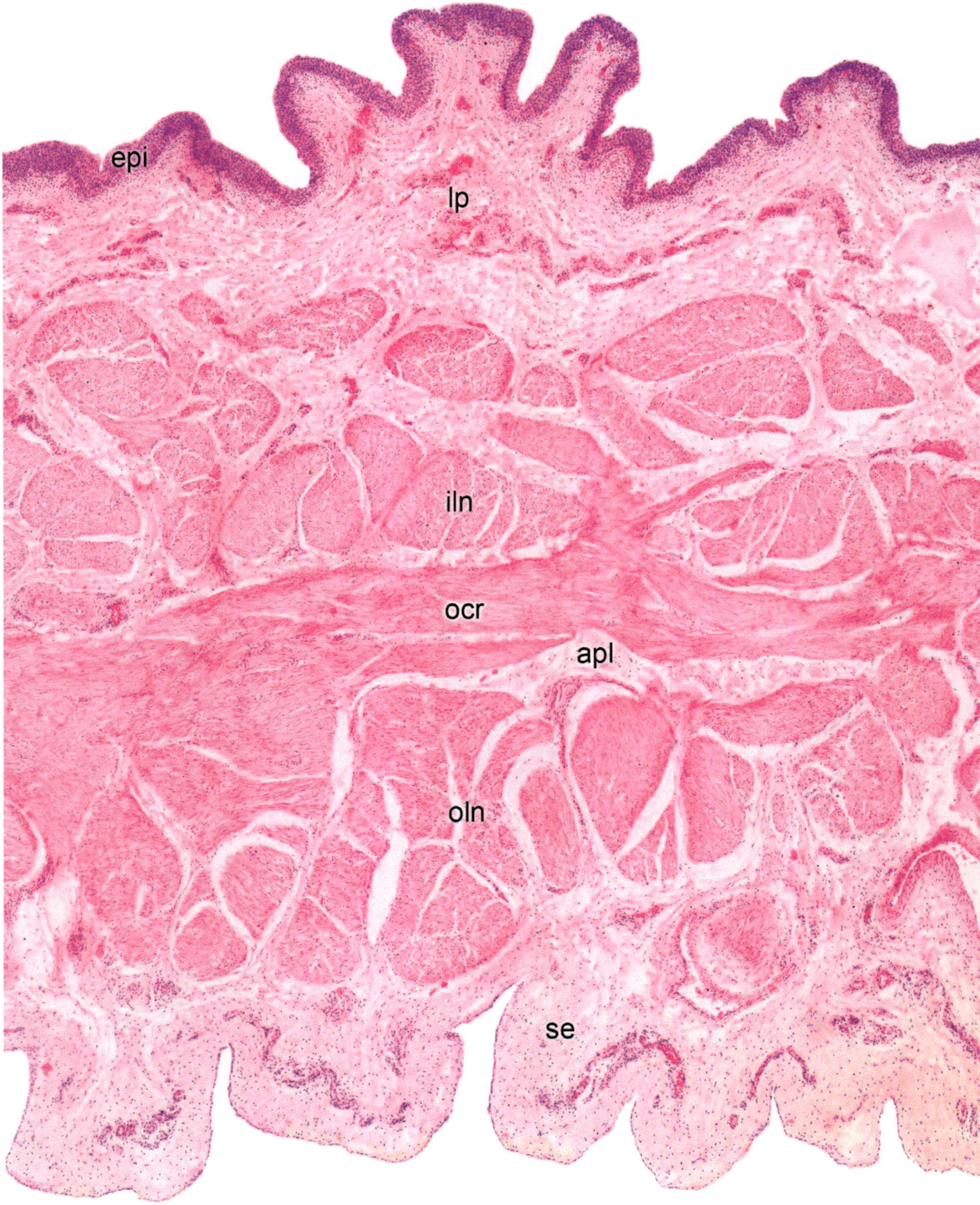

Figure 206. Urinary bladder. The tunica mucosa of the bladder is lined with transitional epithelium (epi) that is characterized with a luminal layer of umbrella cells. Tunica muscularis mucosae and submucosa are missing; lamina propria (lp) rests directly on the tunica muscularis that is composed of vaguely recognizable inner longitudinal (iln), outer circular (ocr) and outermost longitudinal (oln) layers. The upper part of the bladder is covered with tunica serosa (se) that is equivalent to the peritoneal coverage.

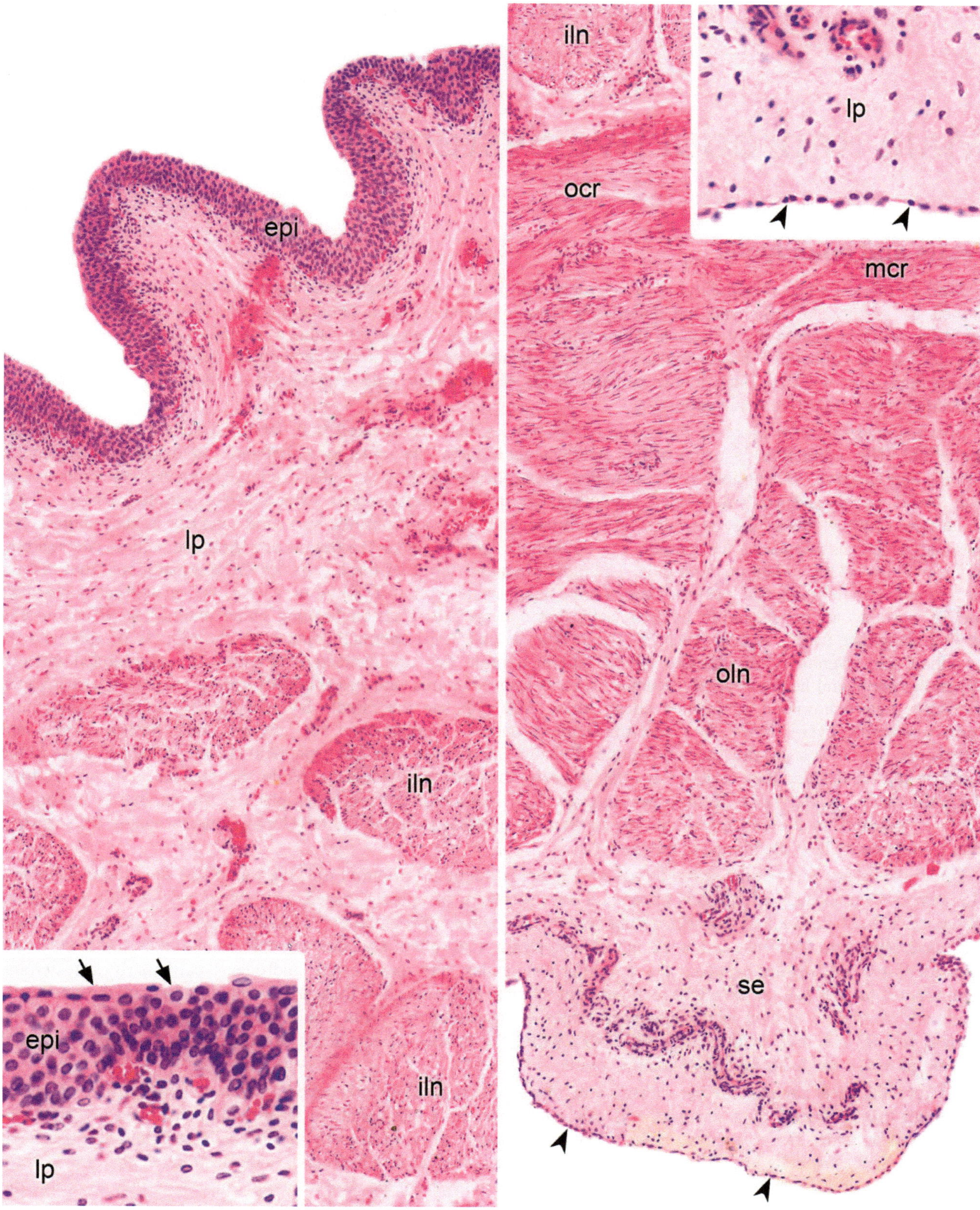

Figure 207. Urinary bladder. The bladder is lined with transitional epithelium (epi) that is characterized with a luminal layer of umbrella cells (arrows, lower inset). Tunica muscularis mucosae and submucosa are missing; lamina propria (lp) rests directly on the tunica muscularis that is composed of vaguely recognizable inner longitudinal (iln), outer circular (ocr) and outermost longitudinal (oln) layers. The upper part of the bladder is covered with tunica serosa (se) that is composed of a loose connective tissue lamina propria serosae (upper inset, lp) covered with mesothelial cells (arrowheads). The inferior surface of the bladder is lined with tunica adventitia.

Chapter 10

Male Reproductive System

Testis and epididymis

Most of the testis and the epididymis are covered with a thin serous membrane layer, the tunica vaginalis, which is formed by mesothelial cells and the underlying loose connective tissue lamina propria serosae. Under the tunica vaginalis, testis is enclosed by a thick dense connective tissue capsule, the tunica albuginea that rests on a thin, vascular, loose connective tissue tunica vasculosa. Tunica albuginea sends incomplete septa to the parenchyma subdividing it into lobules. Each lobule contains loops of extremely long, convoluted seminiferous tubules that are embedded into connective tissue stroma.

Spermatogenesis takes place in the seminiferous tubules that are covered with stratified seminiferous (germinal) epithelium. This epithelium is composed of spermatogenic cells that undergo replication and maturation eventually forming spermatozoa; the epithelium also contains nonreplicating Sertoli cells that are large columnar (triangular) cells with ovoid nuclei and well- developed endoplasmic reticulum. Sertoli cells extend across the full length of the seminiferous epithelium as they surround the spermatogenic cells and provide structural integrity to the tubules. The spermatogenic cells populating the seminiferous epithelium form layers on the basement membrane, and as they mature, they move toward the lumen of the tubule. The most immature forms of these cells are the spermatogonia that rest on the basement membrane and possess small ovoid or spherical nuclei. Type A dark (Ad) spermatogonia have dark nucleus, and they are considered to be stem cells that undergo cell division forming either another set of stem cells or spermatogonia with a lighter nucleus. These type A pale (Ap) spermatogonia enter the process of spermatozoa formation and eventually divide into type B spermatogonia that are characterized with a large, rounded nucleus containing peripheral heterochromatin clumps. Type B spermatogonia undergo mitosis forming primary spermatocytes with large heterochromatic, granulated nuclei; these cells populate the middle layer of the seminiferous epithelium. Primary spermatocytes divide into the transient secondary spermatocytes by meiosis that further divide into spermatids as a result of the second meiotic division. Spermatids with dark, spherical nuclei are located close to the lumen of the tubules and they undergo maturation losing their access cytoplasm, while the nucleus assumes its final elongated shape in the head of the spermatozoa. This maturation process takes place at the luminal compartment, in close proximity to the cell membrane of the Sertoli cells that surround the developing spermatogenic cells as they span from the basement membrane to the lumen.

Under the basal lamina of the seminiferous epithelium, a thin tunica propria forms the wall of the seminiferous tubules. This layer is composed of collagen fibers and several layers of spindle-shaped, peritubular myoid cells that move the sperm toward the hilum by generating peristaltic waves of contraction. Leydig cells are large cells with prominent spherical nucleus and eosinophilic cytoplasm that often contains crystals of Reinke; these cells are embedded into the stroma between the seminiferous tubules, and they secrete androgens, particularly testosterone. Seminiferous tubules lead to the tubuli recti (straight tubules) at the hilum of the testis, eventually forming the rete testis, a network of tubules located in the thickened tunica albuginea at the hilum, the highly vascular mediastinum testis. Straight tubules are lined with Sertoli cells only while the tubules of rete testis are covered with simple cuboidal epithelium with a single apical cilium and microvilli.

Efferent ductules connect the rete testis with the epididymis. They are covered with pseudostratified columnar epithelium that is formed by a series of both tall and shorter cells that results in an irregular, scalloped or cogwheel-shaped lumen; under the epithelium, few layers of circularly arranged smooth muscle cells surround the basement membrane. Epithelial cells often contain brown lipofuscin pigment that is associated with obstructive changes. After a short run, efferent ductules became highly convoluted and partially form the head of epididymis where they enter into the ductus epididymidis, an extremely coiled single tubule forming the bulk of the epididymis. Ductus epididymidis is lined with regularpseudostratified epithelium with stereocilia; occasional basal cells with round nuclei serve as epithelial stem cells. Under the basal membrane of the epithelium, smooth muscle cells form several circular layers in the head; toward the tail of theepididymis this circular layer is gradually accompanied by inner and outer smooth muscle layers that are continuous with the tunica muscularis of the ductus deferens. Epididymis is enclosed into a dense connective tissue capsule that sends trabeculae into the parenchyma.

http://doi.org/10.1016/B978-0-323-91891-6.50010-6,

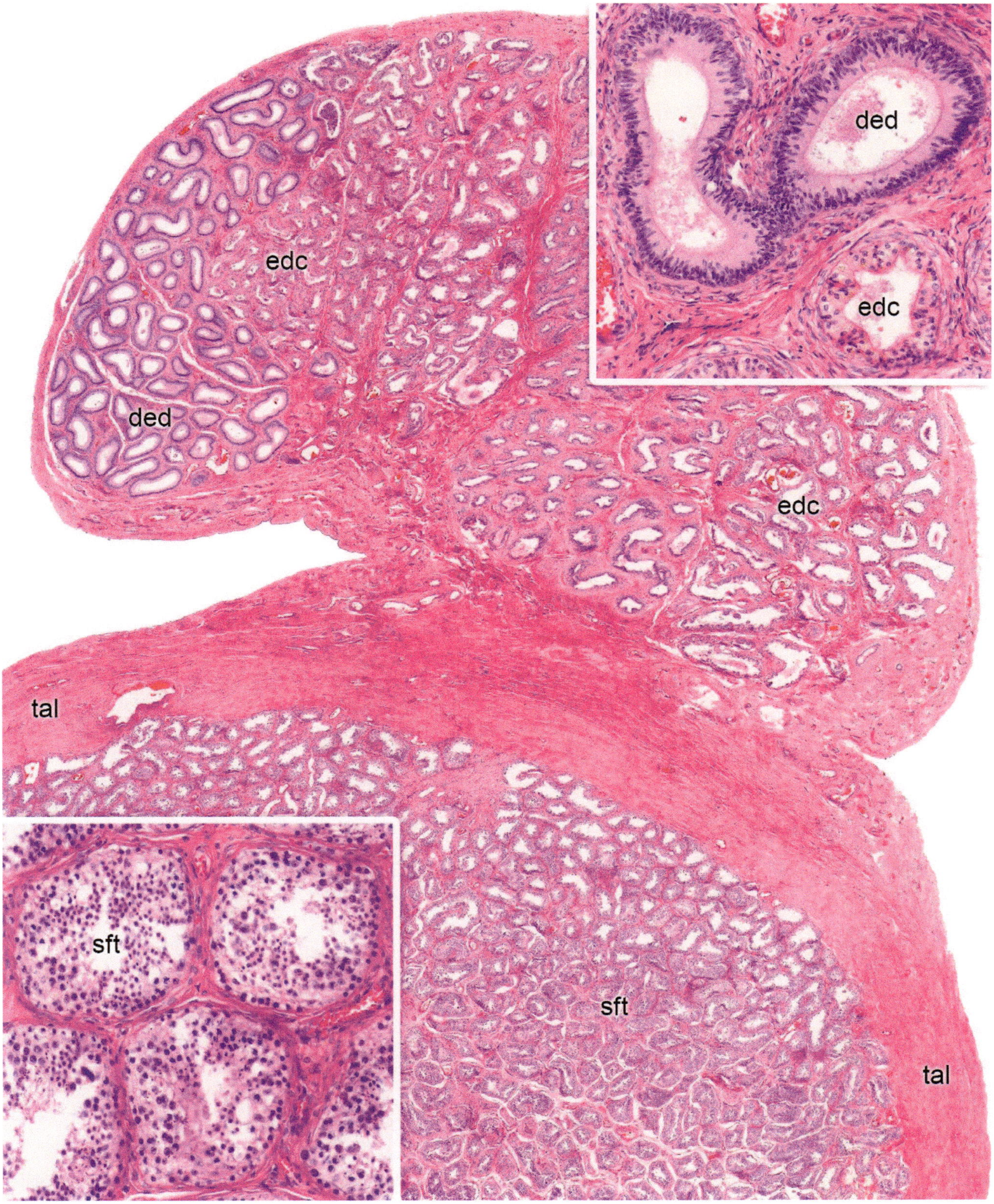

Figure 208. Testis and epididymis. Most of the testis and the epididymis are covered with serous membrane (tunica vaginalis), and the underlying dense connective tissue tunica albuginea (tal). Testis contains convoluted seminiferous tubules (sft) lined with stratified seminiferous epithelium (lower inset). Efferent ductules of the epididymis (edc) are covered with pseudostratified columnar epithelium that is formed by alternating tall and shorter cells resulting in an irregular lumen (upper inset), and they connect the testis with the ductus epididymidis (ded) lined with regular pseudostratified epithelium with stereocilia.

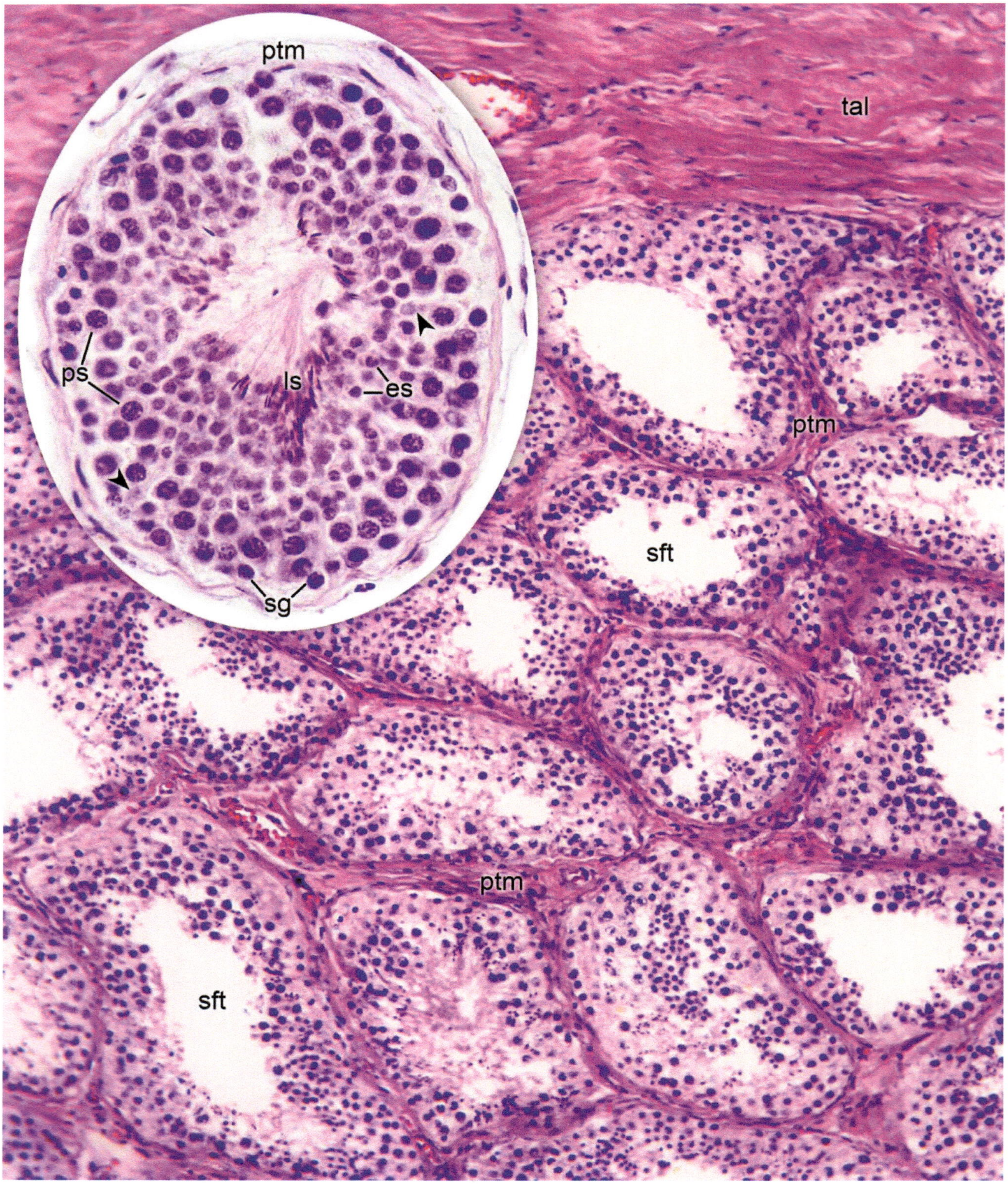

Figure 209. Testis. Testis is covered with dense connective tissue tunica albuginea (tal) and contains convoluted seminiferous tubules (sft) covered with stratified seminiferous epithelium (inset). The epithelium contains differentiating spermatogenic cells embedded into the cytoplasm of Sertoli cells that are characterized with large, light, elongated nuclei (arrowheads). Spermatogonia with spherical nuclei (sg) rest on the basement membrane and they can be divided into several types distinguished from each other according to the morphology of their nuclei. Spermatogonia produce primary spermatocytes (ps) with larger nucleus containing peripheral heterochromatin clumps. These cells undergo meiosis and divide into secondary spermatocytes and then into early spermatids (es) that mature into late spermatids (ls) that eventually transform into spermatozoa. Seminiferous tubules are surrounded by peritubular myoid cells (ptm).

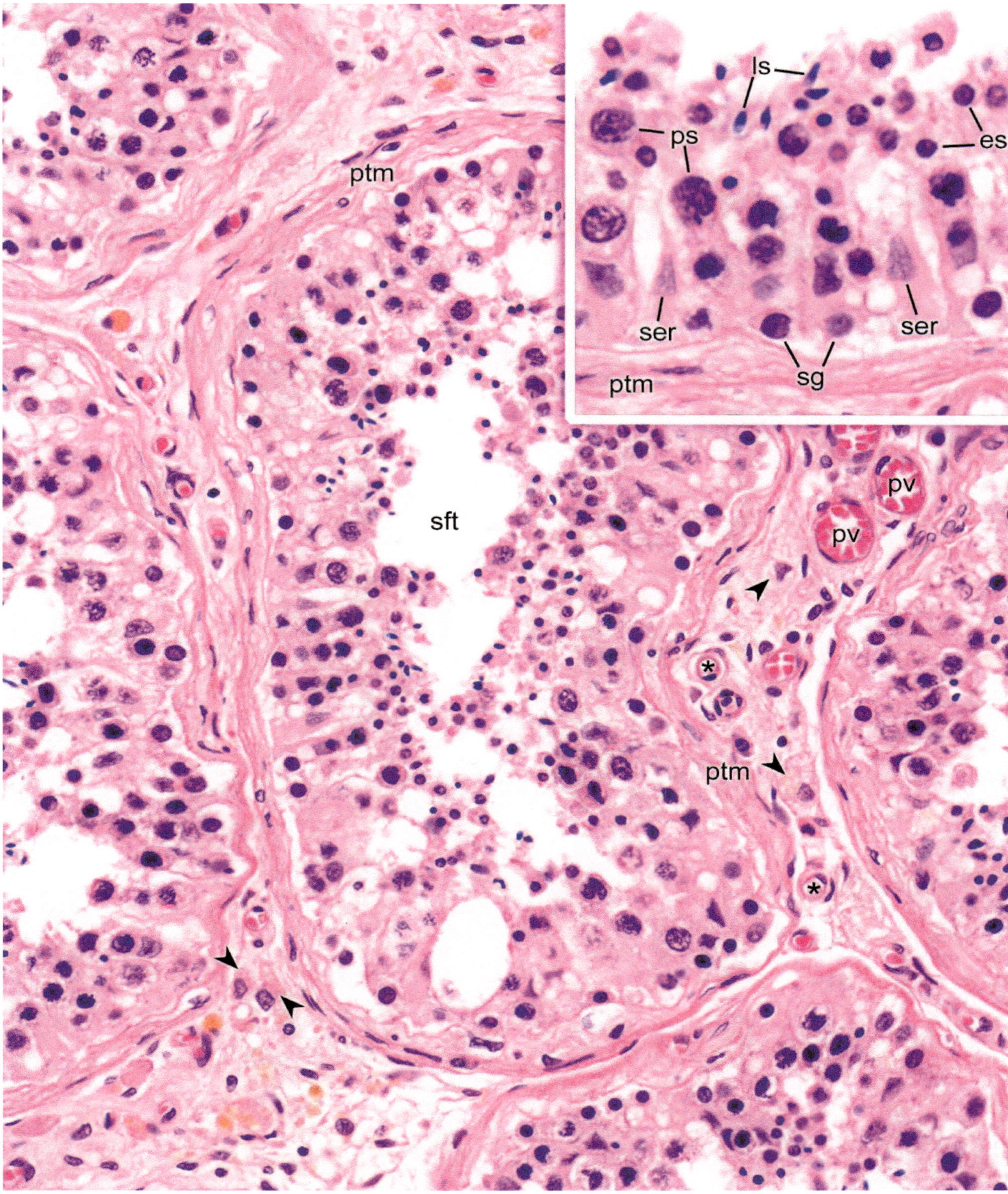

Figure 210. Testis. Seminiferous tubules (sft) are covered with stratified seminiferous epithelium (inset) containing differentiating spermatogenic cells embedded into the cytoplasm of Sertoli cells that are characterized with large, light, elongated nuclei (ser). Dark (Ad) and pale (Ap) spermatogonia with spherical nuclei (sg) rest on the basement membrane Spermatogonia produce primary spermatocytes (ps) with larger nucleus containing peripheral heterochromatin clumps. These cells undergo meiosis and divide into secondary spermatocytes and then into early spermatids (es) that mature into late spermatids (ls) that eventually transform into spermatozoa. Seminiferous tubules are surrounded by contractile peritubular myoid cells (ptm). Intertubular connective tissue is populated by Leydig cells (arrowheads) with eosinophilic cytoplasm often containing lipid droplets as well as by capillaries (asterisks) and postcapillary venules (pv).

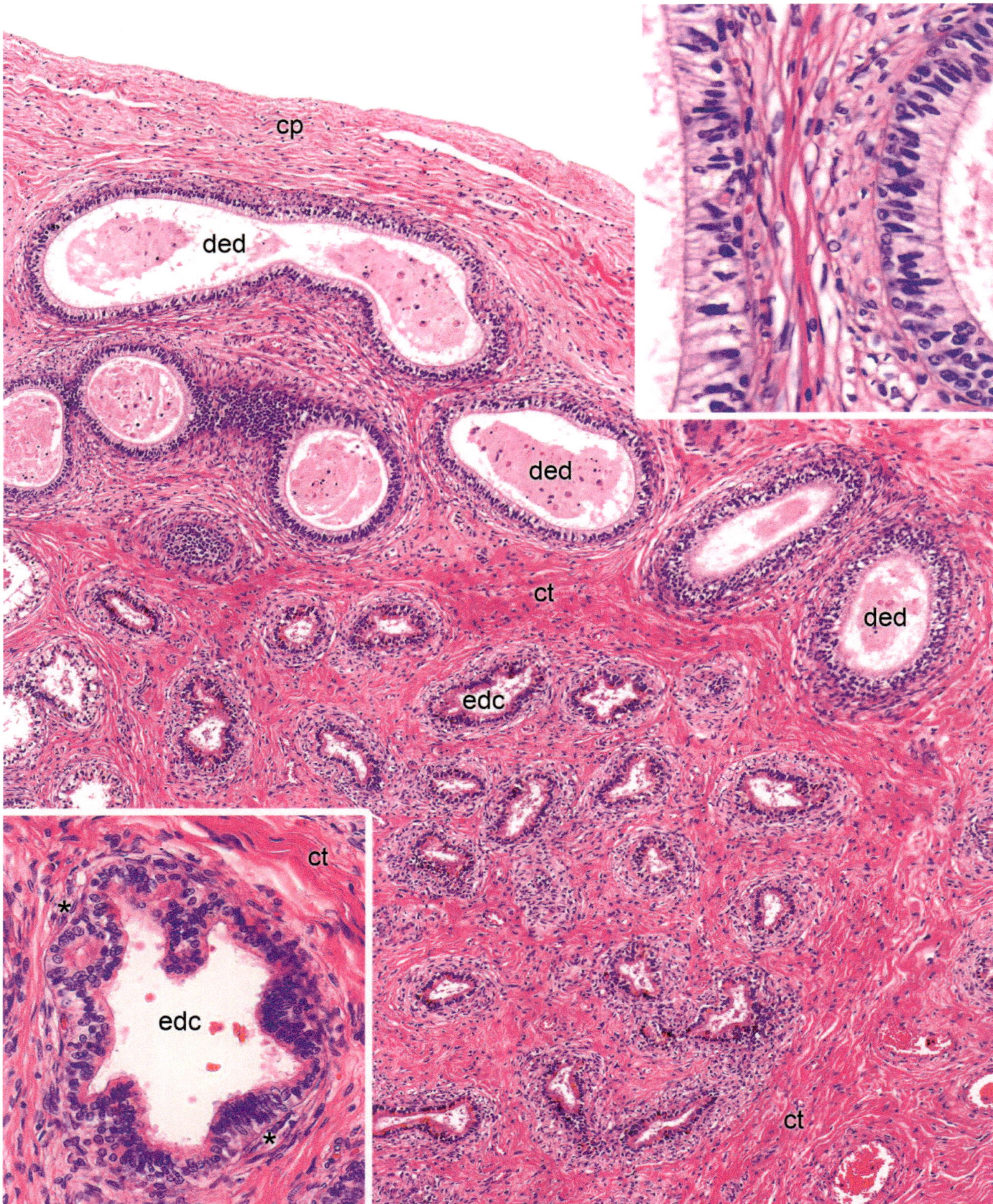

Figure 211. Epididymis. Efferent ductules of the head of epididymis (edc) are covered with pseudostratified columnar epithelium that is formed by alternating tall and shorter cells resulting in an irregular, scalloped lumen (inset). Efferent ductules connect the testis with the ductus epididymidis (ded) lined with pseudostratified columnar epithelium with stereocilia (upper inset). The ductules are embedded into a dense connective tissue stroma (ct) and they have a smooth muscle coverage (asterisks) that becomes gradually thicker towards the distal segments. Epididymis is covered with a dense connective tissue capsule (cp).

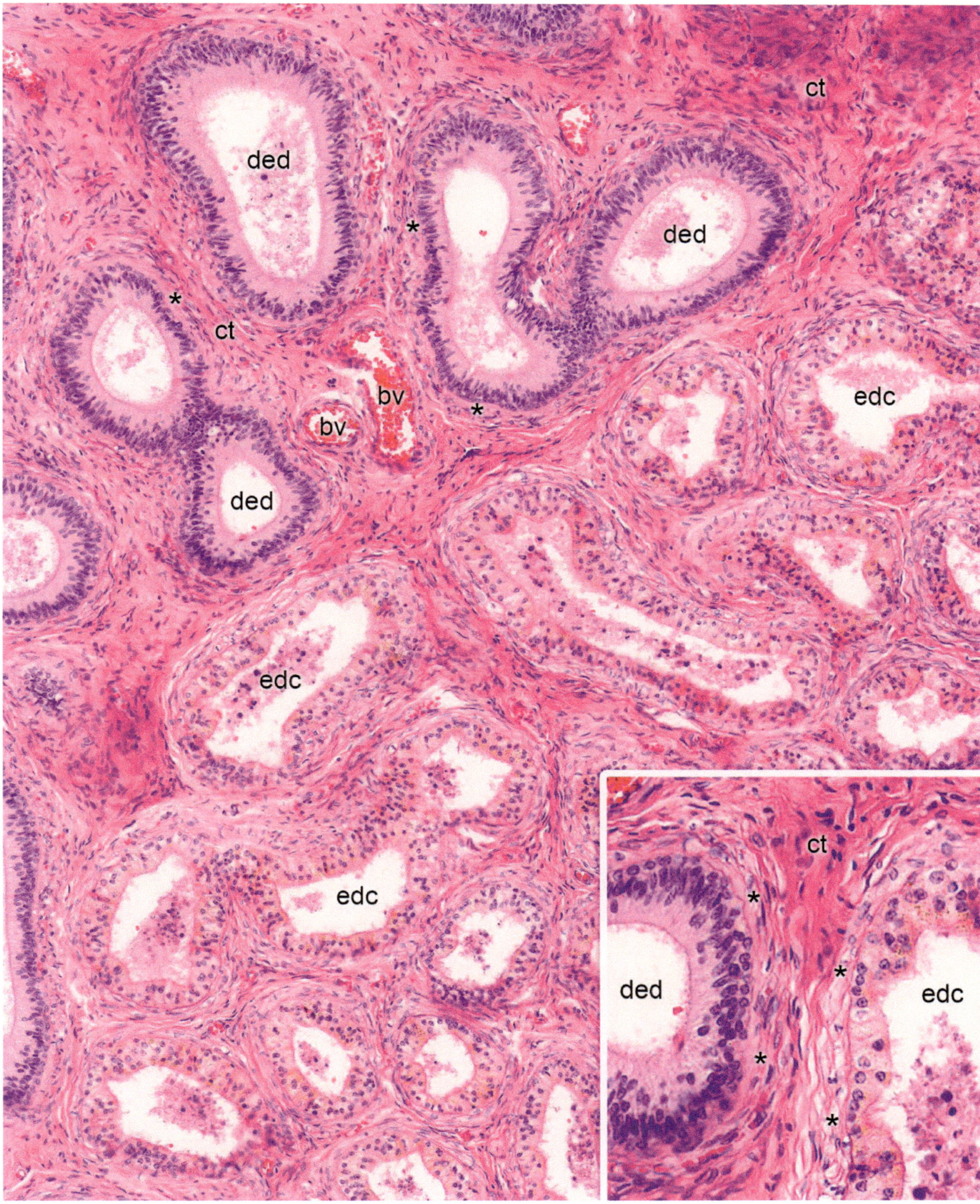

Figure 212. Epididymis. Efferent ductules of the head of epididymis (edc) are covered with pseudostratified columnar epithelium that is formed by alternating tall and shorter cells resulting in an irregular, scalloped lumen (inset). The epithelial cells often contain brown lipofuscin pigment. Efferent ductules connect the testis with the ductus epididymidis (ded) lined with pseudostratified columnar epithelium with stereocilia (inset). Ductules are embedded into a dense connective tissue stroma (ct) and they have a smooth muscle coverage (asterisks) that becomes gradually thicker towards the distal segments.

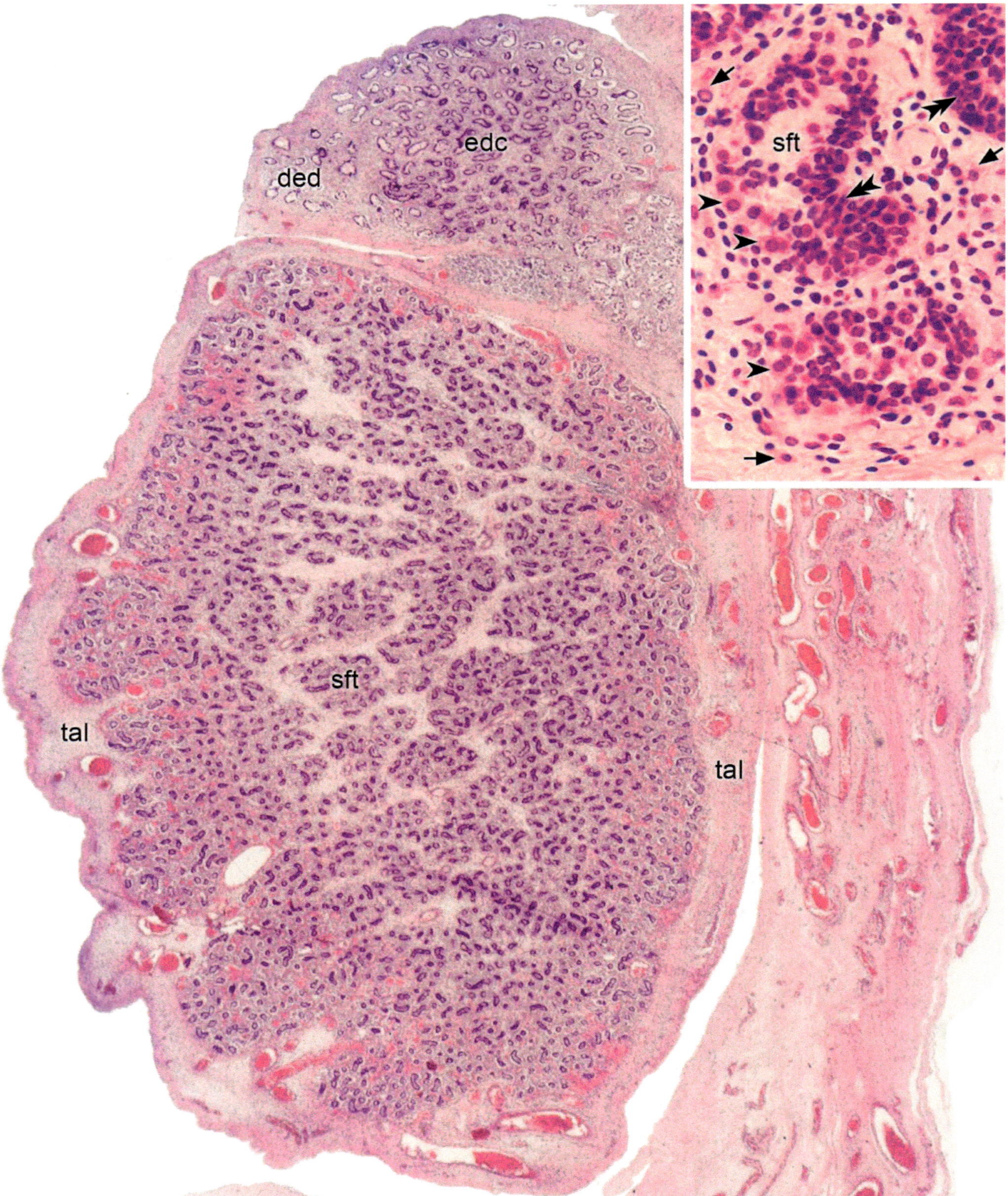

Figure 213. Neonatal testis and epididymis. Dense connective tissue tunica albuginea (tal) surrounds the convoluted seminiferous tubules of the testis (sft) covered with stratified seminiferous (germinal) epithelium (inset). In neonatal testis, germinal epithelium is immature and incapable of sperm production. The epithelium contains mostly Sertoli cells with darker elongated nuclei (double arrowheads) and basally located spermatogonia characterized with pale eosinophilic cytoplasm (arrowheads) and lacks spermatocytes and spermatids. Large, eosinophilic Leydig cells populate the connective tissue between the tubules (arrows). Efferent ductules (edc) and ductus epididymidis (ded) are covered with short pseudostratified columnar epithelium without stereocilia.

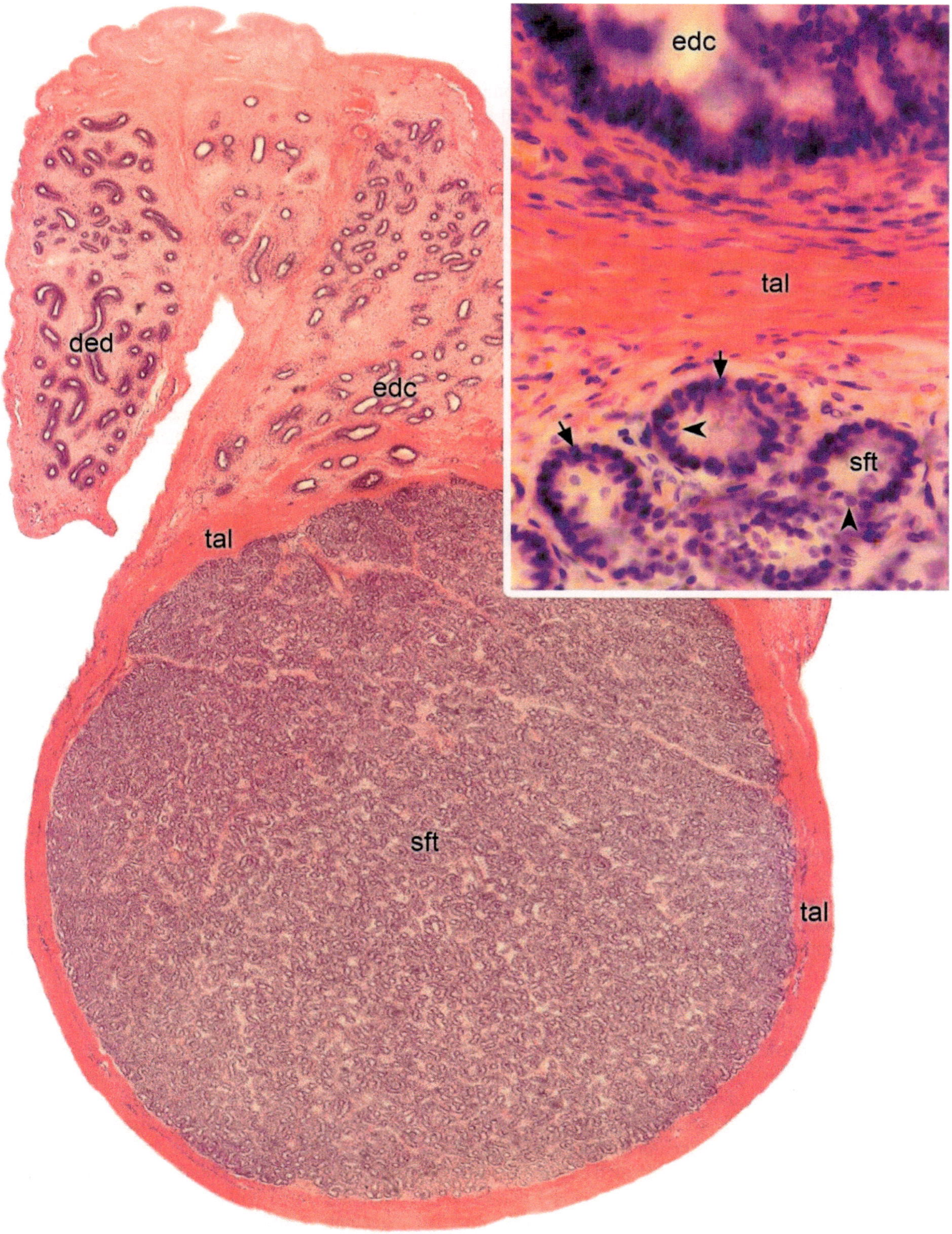

Figure 214. Testis and epididymis of an 8-year-old. Dense connective tissue tunica albuginea (tal) surrounds the convoluted seminiferous tubules of the testis (sft) covered with stratified seminiferous (germinal) epithelium (inset). In prepubertal testis, this germinal epithelium is immature and does not produce spermatozoa. The epithelium contains mostly Sertoli cells with darker elongated nuclei (arrows) and basally located spermatogonia characterized with pale eosinophilic cytoplasm (arrowheads) and lacks spermatocytes and spermatids. Efferent ductules (edc) and ductus epididymidis (ded) are covered with pseudostratified columnar epithelium.

Ductus (vas) deferens

Ductus epididymidis continues into the ductus deferens that is also a component of the spermatic cord when passing through the inguinal canal. Here, ductus deferens is accompanied by several structures, including its own deferential artery, as well as testicular artery, pampiniform plexus and branches of the genitofemoral and sympathetic nerves; these structures are enveloped in a layer formed by skeletal muscle (cremasteric muscle) surrounded by the inner and outer spermatic fasciae. At the posterior side of the urinary bladder, the enlarged part of ductus deferens, the ampulla, joins the duct of the seminal vesicle,forming the ejaculatory duct that opens into the prostatic urethra after piercing the prostate.

The wall structure of ductus deferens is similar to that of the terminal part of ductus epididymidis, containing stereociliated pseudostratified columnar epithelium with basal cells, a thin connective tissue lamina propria and a thick, three-layered muscular coat (inner longitudinal, middle circular and outer longitudinal layers) that becomes thinner in the ampulla and completely disappears in the ejaculatory duct. Proximally, the mucous membrane is thrown into shallow longitudinal folds that are replaced by complex mucosal crests in the ampulla.

Seminal vesicles

Seminal vesicles are coiled tubular glands with complex mucosal folds that are lined with secretory pseudostratified columnar epithelium with basal epithelial stem cells similar to those in the epididymis and vas deferens. Epithelium rests on loose connective tissue lamina propria that forms the core of the mucosal folds. Lamina propria is surrounded by inner circular and outer longitudinal smooth muscle layers; externally, the tubes are covered with a thin layer of connective tissue. The duct of each seminal vesicle merges with the vas deferens forming the ejaculatory ducts that enter the prostate and open into the prostatic urethra. Ejaculatory ducts are lined with secretory pseudostratified columnar epithelium and lack muscular layer.

Prostate gland

Prostate is composed of fibromuscular tissue formed by interwoven layers of collagen fibers and smooth muscle cells. This stroma contains embedded tubuloalveolar glands that are arranged in concentric layers around the urethra. Prostatic urethra is lined with transitional epithelium that is gradually replaced by stratified columnar epithelium as it transitions to the membranous urethra. Posteriorly, the urethra contains a longitudinal fold, the urethral crest, with the seminal colliculus or verumontanum protruding into the urethral lumen. Prostatic utricle is an epithelium lined diverticulum, the remnant of the Müllerian duct that opens in the midline on the verumontanum, slightly distal to the openings of the two ejaculatory ducts. The inner mucosal glands are in close proximity to the urethral mucosa and open directly into the lumen of the urethra and the utricle, while the intermediate submucosal glands and the peripheral main glands open into their ducts that enter the urethral lumen on the sides of the urethral crest.

Prostatic glands are lined with secretory cells that are cuboidal or columnar cells of variable shapes and sizes, often showing pseudostratified columnar epithelium characteristics, resulting in an irregular lumen. With elder age, large eosinophilic concretions (corpora amylacea) commonly appear in the glandular lumina; these structures are composed of concentrically arranged layers of laminated concretions of the glandular secretion.

Penis

Penis is mainly composed of erectile tissue held together by a dense connective tissue tunica albuginea that is enveloped by deep and superficial fasciae covered by skin. Dorsally, the erectile tissueforms the corpora cavernosa, while the corpus spongiosum surrounds the penile (spongy) urethra ventrally. Erectile tissue contains irregular caverns that are covered with endothelium and an underlying smooth muscle layer containing irregularly arranged trabeculae. Smooth muscle cells often form subendothelial cushions under the endothelium of the helicine arteries that are responsible for filling the caverns with blood. The muscle layer is embedded into an interstitial connective tissue layer with collagen and elastic fibers as well as vessels and nerves. Penile urethra is lined with stratified or pseudostratified columnar epithelium that transitions to nonkeratinized stratified squamous epithelium at the urethral orifice. Tubuloalveolar, mucous periurethral glands (glands of Littre) can be often observed opening into the penile urethra and lubricating the lumen.

The glans penis is covered with very thin skin that is firmly attached and lacks adipose tissue, but contains a smooth muscle layer (tunica dartos). Glans is surrounded by the prepuce that is lined with delicate skin with thin stratum corneum.

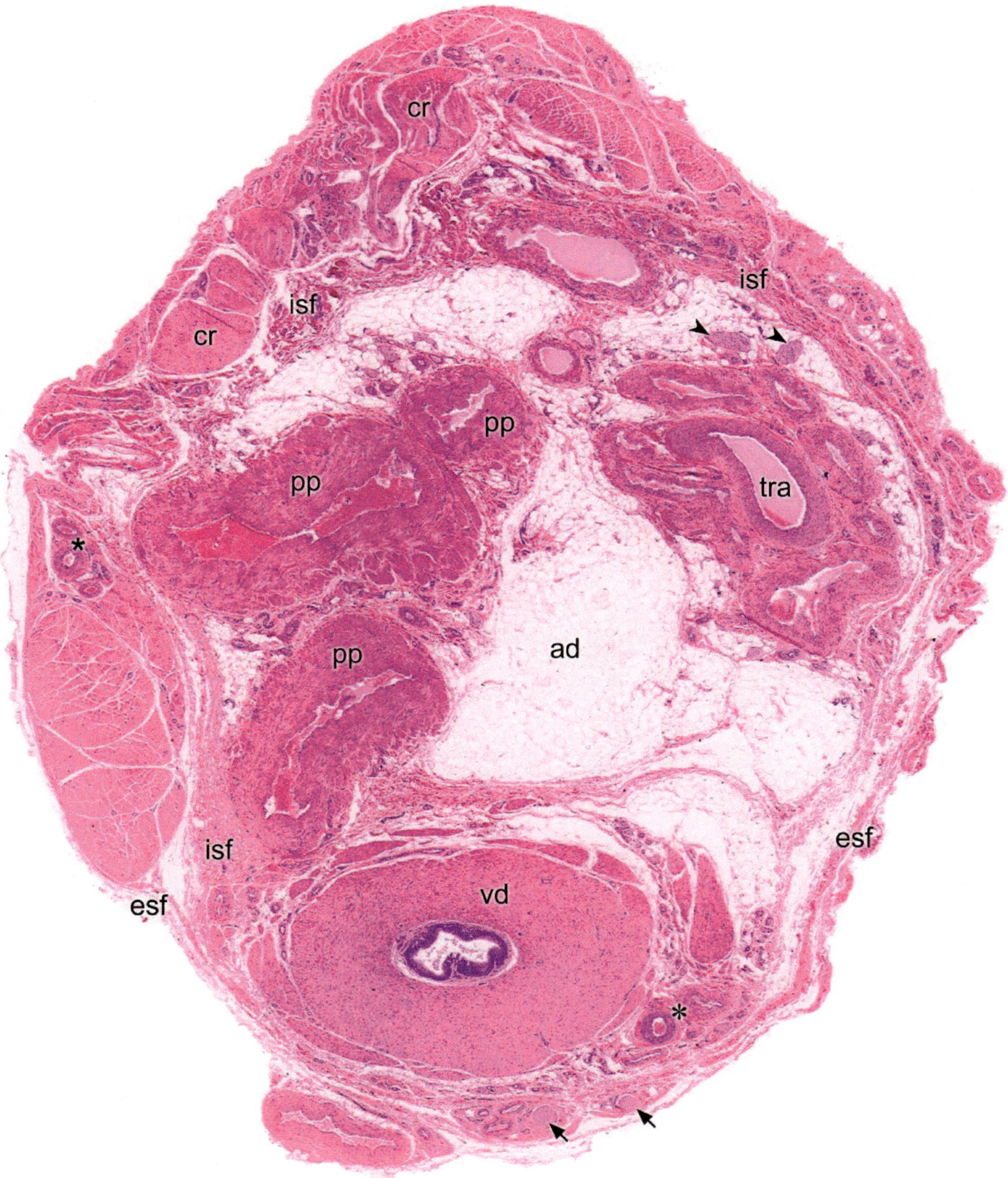

Figure 215. Vas deferens in the spermatic cord. Vas deferens (vd) is the continuation of the ductus epididymidis as it runs in the spermatic cord passing through the inguinal canal where it is supplied by the deferential artery and vein (six-pointed asterisk). Vas deferens is lined with stereociliated pseudostratified columnar epithelium and the underlying lamina propria that rests on inner longitudinal, middle circular and outermost longitudinal layers of smooth muscle. Spermatic cord is covered with the cremasteric muscle (cr) that is a skeletal muscle surrounded by the external (esf) and internal spermatic fasciae (isf) and supplied by the cremasteric vessels (five-pointed asterisks). Within the cord, testicular artery branches (tra) and veins of the pampiniform plexus (pp) are embedded into adipose tissue (ad). Branches of the genital branch of the genitofemoral nerve are located between the external and internal spermatic fasciae (arrows) while testicular autonomic nerves travel within the cord (arrowheads).

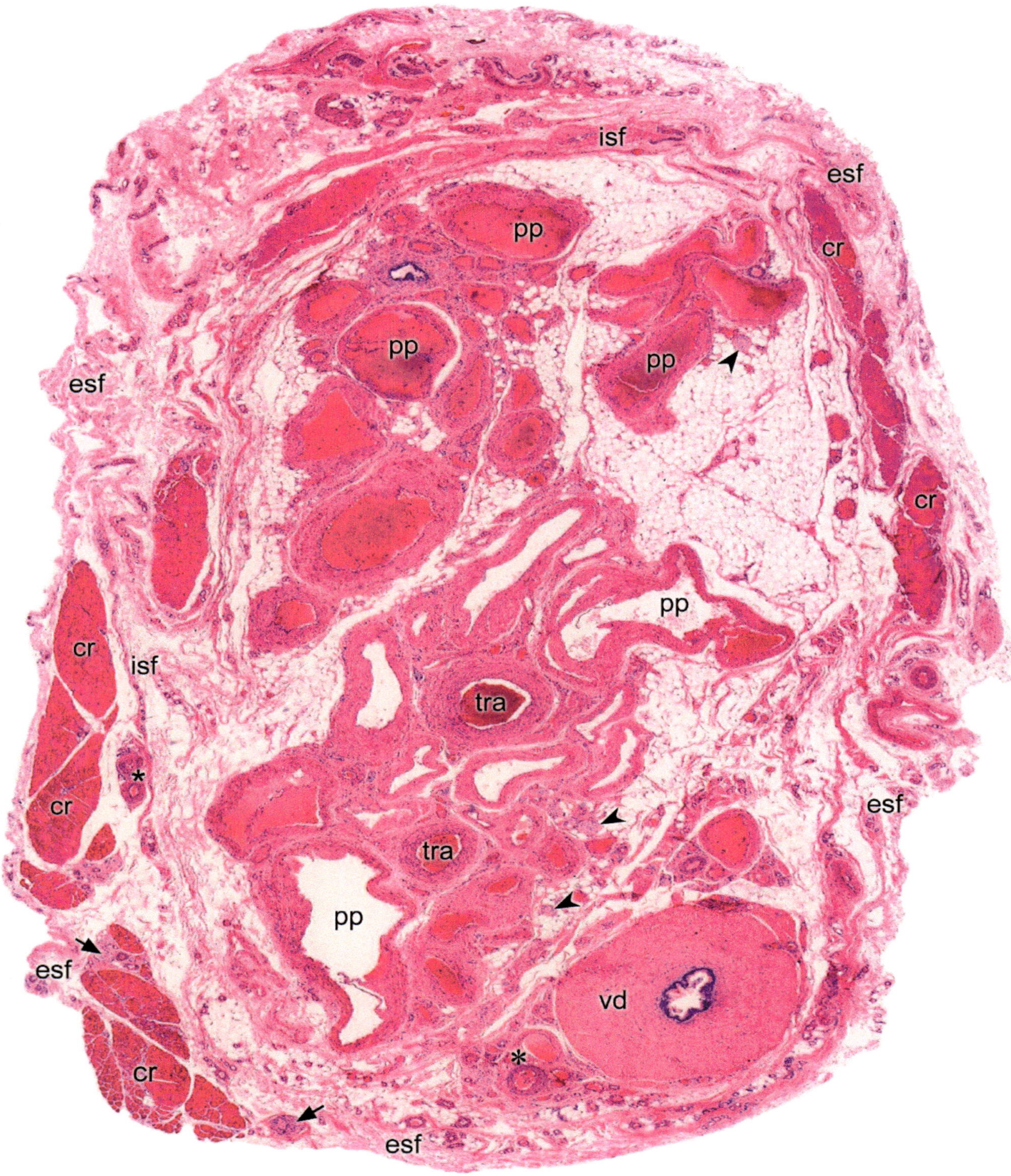

Figure 216. Vas deferens in the spermatic cord. Vas deferens (vd) is the continuation of the ductus epididymidis as it runs in the spermatic cord passing through the inguinal canal where it is supplied by the deferential artery and vein (six-pointed asterisk). Vas deferens is lined with stereociliated pseudostratified columnar epithelium and the underlying lamina propria that rests on inner longitudinal, middle circular and outermost longitudinal layers of smooth muscle. Spermatic cord is covered with the cremasteric muscle (cr) that is a skeletal muscle surrounded by the external (esf) and internal spermatic fasciae (isf) and supplied by the cremasteric vessels (five-pointed asterisks). Within the cord, testicular artery branches (tra) and veins of the pampiniform plexus (pp) are embedded into adipose tissue (ad). Branches of the genital branch of the genitofemoral nerve are located between the external and internal spermatic fasciae (arrows) while testicular autonomic nerves travel within the cord (arrowheads).

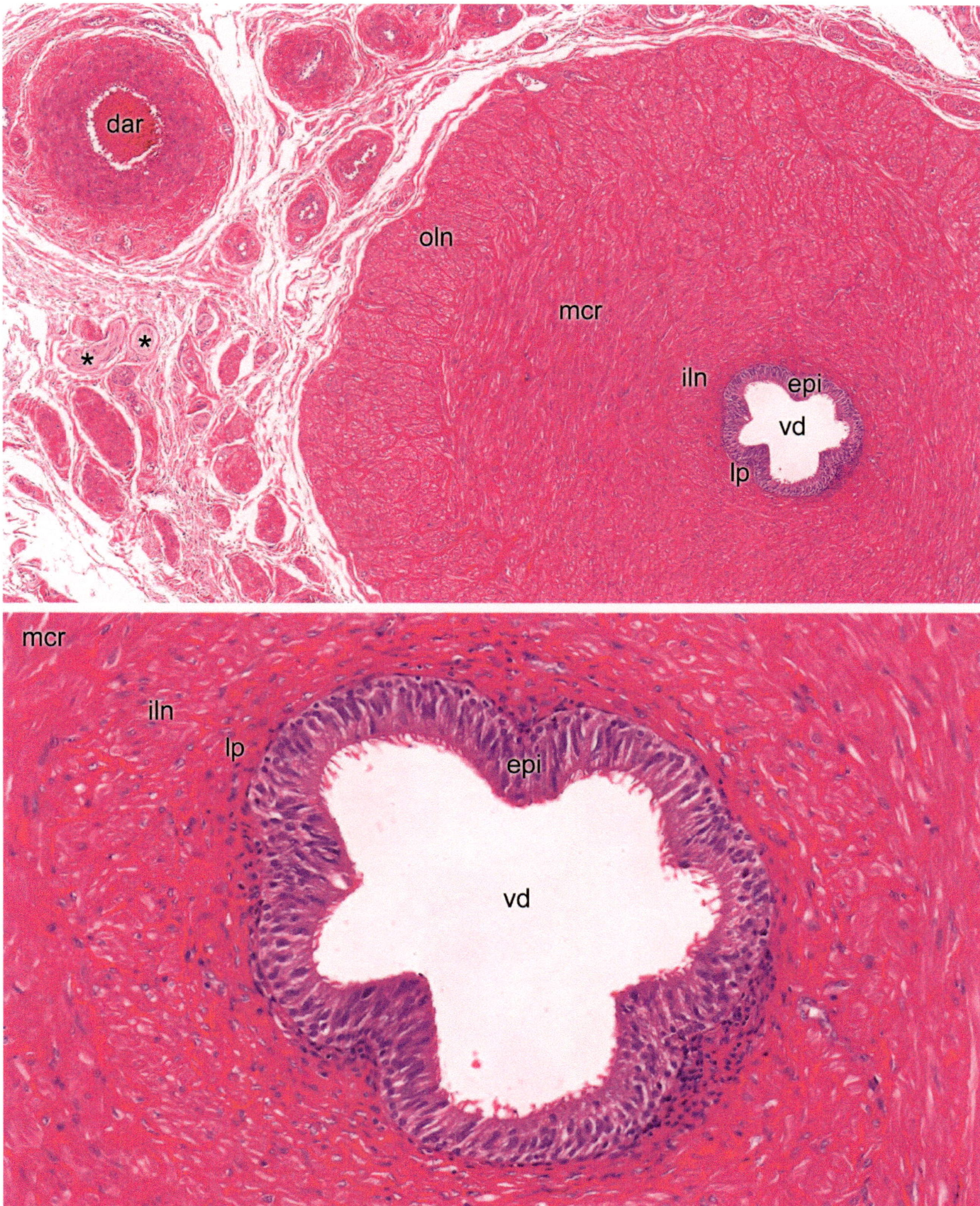

Figure 217. Vas deferens in the spermatic cord. Low (top) and higher magnifications (bottom) of the vas deferens (vd) that is the continuation of the ductus epididymidis and it is supplied by the deferential artery (dar). Vas deferens is lined with stereociliated pseudostratified columnar epithelium (epi) and the underlying lamina propria (lp) that rests on inner longitudinal (iln), middle circular (mcr) and outermost longitudinal layers (oln) of smooth muscle. Testicular autonomic nerves travel within the cord (asterisks).

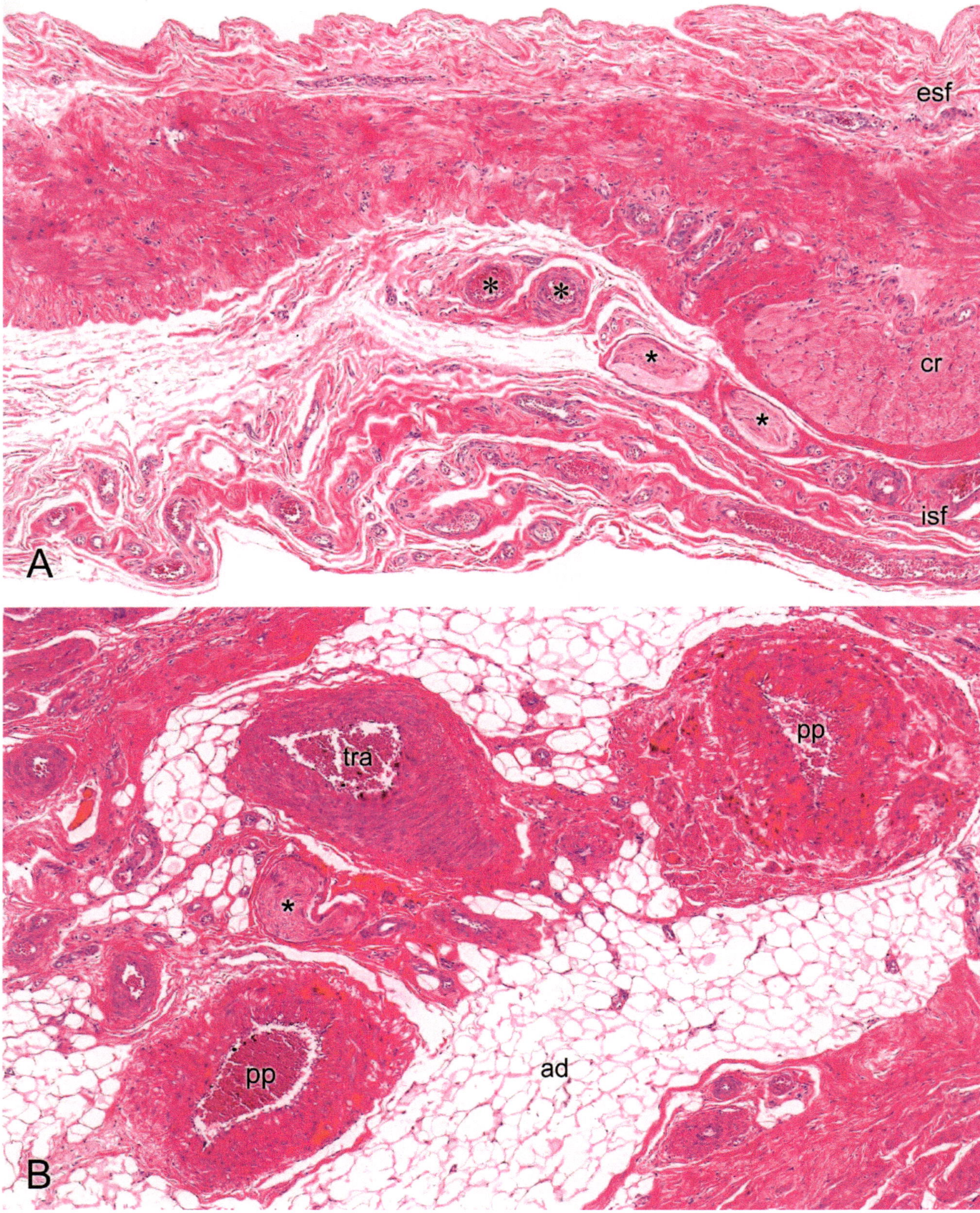

Figure 218. Spermatic cord. (A) Spermatic cord is covered with the cremasteric muscle (cr) that is a skeletal muscle surrounded by the external (esf) and internal spermatic fasciae (isf), supplied by the cremasteric vessels (six-pointed asterisks) and innervated by the fibers of the genital branch of the genitofemoral nerve (five-pointed asterisks). (B) Within the cord, testicular artery branches (tra), veins of the pampiniform plexus (pp), and testicular autonomic nerves (five-pointed asterisks) are embedded into adipose tissue (ad).

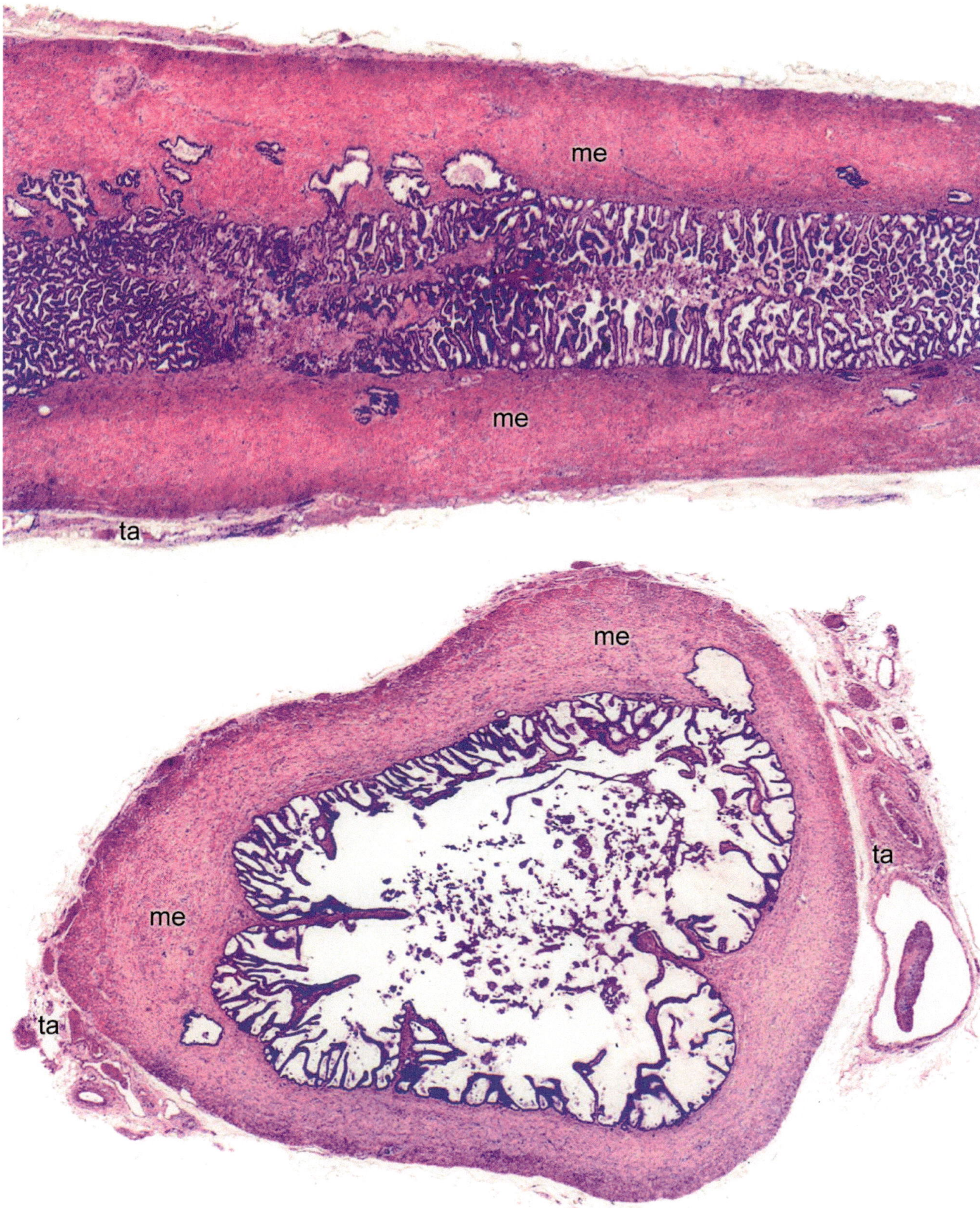

Figure 219. Vas deferens, ampulla. Before joining with the duct of the seminal vesicle, vas deferens forms the distended ampulla. Stereociliated pseudostratified columnar epithelium lines complex mucosal crests that protrude into the lumen. Lamina propria rests on the muscle coat (me) of inner longitudinal, middle circular and outermost longitudinal layers of smooth muscle that is thinner than that of the proximal segments of the vas deferens. The muscle coat is covered with a loose connective tissue tunica adventitia (ta).

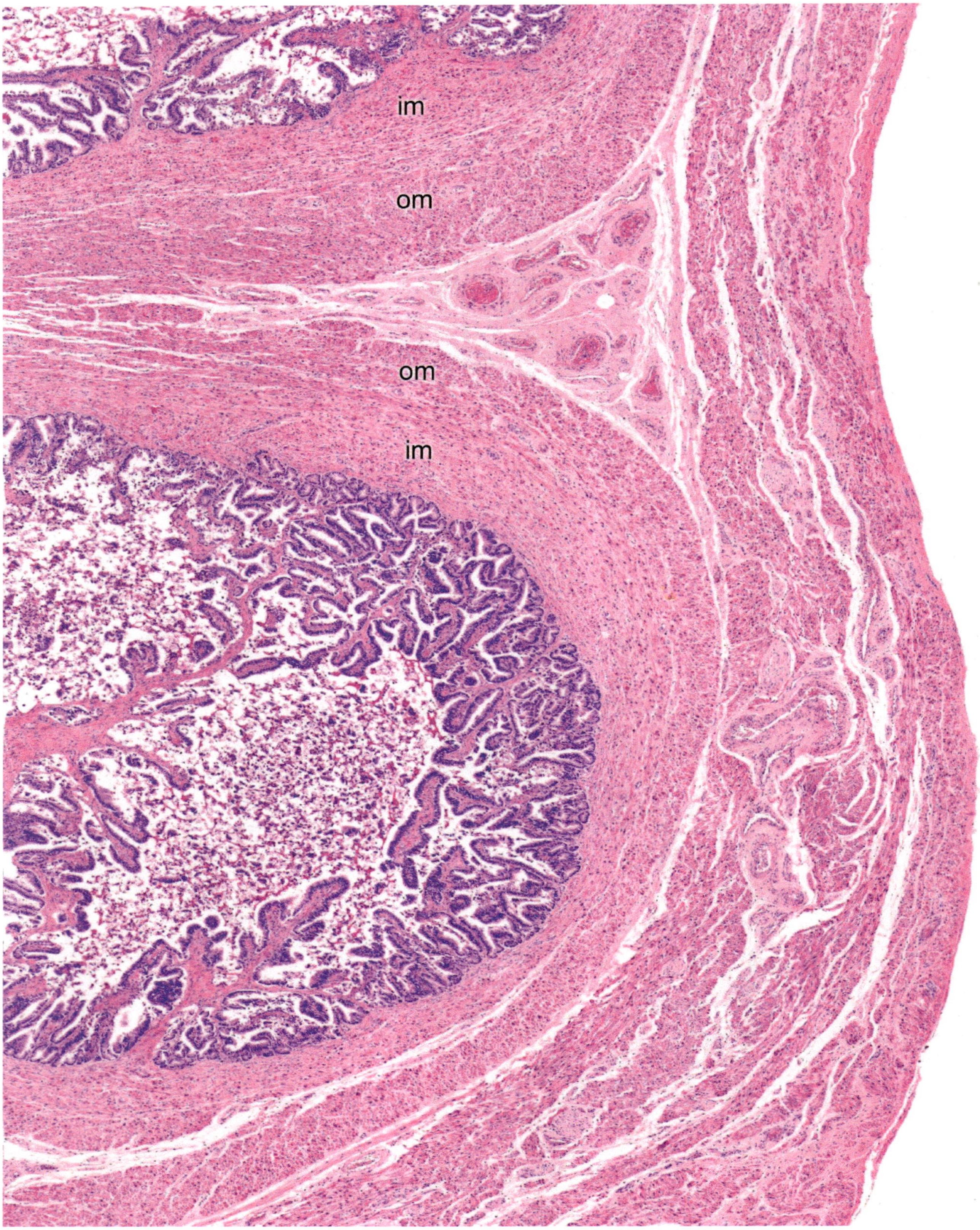

Figure 220. Seminal vesicle. Seminal vesicles are coiled tubular glands with complex mucosal folds that are lined with secretory pseudostratified columnar epithelium with basal epithelial stem cells similar to those in the epididymis and vas deferens. Epithelium is surrounded by inner circular (im) and outer longitudinal smooth muscle layers (om). Outside, the tubes are covered with a thin layer of connective tissue. The duct of the seminal vesicle merges with the vas deferens forming the ejaculatory ducts.

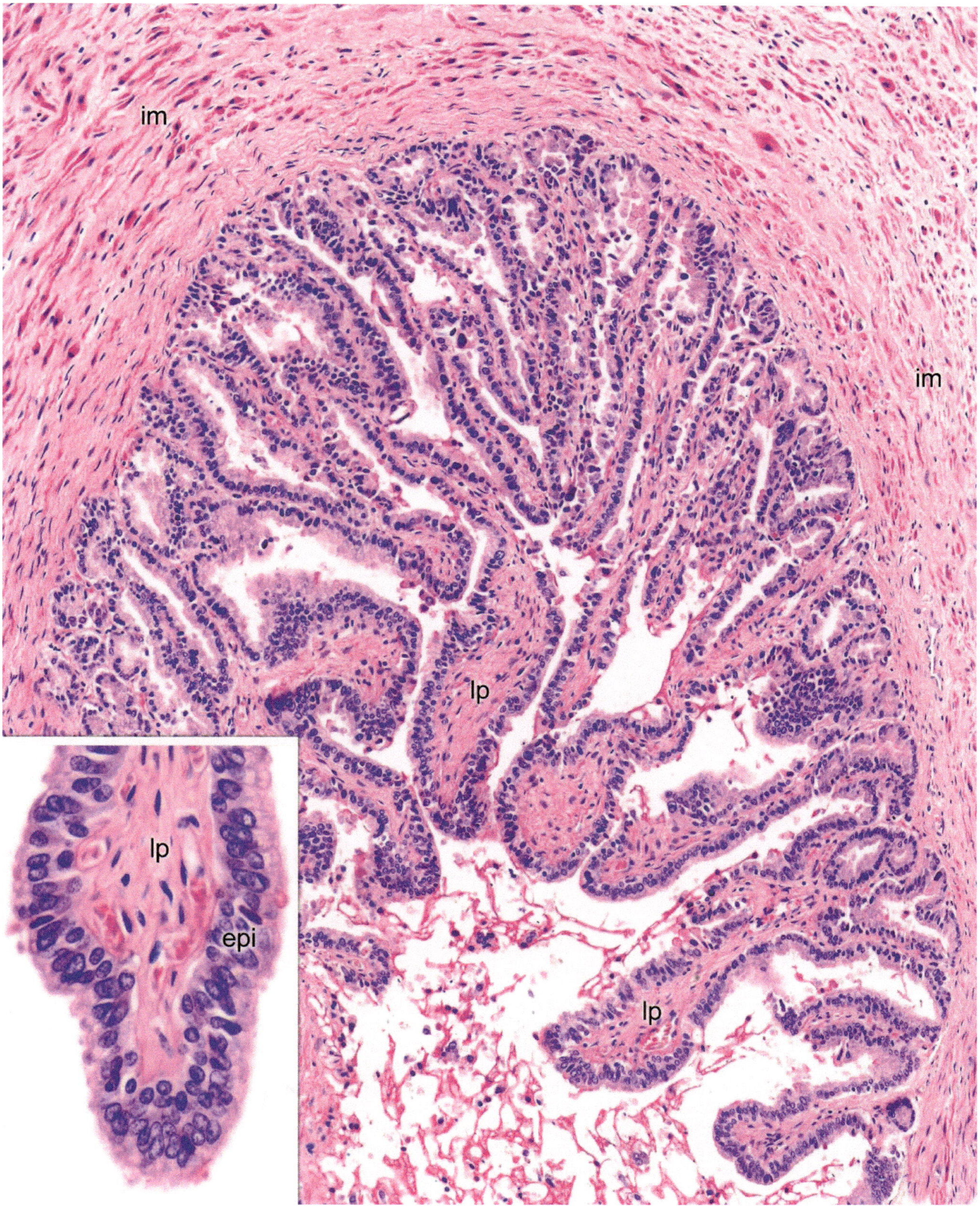

Figure 221. Seminal vesicle. Seminal vesicles are coiled tubular glands with complex mucosal folds that are lined with secretory pseudostratified columnar epithelium (epi) with basal epithelial stem cells similar to those in the epididymis and vas deferens (inset). Epithelium rests on the lamina propria (lp) that forms the core of the folds and it is surrounded by inner circular (im) and outer longitudinal smooth muscle layers. Outside, the tubes are covered with a thin layer of connective tissue.

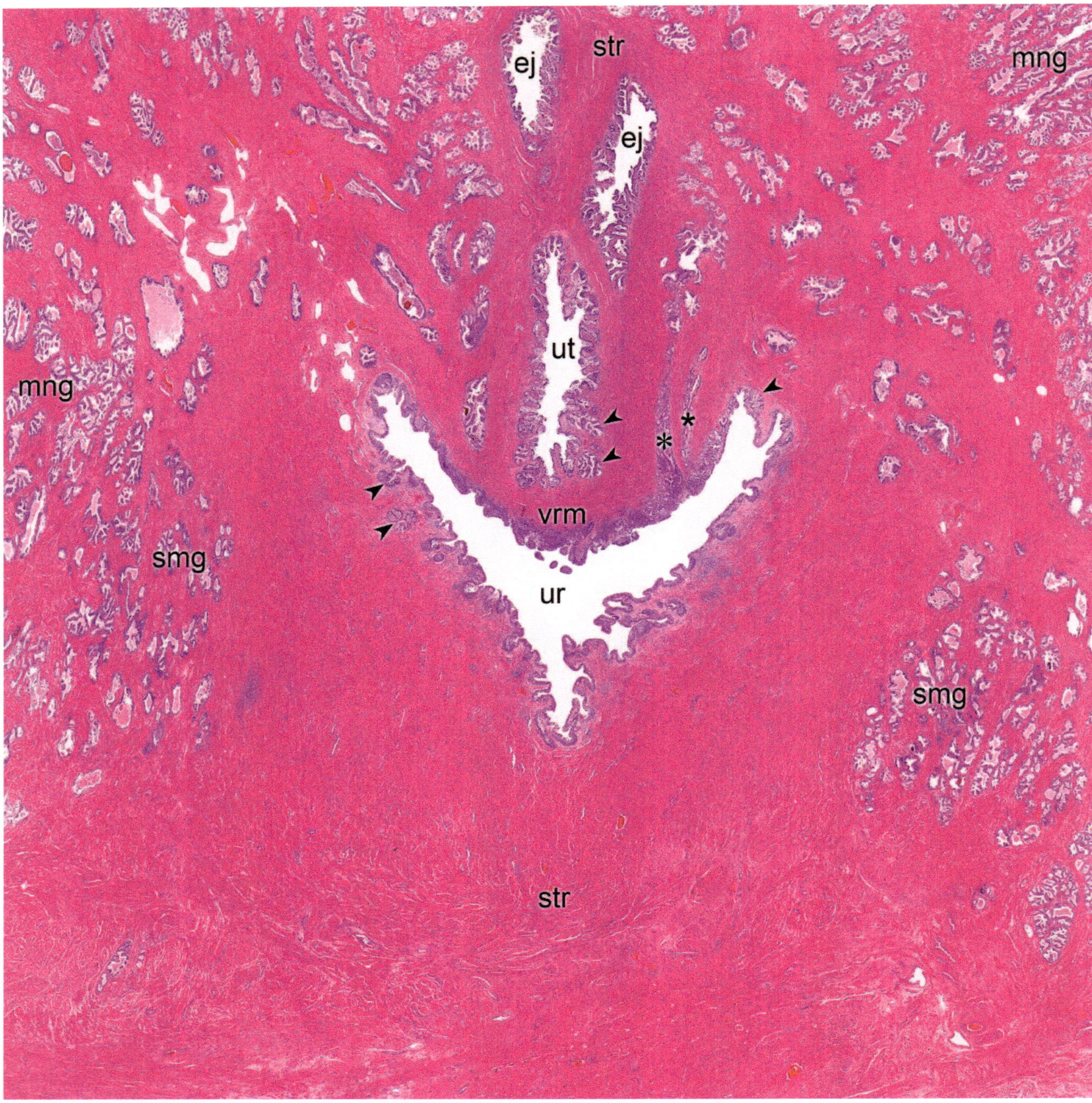

Figure 222. Prostate. Prostate contains tubuloalveolar glands embedded into a fibromuscular stoma (str) formed by collagen fibers and smooth muscle cells. Urethra (ur) perforates the prostate as it receives the two ejaculatory ducts (ej) covered with pseudostratified columnar epithelium identical to the epithelium of the seminal vesicles and the ampulla of the vas deferens. Prostatic utricle (ut) is an epithelium-lined diverticulum, opening between the two ejaculatory ducts on the verumontanum (vrm). Prostatic urethra (ur) is lined with transitional epithelium that is gradually replaced by stratified columnar epithelium distally. Prostatic glands are lined with secretory cuboidal and columnar cells resulting an irregular lumen. Mucosal glands (arrowheads) are adjacent to the urethral epithelium and open directly into the lumen, while the more peripheral submucosal (smg) and main prostatic glands (mng) open into the urethra on the sides of the urethral crest. The ducts of the main glands are usually larger and more medial (six-pointed asterisk) when compared to the smaller and more lateral submucosal ducts (five-pointed asterisk).

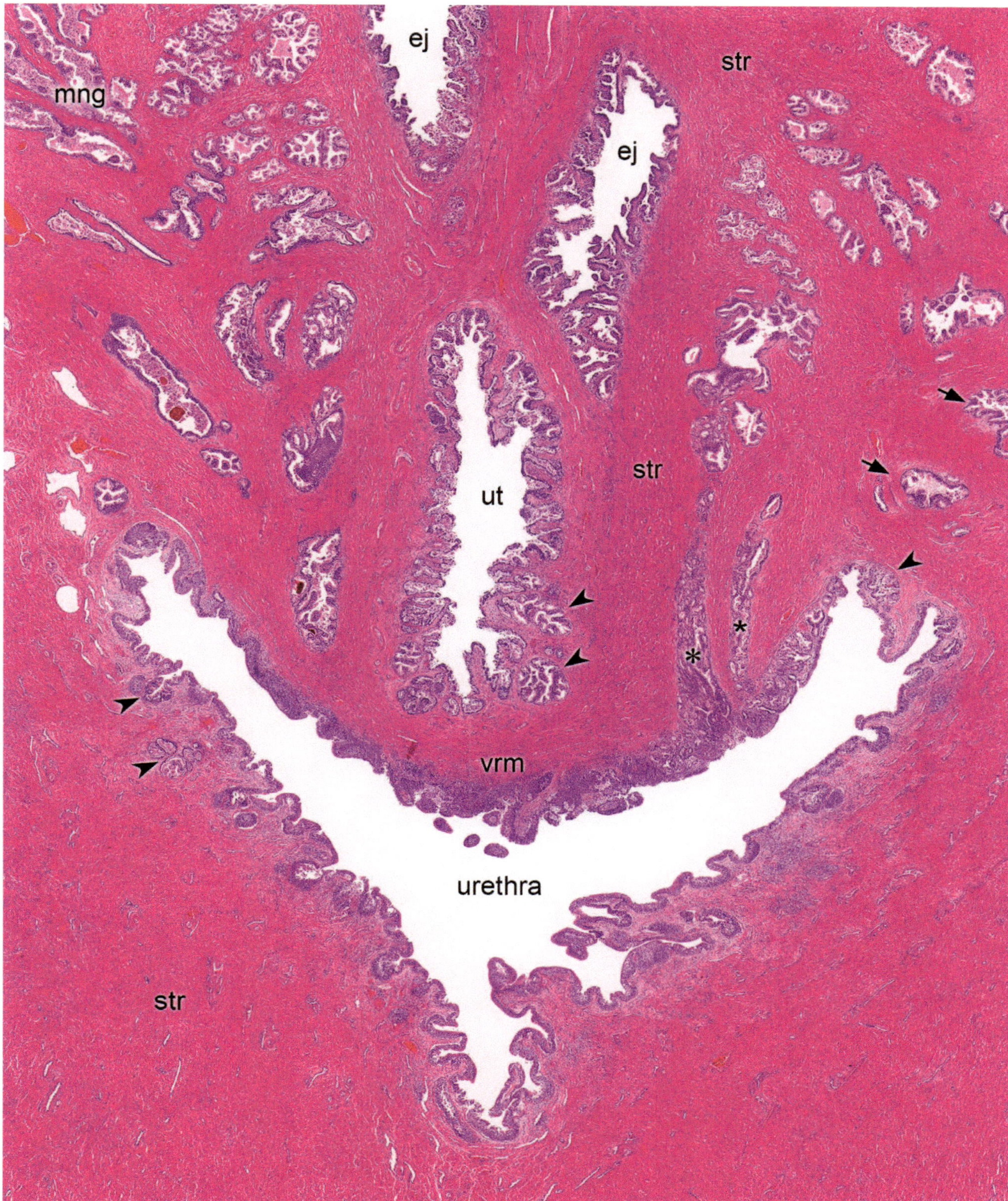

Figure 223. Prostate. Prostate contains tubuloalveolar glands embedded into a fibromuscular stoma (str). Urethra perforates the prostate as it receives the two ejaculatory ducts (ej) covered with pseudostratified columnar epithelium and the prostatic utricle (ut) opening between the two ejaculatory ducts on the verumontanum (vrm). Prostatic glands are lined with secretory cuboidal and columnar cells; mucosal glands (arrowheads) are adjacent to the urethral epithelium and open directly into the lumen, while the more peripheral submucosal (arrows) and main prostatic glands (mng) open on the sides of the urethral crest. The ducts of the main glands are usually larger and more medial (six-pointed asterisk) when compared to the smaller and more lateral submucosal ducts (five-pointed asterisk).

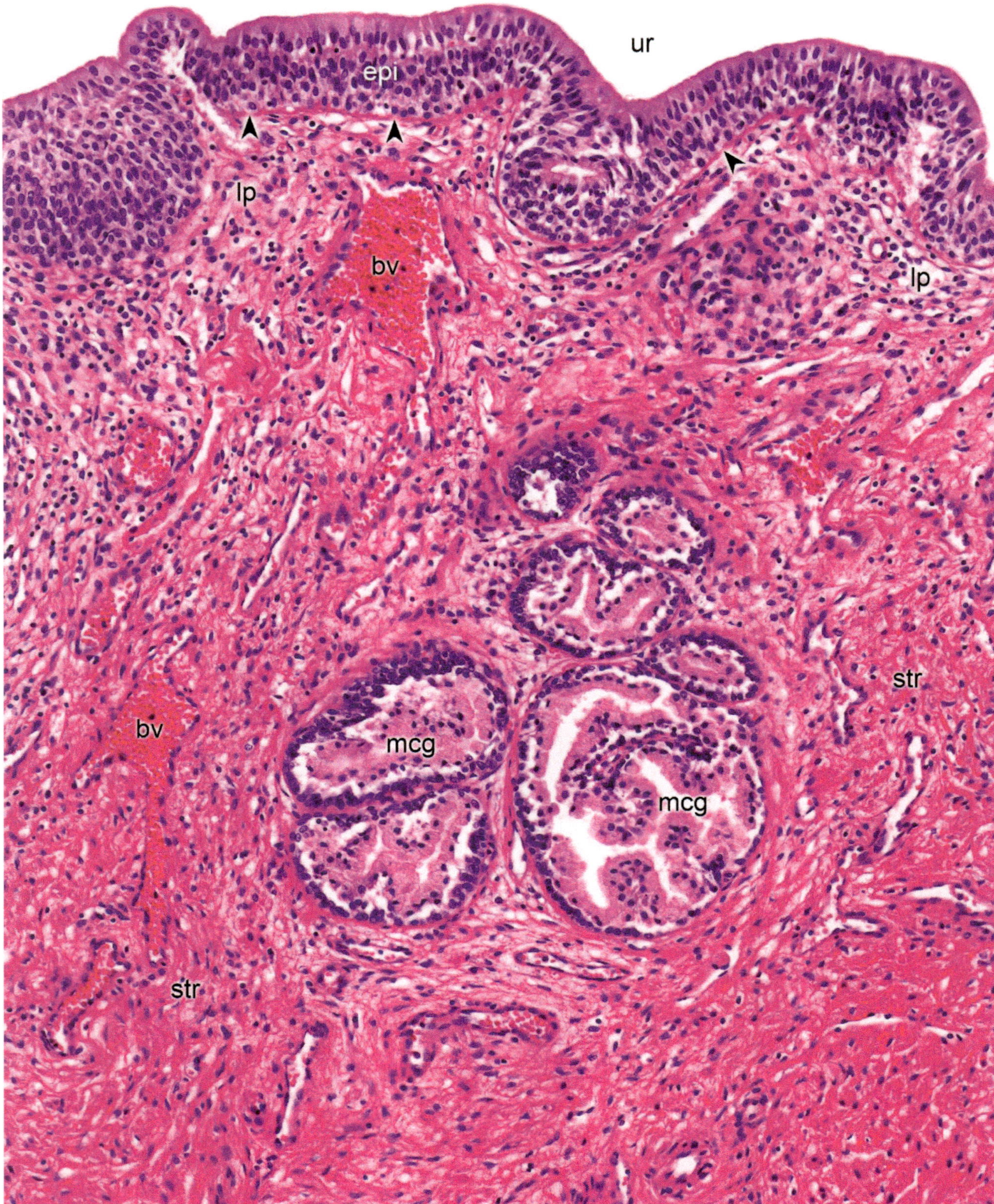

Figure 224. Prostate, mucosal glands. Tubuloalveolar mucosal glands (mcg) are in close proximity to the urethral epithelium (epi) and the underlying basement membrane (arrowheads) and lamina propria (lp), and open directly into the lumen of the urethra (ur). The glands are embedded into a fibromuscular stoma (str) formed by collagen fibers and smooth muscle cells and contain blood vessels, particularly postcapillary and muscular venules. The distal segment of the prostatic urethra (ur) is lined mostly with stratified columnar epithelium (epi) that replaces the transitional epithelium characteristic for the more proximal segments.

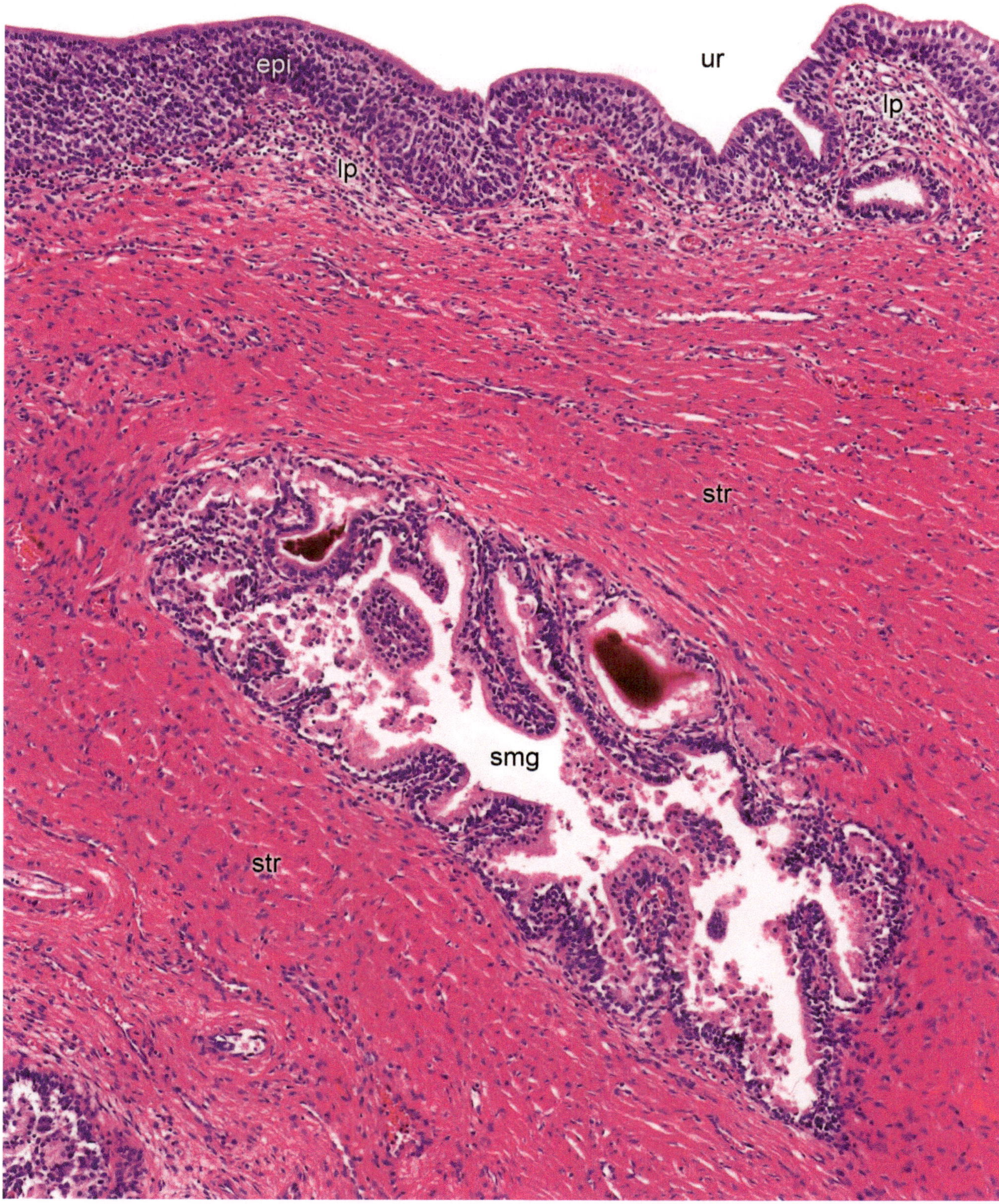

Figure 225. Prostate, submucosal glands. Tubuloalveolar submucosal glands (smg) are embedded into a fibromuscular stoma (str) formed by collagen fibers and smooth muscle cells. Submucosal prostatic glands are more peripheral than the mucous glands and they open into the urethral crest with main ducts. The glands are lined with secretory cuboidal and columnar cells that results in an irregular lumen. The distal segment of the prostatic urethra (ur) is lined mostly with with stratified columnar epithelium (epi) that replaces the transitional epithelium characteristic for the proximal part. Loose connective tissue lamina propria (lp) underlies the epithelium.

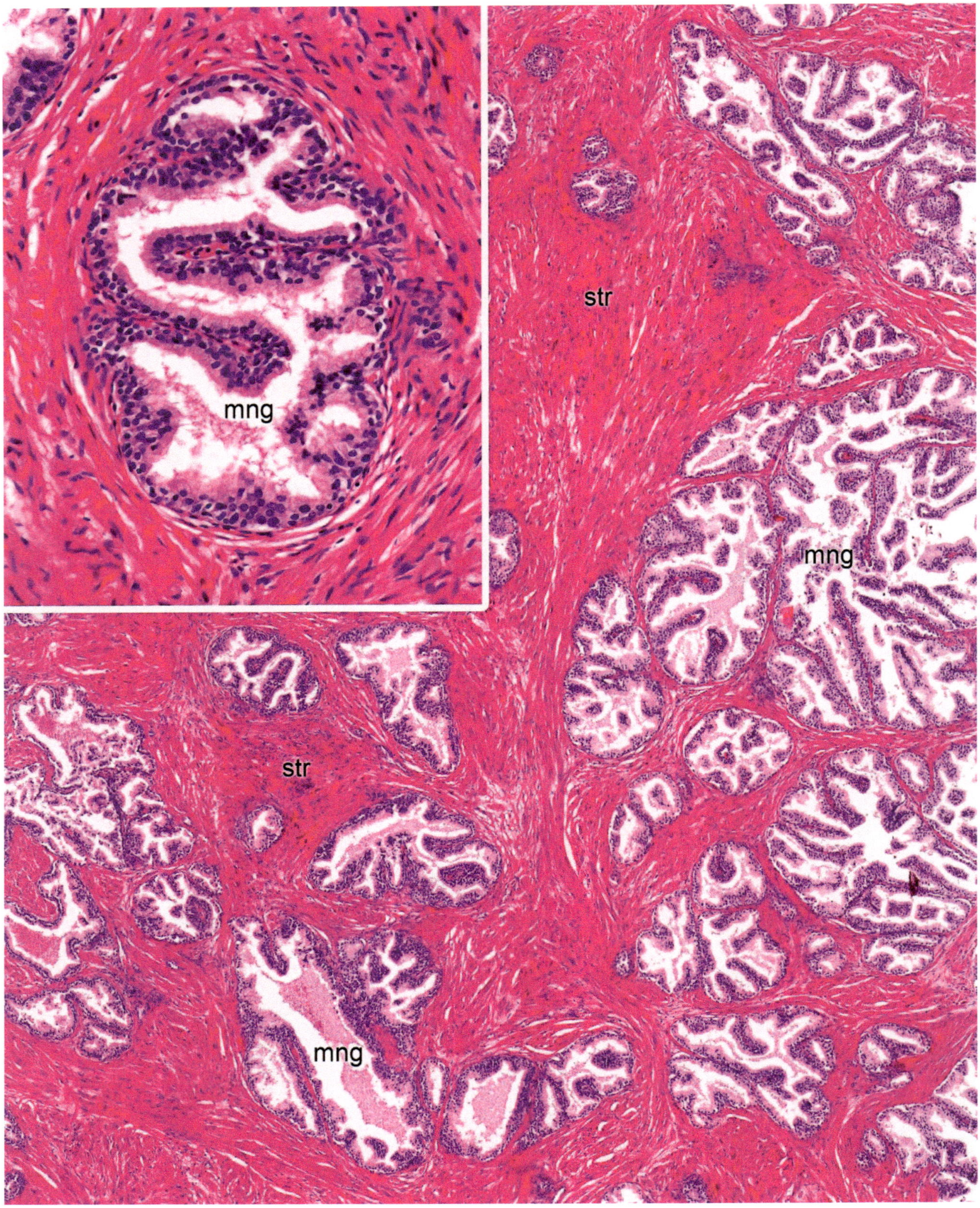

Figure 226. Prostate, main glands. The most peripheral main prostatic glands (mng) are tubuloalveolar glands embedded into a fibromuscular stoma (str) formed by collagen fibers and smooth muscle cells. Main prostatic glands (mng) are lined with secretory cuboidal and columnar cells (inset) and the irregular lumen of their sections resemble walnut kernels. The ducts of the main glands open on the urethral crest.

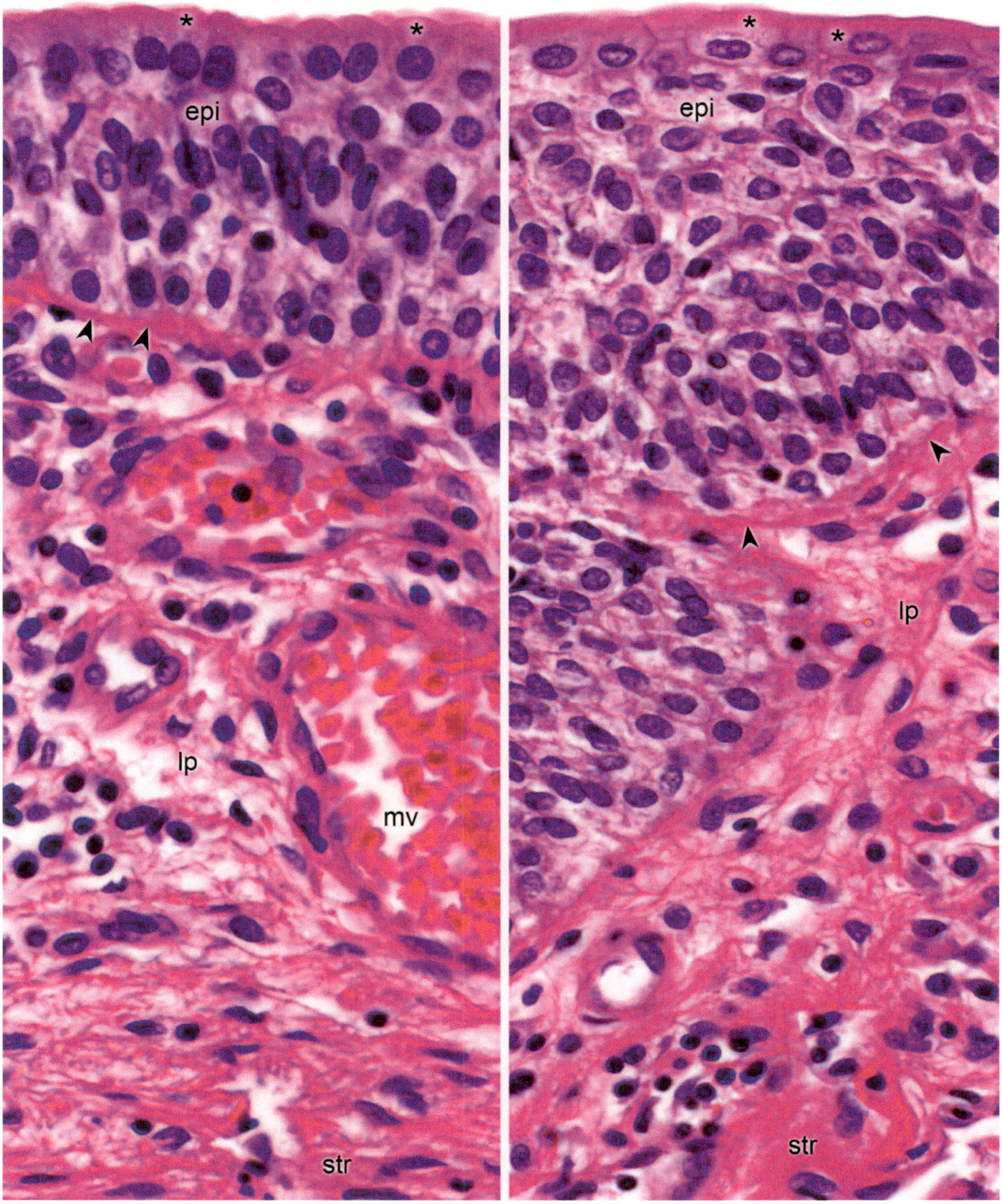

Figure 227. Prostatic urethra, proximal part. The proximal segment of the prostatic urethra is lined with transitional epithelium (epi) that can be in the relaxed (left inset) or in more stretched state (right inset). Consequently, umbrella cells (asterisks), forming the most superficial layer of the epithelium acquire polygonal (left) or more flattened shape (right). The epithelium rests on the basement membrane (arrowheads) and the underlying loose connective tissue lamina propria (lp) that often contains vessels, including muscular venules (mv).Lamina propria covers the fibromuscular stroma (str) formed by collagen fibers and smooth muscle cells.

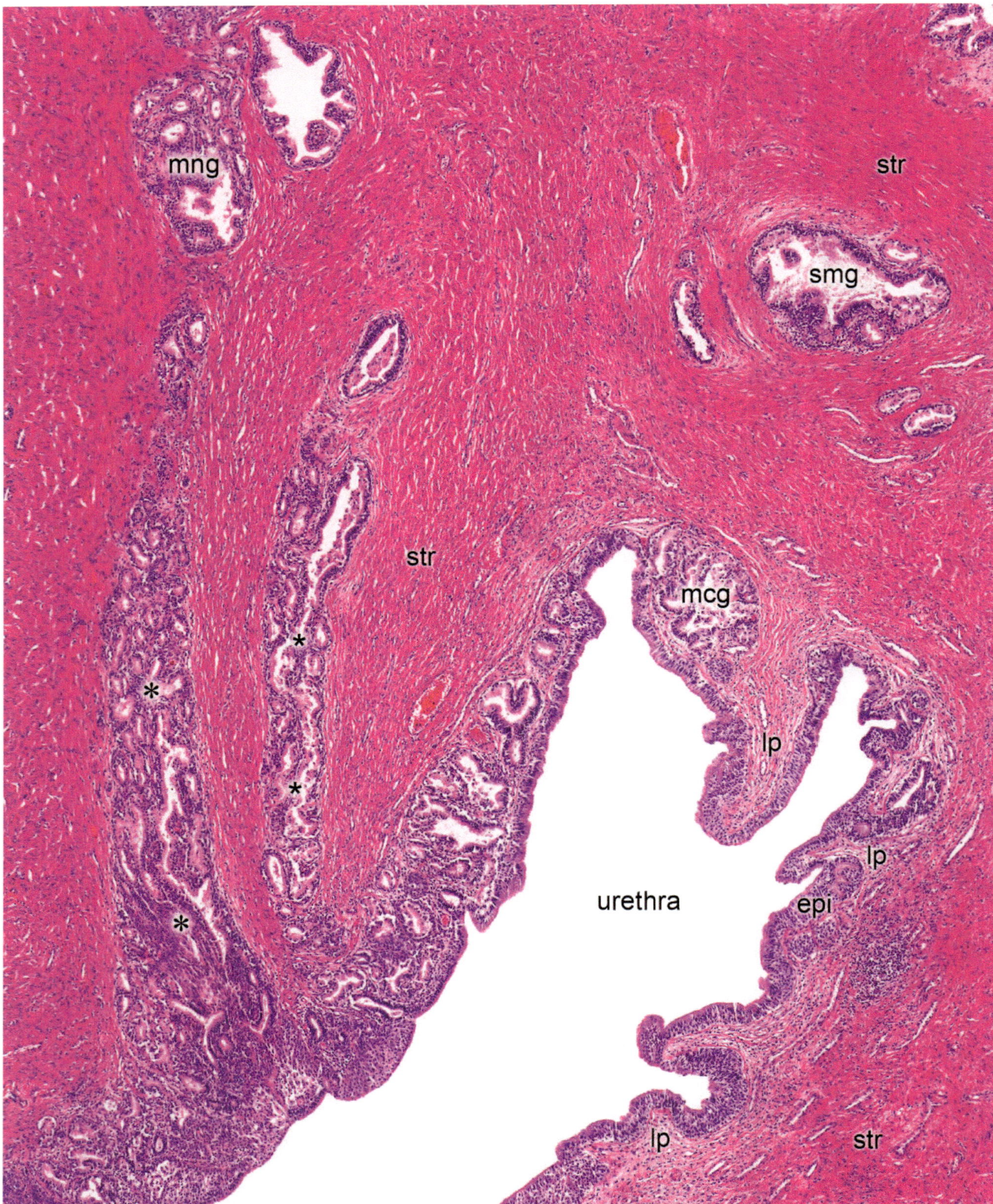

Figure 228. Prostatic glands. Prostate contains tubuloalveolar glands embedded into a fibromuscular stoma (str) formed by collagen fibers and smooth muscle cells. Prostatic glands are lined with secretory cuboidal and columnar cells; mucosal glands (mcg) are adjacent to the urethral epithelium (epi) and the underlying lamina propria (lp) and open directly into the urethral lumen, while the more peripheral submucosal (smg) and main prostatic glands (mng) open on the urethral crest. The ducts of the main glands are usually larger and more medial (six-pointed asterisk) when compared to the smaller and more lateral submucosal ducts (five-pointed asterisk).

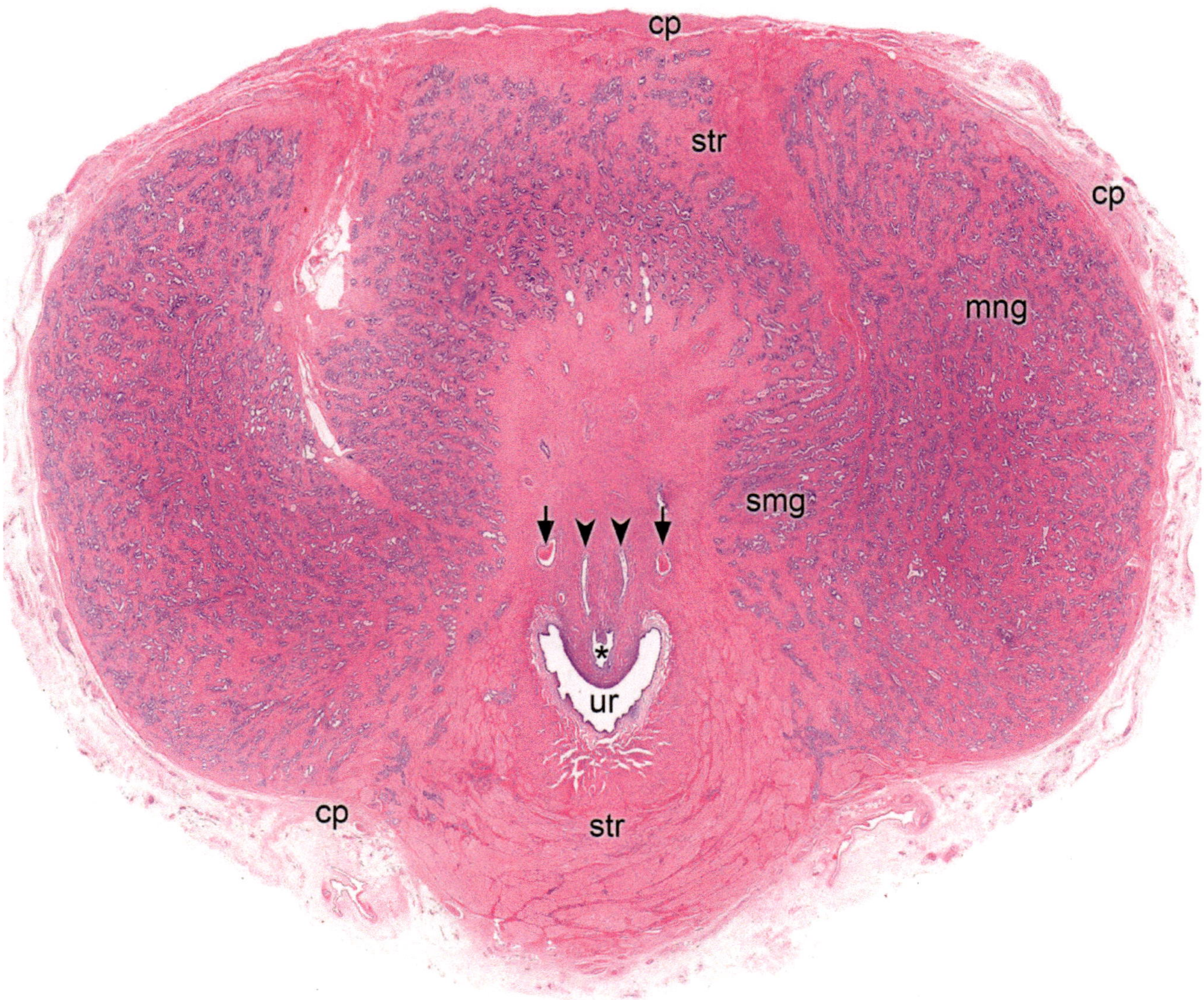

Figure 229. Prostate. Prostate is surrounded by a dense connective tissue capsule (cp) that is covered with the veins of the prostatic plexus. Tubuloalveolar prostatic glands are embedded into a fibromuscular stoma (str) formed by collagen fibers and smooth muscle cells. Urethra (ur) perforates the prostate as it receives the two ejaculatory ducts (arrowheads) covered with pseudostratified columnar epithelium identical to the epithelium of the seminal vesicles and the ampulla of the vas deferens. Prostatic utricle (asterisk) is an epithelium-lined diverticulum, opening between the two ejaculatory ducts on the verumontanum. Prostatic urethra (ur) is lined with transitional epithelium that is gradually replaced by stratified columnar epithelium distally. Prostatic glands are covered with secretory cuboidal and columnar cells resulting an irregular lumen. Mucosal glands are adjacent to the urethral epithelium and open directly into the lumen, while the more peripheral submucosal (smg) and main prostatic glands (mng) open into the urethra on the sides of the urethral crest with ducts (arrows).

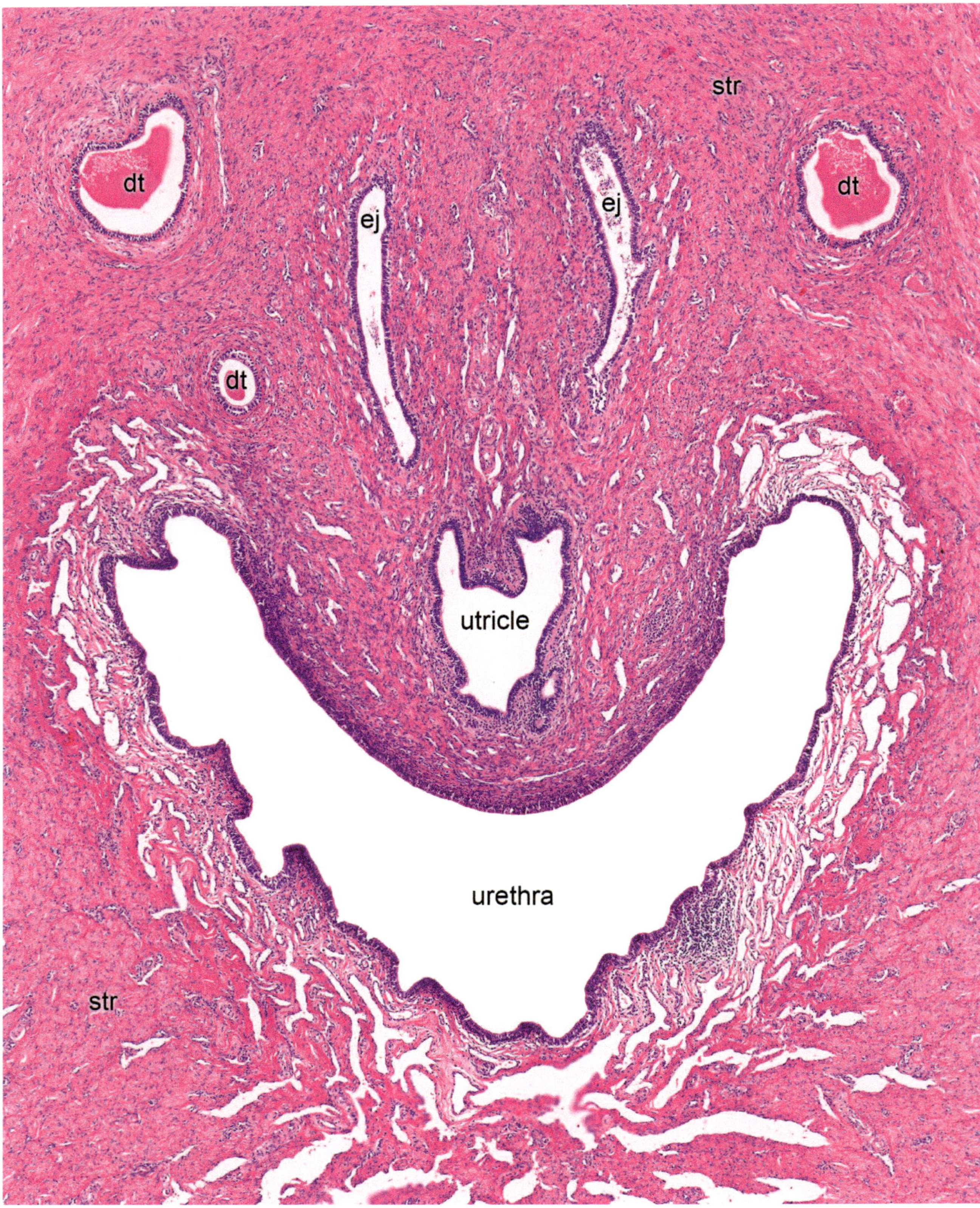

Figure 230. Prostate. Prostate contains tubuloalveolar glands embedded into a fibromuscular stoma (str). Urethra perforates the prostate as it receives the two ejaculatory ducts (ej) covered with pseudostratified columnar epithelium and the prostatic utricle opening between the two ejaculatory ducts on the verumontanum (vrm). Prostatic glands are lined with secretory cuboidal and columnar cells; mucosal glands (arrowheads) are adjacent to the urethral epithelium and open directly into the lumen, while the more peripheral submucosal (arrows) and main prostatic glands (mng) open on the sides of the urethral crest with ducts (dt).

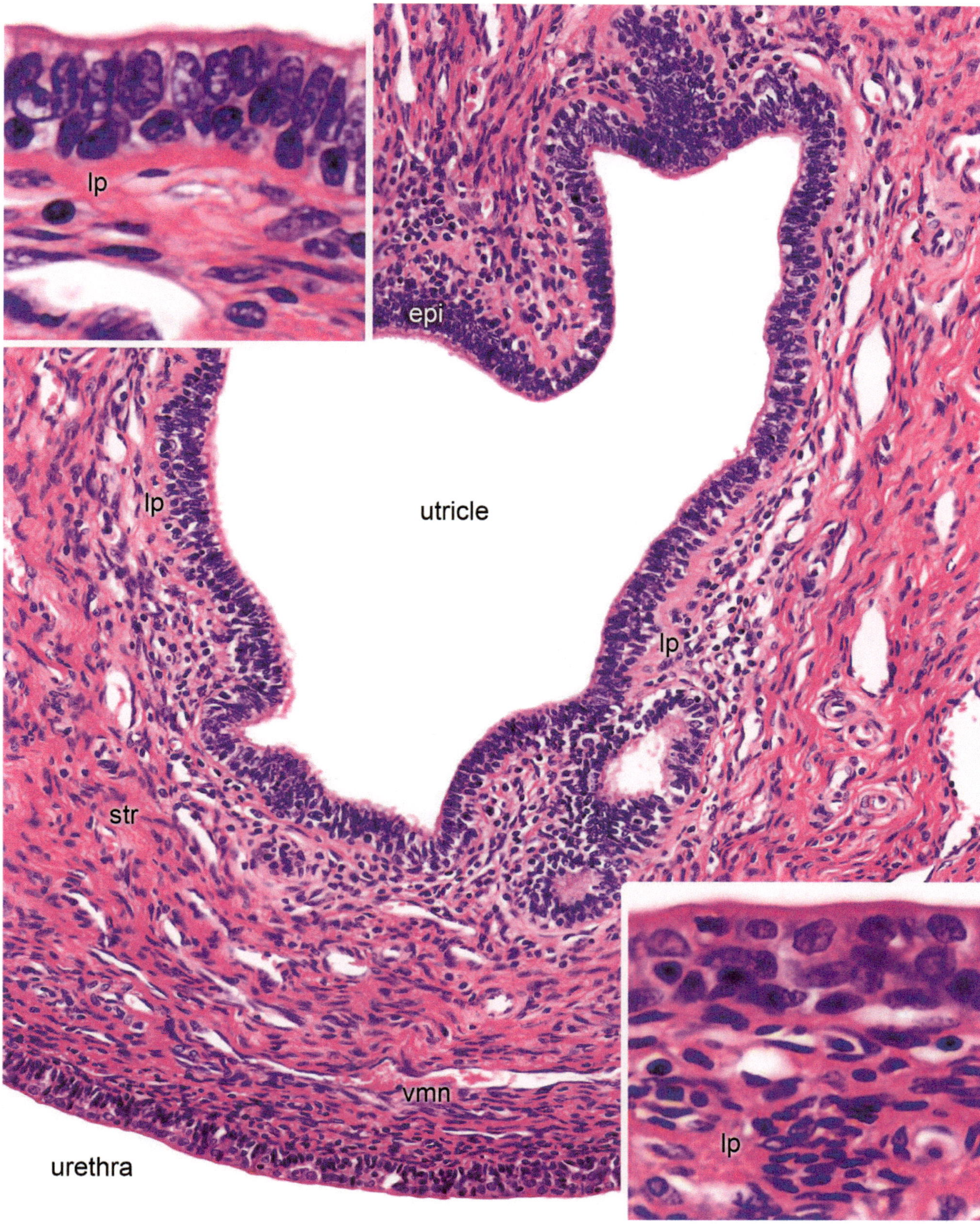

Figure 231. Prostate. Prostatic utricle is a diverticulum, commonly lined with pseudostratified columnar epithelium with basal cells (upper inset) resting on the underlying lamina propria (lp); the utricle opens between the two ejaculatory ducts on the verumontanum (vrm). Prostatic urethra is lined with transitional epithelium (lower inset) that is gradually replaced by stratified columnar epithelium distally. Prostatic tubuloalveolar glands are embedded into a fibromuscular stoma (str). formed by collagen fibers and smooth muscle cells.

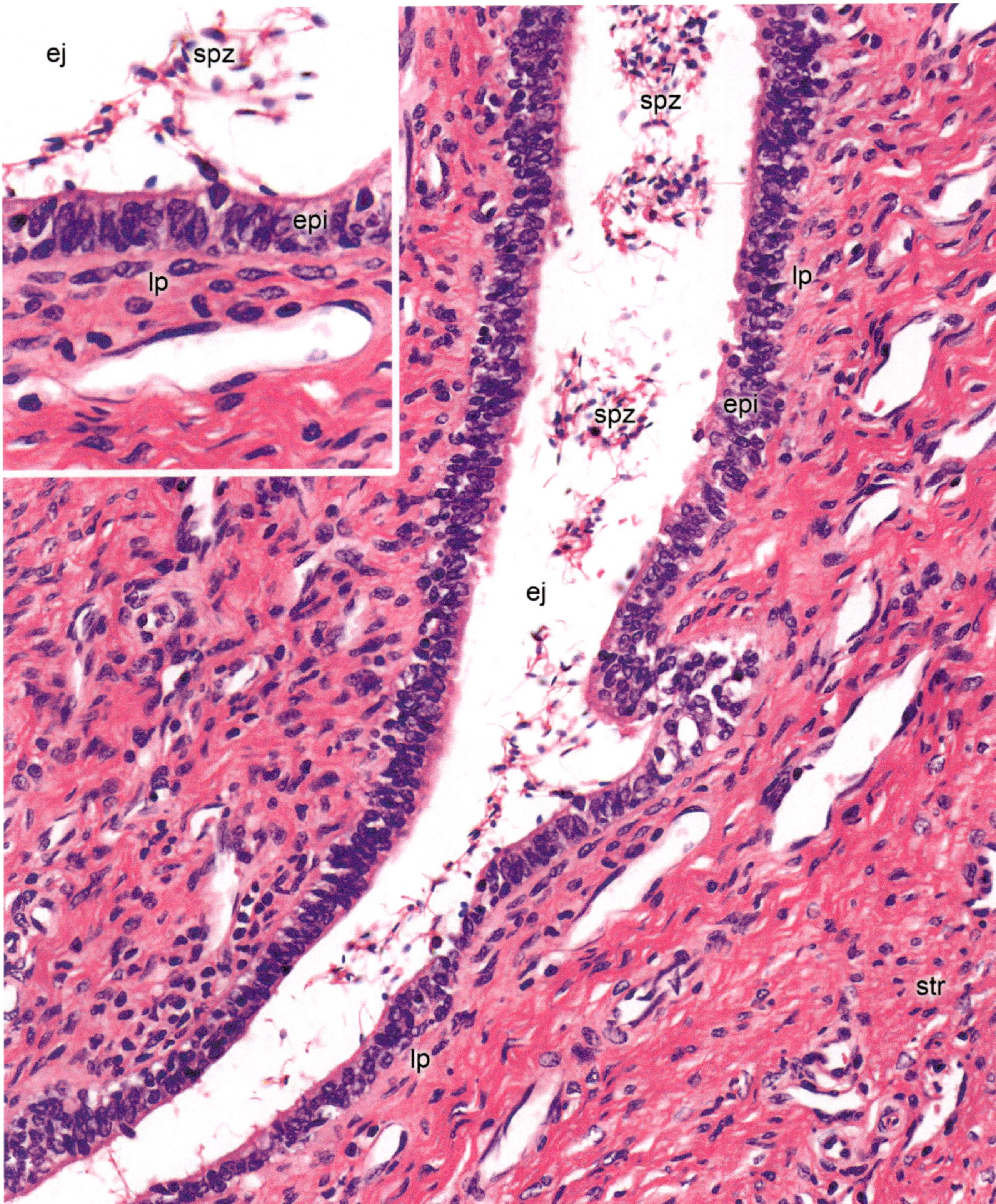

Figure 232. Prostate, ejaculatory duct. The duct of the seminal vesicle merges with the vas deferens forming the ejaculatory ducts (ej) that perforate the prostatic fibromuscular stroma (str) and empty into the prostatic urethra on the sides of the urethral crest. The ejaculatory ducts (ej) are lined with secretory pseudostratified columnar epithelium (epi; inset) and lack a muscular layer under the lamina propria (lp). The lumen of the ducts often contain spermatozoa (spz).

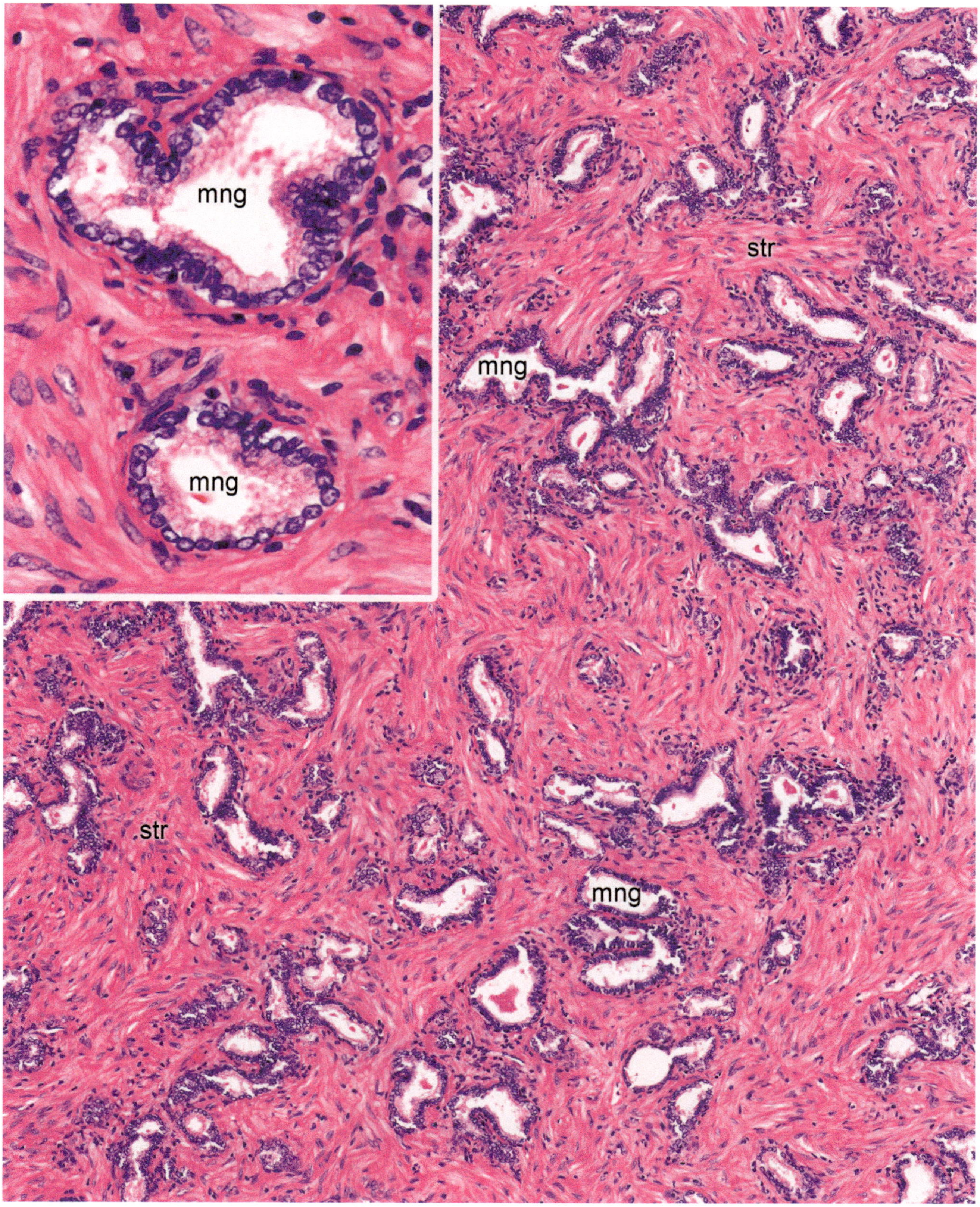

Figure 233. Prostate, main glands. The most peripheral main prostatic glands (mng) are tubuloalveolar glands embedded into a fibromuscular stoma (str) formed by collagen fibers and smooth muscle cells. Main prostatic glands (mng) are lined with secretory cuboidal and columnar cells (inset) and their ducts open on the urethral crest.

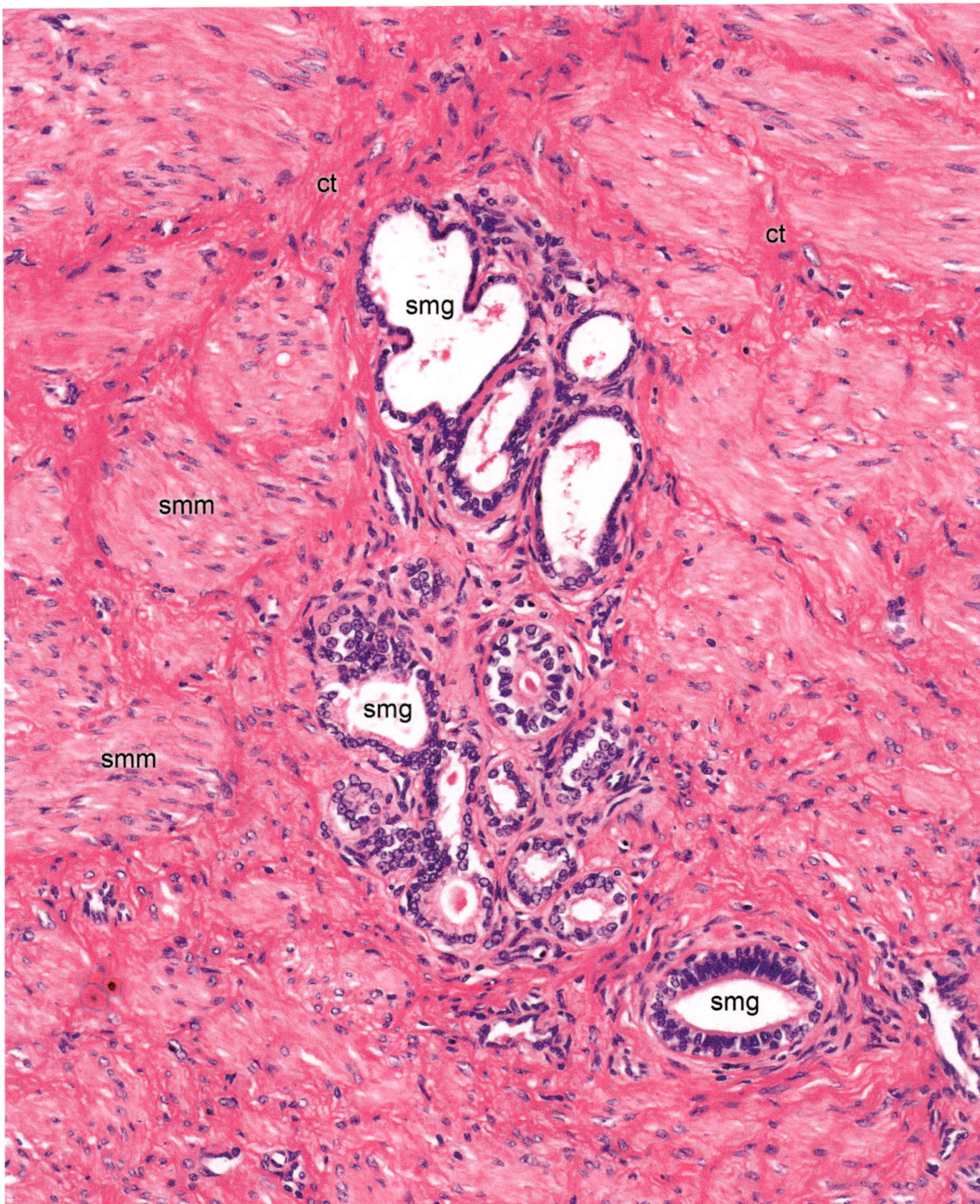

Figure 234. Prostate, submucosal glands. Tubuloalveolar submucosal glands (smg) are embedded into a fibromuscular stoma (str) formed by connective tissue (ct) containing collagen fibers and smooth muscle cells (smm). Submucosal prostatic glands are more peripheral than the mucous glands and they open into the urethral crest with main ducts. The glands lined with secretory cuboidal and columnar cells that results in an irregular lumen.

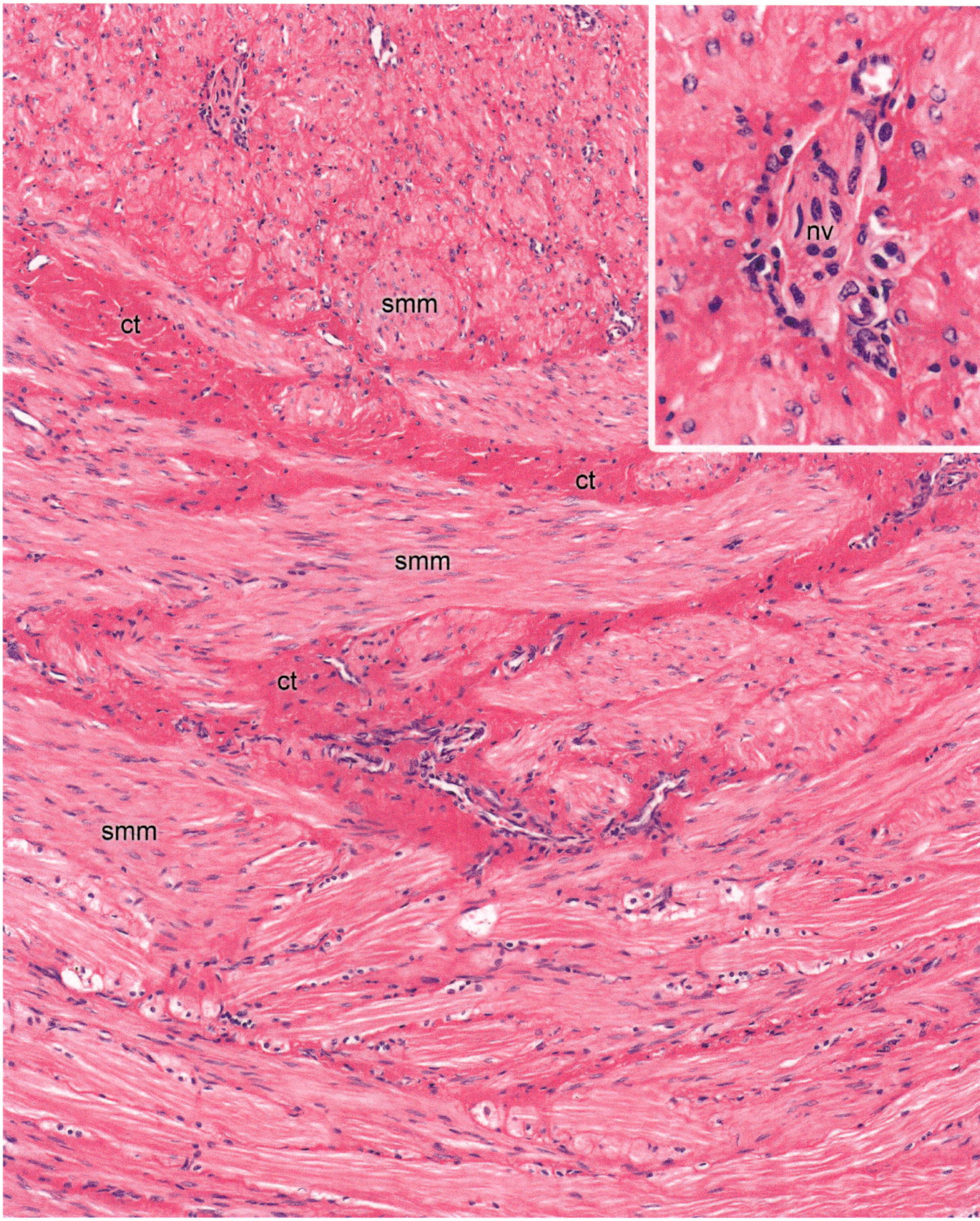

Figure 235. Prostate, fibromuscular stroma. Prostatic tubuloalveolar glands are embedded into a fibromuscular stoma formed by connective tissue (ct) containing collagen fibers and smooth muscle cells (smm). Occasional nerve fibers (nv, inset) and ganglion cells observed in the stroma innervate the smooth muscle cells.

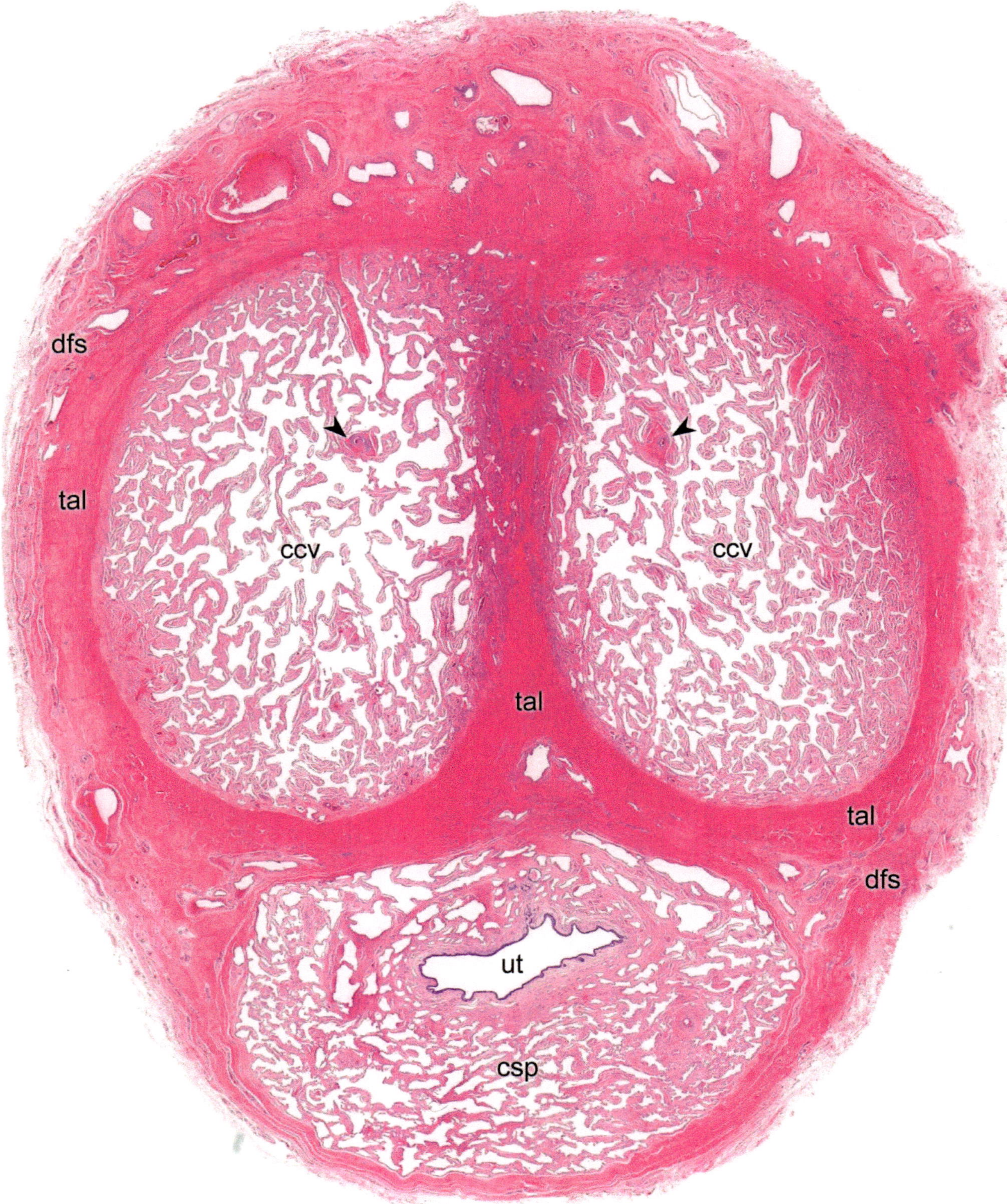

Figure 236. Penis. Cross section of the penis reveals the corpora cavernosa (ccv) and the corpus spongiosum (csp) composed of erectile tissue containing endothelium-lined caverns with underlying smooth muscle trabeculae that is embedded into an interstitial connective tissue layer with collagen and elastic fibers as well as vessels and nerves. Corpora cavernosa contain the deep arteries of the penis (arrowheads) and they are ensheathed by the dense connective tissue tunica albuginea (tal) that is covered by the deep (Buck's) fascia (dfs). Corpus spongiosum contains the penile urethra (ur) that is lined with stratified or pseudostratified columnar epithelium with mucous periurethral glands (glands of Littre) opening into the lumen.

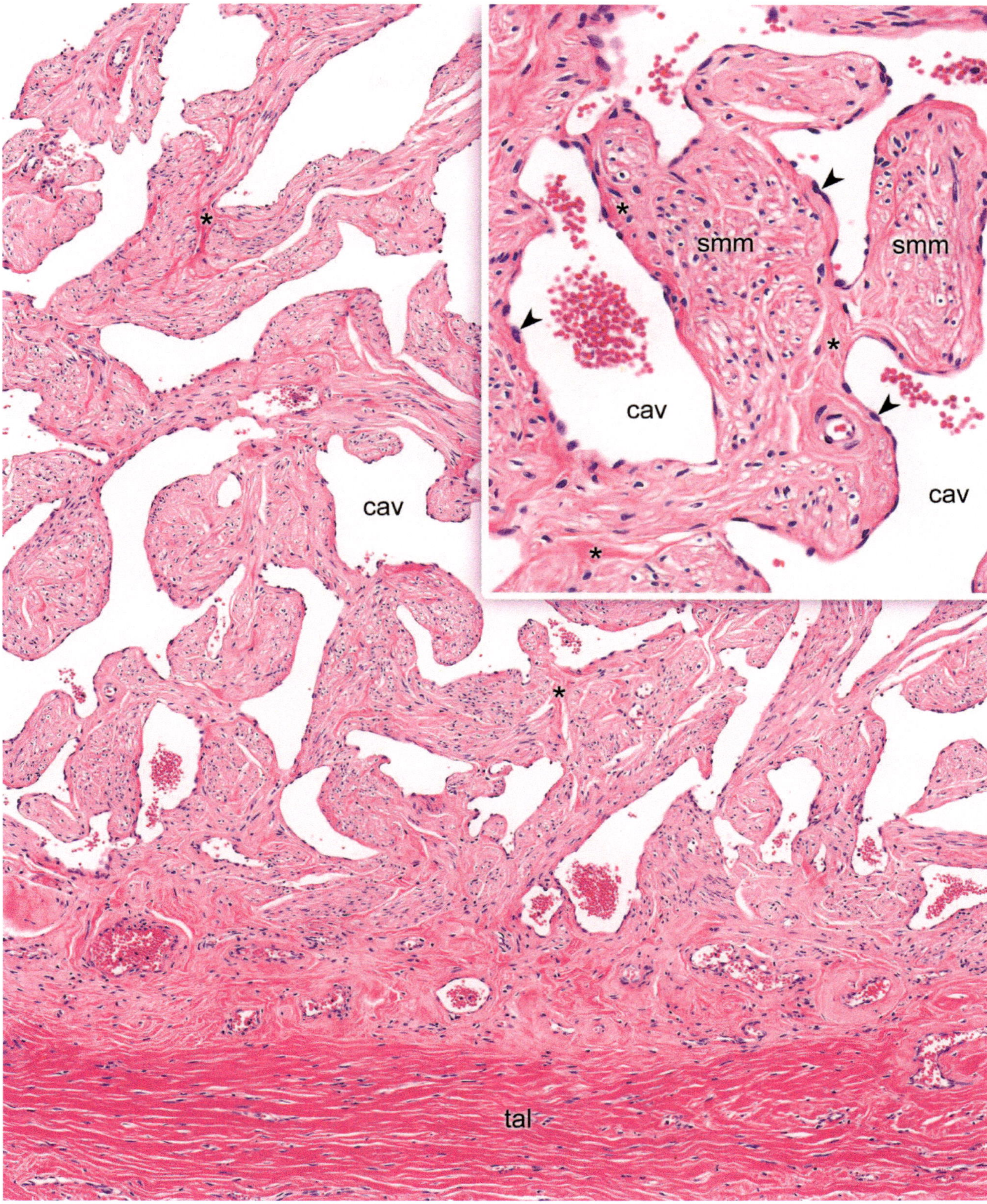

Figure 237. Penis, corpus cavernosum. Corpora cavernosa are composed of erectile tissue containing caverns (cav) lined with endothelial cells (arrowheads, inset). The wall of the caverns is formed by smooth muscle trabeculae (smm) embedded into an interstitial connective tissue layer (asterisks) with collagen and elastic fibers as well as vessels and nerves.

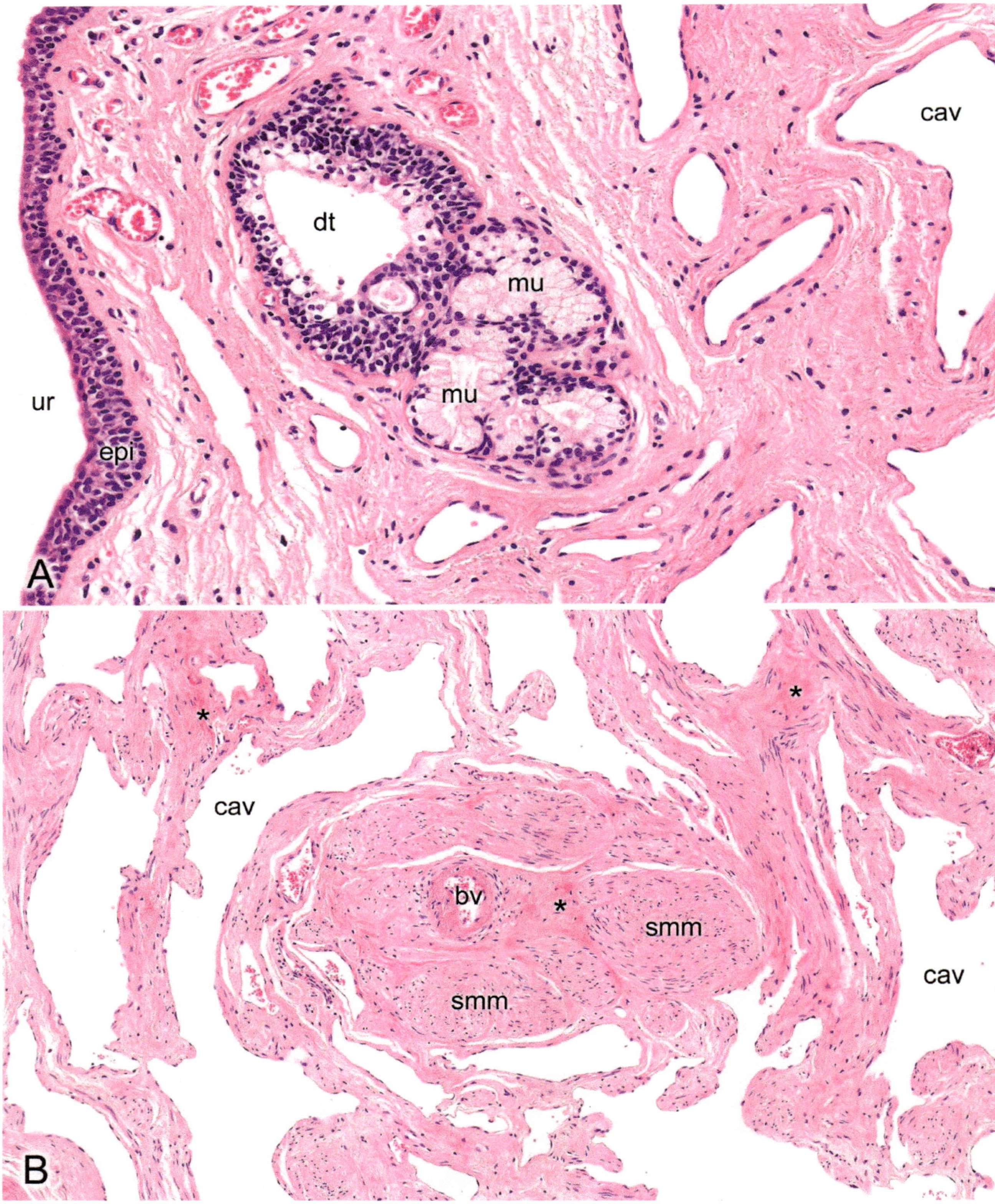

Figure 238. Penis, corpus spongiosum. (A) Penile urethra (ur) is lined with stratified or pseudostratified columnar epithelium that transitions to stratified squamous nonkeratinized epithelium at the urethral orifice. Periurethral glands of Littre are composed of mucous acini (mu) and they open into the penile urethra with their ducts (dt). Urethra is surrounded by the corpus spongiosum that contains caverns (cav) lined with endothelial cells; these caverns are generally smaller than those in the corpora cavernosa. (B) The wall of the caverns is formed by smooth muscle trabeculae (smm) embedded into an interstitial connective tissue layer (asterisks) with collagen and elastic fibers as well as blood vessels (bv) and nerves.

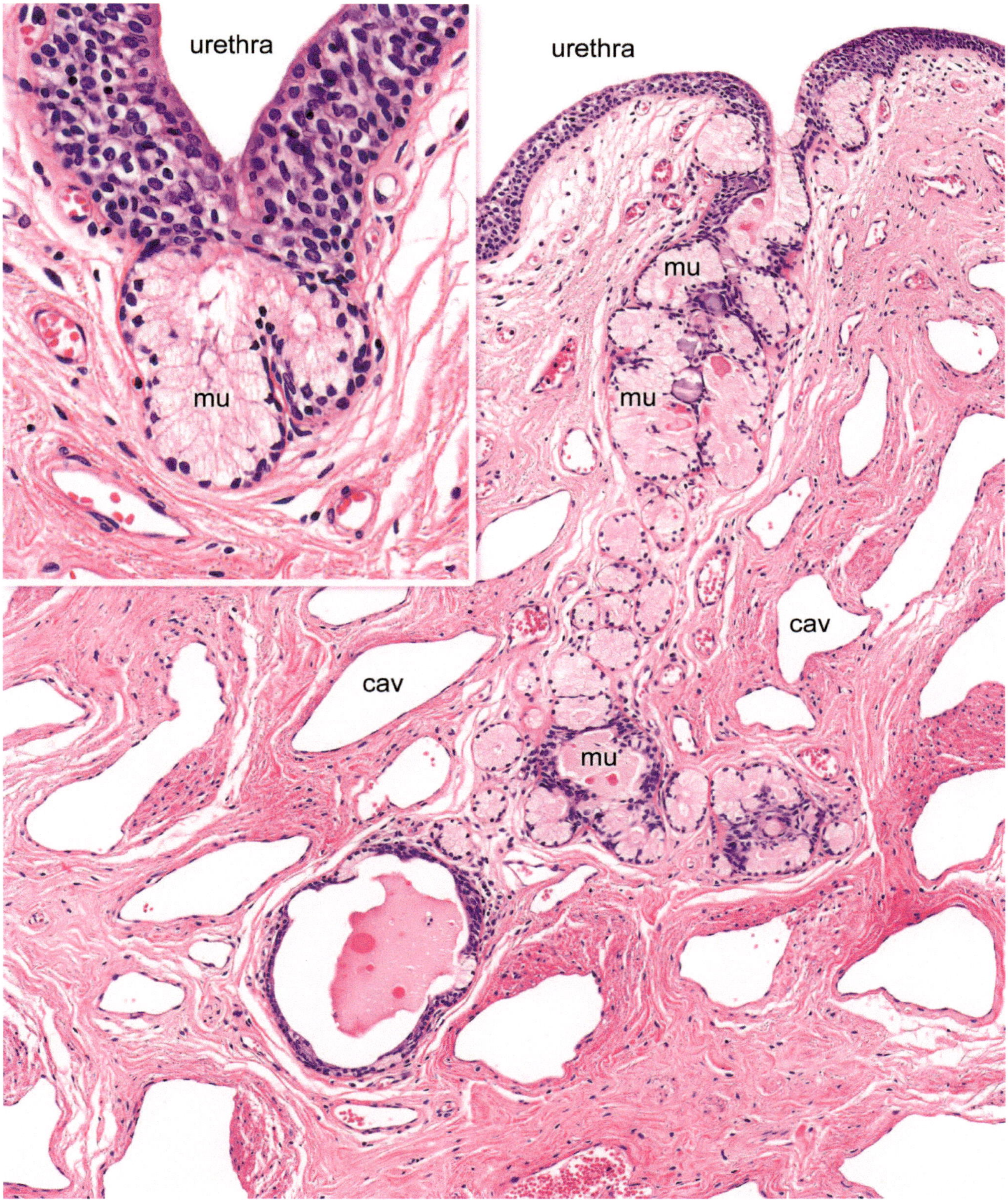

Figure 239. Penile urethra, glands of Littre. Penile urethra is lined with stratified or pseudostratified columnar epithelium that transitions to stratified squamous nonkeratinized epithelium at the urethral orifice (inset). Periurethral glands of Littre are composed of mucous acini (mu) and they open into the penile urethra. Urethra is surrounded by corpus spongiosum that contains caverns (cav) lined with endothelial cells; these caverns are generally smaller than those in the corpora cavernosa. The wall of the caverns is formed by smooth muscle trabeculae embedded into an interstitial connective tissue layer with collagen and elastic fibers.

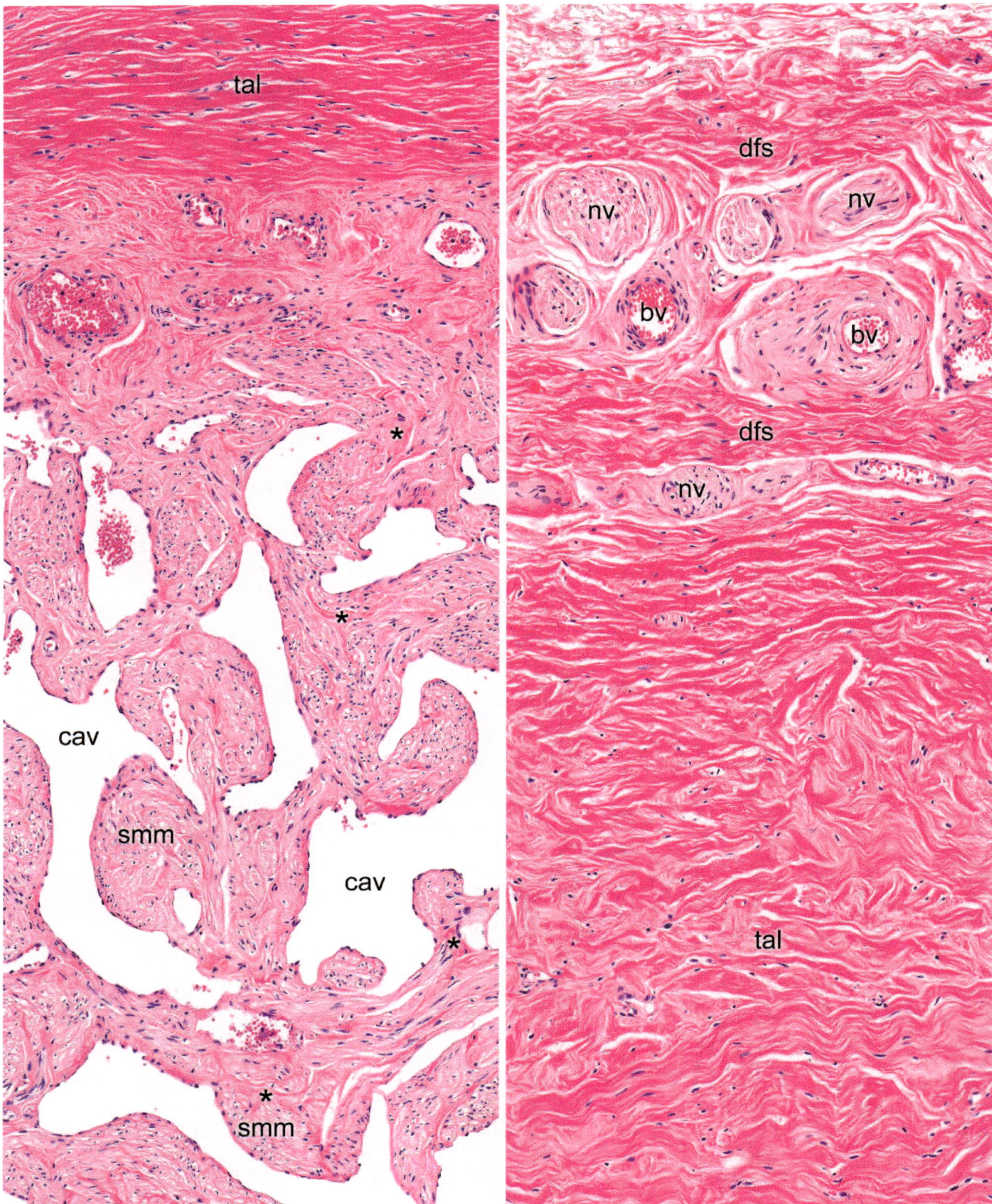

Figure 240. Corpora cavernosa, internal (left) and external layers (right). Corpora cavernosa are composed of erectile tissue containing caverns (cav) lined with endothelial cells. The wall of the caverns is formed by smooth muscle trabeculae (smm) embedded into an interstitial connective tissue layer (asterisks) with collagen and elastic fibers as well as vessels and nerves. Corpora cavernosa are ensheathed by the dense connective tissue tunica albuginea (tal) that is covered by the deep (Buck's) fascia (dfs), enclosing blood vessels (bv) and nerves (nv).

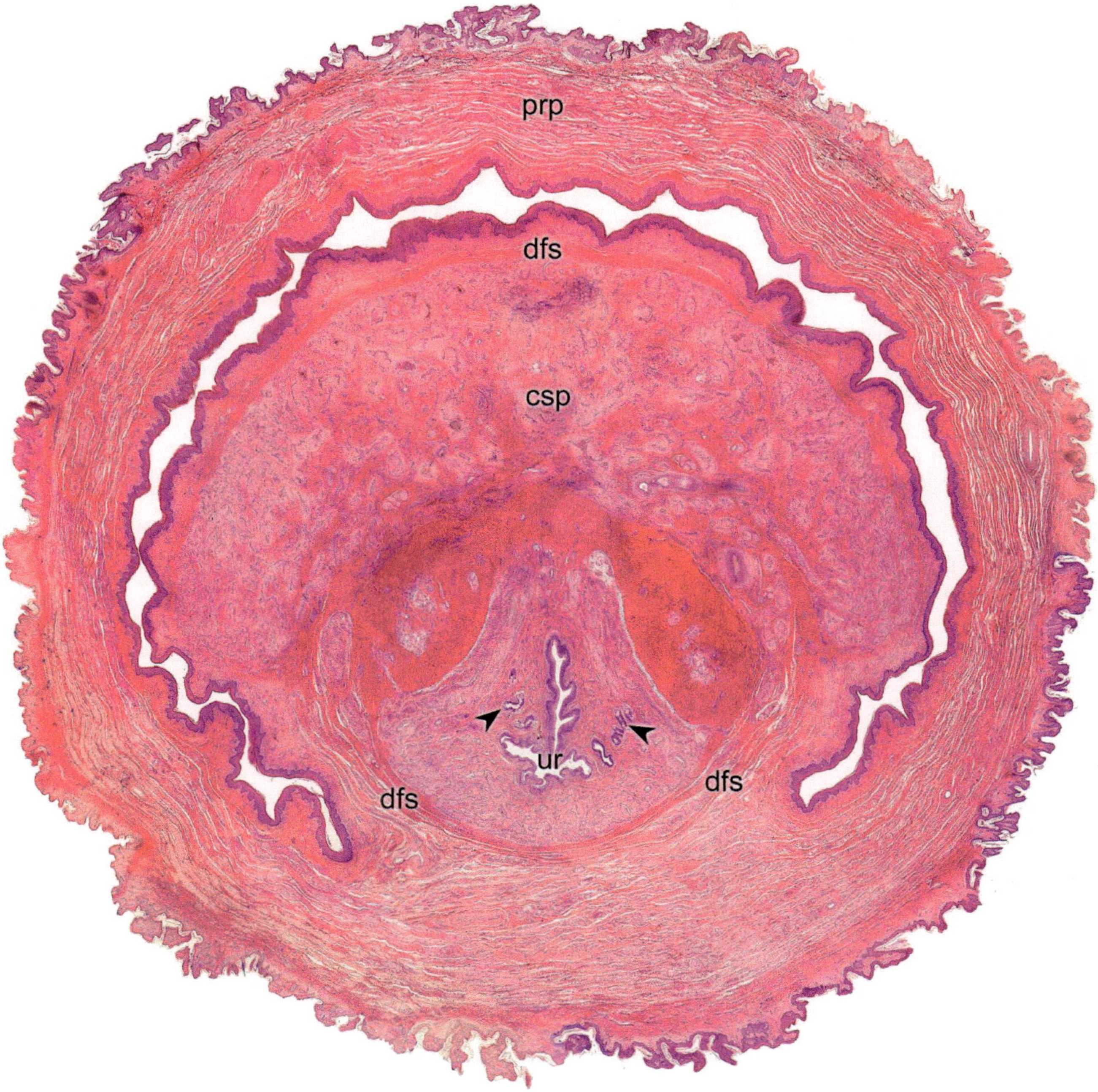

Figure 241. Glans penis, adult. Glans penis is formed by the corpus spongiosum that is surrounded by the deep (Buck's) fascia (dfs). The surface of the glans is lined with thin keratinized stratified squamous epithelium. A fold of the skin, the prepuce (prp) covers the glans; in newborn, the epithelium of the prepuce is fused with the epithelium of the glans, while in adult, they are separated. Glans penis contains the penile urethra (ur) lined with with stratified or pseudostratified columnar epithelium that transitions to stratified squamous nonkeratinized epithelium at the urethral orifice. Periurethral glands of Littre (arrowheads) formed by mucous acini open into the penile urethra.

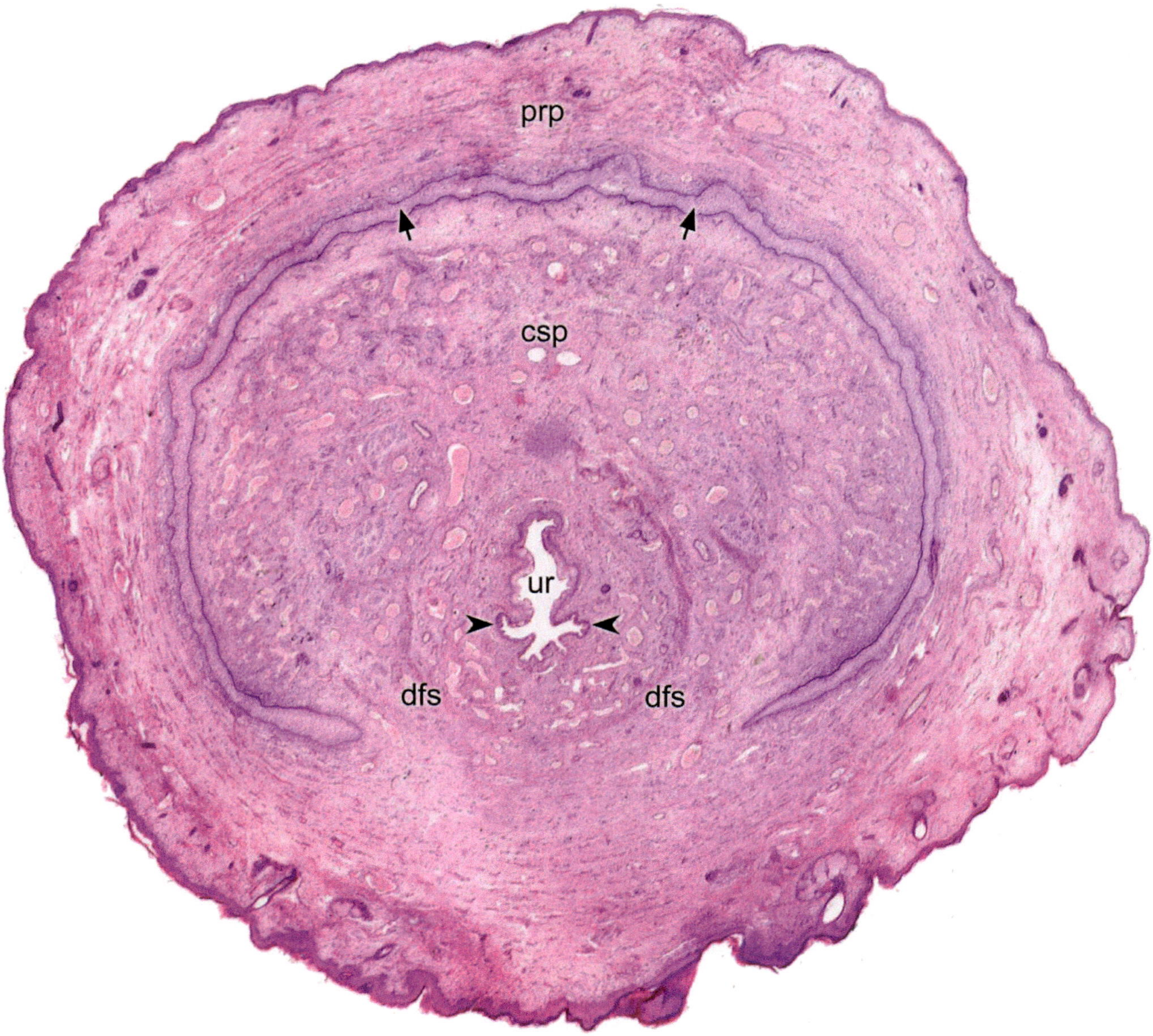

Figure 242. Glans penis, newborn. Glans penis is formed by the corpus spongiosum that is surrounded by the deep (Buck's) fascia (dfs). The surface of the glans is lined with thin keratinized stratified squamous epithelium. A fold of the skin, the prepuce (prp) covers the glans; in newborn, the epithelium of the prepuce is fused with the epithelium of the glans (arrows), while in adult, they are separated. Glans penis contains the penile urethra (ur) lined with with stratified or pseudostratified columnar epithelium that transitions to stratified squamous nonkeratinized epithelium at the urethral orifice. Periurethral glands of Littre (arrowheads) formed by mucous acini open into the penile urethra.

Chapter 11

Female Reproductive System

Ovary

Ovary is covered by a single layer of low cuboidal cells (germinal epithelium) that transitions into a mesothelial cell lining near the mesovarium. The epithelium is attached to a dense connective tissue tunica albuginea that is thinner than the one in the testis. The parenchyma of the ovary is composed of the outer cortex, containing the ovarian follicles embedded into a cell rich stroma, and the inner medulla, formed primarily by loose connective tissue. In the cortex, some of the follicles undergo follicular development; however, most of them regress before the maturation process starts (follicular atresia).

Ovarian follicles contain a single oocyte covered by layers of supporting cells. The first stage of the follicular development is the primordial follicle that appears from the third month of gestation, and it is composed of an oocyte surrounded by a single layer of flattened follicular cells that rest on a basement membrane. Transitioning into the primary follicle, the oocyte enlarges and the follicular cells become cuboidal. Later in this stage, an eosinophilic layer, the zona pellucida starts to appear between the oocyte and the follicular cells. Zona pellucida is rich in glycosaminoglycans and becomes prominent during the follicular development. At later stages of the primary follicles, follicular cells start to proliferate forming multiple layers of cuboidal granulosa cells surrounded by a basal lamina. The connective tissue stroma around the developing follicle forms a sheath of thecal cells that is later separated into an inner vascular theca interna and an outer fibrous theca externa. Theca interna is composed of cuboidal cells that secrete precursors of estrogen in response to luteinizing hormone (LH), while theca externa remains a connective tissue envelope with occasional smooth muscle cells.

As the granulosa cells covering the oocyte proliferate, cavities start to appear between the multiple cell layers that eventually merge and form a single, fluid-filled cavity, the antrum, containing glycosaminoglycan rich liquor folliculi. At this stage, the follicle is categorized as secondary follicle. With the enlargement of the antrum, the location of the oocyte becomes eccentric and it is surrounded by layers of granulosa cells (corona radiata) protruding into the antrum (cumulus oophorus). The mature follicle is at least 10 mm in diameter, typically elevating the surface of the ovary (Graafian follicle). During the ovulation, the thin wall of the follicle facing the ovarian surface ruptures and the pressurized liquor folliculi expels the oocyte with the corona radiata.

Following the ovulation, the walls of the empty follicle collapse and the cavity becomes filled with blood from the surrounding vessels eventually forming a clot (corpus hemorrhagicum). Cells of the theca interna and the granulosa cells dramatically enlarge and accumulate lipid droplets as they turn into lutein cells surrounded by a vascular stroma. This stage is the corpus luteum. Lutein cells secrete progesterone and estrogens that prepare the endometrium of the uterus for implantation of the fertilized oocyte. If the implantation does not take place, corpus luteum eventually turn into corpus albicans where the lutein cells are replaced by scar tissue.

Oviduct

Oviducts (uterine or Fallopian tubes) are tubes extending from the uterus to the ovaries. At the ovary, uterine tube forms funnel-shaped infundibulum that opens into the abdominal cavity adjacent to the ovaries. The infundibulum continues into the ampulla, an enlarged part of the tube that leads into the isthmus with narrow lumen. The intramural part of the oviduct is located in the uterine wall and opens into the uterine cavity.

The mucous membrane of the oviduct is thrown into complex folds that are more elaborate in the ampulla and less complicated at the isthmus. These folds are lined with ciliated simple columnar epithelium containing nonciliated secretory peg cells. The lamina propria of the mucous membrane often contains lymph vessels at the core of the folds. Indistinct layers of inner circular and outer longitudinal bundles of smooth muscle form the tunica muscularis that is surrounded by the connective tissue serosa containing blood vessels.

http://doi.org/10.1016/B978-0-323-91891-6.50011-8,

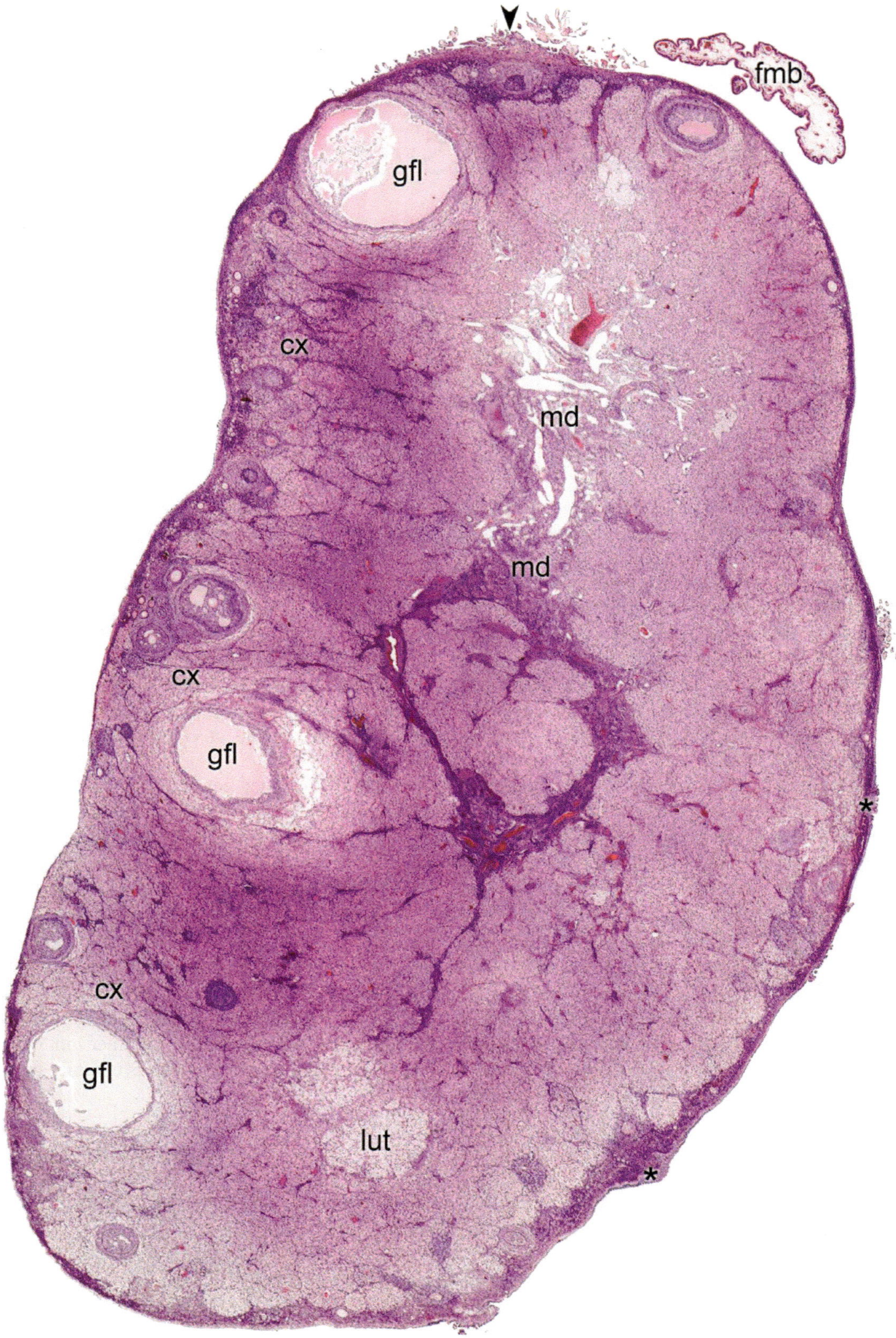

Figure 243. Ovary. The surface of the ovary is lined with a single layer of cuboidal cells, the germinal epithelium and the underlying tunica albuginea (asterisks). Germinal epithelium transitions into a mesothelial cell lining near the mesovarium (arrowhead). Ovarian cortex (cx) contains follicles including the matured Graafian follicles (gfl) that transform into corpus luteum (lut) after the ovulation. Medulla (md) contains loose connective tissue and lacks follicles. Fimbriae of the oviduct (fmb) are adjacent to the ovarian surface.

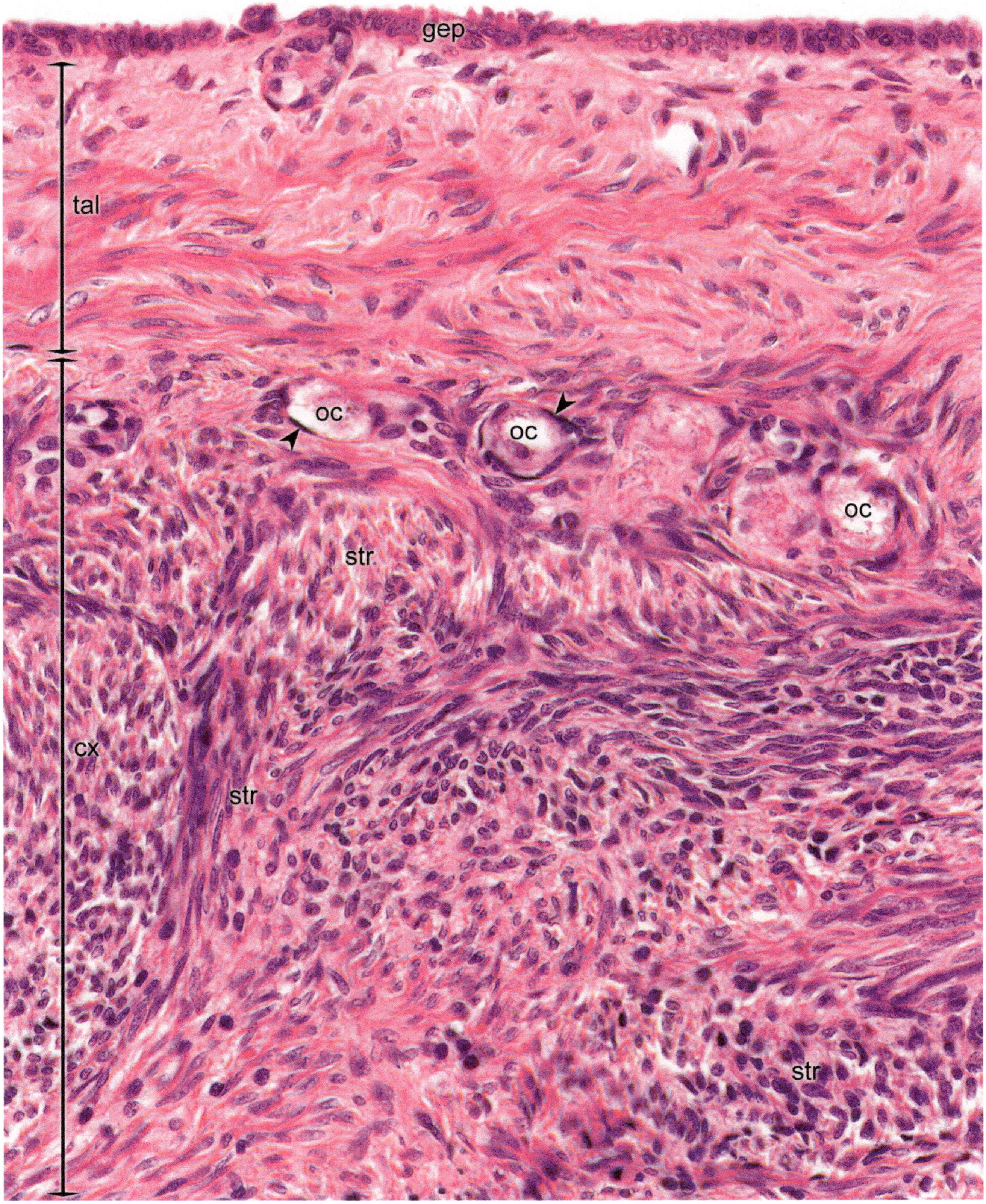

Figure 244. Ovary. The surface of the ovary is lined with a single layer of cuboidal cells, the germinal epithelium (gep) and the underlying tunica albuginea (tal). Germinal epithelium transitions into mesothelial cell lining near the mesovarium. Ovarian cortex (cx) contains follicles including primordial follicles that are composed of an oocyte (oc) surrounded by follicular cells (arrowheads). Follicles are embedded into a cell rich stroma (str).

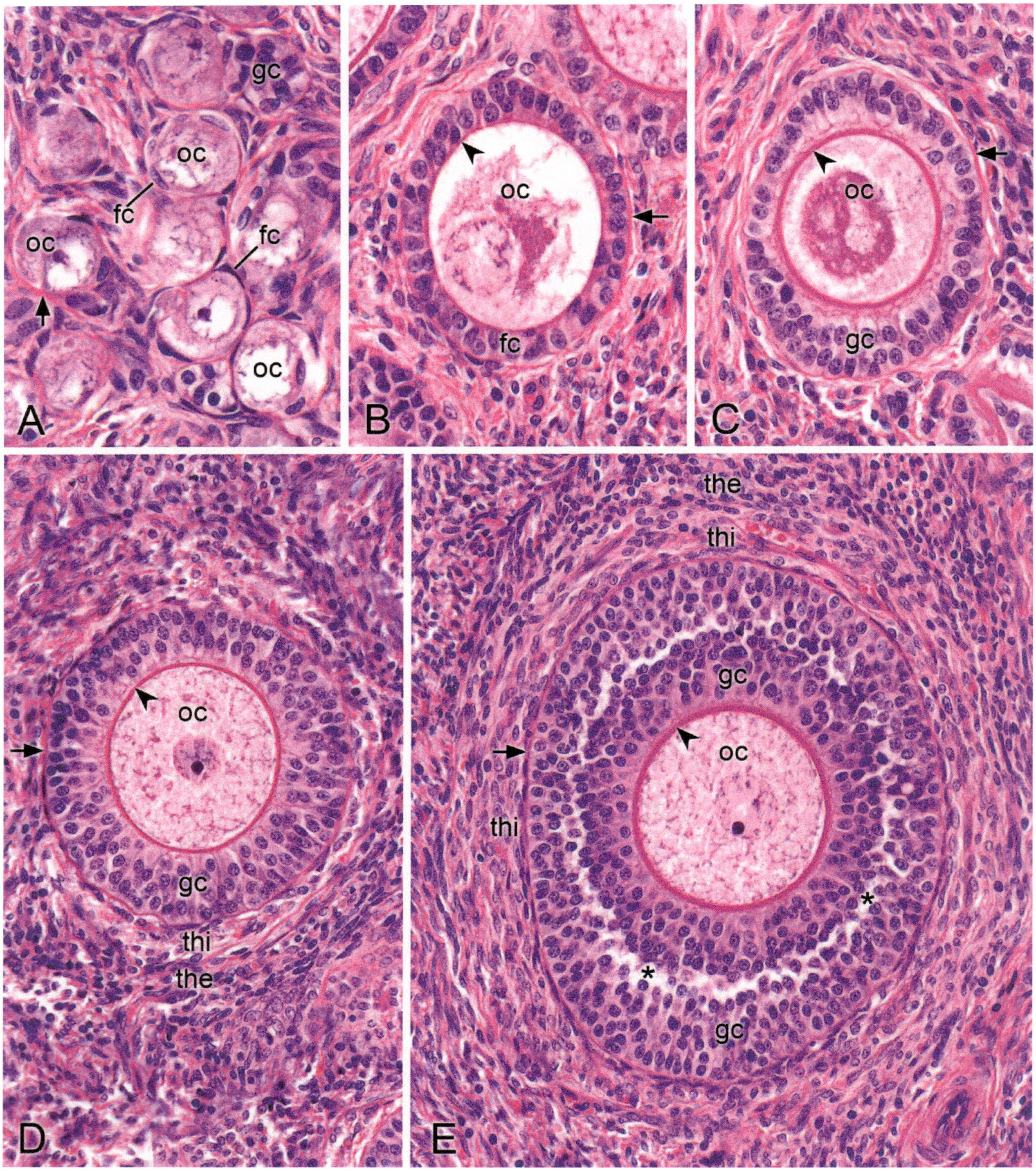

Figure 245. Ovary, follicles. Ovarian cortex contains follicles embedded into a cell rich stroma. (A) Primordial follicles are composed of an oocyte (oc) surrounded by a single layer of flattened follicular cells (fc). (B) In primary follicles, follicular cells (fc) resting on the basement membrane (arrow) become cuboidal, and an eosinophilic layer, the zona pellucida (arrowhead) starts to appear around the oocyte. (C, D) At later stages of the primary follicles, the single layer of follicular cells starts to proliferate forming multiple layers of cuboidal granulose cells (gc) resting on a basal lamina (arrow). Connective tissue stroma around the developing primary follicle forms a sheath of thecal cells that later separates into an inner vascular theca interna (thi) and an outer fibrous theca externa (the). Arrowheads denote the zona pellucida. (E) In secondary follicles, cavities filled with glycosaminoglycan rich liquor folliculi (asterisks) appear between the multiple layers of granulose cells (gc). Zona pellucida (arrowhead) becomes thicker and the secondary follicle is surrounded by a distinct theca interna (thi)and externa (the). Basement membrane is denoted by arrow.

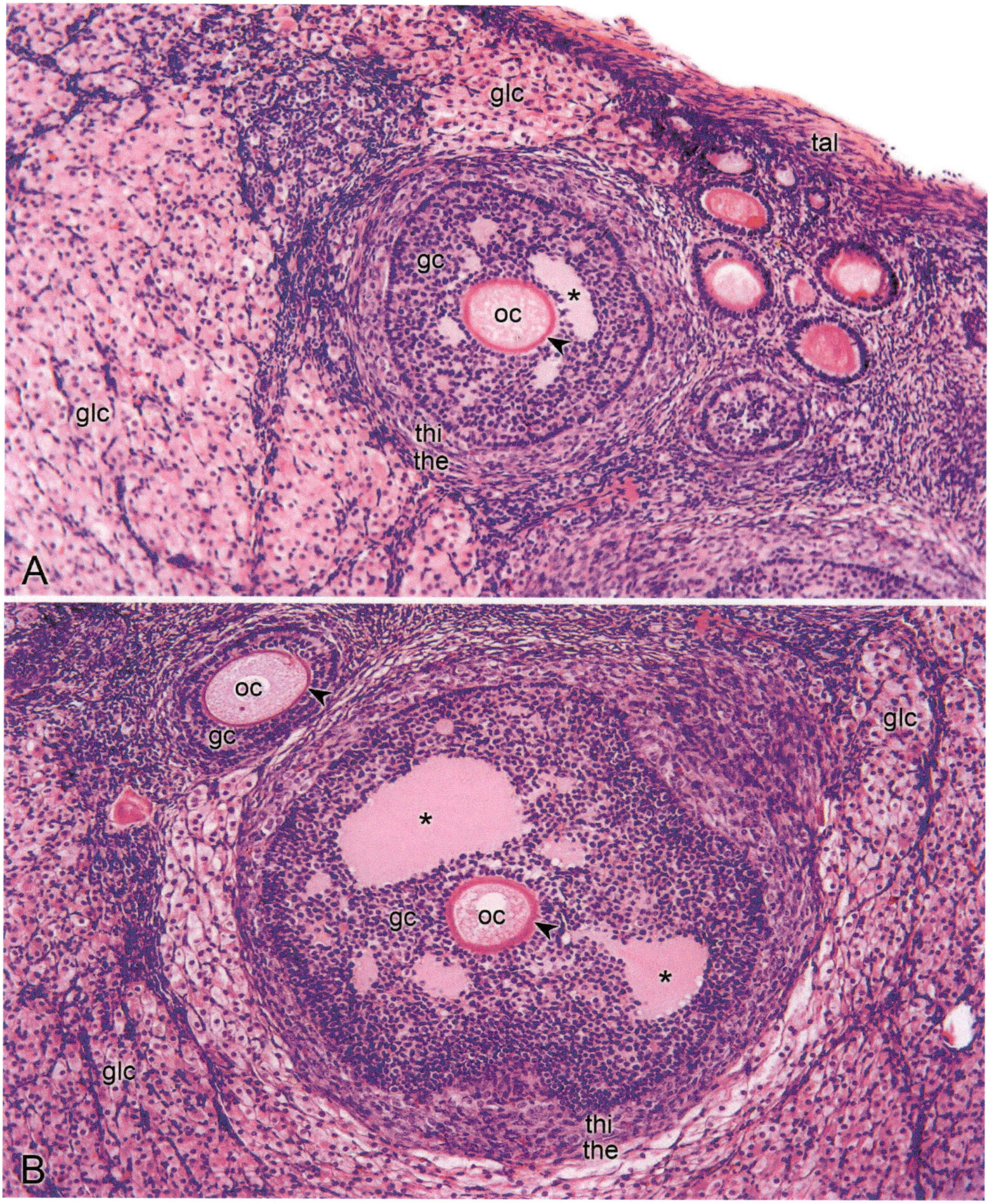

Figure 246. Ovary, secondary follicles. (A) In secondary follicles (center), cavities filled with glycosaminoglycan-rich liquor folliculi (asterisks) appear between multiple layers of granulosa cells (gc). On the right, primary follicles develop. Follicles are embedded into a cell rich stroma that often contains granulosa lutein cells (glc) from a previously formed corpus luteum. Zona pellucida (arrowhead) becomes thicker and the secondary follicle is surrounded by distinct theca interna (thi) and externa (the). (B) As the secondary follicle grows (center), these cavities start to coalesce. At the top left corner, a primary follicle contains few layers of granulosa cells.

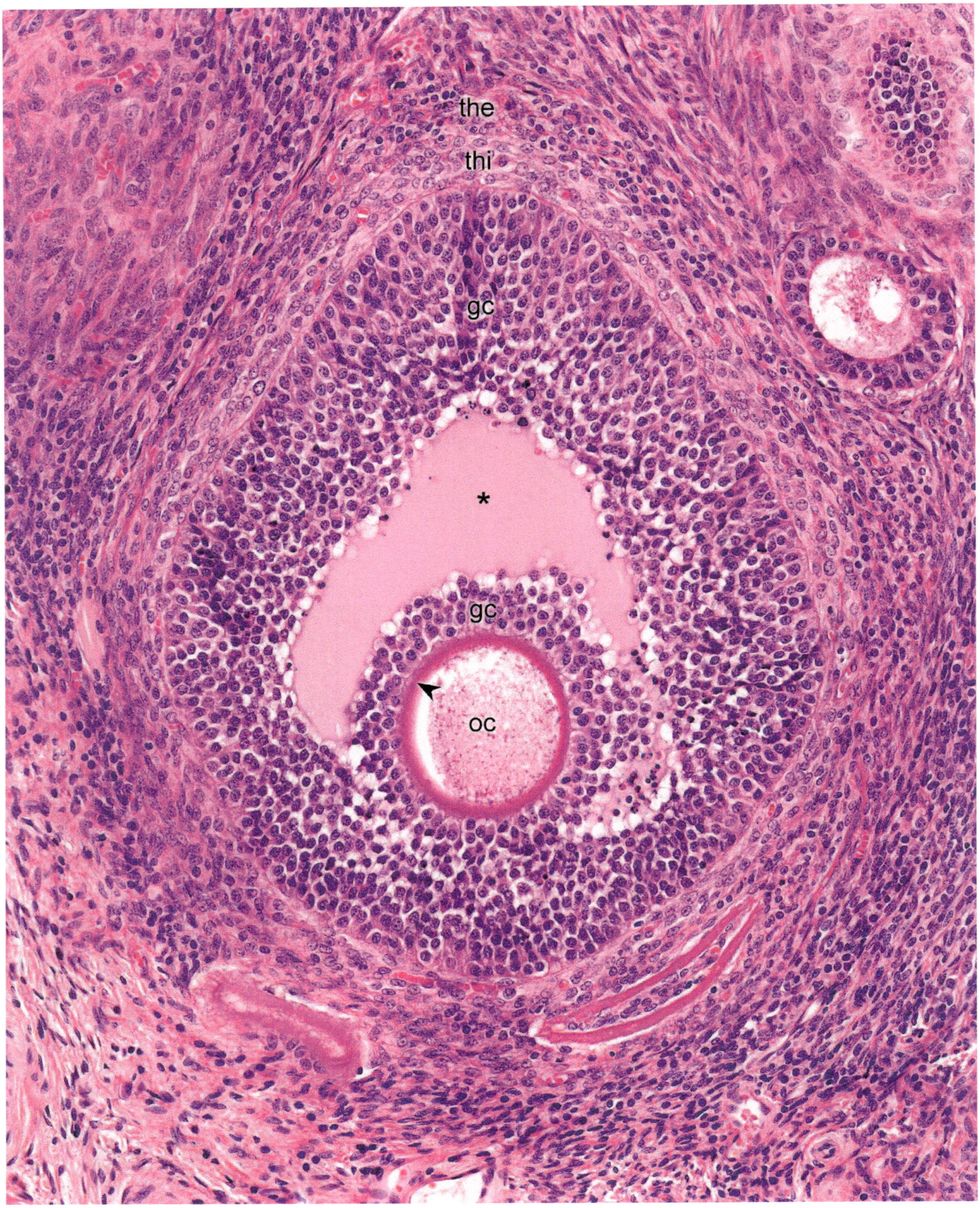

Figure 247. Ovary, secondary follicle. Cavities formed between the layers of granulosa cells merge and form a single, fluid-filled cavity, the antrum, containing glycosaminoglycan rich liquor folliculi (asterisk). The location of the oocyte becomes eccentric, and it is surrounded by layers of granulosa cells (gc) called corona radiata that protrudes into the antrum forming the cumulus oophorus. Zona pellucida (arrowhead) becomes thicker and the secondary follicle is surrounded by a distinct theca interna (thi) and externa (the).

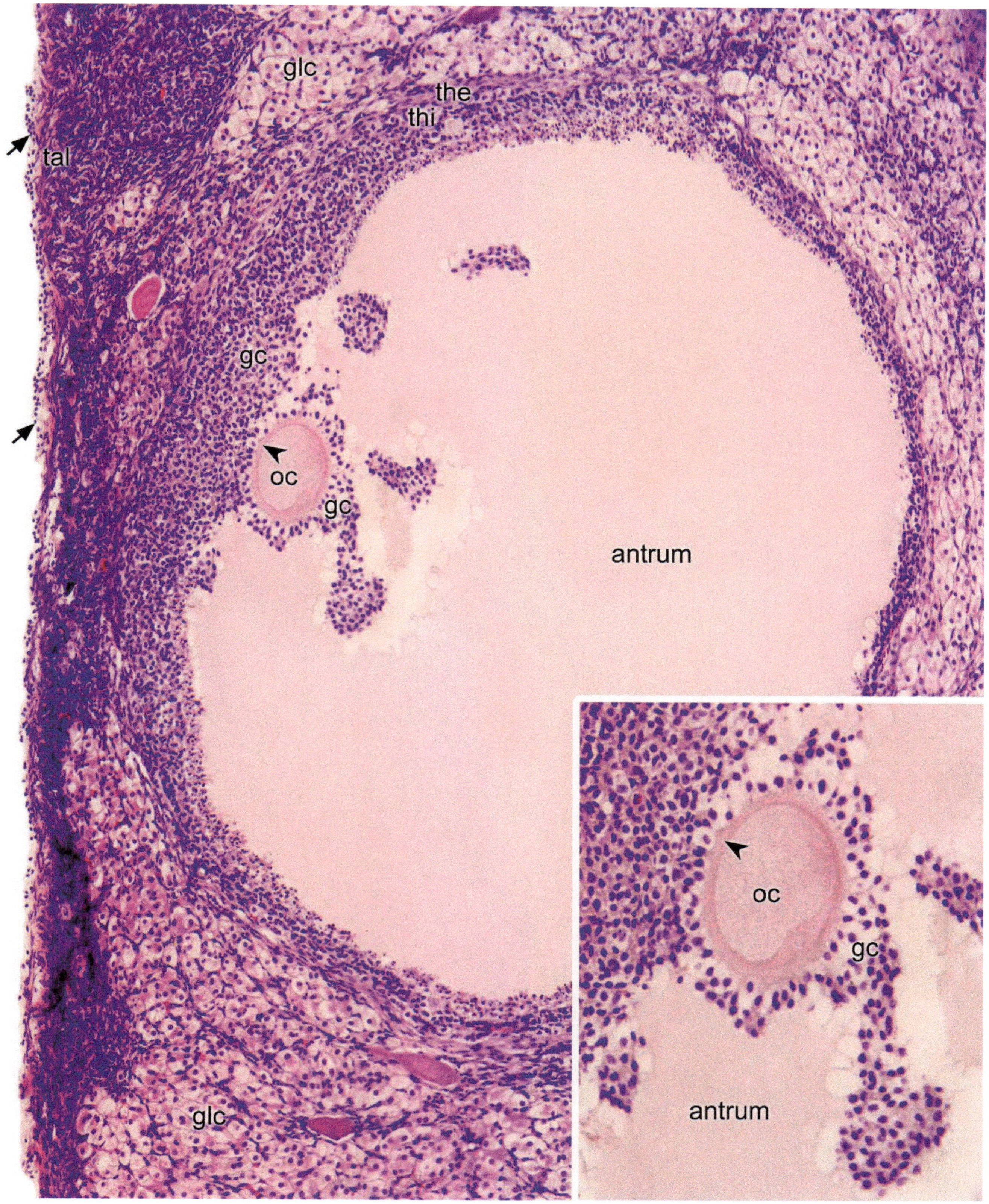

Figure 248. Ovary, Graafian follicle. The mature Graafian follicle is at least 10 mm in diameter located close to the ovarian surface; it is covered with a single layer of cuboidal cells (arrows) and the underlying tunica albuginea (tal). The location of the oocyte is eccentric, and it is surrounded by layers of granulosa cells (gc) called corona radiata that protrudes into the antrum forming the cumulus oophorus (inset). Zona pellucida (arrowheads) is thick and the follicle is surrounded by a distinct theca interna (thi) and externa (the). Follicles are embedded into a cell rich stroma that contains lightly stained granulosa lutein cells (glc) from a previously formed corpus luteum.

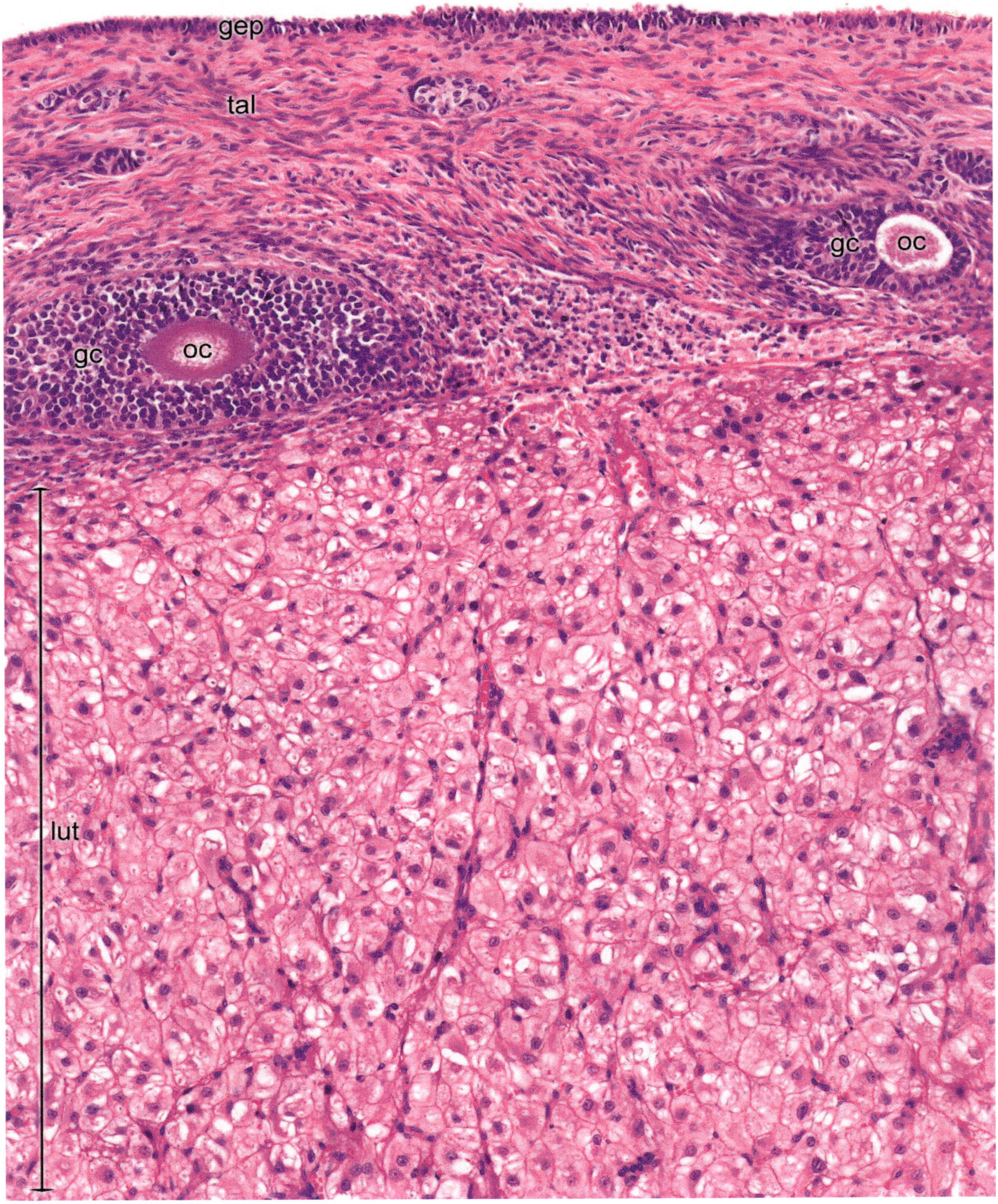

Figure 249. Ovary, corpus luteum. Following the ovulation, the walls of the empty follicle collapse. Cells of the theca interna and the granulosa cells enlarge and accumulate lipid droplets as they turn into lutein cells surrounded by a vascular stroma. This stage is the corpus luteum (lut). The surface of the ovary is lined with a single layer of cuboidal cells, the germinal epithelium (gep) and the underlying tunica albuginea (tal). Ovarian cortex contains follicles that are formed by oocytes (oc) surrounded by granulosa cells (gc).

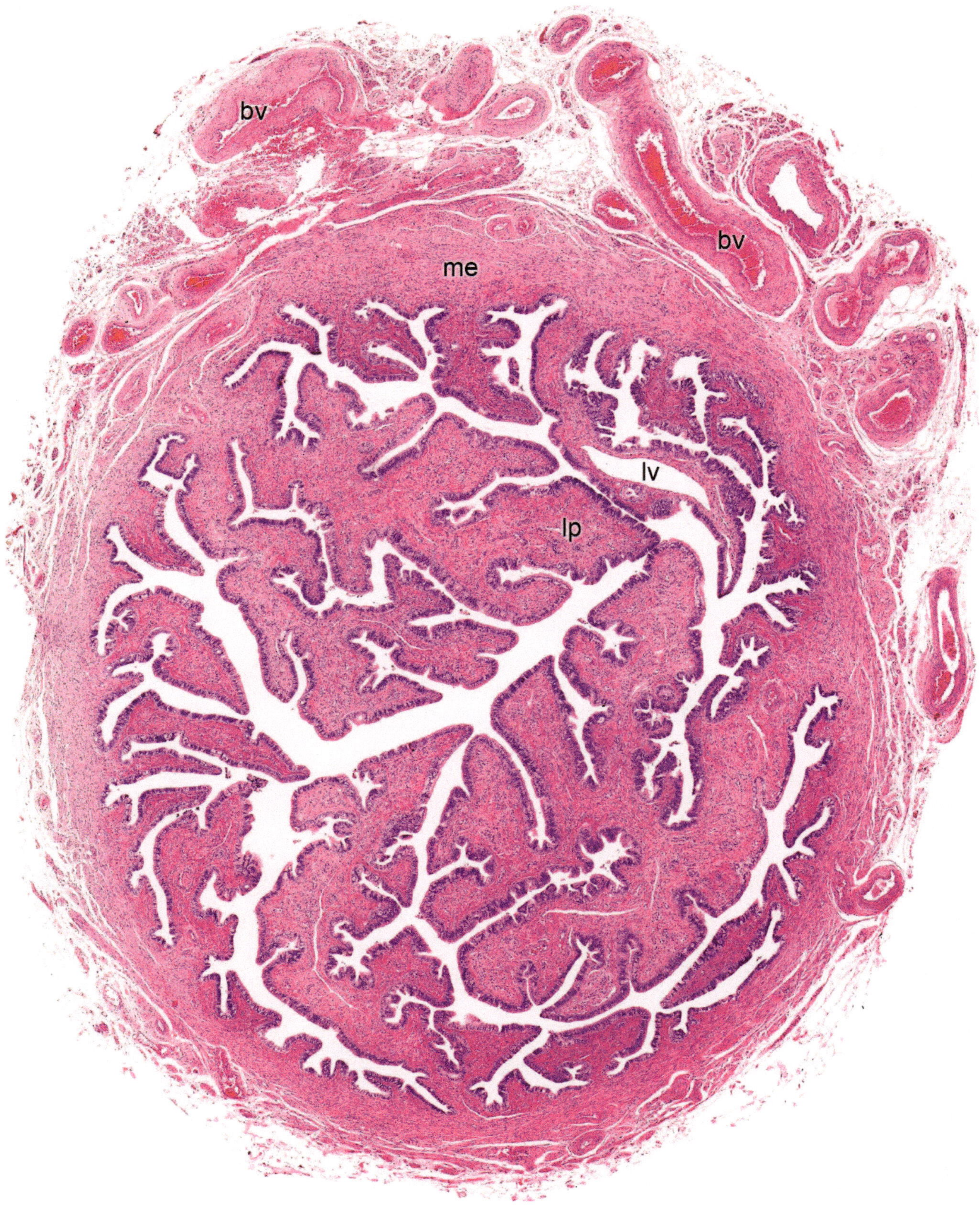

Figure 250. Oviduct, isthmus. Tunica mucosa formed by a simple columnar epithelium lining and an underlying lamina propria (lp) forms complicated folds with occasional lymph vessels (lv) at their axis. Tunica muscularis externa (me) is composed of indistinct layers of inner circular and outer longitudinal bundles of smooth muscle. Tunica muscularis is covered with loose connective tissue containing numerous blood vessels (bv).

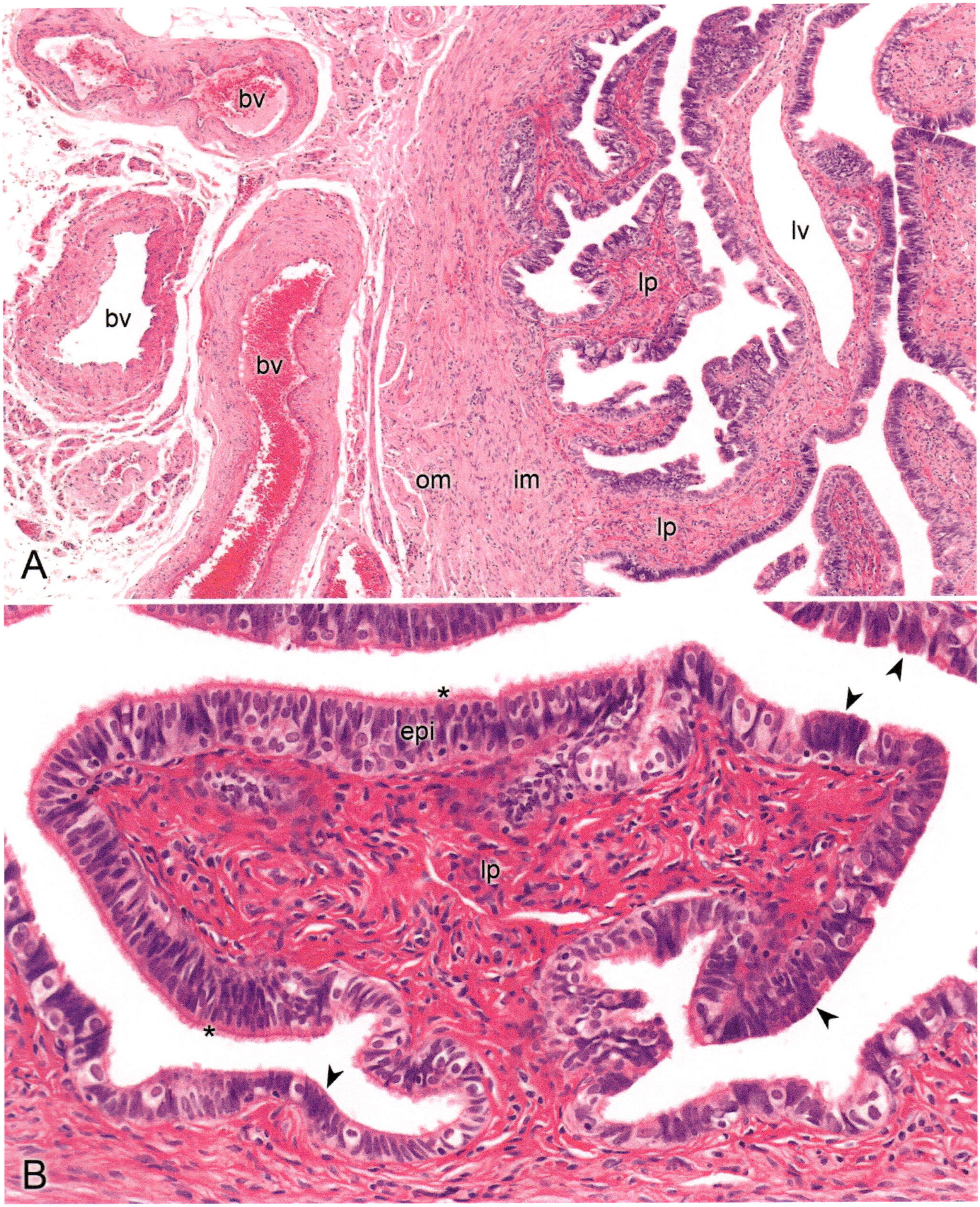

Figure 251. Oviduct, isthmus. (A) Tunica muscularis externa is composed of indistinct layers of inner circular (im) and outer longitudinal (om) bundles of smooth muscle that is covered outside with connective tissue serosa containing blood vessels (bv). Lamina propria (lp) of the mucous membrane contains axial lymph vessels (lv). (B) Loose connective tissue lamina propria (lp) is lined with simple columnar epithelium (epi) with cilia (asterisks). Secretory unciliated peg cells (arrows) produce oviductal fluid.

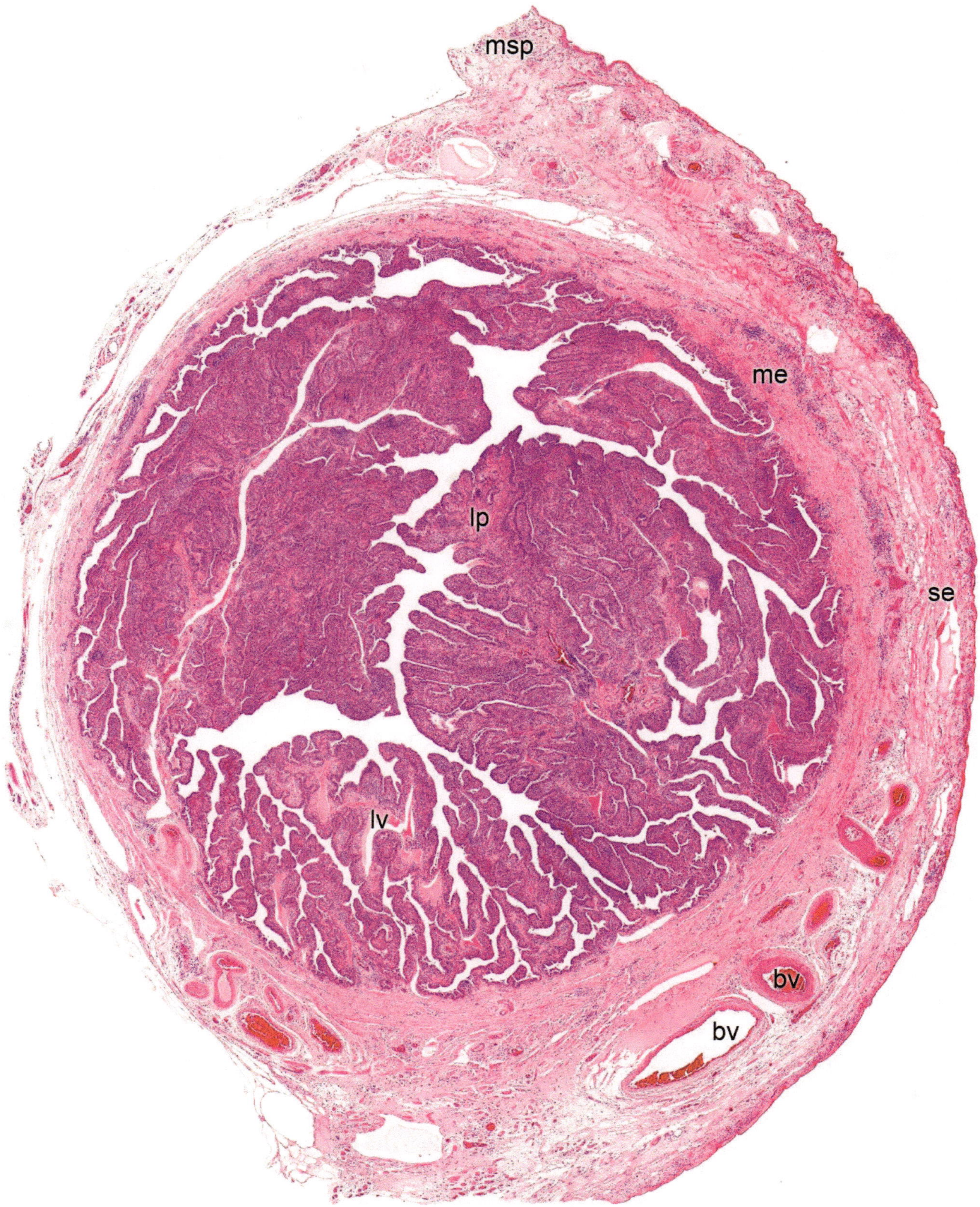

Figure 252. Oviduct, ampulla. Tunica muscularis externa (me) is covered inside by lamina propria (lp) that form folds that are more elaborate in the ampulla than at the ishmus. Lamina propria is lined with simple columnar epithelium and it contains axial lymph vessels (lv). Outside, muscularis externa is surrounded by tunica serosa (se) that contains blood vessels reaching the tuba uterina through the mesosalpinx (msp).

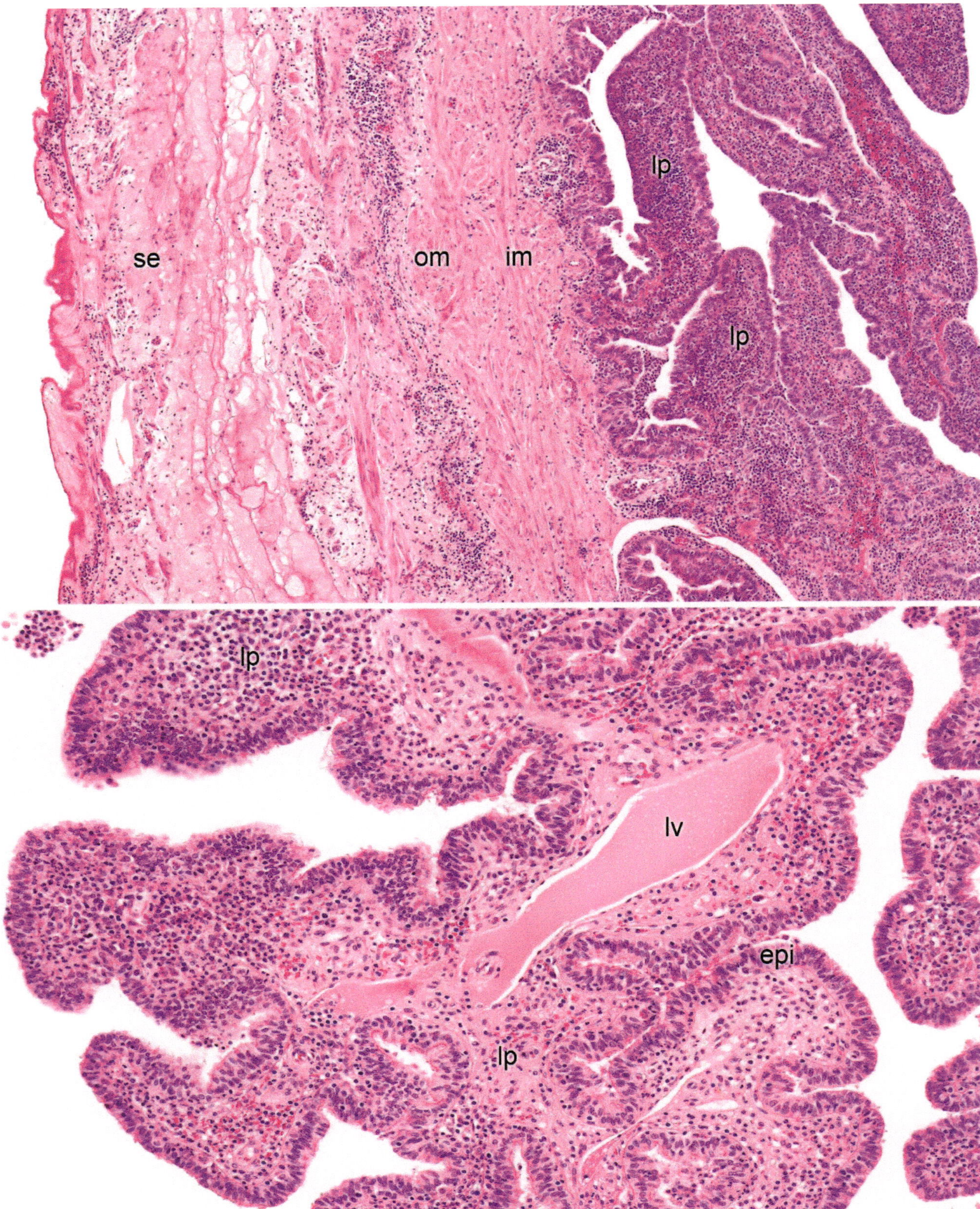

Figure 253. Oviduct, ampulla. Enlarged parts of the previous micrograph. Tunica muscularis is composed of indistinct layers of inner circular (im) and outer longitudinal (om) bundles of smooth muscle (top) that is covered outside with connective tissue tunica serosa (se). Lamina propria (lp) of the tunica mucosa often contains axial lymph vessels (lv, bottom). Mucosa is covered by ciliated simple columnar epithelium (epi) with occasional secretory peg cells.

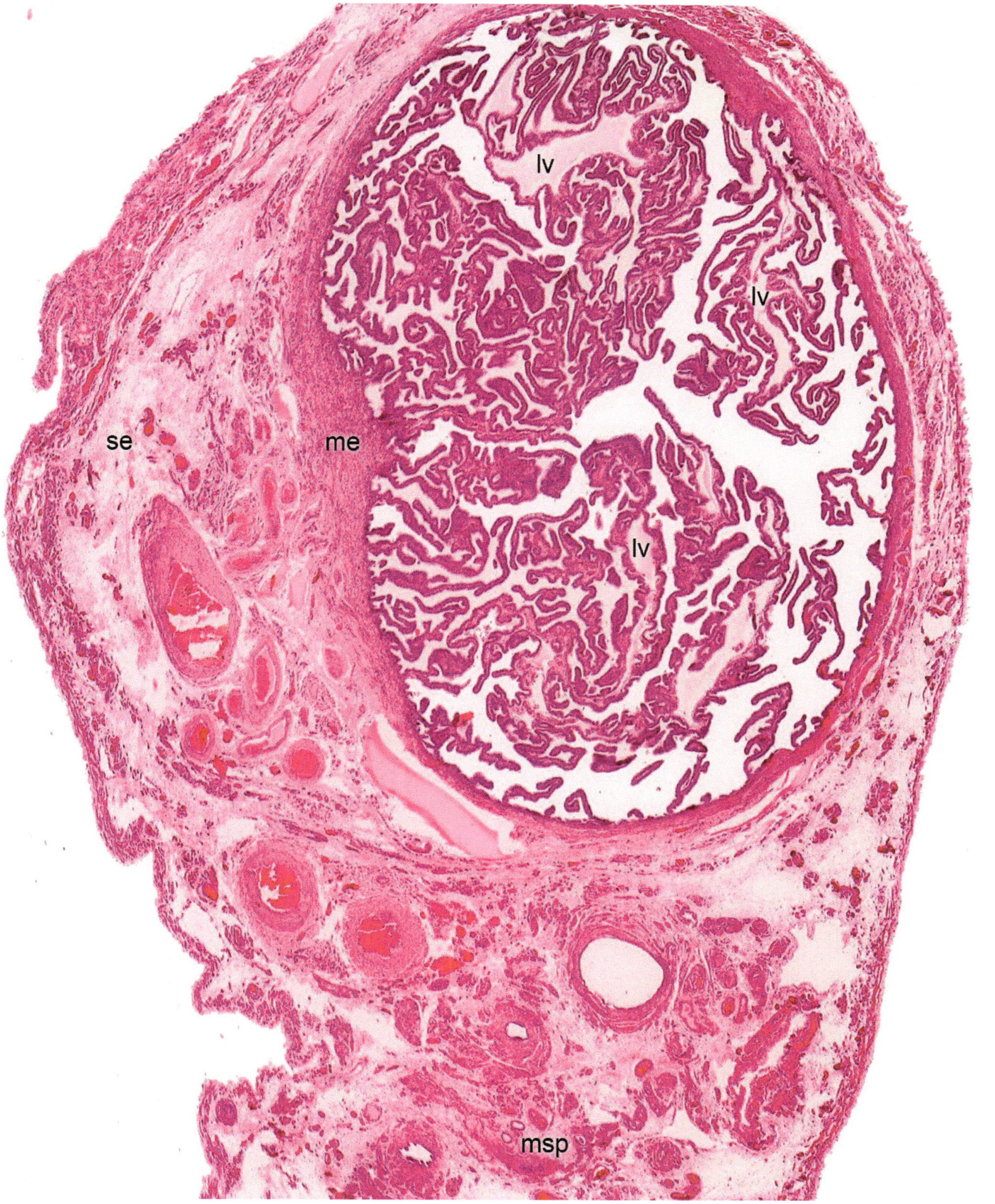

Figure 254. Oviduct, ampulla. Tunica muscularis externa (me) is covered inside by mucous membrane that form folds that are more elaborate in the ampulla than at the ishmus. Lamina propria is lined with simple columnar epithelium and it contains axial lymph vessels (lv). Outside, muscularis externa is surrounded by tunica serosa (se) with mesothelial cell coverage and the underlying lamina propria; duplication of the serosa forms the mesosalpinx (msp).

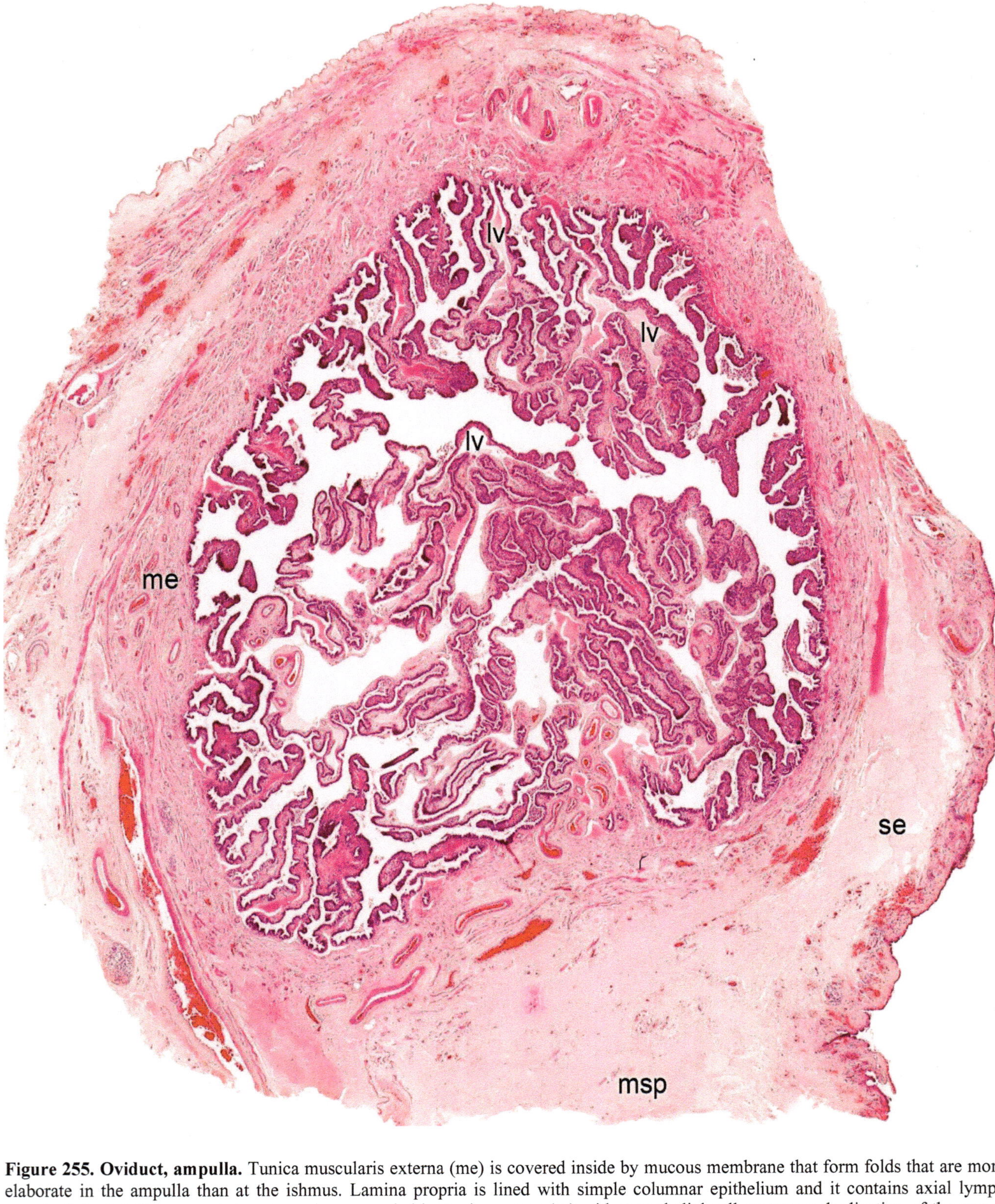

Figure 255. Oviduct, ampulla. Tunica muscularis externa (me) is covered inside by mucous membrane that form folds that are more elaborate in the ampulla than at the ishmus. Lamina propria is lined with simple columnar epithelium and it contains axial lymph vessels (lv). Outside, muscularis externa is surrounded by tunica serosa (se) with mesothelial cell coverage; duplication of the serosa forms the mesosalpinx (msp).

Uterus

The bulk of the uterine wall is formed by the myometrium, a thick layer of interwoven bundles of smooth muscle cells and the supporting dense connective tissue. Myometrium is covered by the endometrium, the mucous membrane of the uterus while the outer perimetrium is the tunica serosa, a peritoneal coverage, formed by mesothelial cells and an underlying layer of loose connective tissue. Uterine cervix is covered by a loose connective tissue envelope (tunica adventitia) instead of serosa. Endometrium is lined with ciliated simple columnar epithelium, and the underlying lamina propria contains tubular glands that may extend into the myometrium, often surrounded by lymphocytes that also infiltrate the lamina propria. These glands are coiled in the secretory phase of the uterus, and they are lined with a single layer of columnar cells. Tunica mucosa reaches its maximal thickness in the secretory phase that is followed by the menstrual phase, when the functional layer (stratum fuctionale) of the mucous membrane is sloughed off and only the bases of the glands remain. Regeneration of the glands follows from the stratum basale, and first the surface epithelium, then the lamina propria with the glands will be restored (proliferative phase). These glands are initially straight with narrow, slightly wavy lumen. Stromal cells between the glands accumulate glycogen in the proliferative phase and they turn into large pale decidual cells with excessive glycogen content after the implantation.

The cervix of the uterus contains less smooth muscle and more connective tissue rich in elastic fibers. The cervical mucosa differs from the endometrium of the rest of the uterus and it does not exhibit dramatic changes during the menstrual cycle. Tunica mucosa contains large branched glands secreting mucus; these glands often develop into large Nabothian cysts as a result of blockage in their ducts. Cervical canal is lined with simple columnar epithelium that abruptly turns into nonkeratinized stratified squamous epithelium of the vagina just outside the external os, the outer cervical orifice of the cervical canal. This part of the uterus protrudes into the vagina (portio vaginalis uteri) and apart from a small area around the external os, it is lined with stratified squamous nonkeratinized epithelium.

Vagina

Vagina is a fibromuscular tube stretching between the cervix of the uterus and the base of the labia minora. Vaginal mucous membrane is lined with stratified squamous nonkeratinized epithelium that shows cyclic changes during the menstrual cycle. During the proliferative phase, superficial cells accumulate glycogen and thus they have light cytoplasm. The mucous membrane of the vagina lacks glands; under the epithelium, lamina propria is composed of loose connective tissue that transitions into a deeper submucosa with denser connective tissue, although the boundaries between these layers are indistinct. Tunica muscularis is formed by vaguely defined inner circular and outer longitudinal layers of smooth muscle. Outside, the vagina is covered by tunica adventitia that is initially denser around the muscle layers and turns into loose connective tissue peripherally.

External genitalia

Vagina leads into the vestibule that is a space between the labia minora that are folds of skin extending to the clitoris. Vestibular mucosa contains small mucous glands (Skene's glands); the larger, mucous Bartholin's glands open into the vestibule laterally. Labia minora are covered with lightly keratinized stratified squamous epithelium, while the vestibular surface also contains nonkeratinized epithelium, often with glycogen-containing apical cells characteristic to the vagina. Labia minora lack hair follicles but contain sebaceous glands that are not associated with hair. Clitoris is also covered with delicate skin and contains erectile tissue with endothelium-lined caverns similar to those of the penis.

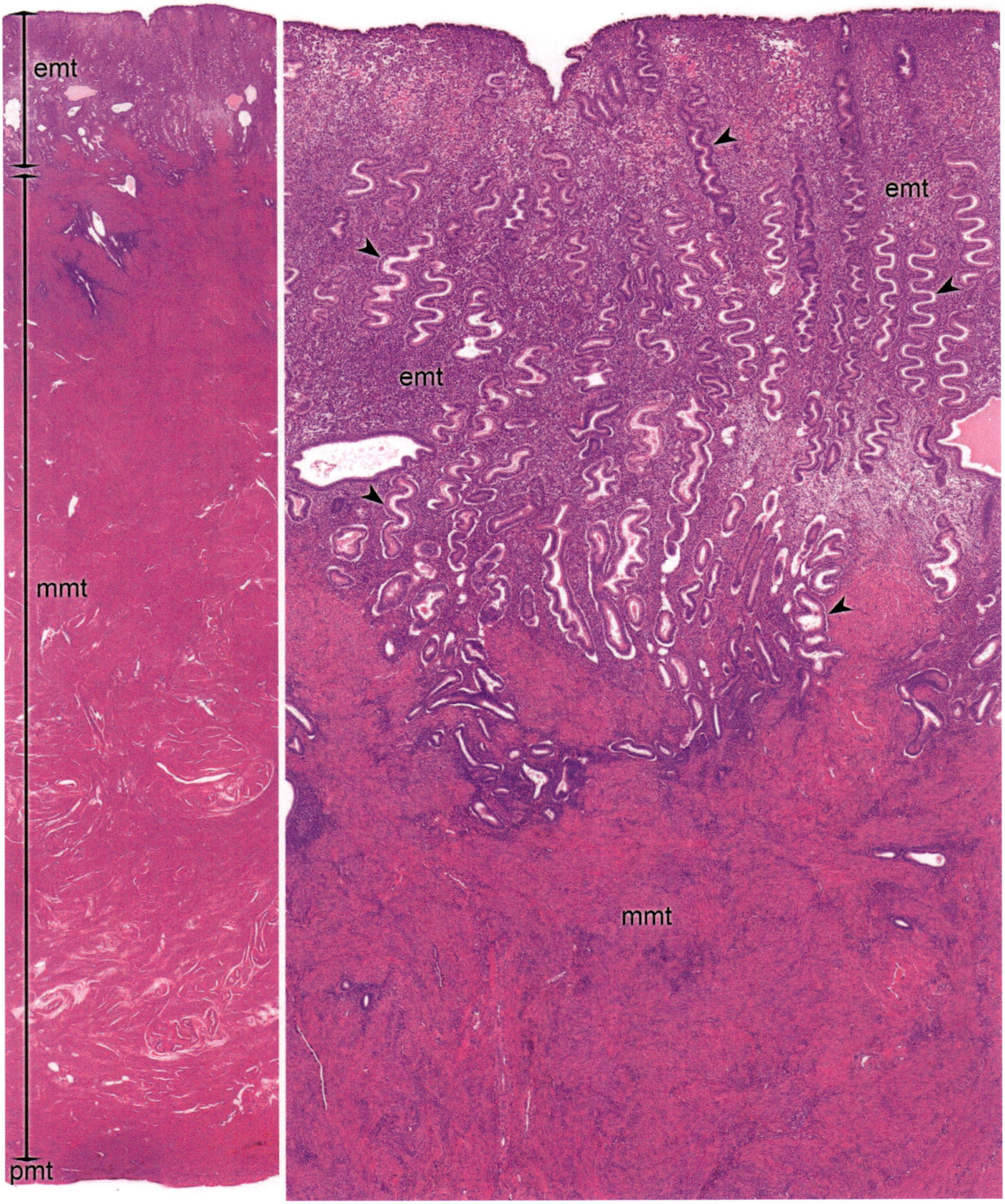

Figure 256. Uterus, secretory phase. Uterus is composed of three layers (left): the mucous membrane, called endometrium (emt), the underlying thick myometrium (mmt) formed by interwoven bundles of smooth muscle cells and supporting dense connective tissue, and the outer perimetrium (pmt) that is identical to the peritoneal coverage. The mucous membrane is lined with ciliated simple columnar epithelium; the underlying lamina propria contains tubular glands (arrowheads) that may extend into the myometrium (right). These glands are coiled in the secretory phase, and they are covered by a single layer of columnar cells.

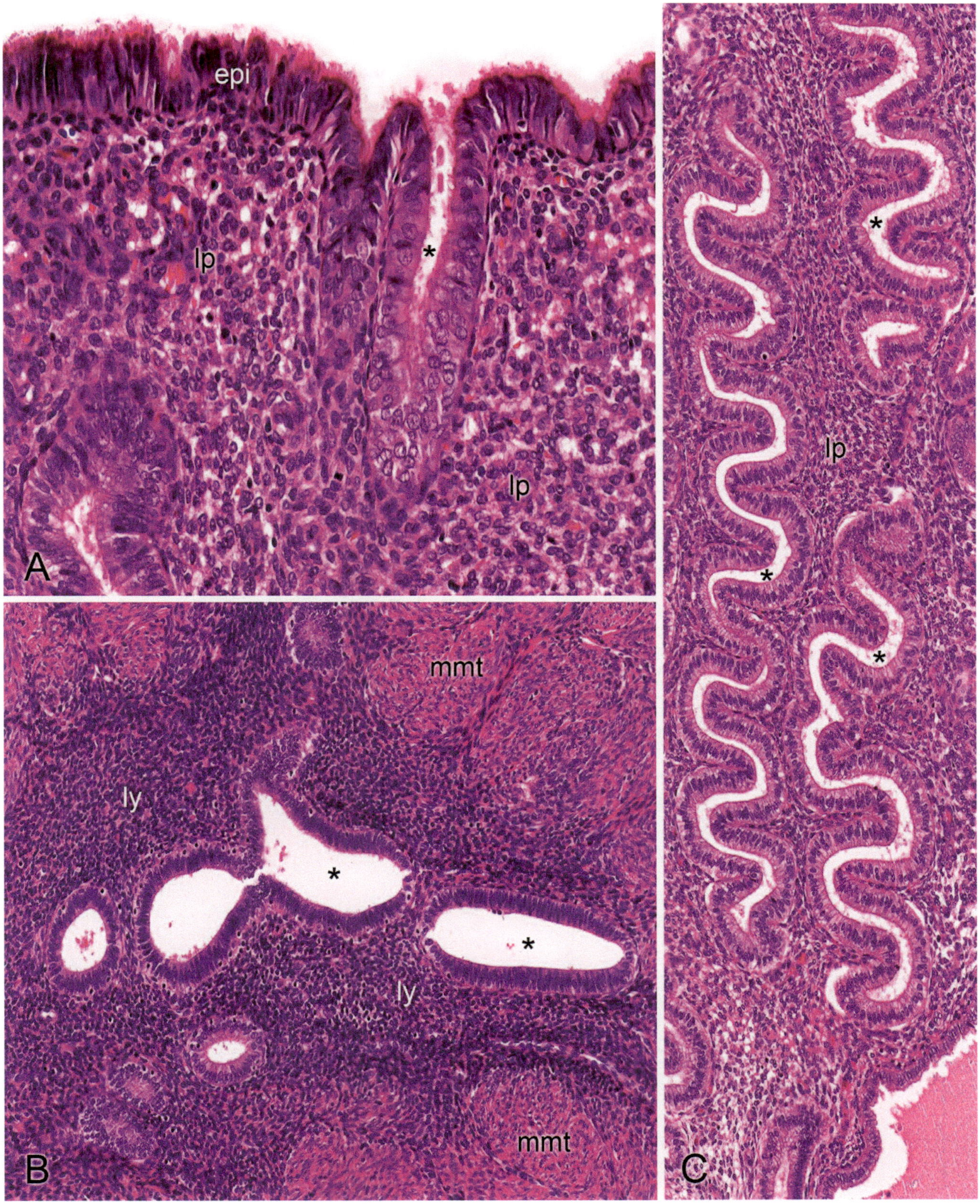

Figure 257. Uterus. (A) Uterine mucosa is lined with ciliated simple columnar epithelium (epi); the underlying lamina propria (lp) contains tubular glands (asterisks). (B) These glands (asterisks) may extend into the myometrium (mmt), often surrounded by lymphocytic aggregations (ly) that also infiltrate the lamina propria. (C) Uterine glands are coiled in the secretory phase, and they are covered by a single layer of columnar cells.

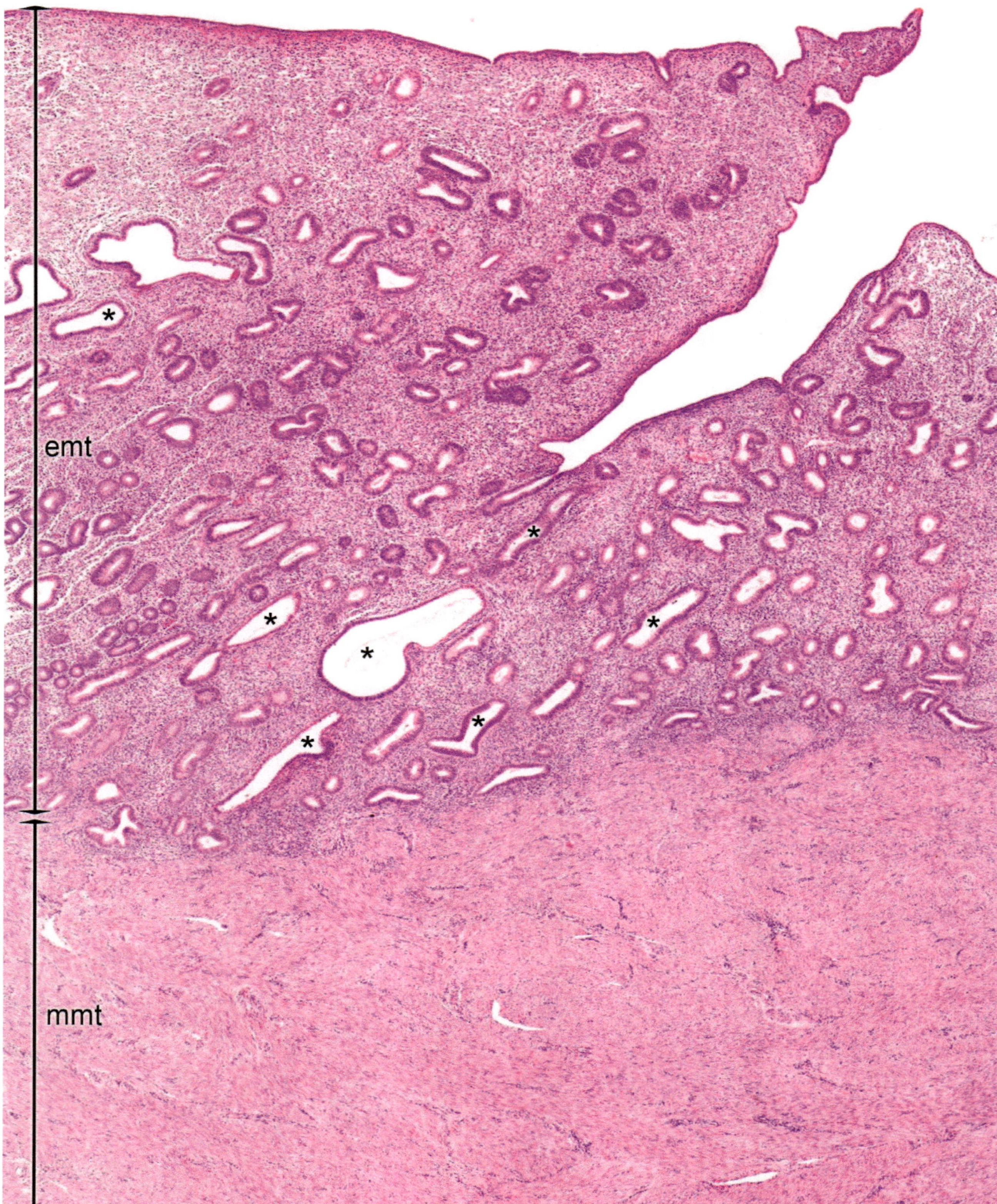

Figure 258. Uterus, proliferative phase. Endometrium (emt) is lined with ciliated simple columnar epithelium; the underlying lamina propria contains tubular glands (asterisks) that are straight in the proliferative phase and may extend into the myometrium (mmt).

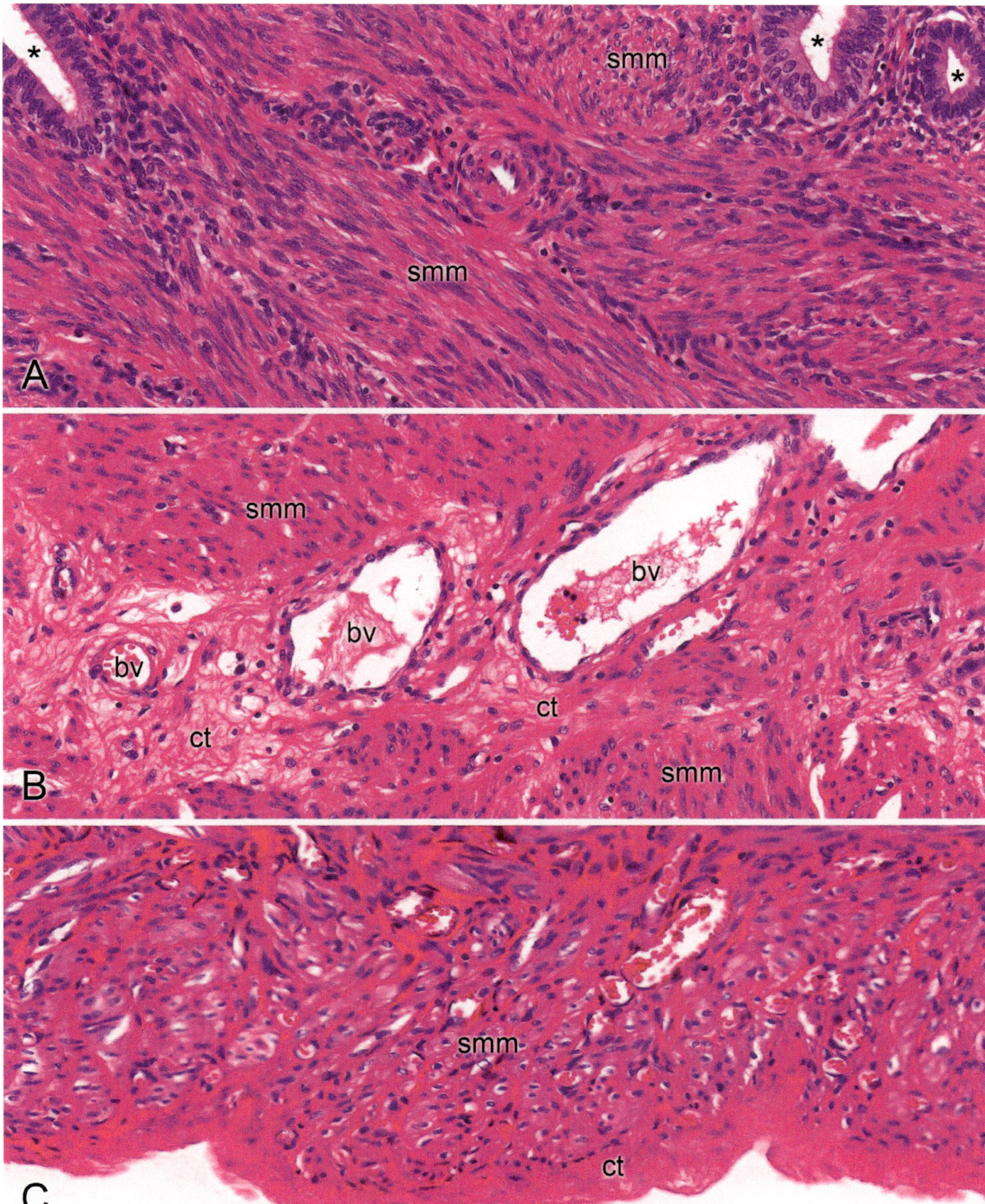

Figure 259. Uterus, myometrium. (A) The base of the uterine glands (asterisks) often extends into the myometrium that is composed of interwoven smooth muscle bundles (smm). The middle layer of the myometrium (B) typically contains dense connective tissue (ct) surrounding blood vessels (bv) and smooth muscle bundles (smm). (C) Myometrium features an outer peritoneal coverage, the perimetrium that contains smooth muscle bundles and a connective tissue layer (ct) lined with mesothelial cells.

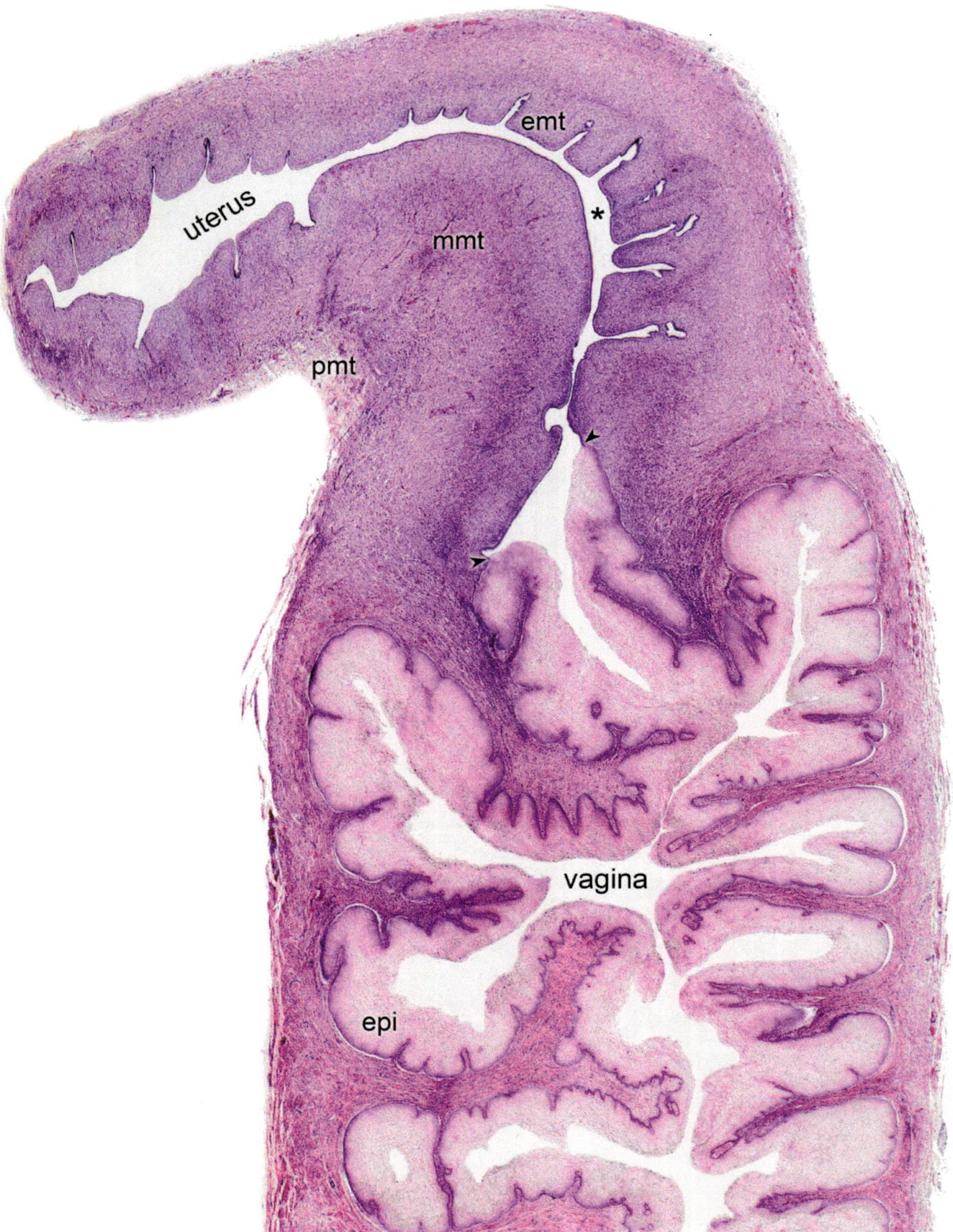

Figure 260. Uterus, vagina (newborn). Uterus is composed of the inner endometrium (emt) that contains uterine glands, the smooth muscle myometrium (mmt) and the outer perimetrium coverage (pmt). The neck of the uterus protrudes into the vagina with the portio vaginalis that contains the cervical canal (asterisks). Simple columnar epithelium of the cervical canal abruptly changes (arrowheads) into the stratified squamous nonkeratinized epithelium of the vagina (epi) at the external os of the cervical canal (arrowheads).

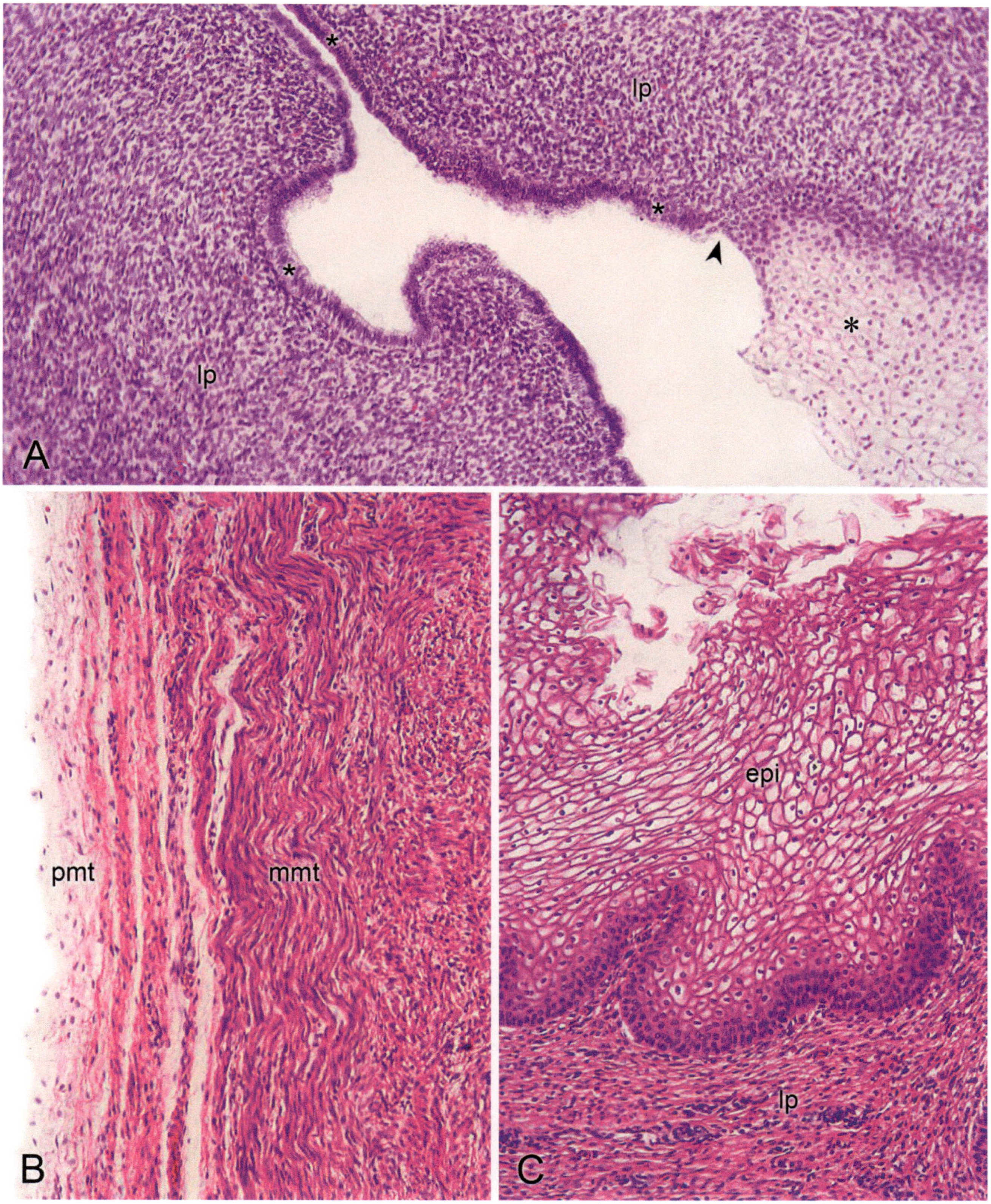

Figure 261. Uterus, vagina. (A) Simple columnar epithelium of the cervical canal (five-pointed asterisks) abruptly changes (arrowheads) into the stratified squamous nonkeratinized epithelium of the vagina (six-pointed asterisks) at the external os. Epithelium and the underlying lamina propria (lp) form the endometrium. (B) Myometrium is composed of smooth muscle mixed with dense connective tissue and it is covered with the perimetrium (pmt). (C) Vagina is lined with stratified squamous nonkeratinized epithelium with lightly stained surface cells containing glycogen (epi). The underlying lamina propria (lp) is formed by loose connective tissue.

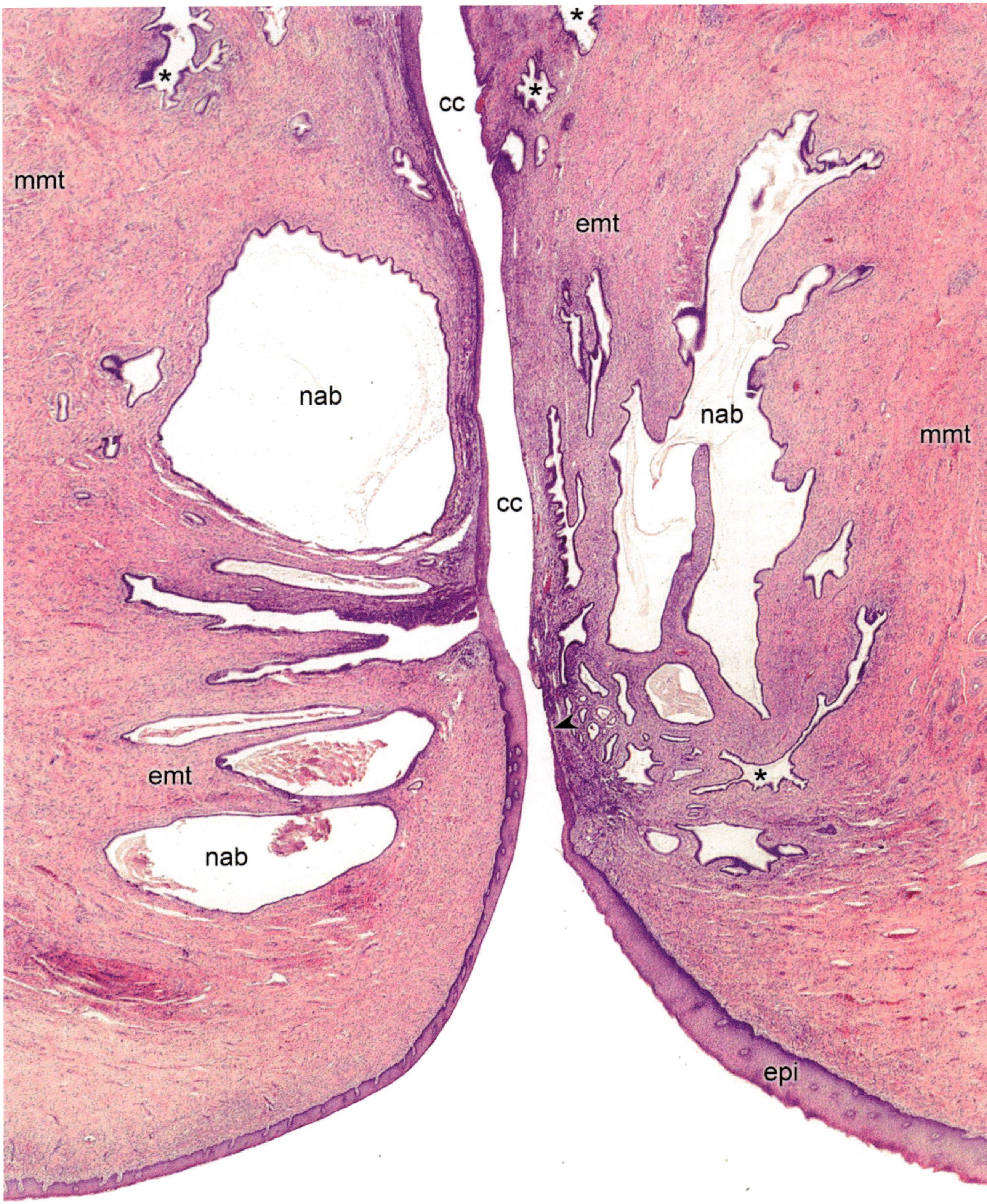

Figure 262. Portio vaginalis uteri. The neck of the uterus protrudes into the vagina with the portio vaginalis that contains the cervical canal (cc). Simple ciliated columnar epithelium of the cervical canal abruptly changes (arrowheads) into the stratified squamous nonkeratinized epithelium of the vagina (epi) at the external os of the cervical canal. The inner endometrium (emt) contains uterine glands (asterisks) that may develop into Nabothian cysts (nab) by occlusion. The underlying myometrium (mmt) is formed by smooth muscle mixed with dense connective tissue.

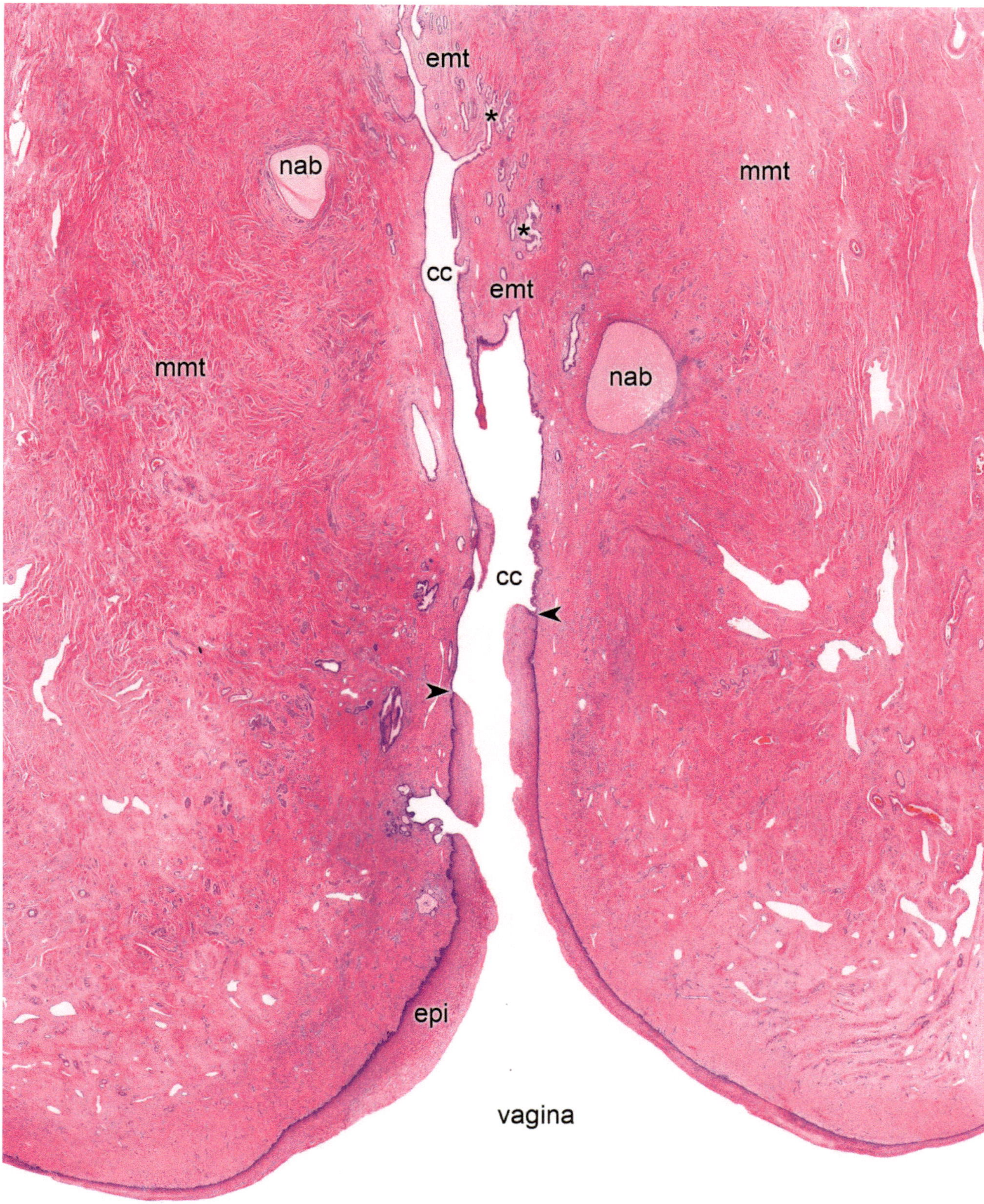

Figure 263. Portio vaginalis uteri. The neck of the uterus protrudes into the vagina with the portio vaginalis that contains the cervical canal (cc). Simple columnar epithelium of the cervical canal abruptly changes (arrowheads) into the stratified squamous nonkeratinized epithelium of the vagina (epi) at the external os of the cervical canal. The inner endometrium (emt) contains uterine glands (asterisks) that may develop into Nabothian cysts (nab) by occlusion. The underlying myometrium (mmt) is formed by smooth muscle mixed with dense connective tissue.

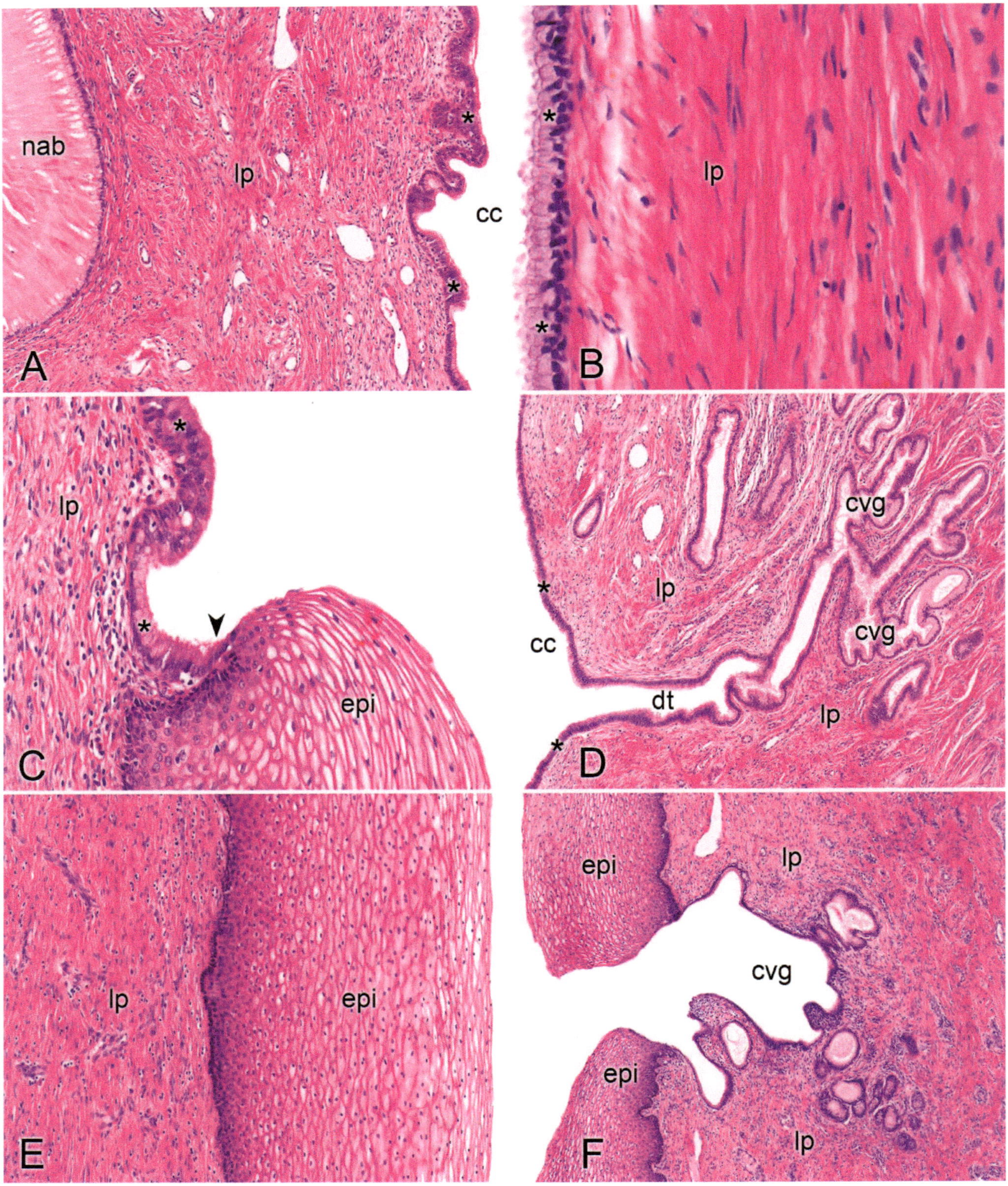

Figure 264. Portio vaginalis uteri. (A) Cervical canal (cc) is lined with simple columnar epithelium (asterisks). Lamina propria often contains occluded cervical glands, the Nabothian cysts (nab). (B) Simple columnar epithelium of the cervical canal lines the lamina propria. (C) Simple columnar epithelium of the cervical canal (asterisks) abruptly changes (arrowheads) into the stratified squamous nonkeratinized epithelium with lightly stained surface cells (epi) at the external os of the cervical canal. (D) Lamina propria (lp) of the endometrium contains cervical glands (cvg) and their ducts (dt) that open on the surface that is lined with simple columnar epithelium (asterisks). (E) Distally to the external os, portio vaginalis is lined with stratified squamous nonkeratinized epithelium. (F) This epithelium may be pierced by ducts of cervical glands (cvg) located in the underlying lamina propria (lp).

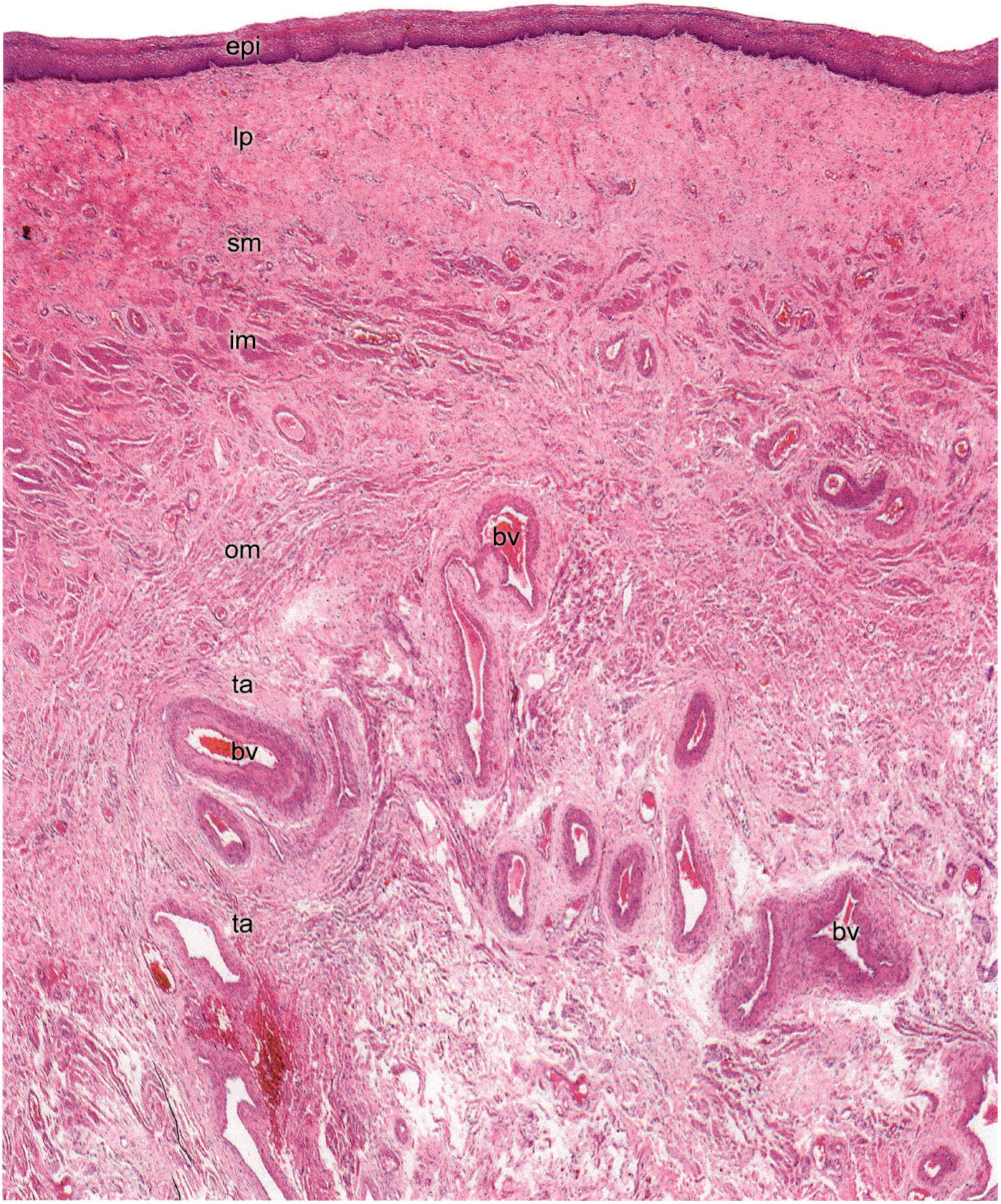

Figure 265. Vagina. Vaginal mucosa is lined with stratified squamous nonkeratinized epithelium (epi) with lightly stained surface cells containing glycogen. The underlying lamina propria (lp) is composed of loose connective tissue that transitions into a deeper submucosa (sm) with more dense connective tissue, although the boundaries between these layers are indistinct. Tunica muscularis is formed by inner circular (im) and outer longitudinal layers (om) of smooth muscle. Outside the vagina is covered by tunica adventitia (ta) containing blood vessels (bv) and nerves.

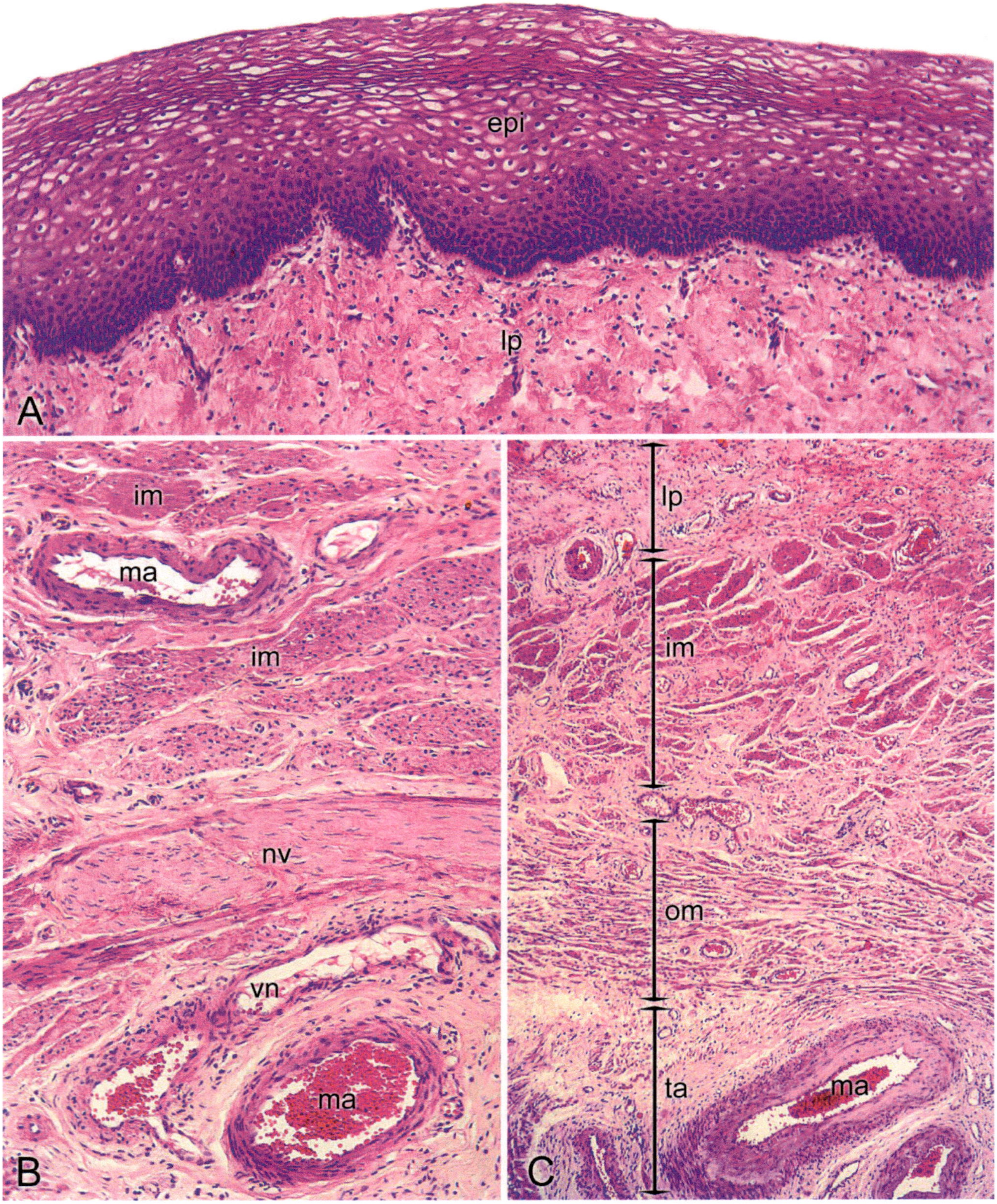

Figure 266. Vagina. (A)Vaginal mucosa is lined with stratified squamous nonkeratinized epithelium (epi) with lightly stained surface cells containing glycogen. The underlying lamina propria (lp) is composed of loose connective tissue. (B, C) Lamina propria transitions into a deeper submucosa (sm) with more dense connective tissue, although the boundaries between these layers are indistinct. Tunica muscularis is formed by an inner circular (im) and outer longitudinal layers of smooth muscle with muscular arteries (ma), veins (vn) and nerves (nv). Outside the vagina is covered by tunica adventitia containing blood vessels and nerves.

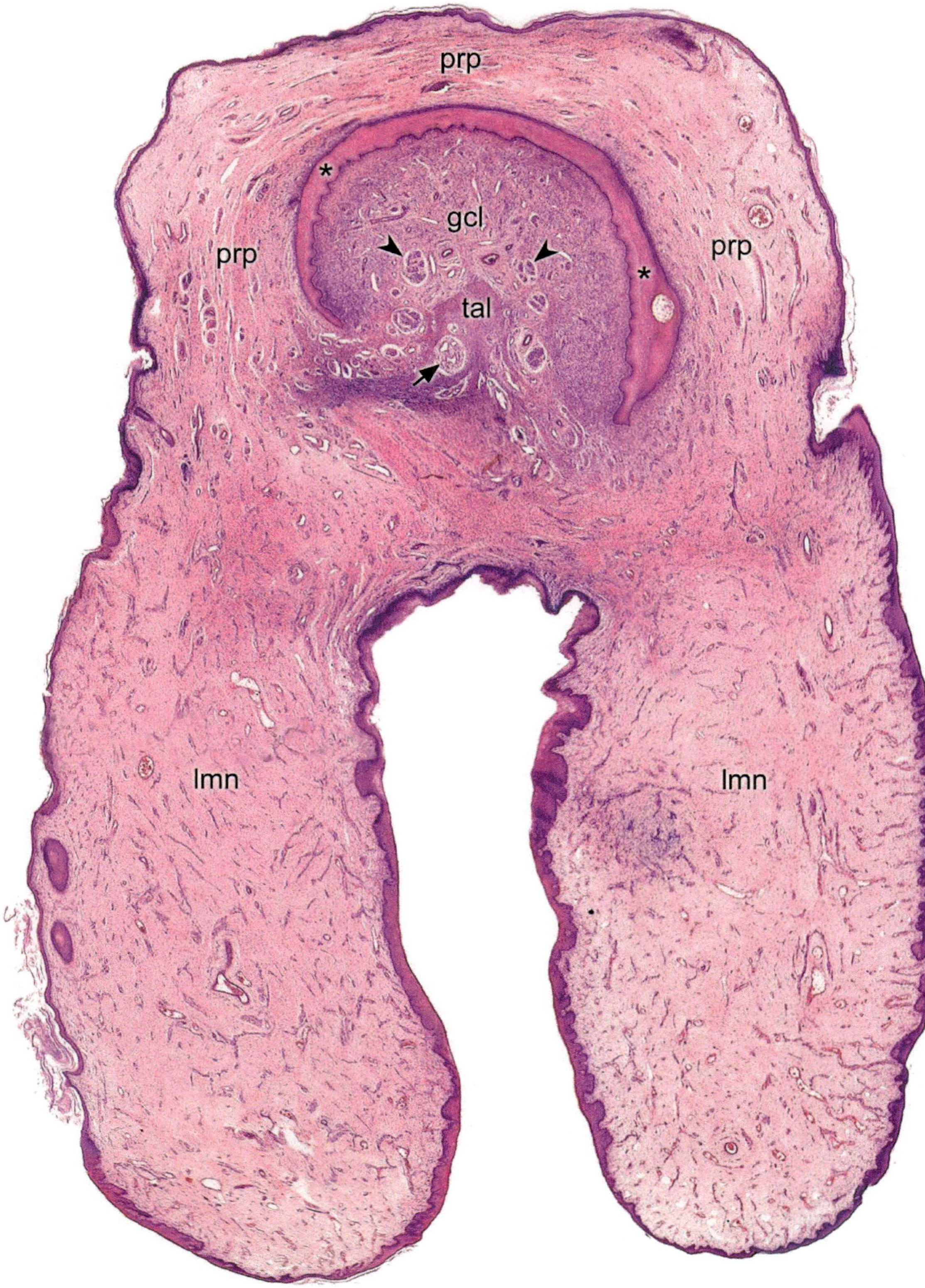

Figure 267. Clitoris and labia minora. The glans of the clitoris (gcl) is covered with thin folds of delicate skin, the prepuce (prp). The epithelium of the glans fuses with the adjacent epithelium of the prepuce in newborn (asterisks). Glans clitoridis contains corpora cavernosa composed of erectile tissue (arrow) that are surrounded by dense connective tissue tunica albuginea (tal) and nerve fibers (arrowheads). Labia minora (lmn) are folds of thin skin that contain a core of dermis rich in blood vessels and lacks subcutaneous adipose tissue. Epidermis on the medial side of the labia minora transitions into the nonkeratinized stratified squamous epithelium of the vagina. Labia minora lack hair follicles but contain sebaceous glands that are not associated with hair.

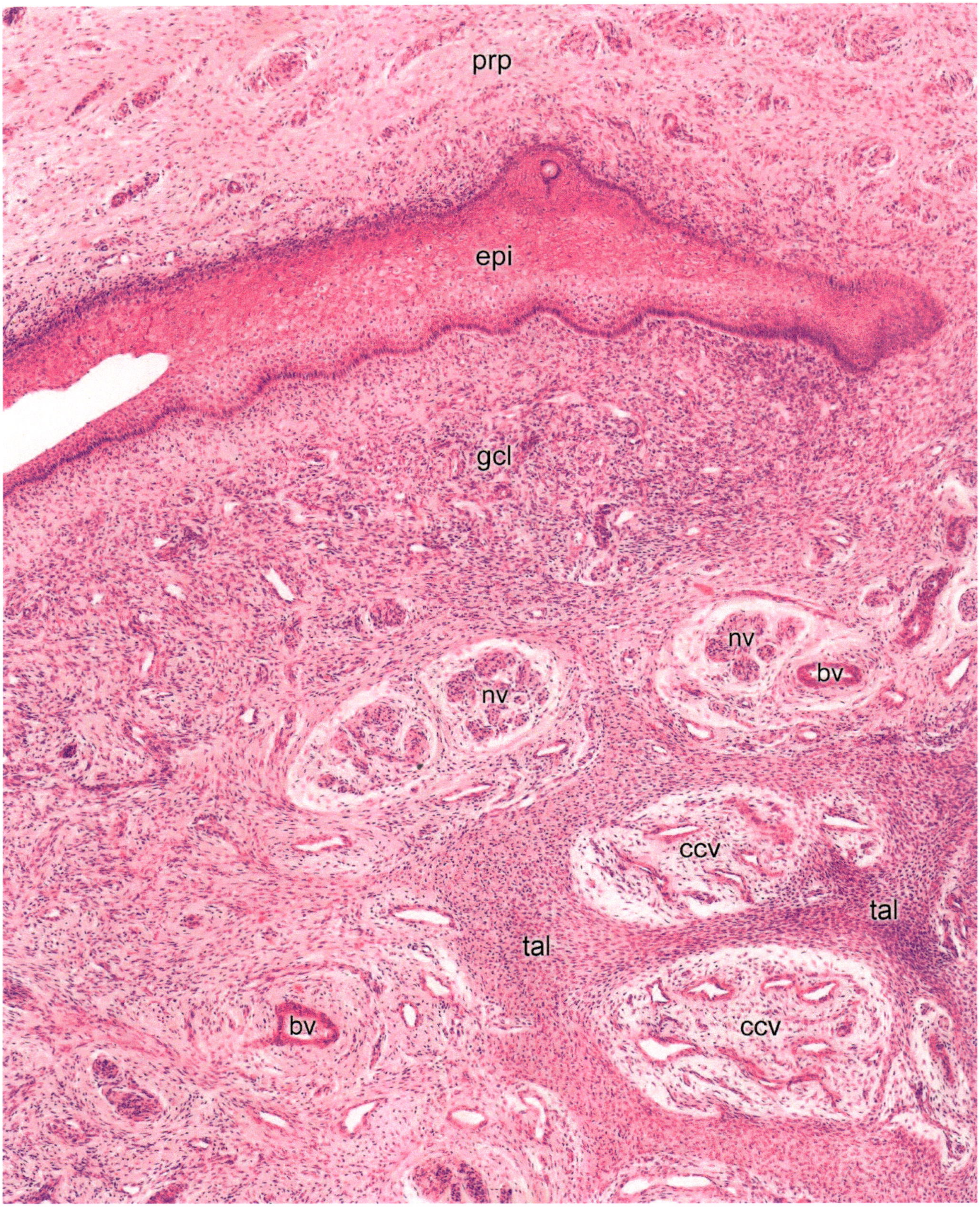

Figure 268. Clitoris. The glans of the clitoris (gcl) is covered with thin folds of delicate skin, the prepuce (prp). The epithelium of the glans fuses with the adjacent epithelium of the prepuce in newborn (asterisks). Glans clitoridis contains corpora cavernosa composed of erectile tissue (ccv) that are ensheathed in dense connective tissue tunica albuginea (tal) surrounded by blood vessels (bv) and nerves (nv).

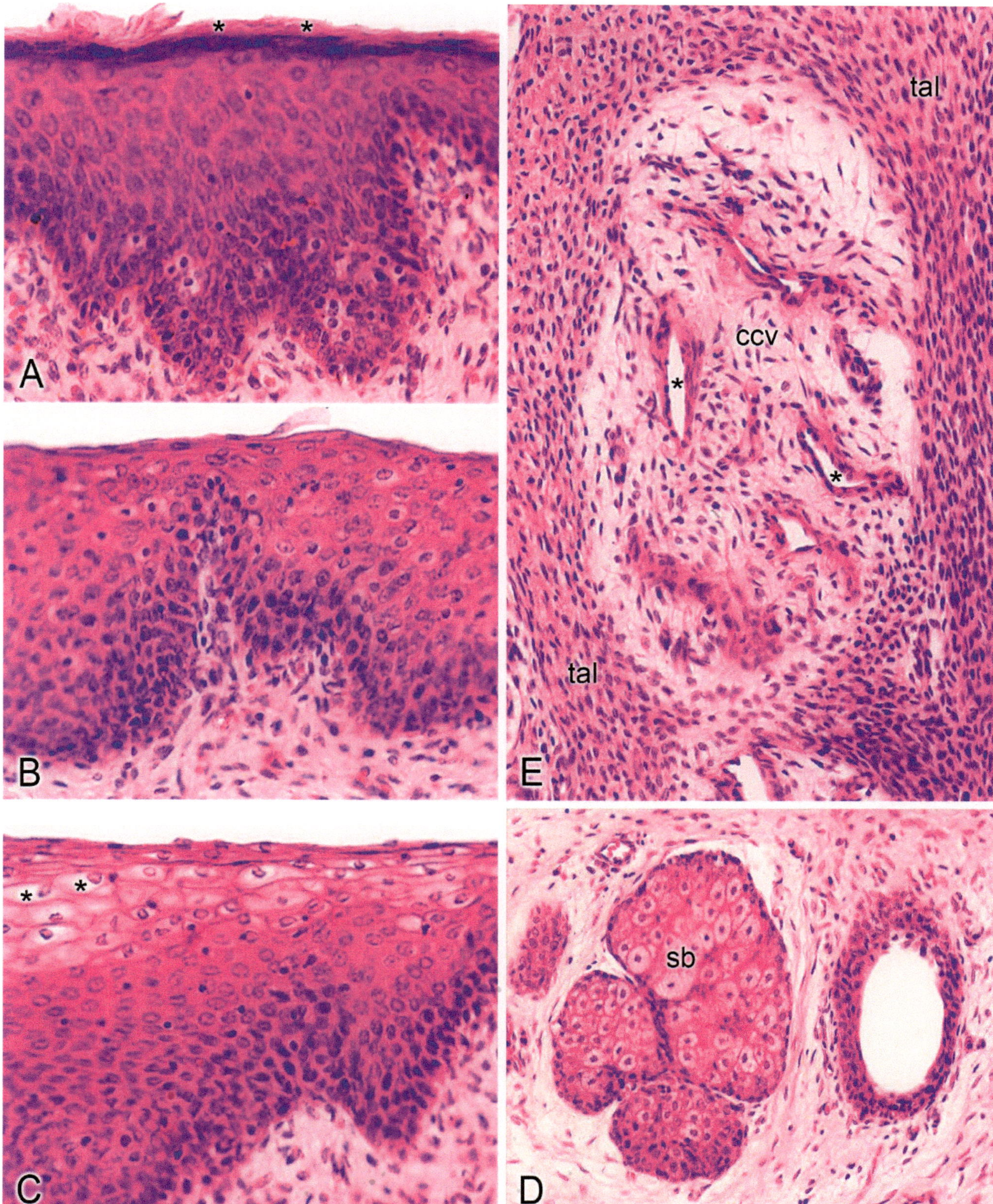

Figure 269. Labia minora and clitoris. (A) Laterally, labia minora are covered with keratinized stratified squamous epithelium with thin stratum corneum (asterisks), while the vestibular surface also contains nonkeratinized epithelium (B), often with glycogen-containing apical cells characteristic to the vagina (C, asterisks). (E) Glans clitoridis contains corpora cavernosa (ccv) composed of erectile tissue with caverns (asterisks); they are surrounded by dense connective tissue tunica albuginea (tal). (D) Labia minora contain sebaceous glands (sb) that are not associated with hair follicles.

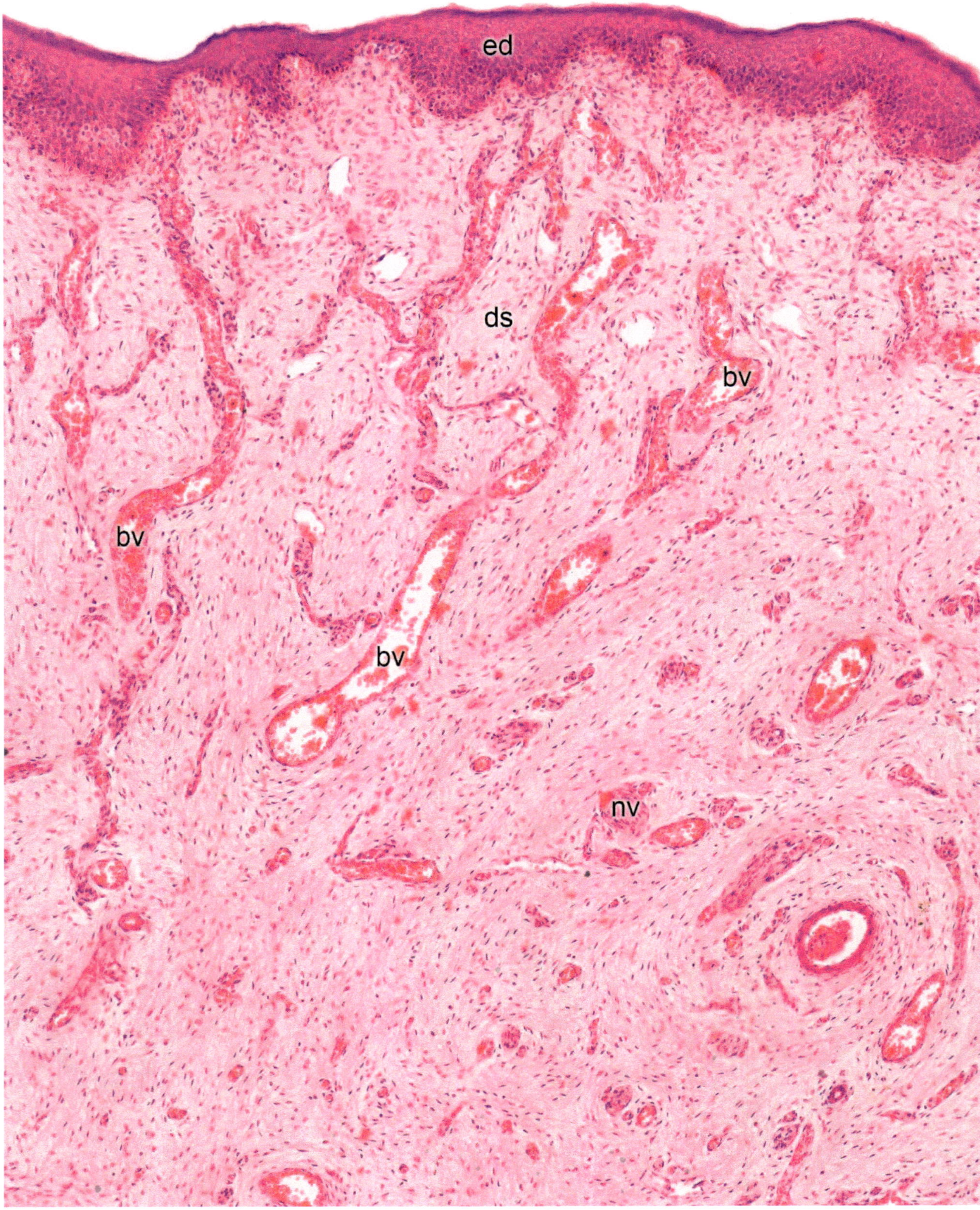

Figure 270. Labium minus, lateral surface. Labia minora are folds of thin skin that are covered laterally with thinly keratinized stratified squamous epithelium epidermis (ed). Under the epithelium, the dense irregular connective tissue dermis (ds) is rich in blood vessels (bv), particularly in venules and nerves (nv). Labia minora lack subcutaneous adipose tissue.

Placenta, umbilical cord

The placenta represents the boundary between the maternal and fetal circulation. Consequently, it is composed of a maternal part, the basal plate (decidua basalis or decidua placentaris) and a fetal part, the chorionic plate covered with amnion at the fetal side. Maternal blood enters into the intervillous space via endometrial spiral arteries penetrating the basal plate and leaves through endometrial veins via the same route. On the opposite side, fetal circulation is accomplished via the vessels of the umbilical cord that is attached to the chorionic plate. Chorionic villi extending into the intervillous space from the chorionic plate (villous chorion or chorion frondosum) are immersed into maternal blood.

In the mature placenta, the fetal side of the chorionic plate is covered by the amnion that is composed of a single layer of cuboidal cells and an underlying basement membrane enveloped by complex layers of stromal cells, vessels and fibers. Mature, tertiary villi of the chorionic plate are covered by a microvillated syncytiotrophoblast layer that is essentially a multinucleated syncytium containing darker zones where the nuclei accumulate and lighter zones where cytoplasm dominates. Under the syncytiotrophoblast layer, a discontinuous layer of cytotrophoblasts cover the core (extraembryonic) mesoderm. This connective tissue stroma contains mesenchymal cells, occasional placental macrophages with extensive cytoplasm (Hofbauer cells) and fetal vessels. Therefore, the barrier between the maternal and fetal blood includes the syncytiotrophoblast and the underlying incomplete cytotrophoblast layers, basal lamina of the trophoblasts, the core mesoderm (mesenchyme), the basal lamina of the endothelium and finally the endothelium of the fetal capillaries. The basal plate contains a large number of decidual cells that are darker than the surrounding connective tissue and possess epithelioid features.

Umbilical cord connects the fetus with the placenta; consequently, it is attached to the fetal side of the chorionic plate and it is lined with cuboidal or flattened cells of the amniotic epithelium. Umbilical cord carries a pair of umbilical arteries and a single umbilical vein. Umbilical vessels are surrounded by mucous connective tissue (Wharton's jelly) that is derived from the extra-embryonic mesoderm. Most of the cells of the mucous connective tissue are fibroblasts with mesenchymal cells and occasional macrophages embedded into a jelly-like matrix that is rich in glycosaminoglycans. Occasionally, the remnant of the allantoic duct can also be observed in the umbilical cord.

Mammary gland

Mammary glands are branched tubuloalveolar glands that are embedded into an extensive stroma with a substantial amount of adipose tissue. The parenchyma of the mammary gland is subdivided into lobules by dense connective tissue septa. The glands enter into the lactiferous ducts opening on the nipple (areola) that also contains sebaceous glands (of Montgomery). When the glands are in the inactive phase, the ducts exhibit a small lumen. During pregnancy glands undergo proliferation and the lumens of the ducts become prominent (secretory phase). Mammary gland also shows cyclic changes with the menstrual cycle, tubular cells become taller with ovulation due to the effects of estrogen. The base of the glands is lined with a single layer of cuboidal cells surrounded by myoepithelial cells, while the lactiferous ducts are covered with nonkeratinized stratified squamous epithelium that is gradually replaced by cuboidal cells in the more proximal parts of the duct system. Milk production involves both merocrine and apocrine secretion; the protein component is formed by merocrine secretion while the lipid component is produced by apocrine mechanism.

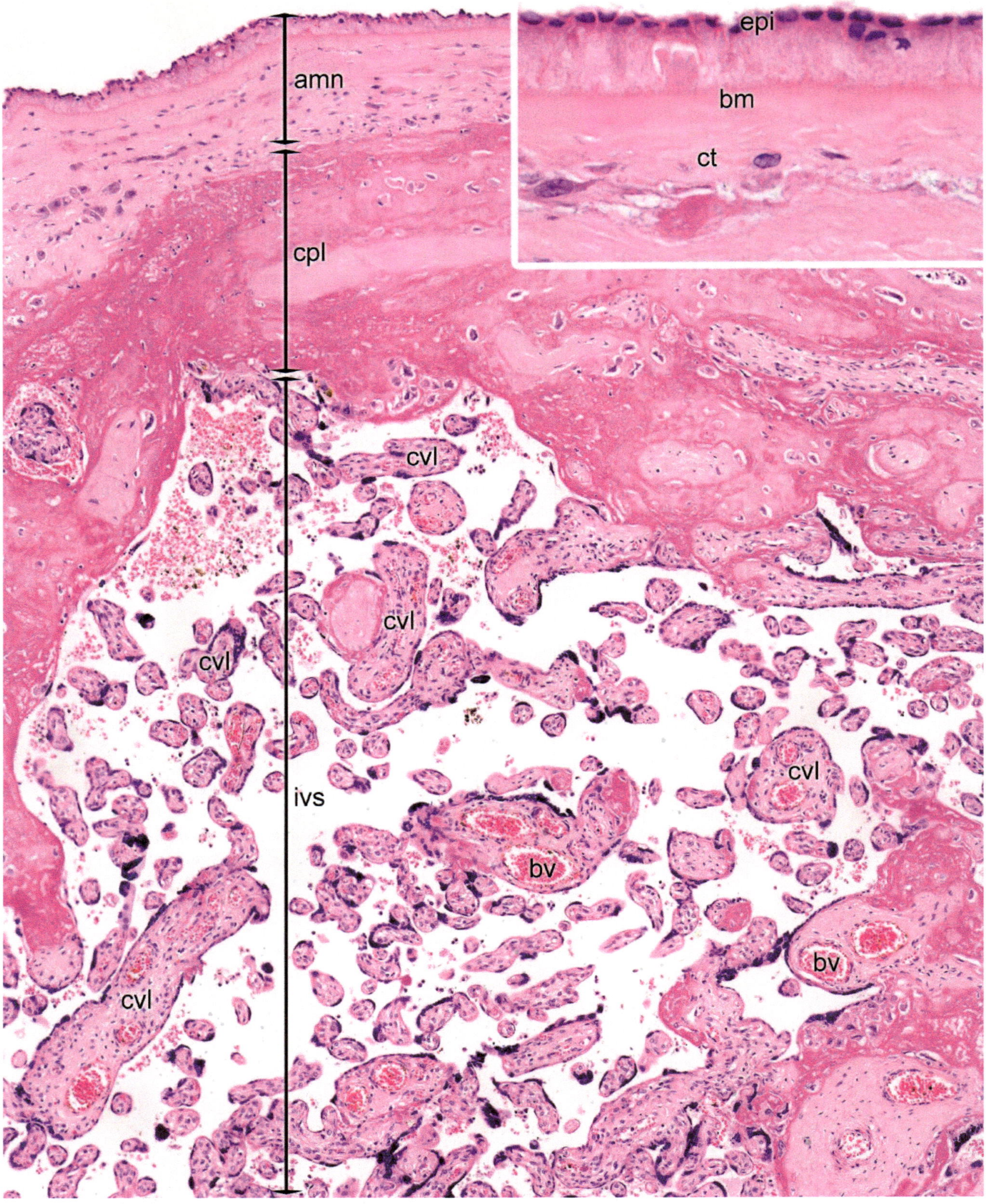

Figure 271. Placenta. At the fetal side of the placenta, chorionic plate is covered with the amnion (amn) that is composed of a single layer of cuboidal cells (epi) and the underlying basement membrane (bm) and connective tissue (ct, inset). Under the chorionic plate (cpl), the intervillous space (ivs) contains chorionic villi (cvl). In mature placenta, these tertiary villi contain blood vessels (bv).

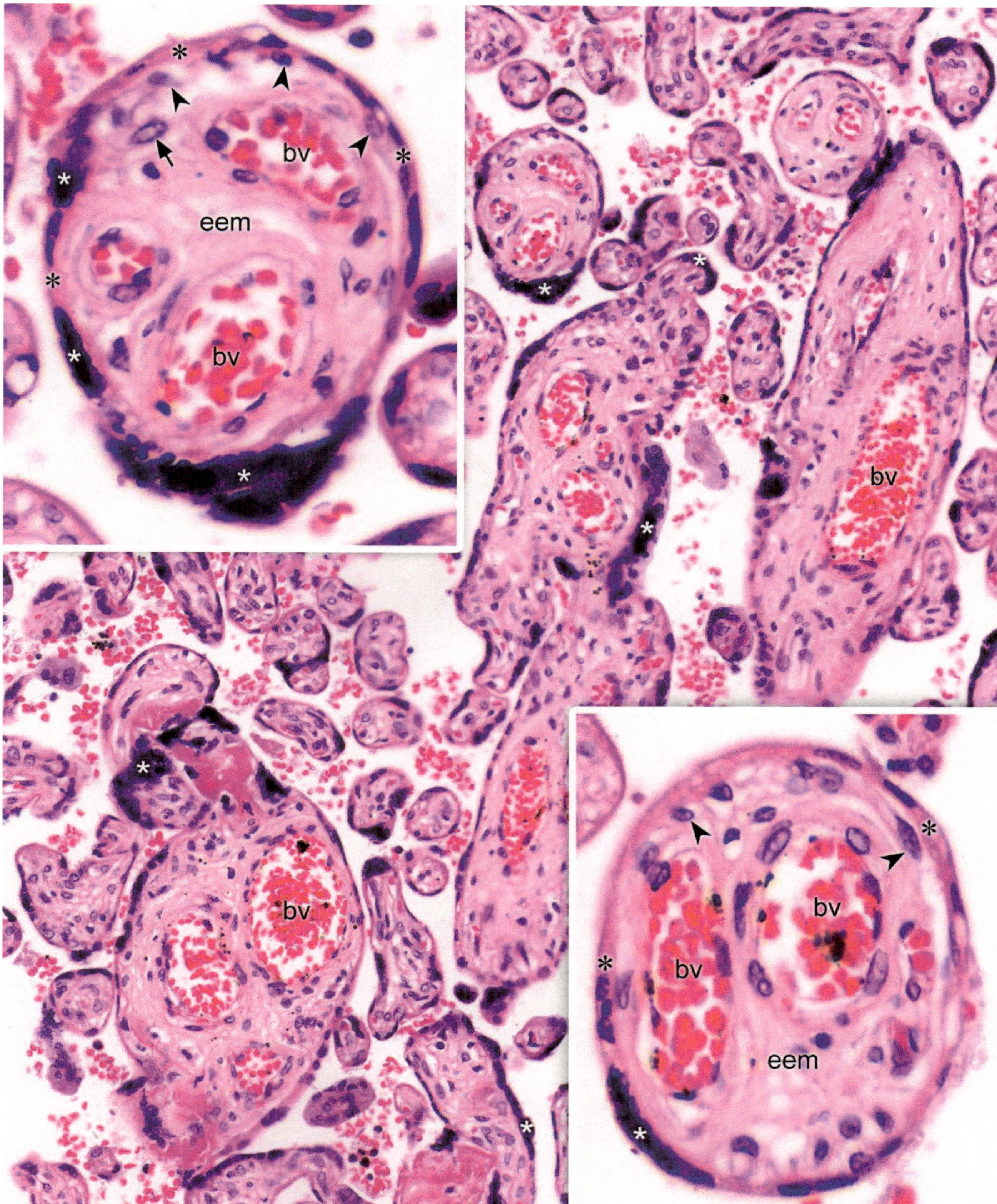

Figure 272. Placenta, chorionic villi. Mature, tertiary villi of the chorionic plate are covered by a layer of multinucleated syncytium formed by microvillated syncytiotrophoblast cells (asterisks). This layer contains darker zones with accumulated nuclei (five-pointed asterisks) and lighter zones where cytoplasm dominates (six-pointed asterisks). Under the syncytiotrophoblast layer, discontinuous layer of cytotrophoblasts (arrowheads) cover the extraembryonic (core) mesoderm (eem). This connective tissue stroma contains mesenchymal cells, occasional placental macrophages with extensive cytoplasm, the Hofbauer cells (arrows; upper inset) and fetal vessels (bv).

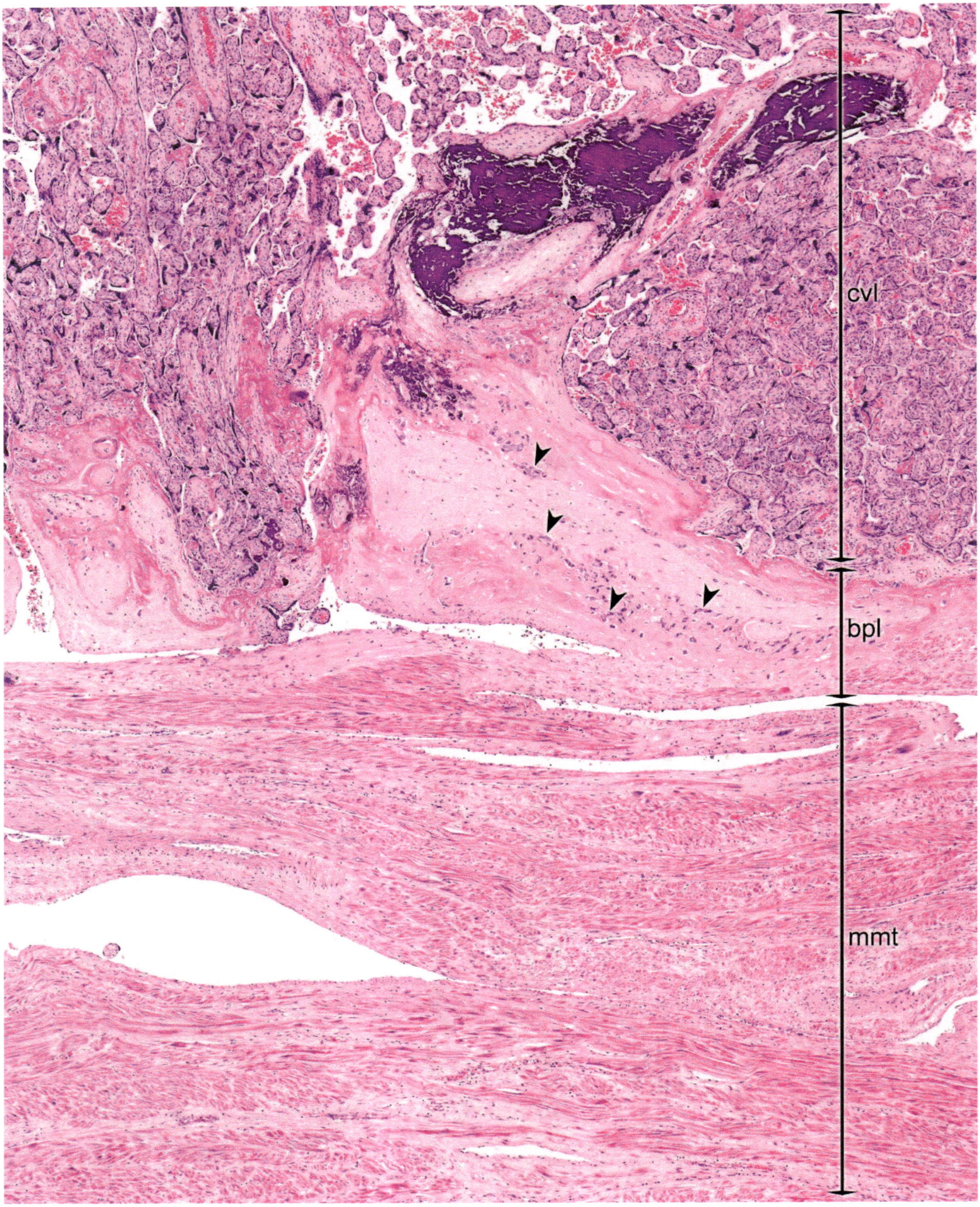

Figure 273. Placenta. At the maternal side of the placenta, the basal plate or decidua basalis (bpl) is anchored to the myometrium (mmt). Basal plate is facing the intervillous space that is filled with the chorionic villi (cvl) immersed into maternal blood. The basal plate contains epitheloid decidual cells (arrowheads) that are darker than the surrounding connective tissue.

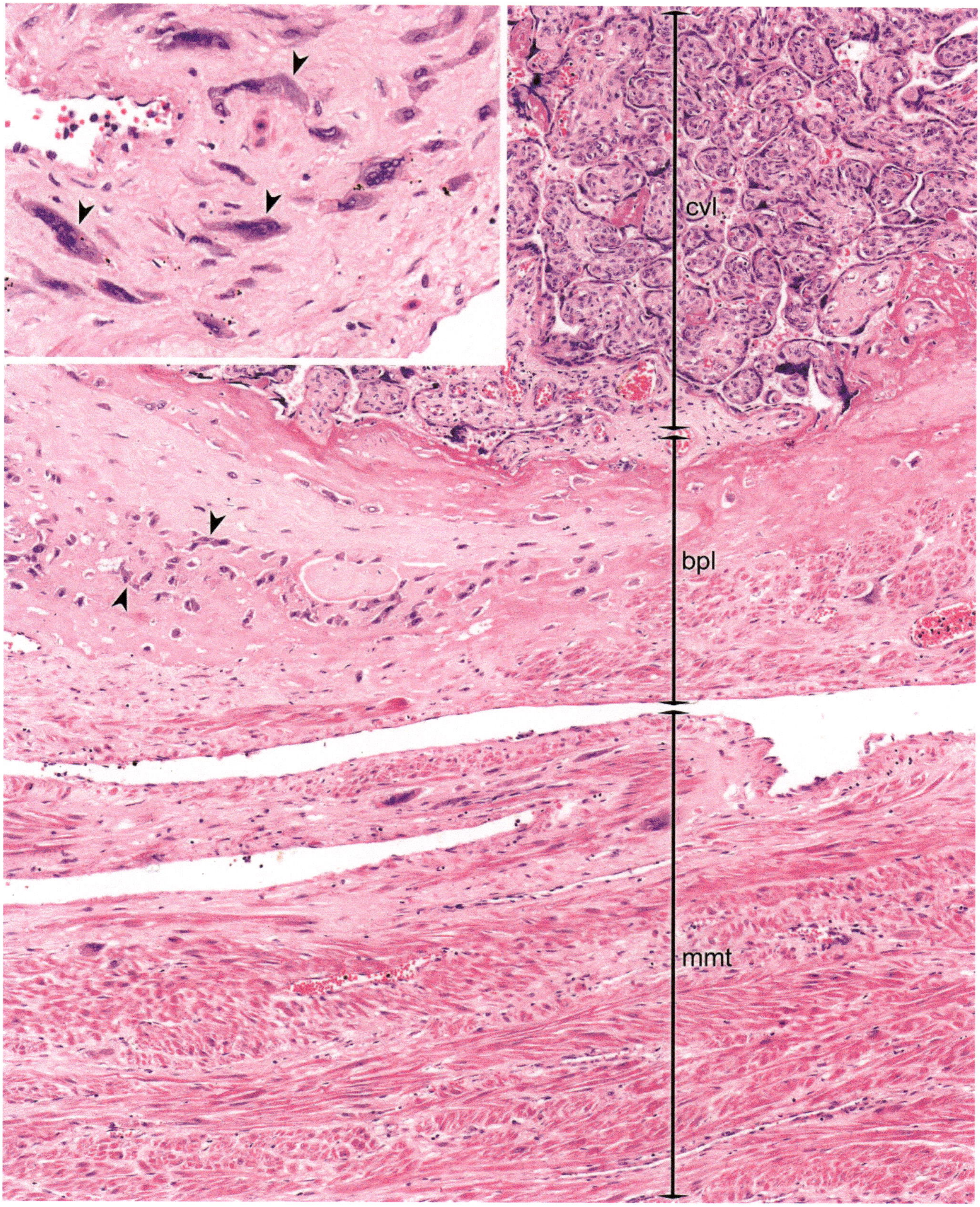

Figure 274. Placenta. At the maternal side of the placenta, the basal plate or decidua basalis (bpl) is anchored to the myometrium (mmt). Above the basal plate, the intervillous space is filled with the chorionic villi (cvl) that are immersed into maternal blood. The basal plate contains epitheloid decidual cells (arrowheads; inset) that are darker than the surrounding connective tissue

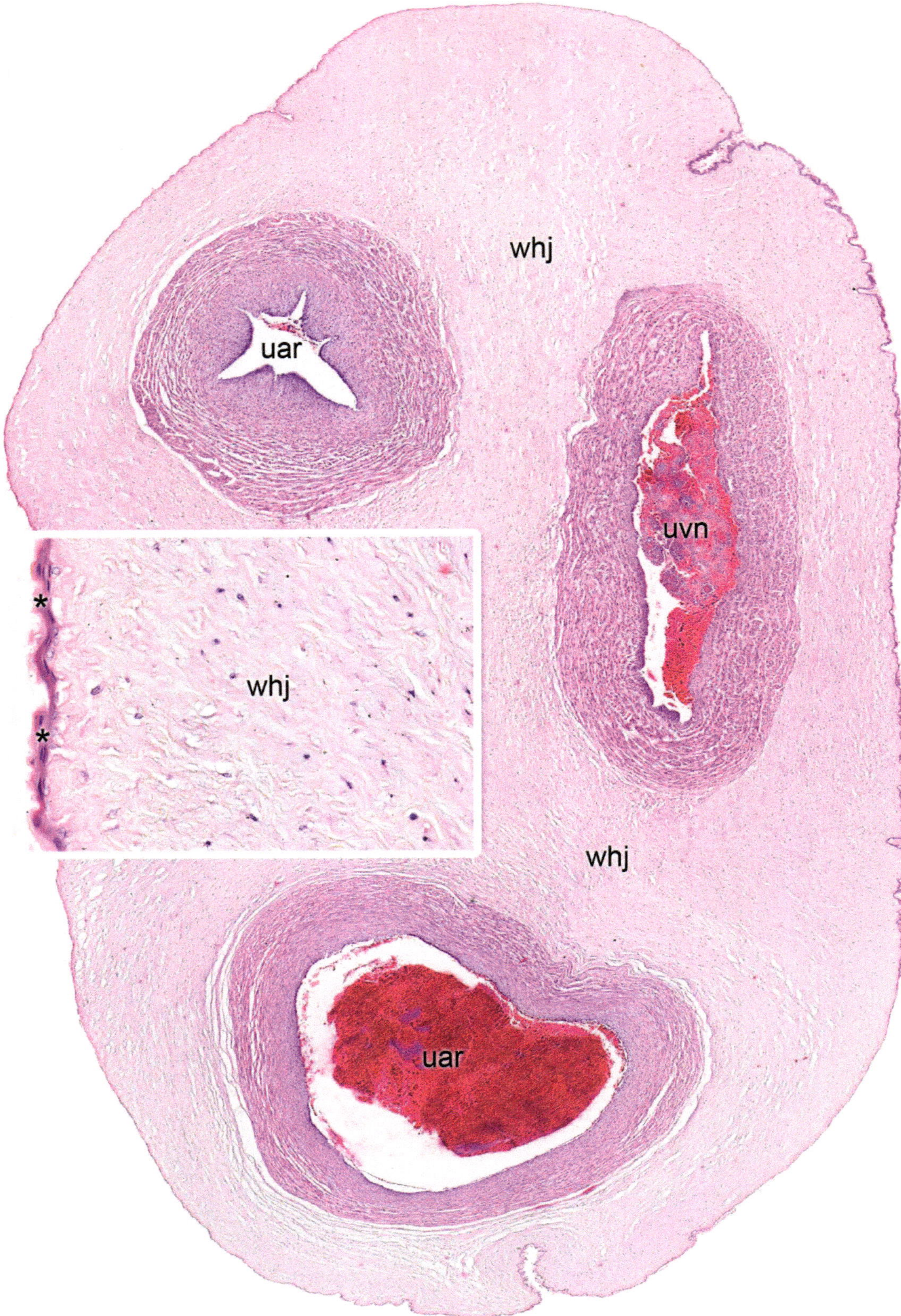

Figure 275. Umbilical cord. The umbilical cord is covered with cuboidal or flattened cells of the amniotic epithelium (asterisks, higher magnification inset). Umbilical cord contains a pair of umbilical arteries (uar) and a single umbilical vein (uvn). Umbilical vessels are surrounded by mucous connective tissue, the Wharton's jelly (whj) that is derived from the extra-embryonic mesoderm and contains mostly fibroblasts with mesenchymal cells and occasional macrophages embedded into a jelly-like matrix.

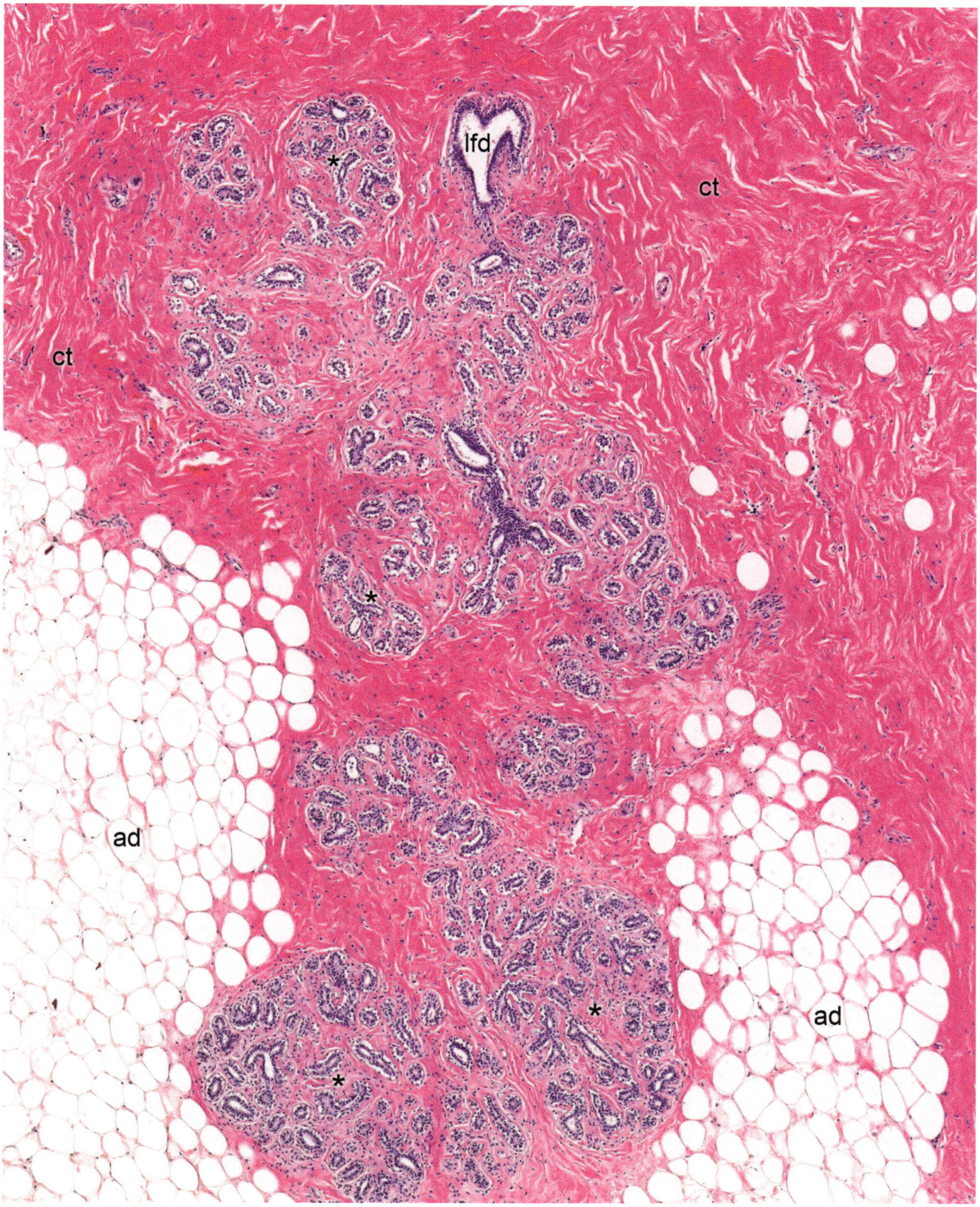

Figure 276. Mammary gland, inactive phase. Mammary glands are branched tubuloalveolar glands that are surrounded by an extensive stroma with adipose tissue (ad) and dense connective tissue septa (ct). When the glands are in the inactive phase, the lobules of the gland (asterisks) are composed of ducts with a narrow lumen surrounded by an eosinophilic layer of myoepithelial cells. The proximal part of the lactiferous ducts (lfd) is covered with stratified or simple cuboidal epithelium.

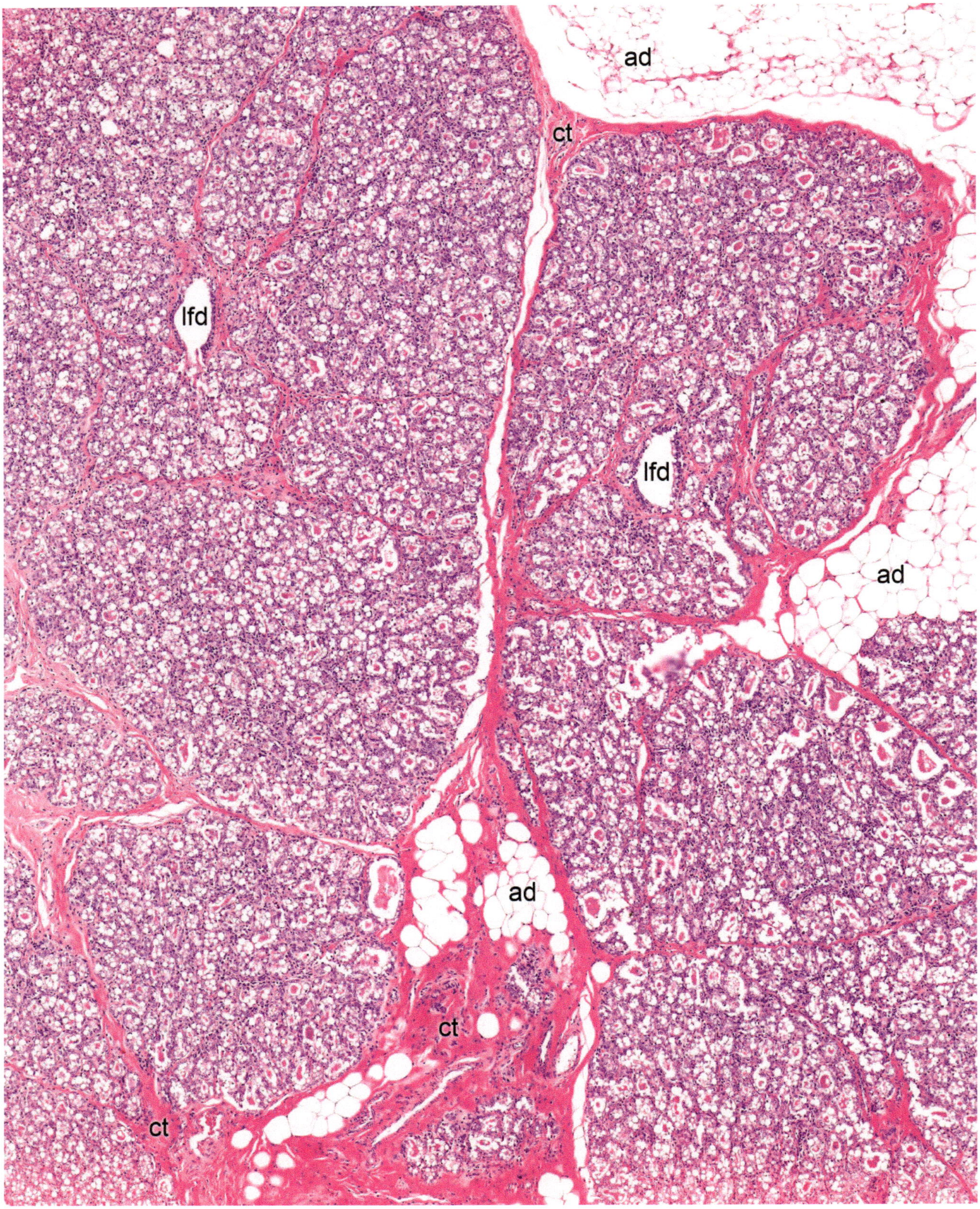

Figure 277. Mammary gland, active phase. Mammary glands are branched tubuloalveolar glands that are surrounded by an extensive stroma with adipose tissue (ad) and dense connective tissue septa (ct). During pregnancy glands undergo proliferation and the lumens of the ducts become larger as the alveoli develop. The proximal part of the lactiferous ducts (lfd) is covered with stratified or simple cuboidal epithelium.

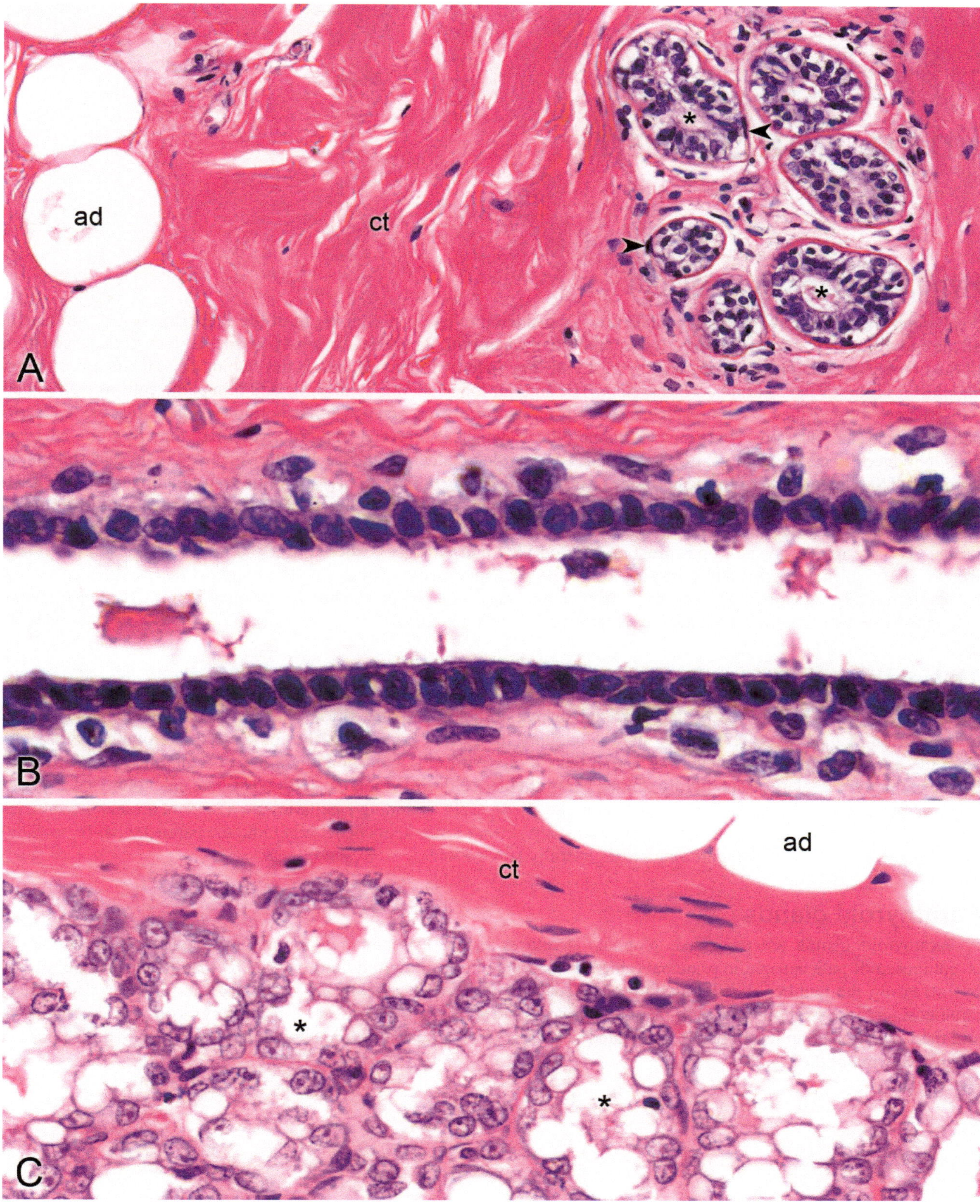

Figure 278. Mammary gland. Mammary glands are branched tubuloalveolar glands that are surrounded by an extensive stroma with adipose tissue (ad) and dense connective tissue septa (ct). (A) When the glands are in the inactive phase, the ducts (asterisks) exhibit a small lumen surrounded by an eosinophilic layer of myoepithelial cells (arrowheads). (B) The proximal part of the lactiferous ducts is covered with stratified or simple cuboidal epithelium. (C) During pregnancy glands undergo proliferation and the lumens of the ducts become larger as the alveoli develop (asterisks).

Chapter 12

Endocrine System

Hypophysis (pituitary gland)

Hypophysis, together with the hypothalamus is considered to be the master organ of the endocrine system influencing all the endocrine functions of the body. It is composed of an anterior lobe (adenohypophysis) that derives from the ectoderm of the oropharynx (Rathke's pouch) and the posterior lobe (neurohypophysis) that is developing from the neuroectoderm, and thus it is part of the central nervous system. The anterior lobe contains acidophilic and basophilic cells as well as cells with pale cytoplasm called chromophobes; these cells produce the tropic and non-tropic hormones regulating other endocrine and non-endocrine functions. Non-tropic hormones, the growth hormone (GH) and prolactin (PRL) directly stimulate target cells to induce effects; in contrast, tropic hormones target other endocrine glands, and they include the thyroid-stimulating hormone (TSH), adrenocorticotropic hormone (ACTH), luteinizing hormone (LH) and follicle-stimulating hormone (FSH). Acidophils populating the lateral wedge of the adenohypophysis produce GH and PRL (somatotropes and lactotropes, respectively), while basophils producing ACTH and TSH (corticotropes and thyrotropes, respectively) populate the central wedge. The basophilic gonadotropes produce LH and FSH and they are scattered over the lateral wings and the central wedge. Between the anterior and posterior lobes, the pars intermedia contains cystic, colloid-filled cavities as well as basophils and chromophobes. The posterior lobe contains non-myelinated axon bundles originating from the hypothalamic magnocellular system as well as dilated axon terminals (Herring bodies).

Thyroid gland

Thyroid gland is composed of large, round-shaped follicles that are filled with eosinophilic colloid containing thyroglobulin that serves as an inactive storage medium for thyroid hormones. The follicles are surrounded by a single layer of cuboidal, follicular (principal) cells that secrete thyroxine (T_4) and triiodothyronine (T_3) into the follicular lumen. Between the follicles, larger, and lightly stained parafollicular cells often form clusters and secrete calcitonin that regulates calcium metabolism by suppressing blood calcium levels. Vessels, including capillaries are abundant around the follicles. Thyroid gland is surrounded by a connective tissue capsule that forms septa penetrating into the parenchyma. In newborn, the follicles of the thyroid gland are small and generally lack colloid.

Parathyroid glands

Parathyroid glands are pea-sized, globular structures embedded into the thyroid parenchyma. Most of the gland is composed of pale basophilic principal (chief) cells producing parathyroid hormone (PTH) that activates bone resorption by increasing the number and the activity of osteoclasts, thus elevating calcium and phosphate blood levels. Clusters of large eosinophilic, oxyphil cells with numerous mitochondria can also be observed; they are probably not secretory cells, and their exact function has not been elucidated yet. Parathyroid gland is surrounded by adipose tissue that also penetrates the parenchyma of the gland.

http://doi.org/10.1016/B978-0-323-91891-6.50012-X,

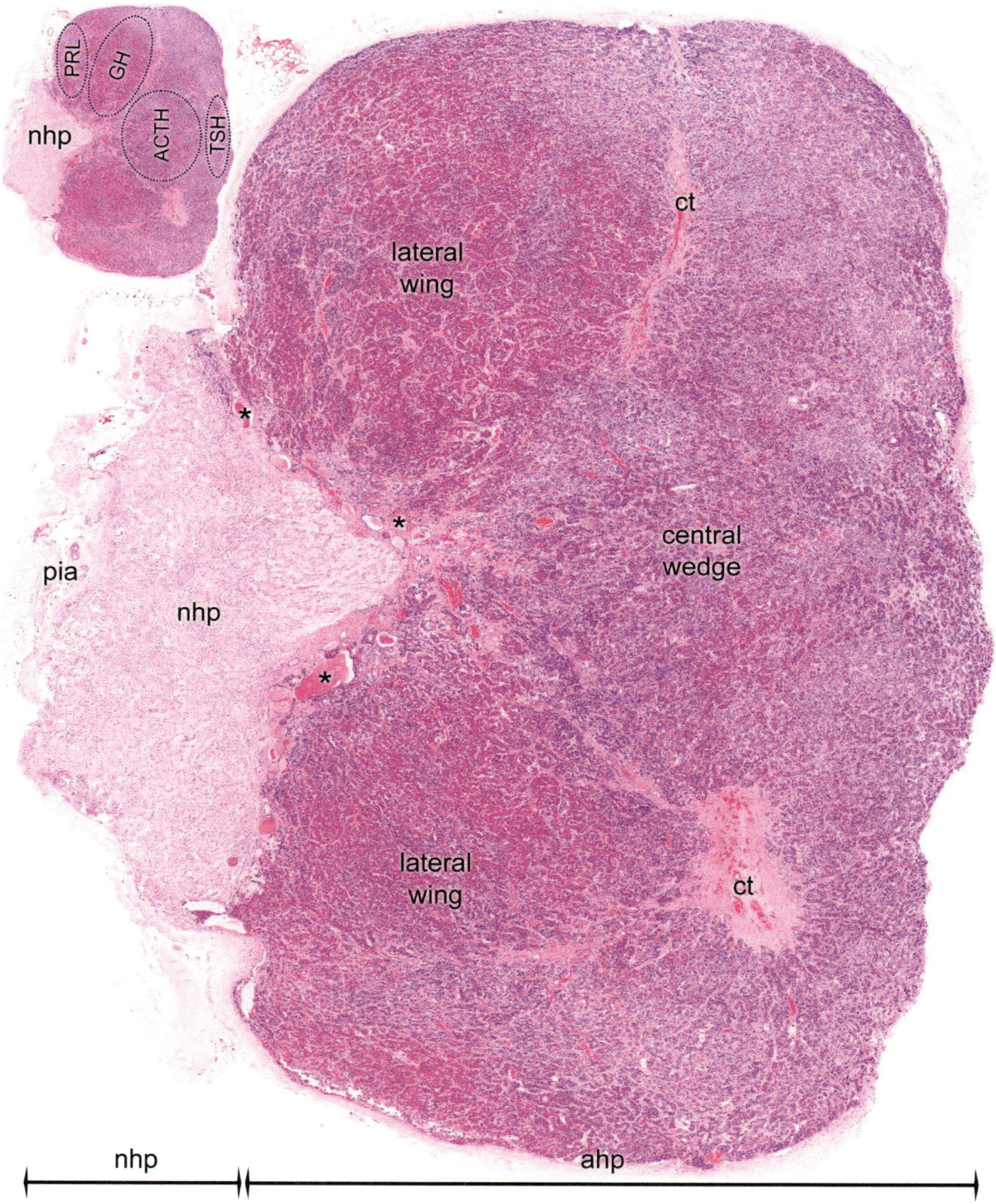

Figure 279. Hypophysis, horizontal section. Hypophysis is formed by the anterior adenohypophysis (ahp) and the posterior neurohypophysis (nhp) that is covered by pia mater (pia) and contains unmyelinated axon bundles. Adenohypophysis is composed of the central wedge and the two lateral wings separated by connective tissue septa (ct), and it contains acidophilic and basophilic cells as well as lightly stained cells, the chromophobes; these cells produce tropic and non-tropic hormones, organized in clusters (inset). Between the anterior and posterior lobes, the pars intermedia contains cystic, colloid-filled cavities (asterisks) as well as basophils and chromophobes.

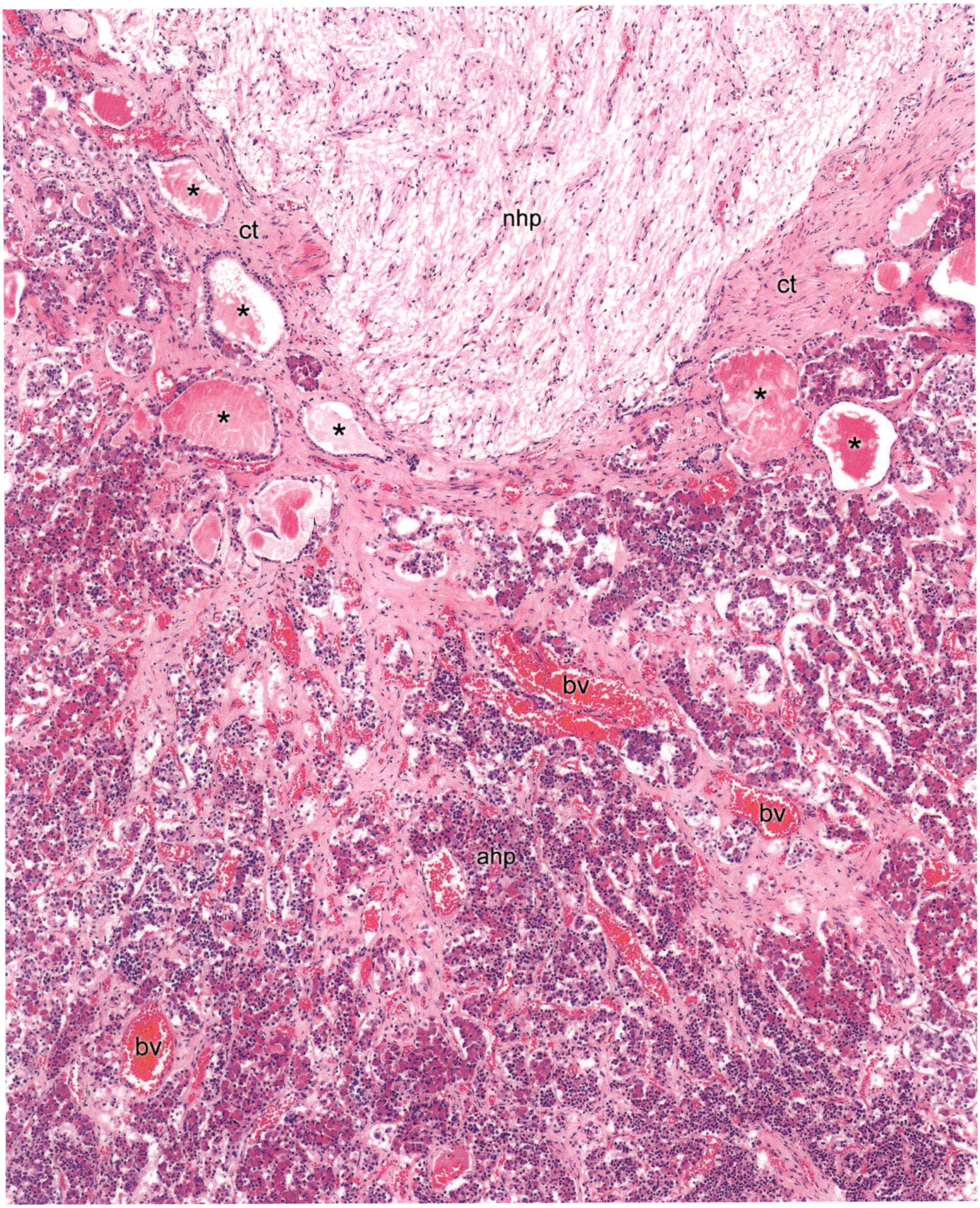

Figure 280. Hypophysis, central wedge. Hypophysis is formed by the anterior adenohypophysis (ahp) and the posterior neurohypophysis (nhp) that contains unmyelinated axon bundles. The central wedge of the adenohypophysis (ahp) contains blood vessels (bv) surrounded by a mixture of acidophilic and basophilic cells as well as lightly stained chromophobes. Between the adeno- and neurohypophysis, the pars intermedia contains cystic, colloid-filled cavities (asterisks) embedded into dense connective tissue septa (ct).

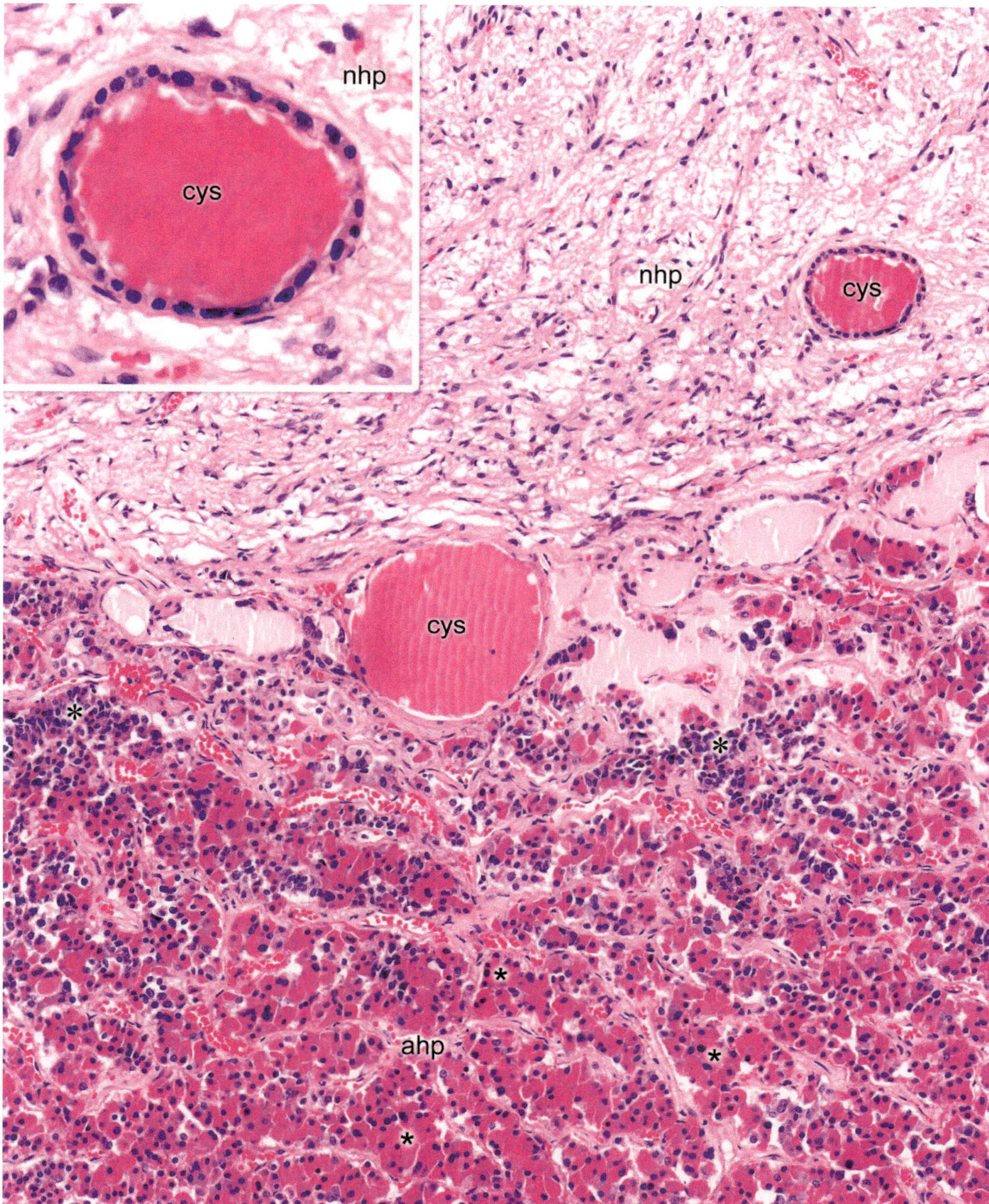

Figure 281. Hypophysis, lateral wing. Hypophysis is formed by the anterior adenohypophysis (ahp) and the posterior neurohypophysis (nhp) that contains unmyelinated axon bundles. The lateral wings of the adenohypophysis contain clusters acidophilic cells (five-pointed asterisks) mixed with groups of basophils (six-pointed asterisks) and lightly stained chromophobes that can be reliably identified with higher magnification. Between the adeno- and neurohypophysis, the pars intermedia contains cystic, colloid-filled cavities that are lined with flattened or cuboidal cells (cys; inset).

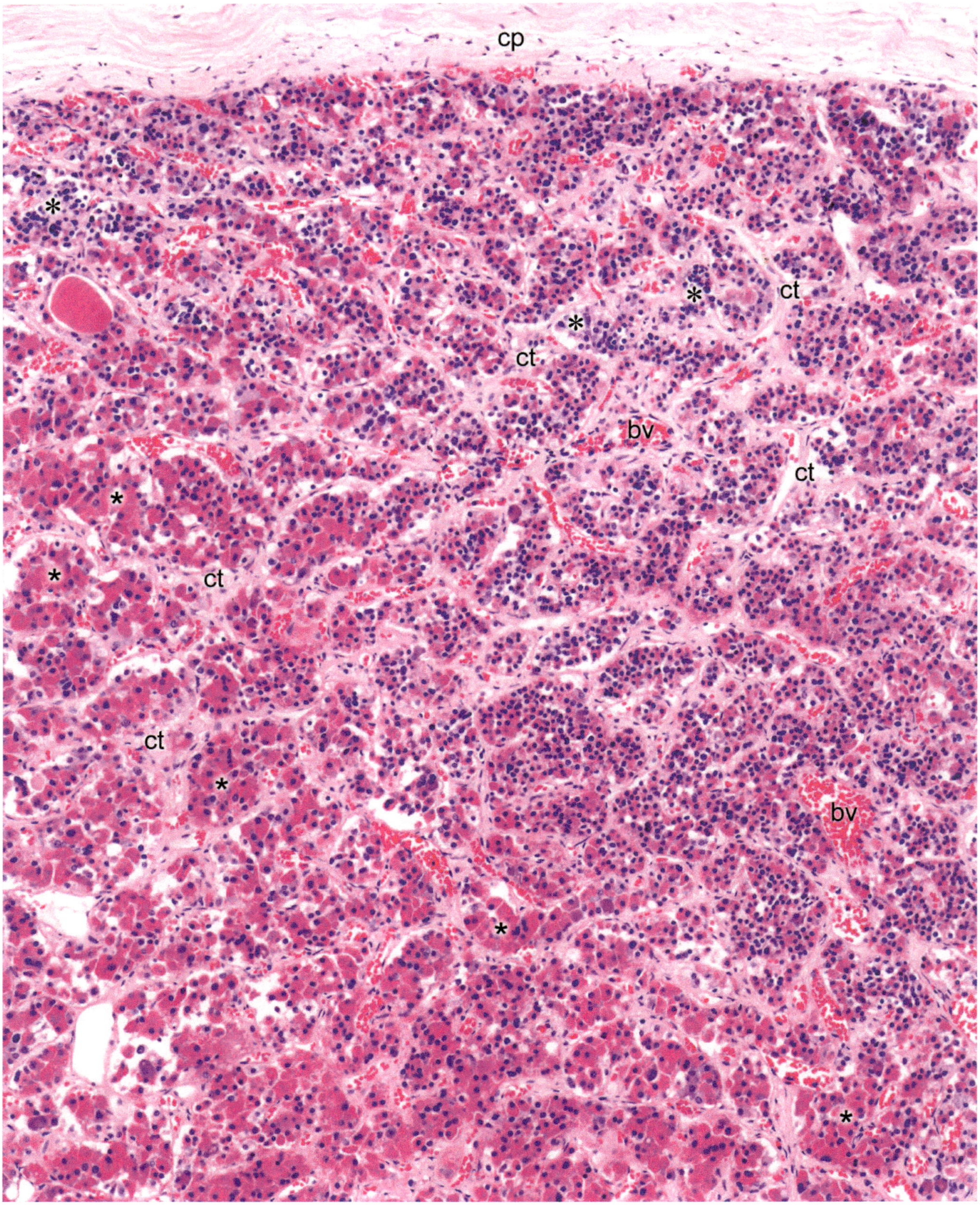

Figure 282. Hypophysis, lateral wing. Adenohypophysis is covered with a thin loose connective tissue capsule (cp). The lateral wings of the adenohypophysis contain clusters acidophilic cells (five-pointed asterisks) mixed with groups of basophils (six-pointed asterisks) and lightly stained chromophobes that can be reliably identified with higher magnification. The clusters of cells are surrounded by connective tissue septa (ct) and blood vessels (bv).

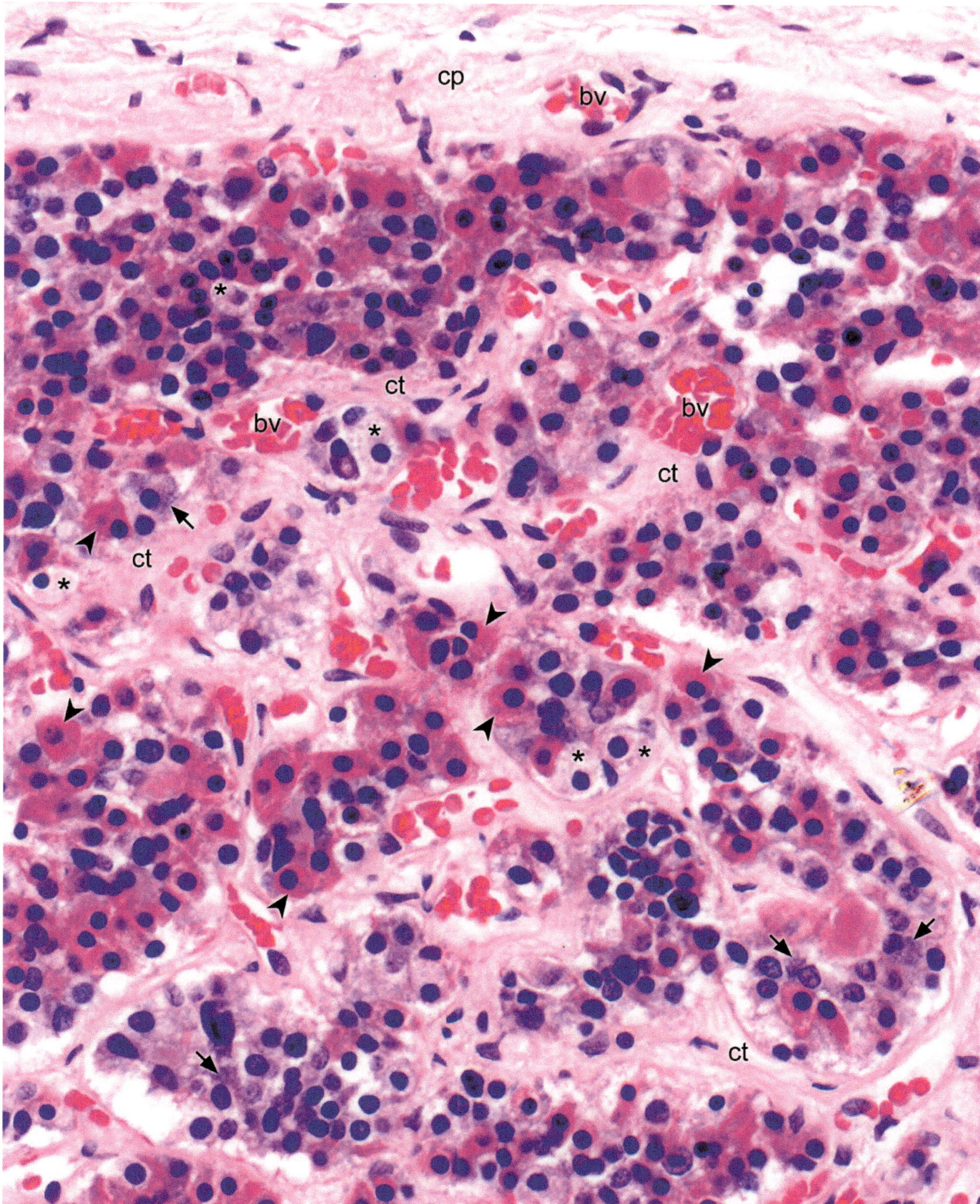

Figure 283. Hypophysis, lateral wing. Adenohypophysis is covered with a thin loose connective tissue capsule (cp). The lateral wings of the adenohypophysis contain clusters acidophilic cells (arrowheads) mixed with groups of basophils (arrows) and lightly stained chromophobes (asterisks). The clusters of cells are surrounded by connective tissue septa (ct) and blood vessels (bv).

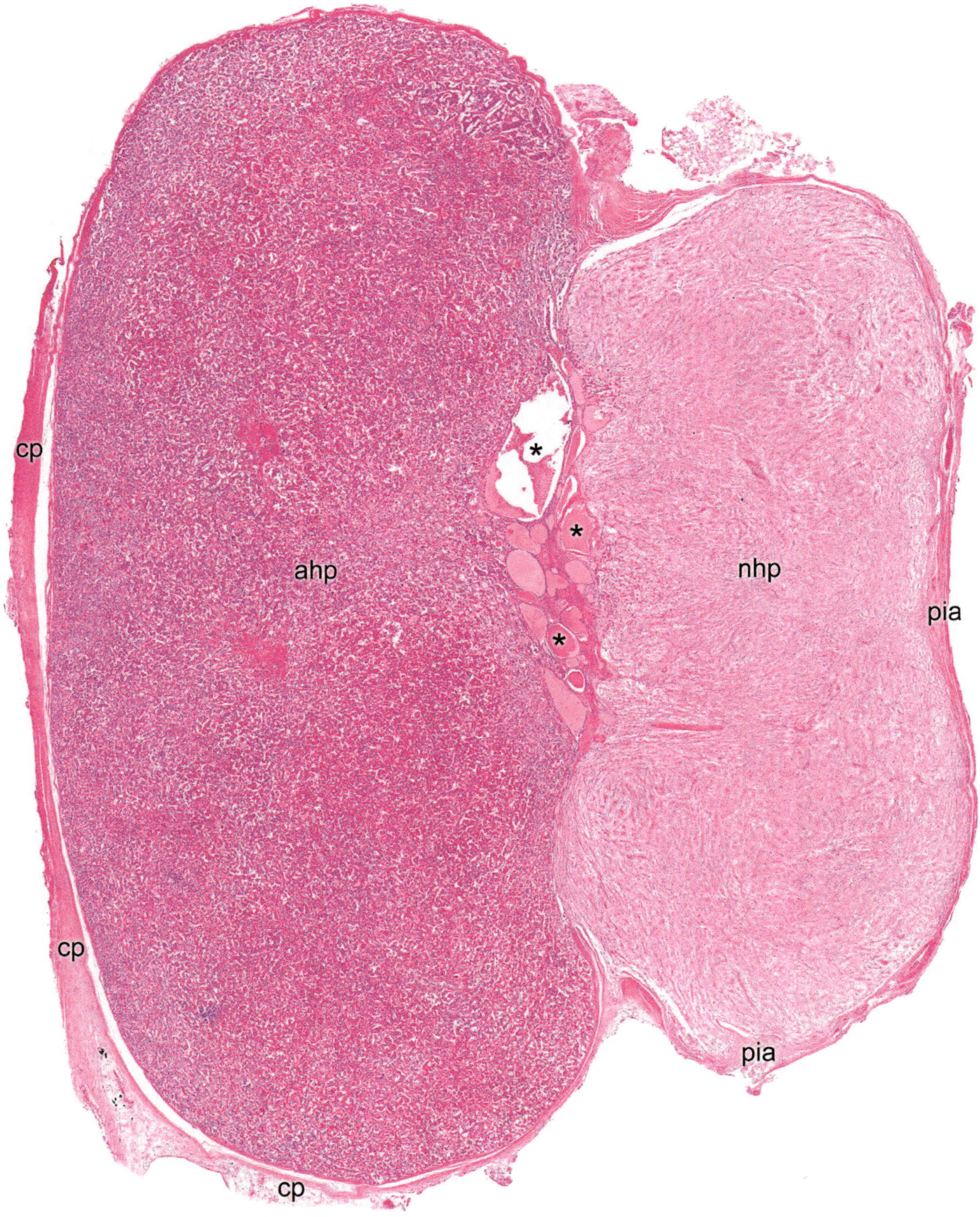

Figure 284. Hypophysis, horizontal section. Hypophysis is formed by the anterior adenohypophysis (ahp) and a posterior neurohypophysis (nhp) that is covered by pia mater (pia) and contains unmyelinated axon bundles. Adenohypophysis covered by a connective tissue capsule (cp) and it contains acidophilic and basophilic cells as well as lightly stained cells, the chromophobes. Between the anterior and posterior lobes, the pars intermedia contains cystic, colloid-filled cavities (asterisks).

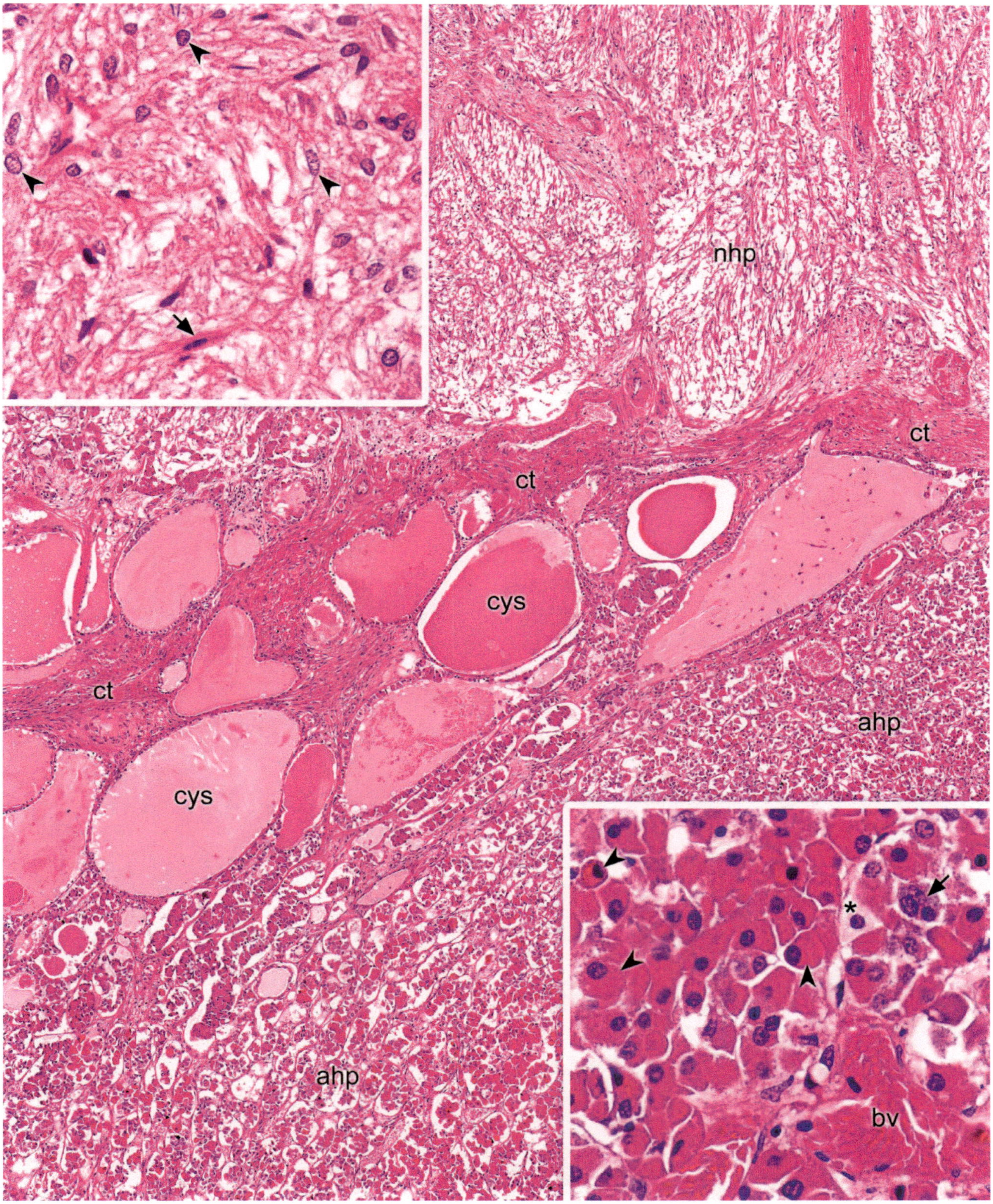

Figure 285. Hypophysis, horizontal section. Hypophysis is formed by the anterior adenohypophysis (ahp) and a posterior neurohypophysis (nhp) that is formed by unmyelinated axon bundles as well as fibroblasts (arrow, upper inset) and branched glial cells, the pituicytes (arrowheads, upper inset). Adenohypophysis contains clusters of acidophils (arrowheads, lower inset) mixed with basophils (arrow, lower inset) and lightly stained chromophobes (asterisk). Cystic, colloid-filled cavities (cys) lined with flattened or cuboidal cells represent the pars intermedia between the adeno- and neurohypophysis.

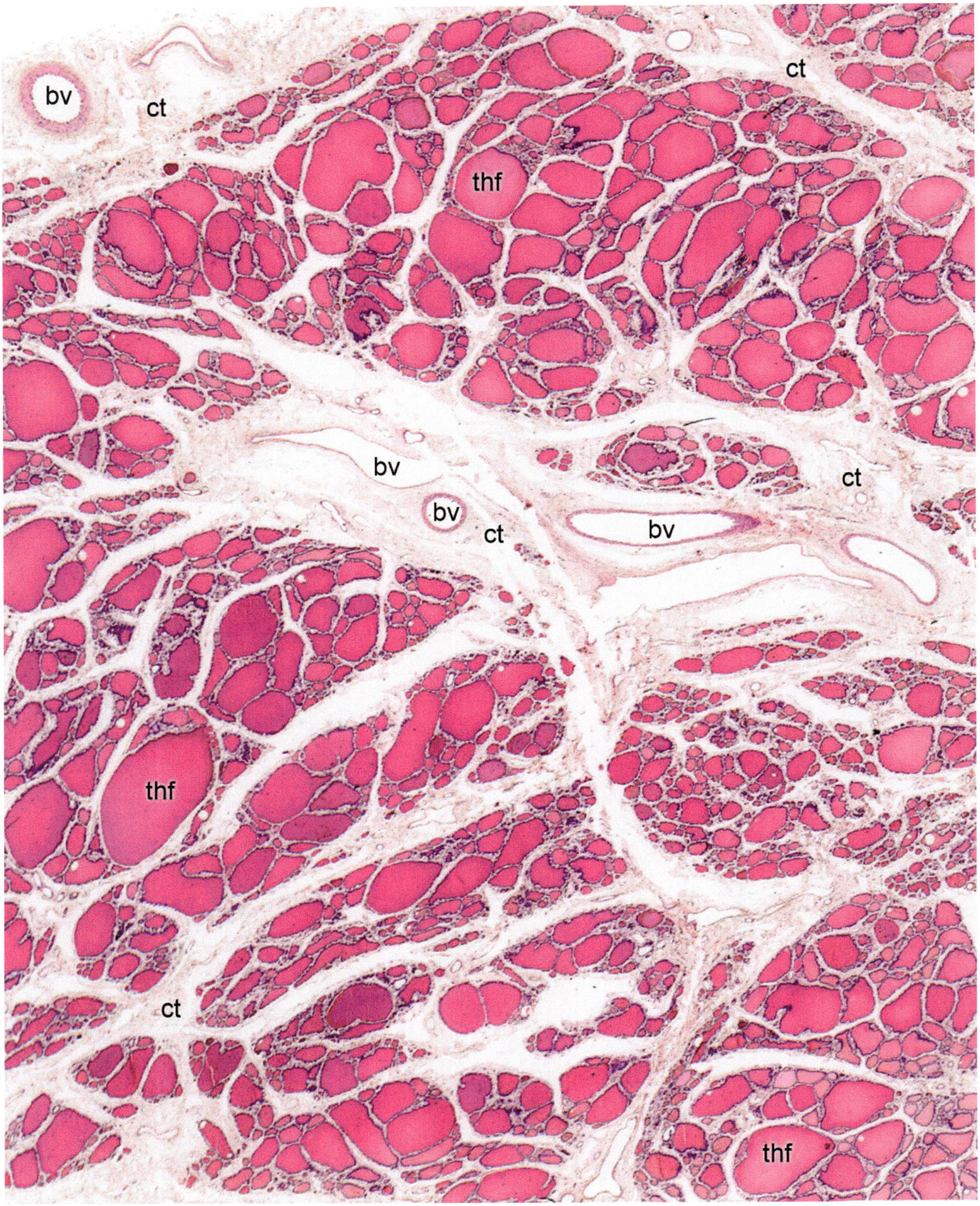

Figure 286. Thyroid gland. Thyroid gland is formed by spherical follicles (thf) filled with eosinophilic colloid containing thyroglobulin. The follicles are lined by a single layer of cuboidal, follicular (principal) cells. Between the follicles, connective tissue septa (ct) contain blood vessels (bv).

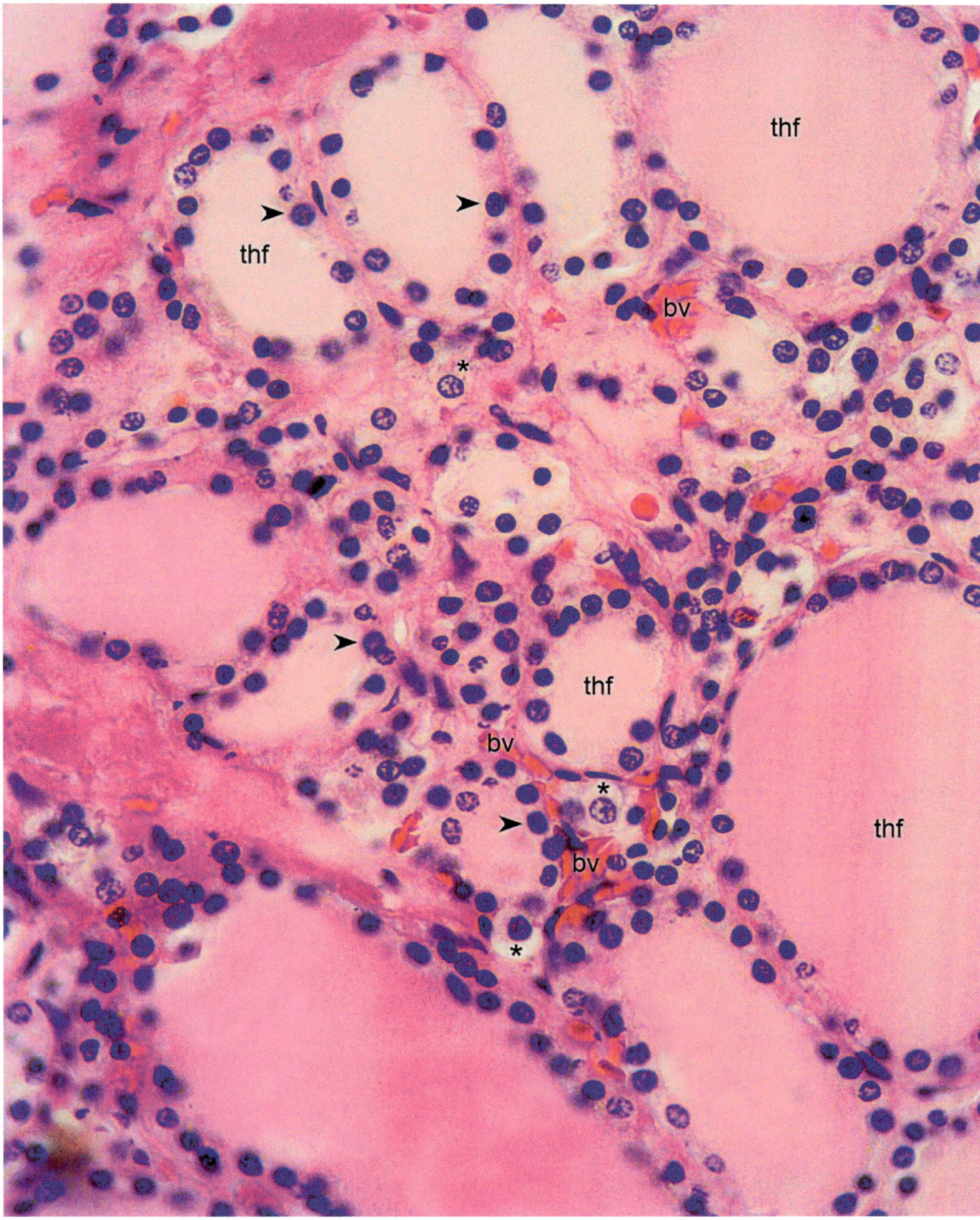

Figure 287. Thyroid gland. Thyroid gland contains spherical follicles (thf) filled with eosinophilic colloid containing thyroglobulin. The follicles are lined by a single layer of cuboidal, follicular (principal) cells (arrowheads). Between the follicles, larger and lightly stained parafollicular cells (asterisks) often form clusters. Parafollicular cells secrete calcitonin that regulates calcium metabolism by suppressing blood calcium levels. Blood vessels (bv) are abundant between the follicles.

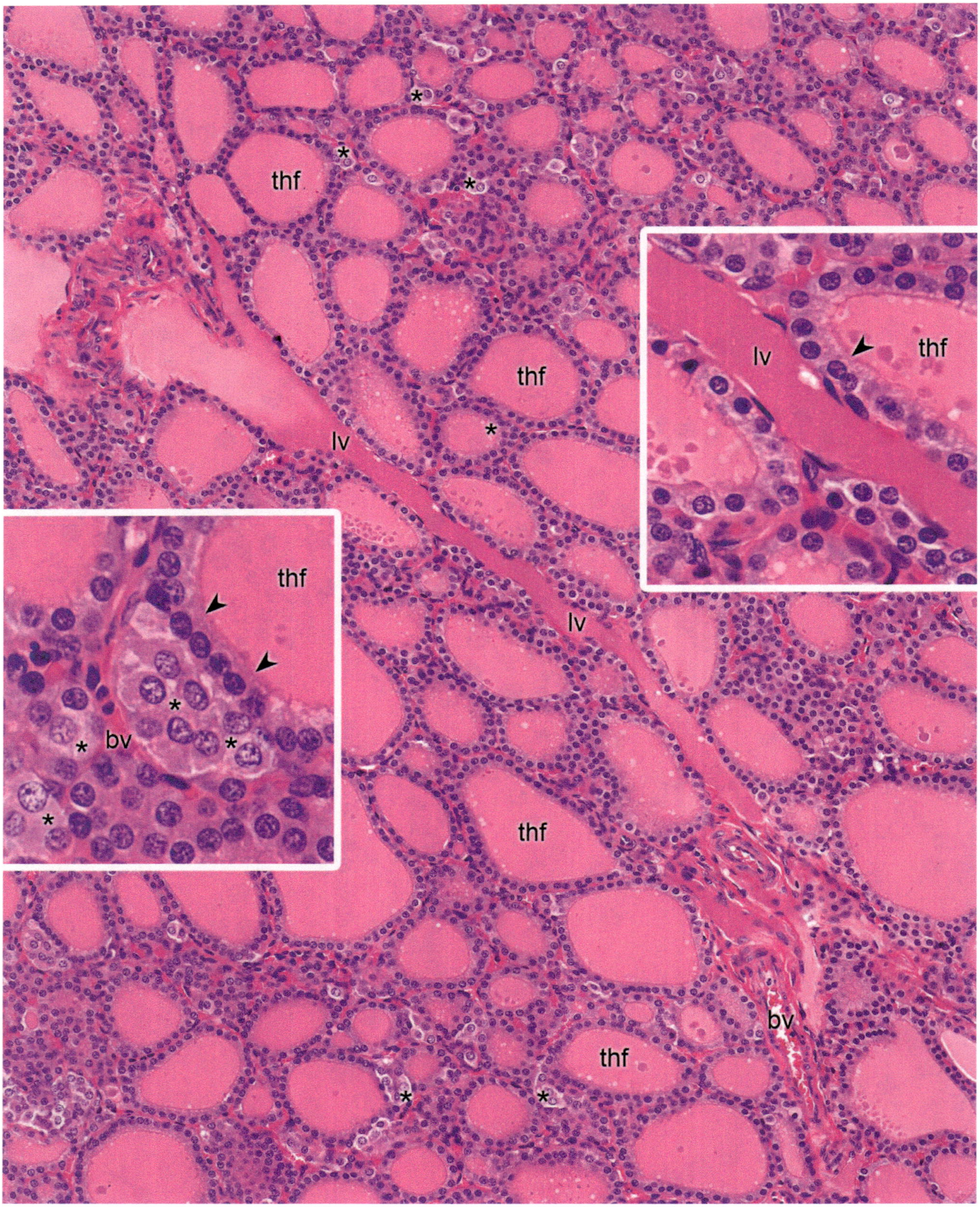

Figure 288. Thyroid gland. Thyroid gland contains spherical follicles (thf) filled with eosinophilic colloid containing thyroglobulin. The follicles are lined by a single layer of cuboidal, follicular (principal) cells (arrowheads; insets). Between the follicles, larger and lightly stained parafollicular cells (asterisks; left inset) often form clusters. Parafollicular cells secrete calcitonin that regulates calcium metabolism by suppressing blood calcium levels. Blood vessels (bv) and lymph vessels (lv; right inset) are abundant between the follicles.

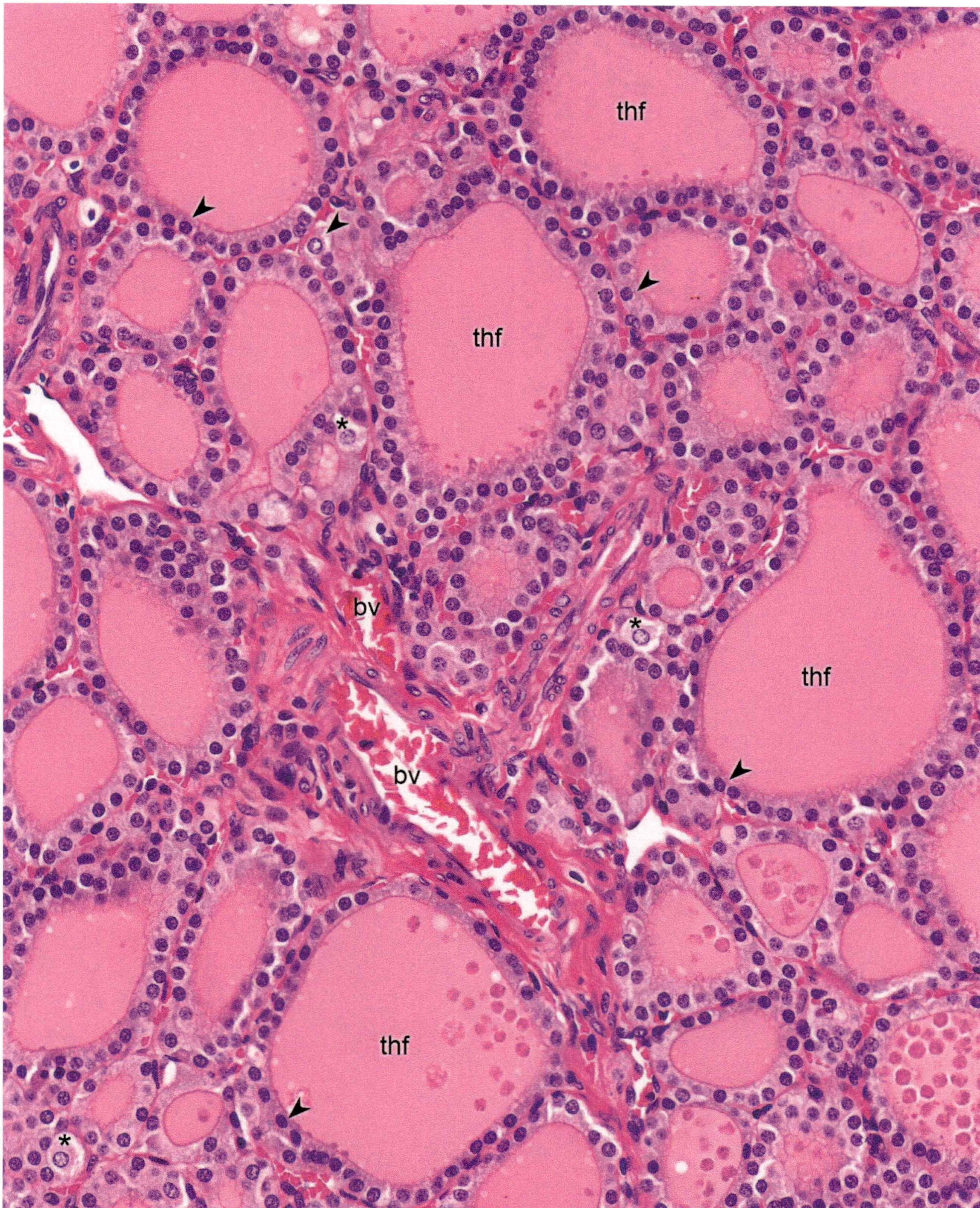

Figure 289. Thyroid gland. Thyroid gland contains spherical follicles (thf) filled with eosinophilic colloid containing thyroglobulin. The follicles are lined by a single layer of cuboidal, follicular (principal) cells (arrowheads). Between the follicles, larger and lightly stained parafollicular cells (asterisks) often form clusters. Parafollicular cells secrete calcitonin that regulates calcium metabolism by suppressing blood calcium levels. Blood vessels (bv), including capillaries are abundant between the follicles.

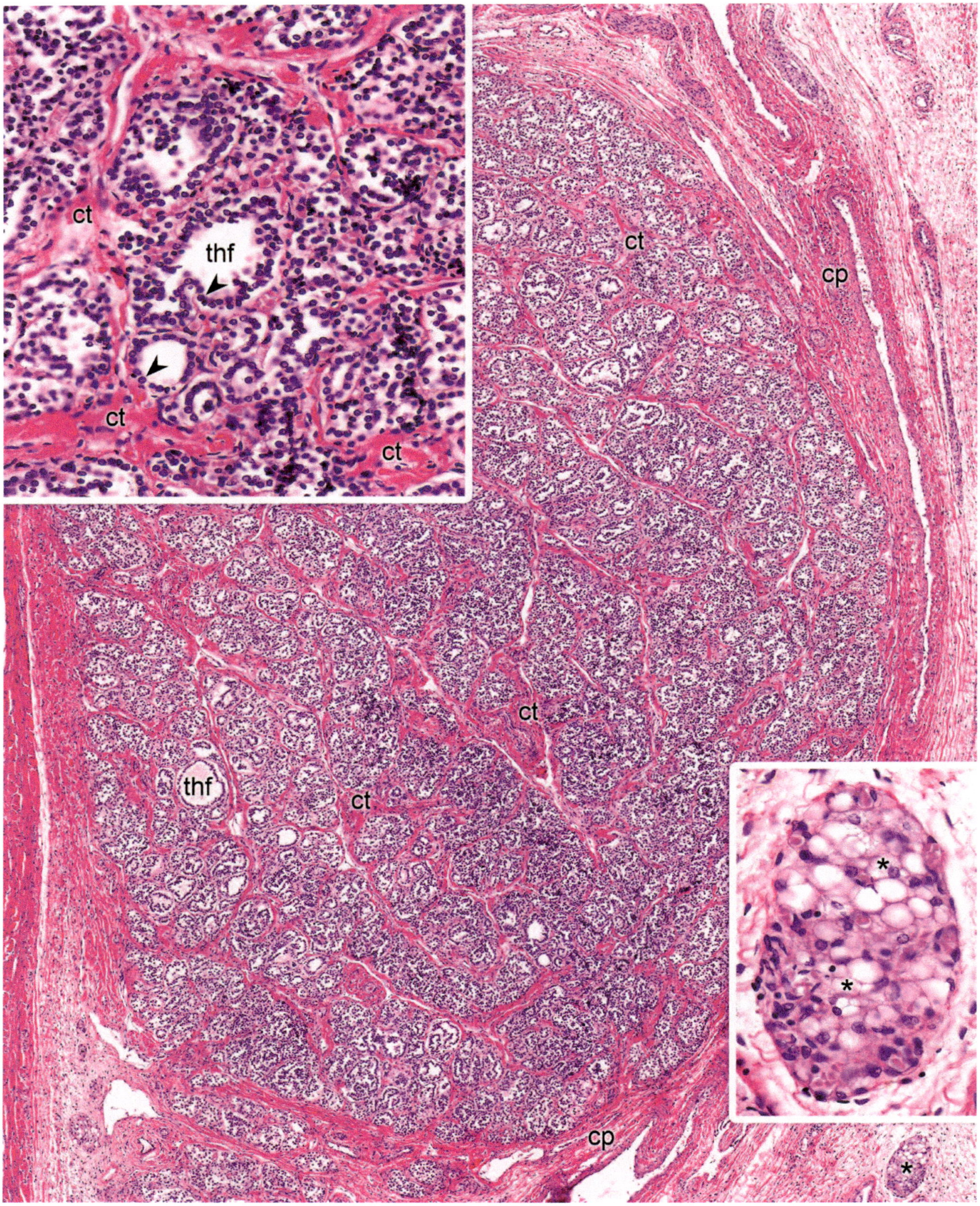

Figure 290. Thyroid gland, newborn. In newborn, thyroid gland contains spherical follicles (thf; upper inset) that are small and generally lack colloid. The follicles are surrounded by a single layer of cuboidal, follicular (principal) cells (arrowheads; upper inset). The gland is covered by a dense connective tissue capsule (cp) forming connective tissue septa (ct) that penetrate the parenchyma and subdivide it into lobules. The connective tissue stroma around the thyroid often contains brown adipose tissue (asterisks; lower inset).

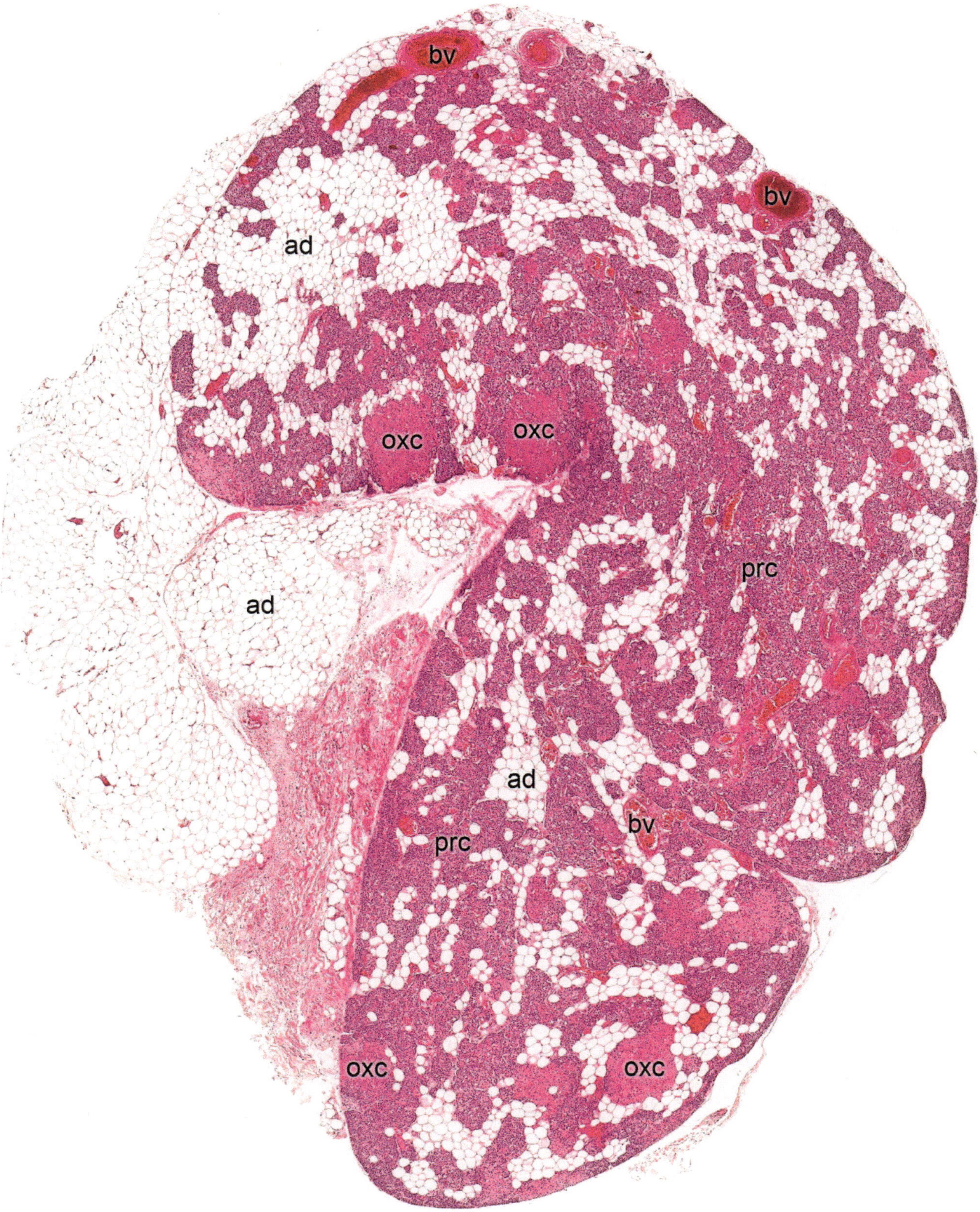

Figure 291. Parathyroid gland. Parathyroid glands are spherical, pea-sized structures surrounded by adipose tissue (ad) that penetrates the parenchyma of the gland. Most of the parathyroid gland is composed of basophilic principal (chief) cells (prc) producing parathyroid hormone (PTH) that activates bone resorption and elevates blood calcium and phosphate levels. Clusters of eosinophilic oxyphil cells (oxc) can also be observed; their exact function has not been elucidated yet. Blood vessels (bv) are abundant in the parathyroid gland.

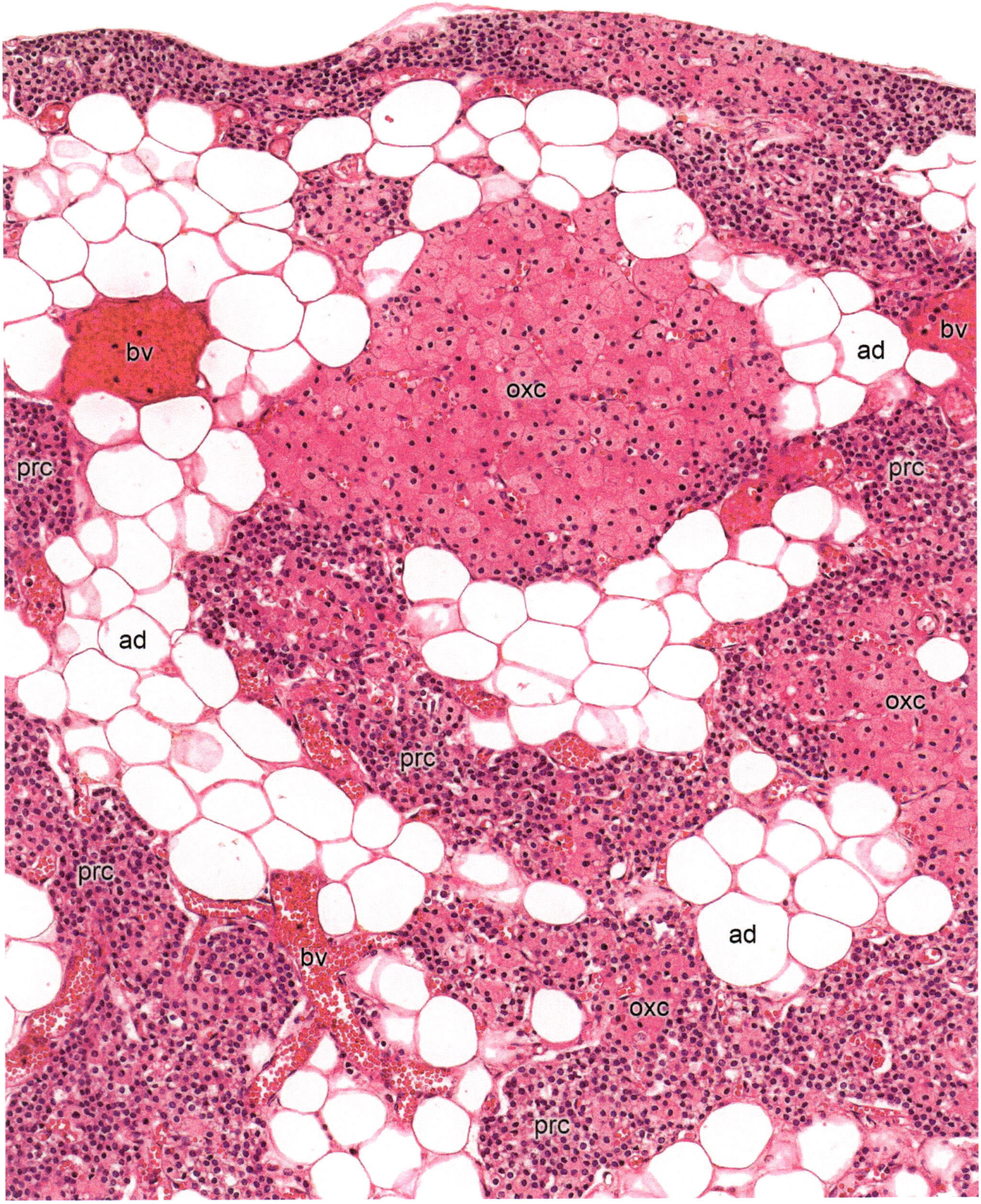

Figure 292. Parathyroid gland. Parathyroid glands are spherical, pea-sized structures surrounded by adipose tissue (ad) that penetrates the parenchyma of the gland. Most of the parathyroid gland is composed of basophilic principal (chief) cells (prc) producing parathyroid hormone (PTH) that activates bone resorption and elevates blood calcium and phosphate levels. Clusters of eosinophilic oxyphil cells (oxc) can also be observed; their exact function has not been elucidated yet. Blood vessels (bv) are abundant in the parathyroid gland.

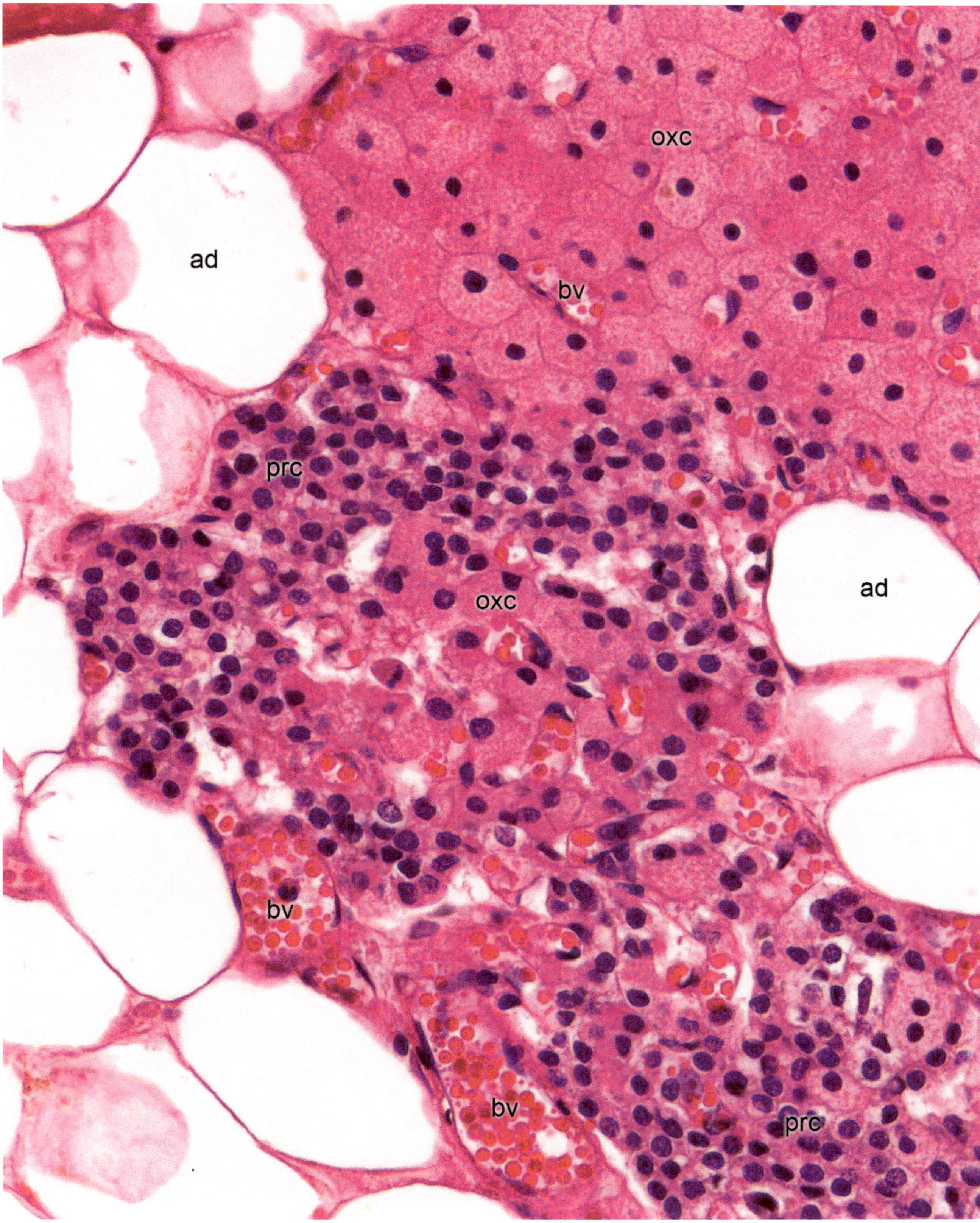

Figure 293. Parathyroid gland. Parathyroid glands are spherical, pea-sized structures surrounded by adipose tissue (ad) that penetrates the parenchyma of the gland. Most of the parathyroid gland is composed of basophilic principal (chief) cells (prc) producing parathyroid hormone (PTH) that activates bone resorption and elevates blood calcium and phosphate levels. Clusters of eosinophilic oxyphil cells (oxc) can also be observed; their exact function has not been elucidated yet. Blood vessels (bv) are abundant in the parathyroid gland.

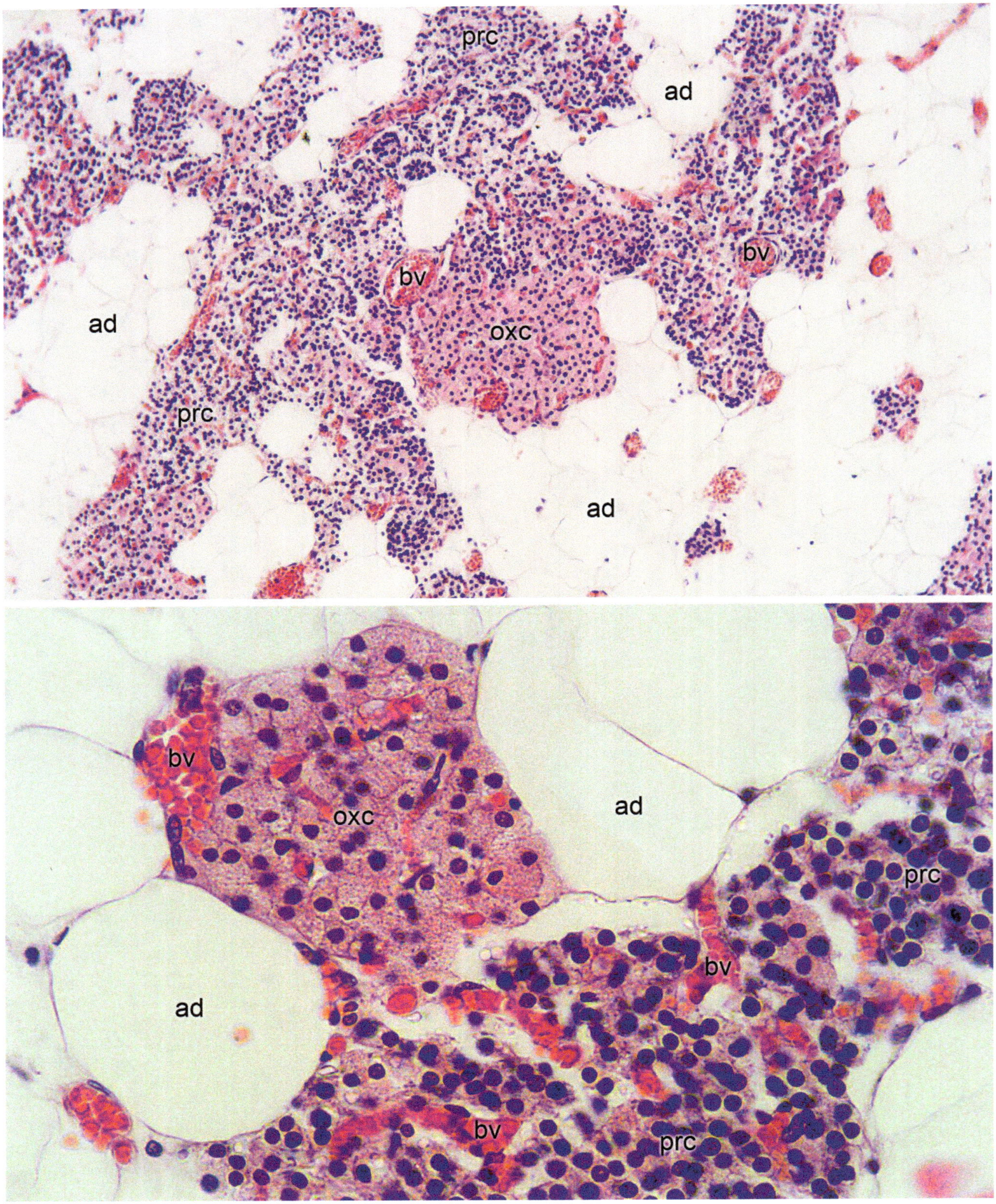

Figure 294. Parathyroid gland. Low magnification (top) and high magnification (bottom) of the parathyroid parenchyma. Parathyroid glands are spherical, pea-sized structures surrounded by adipose tissue (ad) that penetrates the gland. Most of the parathyroid gland is composed of basophilic principal (chief) cells (prc) producing parathyroid hormone (PTH) that activates bone resorption and elevates blood calcium and phosphate levels. Clusters of eosinophilic oxyphil cells (oxc) can also be observed; their exact function has not been elucidated yet. Blood vessels (bv) are abundant in the parathyroid gland.

Adrenal (suprarenal) gland

Adrenal gland is composed of the cortex that originates from mesodermal mesenchyme, and the medulla that develops from the neural crest. Under the connective tissue capsule, cortex forms three distinct layers. The outermost layer, the zona glomerulosa is composed of polygonal cells forming globular clusters that secrete primarily mineralocorticoids. Under the zona glomerulosa, the zona fasciculata is formed by straight cords of polygonal cells that produce glucocorticoids. Finally, the innermost layer of the cortex, the zona reticularis consists of cells that form a network of cords; these cells secrete androgens and glucocorticoids. Cortical cells are surrounded by rich capillary network. Adrenal medulla contains large cells with pale staining, forming spherical clusters with capillaries between; noradrenaline and adrenaline is secreted by different populations of cells in the medulla. The central adrenomedullary vein with unusually thick longitudinal smooth muscle bundles (smm) in the tunica media is also located in the medulla.

Pineal gland

Pineal gland is a neuroendocrine gland that plays a key role in regulating circadian rhythm. The thin layer of capsule that is identical to the pia mater, sends trabeculae into the parenchyma that surround globular clusters of pinealocytes and interstitial (glial) cells. Pinealocytes are neuroendocrine cells with lightly stained spherical or oval nuclei, and they constitute approximately 95% of the pineal cell population, while the less common interstitial glial cells are similar to astrocytes and possess darker, elongated nuclei. Melatonin is released by the pinealocytes at night and it is associated with the regulation of circadian rhythm; it also inhibits gonadal functions. Large calcified concretions (corpora aranacea, brain sand) can be often observed in the pineal gland in elder age.

Endocrine pancreas

The endocrine pancreas is represented by the islets of Langerhans that are well defined globular clusters of polygonal cells with pale staining and rounded, central nucleus; these clusters are embedded into the exocrine pancreas that is mostly composed of serous acini and their duct system. Capillaries are abundant around the alpha, beta, delta, epsilon and gamma (PP) cells that secrete primarily glucagon, insulin, somatostatin, ghrelin and pancreatic polypeptide, respectively; these cells are morphologically indistinguishable with routine H&E staining.

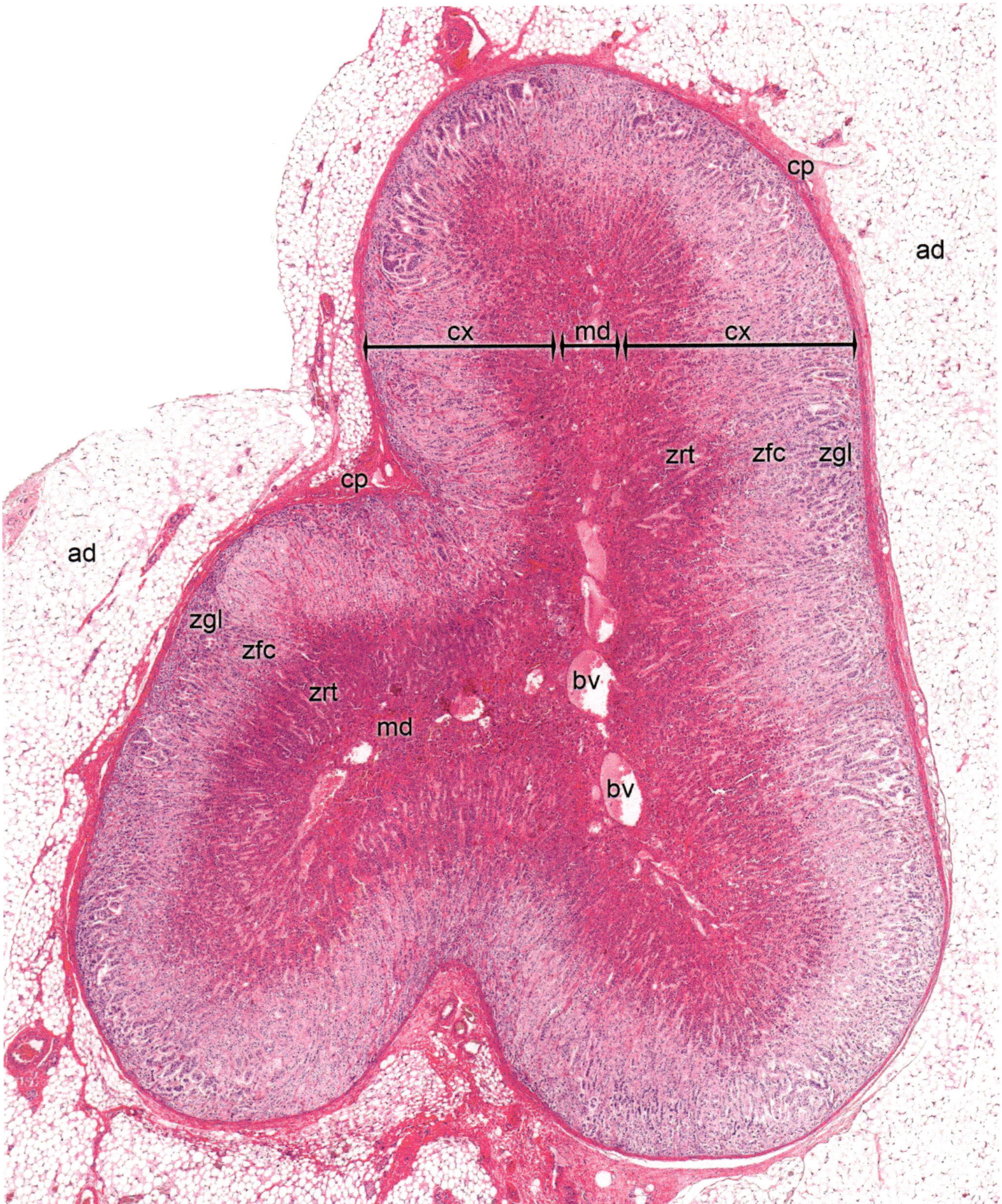

Figure 295. Adrenal gland. Under the connective tissue capsule (cp) that is covered by the perirenal adipose tissue (ad), the adrenal cortex (cx) is composed of three layers. The zona glomerulosa (zgl) is formed by polygonal cells accumulating into globular clusters that secrete primarily mineralocorticoids. The underlying zona fasciculata (zfc) is composed of cords of polygonal cells that produce glucocorticoids, and the innermost zona reticularis (zrt) consists of cells that form a network of cords; these cells secrete androgens and glucocorticoids. Adrenal medulla (md) contains blood vessels (bv) and large pale cells secreting noradrenaline and adrenaline.

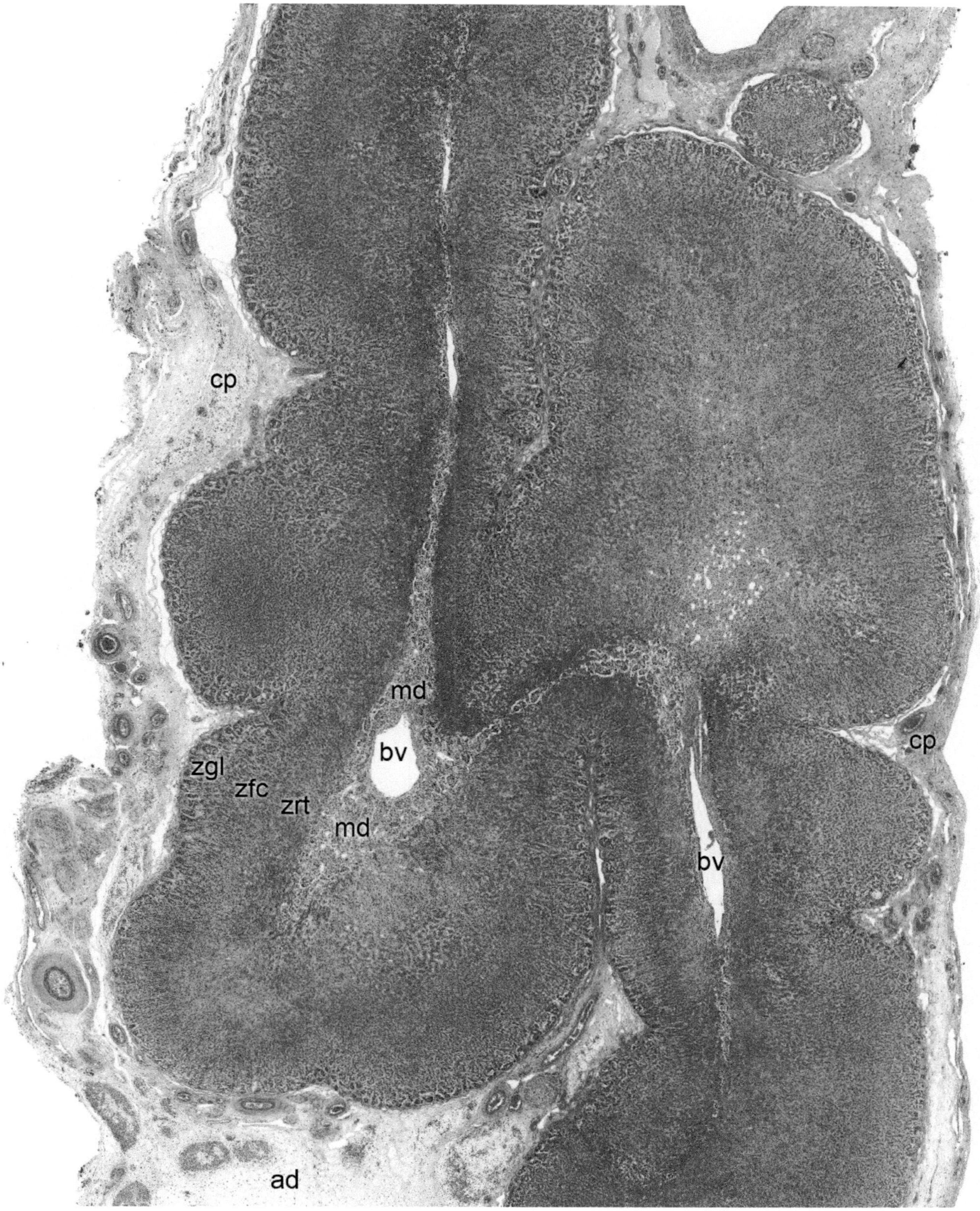

Figure 296. Adrenal gland. Under the connective tissue capsule (cp) that is surrounded by the perirenal adipose tissue (ad), the adrenal cortex is composed of three layers. The zona glomerulosa (zgl) is formed by polygonal cells accumulating into globular clusters that secrete primarily mineralocorticoids. The underlying zona fasciculata (zfc) is composed of cords of polygonal cells that produce glucocorticoids, and the innermost zona reticularis (zrt) consists of cells that form a network of cords; these cells secrete androgens and glucocorticoids. Adrenal medulla (md) contains blood vessels (bv) and large pale cells secreting noradrenaline and adrenaline.

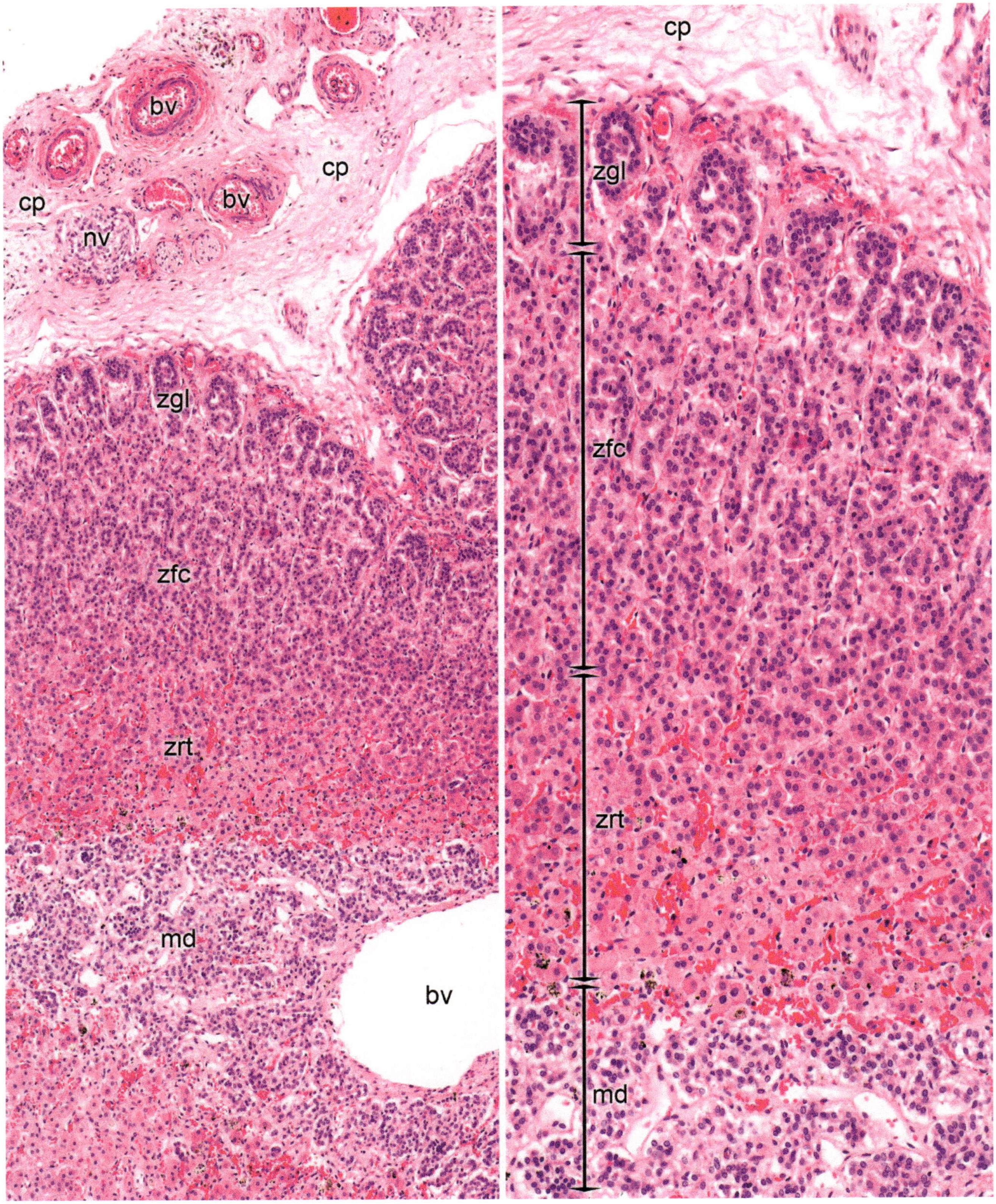

Figure 297. Adrenal gland. Under the connective tissue capsule (cp), carrying blood vessels (bv) and nerves (nv), the adrenal cortex is composed of three layers. The zona glomerulosa (zgl) is formed by polygonal cells of globular clusters and secretes primarily mineralocorticoids. The underlying zona fasciculata (zfc) is composed of cords of polygonal cells that produce glucocorticoids, and the innermost zona reticularis (zrt) consists of cells that form a network of cords; these cells secrete androgens and glucocorticoids. Adrenal medulla (md) contains blood vessels (bv) and large pale cells secreting noradrenaline and adrenaline.

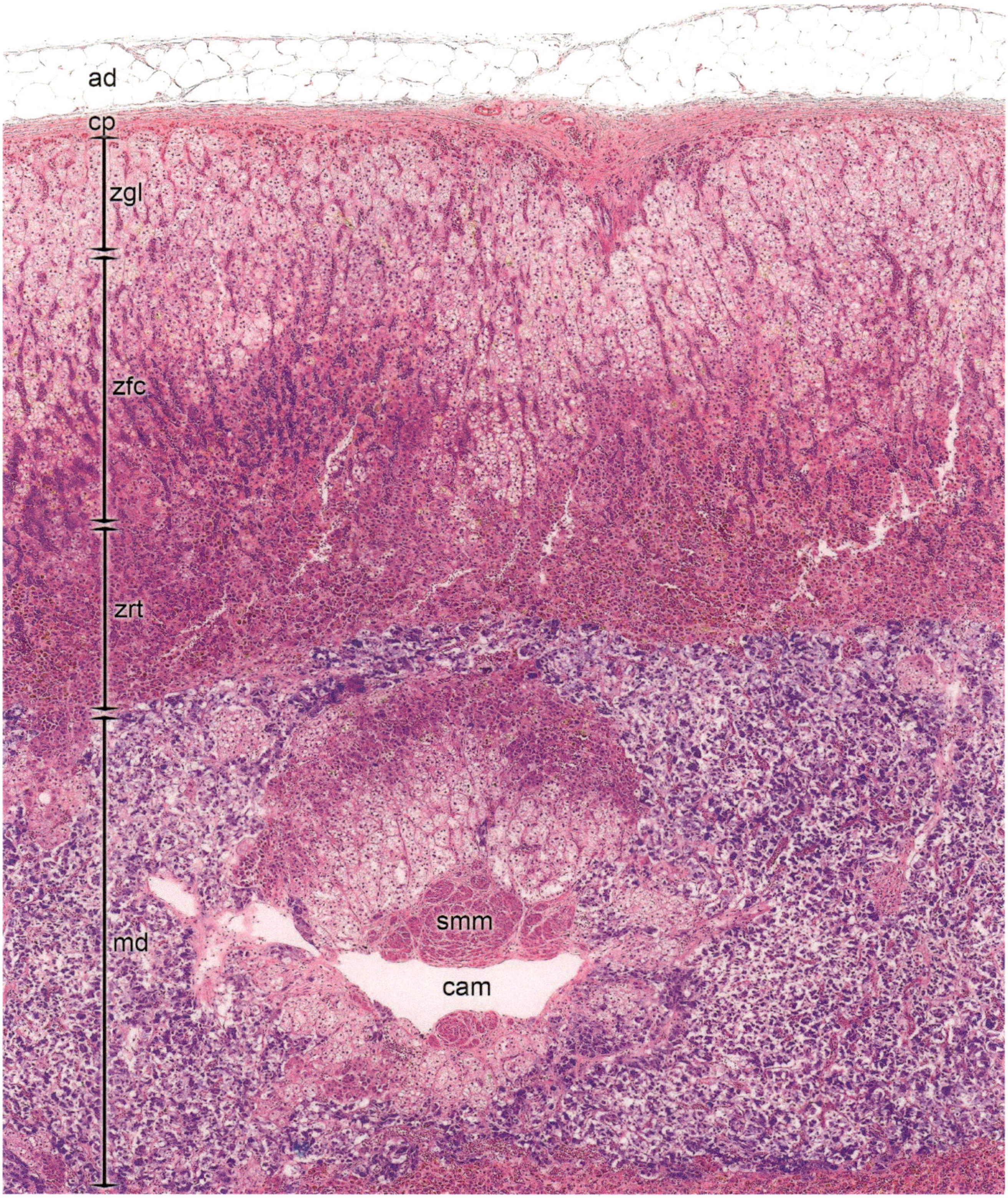

Figure 298. Adrenal gland. Under the connective tissue capsule (cp), surrounded by the perirenal adipose tissue (ad), the adrenal cortex is composed of three layers. The zona glomerulosa (zgl) is formed by polygonal cells accumulating into globular clusters that secrete primarily mineralocorticoids. The underlying zona fasciculata (zfc) is composed of cords of polygonal cells that produce glucocorticoids, and the innermost zona reticularis (zrt) consists of cells that form a network of cords; these cells secrete androgens and glucocorticoids. Adrenal medulla (md) contains cells secreting noradrenaline and adrenaline. The central adrenomedullary vein (cam) with unusually thick longitudinal smooth muscle bundles (smm) in the tunica media is also located in the medulla.

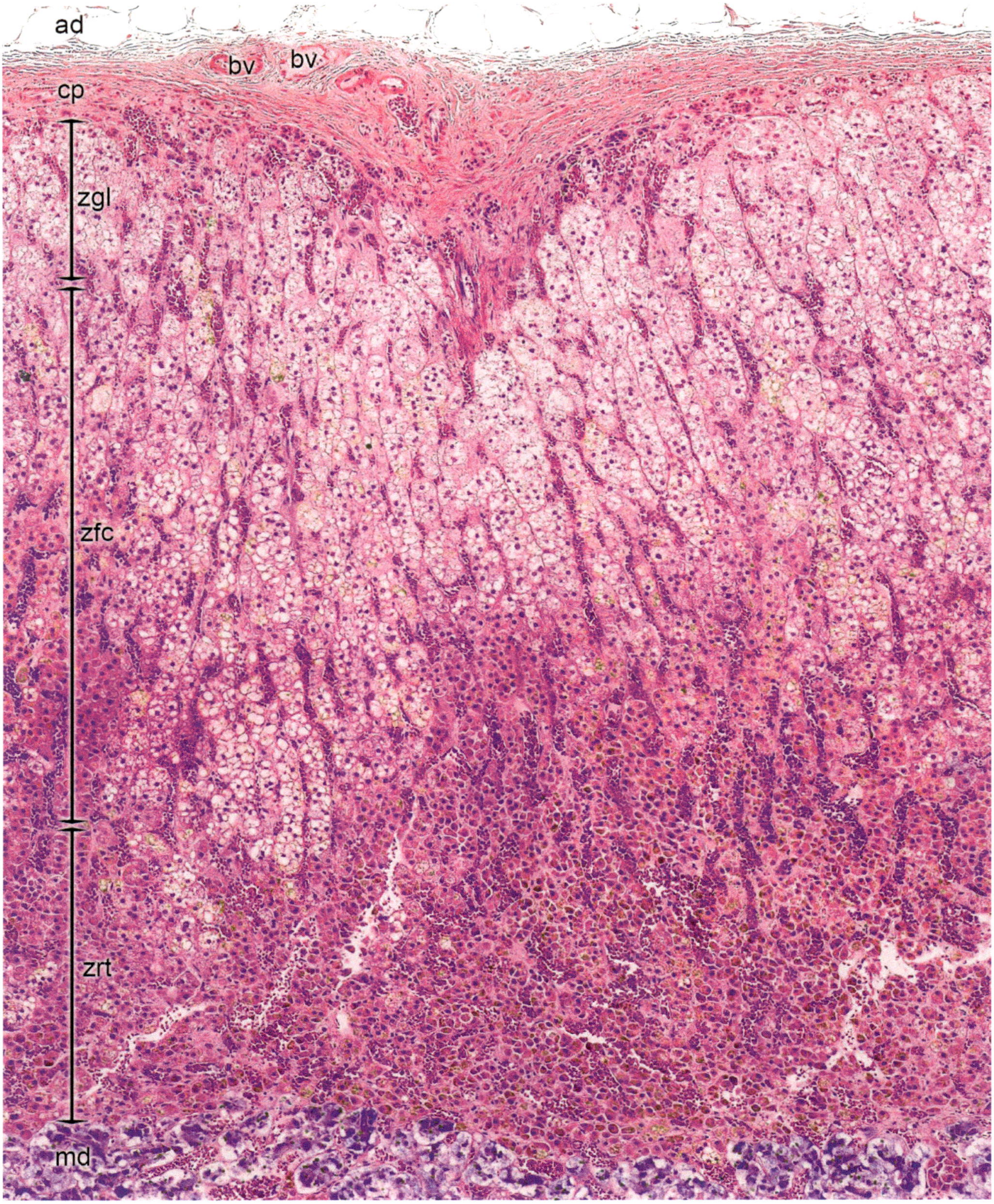

Figure 299. Adrenal gland. Under the connective tissue capsule (cp), carrying blood vessels (bv) and surrounded by the perirenal adipose tissue (ad), the adrenal cortex is composed of three layers. The zona glomerulosa (zgl) is formed by polygonal cells of globular clusters and secretes primarily mineralocorticoids. The underlying zona fasciculata (zfc) is composed of cords of polygonal cells that produce glucocorticoids, and the innermost zona reticularis (zrt) consists of cells that form a network of cords; these cells secrete androgens and glucocorticoids. Adrenal medulla (md) contains cells secreting noradrenaline and adrenaline.

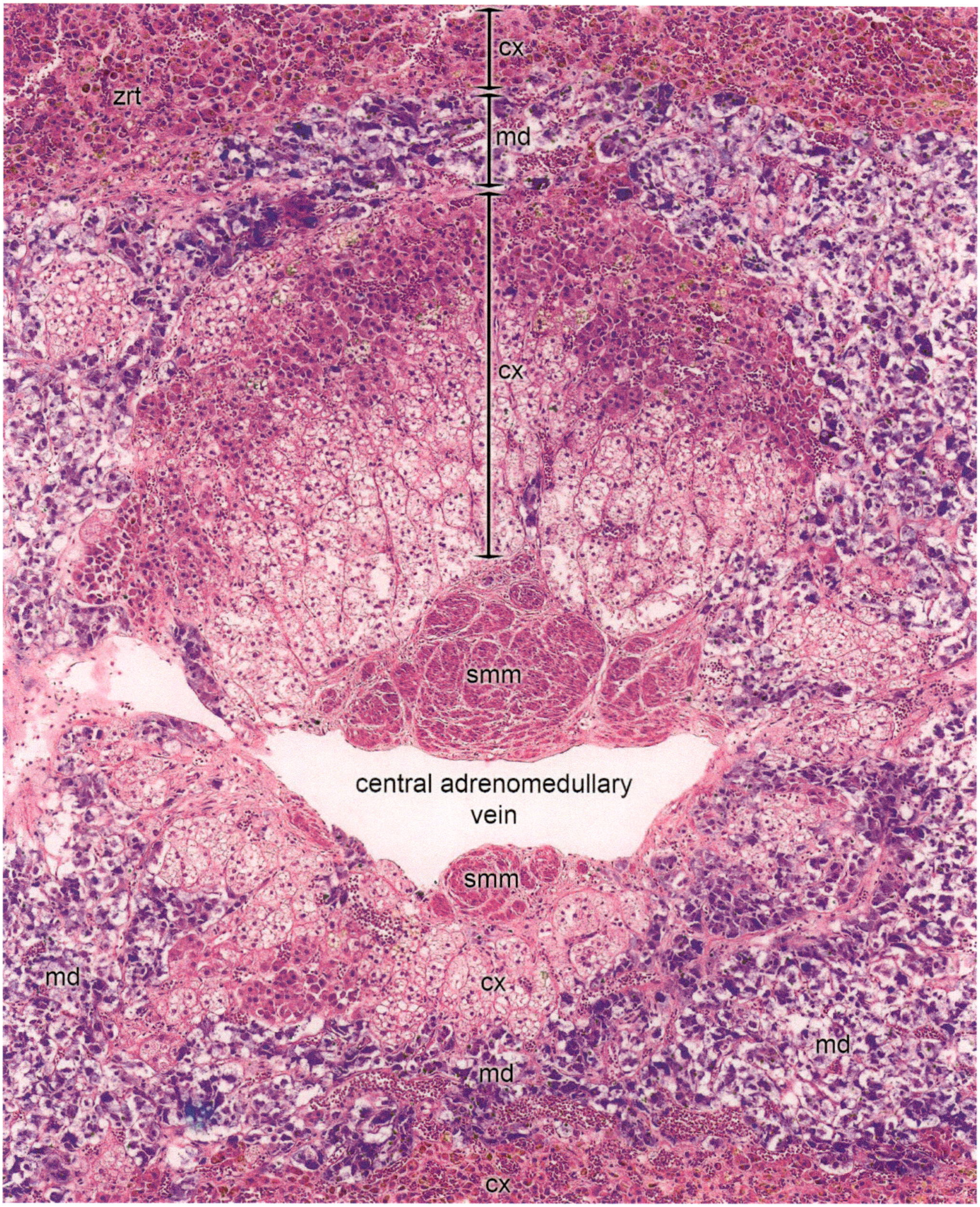

Figure 300. Adrenal gland, adrenomedullary vein. Adrenal medulla (md) is surrounded by the adrenal cortex (cx) and it contains large cells secreting noradrenaline and adrenaline. The central adrenomedullary vein, with unusually thick longitudinal smooth muscle bundles (smm) in the tunica media, is located in the medulla, adjacent to the cortical border.

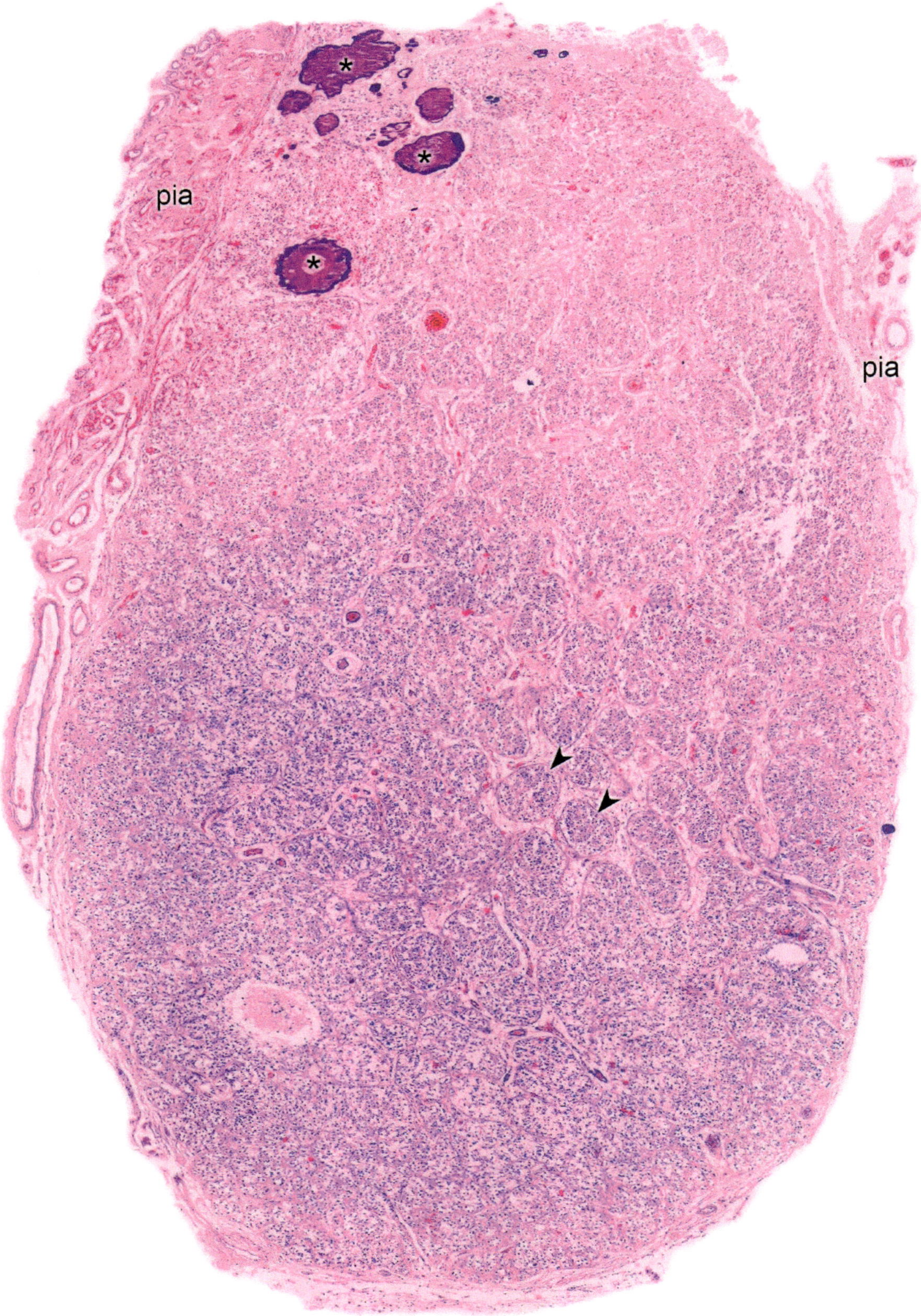

Figure 301. Pineal gland. The pineal gland is surrounded by pia mater (pia) that sends septa composed of fibroblasts and connective tissue fibers into the gland. Between the septa, the parenchyma forms globular clusters or cords of neuroendocrine pinealocytes (arrowheads). These cells have lightly stained spherical or oval nuclei and they are accompanied by few interstitial glial cells that resemble astrocytes and possess darker, elongated nuclei. Large calcified concretions (corpora aranacea, brain sand) can be observed in the pineal gland in elder age (asterisks).

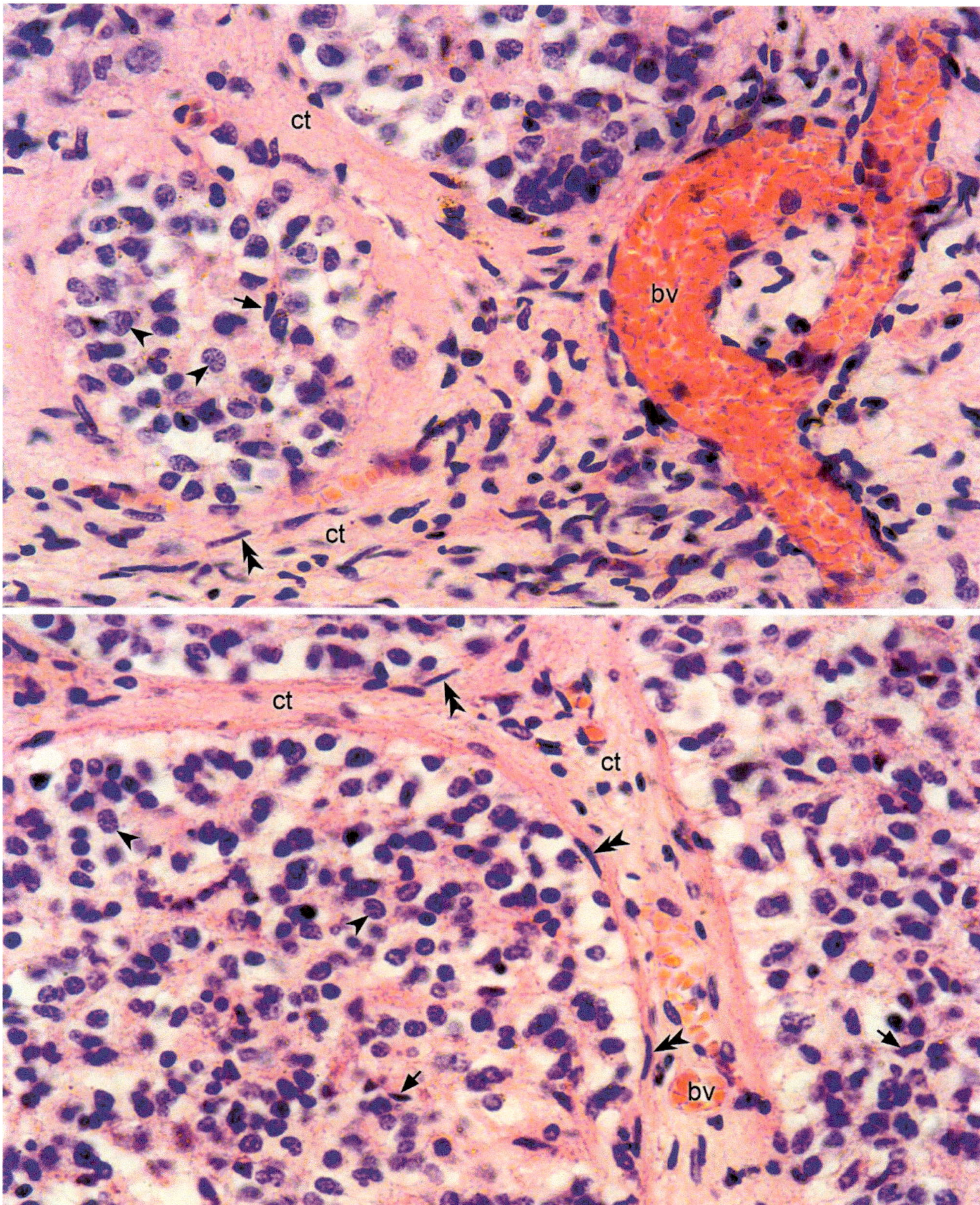

Figure 302. Pineal gland. Connective tissue septa (ct), containing fibroblasts (double arrowheads), connective tissue fibers and blood vessels (bv) divide the parenchyma into globular clusters or cords. These clusters are primarily composed of neuroendocrine pinealocytes with lightly stained spherical or oval nuclei (arrowheads). Interstitial glial cells with darker, elongated nuclei (arrows) are less common; these cells resemble astrocytes.

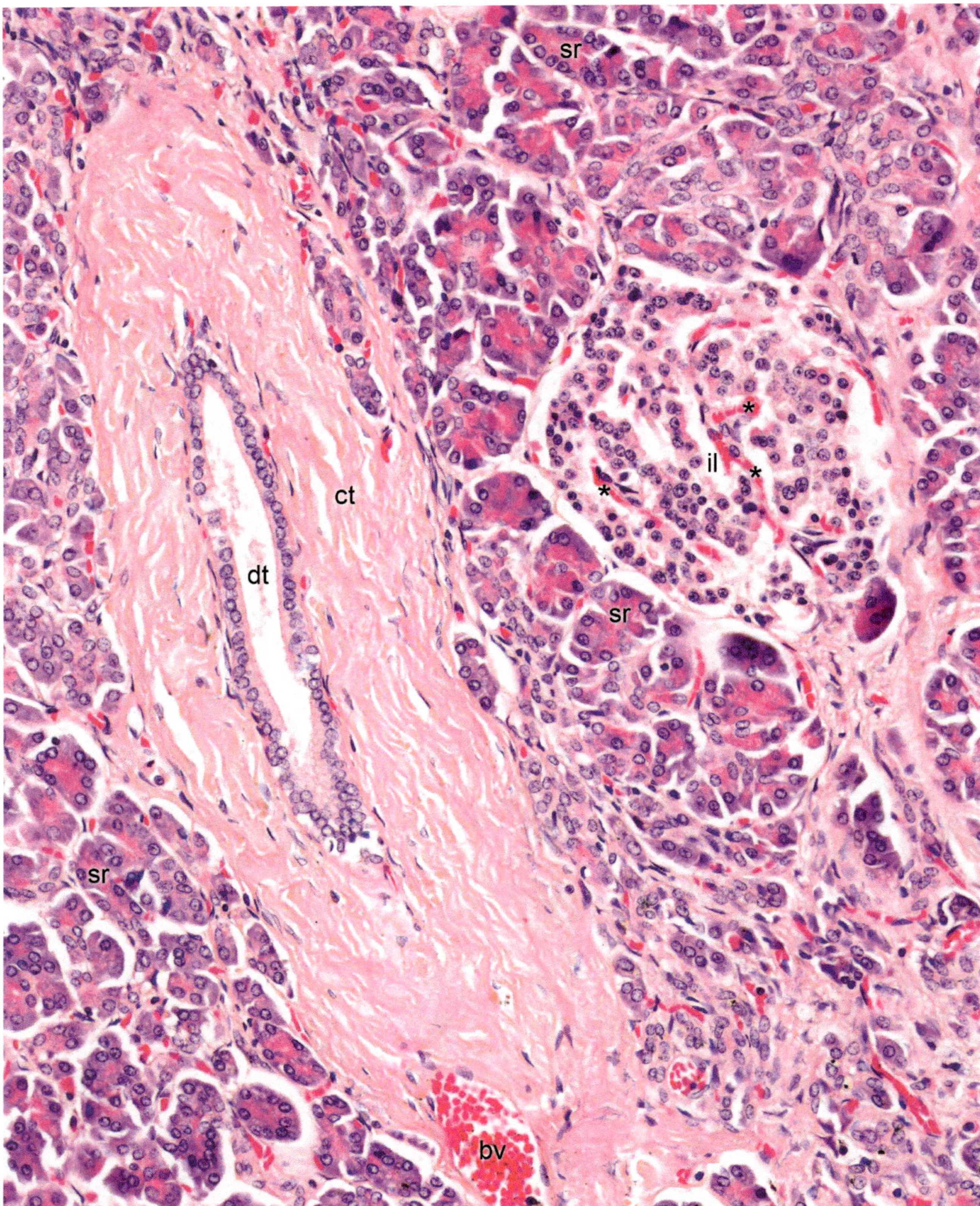

Figure 303. Pancreas, islets of Langerhans. Endocrine pancreas is represented by islets of Langerhans (il) that are globular clusters of polygonal cells. These clusters are embedded into the exocrine pancreas that is a tubuloalveolar gland containing serous acini (sr) and connective tissue trabeculae (ct) with ducts (dt) and blood vessels (bv). Islets of Langerhans contain capillaries (asterisks) surrounded by cells that produce primarily insulin, glucagon and somatostatin.

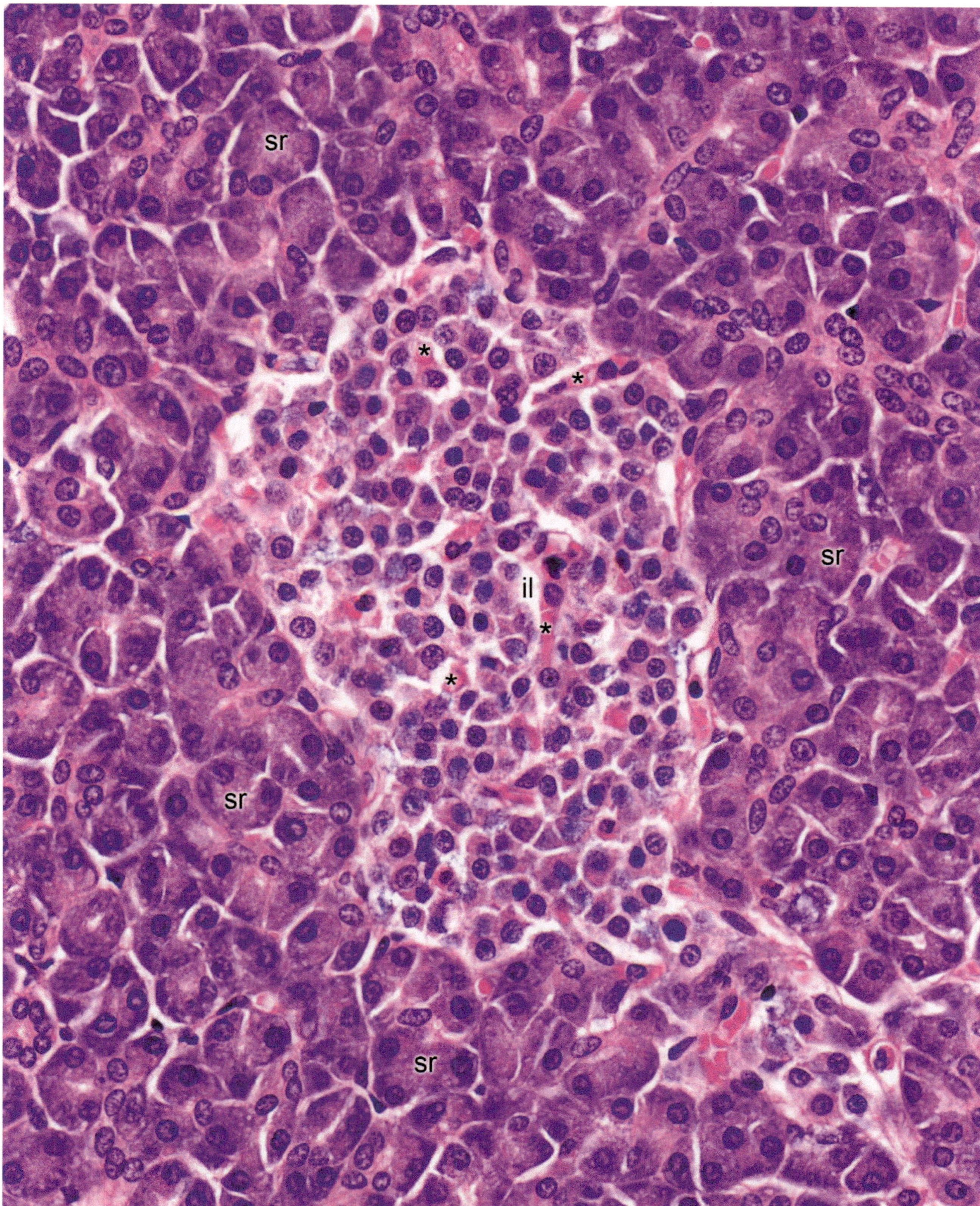

Figure 304. Pancreas, islets of Langerhans. Endocrine pancreas is represented by islets of Langerhans (il) that are globular clusters of polygonal cells. These clusters are embedded into the exocrine pancreas that is a tubuloalveolar gland containing serous acini (sr) and their ducts. Islets of Langerhans contain capillaries (asterisks) surrounded by cells that produce primarily insulin, glucagon and somatostatin.

Chapter 13

Nervous System

Nervous tissue

Nervous tissue forms the parenchyma of the nervous system, and it is composed of neurons and glial cells. The major glial cell types of the central nervous system (CNS) are the astrocytes (astroglia), oligodendrocytes (oligodendroglia) microglia and ependymal cells, while peripheral nervous system (PNS) contains Schwann cells and satellite cells. Detailed description of the cells populating the nervous system is given in Chapter I (basic tissues). It is not the primary goal of histology to elaborate on neuroanatomical details, including the complex circuitry of the central nervous system; therefore, only the basic microanatomy of the cerebrum, cerebellum, and the spinal cord (CNS), along with the histology of ganglia and peripheral nerves (PNS) are described below.

Peripheral nerve

Peripheral somatic nerves are composed of axons of neurons whose cell bodies are located in the central or peripheral nervous systems. These axons are surrounded by myelin sheath formed by numerous layers of the cell membrane duplications of Schwann cells that also envelope the myelin sheath itself. This Schwann sheath or neurilemma contains the cytoplasm and the elongated nuclei of Schwann cells. Indeed, the vast majority of nuclei that populate the parenchyma of the peripheral nerve are nuclei of Schwann cells.

The myelin sheath is formed by multiple sections that represent the myelinization process carried out by a single Schwann cell; the gaps between the sections are covered with Schwann sheath and contain an axonal swelling, the node of Ranvier. Circularly arranged cytoplasmatic remains between the myelin sheath layers form the Schmidt-Lanterman clefts as well as the perinodal cytoplasmatic remnants at the Ranvier nodes.

Peripheral nerves are surrounded by several succeeding layers of connective tissue envelopes. Endoneurium is formed around individual Schwann cells surrounding the myelin and Schwann sheaths, and it is composed of thin, type III collagen fibers that are probably secreted by the Schwann cells. Occasional fibroblasts can also be observed among the myelinated neurites. Fascicles of nerve fibers are enveloped by the perineurium formed by fibroblasts and collagen fibers along with flattened contractile cells. The outermost envelope, the epineurium encloses the entire nerve into a dense connectivetissue sheath that often extends into the nerve between the fascicles and is continuous with the dura mater and the underlying arachnoid. Epineurium contains vessels (vasa nervorum) and nerves (nervi nervorum); the latter providing the intrinsic innervation of the nerve sheaths.

Ganglia

Ganglia represent accumulation of perikarya in the peripheral nervous system. Sensory ganglia contain pseudounipolar neurons while autonomic ganglia are formed by multipolar perikarya of nerve cells. Ganglion cells are large neurons with pale nucleus, prominent nucleolus and well developed rER that forms basophilic clusters in the cytoplasm (Nissl substance). Ganglia are surrounded by dense connective tissue capsule derived from the dura mater that transitions into the epineurium of the spinal nerves. Perikarya of the ganglion cells are surrounded by satellite cells that are small glial cells whose primary role is providing insulation and support to the ganglionic perikarya. Ganglia also contain bundles of myelinated and unmyelinated axons.

Intramural ganglia are formed by autonomic ganglion cells located in the wall of hollow organs, particularly in the gastrointestinal tract. These cells are located between the smooth muscle layers of the tunica muscularis externa, where they form the myenteric (Auerbach's) plexus while the Meissner's plexus is composed of submucosal ganglion cells and nerve fibers.

http://doi.org/10.1016/B978-0-323-91891-6.50013-1,

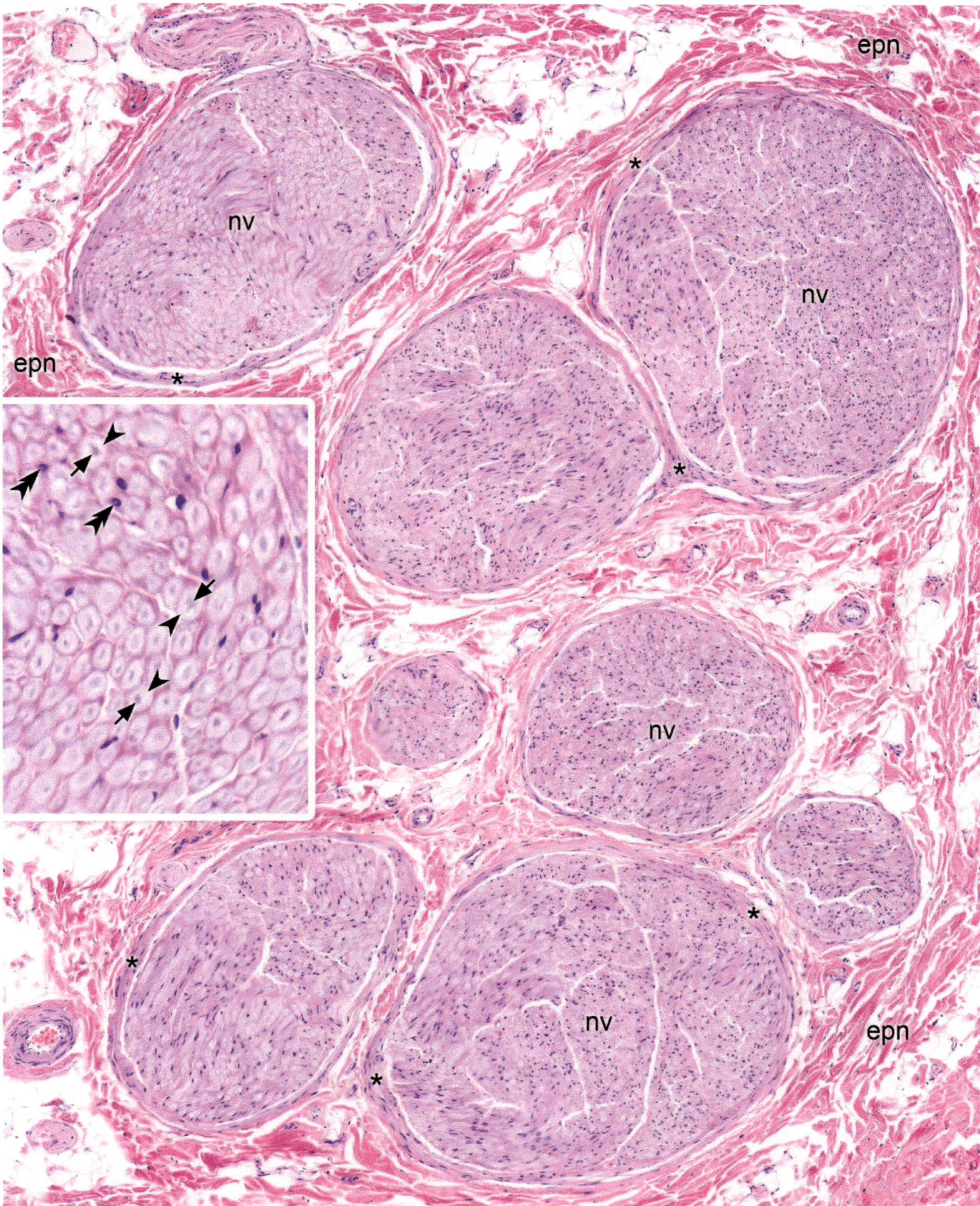

Figure 305. Peripheral nerve, cross section. Peripheral nerves are surrounded by connective tissue envelopes. Epineurium (epn) envelopes the entire nerve, perineurium (asterisks) surrounds nerve bundles (nv), while endoneurium is located around individual axons. Myelin sheath (arrowheads, inset) is formed around the axons (arrows, inset) by Schwann cells that envelope the myelin sheath as Schwann sheath (neurilemma) containing the cytoplasm and the elongated nuclei of Schwann cells (double arrowheads, inset).

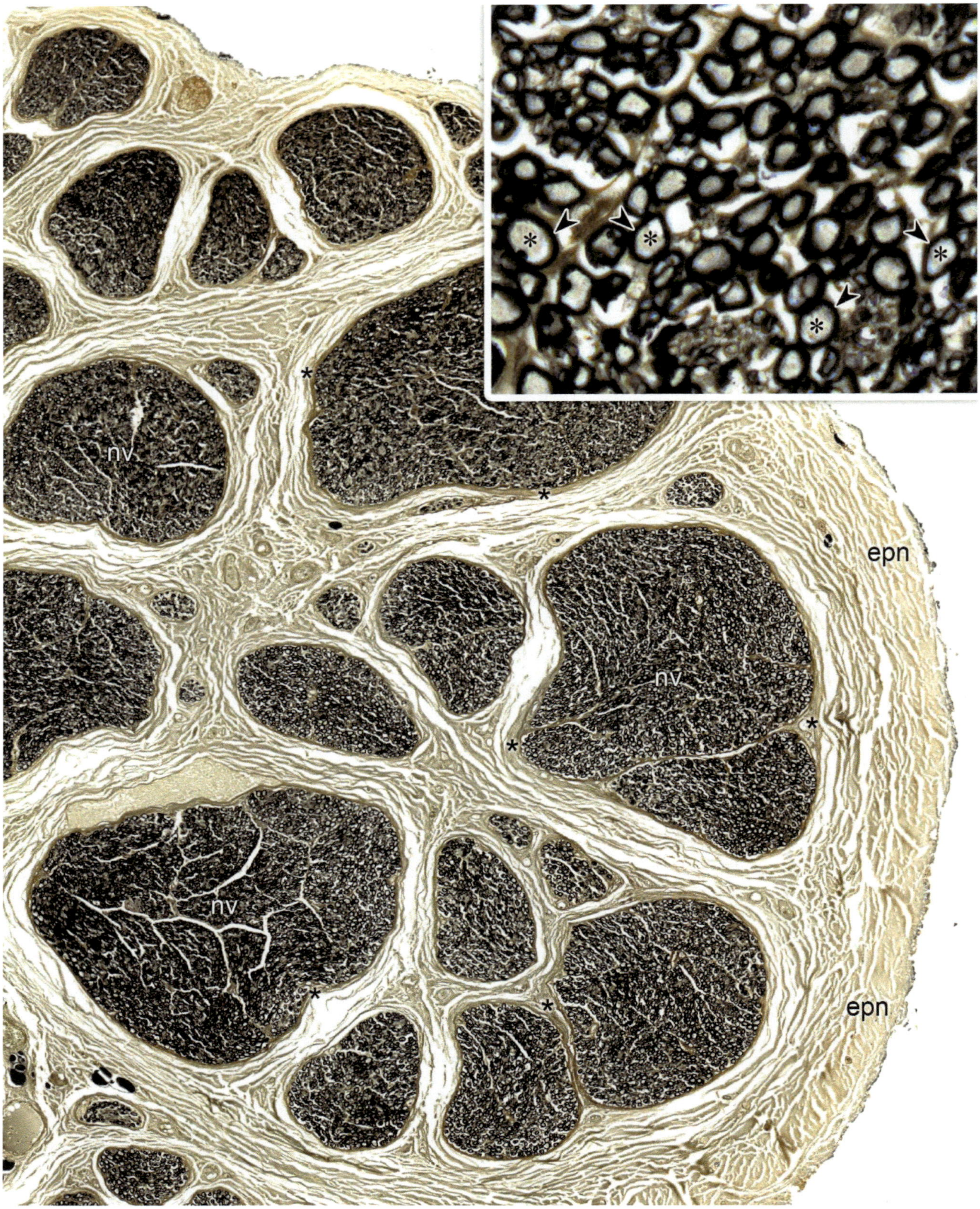

Figure 306. Peripheral nerve, cross section, myelin staining. Peripheral nerves are surrounded by connective tissue envelopes. Epineurium (epn) envelopes the entire nerve, perineurium (five-pointed asterisks) surrounds nerve bundles (nv), while endoneurium is located around individual axons. Myelin sheath (arrowheads, inset) is formed around the axons (six-pointed asterisks) by Schwann cells that envelope the myelin sheath as Schwann sheath (neurilemma) containing the cytoplasm and the nuclei the of Schwann cells.

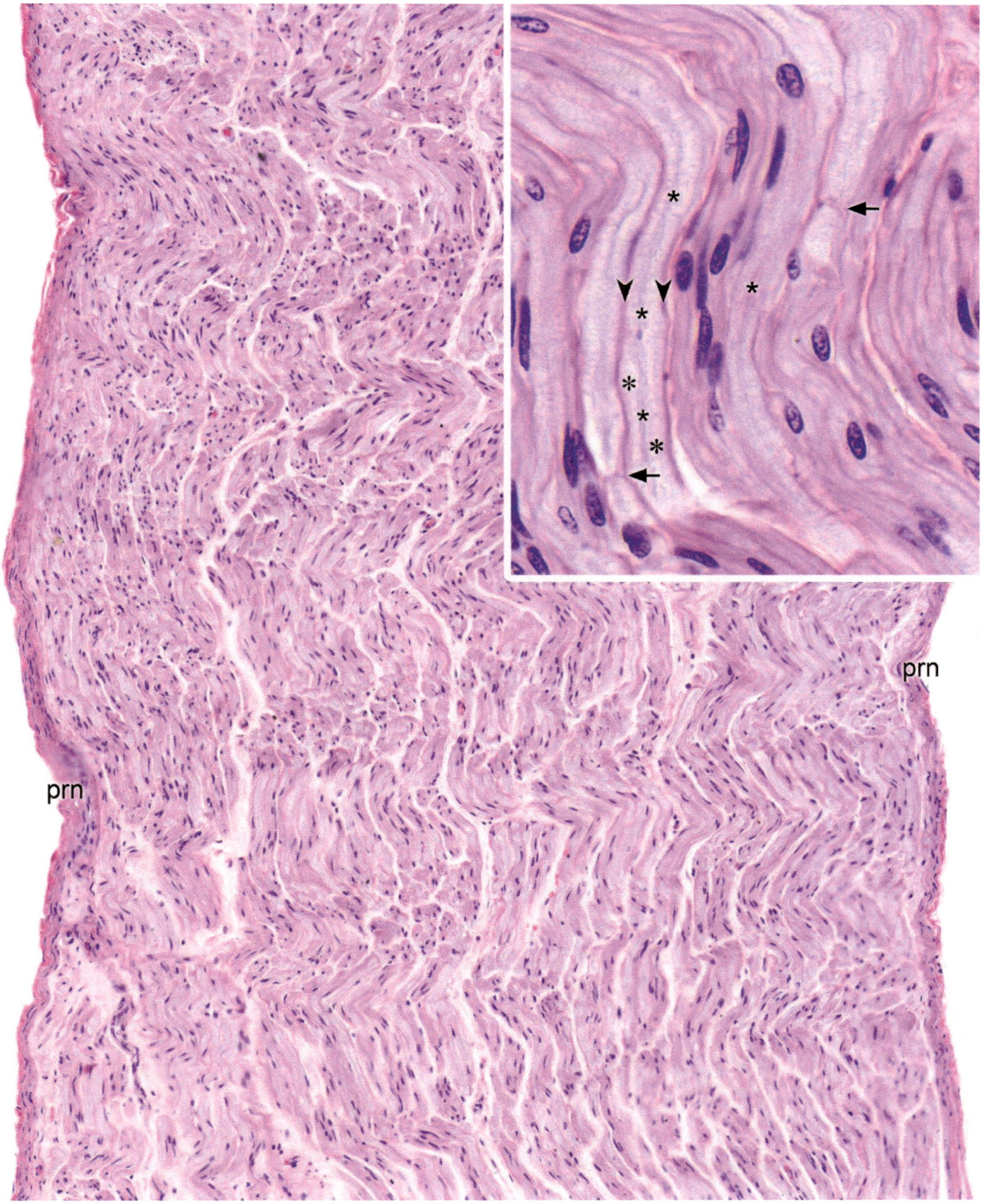

Figure 307. Peripheral nerve, longitudinal section. Perineurium (prn) surrounds bundles of nerve fibers while endoneurium is located around individual axons. Myelin sheath (six-pointed asterisks, inset) is formed around the axons (five-pointed asterisks, inset) by Schwann cells that envelope the myelin sheath as Schwann sheath (neurilemma) containing the cytoplasm (arrowheads) and the elongated nuclei of Schwann cells. Myelin sheath is formed by multiple sections carried out by a single Schwann cell; the gaps between the sections of myelin sheath is covered with neurilemma and contains an axonal swelling, the node of Ranvier (arrows).

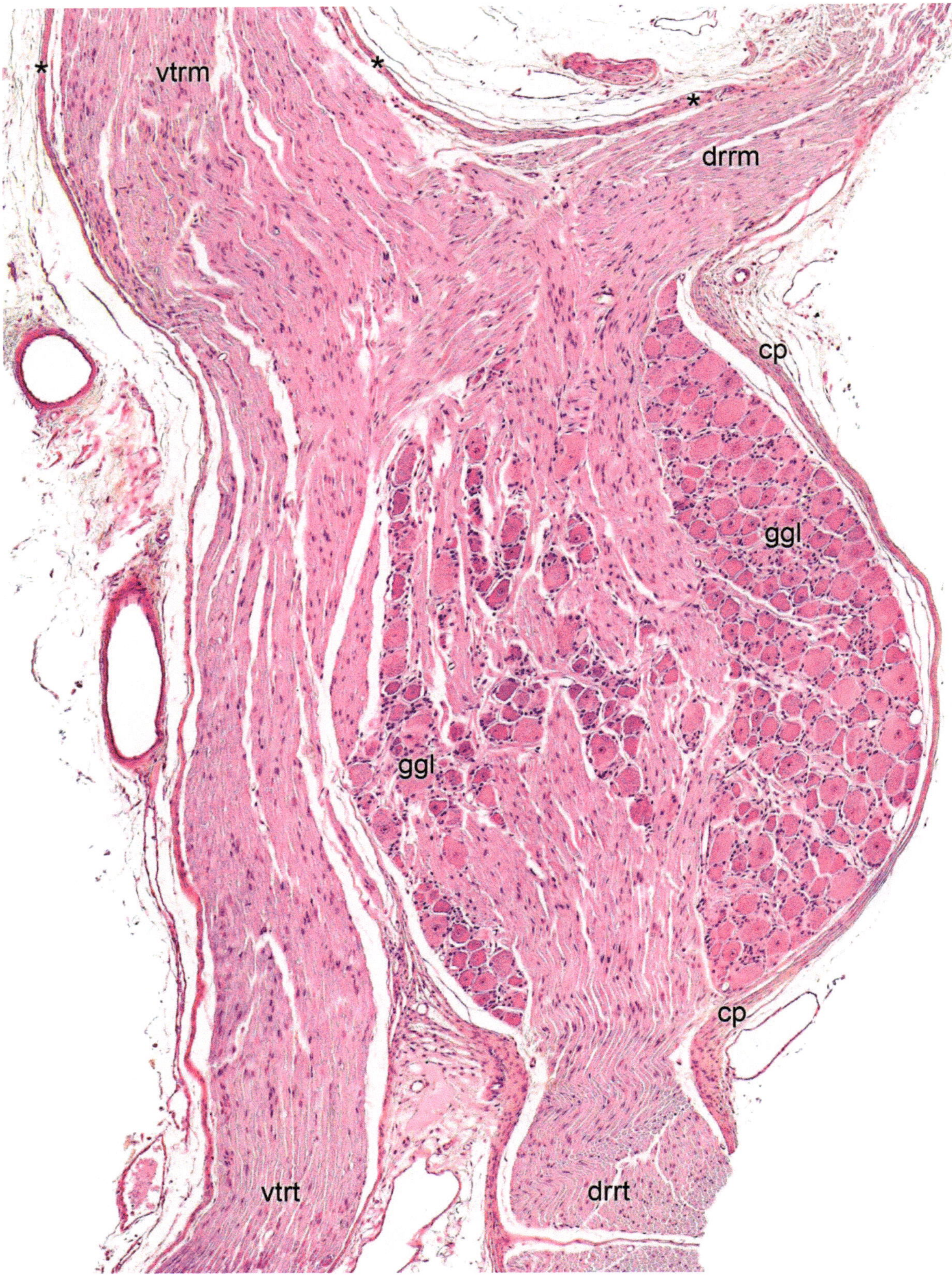

Figure 308. Spinal ganglion. Sensory ganglia are composed of pseudounipolar neurons (ggl) with pale nucleus, prominent nucleolus and well developed rER that is represented by the Nissl substance. Spinal ganglia are surrounded by dense connective tissue capsule (cp) derived from the dura mater; the capsule transitions into the epineurium (asterisks) of the ventral ramus (vtrm) and the dorsal ramus (drrm) of the spinal nerves. The central axons of the sensory ganglion cells form the dorsal root (drrt) that joins the ventral root (vtrt) forming the spinal nerve. Ganglionic perikarya are surrounded by satellite cells. Sensory ganglia contain bundles of myelinated axons.

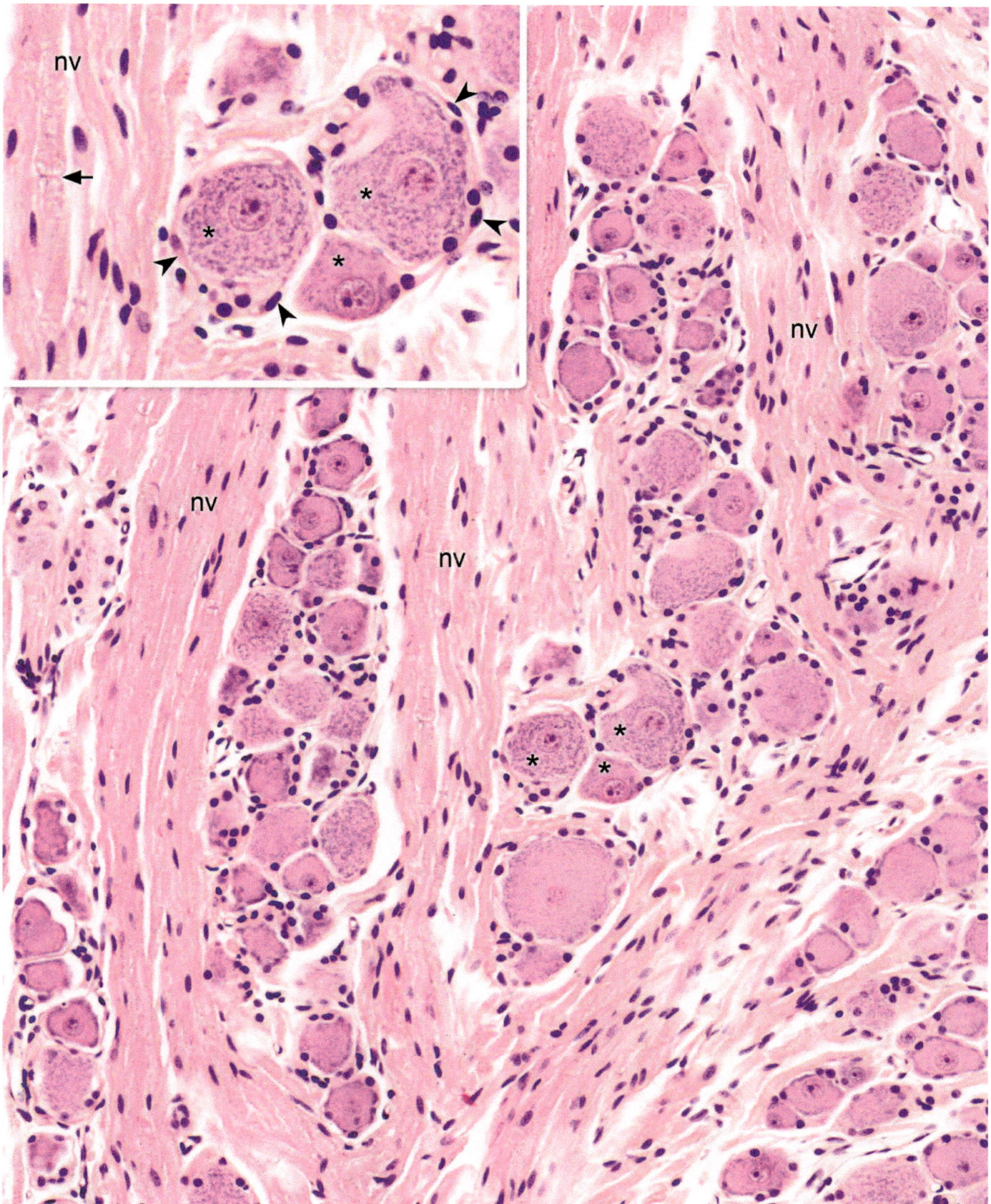

Figure 309. Spinal ganglion. Sensory ganglia are composed of pseudounipolar neurons (asterisks) with pale nucleus, prominent nucleolus, and well developed rER that forms basophilic clusters in the cytoplasm called Nissl substance (inset). Ganglionic perikarya are surrounded by satellite cells (arrowheads, inset). Sensory ganglia contain bundles of myelinated axons (nv) with nodes of Ranvier that lack myelin sheath (arrow).

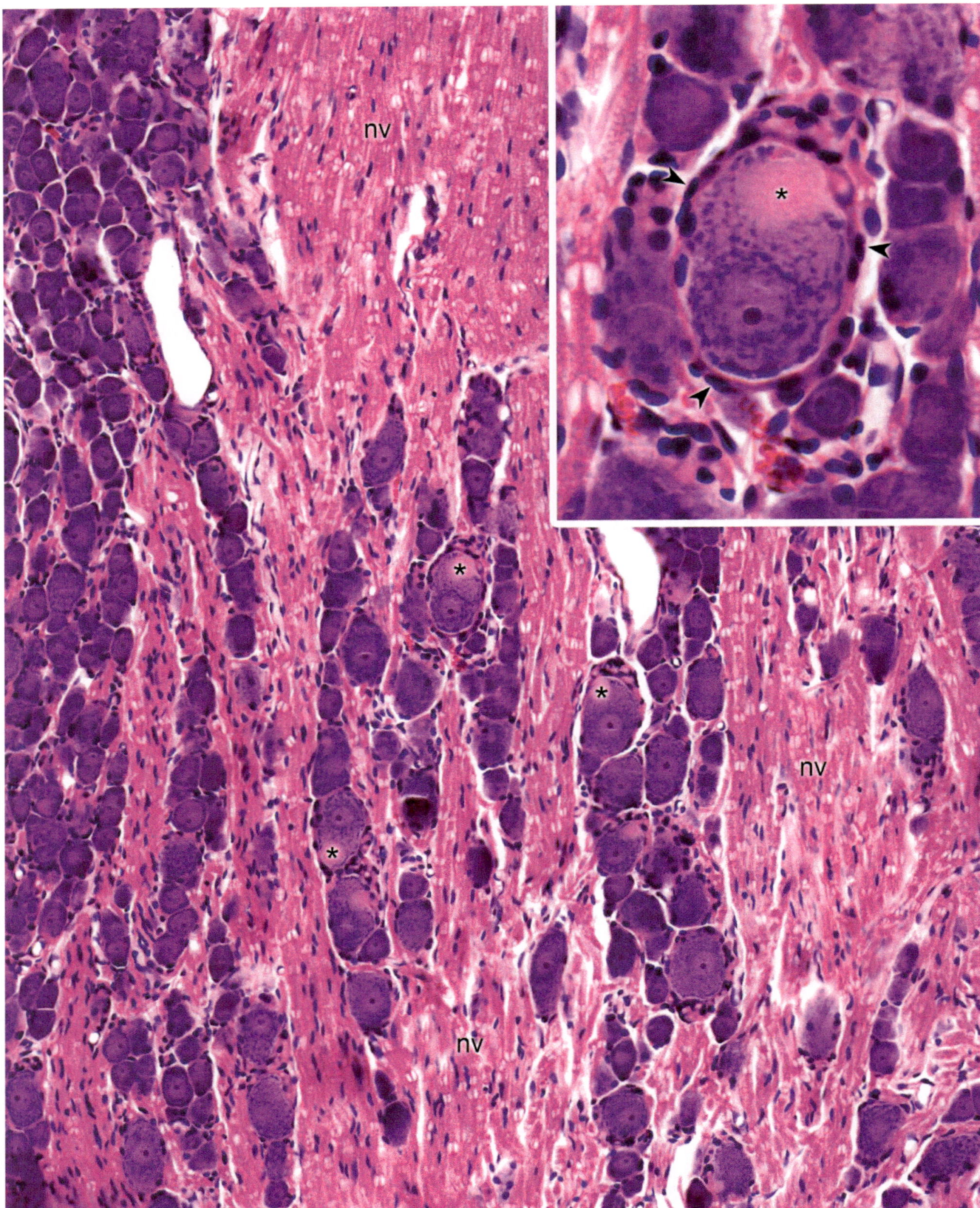

Figure 310. Spinal ganglion. Sensory ganglia are composed of pseudounipolar neurons (asterisks) with pale nucleus, prominent nucleolus and well developed rER that is represented by Nissl substance (inset). Ganglionic perikarya are surrounded by satellite cells (arrowheads, inset). Sensory ganglia contain bundles of myelinated axons (nv).

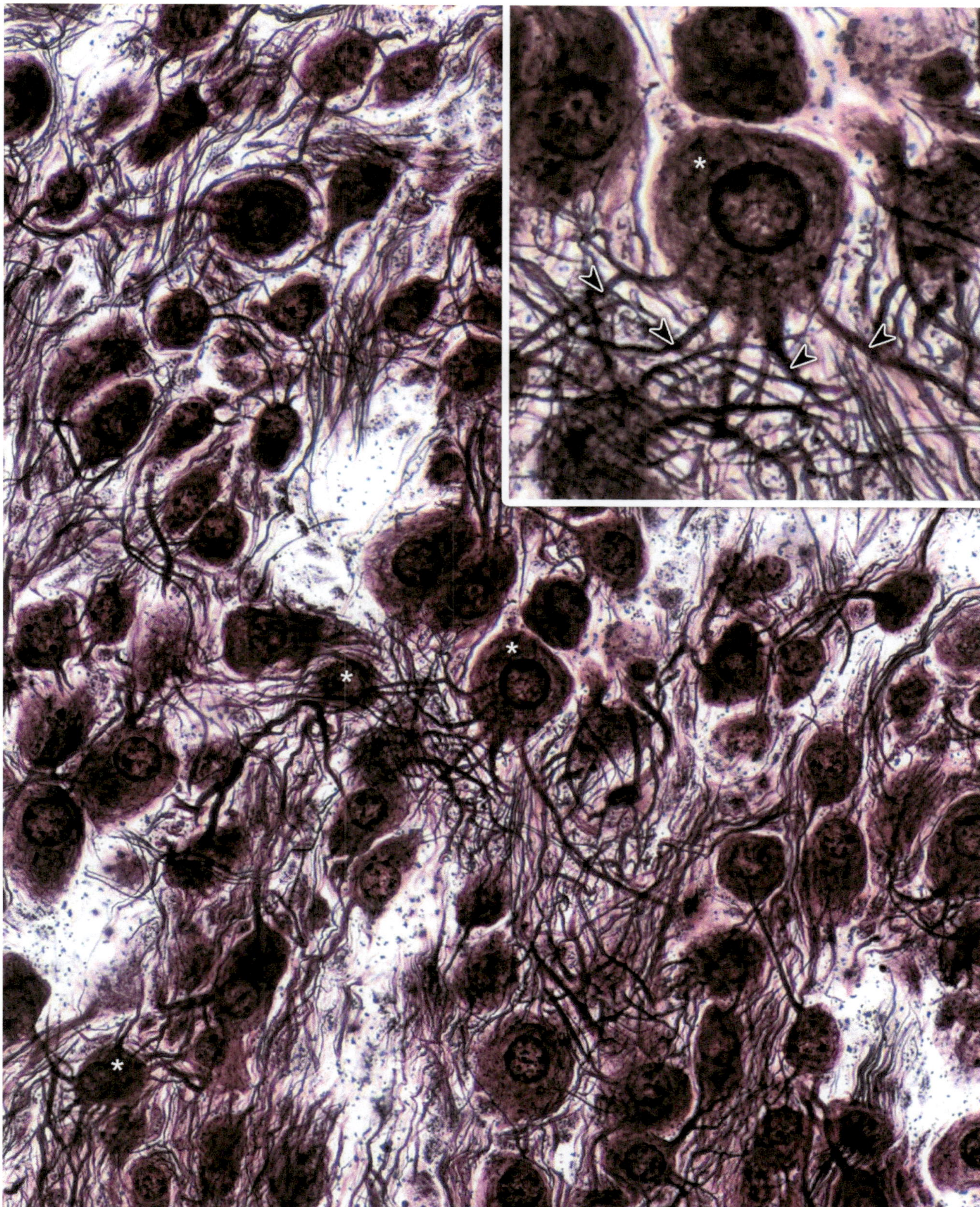

Figure 311. Autonomic ganglion, silver impregnation. Autonomic ganglia are composed of multipolar neurons (asterisks) with large, spherical nucleus (inset), prominent nucleolus, well developed rER and multiple processes (arrowheads, inset). Ganglionic perikarya are surrounded by satellite cells that are hard to identify with silver impregnation.

Meninges

Central nervous system is surrounded by the meninges that are connective tissue envelopes. The outermost meninx is the dura mater (pachymeninx, "thick" meninx) that is composed of dense irregular connective tissue. Dura mater receives rich arterial supply and forms venous sinuses that are duplications of the dura lined with endothelium. The internal surface of the dura is loosely attached to the arachnoid, primarily held in position by the pressure of cerebrospinal fluid that fills the subarachnoid space. Arachnoid sends fine spiderweb-like trabeculae to the innermost meninx, the pia mater, hence the name; these trabeculae and the arachnoid itself is composed of avascular tissue rich in collagen and elastic fibers. Pia mater is a thin connective tissue layer that closely follows the surface of the brain and the spinal cord, and it is continuous with the perivascular connective tissue as the vessels penetrate the white matter. Arachnoid and pia mater are considered together as a functional and morphological unit, the leptomeninx ("thin" meninx).

Spinal Cord

Spinal cord is part of the CNS, and therefore, it is composed of an outer, white matter shell and an inner, butterfly-shaped grey matter core. The central canal, lined with ependymal cells and filled with cerebrospinal fluid is located at the center of the grey matter. White matter is composed of neuroglia as well as myelinated and unmyelinated axons that form ascending and descending tracts in the lateral funiculus (between the ventral and dorsal horns), as well as in the posterior (between the dorsal horns) and in the anterior funiculi (between the ventral horns). The butterfly-shaped grey matter core contains the neuronal perikarya and it is organized into ventral and dorsal horns that have a laminated structure; these laminae (of Rexed) correspond to the nuclei of the CNS and represent groups of neuronal cell bodies. Ventral horns contain large multipolar motor neurons with pale granular nucleus and prominent nucleolus. Cytoplasm contains large amount of basophilic, clumped rER (Nissl substance) that is missing from the area of the cell body that serves for the origination of the myelinated axon (axon hillock). The dorsal horn and the central area of the grey matter is composed of interneurons and afferent sensory axons whose pseudounipolar perikarya are located in the sensory spinal ganglia.

The surface of the spinal cord is covered with the delicate pia mater that is composed of thin connective tissue. In contrast, dura mater and the underlying arachnoid loosely surround the spinal cord as well as the ventral and dorsal roots and rootlets forming the subarachnoid space between the outer meninges and the pia mater. Dura mater and the arachnoid follows the initial segment of the spinal nerves where they transition into the epineurium.

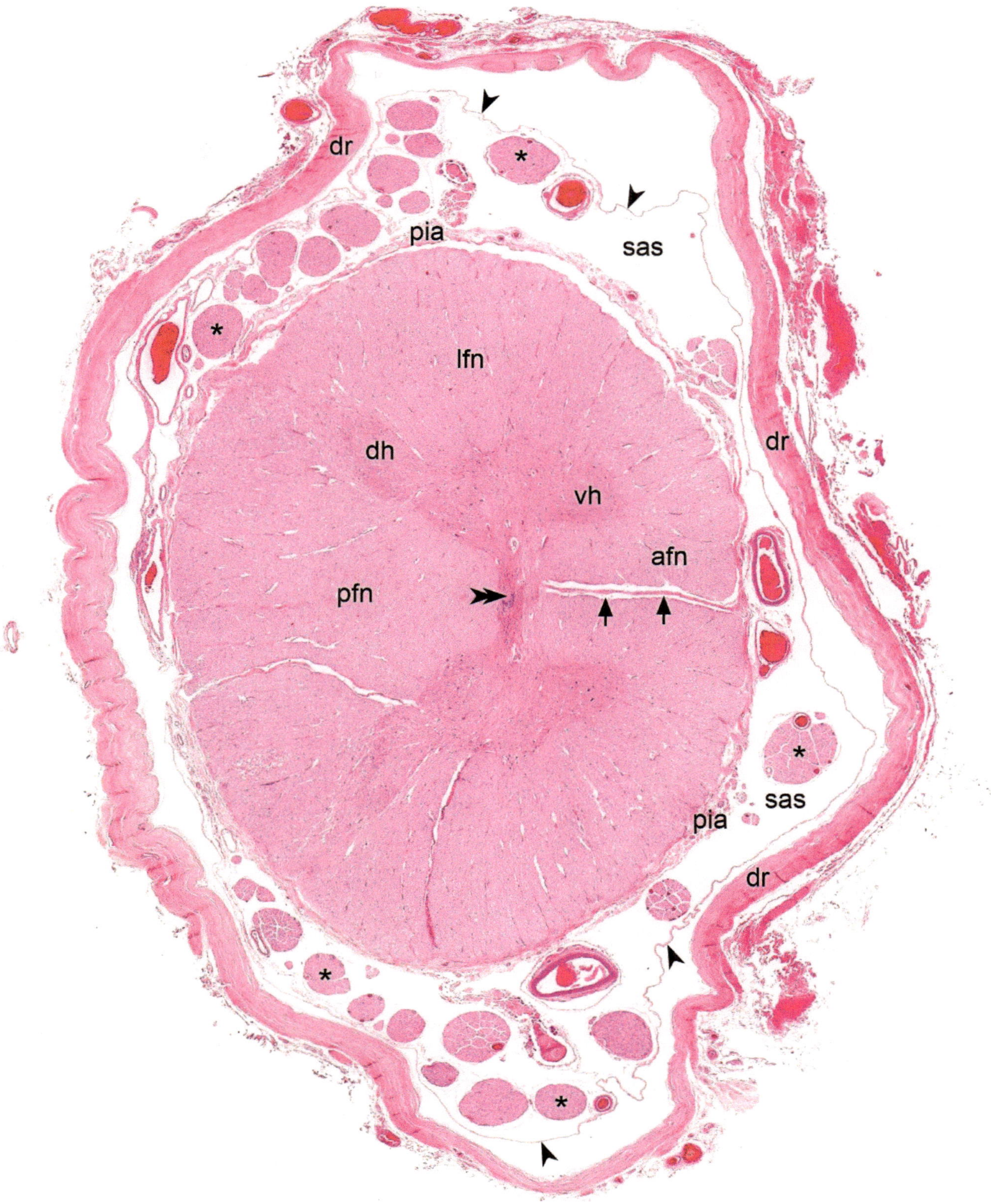

Figure 312. Spinal cord. White matter is composed of the lateral (lfn), posterior (pfn) and anterior funiculi (afn); the latter is split by the anterior fissure (arrows). The butterfly-shaped grey matter core contains the neuronal perikarya and it is organized into ventral (vh) and dorsal horns (dh) as well as the central area with the central canal lined with ependymal cells (double arrowheads). The surface of the spinal cord is covered with a pia mater (pia). In contrast, dura mater (dr) and the underlying arachnoid (arrowheads) loosely surround the spinal cord as well as the roots and rootlets (asterisks) forming the subarachnoid space (sas) between the outer meninges and the pia mater.

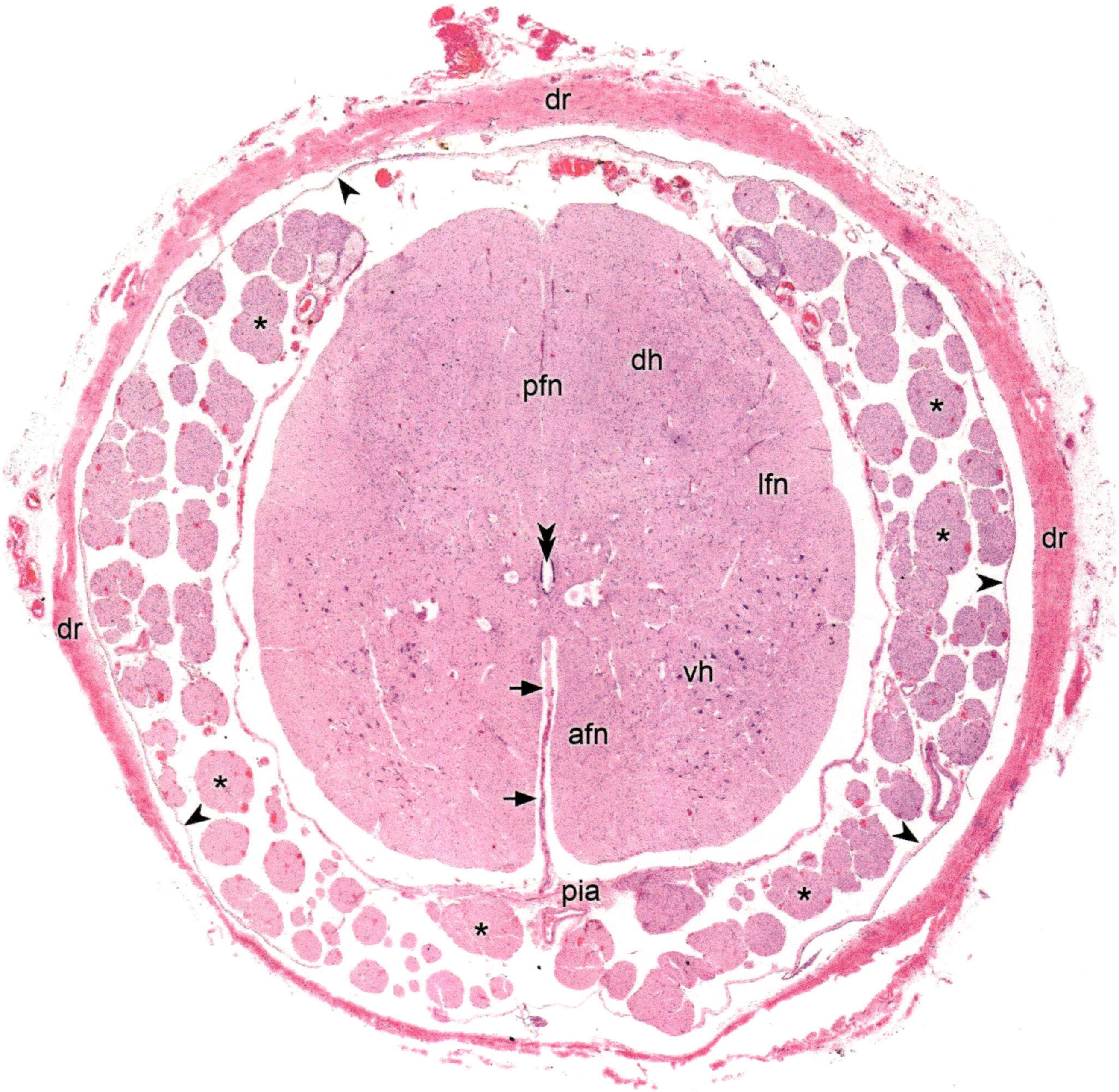

Figure 313. Spinal cord, newborn. White matter is composed of the lateral (lfn), posterior (pfn) and anterior funiculi (afn); the latter is split by the anterior fisure (arrows). The butterfly-shaped grey matter core contains the neuronal perikarya and it is organized into ventral (vh) and dorsal horns (dh) as well as the central area with the central canal lined with ependymal cells (double arrowheads). The surface of the spinal cord is covered with a pia mater (pia). In contrast, dura mater (dr) and the underlying arachnoid (arrowheads) loosely surround the spinal cord as well as the roots and rootlets (asterisks) forming the subarachnoid space (sas) between the outer meninges and the pia mater.

Cerebrum

Cerebrum is formed by an outer cortex an inner medulla; the cortex is mostly composed of grey matter containing neuronal perikarya, while medulla is mostly formed by myelinated or nonmyelinated fibers. In the deeper regions of the brain, medulla surrounds nuclei that are islands of clustered neuronal cell bodies. Cerebral cortex contains perikarya of neurons surrounded by the neuropil that is composed of axons, dendrites and glial processes. In the neocortex, these structures are organized into six layers that are histologically distinguishable with H&E staining. The most characteristic cell types of these layers are the pyramidal cells that possess a pyramid- (triangular)-shaped cell body with dendrites emanating from the apex (apical dendrite) and the sides, while the single axon originates from the base of the cell. The spherical nucleus of the pyramidal cell is pale, granular, with prominent nucleolus; cytoplasm contains well-defined clumped rER (Nissl substance). The axis of the pyramidal cells is perpendicular to the cortical surface. Apical dendrites ascend and form tufts in the *molecular layer* (layer I) that contains few scattered, star-shaped stellate neurons. The underlying layer is the *external granular layer* (layer II) that appears to be more dense and therefore darker, since it is more cellular, containing small pyramidal and stellate cells. Layer III, the *external pyramidal layer*, is mostly composed of scattered small and medium sized pyramidal cells; this layer rests on the *internal granular layer* (layer IV) that is similar to the external one in composition. *Internal pyramidal layer* (layer V) is under the internal granular layer, and it is very characteristic with its large pyramidal cells that are especially prominent in the motor cortex (Betz motor neurons). Finally, the underlying *polymorph layer* (layer VI) contains a wide variety of multiform neurons along with few pyramidal cells and rests on the white matter that is primarily composed of axons and glial cells.

Cerebellum

Cerebellar cortex is composed of three layers. The most characteristic layer is the *Purkinje cell layer* (layer II) that contains the cell bodies of the Purkinje cells that are one of the largest multipolar neurons in humans, possessing spherical granular nucleus with prominent nucleolus as well as basophilic Nissl substance in the cytoplasm. Purkinje cells have large, distinct, pear-shaped perikarya and elaborate, flattened dendritic arborization emanating from the apical dendrite of the neuron. These processes are located in the superficial *molecular layer* (layer I) and contact with the parallel fibers at right angles. Molecular layer also contains stellate and basket neurons; the former possesses star-shaped perikaryon while the latter forms basket-like extensions around the perikarya of the Purkinje cells. The deepest layer, the *granular layer* (layer III) rests on the cerebellar white matter and it contains several types of neurons including Golgi neurons and small granular cells with small cell body and claw-shaped terminals.

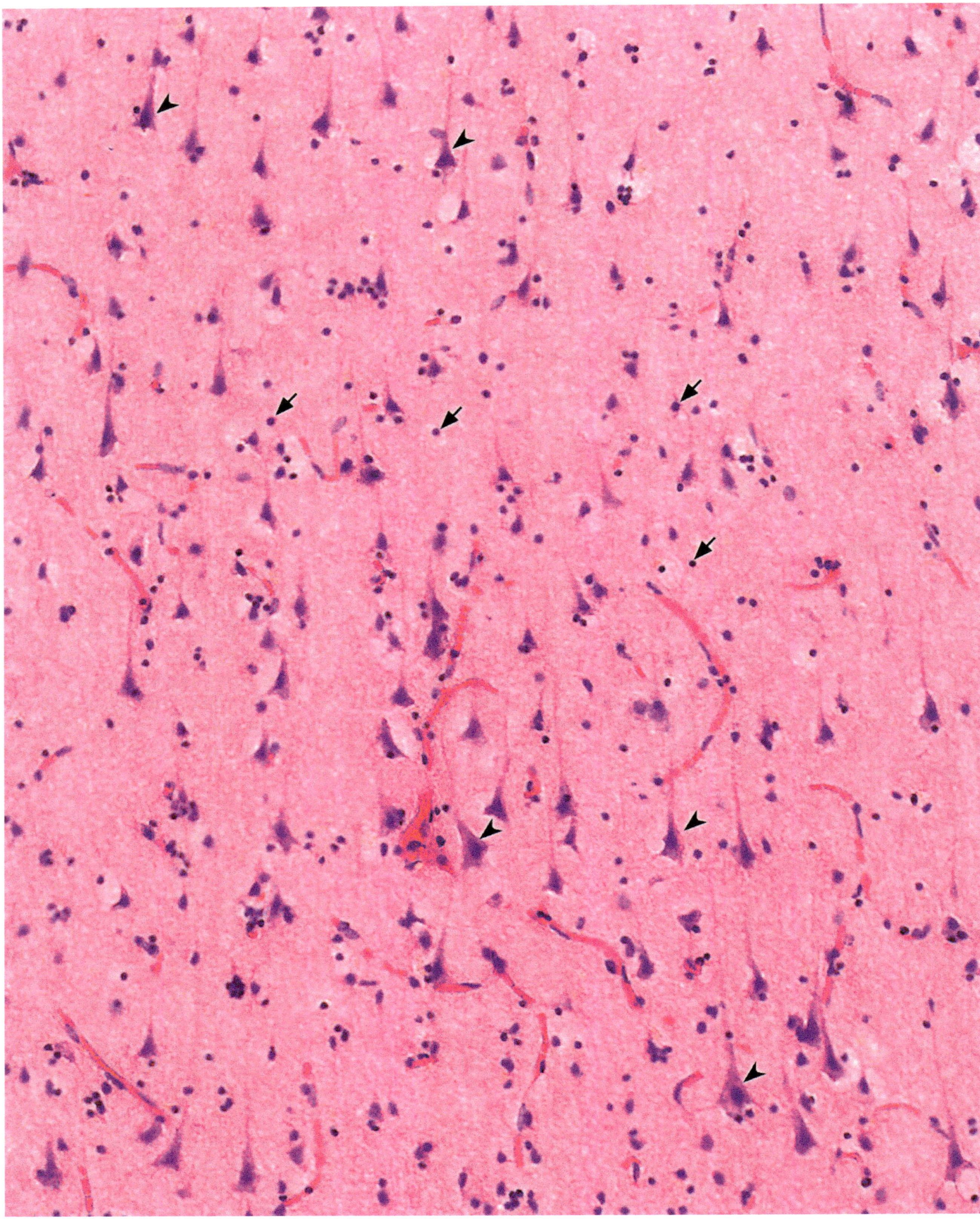

Figure 314. Cerebral cortex. The characteristic cell types of the cerebral cortex are the pyramidal cells (arrowheads) that possess a pyramid (triangular)-shaped cell body with dendrites emanating from the apex (apical dendrite) and the sides, while the single axon originates from the base. Cytoplasm is dark due to the well-defined clumped rER (Nissl substance). Oligodendrocytes can be easily identified from their small, darkly stained, spherical nucleus (arrows). *(Brain sample provided by my father, Dr. I.J.D.)*

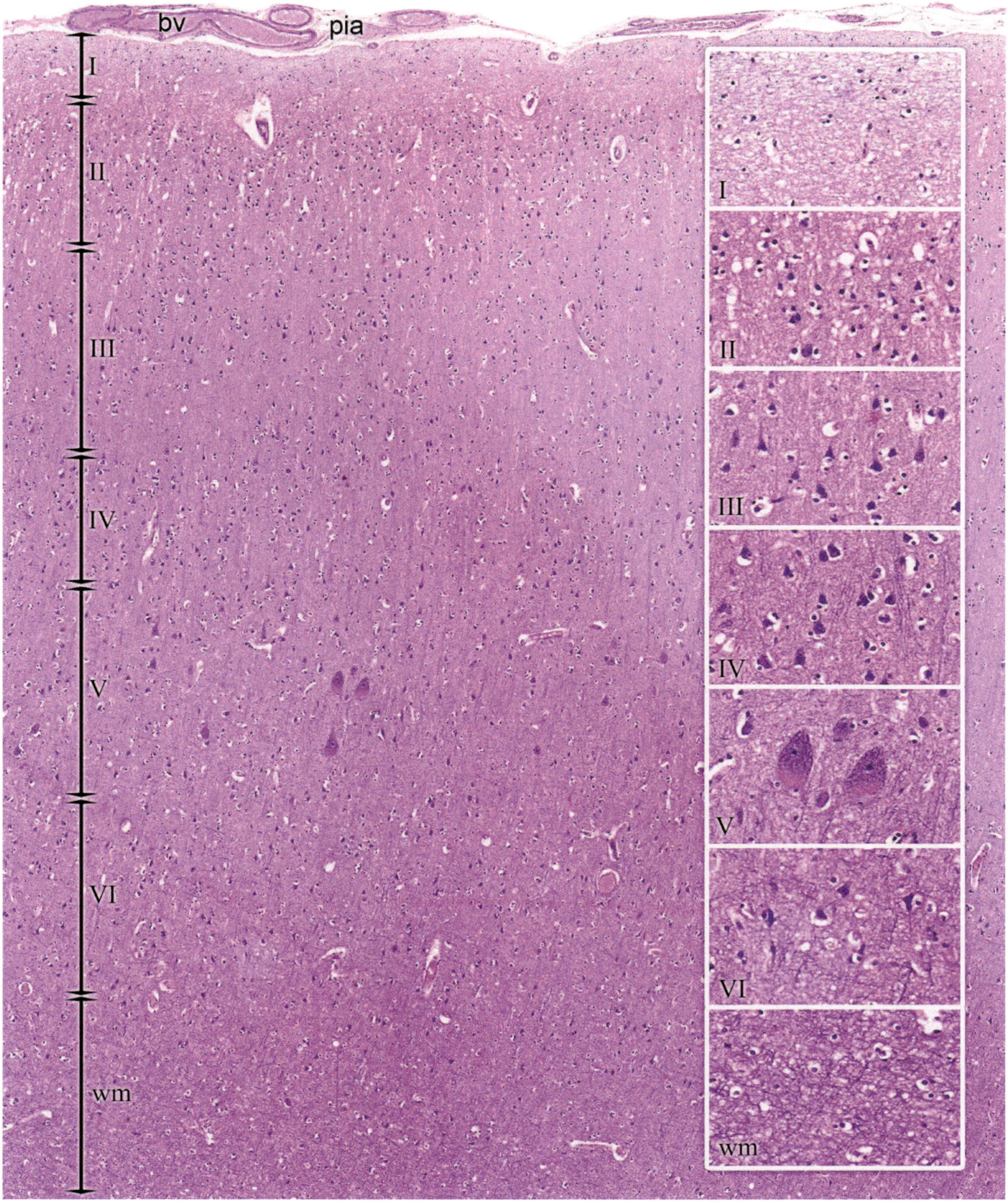

Figure 315. Layers of the cerebral cortex. Cerebral cortex is covered by pia mater (pia) carrying blood vessels (bv). The most superficial *molecular layer* (layer I) is located under the pia mater and it contains few, scattered, star-shaped stellate neurons. E*xternal granular layer* (layer II) appears to be denser and contains small pyramidal and stellate cells. Layer III, the *external pyramidal layer*, is mostly composed of small and medium sized pyramidal cells; this layer rests on the *internal granular layer* (layer IV) that is similar to the external one in composition. *Internal pyramidal layer* (layer V) contains large pyramidal cells. The underlying *polymorph layer* (layer VI) contains a wide variety of multiform neurons along with few pyramidal cells and rests on the white matter (vm).

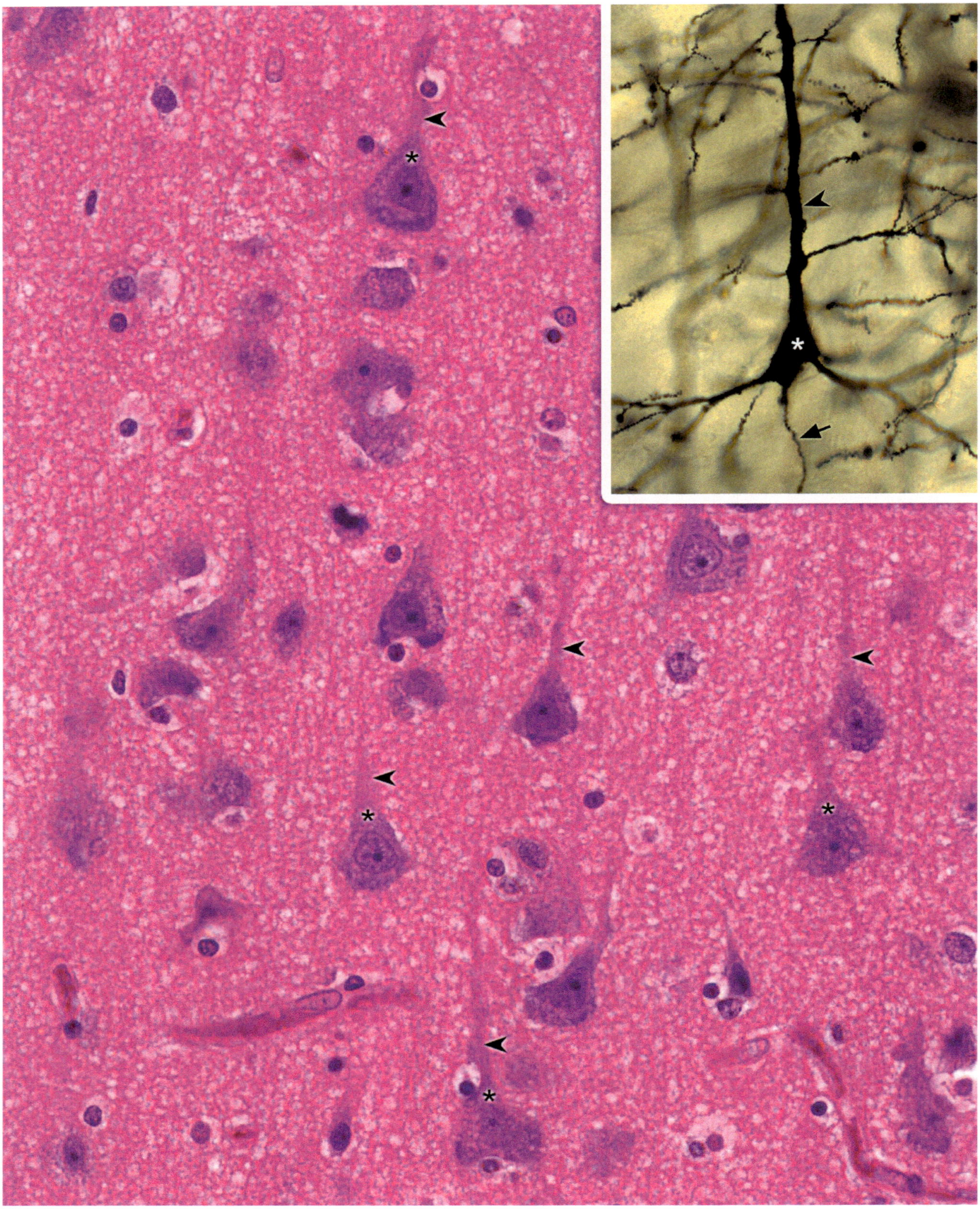

Figure 316. Cerebral cortex, layer V, H&E staining. The inset depicts a pyramidal cell with silver impregnation. Pyramidal cells (asterisks) have a pyramid (triangular)-shaped cell body with dendrites emanating from the apex (apical dendrite, arrowheads) and the sides, while the single axon originates from the base (arrow, inset). The spherical nucleus of the pyramidal cell is pale, granular, with prominent nucleolus; cytoplasm contains well-defined clumped rER (Nissl substance). Apical dendrites (arrowheads) ascend perpendicular to the cortical surface and form tufts in the *molecular layer* (layer I).

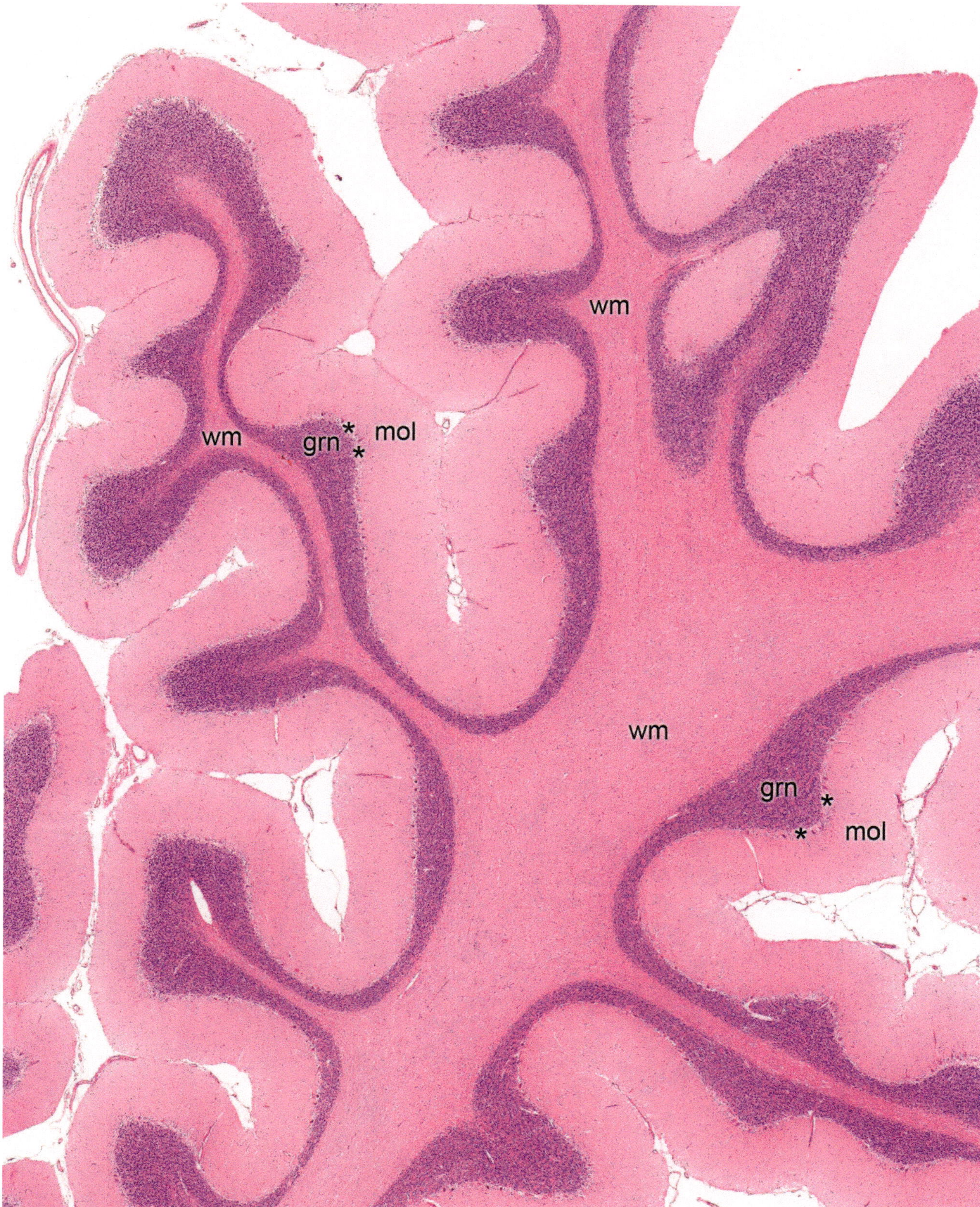

Figure 317. Cerebellum. Cerebellar cortex is composed of three layers. *Purkinje cell layer* (layer II; asterisks) contains the cell bodies of the Purkinje cells that are large multipolar neurons, with pear-shaped perikarya and elaborate dendritic arborization emanating from the apical dendrite of the neuron. These processes are located in the superficial *molecular layer* (layer I; mol) that contains parallel fibers and stellate and basket neurons. The underlying *granular layer* (layer III; grn) rests on the cerebellar white matter (wm).

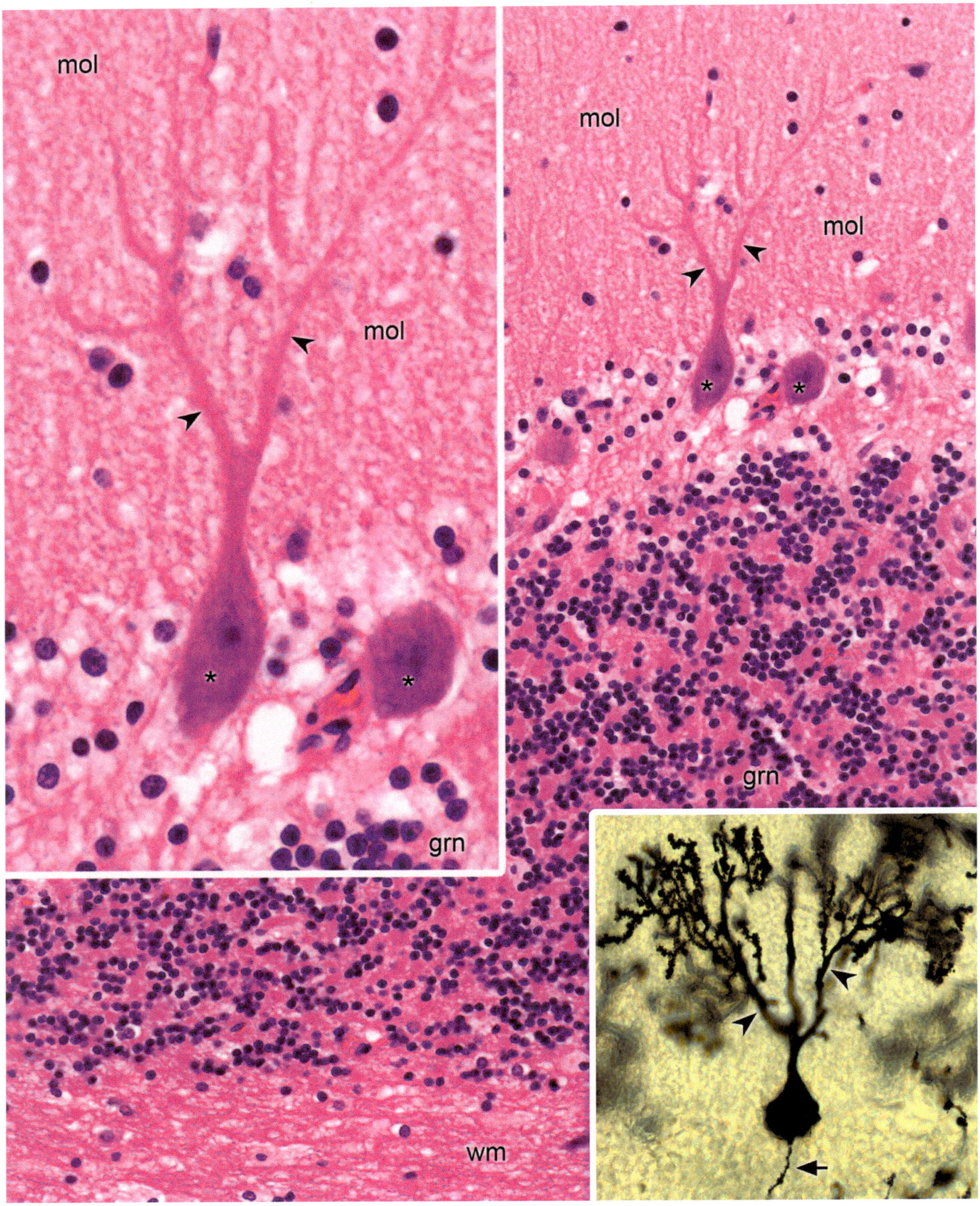

Figure 318. Cerebellum, H&E staining. Lower inset shows a Purkinje cell with silver impregnation. Cerebellar cortex is composed of three layers. *Purkinje cell layer* (layer II) contains the cell bodies of the Purkinje cells (asterisks) that are large multipolar neurons, with pear-shaped perikarya, elaborate dendritic arborization emanating from the apical dendrite of the neuron (arrowheads) and a single axon projecting from the base of the cell body (arrow, lower inset). The apical dendrites are located in the superficial *molecular layer* (layer I; mol) that contains parallel fibers and stellate and basket neurons. The underlying *granular layer* (layer III; grn) rests on the white matter (wm).

Chapter 14

Sensory Organs

Eye

The eyeball is formed by three distinct layers. The outermost fibrous corneoscleral layer comprises the translucent cornea and the nontransparent sclera. The middle layer is the uvea, a vascular layer that also includes the choroid (choroidea) as well as the stroma of the ciliary body and the iris. The innermost coat is the retina that surrounds the vitreous chamber.

Cornea is composed of a corneal stroma, covered externally by the corneal epithelium and the underlying basement membrane (Bowman's membrane). Corneal epithelium is continuous with the epithelium of the conjunctiva and it is composed of thin (approximately five cell layers thick), nonkeratinized stratified squamous epithelium without penetrating papillae. Internally, the stroma is lined with simple squamous epithelium that rests on the basement membrane adjacent to the stroma (Descemet's membrane). The stroma of the cornea is avascular, and it is composed of several layers of regularly arranged collagen fibrils and fibroblasts. Cornea transitions to the sclera that is formed by irregular connective tissue containing vessels, and due to the random arrangement of the collagen fibrils, it is opaque. The inner part of the sclera, adjacent to the choroid, is composed of a delicate network of collagen and elastic fibers (lamina fusca). The outer part of the sclera that is facing the eyelid (palpebra) is covered outside with the conjunctiva, while the outer surface of the posterior part is coated with a loose connective tissue envelope.

Iris is the anterior part of the uvea, surrounding a central aperture, the pupil. Iris is composed of vascularized stroma formed by a network of fibroblast and melanocytes with deep spaces between the interconnecting bundles. At the pupillary margin, stroma contains circularly arranged smooth muscle cells forming the sphincter pupillae muscle that is under parasympathetic control. The stroma of the iris is covered posteriorly by two layers of pigmented epithelium; the anterior layer is composed of melanin-containing myoepithelial cells, forming the dilator pupillae muscle that is controlled by sympathetic nerves. This layer is lined with a layer of extremely pigmented epithelial cells (posterior pigmented epithelium) that rest on a basement membrane facing the lens while the basement membrane of the anterior layer faces the stroma of the iris. Melanin pigments in the cells of the posterior layer obscure the nucleus and the cell boundaries. Anteriorly, iris is lined with a discontinuous layer of melanocytes and fibroblasts.

Ciliary body is part of the uvea located between the iris and the posterior choroid that covers the inner surface of the sclera. As the continuation of the iris, the ciliary body is also composed of an inner epithelial layer and an outer vascular stroma that accommodates the ciliary muscle forming the bulk of the ciliary body. Ciliary muscle consists of smooth muscle cells that are organized circularly (inner zone), radially (middle zone) and longitudinally (outer zone); these fibers are responsible for reducing the tension on the lens, making it thicker (circular and longitudinal fibers) or flatten the lens (radial fibers). From the vascular layer of the ciliary body, ciliary processes are protruding centripetally. The movement of the ciliary muscles is transmitted to the lens by zonular fibers that originate from the basal lamina of the nonpigmented epithelial cells covering the ciliary processes, and they are attached to the margin of the lens.

The vascular stroma containing the ciliary muscles is covered inside by a double layer of pigmented epithelial cells; this layer is continuous with the anterior and posterior pigmented epithelium of the iris. An additional, nonpigmented layer of cuboidal epithelial cells covers the ciliary body inside; this layer is the continuation of the non-photosensitive portion of the retina, and it is responsible for the production of aqueous humor. This liquid is reabsorbed at the iridocorneal angle through the vascular spaces of Fontana that lead into the canal of Schlemm running circularly at the iridocorneal angle. Schlemm's canal is a vessel covered by endothelium and opens into the aqueous veins of the sclera that drain the aqueous humor into the blood.

The posterior part of the uvea is the choroid that covers the internal surface of the sclera. The diameter of the vessels forming the bulk of choroid decreases centrifugally; the outermost lamina choriocapillaris contains mostly capillaries. The vascular layer of the choroid is covered by the vitreous Bruch's membrane that is composed of the basement membrane formed by the lamina choriocapillaris, the basement membrane of the retinal pigment epithelium and a lamina of collagen and elastic fibers between.

http://doi.org/10.1016/B978-0-323-91891-6.50014-3,

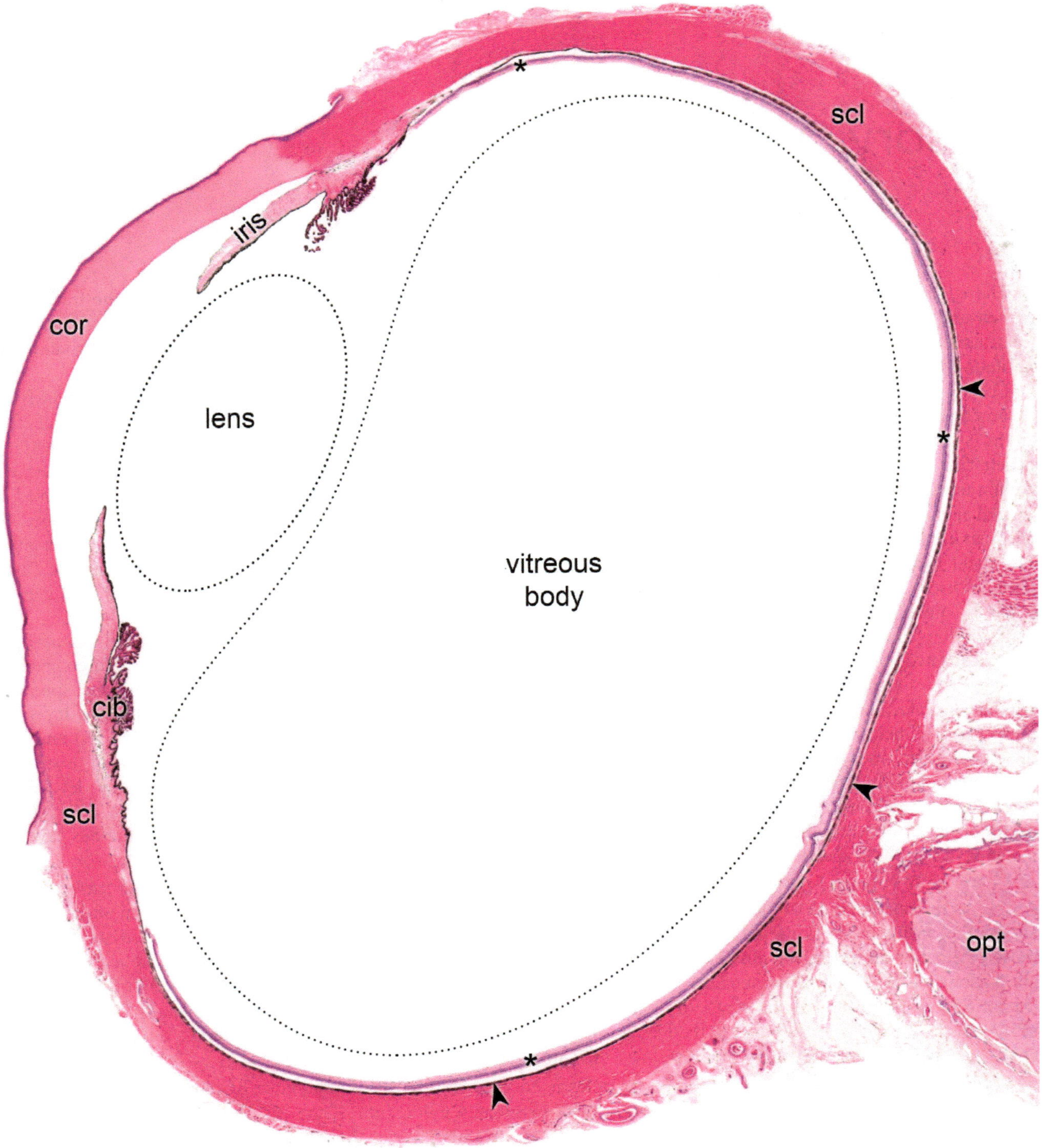

Figure 319. Eye. The eyeball is formed by three distinct layers. The outermost fibrous corneoscleral layer comprises the anterior, translucent cornea (cor) that continues into the posterior, nontransparent sclera (scl). The middle layer is the uvea, a vascular layer that also includes the posterior choroid (choroidea, arrowheads) as well as the stroma of the ciliary body (cib) and the iris anteriorly. The innermost coat is the retina (asterisks) that surrounds the vitreous chamber. Axons of the retinal ganglion cells form the optic nerve (opt). The chambers of the eye contain the anterior lens and the posterior vitreous body that are missing from this specimen, only their contours are shown; the eye is compressed anteroposteriorly due to the tissue preparation. Lens is an avascular structure and it is anchored to the ciliary body by the zonular fibers that end in the capsule of the lens. The capsule is a thick basal lamina that is formed by the underlying cuboidal epithelial cells (lens epithelium) that are located only on the anterior side of the lens. Similar to the lens, the vitreous body is a transparent structure that fills the space between the lens and the retina (vitreous chamber) and it is composed of a gelatinous substance rich in glycosaminoglycans that is surrounded by a delicate vitreous membrane formed by thin collagen fibers.

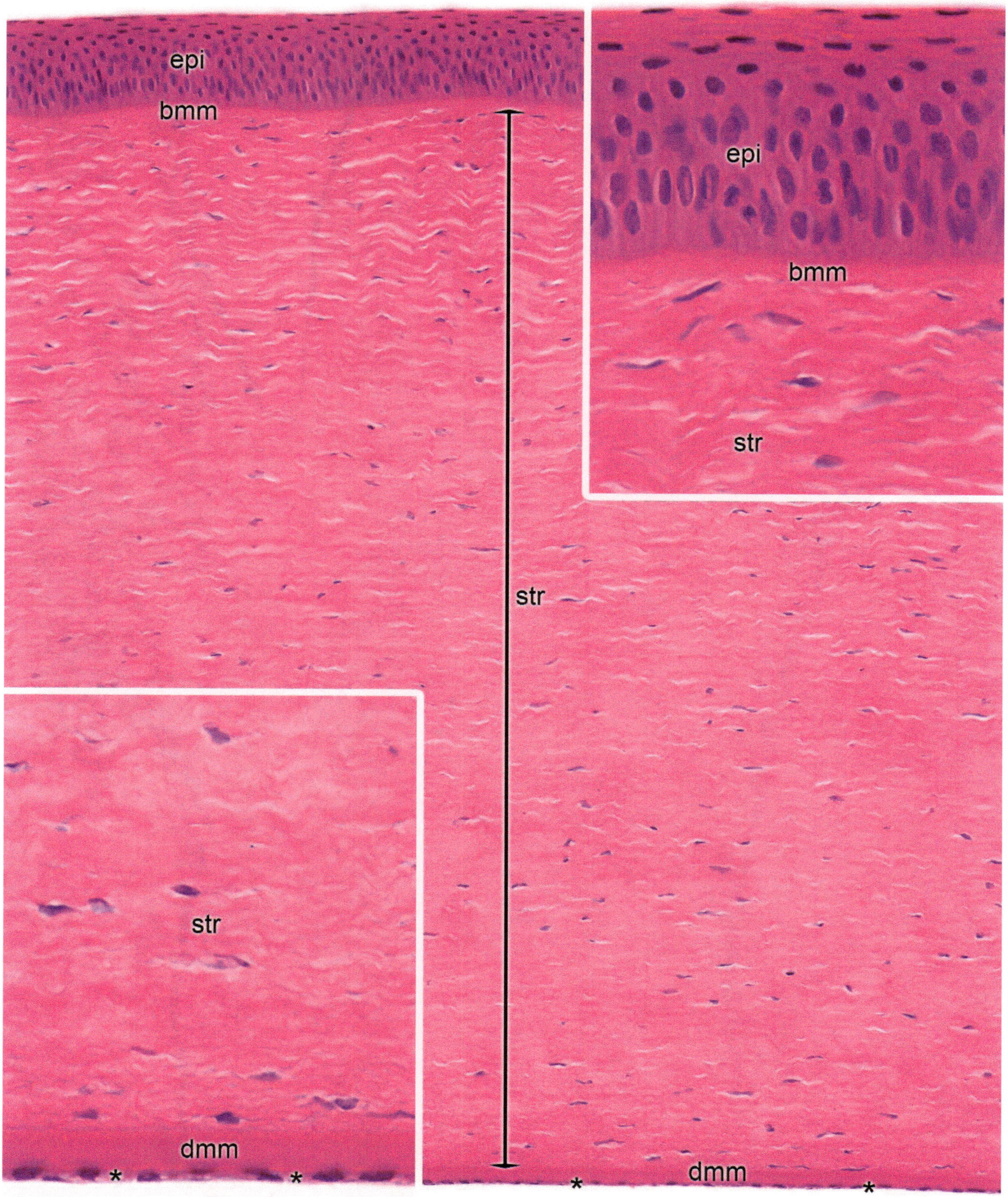

Figure 320. Cornea. The cornea is composed of a corneal stroma (str) covered externally by the corneal epithelium (epi) and the underlying basement membrane called Bowman's membrane (bmm, upper inset). Corneal epithelium (epi) is continuous with the epithelium of the conjunctiva and it is composed of thin (approximately five cell layers thick) stratified squamous epithelium without penetrating papillae (upper inset). Internally, the stroma is lined with simple squamous epithelium (asterisks, lower inset) that rests on the basement membrane facing the stroma (Descemet's membrane, dmm). The stroma of the cornea (str) is avascular and it is composed of several layers of regularly arranged collagen fibrils and fibroblasts.

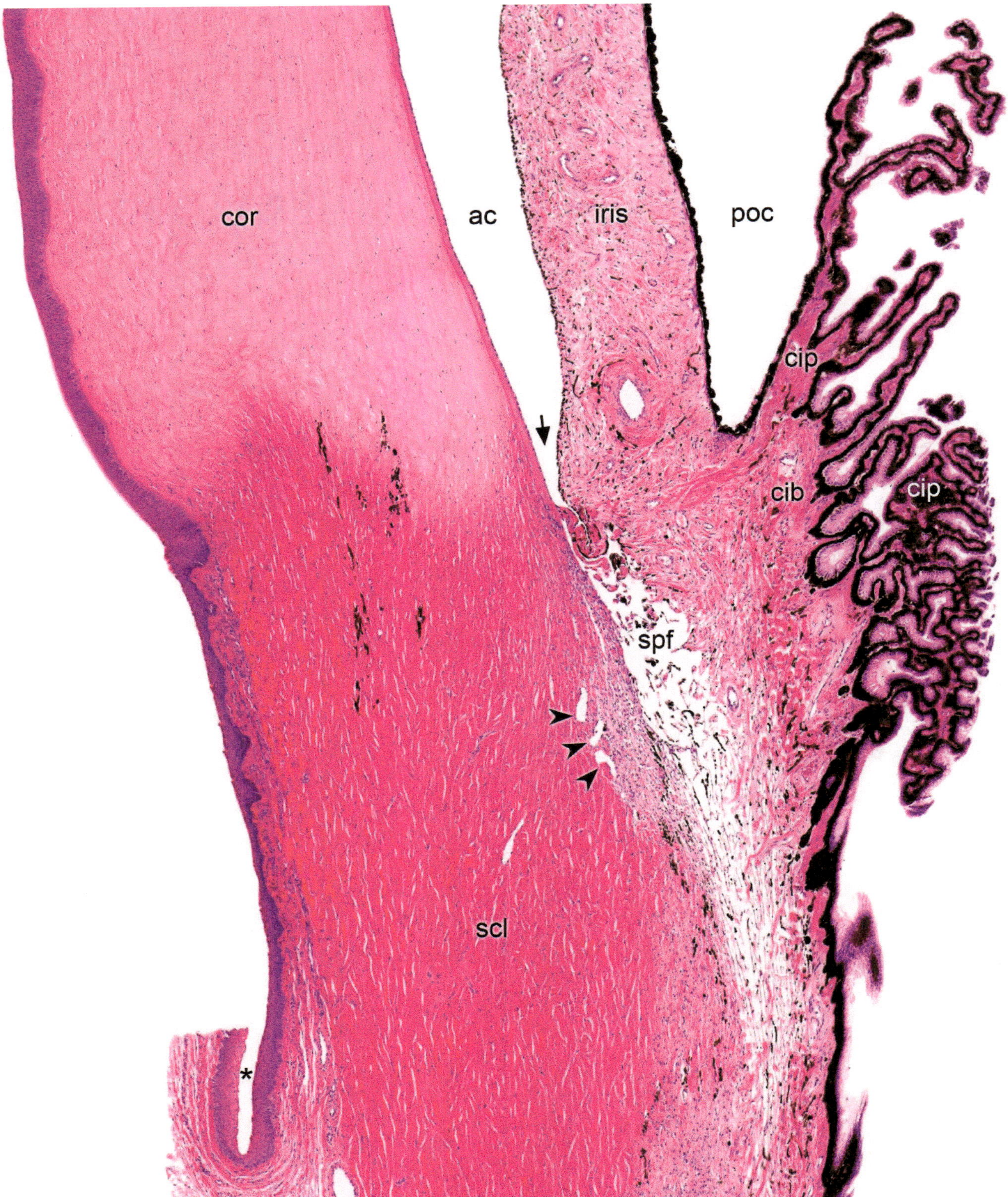

Figure 321. Iridocorneal angle, small magnification. The anterior, translucent cornea (cor) continues into the posterior, nontransparent sclera (scl) covered anteriorly with the bulbar conjunctiva that transitions into the palpebral conjunctiva at the conjunctival fornix (asterisk) covered with stratified columnar epithelium with goblet cells. The choroid continues anteriorly into the ciliary body (cib) with the ciliary processes (cip) and then into the iris. Aqueous humor is reabsorbed at the iridocorneal angle (arrow) through the vascular spaces of Fontana (spf) that lead into the canal(s) of Schlemm (arrowheads) running circularly at the iridocorneal angle. Between the cornea and the iris, the anterior chamber (ac) communicates with the posterior chamber (poc) through the pupil.

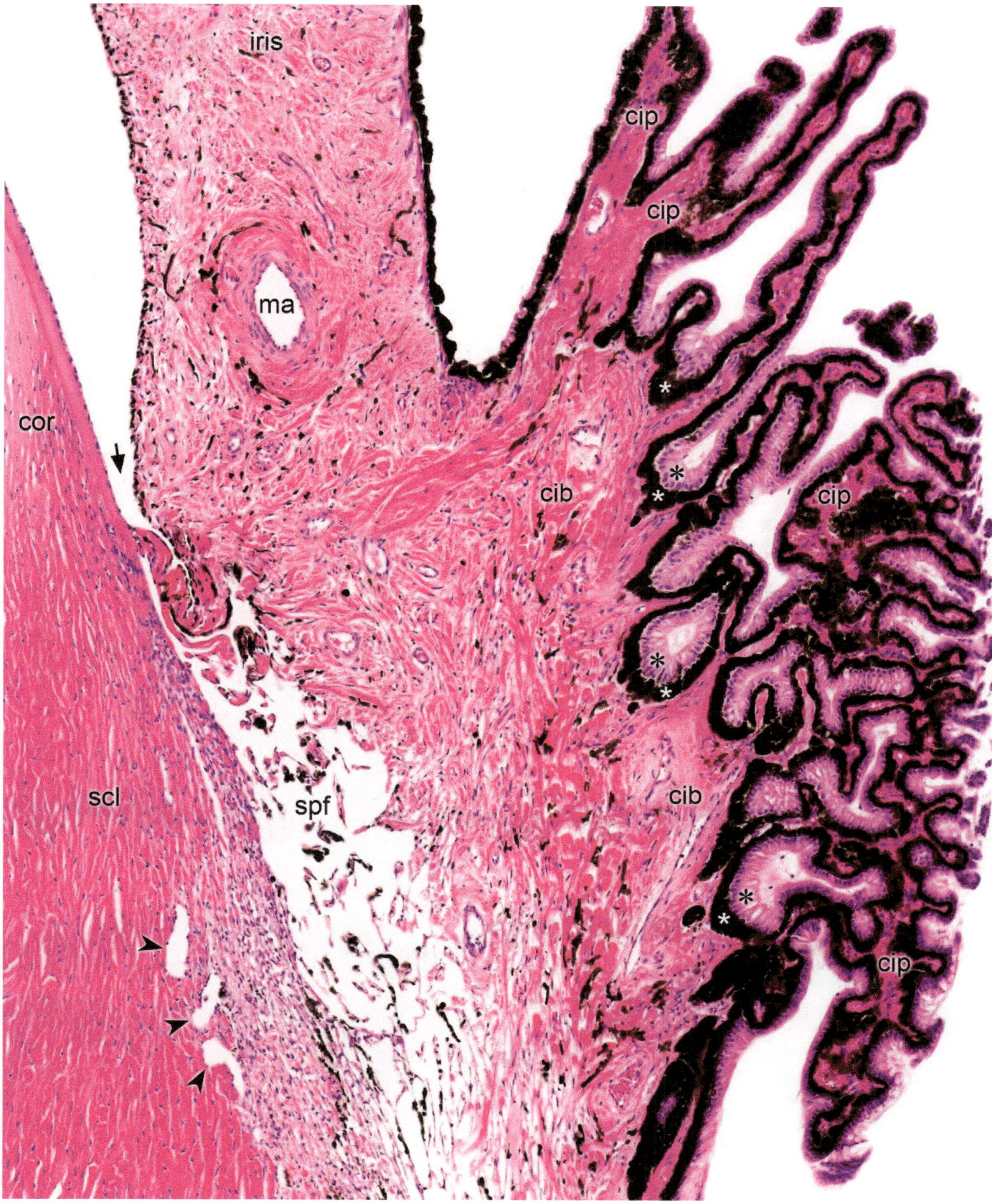

Figure 322. Iridocorneal angle, large magnification. The anterior, translucent cornea (cor) continues into the posterior, nontransparent sclera (scl). Ciliary processes are lined with an inner pigmented (five-pointed asterisks) and an outer nonpigmented layer of cuboidal cells (six-pointed asterisks); the latter is the continuation of the non-photosensitive portion of the retina. This layer is responsible for the production of aqueous humor that is reabsorbed at the iridocorneal angle (arrow) through the vascular spaces of Fontana (spf) that lead into the canal(s) of Schlemm (arrowheads) running circularly at the iridocorneal angle. Muscular arteries (ma) form an arterial circle at the junction of the iris and the ciliary body.

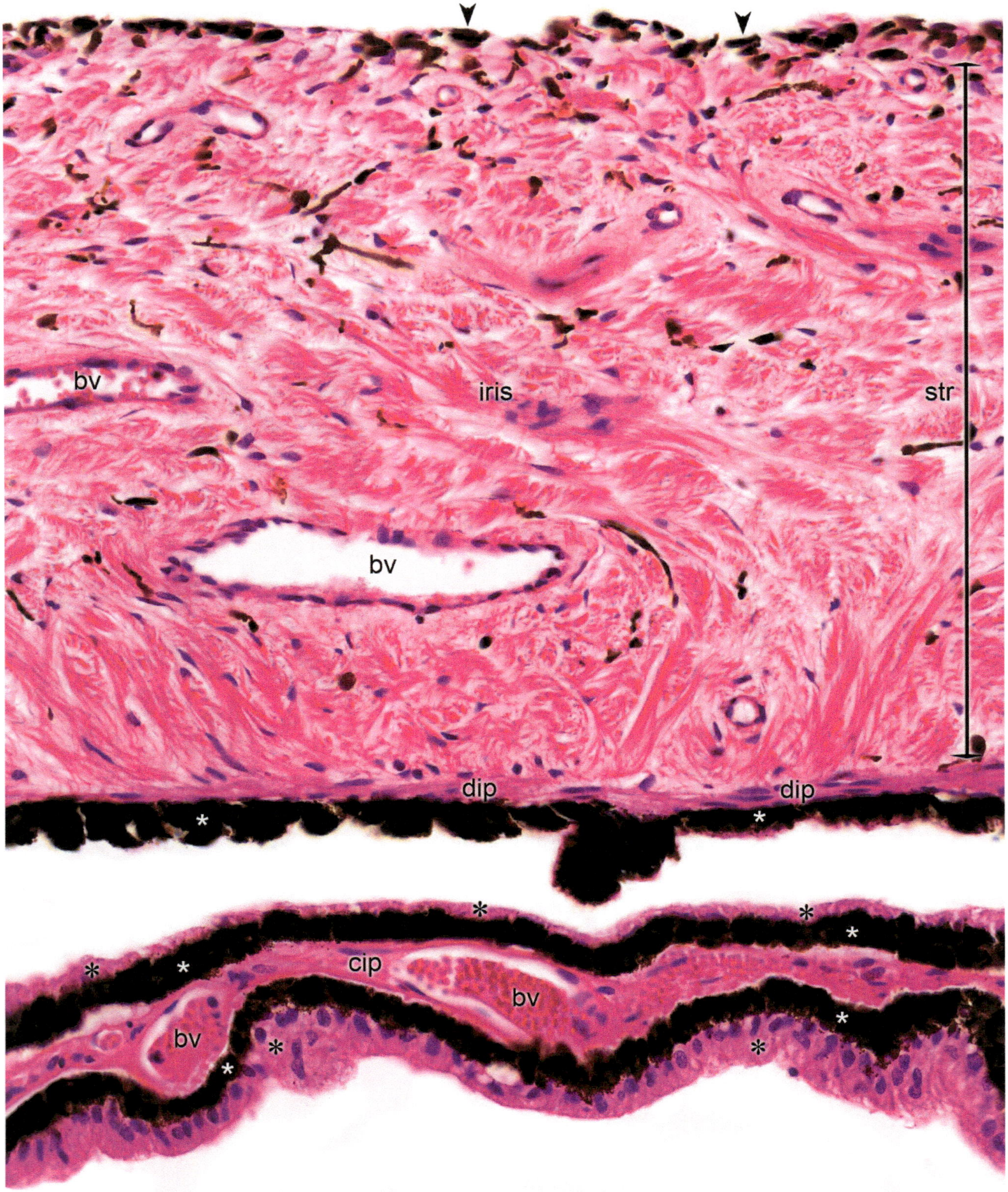

Figure 323. Iris with a ciliary process (cip). Iris is composed of vascularized stroma (str) formed by fibroblast and melanocytes with blood vessels. The stroma of the iris is covered posteriorly by melanin-containing, myoepithelial cells forming the dilator pupillae muscle (dip) and a layer of extremely pigmented epithelial cells (five-pointed asterisks) that also lines the ciliary processes. Anteriorly, iris is lined with a discontinuous layer of melanocytes (arrowheads) and fibroblasts. Ciliary processes (cip) contain vessels (bv) and they are lined with an inner pigmented (five-pointed asterisks) and an outer nonpigmented layer of cuboidal cells (six-pointed asterisks); the latter is the continuation of the non-photosensitive portion of the retina and produces the aqueous humor.

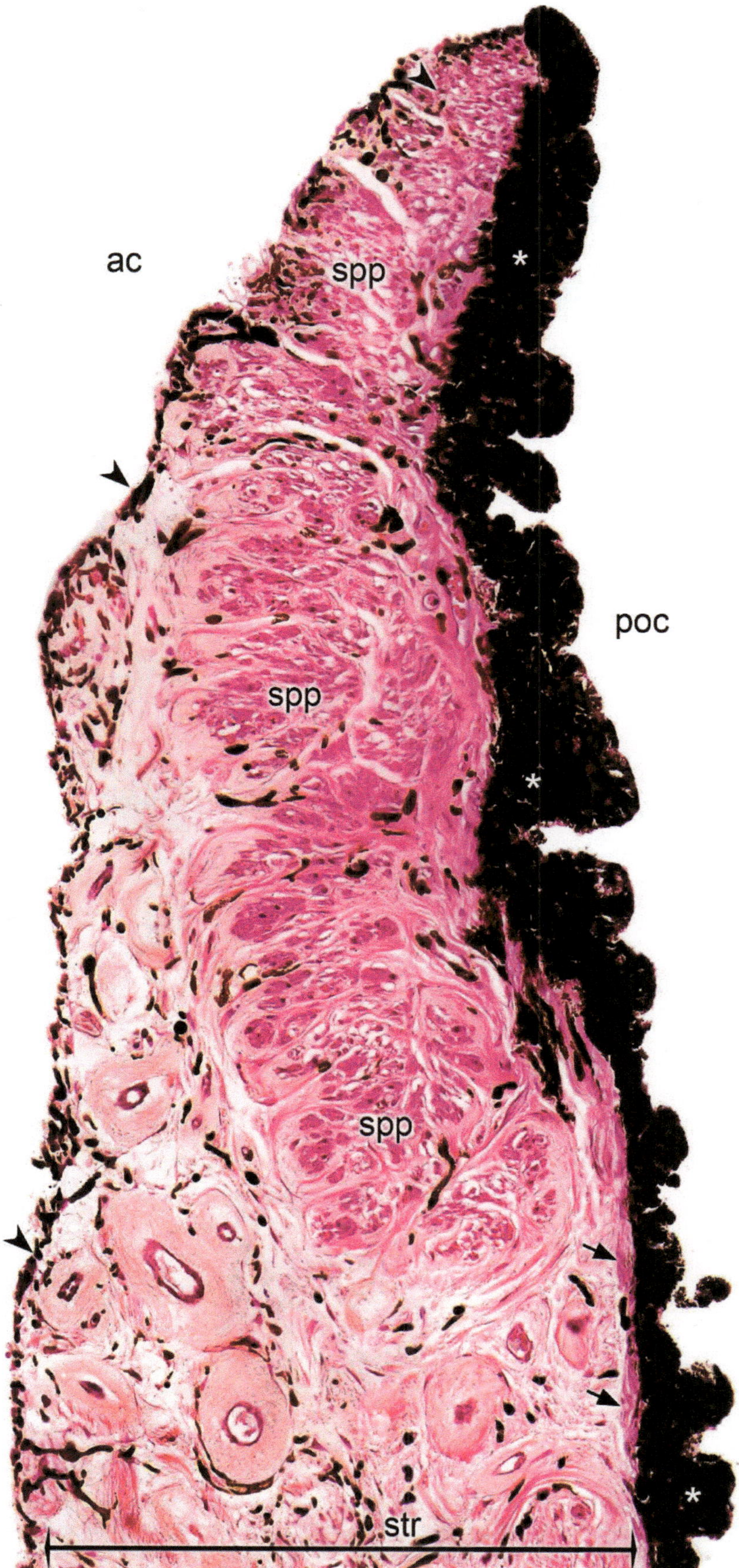

Figure 324. Iris, pupillary margin. Iris is composed of vascularized stroma (str) formed by fibroblast and melanocytes with deep spaces between the interconnecting bundles. At the pupillary margin, stroma contains circularly arranged smooth muscle cells forming the sphincter pupillae muscle (spp) that is under parasympathetic control. Facing the posterior chamber (poc), the stroma of the iris is covered posteriorly by myoepithelial cells forming the dilator pupillae muscle (arrows) and a layer of extremely pigmented epithelial cells (five-pointed asterisks) that also line the ciliary processes. Anteriorly, facing the anterior chamber (ac), iris is lined with a discontinuous layer of scattered melanocytes (arrowheads) and fibroblasts.

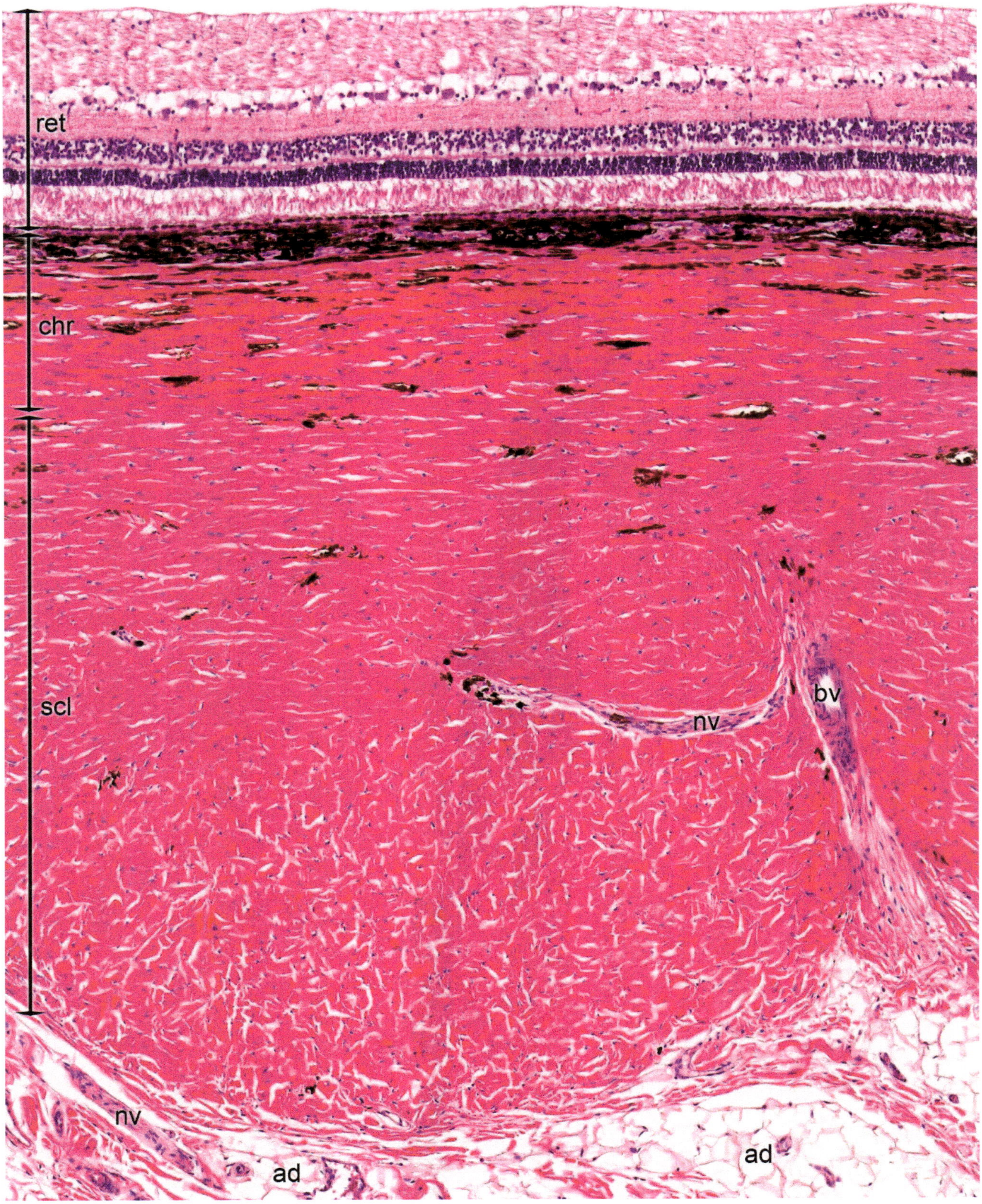

Figure 325. Retina, choroid and sclera. Posteriorly, the eyeball is formed by three distinct layers. The outermost nontransparent sclera (scl) is composed of dense connective tissue with collagen fibers, while the middle layer, the choroid (chr) is a vascular layer. The innermost coat is the retina (ret) that surrounds the vitreous chamber. Outside, the sclera is covered with the loose connective tissue of the orbit, containing adipose tissue (ad) with nerves (nv) and blood vessels (bv) that also penetrate the sclera.

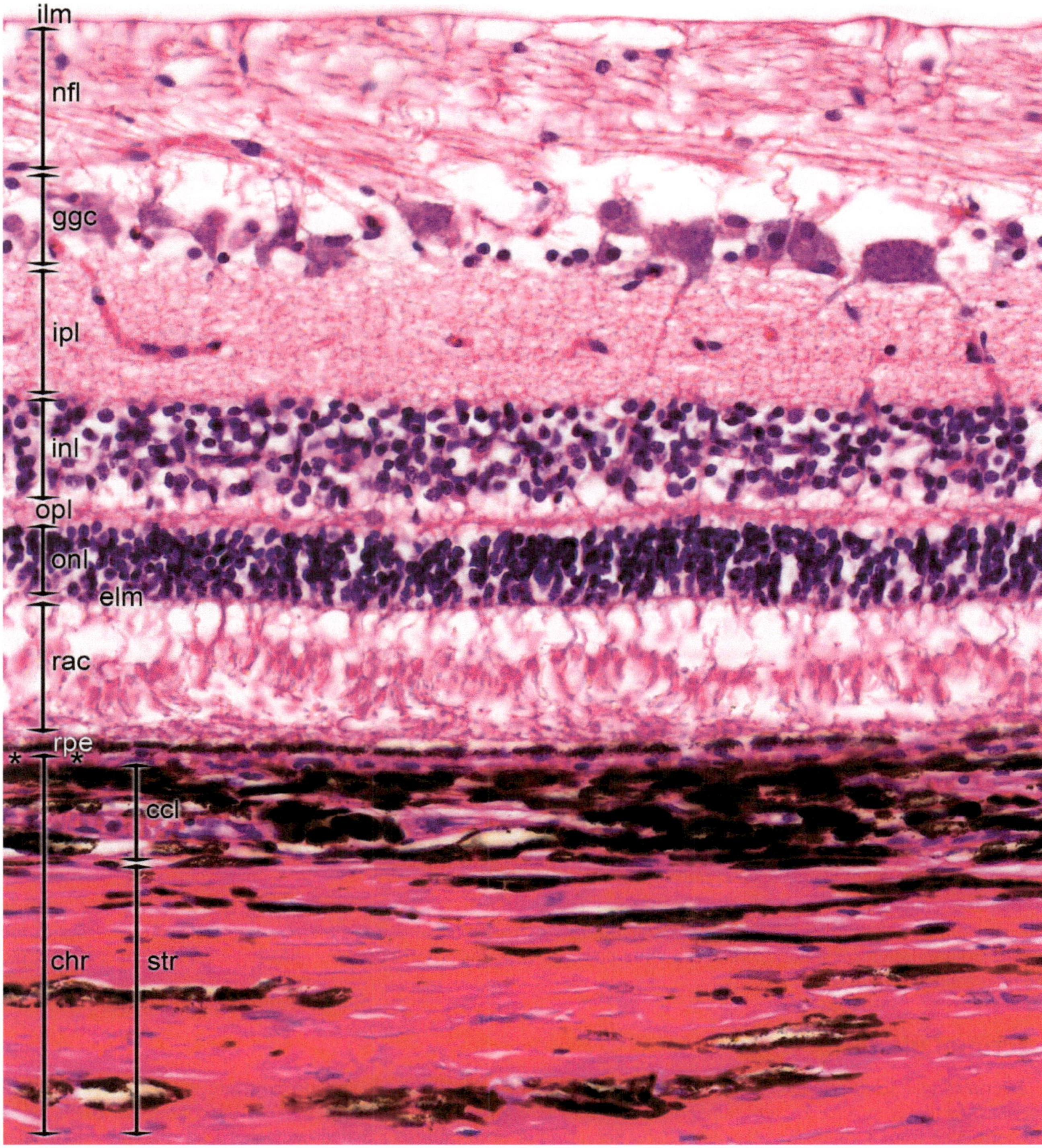

Figure 326. Layers of the retina. The innermost layer, the *inner limiting membrane* (ilm) is formed by the basement membrane of the radial glial (Müller) cells. The processes of the ganglion cells that eventually form the optic nerve are collected in the *layer of nerve fibers* (nfl), while the perikarya of the ganglion cells occupy the underlying *ganglion cell layer* (ggc). The *inner plexiform layer* (ipl) contains the synapses between the bipolar cells and the processes of the amacrine and ganglion cells. The *inner nuclear layer* (inl) is tightly packed with spherical nuclei of the bipolar horizontal, amacrine and Müller cells. The inner and outer segments of the photoreceptor cells are located in the *layer of rods and cones* (rac) that rests on the layer of the retinal pigmented epithelium (rpe), while the *outer nuclear layer* (onl) contains the spherical nuclei of the photoreceptor cells, whose projections, along with the processes of the connecting bipolar and horizontal cells are located in the *outer plexiform layer* (opl). Between the outer nuclear layer and the layer of rods and cones, the border of the Müller cells forms the *external limiting membrane* (elm). Retina rests on the vitreous Bruch's membrane (asterisks) and the underlying lamina choriocapillaris (ccl) of the choroid (chr). Choroid stroma (str) contains vessels with increasing diameters toward the sclera.

The innermost layer of the eye is the retina that is composed of the inner neural retina and an outer single layer of pigmented cuboidal cells that are attached to the choroid. There is a virtual space between these layers that can lead to detachment of the neural retina from the *pigmented epithelium*, the deepest layer of the retina. Internal to the pigmented epithelium, retina has nine additional layers. The outermost, the *layer of rods and cones* contains the inner and outer segments of the photoreceptor cells. The *external limiting membrane* marks the border of the Müller cells that are large, supporting glial cells extending to the retinal surface. The next layer, the *outer nuclear layer* contains the spherical nuclei of the rods and cones, while the projections of the photoreceptor cells along with the processes of the connecting bipolar and horizontal cells are located in the *outer plexiform layer*. The *inner nuclear layer* is tightly packed with spherical nuclei since it is composed of the perikarya of the bipolar horizontal and amacrine cells; this layer also contains the nuclei of the Müller cells. The *inner plexiform layer* contains the synapses between the bipolar cells and the processes of the amacrine and ganglion cells, while the cell bodies of the large ganglion cells are located in the *ganglion cell layer*. The processes of the ganglion cells that eventually form the optic nerve are collected in the *layer of nerve fibers*. Finally, the innermost layer, the *inner limiting membrane* is formed by the basement membrane of the Müller cells.

Lens is an avascular structure and it is anchored to the ciliary body by the zonular fibers that end in the capsule of the lens. The capsule is a thick basal lamina that is formed by the underlying cuboidal epithelial cells (lens epithelium) that are located only on the anterior side of the lens. These subcapsular epithelial cells form the lens fibers that make up the bulk of the lens extending from the epithelium to the capsule covering the posterior surface of the lens. Similar to the lens, the vitreous body is a transparent structure that fills the space between the lens and the retina (vitreous chamber). Vitreous body is composed of a gelatinous substance rich in glycosaminoglycans that is surrounded by a delicate vitreous membrane formed by collagen fibers. Occasional vitreous cells, the hyalocytes can be observed at the periphery of the vitreous body; they are believed to produce the gelatinous matrix.

Palpebra

The anterior surface of the eye is protected by the palpebra (eyelid). The internal surface of the palpebra facing the eye is lined with conjunctiva that extends to the anterior surface of the eye covering the sclera, and terminates circularly around the cornea at the corneoscleral junction. The transition between the palpebral and bulbar (scleral) conjunctiva is the conjunctival fornix that receives the opening of the accessory lacrimal gland (of Krause). The conjunctiva is lined with stratified columnar epithelium with goblet cells and the underlying loose connective tissue lamina propria.

The core of the palpebra is formed by the palpebral part of the orbicularis oculi muscle covered by the tarsus or tarsal plate posteriorly. The former is composed of skeletal muscle while the latter is a dense connective tissue plate that also contains aggregated sebaceous glands (tarsal glands or Meibomian glands). These glands are extremely long and open directly into the palpebral margin. Accessory lacrimal glands of Wolfring are embedded in the superior part of the tarsus. The outer surface of the palpebra is covered with skin that extends to the palpebral margin. Here, sebaceous glands of Zeis as well as apocrine sweat glands of Moll open into the hair follicles of the eyelashes. Between the eyelashes, the ducts of the Meibomian glands open at the palpebral margin.

Palpebral movements are controlled by several layers of muscles. The palpebral part of the orbicularis oculi is the most anterior; behind it, the aponeurosis of the levator palpebrae superioris radiates into the eyelid in front of the tarsus. Both muscles are striated, while the more posterior superior tarsal muscle, originating from the upper part of the tarsus is a smooth muscle.

Lacrimal gland

Lacrimal gland is a compound tubuloalveolar serous gland that produces tears to lubricate the ocular surface and open at the inner surface of the palpebra. The gland is composed of lobules separated by connective tissue septa that originate from a connective tissue capsule. Serous acini open into intercalated ducts that lead into the intralobular ducts forming interlobular ducts surrounded by connective tissue; these ducts are covered with cuboidal epithelium that becomes taller distally and transitions into the pseudostratified epithelium of the largest excretory ducts. Lacrimal glands lack striated ducts. Myoepithelial cells surround the base of the acini; they are enclosed within the basement membrane and facilitate the release of tear from the ducts. Lacrimal glands contain scattered lymphocytes with occasional lymph follicles as well as adipocytes dispersed among the acini. Accessory lacrimal glands open into the conjunctival fornix (glands of Krause) or at the inner surface of the palpebra (glands of Wolfring).

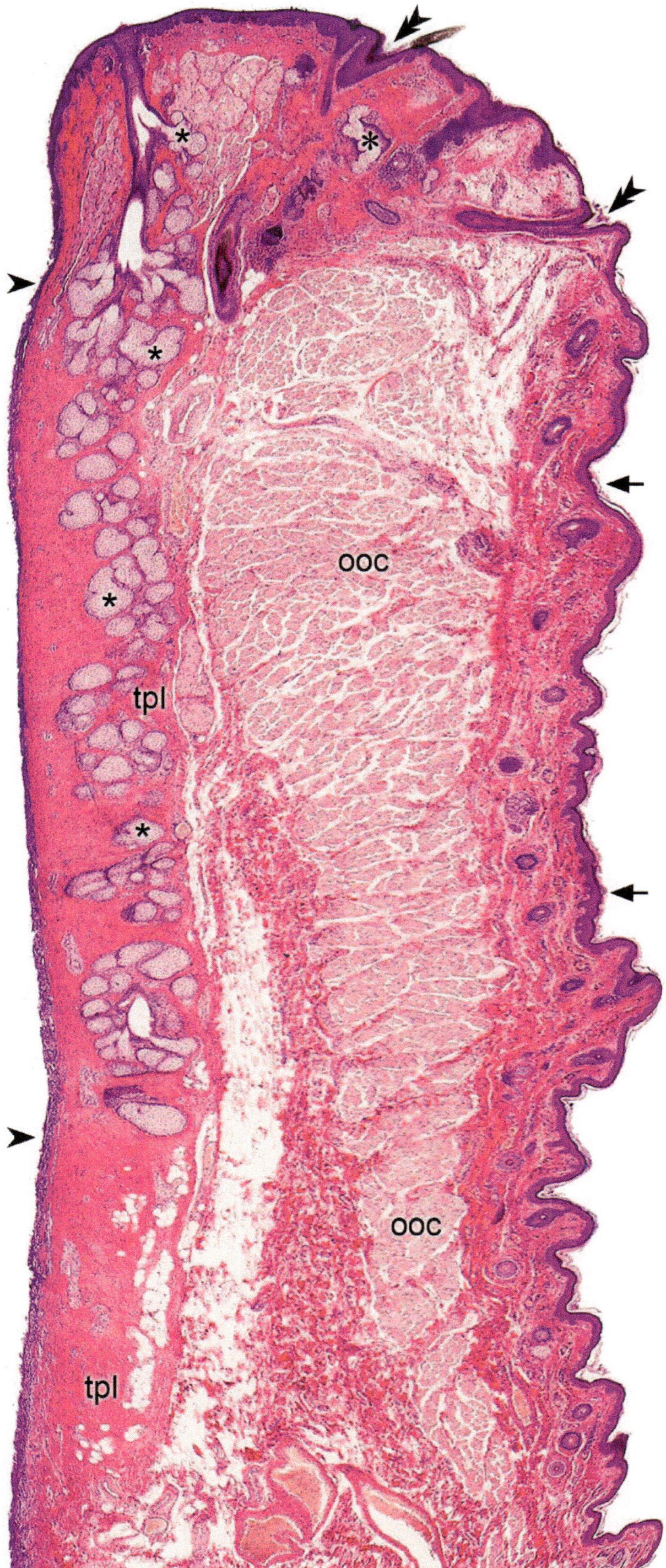

Figure 327. Eyelid. The internal surface of the palpebra facing the eye is lined with conjunctiva (arrowheads) that is composed of stratified columnar epithelium with goblet cells and the underlying loose connective tissue lamina propria. The core of the palpebra is formed by the palpebral part of the orbicularis oculi muscle (ooc) anteriorly, and the dense connective tissue tarsal plate posteriorly (tpl). Tarsal plate contains extremely long sebaceous glands, the Meibomian glands (five-pointed asterisks) opening directly into the palpebral margin. The outer surface of the palpebra is covered with skin (arrows) that extends to the palpebral margin. Here, sebaceous glands of Zeis (six-pointed asterisks) as well as apocrine sweat glands of Moll open into the hair follicles of eyelashes (double arrowheads).

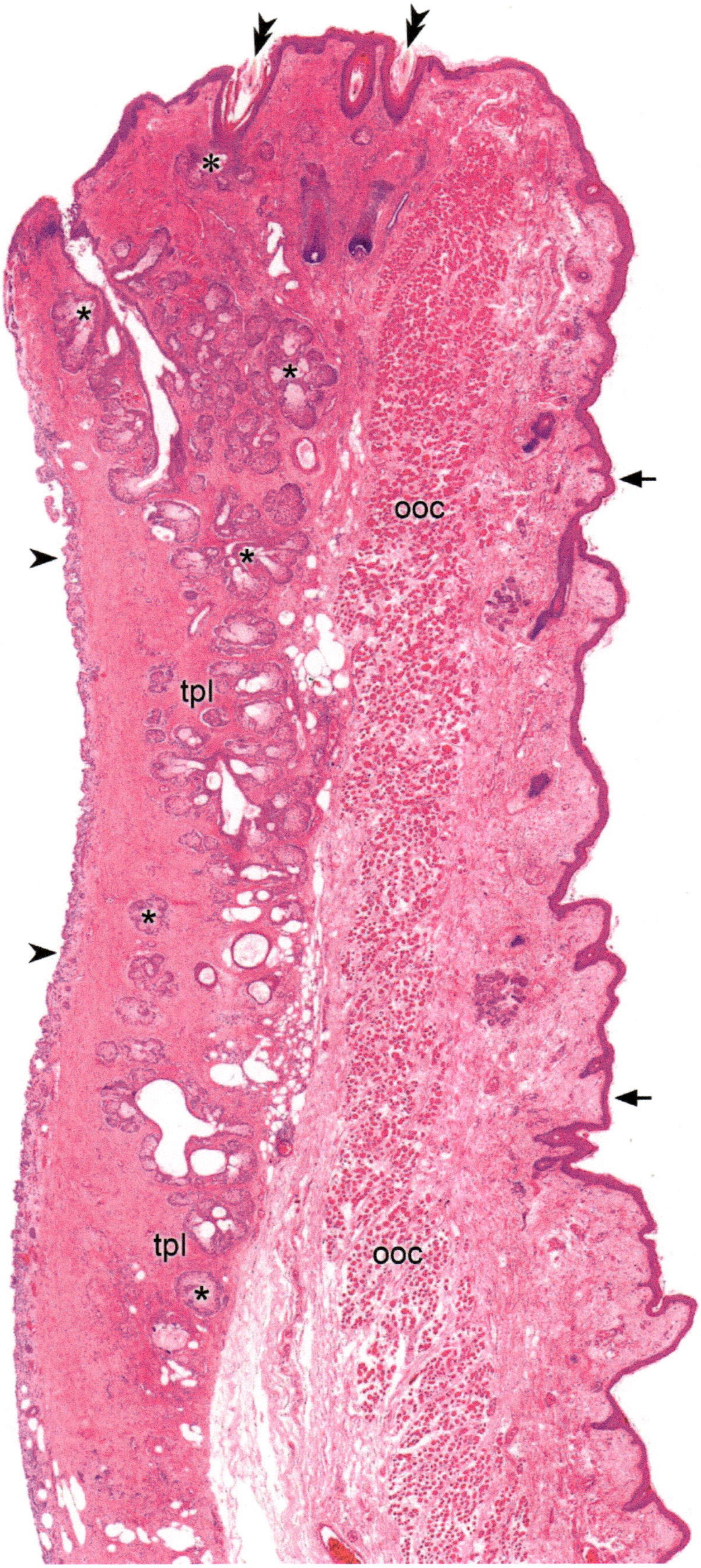

Figure 328. Eyelid. The internal surface of the palpebra facing the eye is lined with conjunctiva (arrowheads) that is composed of stratified columnar epithelium with goblet cells and the underlying loose connective tissue lamina propria. The core of the palpebra is formed by the palpebral part of the orbicularis oculi muscle (ooc) anteriorly, and the dense connective tissue tarsal plate posteriorly (tpl). Tarsal plate contains extremely long sebaceous glands, the Meibomian glands (five-pointed asterisks) opening directly into the palpebral margin. The outer surface of the palpebra is covered with skin (arrows) that extends to the palpebral margin. Here, sebaceous glands of Zeis (six-pointed asterisks) as well as apocrine sweat glands of Moll open into the hair follicles of eyelashes (double arrowheads).

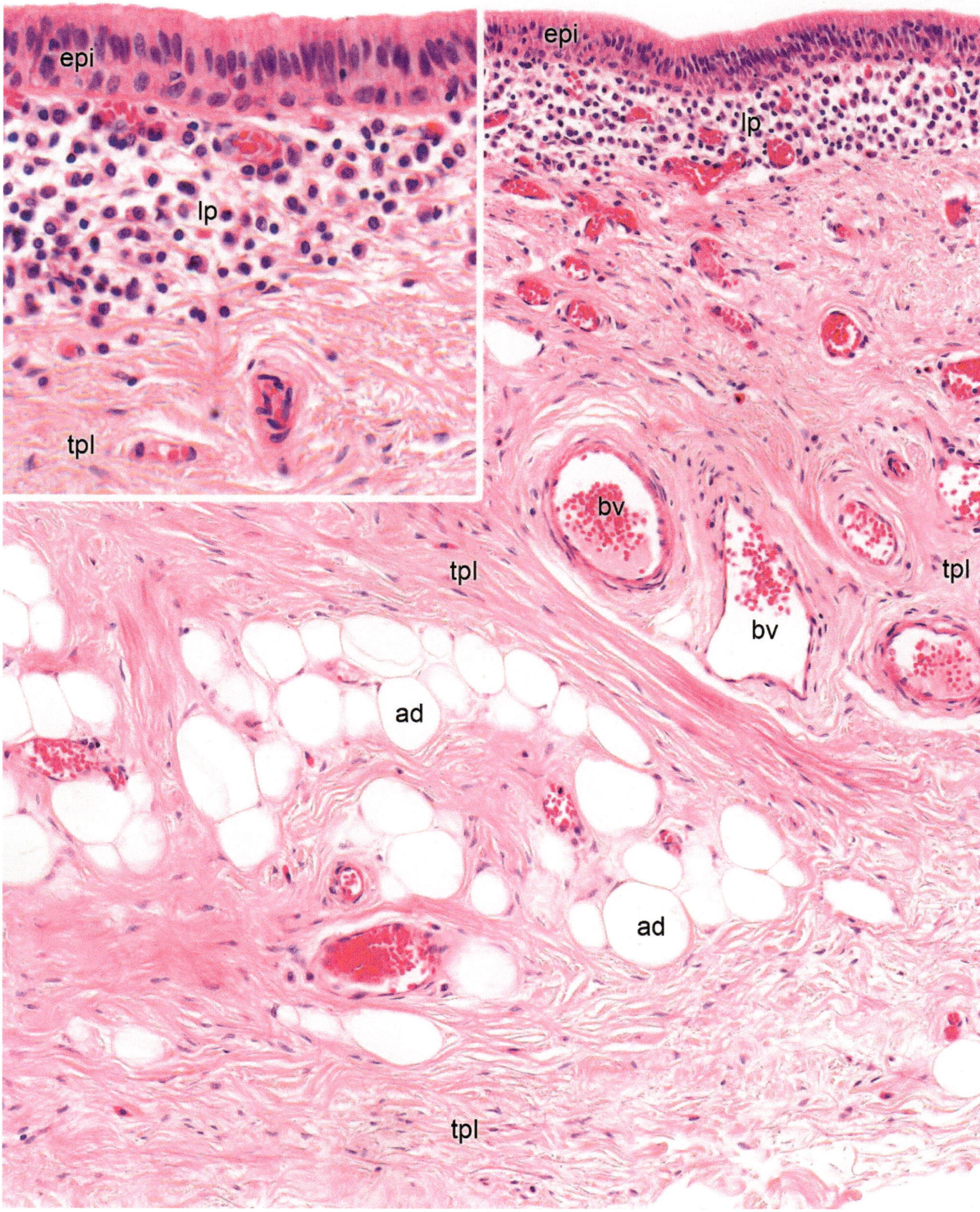

Figure 329. Conjunctiva. The internal surface of the palpebra facing the eye is lined with conjunctiva that is composed of stratified columnar epithelium (epi; inset) with occasional goblet cells and the underlying loose connective tissue lamina propria (lp). Under the conjunctiva, the core of the palpebra is formed by the dense connective tissue tarsal plate (tpl) that also contains adipocytes and blood vessels surrounded by collagen fibers.

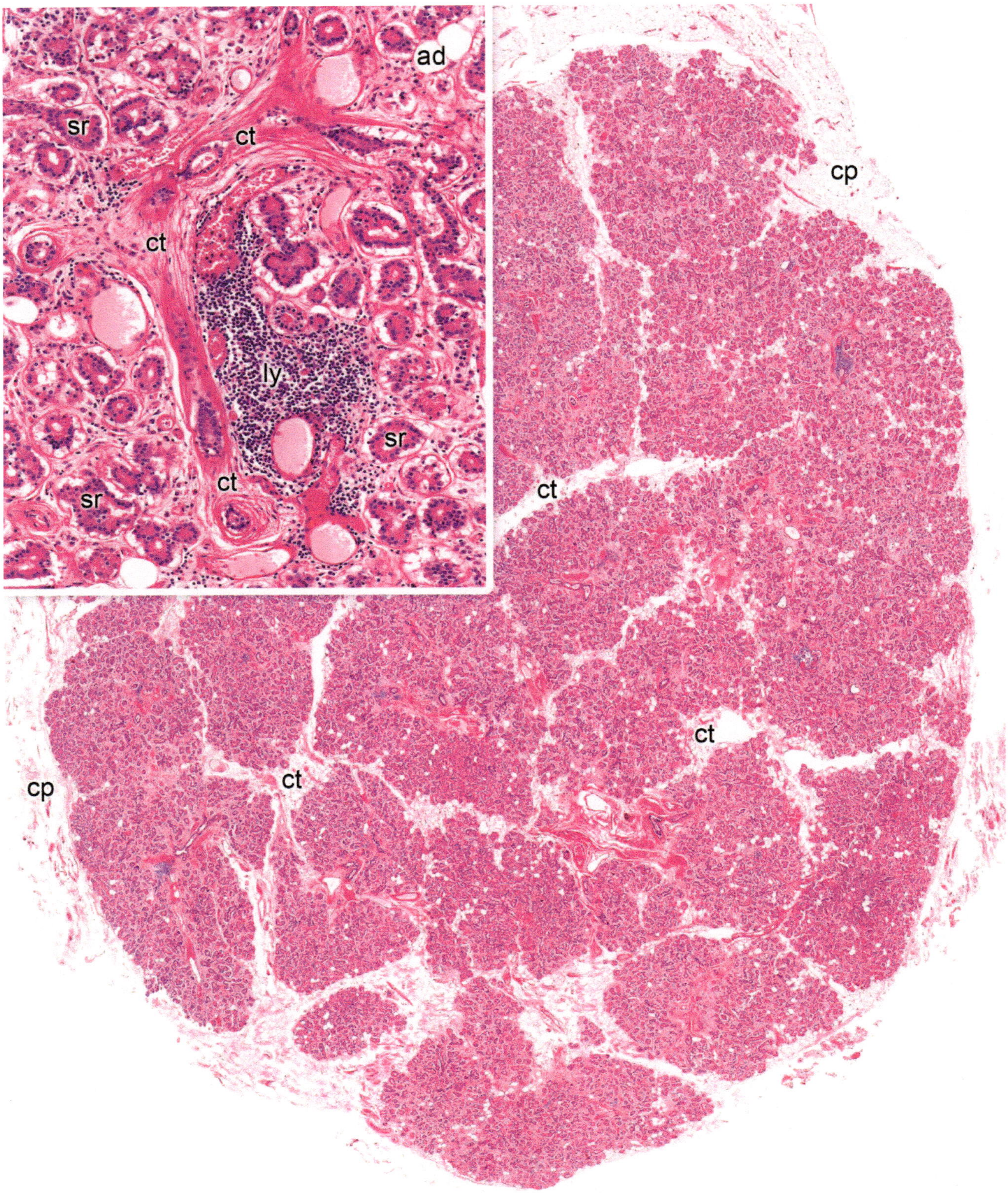

Figure 330. Lacrimal gland. Lacrimal gland is a serous, compound tubuloalveolar gland that is composed of lobules separated by connective tissue septa (ct) that originate from a connective tissue capsule (cp). Serous acini (sr, inset) open into intercalated ducts that lead to intralobular ducts forming interlobular ducts surrounded by connective tissue; these ducts are covered with cuboidal epithelium that becomes gradually higher. Lacrimal gland lacks striated ducts. Myoepithelial cells surround the base of the acini, and they are enclosed within the basement membrane. Lacrimal gland contains lymphocytic aggregations (ly) as well as adipocytes (ad) dispersed among the acini.

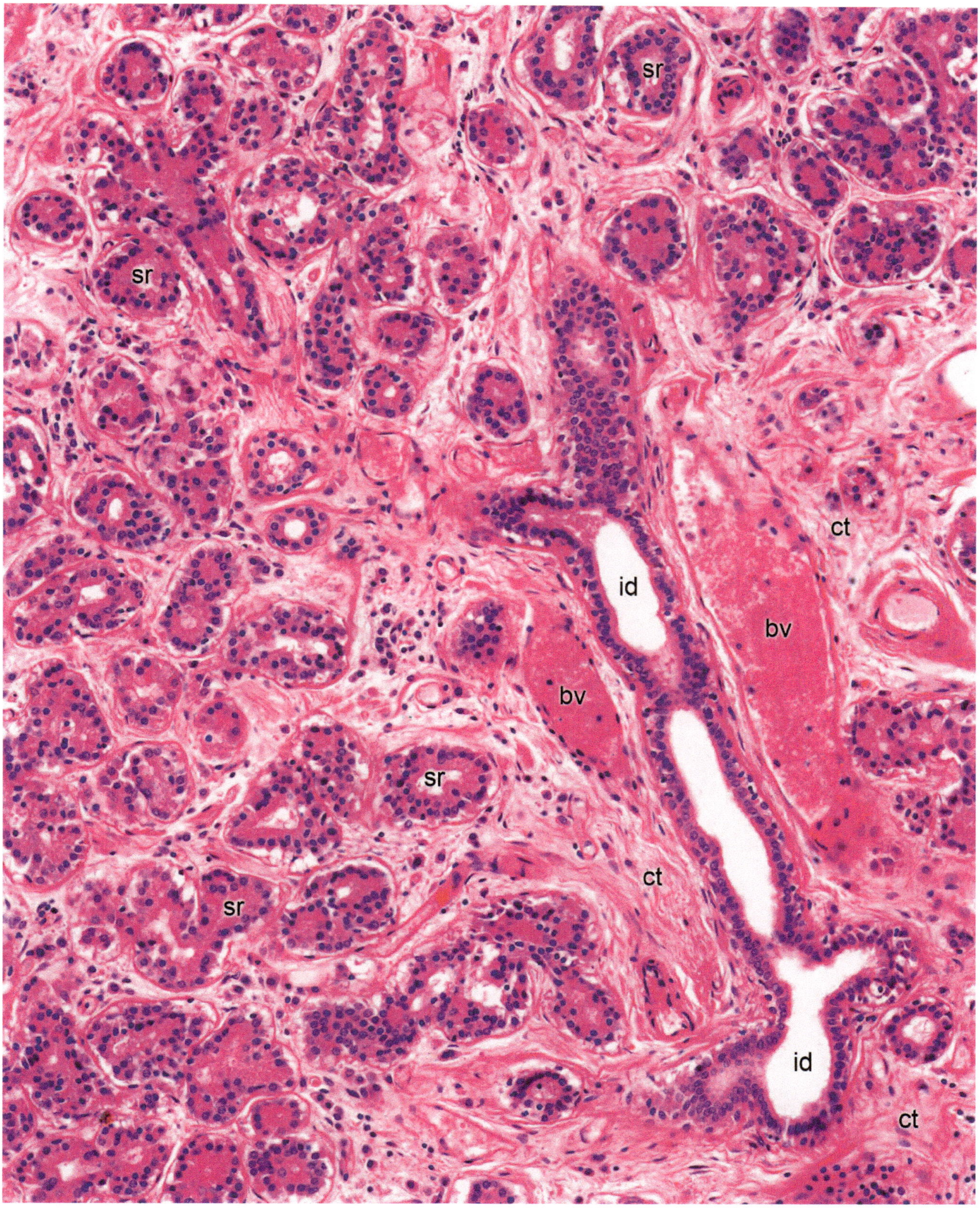

Figure 331. Lacrimal gland. Lacrimal glands contain serous acini (sr) open into intercalated ducts that lead to intralobular ducts forming interlobular ducts (id), surrounded by connective tissue (ct) that distributes blood vessels (bv). These ducts are covered with cuboidal epithelium that becomes gradually higher distally. Lacrimal glands lack striated ducts. Myoepithelial cells surround the base of the acini, and they are enclosed within the basement membrane.

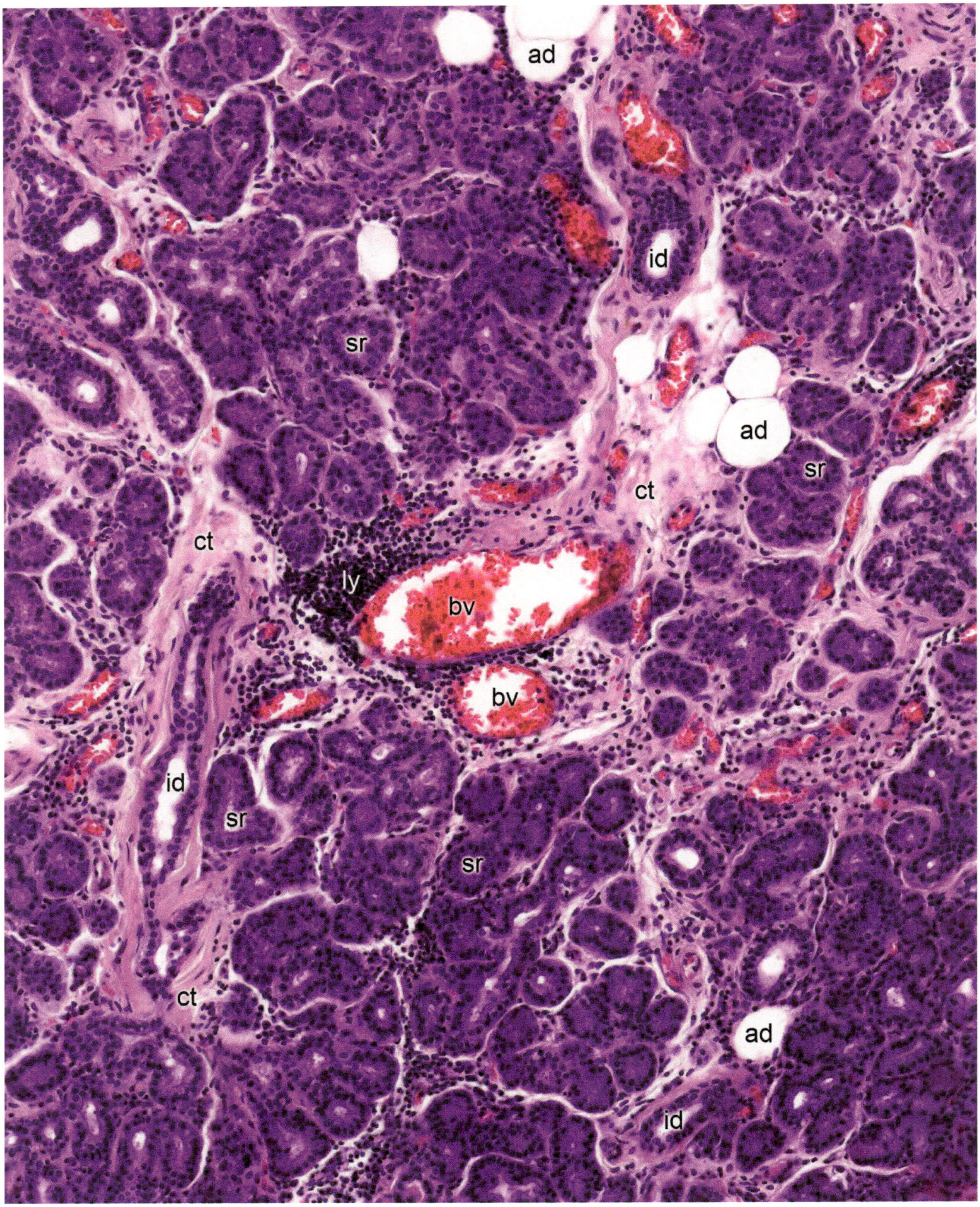

Figure 332. Lacrimal gland. Lacrimal glands contain serous acini (sr) opening into intercalated ducts that lead to intralobular ducts forming interlobular ducts (id), surrounded by connective tissue (ct) with blood vessels (bv). These ducts are lined with cuboidal epithelium that becomes gradually higher distally. Lacrimal glands lack striated ducts. Myoepithelial cells surround the base of the acini, and they are enclosed within the basement membrane. Lacrimal gland contains lymphocytic aggregations (ly) as well as adipocytes (ad).

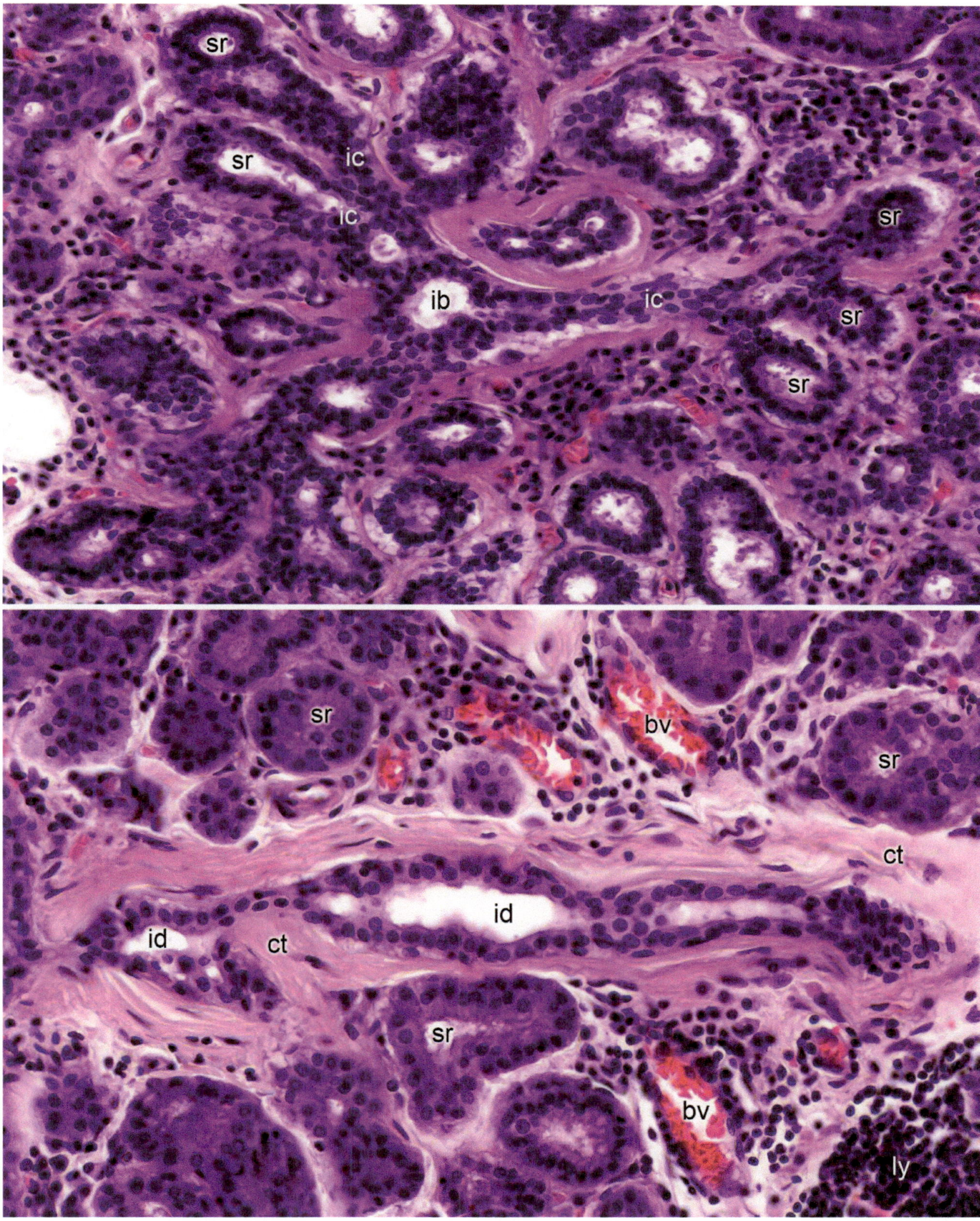

Figure 333. Lacrimal gland. Lacrimal gland contains serous acini (sr) opening into intercalated ducts (ic) that lead to intralobular ducts (ib) forming interlobular ducts (id), surrounded by connective tissue (ct) with blood vessels (bv). These ducts are lined with cuboidal epithelium that becomes gradually higher distally. Lacrimal glands lack striated ducts. Myoepithelial cells surround the base of the acini, and they are enclosed within the basement membrane. Lacrimal gland contains lymphocytic aggregations (ly) as well as adipocytes.

Cochlea

Cochlea is a helical chamber of the bony labyrinth, a complex space inside the temporal bone. The membranous cochlea that is part of the membranous labyrinth is suspended inside the bony cochlea as the cochlear duct that subdivide the bony cochlea into three distinct spaces. The middle scala media is the cochlear duct itself, bordered by the superior scala vestibuli and the inferior scala tympani that join via the helicortrema at the apex of the bony cochlea. Scala media is filled by endolymph, while scala vestibuli and tympani contain perilymph. Scala media is separated from the scala tympani by the acellular basilar membrane that also serves as a base for the spiral organ of Corti, located within the cochlear duct. The vestibular (Reissner's) membrane stretches between the scala vestibuli and the scala media. The lateral wall of the scala media is formed by the stria vascularis that lines the dense connective tissue spiral ligament attached to the bony wall. Stria vascularis is covered by pseudostratified columnar epithelium.

The spiral organ of Corti occupies the floor of the scala media. It is a complex structure that is composed of hair (sensory) cells, phalangeal (supporting) cells and pillar cells that line tunnel-like structures. The single row of inner hair cells along with the inner phalangeal cells are located closer to the modiolus (axis of the cochlea) while the three rows of outer hair cells with their corresponding outer phalangeal cells are located farther from the modiolus. Between these sets of hair and phalangeal cells, the inner and outer pillar cells line the inner spiral tunnel. The periphery of the organ of Corti is formed by outer border cells (of Hansen) as well as the cuboidal cells of Claudius and Böttcher. Closer to the modiolus, the cuboidal inner border cells transition to the acellular, gelatinous tectorial membrane that extends from the modiolus over the organ of Corti and its lower surface lies on the stereocilia covering the apical surface of the hair cells. Vibration of the hair cells against the tectorial membrane at specific sites of the cochlear duct results in action potentials carried by the cochlear nerve into the brain where it translates as sound perception.

Carotid body

At the bifurcation of carotid arteries, the adventitia of the vessels contains encapsulated, extremely vascular clusters of cells that act as chemoreceptors primarily sensitive to changes in the partial pressure of oxygen in the perfusing blood; their sensitivity is increased with low pH (high partial pressure of CO_2). These cell clusters contain type I glomus cells that are associated with capillaries and they are innervated by nerve fibers releasing various neurotransmitters upon the stimulus. Type II glomus cells are supporting (sustentacular) glial cells dispersed between the type I cells. Blood vessels in the carotid body receive both a sympathetic and parasympathetic innervation; parasympathetic ganglion cells are commonly located in the carotid body while sympathetic postganglionic fibers come from the superior cervical ganglion.

Cutaneous receptors

In addition to the barrier function, skin is also a sensory organ containing several cutaneous receptors. The structure and function of Merkel's cells (disks), Pacinian corpuscles and the Meissner's corpuscles have been discussed in detail in the chapter of Integumentary system (Chapter 4).

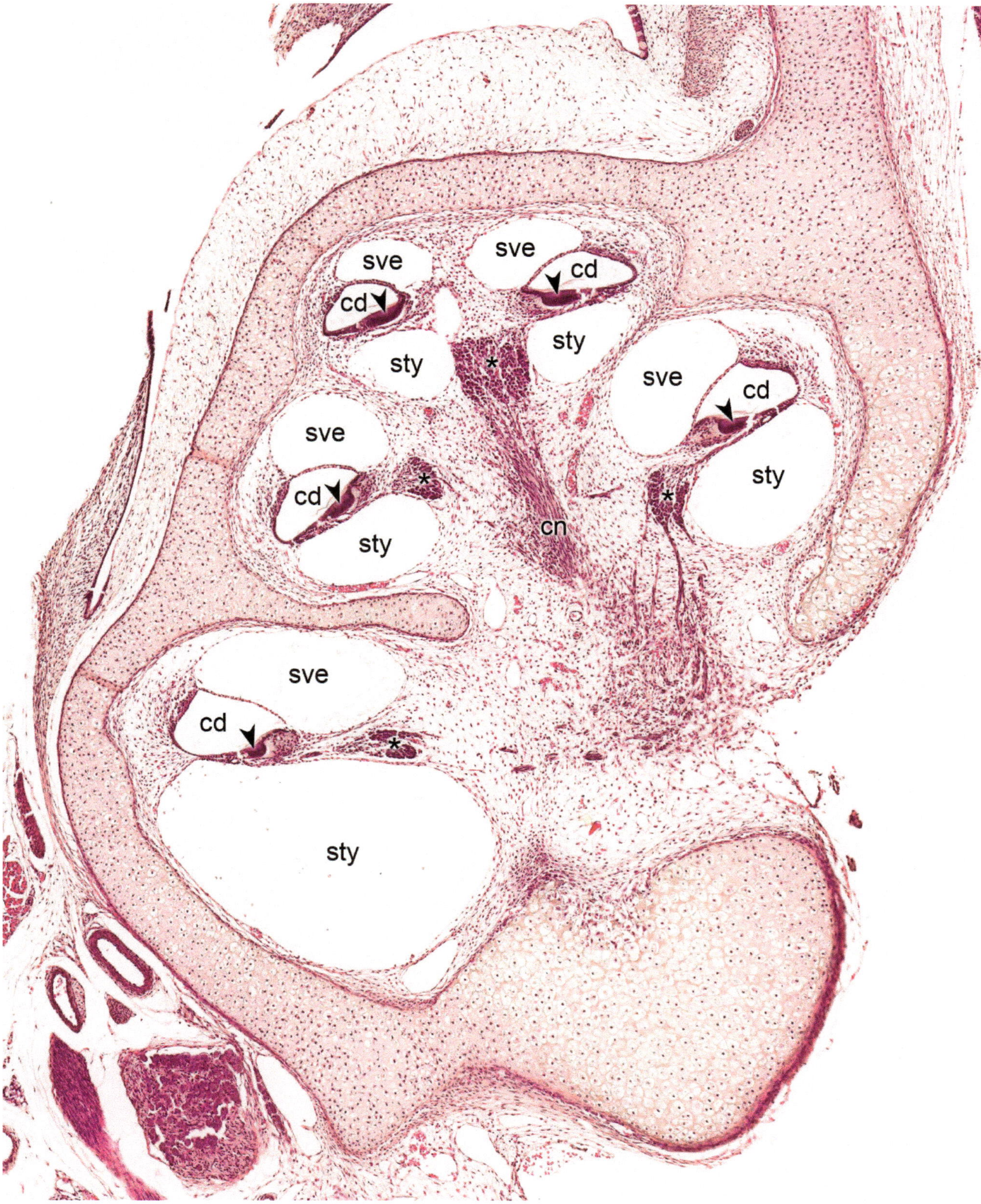

Figure 334. Cochlea. Cochlear duct (cd) is bordered by the superior scala vestibuli (sve) and the inferior scala tympani (sty) that join via the helicortrema at the apex of the bony cochlea. Scala media is filled by endolymph while scala vestibuli and tympani contain perilymph. Scala media is separated from the scala tympani by the acellular basilar membrane that also serves as a base for the spiral organ of Corti (arrowheads), located within the cochlear duct. The vestibular (Reissner's) membrane stretches between the scala vestibuli and the scala media. Organ of Corti is innervated by the cochlear nerve (cn) through the spiral ganglion (asterisks).

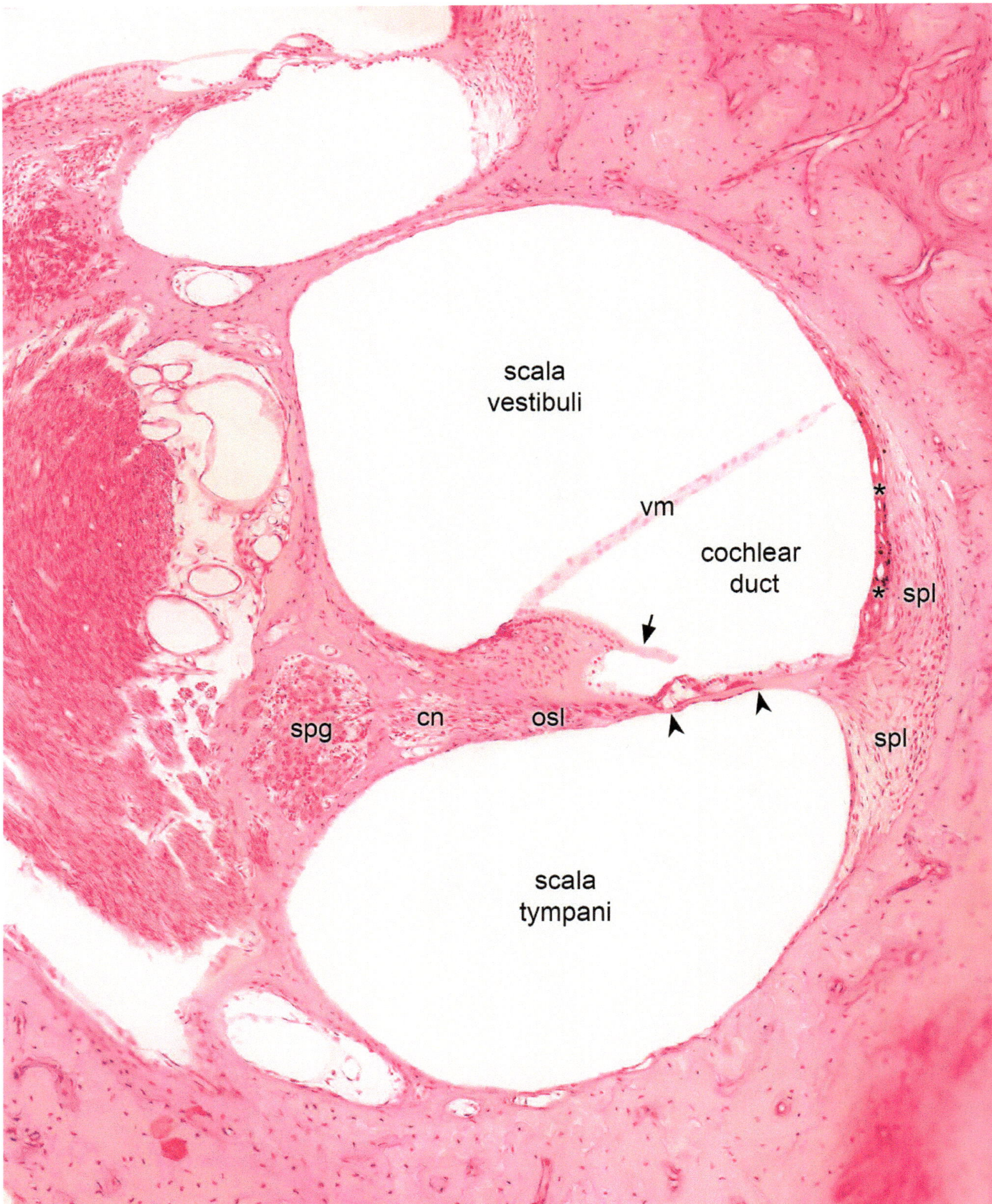

Figure 335. Cochlea. Cochlear duct is bordered by the superior scala vestibuli (sve) and the inferior scala tympani (sty) that join via the helicortrema. Cochlear duct is separated from the scala tympani by the acellular basilar membrane (arrowheads) that also serves as a base for the spiral organ of Corti (arrowheads) with the overhanging acellular, gelatinous tectorial membrane (arrow). Basilar membrane attaches to the spiral ligament (spl) lined by the stria vascularis (asterisks). The vestibular (Reissner's) membrane (vm) stretches between the scala vestibuli and the cochlear duct. The organ of Corti is innervated by the spiral ganglion (spg), through the cochlear nerve (cn) running in the osseous spiral lamina (osl).

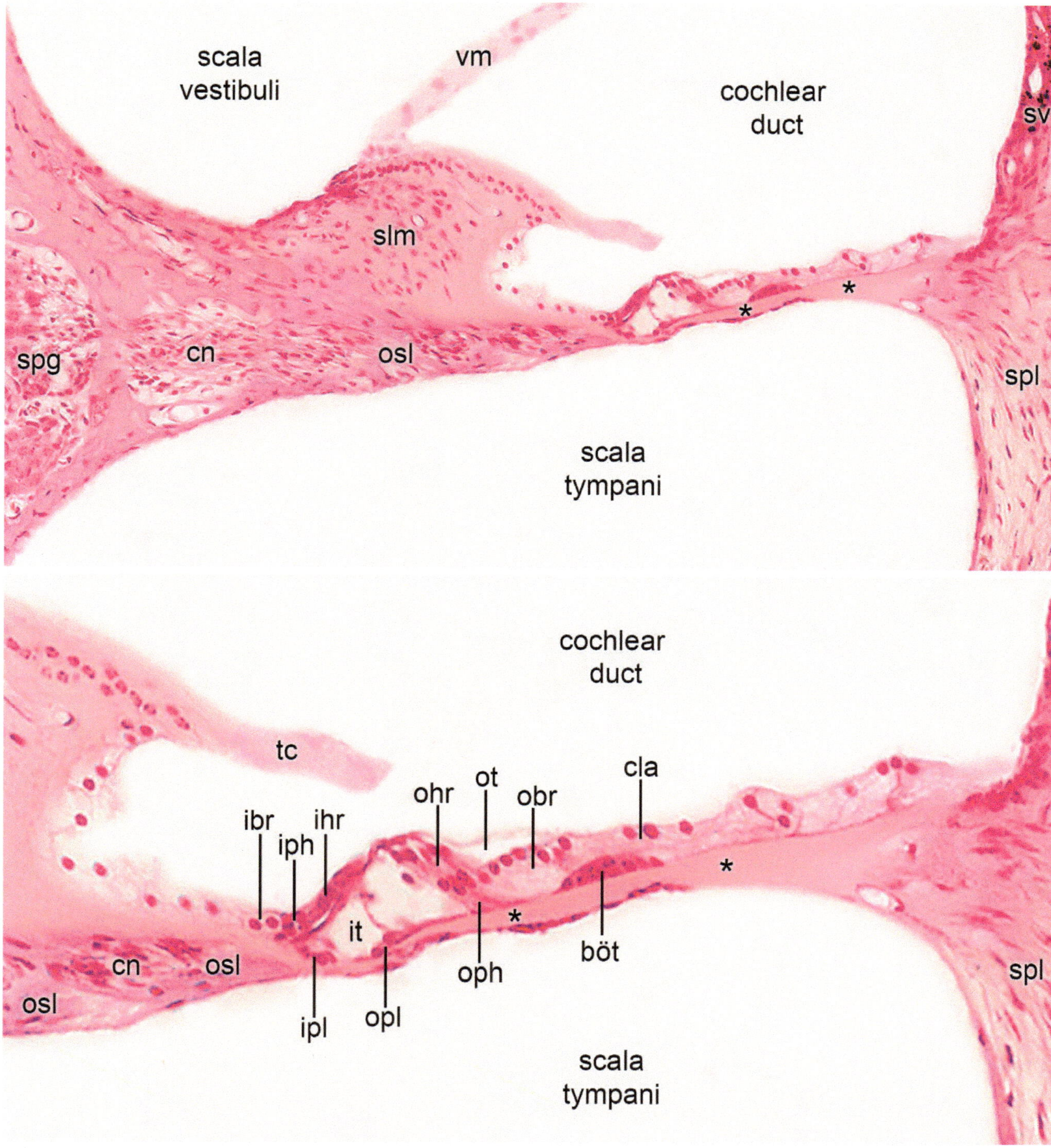

Figure 336. Cochlear duct, lower (top) and higher magnifications (bottom). The spiral organ of Corti occupies the floor of the cochlear duct resting on the acellular basilar membrane (bm) that stretches from the osseous spiral lamina (osl) to the spiral ligament (spl). It is a complex structure that is composed of hair (sensory) cells, phalangeal (supporting) cells and pillar cells that line tunnel-like structures. The single row of inner hair cells (ihr) along with the inner phalangeal cells (iph) are located closer to the modiolus (axis of the cochlea) while the three rows of outer hair cells (ohr) with their corresponding outer phalangeal cells (oph) are located farther away from the modiolus. Between these sets of hair and phalangeal cells, inner (ipl) and outer pillar cells (opl) line the inner spiral tunnel (it) while the outer spiral tunnel (ot) is formed by the outer border cells (obr). The periphery of the organ of Corti is formed by outer border cells of Hansen (obr) as well as the cuboidal cells of Claudius (cla) and Böttcher (böt). Closer to the modiolus, the cuboidal inner border cells (ibr) transition to the cells forming the acellular gelatinous tectorial membrane (tc) that extends from the modiolus over the organ of Corti and its lower surface lies on the stereocilia covering the apical surface of the hair cells. Vibration of the hair cells against the tectorial membrane at specific sites of the cochlear duct results in action potentials carried by the cochlear nerve (cn) located in the osseous spiral lamina (osl) into the brain where it translates as sound perception.

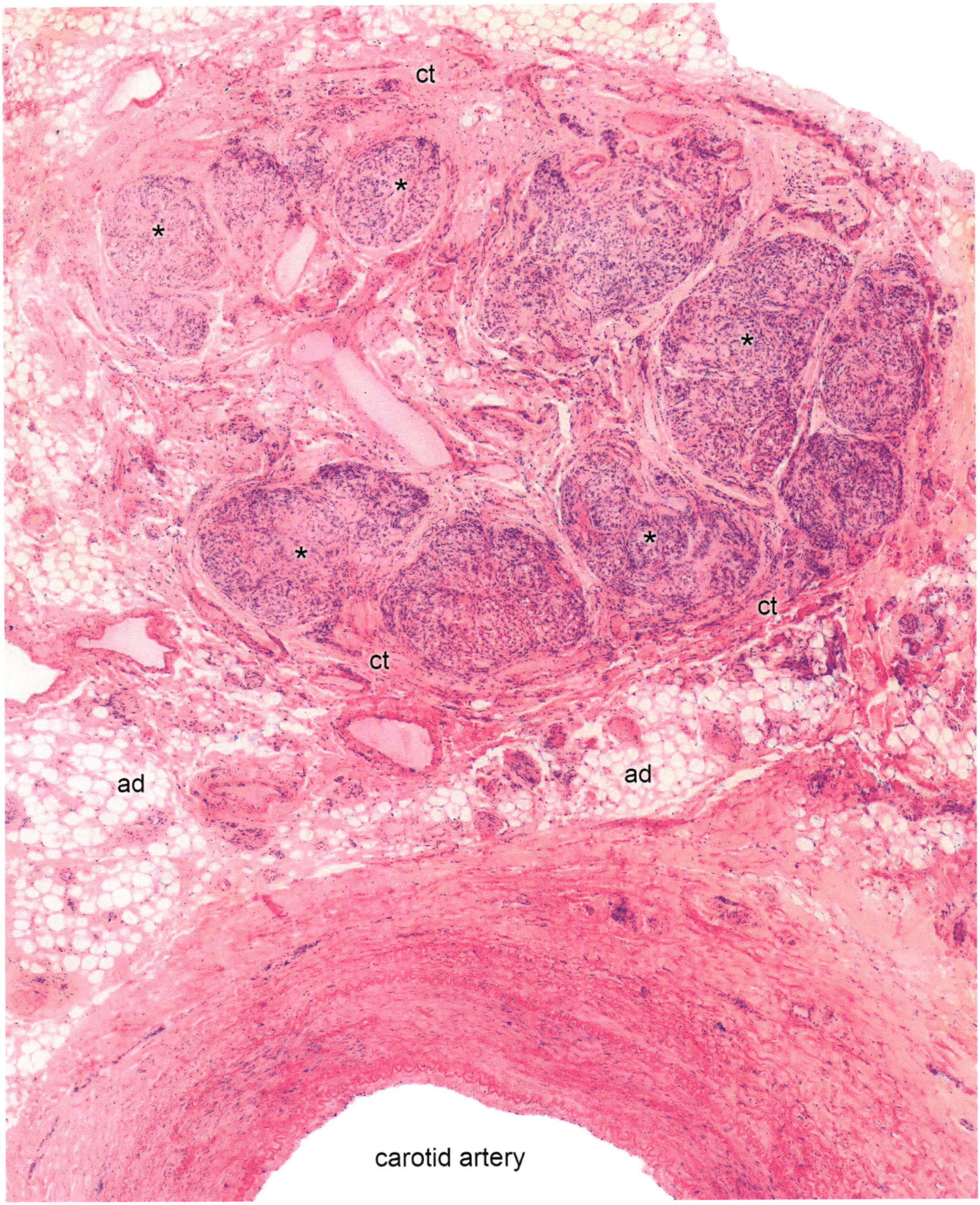

Figure 337. Carotid body, low magnification. At the bifurcation, the adventitia of the carotid arteries contains extremely vascular clusters of cells (asterisks) surrounded by connective tissue (ct) and adipocytes (ad). These cell clusters act as chemoreceptors and contain type I glomus cells that are associated with small vessels. Type I cells are innervated by nerve fibers and they are sensitive to changes in the partial pressure of oxygen in the perfusing blood. Type II glomus cells are supporting (sustentacular) glial cells.

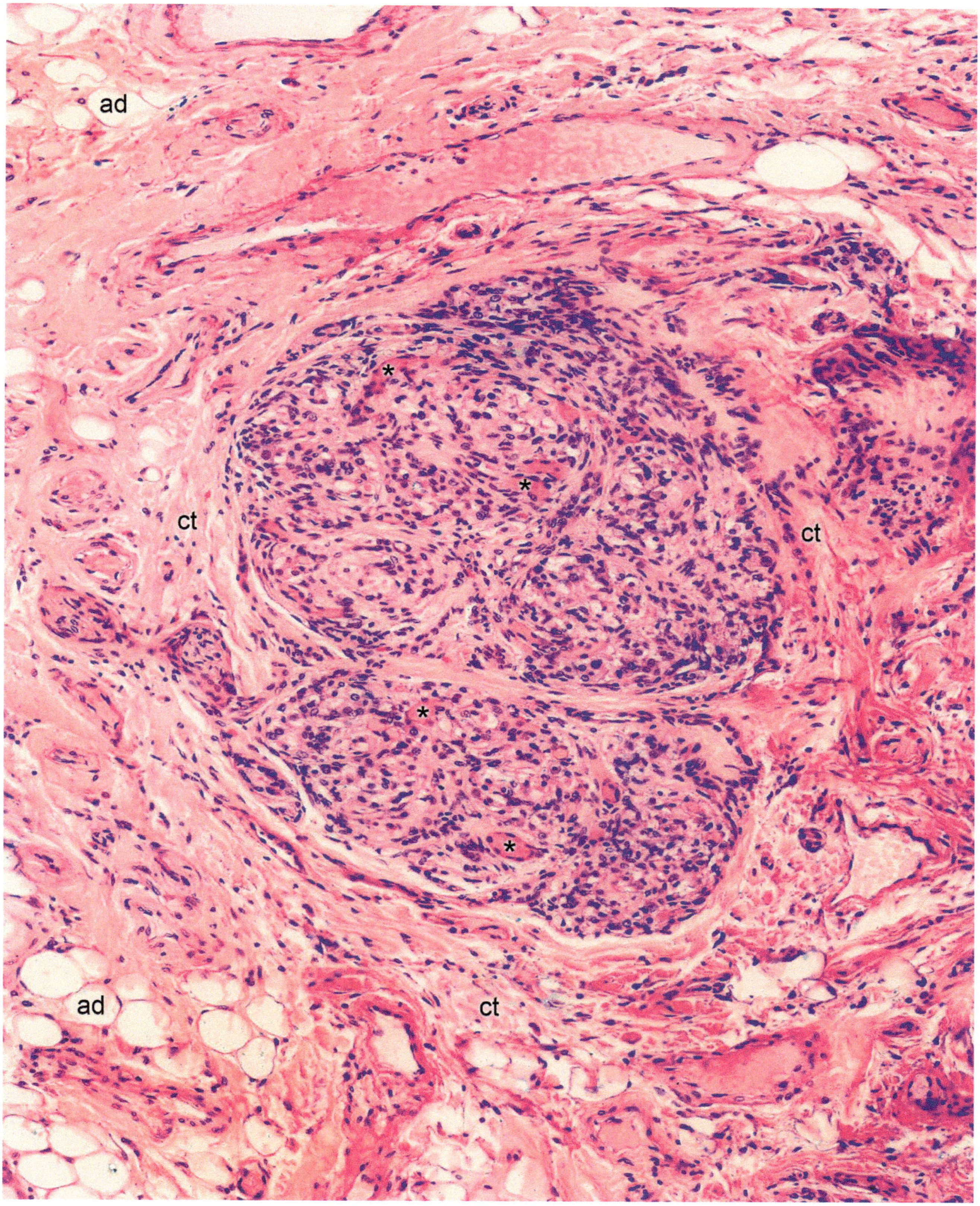

Figure 338. Carotid body, high magnification. At the bifurcation, the adventitia of the carotid arteries contains extremely vascular clusters of cells surrounded by connective tissue (ct) and adipocytes (ad). These cell clusters act as chemoreceptors and contain type I glomus cells that are associated with small vessels (asterisks). Type I cells are innervated by nerve fibers and they are sensitive to changes in the partial pressure of oxygen in the perfusing blood. Type II glomus cells are supporting (sustentacular) glial cells.

Acknowledgement

The author is indebted to Prof. Antal Nógrádi, Head of Department of Anatomy at University of Szeged, Hungary, for providing access to the department's slide scanner and to the histological specimens supplementing the author's personal collection. Similarly, the author is grateful to Zoltán Fekécs M.Sc. and László Gál M.Sc., at University of Szeged, as well as to Toshi Kinoshita at University of Wisconsin, Madison, WI, for the valuable technical support. Finally and most importantly, the present work could not have been achieved without the long line of dedicated histology technicians who prepared the quality sections the book is based on during the past eight decades. Beyond the educational purpose, the present volume is intended to commemorate their accomplishment.

Index

A
Adenohypophysis, 304–311
Adipocytes, 5, 14, 21, 31, 46, 47, 62, 63, 73, 74, 79, 80, 82, 88–91, 93, 106, 110, 111, 130, 137, 141, 157, 159, 188, 197, 206, 214, 358, 361, 362, 364, 365, 370, 371
Adipose tissue, 5, 14, 43, 50, 51, 54, 56, 66–68, 72, 84, 86, 87, 99, 100, 106, 109, 118, 194, 213, 236–238, 240, 295, 302, 304, 317–320, 322, 323, 325, 326, 356
Adrenal gland, 321–327
Alveolar bone, 101–104
Alveolar duct, 199
Alveolar sacs, 199, 209
Alveolus, 88, 208–210, 302, 303
Amnion, 16, 295, 296
Anal canal, 141, 167–174
Anal submucosal muscle, 141, 168, 169, 171
Anal transitional zone, 141, 167, 168, 170–172
Anterior chamber of the eye, 352, 355
Anterior funiculus, 340–342
Apocrine sweat gland, 1, 4, 71, 73, 80, 173, 174, 358–360
Arteries, 56
Arteriole, 43, 52, 53, 56, 62, 63, 65, 116, 206, 212, 218
Arytenoid cartilage, 193, 195, 196, 198
Arytenoid muscle, 193
Autonomic plexus, 106, 110, 121, 123, 148, 159, 163

B
Basal plate or decidua basalis, 295, 298, 299
Basilar membrane, 366–369
Blood, 32, 33
Blood vessel, 14, 16, 26, 35, 41, 42, 55, 57–60, 67, 69, 72, 79, 84, 87, 90, 91, 98, 99, 112, 116, 123, 126, 159–161, 163–165, 203, 208–210, 223, 224, 246, 260, 262, 265, 273–275, 283, 289–292, 294, 296, 297, 306, 308, 309, 312–315, 317–320, 322–324, 326, 329, 330, 345, 354, 356, 361, 363–366
Bone, 17, 23–27
Bone marrow, 26, 27, 31–33, 56
Bone trabeculae, 31, 192
Cuboidal cells of Böttcher, 366, 369
Bowman's capsule, 212, 215, 218, 219
Bowman's membrane, 349, 351
Bronchiolus, 210
Bronchus, 209
Brunner's gland, 141–144

C
Caecum, 149, 150, 152–154
Minor calyx, 213, 214, 220
Capillaries, 56
Capsule, 5, 35, 38–44, 46–48, 52, 66, 82, 106, 115, 179, 212, 217, 221, 228, 232, 251, 304, 308–310, 316, 321–326, 332, 336, 350, 358, 362
Cardiac gland, 118, 124–126
Cardiac muscle, 9–11
Cardiovascular system, 54–65
Carotid body, 366, 370, 371
Cartilage, 17
Cavern, 236, 258–262, 279, 293
Cementum, 88, 101, 104
Central adrenomedullary vein, 321, 325, 327
Central nervous system, 32, 34, 304, 332, 340
Central vein, 175–178
Cerebellum, 332, 343, 347, 348
Cerebrum, 332, 343
Cervical canal, uterus, 279, 284–288
Cervical gland, uterus, 288
Choriocapillary layer, 349, 357
Chorionic plate, 295–297
Chorionic villi, 295–299
Choroid of the eye, 349, 356, 357
Ciliary body, 349, 350, 352, 353, 358
Ciliary process, 349, 352–355
Circular skeletal muscle, 110
Circumvallate papillae, 88, 92–95
Clitoris, 279, 291–293
Cochlea, 366–369
Cochlear duct, 366–369
Cochlear nerve, 366–369
Collagen fiber, 5, 15, 17, 36, 37, 54, 85, 87, 88, 91, 172, 209, 210, 228, 236, 244, 246–251, 253, 255–257, 332, 350, 356, 358, 361
Colorectal zone, 141, 167–170
Conjunctiva, 349, 351, 352, 358–361
Connective tissue, 4–6, 11, 13–18, 20–22, 26–28, 30, 35–48, 52, 54–60, 66–73, 76–78, 82, 83, 85,

87, 88, 98, 101–108, 110, 112–123, 125, 130, 131, 134–139, 141, 143, 144, 146, 157, 159, 161–166, 172–179, 184–189, 191, 192, 196, 199–207, 212–214, 221–225, 227–236, 241–243, 247, 249, 251, 256–262, 265, 266, 268, 273, 274, 276, 279, 280, 283, 285–287, 289–306, 308–310, 312, 316, 321–326, 328–330, 332–334, 336, 340, 349, 356, 358–365, 370, 371
Cornea, 349–353, 358
Corpora cavernosa, 236, 258–260, 262, 291–293
Corpus cavernosum, 258, 259, 292, 293
Corpus of the stomach, 127–129
Corpus spongiosum, 236, 258, 260, 261, 263, 264
Cortex, 35, 38–51, 71, 76–78, 212–216, 220, 221, 265–268, 272, 321–327, 343–348
Cremasteric muscle, 236–238, 240
Cricoid cartilage, 193, 196
Crypt, 35–37, 88, 96–99
Crypts of Lieberkühn, 141–158, 160–162, 164–170
Cuboidal cells of Claudius, 366, 369
Cutaneous receptors, 66, 366
Cutaneous zone, 167, 171–174
Cystic cavity, 304–307, 310, 311

D

Deep (Buck's) fascia of penis, 258, 262–264
Deferential artery, 236–239
Dense connective tissue, 15
Dentin, 88, 101–105
Dermal papilla, 66, 67, 69–71, 76, 81
Dermal sheath, 71, 76–78
Dermis, 5, 66–73, 75, 79, 84–87, 172–174, 291
Descemet's membrane, 349, 351
Dilator pupillae muscle, 349, 354, 355
Distal convoluted tubule, 212, 217–219
Dorsal horn, 340–342
Dorsal ramus, spinal nerve, 336
Dorsal root, 336, 340
Dorsum of the tongue, 91
Duct, 1, 4, 67, 71, 79, 88–90, 92–96, 98, 99, 106, 109, 179, 200, 202, 203, 206, 207, 212, 236, 244, 245, 247, 248, 250, 252, 254–256, 260, 279, 288, 295, 301–303, 330, 331, 358, 362–365
Ductus epididymidis, 228, 229, 232–239
Duodenum, 118, 141–144
Dura mater, 332, 336, 340–342

E

Early spermatid, 230, 231
Efferent ductules, epididymis, 228, 229, 232–235
Ejaculatory duct, 236, 242, 244, 245, 251–254
Elastic artery, 5, 56–60
Elastic cartilage, H &E staining, 20
Elastic cartilage orcein staining, 21
Embryonic connective tissue, 16
Enchondral ossification, 17, 28, 29
Endocrine pancreas, 321, 330, 331
Endocrine system, 304–331
Endometrium, 265, 279, 280, 282, 284–288
Endosteal lamellae, 23, 24
Epicardium, 54, 55
Epidermis, 1, 66–75, 78, 81, 85, 87, 291, 294
Epididymis, 228, 229, 232–236, 242, 243
Epiglottic vallecula, 193
Epiglottis, 17, 21, 190, 193
Epineurium, 332–334, 336, 340
Epithelial glands, 71
Epithelium, 1, 21, 36, 37, 85, 87, 89, 96–100, 103, 108–110, 114, 119–125, 129, 132, 134–136, 138–140, 143, 144, 151, 158, 162, 166, 168, 169, 171–173, 181–184, 192, 197, 202, 203, 205–207, 209, 210, 214, 223–227, 239, 243, 246, 247, 249, 250, 253, 274, 276, 281, 284, 286–290, 294, 296, 351, 361
Esophageal-cardiac junction, 124, 125
Esophagus, 3, 118–125, 199–201
Excretory duct, 3, 106, 111, 112, 114, 116, 117, 197, 198, 358
Exocrine pancreas, 179, 321, 330, 331
External anal sphincter, 167
External elastic lamina, 56, 61
External genitalia, 279
External limiting membrane (retina), 357, 358
External root sheath, 76–78
External spermatic fascia, 237, 238, 240
Extraembryonic mesoderm, 295, 297
Eye, 349, 350
Eyelid, 71, 358–360

F

Female reproductive system, 265–303
Fibrocartilage, 22
Fibrous periosteum, 27
Filiform papilla, 88, 89, 91, 92
Fimbria, 266
Follicular cells, 265, 267, 268
Fundic glands, 127–132, 134–136
Fundic mucosa of the stomach, 131, 135
Fundus of the stomach, 130, 134
Fungiform papilla, 88, 89, 91, 92

G

Gallbladder, 2, 179–184
Ganglia, 32, 332, 336–339
Ganglion cell layer (retina), 357, 358
Ganglion cells, 32, 83, 106, 110, 118, 119, 121, 133, 141, 163, 212, 257, 332, 336, 350, 357, 358, 366
Gastric cardiac glands, 124–126
Gastric pit, 118, 124–132, 134–140
Germinal epithelium, 228, 234, 235, 265–267, 272

Gingiva, 88, 101–103
Glandular epithelium, 1, 4
Glans clitoridis, 291–293
Glans penis, 236, 263, 264
Glial cells, 34
Glomerulus, 212, 215, 218, 219
Graafian follicle, 265, 266, 271
Granular layer, 343, 345, 347, 348
Granulosa cells, 265, 268–272
Granulosa lutein cells, 269, 271

H

Hair, 71, 73, 74, 76–78, 279, 291, 366, 369
Hair follicle, 4, 66, 71–78, 80, 83–87, 141, 173, 174, 279, 291, 293, 358–360
Papilla of the hair follicle, 74, 76, 78, 174
Haversian canal, 17, 23–25
Haversian lamellae, 17, 23–25, 214
Heart, 54
Hyalin cartilage, 17–19, 208–210
Hypopharynx, 106, 110, 195, 196
Hypophysis, 304–311

I

Ileocaecal valve, 141, 149–151
Ileum, 35, 141, 147–151
Inferior nasal concha, 192
Inner border cells, 366, 369
Inner hair cells, 366, 369
Renal capsule, inner layer, 217
Inner limiting membrane (retina), 357, 358
Inner longitudinal layer, 212, 222–227, 236–239, 241
Inner muscularis externa, 118–122, 125, 127, 141–143, 145, 147–149, 152, 154–157, 159, 160, 163, 164, 242, 243, 273, 274, 276, 289, 290
Inner nuclear layer (retina), 357, 358
Inner phalangeal cells, 366, 369
Inner pillar cells, 366, 369
Inner plexiform layer (retina), 357, 358
Inner spiral tunnel, 366, 369
Integumentary system, 66–82
Interarythenoid notch, 195
Intercalated duct, 4, 5, 9, 10, 54, 106, 111, 116, 117, 179, 187, 358, 362–365
Interlobar sulcus, 220, 221
Interlobular duct, 106, 175, 179, 189, 358, 362–365
Internal anal sphincter, 141, 167
Internal elastic lamina, 56, 58, 61–63
Internal root sheath, 71, 76–78
Internal spermatic fascia, 237, 238, 240
Interstitial lamellae, 17, 23–25
Intestinal glands, 141–143, 151, 162, 165, 166
Intralobular duct, 106, 179, 185–188, 358, 362–365
Intramembranous ossification, 17, 30
Iridocorneal angle, 349, 352, 353
Iris, 349, 350, 352–355
Islets of Langerhans, 179, 185–189, 330, 331

J

Jejunum, 145, 146

K

Kidney, 2, 212–214, 217, 220, 221

L

Labia minora, 279, 291, 293, 294
Labium minus, 294
Lacrimal gland, 358, 362–365
Lactiferous duct, 295, 301–303
Lamina propria, 5, 32, 35–37, 83–91, 93–99, 106–110, 118–131, 134–143, 146–148, 151, 153, 157, 158, 160–162, 164–166, 168–171, 179–184, 190–192, 197–207, 209–212, 214, 222–228, 236–239, 241, 243, 246, 247, 249, 250, 253, 254, 265, 273–282, 285, 288–290, 358–361
Large intestine, 141, 149–153, 161, 162, 165–170
Larynx, 17, 190, 193–198
Lateral cricoarytenoid muscle, 194, 196
Lateral funiculus, 340–342
Late spermatid, 230, 231
Layers of the gastrointestinal tract, 83
Lingual tonsil, 35, 88, 96, 98–100
Lip, 83–87
Liver, 175–178
Longitudinal skeletal muscle, 110
Loop of Henle, 212, 214, 216
Loose connective tissue, 13
Lower alimentary tract, 118–174
Lung, 199, 208
Corpus luteum, 265, 266, 269, 271, 272
Lymphatic aggregation, 43, 83, 106, 141, 213, 281, 362, 364, 365
Lymphatic follicle, 35–43, 52, 53, 88, 95–100, 106, 110, 118, 126–130, 137–139, 141, 147–150, 152–157, 159–162, 164, 165, 169, 171, 190, 194, 214, 358
Lymphatic system, 35–53
Lymphatic vessels, 56
Lymph node, 35, 38–42, 56, 119, 125, 149
Lymph vessel, 35, 38–42, 65, 98, 99, 141, 165, 168, 169, 175–178, 265, 273–278, 314

M

Prostatic main glands, 244, 245, 247, 248, 250–252, 255, 256
Male reproductive system, 228–264
Malpighian (splenic) corpuscles, 43, 52, 53
Mammary gland, 1, 295, 301–303
Matrix, 16–22, 26, 27, 30, 32, 71, 76–78, 175, 295, 300, 358
Medulla, 35, 38–51, 71, 76–78, 212–214, 220, 221, 265, 266, 321–327, 343

Medullary cords, 40
Medullary rays of Ferrein, 212–216, 220, 221
Medullary sinus, 35, 38–42
Meissner's corpuscle, 66, 81, 366
Meissner's plexus, 83, 118, 332
Meninges, 340–342
Merocrine sweat gland, 4, 71, 79
Mesosalpinx, 275, 277, 278
Microvasculature, 62–65
Middle circular layer, 118, 133, 236–239, 241
Molecular layer, 343, 345–348
Esophageal mucosal gland, 118, 124
Prostatic mucosal glands, 246, 250
Mucous acinus, 1, 4, 90, 106, 112–116, 124, 197–199, 202, 203, 206, 260, 261, 263, 264
Muscle fiber, 5–8, 88–90, 106, 110
Muscle tissue, 5
Muscular artery, 50, 56, 61–63, 108, 110, 114, 180, 181, 183, 290, 353
Caecal muscularis externa, 149
Ileal muscularis externa, 149
Tunica muscularis externa, 83, 118–121, 123, 127, 130, 133–135, 137–139, 141–143, 145, 147–150, 152–157, 160, 163, 164, 180–184, 273–275, 277, 278, 332
Tunica muscularis mucosae, 83, 110, 119–127, 129–132, 134–146, 151, 153, 157–159, 162, 166, 168–171, 179–184, 199, 209, 210, 222–227
Muscular venule, 56, 62, 63, 65, 116, 206, 246, 249
Myocardium, 5, 54, 55
Myometrium, 54, 279–287, 298, 299

N
Nabothian follicle, 279, 286–288
Nasal cavity, 190, 192
Nasopharynx, 106, 190
Neonatal testis, 234
Nerve, 9, 17, 26, 54, 56–63, 66, 67, 69, 72, 79, 83, 88, 106, 110, 112, 116, 119, 123, 159–161, 194, 236, 258–260, 262, 289–292, 294, 324, 332–334, 356, 358
Layer of nerve fibers (retina), 357, 358
Nervous system, 332–348
Nervous tissue, 32, 332
Neurohypophysis, 304–307, 310, 311
Neuromuscular junctions, 8

O
Odontoblast cells, 88, 102–105
Oocyte, 265, 267, 268, 270–272
Optic nerve, 350, 357, 358
Orbicularis oculi muscle, 358–360
Osseous spiral lamina, 368, 369
Ossification center, 17, 28
Osteoclast, 17, 26
Osteogenic periosteum, 17, 27
Osteon, 17, 23–25, 279, 284–288
Outer border cells of Hansen, 366, 369
Outer circular layer, 212, 222–227
Outer hair cells, 366, 369
Outermost longitudinal layer, 222–227, 237–239, 241
Outer muscularis externa, 118–122, 125, 127, 133, 141–143, 145, 147–149, 152, 154–157, 159, 160, 163, 164, 273, 274, 276, 289
Outer nuclear layer (retina), 357, 358
Outer phalangeal cells, 366, 369
Outer pillar cells, 366, 369
Outer plexiform layer (retina), 357, 358
Outer spiral tunnel, 369
Ovary, 265–272
Oviducts, 265
Oxyphil cells, parathyroid, 304, 317–320

P
Pacinian corpuscle, 66–68, 82, 366
Palatine tonsil, 35–37
Palpebra, 349, 358–361
Pampiniform plexus, 236–238, 240
Pancreas, 179, 185–189, 321, 330, 331
Papillae of the tongue, 92
Paracortex, 35, 38–42
Parathyroid glands, 304, 317–320
Parotid gland, 106, 111
Pars cutanea, 83–87
Pars mucosa, 83–87
Penile urethra, 236, 258, 260, 261, 263, 264
Penis, 236, 258–260, 279
Perichondrium, 17–22, 190, 197, 198, 200–203, 205, 206
Perimetrium, 279, 280, 283–285
Perimysium, 5–7
Perineurium, 332–335
Periodontal ligament, 88, 101–104
Periosteal lamellae, 23, 24
Peripheral nerve, 332–335
Peripheral nervous system, 32, 332
Peritubular myoid cells, 228, 230, 231
Pharynx, 106, 190
Pia mater, 305, 310, 321, 328, 340–342, 345
Pineal gland, 321, 328, 329
Placenta, 295–299
Portal vein, 175–178
Portio vaginalis uteri, 279, 284, 286–288
Postcapillary venule, 4, 35, 56, 62, 64, 65, 85, 94, 172, 231
Posterior chamber of the eye, 352, 355
Posterior cricoarytenoid muscle, 196
Posterior funiculus, 340–342
Prepuce (foreskin), 236, 263, 264, 291, 292
Primary spermatocyte, 228, 230, 231
Principal cells, parathyroid, 304, 317–320

Prostatatic urethera, 249
Prostate gland, 236, 244–248, 250–257
Proximal convoluted tubule, 212, 217–219
Dental pulp, 88, 101–105
Pyloric glands, 118, 137–141
Pylorus of the stomach, 137
Pyriform recess, 194

R

Layer of rods & cones (retina), 357, 358
Renal capsule, outer layer, 217
Renal column, 220, 221
Renal corpuscle, 212–216, 218, 219, 221
Renal cortex, 213–217, 221
Renal papilla, 212–214, 220, 221
Renal pelvis, 212–214, 220, 221
Respiratory epithelium, 35, 106, 190, 191
Respiratory system, 190–211
Retina, 349, 350, 353, 354, 356–358
Rokitansky-Aschoff sinuses, 179–183
Root of the tongue, 96–100
Root of the tooth, 101–105

S

Salivary gland, 1, 3, 4, 84–90, 96, 98–100, 106–109, 111, 114, 120, 121, 123, 193–201, 204, 205, 207, 208
Scala tympani, 366–368
Scala vestibuli, 366–368
Sclera, 349, 350, 352, 353, 356–358
Sebaceous gland, 1, 4, 71–75, 80, 83, 85, 141, 173, 174, 279, 291, 293, 295, 358–360
Seminal vesicles, 236, 241–244, 251, 254
Seminiferous tubules, 228–231, 234, 235
Sensory organs, 349–371
Tunica serosa, 83, 127, 134, 137, 141, 145, 147–149, 152, 154–157, 159, 160, 163, 164, 180, 183, 184, 226, 227, 275–279
Serous acinus, 106, 111–113, 115–117, 197, 198, 202, 203, 321, 330, 331, 358, 362–365
Sertoli cell, 228, 230, 231, 234, 235
Sigmoid colon, 164–166
Sinusoid, 26, 27, 31, 43, 56, 175–178
Skeletal muscle, 5–8, 36, 84, 86, 88–90, 93, 96, 98–100, 106–110, 118, 123, 125, 167, 190, 196, 236–238, 240, 358
Skin, 66–69, 72, 73
Small intestine, 141, 151
Smooth muscle, 5, 12, 54, 56–60, 62, 63, 65, 66, 83, 110, 118–122, 125, 127, 141–143, 145, 147–149, 166–169, 171, 179–184, 199–201, 203, 204, 207, 212, 228, 232, 233, 236–239, 241–244, 246–251, 253, 255–262, 265, 273, 274, 276, 279, 280, 283–287, 289, 290, 321, 325, 327, 332, 349, 355, 358
Soft palate, 106–108, 190
Spaces of Fontana, 349, 352, 353
Spermatic cord, 236–240
Spermatogonia, 228, 230, 231, 234, 235
Sphincter pupillae muscle, 349, 355
Spinal cord, 32, 332, 340–342
Spiral ganglion, 367, 368
Spiral ligament, 366, 368, 369
Spleen, 43, 52, 56
Stomach, 118, 124–130, 141
Stratum basale, 66, 70, 71, 78, 88, 94, 172, 279
Stratum corneum, 1, 3, 66, 69–71, 81, 87, 88, 172, 236, 293
Stratum granulosum, 66, 70
Stratum lucidum, 66, 70, 88
Stratum spinosum, 66, 70, 78
Striated duct, 106, 111–113, 115–117, 358, 362–365
Stroma, 54, 83, 106, 118, 175, 228, 232, 233, 236, 244–257, 265, 267–269, 271, 272, 295, 297, 301–303, 349–351, 354, 355, 357
Subarachnoidal space, 340–342
Subcapsular sinus, 35, 39–42
Subcutis, 66–68, 72, 73, 79, 80, 82, 84, 86
Sublingual glands, 106, 114–117
Submandibular glands, 106, 112, 113
Caecal submucosa, 149, 150, 152, 153
Ileal submucosa, 149, 150
Prostatic submucosal glands, 244, 247, 250, 251, 256
Tunica submucosa, 15, 83, 88, 110, 118, 119, 122, 130, 131, 133–139, 142–148, 151–166, 168, 169, 171, 179–184, 190, 193, 194, 196–199, 203, 204, 207, 209, 210, 223–227
Superior thyroid notch, 195
Sweat gland, 1, 3, 4, 66–69, 71–73, 79–84, 86, 173, 174, 358–360

T

Tarsal plate, 358–361
Tectorial membrane, 366, 368, 369
Teeth, 88
Terminal bronchiolus, 199, 208, 210, 211
Testicular artery, 236–238, 240
Testis, 228–231, 234, 235, 265
Theca externa, 265, 268–271
Theca interna, 265, 268–272
Thymic lobule, 45
Thymus, 43, 44, 46–49
Thymus persistens, 50, 51
Thyroarytenoid muscle, 194, 196
Thyroepiglottic muscle, 194
Thyroid cartilage, 193–197
Thyroid follicles, 312–316
Thyroid gland, 193, 304, 312–316
Tip of the tongue, 89, 90
Tongue, 88
Tonsils, 35
Tooth, 101–103

Trabecular sinus, 35, 39–42
Trabecule, 17, 29, 31, 35, 38–43, 52, 228, 236, 321, 340
Trachea, 17, 199–207
Tracheal cartilage, 193, 200–207
Transverse colon, 160–164
Tube uterina, 273–278
Tunica adventitia, 41, 42, 54, 56–61, 83, 106, 110, 118–123, 125, 142, 179–181, 183, 184, 200, 201, 203–206, 212, 222–225, 227, 241, 279, 289, 290
Tunica albuginea, 228–230, 234–236, 258, 262, 265–267, 271, 272, 291–293
Tunica intima, 54, 56–61, 63
Tunica media, 5, 54, 56–61, 63, 321, 325, 327

U
Umbilical artery, 295, 300
Umbilical cord, 5, 16, 295, 300
Umbilical vein, 295, 300
Umbrella cells, 1, 3, 222–227, 249
Upper alimentary tract, 83–117
Ureter, 212, 222–225
Urinary bladder, 212, 226, 227, 236
Urinary system, 212–227
Uterus, 265, 279–285
Prostatic utricle, 236, 244, 245, 251–253
Uvula, 106–109
Musculus uvulae, 106–109

V
Vagina, 279, 284–287, 289–291, 293
Vas deferens, 236–239, 241–244, 251, 254
Vein, 35, 50, 52, 54, 61, 84, 108, 110, 167–169, 171, 180, 181, 183, 237, 238, 240, 251, 290, 349
Venous sinus, 192, 340
Ventral horn, 32, 340–342
Ventral ramus, spinal nerve, 336
Ventral root, 336
Vermiform appendix, 35, 141, 154–159
Vermillion zone, 83–87
Verumontanum, 236, 244, 245, 251–253
Vessels, general structure, 54
Vestibular fold, 190, 194
Vestibular membrane, 368
Villi, 141–151, 161, 162, 165, 166, 295–297, 373
Vocal fold, 190, 193, 194, 196
Vocalis muscle, 194, 196
Vocal ligament, 194, 196
Vocal process, 196
Von Ebner's gland, 88, 92–95

W
Wharton's jelly, 5, 16, 295, 300
White matter, 32, 340–343, 345, 347, 348

Z
Zona fasciculata, 321–326
Zona glomerulosa, 321–326
Zona reticularis, 321–326
Zone of hypertrophy, 17, 28, 29
Zone of ossification, 17, 28, 29
Zone of proliferation, 17, 28, 29
Zone of reserve cartilage, 28, 29